MONITORING POLYMERIZATION REACTIONS

MONITORING POLYMERIZATION REACTIONS

From Fundamentals to Applications

Edited by

WAYNE F. REED
ALINA M. ALB

Tulane University,
New Orleans, Louisiana

Published by John Wiley & Sons, Inc., Hoboken, New Jersey.
Published simultaneously in Canada.

For general information on our other products and services or for technical support, please contact our Customer Care Department within the United States at (800) 762–2974, outside the United States at (317) 572–3993 or fax (317) 572–4002.

Wiley also publishes its books in a variety of electronic formats. Some content that appears in print may not be available in electronic formats. For more information about Wiley products, visit our web site at www.wiley.com.

Library of Congress Cataloging-in-Publication Data:

Monitoring polymerization reactions : from fundamentals to applications / edited by Wayne F. Reed, Tulane University, New Orleans, Louisiana, USA, Alina M. Alb, Tulane University, New Orleans, Louisiana, USA.
    pages   cm
  Includes index.
  ISBN 978-0-470-91738-1 (cloth)
  1. Polymerization. 2. Chemical reactions. I. Reed, Wayne F. II. Alb, Alina M.
  TP156.P6M585 2014
  668.9′2–dc23
                            2013018919

10 9 8 7 6 5 4 3 2 1

CONTENTS

PREFACE

Monitoring polymerization reactions has both fundamental and applied motivations. At the basic level, simultaneous monitoring of the various reaction characteristics, such as the evolution of copolymer composition and molecular weight, can reveal the underlying kinetics and mechanisms involved in reactions and also illuminate ways in which the reaction may deviate from what is ideally thought to occur. The ability to monitor reactions will become increasingly important as new, stimuli-responsive, and "intelligent" polymers are developed at the frontiers of twenty-first century materials science.

From the practical viewpoint, optimization of reaction conditions at the laboratory and pilot plant level can be an important step in process development. Ultimately, application of monitoring and subsequent feedback control to full-scale industrial reactors will lead to improved efficiency in the use of energy, feedstocks, and other nonrenewable resources, as well as plant and labor time, enhanced product quality and reliability, and the reduction of emissions.

While it is not possible to be all encompassing in scope, this book attempts to be as self-contained as possible. Hence, the basic elements of different types of polymer reactions and their kinetics are reviewed in Section 1, including a brief overview of stimuli-responsive polymers. This is followed in Section 2 by a description of the most relevant polymer and reaction characteristics to be monitored. The section then explains the principles and applications of some of the most important polymer characterization tools, such as light scattering, gel permeation chromatography, calorimetry, rheology, and various types of spectroscopy. Section 3 is devoted to Automatic Continuous Online Monitoring of Polymerization (ACOMP) reactions, which is a flexible platform that allows characterization tools discussed in Section 2 to be employed simultaneously during reactions in order to obtain a comprehensive and quantitative record of multiple reaction features. The chemical engineering community has devoted considerable effort to the modeling of polymerization reactions, and Section 4 provides an account of the theoretical and experimental bases for these efforts, as well as a treatment of the type of modeling and numerical approaches used. Finally, Section 5 gives a short selection of industrially important polymers and perspectives on their production and how monitoring can be employed to optimize their manufacture.

INTRODUCTION

Observing processes as they occur is a widespread approach in many areas of inquiry and industry because it allows fundamental understanding of the mechanisms involved in the process and offers the possibility for control. The many areas of process monitoring are striking: time-lapse photography of plant growth, embryology, crystallization processes; electrophysiological monitoring of vital signs; using the electromagnetic wave spectrum to detect and monitor the expansion of the universe; the dynamic flows of ocean currents; and the paths of hurricanes.

As the focus turns to chemistry, science and engineering have developed rich and deep layers of useful theory and modeling concerning reaction kinetics and processes. In the realm of the vast chemical industry, monitoring temperature, pressure, and other parameters, the presence of specific small molecules, particulates, and pollutants has allowed enormous gains in efficiency, safety, product quality, and reduction of negative environmental impact.

When it comes to monitoring polymer reactions, however, the nature of the task shifts. This is because polymer characterization is a complex and challenging field even "offline." In principle, polymers have definable molecular weight distributions, conformations, and interactions in different solvents and at different concentrations and temperatures, architectures, hydrodynamic properties, comonomer compositions, sequence length distributions or comonomer blockiness, propensities to aggregate, and so forth. The experimental measurement of these quantities and behaviors requires many instrumentational and methodological approaches. Hence, polymer characterization is a great interdisciplinary undertaking, which has united efforts from diverse areas of the physical and biological sciences and engineering. It has rapidly incorporated the advances in fields such as light, x-ray, and neutron scattering, nuclear and electron magnetic resonance, ultraviolet, visible, and infrared spectroscopies, thermal measurements, ultrasonics, viscometry, time-of-flight mass spectroscopy, and microscopies.

Given the challenges that exist for the characterization of polymers in equilibrium, taking the next step to provide characterization while polymer reactions occur requires new concepts and strategies but is rewarding for both fundamental science and improved industrial manufacturing.

In this book, all characterization will be of the type that can be made while polymers are in solutions, bulk, or melts, but not in the solid state. Hence, issues of crystallinity, and mechanical, electrical, optical, and other properties of polymers used in their solid state are not treated. There are possibilities for incorporating and automating such measurements into the manufacturing process; however, these will require certain delay times and conditioning steps.

The term "polymer reactions" is used in a very broad sense. It refers not only to the usual step growth and chain growth reactions that produce polymers, but also to postpolymerization modifications, conjugation, degradation processes, noncovalent polymers, physical reactions producing aggregates and gels, and supramacromolecular assemblies.

As regards the building of covalently bonded polymer chains, attention will be paid to both chain growth and step growth methods. In both these categories, some of the newer "living"-type approaches will be presented as well as the features that can make these advantageous. In the area of copolymers, block, statistical, and gradient copolymers will be treated, along with the extra dimensions of characterization work required. Copolymers produced by postpolymerization modifications, such as hydrolysis, functionalization, grafting, and conjugation are an interesting alternate route to forming copolymers, where architecture can also be controlled.

While the differences between chemists' and chemical engineers' approaches may not be obvious to people outside these fields, they become readily apparent in the polymer industry, to the point where a common adage suggests that if a manufacturer is producing polymers in batch reactors then the company founders were likely chemists, whereas if the manufacture is via continuous reactors the founders were likely chemical engineers. Likewise, in their approach to understanding and controlling polymer properties, one will often see the chemist focused on the types of detailed organic reactions and side reactions used, while the chemical engineer focuses more on mixing, spatial inhomogeneity, heat and mass transfer, shear effects, viscosity, and the like.

Although polymer reaction monitoring is of great interest to both the polymer scientist in a research laboratory and the process engineer in a production plant, their motivations differ and the approaches can differ. Whereas the laboratory scientist has the knowledge and means to both understand and run complex analytical instrumentation yielding information-rich data, and to invent and change chemistry, the process engineer wants to be able to produce a fixed range of products reliably and efficiently with minimum amount of instrumentation, such as monitoring. Hence, the laboratory monitoring platform can be complex, multifaceted, delicate, and rather challenging to interpret, whereas factory monitoring is simplified to follow the synthesis process with the least amount of operator intervention and interpretation possible.

There are a vast number of polymers produced in university and industrial research laboratories, but far fewer in industrial production. This book does not intend to cover all types of polymers; rather it will use some of the best known classes to illustrate reaction monitoring. It is important to put into perspective the height of the "macromolecular pyramid." At the base of the pyramid are the many billions of pounds per year of commodity plastics, such as polyolefins.

Working up the pyramid are the highly developed markets for both engineering plastics, such as polycarbonates and polysulfones, and specialty polymers, such as customized water-soluble copolymers, used in water purification, secondary oil recovery, personal care, etc. They have higher performance specifications than commodity plastics and are under more strict regulations.

Intertwined throughout the base, middle, and peak of the pyramid are the many natural products used in industries such as food, flavoring and fragrances, pharmaceuticals, biotechnology, personal care, nanostructures, hybrid materials, and others. In keeping with the book's broad definition of polymer reactions, fermentation processes involving natural products, as well as degradation reactions leading to biofuels, are open and exciting areas for online monitoring.

At the tip of the pyramid is the vibrant activity in synthetic polymer chemistry. Creative concepts are used to produce a new generation of twenty-first century polymeric materials, including organic electronic and optical devices, self-healing polymers, drug delivery, device interfaces, nanocomposites, and polymers that react to environmental stimuli. Among these latter "stimuli-responsive polymers" are those that can change conformation, go through phase transitions, micellize, and associate with other substances when conditions such as irradiation, presence of specific agents, and changes in temperature, pH, and ionic strength occur. And, of course, to accompany the ferment of activity at all levels of the pyramid are the evolving techniques and methodologies necessary to characterize polymers of all types.

In the following chapters, the experience and thoughts of many leading scientists and engineers from very different parts of the great interdisciplinary field of macromolecules are brought together around the theme of polymer reaction monitoring.

WAYNE F. REED
ALINA M. ALB

CONTRIBUTORS

Brooks A. Abel, University of Southern Mississippi, Hattiesburg, MS, USA

Alina M. Alb, Tulane University, New Orleans, LA, USA

José M. Asua, University of the Basque Country, Donostia-San Sebastián, Spain

Guy C. Berry, Carnegie Mellon University, Pittsburgh, PA, USA

Mark L. Brader, Biogen Idec Corp, Cambridge, MA, USA

José C. de la Cal, University of the Basque Country, Donostia-San Sebastián, Spain

Michael F. Drenski, Advanced Polymer Monitoring Technologies, Inc., New Orleans, LA, USA; Tulane University, New Orleans, LA, USA

Joel D. Flores, University of Southern Mississippi, Hattiesburg, MS, USA

Daniela Held, PSS Polymer Standards Service GmbH, Mainz, Germany

Matthew Kade, University of Chicago, IL, USA

Peter Kilz, PSS Polymer Standards Service GmbH, Mainz, Germany

Jose R. Leiza, University of the Basque Country, Donostia-San Sebastian, Spain

Marcelo K. Lenzi, Universidade Federal do Paraná, Curitiba, Brasil

Andrew J. D. Magenau, Carnegie-Mellon University, Pittsburgh, PA, USA

Robert T. Mathers, Carnegie-Mellon University, Pittsburgh, PA, USA

Krzysztof Matyjaszewski, Carnegie-Mellon University, Pittsburgh, PA, USA

Charles L. McCormick, University of Southern Mississippi, Hattiesburg, MS, USA

Timothy McKenna, Queens University, Kingston, ON, Canada

Aline Nicolau, Universidade Federal do Rio Grande do Sul, Porto Alegre, Rio Grande do Sul, Brasil

José C. Pinto, Universidade Federal do Rio de Janeiro, Rio de Janeiro, Brasil

Alex W. Reed, Advanced Polymer Monitoring Technologies, Inc., New Orleans, LA, USA; Tulane University, New Orleans, LA, USA

Wayne F. Reed, Tulane University, New Orleans, LA, USA

José A. Romagnoli, Louisiana State University, Baton Rouge, LA, USA

Miriam B. Roza, Universidade Federal do Rio Grande do Sul, Porto Alegre, Rio Grande do Sul, Brasil

Dimitrios Samios, Universidade Federal do Rio Grande do Sul, Porto Alegre, Rio Grande do Sul, Brasil

Alexandre F. Santos, Universidade Tiradentes, Aracaju, Brasil.

F. Joseph Schork, School of Chemical & Biomolecular Engineering, Georgia Institute of Technology, Atlanta, GA, USA

Kristin Schröder, Carnegie-Mellon University, Pittsburgh, PA, USA

Fabricio M. Silva, Universidade de Brasília, Brasilia, Brasil

DeeDee Smith, University of Southern Mississippi, Hattiesburg, MS, USA

Jorge Soto, Lion Copolymer LLC, Baton Rouge, LA, USA

Matthew Tirrell, University of Chicago, Chicago, IL, USA

Radmila Tomovska, University of the Basque Country, Donostia-San Sebastián, Spain

Wesley L. Whipple, Nalco, An Ecolab Company, Naperville, IL, USA

Joseph Zeaiter, American University of Beirut, Beirut, Lebanon

Hua Zheng, Nalco, An Ecolab Company, Naperville, IL, USA

Zifu Zhu, Tulane University, New Orleans, LA, USA

SECTION 1

OVERVIEW OF POLYMERIZATION
REACTIONS AND KINETICS

1

FREE RADICAL AND CONDENSATION POLYMERIZATIONS

MATTHEW KADE AND MATTHEW TIRRELL

1.1 INTRODUCTION

Polymers are macromolecules composed of many monomeric repeat units and they can be synthetic or naturally occurring. While nature has long utilized polymers (DNA, proteins, starch, etc.) as part of life's machinery, the history of synthetic polymers is barely 100 years old. In this sense, man-made macromolecules have made incredible progress in the past century. While synthetic polymers still lag behind natural polymers in many areas of performance, they excel in many others; it is the unique properties shared by synthetic and natural macromolecules alike that have driven the explosion of polymer use in human civilization. It was Herman Staudinger who first reported that polymers were in fact many monomeric units connected by covalent bonds. Only later we learned that the various noncovalent interactions (i.e., entanglements, attractive or repulsive forces, multivalency) between these large molecules are what give them the outstanding physical properties that have led to their emergence.

In recent years, the uses of synthetic polymers have expanded from making simple objects to much more complex applications such as targeted drug delivery systems and flexible solar cells. In any case, the application for the polymer is driven by its physical and chemical properties, notably bulk properties such as tensile strength, elasticity, and clarity. The structure of the monomer largely determines the chemical properties of the polymer, as well as other important measurable quantities, such as the glass transition temperature, crystallinity, and solubility. While some impor-

tant determinants of properties, such as crystallinity, can be affected by polymer processing, it is the polymerization itself that determines other critical variables such as the molecular weight, polydispersity, chain topology, and tacticity. The importance of these variables cannot be overstated. For example, a low-molecular-weight stereo-irregular polypropylene will behave nothing like a high-molecular-weight stereo-regular version of the same polymer. Thus, it is easy to see the critical importance the polymerization has in determining the properties and therefore the potential applications of synthetic polymers. It is therefore essential to understand the polymerization mechanisms, the balance between thermodynamics and kinetics, and the effect that exogenous factors (i.e., temperature, solvent, and pressure) can have on both.

1.1.1 Structural Features of Polymer Backbone

1.1.1.1 *Tacticity* Tacticity is a measure of the stereochemical configuration of adjacent stereocenters along the polymer backbone. It can be an important determinant of polymer properties because long-range microscopic order (i.e., crystallinity) is difficult to attain if there is short-range molecular disorder. Changes in tacticity can affect the melting point, degree of crystallinity, mechanical properties, and solubility of a given polymer. Tacticity is particularly important for a, a'-substituted ethylene monomers (e.g., propylene, styrene, methyl methacrylate). For a polymer to have tacticity, it is a requirement that a does not equal a' because otherwise the carbon in question would not be a stereocenter. The tacticity is determined

Monitoring Polymerization Reactions: From Fundamentals to Applications, First Edition. Edited by Wayne F. Reed and Alina M. Alb.
© 2014 John Wiley & Sons, Inc. Published 2014 by John Wiley & Sons, Inc.

during the polymerization and is unaffected by the bond rotations that occur for chains in solution. The simplest way to visually represent tacticity is to use a Natta projection, as shown in Figures 1.1–1.3 using poly (propylene) as a representative example.

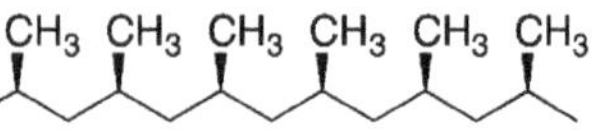

FIGURE 1.1 Isotactic polypropylene.

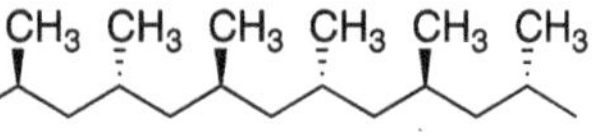

FIGURE 1.2 Syndiotactic polypropylene.

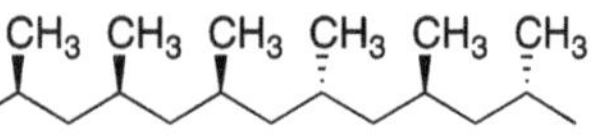

FIGURE 1.3 Atactic polypropylene.

An isotactic chain is one in which all of the substituents lie in the same plane (i.e., they have the same stereochemistry). Isotactic polymers are typically semicrystalline and often adopt a helical configuration. Polypropylene made by Ziegler–Natta catalysis is an isotactic polymer.

A syndiotactic chain is the one where the stereochemical configuration between adjacent stereocenters alternates.

An atactic chain lacks any stereochemical order along the chain, which leads to completely amorphous polymers.

1.1.1.2 *Composition* Copolymer composition influences a number of quantities, including the glass transition temperature. One commercially relevant example of this effect is with Eastman's copolymer Tritan™, which has been replacing polycarbonate in a number of applications due to concerns over bisphenol-A's (BPA's) health effects. Tritan™ can be considered poly(ethylene terephthalate) (PET), where a percentage of the ethylene glycol is replaced by 2,2,4,4-tetramethyl-1,3-cyclobutane diol (TMCBDO). In the case of beverage containers, T_g must be greater than $100\,°C$ so they can be safely cleaned in a dishwasher or autoclave. The T_g of Tritan is engineered to be ~110 °C by tuning the relative incorporation of the ethylene glycol (low T_g) and TMCBDO (T_g-increasing) diol monomers.

Altering the glass transition temperature is by no means the only reason to include comonomers in a polymerization. In designing copolymers with specialized applications, comonomers can be included for specific functions, or as sites for further functionalization or initiation of a secondary polymerization (e.g., to make graft copolymers in a graft-from approach). In more broadly used commercial polymers, comonomers can be included to alter different properties, including swelling in particular solvents, stability, viscosity, or to induce self-assembly (e.g., styrene-butadiene-styrene rubbers where styrene domains within the butadiene matrix provide mechanical integrity). While block copolymers produced in sequential polymerizations are not confronted with the problem of unequal reactivity, monomers often have different reactivities within a polymerization. Such discrepancies lead to differences between the composition of monomer feed and the composition of the final polymer.

1.1.1.3 *Sequence* The difference in reactivity between comonomers affects the composition and also alters the placement of the monomer units along the chain. In the case of living polymerization, sequential monomer addition leads to the formation of block copolymers. However, when a random copolymer is targeted, reactivity differences can lead to nonrandom distribution of monomer units. If the incorporation of a comonomer B is intended to disrupt crystallinity of poly(A), uninterrupted sequences of A can lead to domains of crystallinity. For example, block copolymers of ethylene–propylene are highly crystalline, while random copolymers are completely amorphous.

1.1.1.4 *Regioselectivity* The issue of regioselectivity is most relevant here to vinyl monomers undergoing free radical polymerization, but also applies to other polymerization mechanisms discussed (particularly the synthesis of conducting polymers, which often entails the use of monomers bearing alkyl chains designed to improve solubility). The example of 1-substituted ethylene derivatives (e.g., styrene) is shown in Scheme 1.1. When a propagating chain adds a monomer unit, the radical can add to either C^1 or C^2. If each successive addition occurs in the same fashion, the result is an isoregic chain, typically referred to as a head-to-tail arrangement.

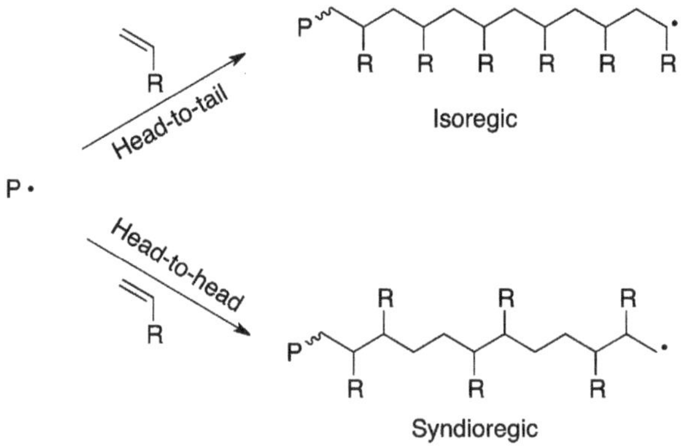

SCHEME 1.1 Regioselectivity in free radical polymerization.

The alternate configuration is achieved when each successive monomer addition alternates between C^1 and C^2 additions, giving a syndioregic chain, commonly called a head-to-head arrangement. For free radical polymerizations, isoregic addition is overwhelmingly favored. This is due jointly to resonance and/or inductive stabilization of the resulting radical, which favors head-to-tail addition, and steric constriction around the R group, which discourages head-to-head addition.

1.1.2 The Chain Length Distribution

It is evident that the molecular weight of a polymer chain determines important properties such as viscosity and mechanical strength. Because synthetic polymers do not have a single chain length i and are instead polydisperse, any measure of molecular weight is an average. The chain length distribution is typically characterized by the first three moments of the distribution, where the kth moment is described as follows:

$$\mu^{(k)} = \sum_{P=1}^{\infty} P^k \cdot \chi_P \qquad (1.1)$$

where P is the length of an individual polymer chain and χ_P is the number of chains of length P.

The weighted degrees of polymerization are defined as the ratio of successive moments, as seen in Equations 1.2 through 1.4:

$$dp_n = \overline{P_n} = \frac{\mu^1}{\mu^0} = \frac{\sum_{P=1}^{\infty} P \cdot \chi_P}{\sum_{P=1}^{\infty} \chi_P} \qquad (1.2)$$

$$dp_w = \overline{P_w} = \frac{\mu^2}{\mu^1} = \frac{\sum_{P=1}^{\infty} P^2 \cdot \chi_P}{\sum_{P=1}^{\infty} P \cdot \chi_P} \qquad (1.3)$$

$$dp_z = \overline{P_z} = \frac{\mu^3}{\mu^2} = \frac{\sum_{P=1}^{\infty} P^3 \cdot \chi_P}{\sum_{P=1}^{\infty} P^2 \cdot \chi_P} \qquad (1.4)$$

The number-average degree of polymerization is the number of polymerized units divided by the number of polymer chains, obtained by end-group analysis (e.g., NMR). The weight-average degree of polymerization determines most important properties of a polymer:

$$\text{PDI} = \frac{dp_w}{dp_n} = \frac{\mu^{(2)} \mu^{(0)}}{\mu^{(1)} \mu^{(1)}} \qquad (1.5)$$

The polydispersity of a polymer sample is described by the polydispersity index (PDI), which is a ratio of the weight-average and number-average degrees of polymerization. For monodisperse polymers, such as a proteins, the PDI will equal 1, while synthetic polymers have PDIs that can approach 1, or conversely go to values higher than 10.

1.1.3 Polymerization Mechanisms

It is useful in the classification of polymerizations to define several mechanisms of polymer growth, each one with distinctive and defining features. In the context of this chapter, three mechanisms are considered: step growth, chain growth, and "living" polymerization. Carothers initially classified polymers into condensation and addition, and while these terms are often used interchangeably with step and chain polymerizations, it must be stressed that this is not entirely accurate.

Step polymerization indicates a mechanism of growth where monomers combine with each other to form dimers, the dimers combine with each other or other monomer units to form tetramers or trimers, respectively, the process continuing until polymer is formed. While each coupling step in a step polymerization is often accompanied by the elimination of a small molecule (e.g., water), making it a condensation polymerization, this is not always the case (e.g., isocyanates and alcohols reacting to make polyurethanes). Furthermore, not all polymerizations in which a condensate is formed follow a stepwise mechanism.

Step polymerization leads to high-molecular-weight polymer when monomer conversion is very high (see Table 1.1). In comparison, the chain growth mechanism immediately leads to high-molecular-weight polymer regardless of monomer conversion. In this case, there is an active chain end, which adds monomer units one by one until the chain is rendered inactive by termination or transfer. In a normal chain growth process, a chain lifetime is short compared to the polymerization process, new chains being constantly initiated and terminated.

Living polymerization is a chainwise mechanism where transfer and termination reactions have been eliminated. Therefore, all polymer chains are active throughout the entire polymerization and grow at similar rates. A major consequence of living polymerization is that PDIs are much lower (≤ 1.1) than for the usual chainwise mechanism. Table 1.1 highlights some salient features of each mechanism.

1.2 FREE RADICAL POLYMERIZATION

Free radical polymerization is a globally important method for the production of polymers, both academic and industrial. In fact, free radical polymerization is used to produce a significant percentage of the polymers made worldwide, including 45% of manufactured plastics and 40% of synthetic rubber, which amounts to 100 and 4.6 million tons, respectively.

Despite its widespread use, gaining a full understanding of the polymerization process is not a straightforward task. Free radical polymerization is controlled by a number of

TABLE 1.1 Distinctions between Stepwise, Chainwise and Living Polymerization

Characteristic	Stepwise	Chainwise	"Living"
Number and type of reactions	Only one: reaction between two (usually dissimilar) functional groups	Initiation Propagation Termination Also: Transfer Inhibition	Initiation Propagation
Convention as to what is considered polymer	All species considered to be polymer	Unreacted monomer is distinct from polymer	Unreacted monomer is distinct from polymer
Polymer concentration with conversion p			
Degree of polymerization with conversion p			

different processes, each of which has its own kinetics and thermodynamics. If each of these individual processes is fully understood and its rate coefficient determined, the kinetics of the overall polymerization can be determined and the full molecular weight distribution can be accurately predicted. A major complicating factor is that all these processes are closely related to each other, making it challenging to separate and determine kinetic rate coefficients. However, much effort has been devoted to study the processes that constitute a free radical polymerization and modern experimental techniques have improved their understanding.

This section will focus on the kinetics of free radical polymerization, but will also address the effect of kinetics on molecular weight distributions of commonly used monomers. This chapter will not address controlled free radical polymerization (CRP) since it is covered at length in Chapter 2.

1.2.1 Initiation

Initiation is the process by which radicals are formed and then subsequently initiate polymerization by reacting with a monomer molecule. The prerequisite step is for an initiator molecule to decompose into a radical species. While not always the case, the most common scenario is for an initiator molecule to decompose into two radical species:

$$\text{Initiator} \xrightarrow[hv, \Delta]{k_d} I_1^\bullet + I_2^\bullet \qquad (1.6)$$

Initiator decomposition can be triggered in a variety of ways. The most common method for industrial free radical polymerization is thermal initiation (typically using azo or peroxy initiating species), while photoinitiation is more popular for laboratory scale kinetic studies. In either case, Equation 1.6 describes the decomposition of initiator into two radical species, which may or may not have equal reactivities, depending on the choice of initiator [1, 2]. The concentration of initiator can then be calculated by:

$$-\frac{d[I]}{dt} = k_d[I] \qquad (1.7)$$

An important consideration is that initiator decomposition is not equivalent to chain initiation because of the various side reactions that can take place before reaction of the radical species with a monomer unit. To achieve initiation of a growing polymer chain, the radical species must escape the solvent cage [3] before undergoing deleterious side reactions that reduce chain initiation efficiency. The quantity f represents the fraction of produced radicals that can initiate polymerization, typically between 0.5 and 0.8 for most free radical polymerization initiators. Odian has demonstrated, using benzoyl peroxide as initiator, that initiating radicals can undergo side reactions which decrease the initiator efficiency, f, before escaping the solvent cage [4].

The first-order rate law, R_d, for the production of radicals that can initiate polymerization is conveyed by Equation 1.8. The f term accounts for all of the various inefficiencies in initiating polymerization:

$$R_d = \frac{d[I^\bullet]}{dt} = -2f\frac{d[I]}{dt} = 2fk_d[I] \qquad (1.8)$$

This leads directly to the concentration of initiator molecules as a function of time:

$$[I] = [I]_0 e^{-k_d t} \qquad (1.9)$$

Another complication is that for many initiators, decomposition leads to two radicals of different structures and reactivities [1, 2]. The difference in reactivities between the radicals produced in an unsymmetrical decomposition is addressed in the following equations:

$$I^\bullet_{(1)} + M \xrightarrow{k_i^{(1)}} R^\bullet_1 \qquad (1.10)$$

$$I^\bullet_{(2)} + M \xrightarrow{k_i^{(2)}} R^\bullet_1 \qquad (1.11)$$

This means that in the case of unsymmetrical decomposition of initiator into two radicals with differing reactivity, the expression for the overall rate of initiator is actually a composite of two different reactions (Eq. 1.12). However, for the sake of simplicity, the two different initiation rate coefficients will be combined into an average rate constant to give the overall rate of initiation, R_i:

$$R_i = \frac{d[R^\bullet]}{dt} = -\frac{d[I^\bullet_1]}{dt} - \frac{d[I^\bullet_2]}{dt} \qquad (1.12)$$

$$R_i = k_i^{(1)}[M]\left[I^\bullet_{(1)}\right] + k_i^{(2)}[M]\left[I^\bullet_{(2)}\right] \qquad (1.13)$$

$$R_i = k_i[M][I^\bullet], \quad \text{where } k_i = \frac{k_i^{(1)} + k_i^{(2)}}{2} \qquad (1.14)$$

1.2.1.1 Thermal Initiation Thermal initiators are very common and typically decay following a first-order rate law, as shown in Equation 1.9. Most common thermal initiators are peroxides or diazo compounds, such as azobisisobutyronitrile (AIBN) [5]. Initiators are chosen so that at polymerization temperature, decomposition is slow with typical values for k_d ranging from 10^{-6} to 10^{-4} s^{-1}. Commonly, the rate at which a thermal initiator decomposes is reported as the temperature at which the half-life (Eq. 1.15) is equal to 10 h:

$$t_{1/2} = \frac{\ln 2}{k_d} \qquad (1.15)$$

Table 1.2 shows temperature for a 10 h half-life for several common thermal initiators. Their slow decomposition allows initiators concentration to be considered constant over the

TABLE 1.2 Decomposition Rate and 10 h $t_{1/2}$ for Common Thermal Initiators

Initiator	Solvent	10 h Half-life °C
4,4-Azobis(4-cyanovaleric acid)	Water	69
2,2′-Azobisisobutyronitrile (AIBN)	Toluene	65
tert-Amyl peroxybenzoate	Benzene	99
Benzoyl Peroxide	Benzene	70
tert-Butyl peracetate	Benzene	100
tert-Butyl peroxide	Benzene	125
Dicumyl peroxide	Benzene	115
Peracetic acid	Toluene	135
Potassium persulfate	Water	60

course of polymerization, particularly when compared to the average lifetime of an active chain.

1.2.1.2 Photoinitiation Photoinitiation [6] takes advantage of initiators that can form radical species upon UV irradiation. Unlike thermal initiation, which produces a relatively small supply of radicals throughout the course of a polymerization, photoinitiation can provide a burst of radicals when desired. This makes photoinitiation an ideal candidate for kinetic experiments or surface-initiated polymerization because the production of radicals is limited to the area that is irradiated at the time of irradiation. Furthermore, the concentration of radicals, ρ, produced by a given number of photons can be easily calculated as follows:

$$\rho = 2\Phi\frac{n_{abs}}{V} \qquad (1.16)$$

where Φ is the primary quantum yield, n_{abs} is the number of absorbed photons, and V is the irradiated volume.

Rearrangement of Beer's law and combination with Equation 1.16 gives a final expression for the concentration of radicals produced by an irradiation event:

$$\rho = 2\Phi\frac{(E_{tot}/E_\lambda)\cdot(1-10^{-\varepsilon bc})}{V} \qquad (1.17)$$

1.2.1.3 Self-Initiation A free radical polymerization can be started by self-initiation of the monomer species. In fact, true self-initiation is very rare and some of the cases reported in the literature are actually due to oxygen producing peroxide species that can act as initiators, or other impurities that lead to radical formation [7].

One monomer that is known to self-initiate, even at high purity is styrene [8–10]. As shown in Scheme 1.2, styrene undergoes a Diels–Alder reaction to give a styrene dimer. This dimer can then react with another styrene monomer to give a styrene radical or $R^\bullet_1$. Significantly, the activation energy for the self-initiation is rather large.

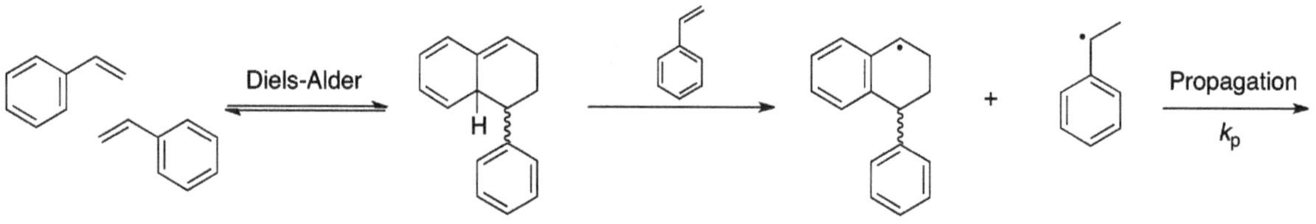

SCHEME 1.2   Initiation mechanism in the auto-polymerization of styrene. With permission from Odian G. *Principles of Polymerization.* 4th ed. © 2004 Hoboken (NJ): John Wiley & Sons, Inc.

The half-life for 50% monomer conversion is only 4 h at 127 °C, but it is 400 days at 29 °C.

1.2.2   Propagation

Propagation is the step most closely associated with the actual polymerization reaction as it is the addition of a monomer unit to the propagating macroradical. Writing a rate law for the propagation reaction is somewhat complicated by the fact that the rate of propagation is chain length-independent [11–16]. For example, for the polymerization of methyl methacrylate at 60 °C, the first propagation step is 16 times faster than the long chain propagation reaction.

This can be accounted for by a simple summation of the propagation for each chain length *i*:

$$-\frac{dM}{dt} = \sum_i k_p^i [R_i][M] \tag{1.18}$$

where k_p^i is the propagation rate constant for a macroradical with chain length *i* and $[R_i]$ is the concentration of polymers with chain length *i*.

It is important to review the stringent requirements that lead to successful propagation. For a typical free radical polymerization, a successful propagation reaction can be expected to occur with frequency of $10^3 s^{-1}$, while the collision frequency in a liquid near room temperature is much higher: $10^{12} s^{-1}$. Given the high monomer concentration in a polymerization, this effectively means that only one in every 10^9 collision events leads to a successful propagation step [17] (i.e., addition of one monomer molecule to the growing macroradical). These values highlight the fine balance between the reactivity and stability of the propagating macroradical. The radical must be reactive enough to produce a polymer in a matter of seconds but also must be stable enough to survive the 10^9 nonproductive collisions that occur for every successful propagation reaction.

Furthermore, there is a fine balance between the reactivity of the monomer and the stability of the macroradical, quantities typically inversely related. For example, styrene is a very reactive monomer but produces a more stable (i.e., less reactive) chain end in the form of a resonance-stabilized

secondary benzyl radical. The other extreme would be ethane, which is a very nonreactive monomer that leads to an extremely reactive primary radical chain end.

The kinetic rate coefficient for propagation, k_p, is chain length and monomer concentration dependent. Solvent choice normally does not have a significant effect on k_p [18–20], although this is not the case when ionic liquids are used as solvents [21–23]. However, the dependence of k_p on monomer concentration is not nearly as significant as dependence on pressure. Free radical polymerizations have large negative activation volumes (Eq. 1.14), meaning that at higher pressures the rate of propagation increases [24–26]:

$$\frac{d \ln k}{dP} = -\frac{\Delta V}{RT} \tag{1.19}$$

1.2.3   Transfer

Transfer reactions involve the transfer of the radical from a growing polymer chain to another molecule, T, typically by the donation of a hydrogen atom to the macroradical, $R_i^{\bullet}$, to produce an inactive polymer chain, P_i, and another radical T:

$$-\frac{dT}{dt} = k_{tr}[R^{\bullet}][T] \tag{1.20}$$

Each molecule involved in radical transfer reactions is characterized by a transfer constant, *C*, which is a ratio of the rate constant for transfer and the rate constant for propagation:

$$C = \frac{k_{tr}}{k_p} \tag{1.21}$$

Both monomer and solvent can act as transfer agents; often chain transfer agents (CTAs) are intentionally added to polymerization reactions. While such a transfer reaction renders the propagating chain inactive and thus affects the molecular weight of the chain, it does not affect the kinetic chain, which is a measure of how long a given radical persists. Thus, in most cases, transfer reactions do not affect the rate of polymerization but do alter the molecular weight distribution.

There are a number of different possible cases for transfer reactions, the relative rates of propagation, k_p, transfer, k_{tr}, and reinitiation, k_{re-in}, determining the effects of the transfer reactions on the overall rate of polymerization, as well as on the molecular weight distribution [17]. The first case is when k_p is much greater than k_{tr} and k_{re-in} is much greater than k_{tr}, which is considered normal chain transfer. In this scenario, because there is a relatively low amount of chain transfer and the small molecule radical formed quickly reinitiates polymerization, normal chain transfer does not affect the overall rate of polymerization, R_p, but leads to a decrease in the molecular weight. The next case is where k_p is much smaller than k_{tr}, but comparable to k_{re-in}. This type of transfer leads to a high percentage of active radicals existing on the transfer agent, T, but again does not decrease the overall rate of polymerization. It does drastically decrease the molecular weight of the resulting polymers, leading to telomerization, or the production of mostly dimers and trimers. The third case of chain transfer is when propagation is much faster than transfer ($k_p >> k_{tr}$), but reinitiation is slow relative to propagation ($k_{re-in} < k_p$). Here, both the rate of polymerization and the molecular weight decrease, but not enough that the polymerization would be completely stopped; this is called retardation [27, 28]. Finally, there is the case of inhibition [27, 28], which occurs when the rate of transfer is much higher than propagation ($k_{tr} >> k_p$) and reinitiation is slower than propagation ($k_{re-in} < k_p$). Inhibition occurs when the transfer agent efficiently traps radicals and the resultant transfer radical is very stable. Examples of radical inhibitors include BHT, nitrobenzene, and diphenyl picryl hydrazyl (DPPH), which are useful for preventing autopolymerization of vinyl monomers stored over long periods of time.

Radical transfer could greatly complicate the kinetics of polymerization, particularly because a wide variety of molecules can act as transfer agents, including but not limited to monomer, solvent, initiator, polymer [29], and added CTAs. Even molecular oxygen can be a radical transfer agent [30], which, if present in significant amounts, acts as an inhibitor in most free radical polymerizations. While the possibilities for transfer seem endless, careful planning of the reaction conditions can control most transfer reactions. For example, a decrease in temperature will generally lower the transfer constant C for all species. Furthermore, a judicious choice of initiator or simply a decrease in initiator concentration can significantly reduce transfer to the initiating species. The only species to which transfer cannot be avoided is the monomer, which in fact is often a limiting factor for the molecular weight. Table 1.3 lists the values for the monomer transfer constant, C_M, for various common monomers. Another important transfer reaction is to the solvent, which can be problematic because of the high solvent concentrations used in industrial polymerizations (Table 1.2).

TABLE 1.3   Transfer Constants to Monomers, $C_M \times 10^4$

Monomers	$T\,(°C)$	$C_M \times 10^4$
Methyl methacrylate	0	0.128
	60	0.18
	120	0.58
Acrylonitrile	60	0.26
Styrene	0	0.108
	60	0.75
	117	1.40
Methyl acrylate	60	0.036
Ethylene	60	0.40
Methacrylamide	60	10×10^5
Vinyl acetate	60	1.75

Despite the tendency for radical transfer reactions to slow polymerization kinetics, decrease or limit molecular weight, and complicate the kinetic picture of a given polymerization, the transfer process can also be very useful for the process engineer. For example, a simple way to achieve lower molecular weight polymers is to increase the initiator concentration. As consequence, the rate of polymerization would increase, which could, on the other hand, lead to the loss of control and exothermicity. The addition of a CTA can regulate molecular weight without affecting the rate of polymerization, avoiding the associated problems. Furthermore, if CTAs chosen have high chain transfer constants, they can be used in relatively low concentrations.

1.2.4   Termination

Termination is probably the most complex step in the free radical process, owing to the fact that k_t depends on monomer conversion, pressure, temperature, system viscosity, and the chain length of the terminating macroradicals [31, 32]. The complexity of termination is manifested in the widely spread k_t values found in the literature for any given system [33, 34]:

$$-\frac{d[R^\bullet]}{dt} = \sum_i \sum_i 2k_t^{i,j}[R_i^\bullet][R_j^\bullet] \qquad (1.22)$$

There are different modes of termination: combination and disproportionation. Active chains terminated by disproportionation will have the same molecular weight, where one of the chains will have an unsaturation and the other will be fully saturated. When chains are terminated by combination, because two propagating chains combine, the number of chains decreases by one, and the resultant molecular weight is the sum of the two macroradicals, thereby increasing the final molecular weight distribution.

The relative contribution of each mode of termination is described by δ in the following equation:

$$\delta = \frac{k_{t,d}}{k_{t,d} + k_{t,c}} \tag{1.23}$$

Disproportionation is generally favored slightly over combination at increased temperatures, but other factors such as monomer choice can have a greater impact on δ.

Looking at the rate and activation energy for termination in comparison with the other steps in a polymerization, it might seem surprising that polymers can be produced at all. The rate constant for termination is always very high and the activation energy for the chemical reaction can be considered 0 [35]. Indeed, the reason that termination is not the dominating reaction in a given polymerization is because two propagating macroradicals (i.e., polymer chain ends) must first find each other before they can react. To better understand chain termination, the process can be broken into three stages [36–38]:

1. Translational diffusion of the macroradical coils toward each other within the reaction medium.
2. Segmental diffusion of the chains ends toward each other, putting them in a position to react.
3. The chemical reaction between the two radicals that leads to termination.

As it is always the case, the slowest process will be the rate-determining step. Because the chemical reaction rate is very high (on the order of $10^{10}\,1\,mol^{-1}\,s^{-1}$), the rate-determining step will always be either translational (i.e., center-of-mass) diffusion or segmental diffusion [39]. At low conversion, segmental diffusion is the rate-limiting step, while at high conversion, center-of-mass diffusion controls the rate of termination. This phenomenon occurs because at high conversion the polymer chains become entangled and translational diffusion becomes difficult. Polymer chains must undergo translational diffusion by reptation, significantly slowing this mode of diffusion. At very high conversion (>80%), diffusion can actually be controlled by reaction of monomer [40] (i.e., the position of the chain end moves by addition of a monomer unit). However, the case of reaction-controlled diffusion will not be treated in great detail here.

Because both rate-controlling termination processes are diffusion controlled, it should follow that both processes will be chain length dependent. However, segmental diffusion and translational diffusion show very different dependencies on molecular weight. A facile way to envision this is to consider a macroscopic termination rate constant, k_t, which is a weighted summation of the microscopic termination reactions. The molecular weight dependence of this macroscopic rate constant is described in Equation 1.24. The value for α is empirically known for both translation diffusion and segmental diffusion:

$$\langle k_t \rangle = k_t^0 \cdot \bar{P}^{-\alpha} \tag{1.24}$$

For translational diffusion, α is between 0.5 and 0.6, depending on the solvent quality, while segmental diffusion shows much less of a molecular dependence, with $\alpha \sim 0.16$ [41–46].

While the chain length dependence of termination was discussed earlier, the reality is that termination is much more strongly dependent on pressure [47, 48] than on chain length. The large negative activation volumes typical for termination describe this effect. Because increased pressure not only decreases the rate of termination but also increases k_p, pressure can lead to a marked increase in the final molecular weight.

1.2.5 Rate of Polymerization

The overall rate of polymerization is determined by the contributions of the various processes discussed in the aforementioned sections: initiation, propagation, transfer, and termination. It is instructive to separate a polymerization into different regimes and to understand their kinetics. Thus, at the beginning of the polymerization, when the concentration of radicals is increasing (this phase lasts only a few seconds [49]), a stationary phase is observed, where the concentration of radicals can be considered constant; dead-end polymerization [50, 51] occurs if the initiator is completely consumed before monomer conversion is complete. The latter scenario can be easily avoided by carefully choosing the concentration and type of initiator (half-life time, $t_{1/2}$), so that the polymerization can be completed before the initiator is consumed.

1.2.5.1 *Stationary Polymerization* The most classic kinetic treatment for the rate of polymerization is the quasi steady-state polymerization, which assumes a constant free radical polymerization throughout the course of the polymerization [52]:

$$\frac{d[R^\bullet]}{dt} = 0 \tag{1.25}$$

A number of assumptions are made to derive the overall rate of polymerization, R_p, in a straightforward way. These assumptions are as follows:

1. The concentration of initiator-derived radicals remains constant throughout the polymerization.
2. Instantaneous establishment of a steady-state free radical concentration.
3. Chain length and conversion-independent rate coefficients, k_t and k_p.

4. Monomer is only consumed by chain propagation (which allows the loss of monomer to be directly associated with R_p).
5. All reactions are irreversible.

The central tenet of the steady-state (or stationary) polymerization is that the concentration of radicals is constant. It closely follows that the rate of formation of radicals must equal the rate of radical termination.

Combining Equations 1.8 and 1.22 gives the following:

$$2fk_d[I] = 2k_t[R_i^\bullet][R_j^\bullet] \tag{1.26}$$

$$2fk_d[I] = 2k_t[R^\bullet]^2 \tag{1.27}$$

The right half of the equation can be simplified using assumption 3 to give $[R^\bullet]^2$ instead, because there is no need to distinguish between different chain lengths of the macroradicals. Furthermore, when Equation 1.17, which describes the disappearance of monomer, is simplified by assumption 3, it can be directly correlated with the rate of polymerization:

$$R_p = -\frac{d[M]}{dt} = k_p[M][R^\bullet] \tag{1.28}$$

By solving for $[R^\bullet]^2$ in Equation 1.27, and substituting into Equation 1.28, an expression for the rate of polymerization is obtained. Integration of Equation 1.28 with respect to time and combination of the various rate constants into a single empirical rate constant, k_{obs}, give an expression for the rate of polymerization, in terms of monomer conversion, p:

$$k_{obs}t = \ln\left(\frac{1}{1-p}\right), \quad \text{where } k_{obs} = k_p\left(f\frac{k_d}{k_t}[I]\right)^{0.5} \tag{1.29}$$

1.2.6 The Chain Length Distribution

The chain length distribution for a given monomer determines numerous properties of the resulting polymer; therefore, understanding how different polymerization parameters affect the distribution is of paramount importance. Here, the focus is on calculating the chain length distribution rather than the molecular weight distribution, even though molecular weights are reported often.

The chain length distribution can easily be converted to a molecular weight distribution considering that a chain of length i has a molecular weight of i times the mass of the repeat unit plus the mass of the two end-groups. In the case of unknown end-groups (e.g., polymers initiated by benzoyl peroxide, which can initiate through a number of different radical species), it may be difficult to calculate the exact mass of the polymer chain. Fortunately, the mass of the end-groups becomes insignificant for longer polymers.

Typically, the chain length distribution is characterized by the moments of the distribution. It is also possible to gain an understanding of the distribution by focusing on the microscopic distribution. By knowing the concentration of every macroradical species, one can build a picture of the entire distribution.

For example, Equation 1.30 shows the solution for the rate of change in concentration of macroradicals with chain length i; that is, the production by addition of one monomer unit from macroradicals of length $i - 1$, subtracted by the combined loss through transfer and termination reactions, or the addition of another monomer unit to make a macroradical of chain length $i+1$. However, solving this set of differential equations becomes increasingly complex mathematically:

$$\frac{d[R_i^\bullet]}{dt} = k_p^{i-1}[M][R_{i-1}^\bullet]$$
$$- \left(k_p^i[M] + k_{tr}^M[M] + k_{tr}^T[T] + 2\sum_{j=1}^{\infty} k_t^{i,j}[R_j^\bullet]\right)[R_i^\bullet] \tag{1.30}$$

An alternate starting point involves the use of the kinetic chain length, defined as the total number of monomer units added divided the total number of initiation steps:

$$\text{Kinetic chain length } \nu = \frac{\text{total number of polymerized units}}{\text{total number of initiation steps}}$$
$$= \frac{\int_0^t (d[M]/dt)\,dt}{\int_0^t (d[I^\bullet]/dt)\,dt} \tag{1.31}$$

The kinetic chain deviates from dp_n because of transfer reactions and termination by combination but remains a good starting place. In the absence of all transfer reactions and for termination occurring exclusively by disproportionation, the kinetic chain length will equal dp_n. In the analogous case (no transfer reactions) where combination is the only termination method, dp_n will equal twice the kinetic chain length. The relation between the kinetic chain length and dp_n when there is no chain transfer is shown in Equation 1.32:

$$dp_n = \left(\frac{2}{1+\delta}\right)\nu \tag{1.32}$$

A more useful simplification is to assume a steady-state polymerization, which means that the radical concentration (and the monomer and initiator concentrations) and the relevant rate constants will remain constant over the course of the polymerization. By adopting a steady-state model, one can substitute the rate of polymerization R_p (Eq. 1.26) and the rate of dissociation R_d (Eq. 1.23) into Equation 1.31 to give an expression for the kinetic chain length [53, 54]:

$$\nu = \frac{R_p}{R_d} = \frac{k_p[R^\bullet][M]}{2fk_d[I]} \qquad (1.33)$$

In the steady-state model, the simplified expression for $[R^\bullet]$ can be substituted to give an expression for the kinetic chain length in terms of only rate constants and concentrations, which can be controlled by the polymerization engineer:

$$[R^\bullet] = \left(\frac{fk_d[I]}{k_t}\right)^{0.5} \qquad (1.34)$$

$$\nu = \frac{R_p}{R_d} = \frac{k_p k_t^{0.5}[M]}{2(k_d f[I])^{0.5}} \qquad (1.35)$$

While Equation 1.35, in combination with Equation 1.32, can give the number-average degree of polymerization, it is important not to ignore the role of the transfer reactions. Even in the case where transfer to initiator and solvent is nonexistent (presumably by careful initiator choice and a solvent-free polymerization), transfer to monomer can never be avoided entirely. Another way to approach the problem is to consider the simplest definition of dp_n; that is, the total number of polymerized monomers units divided by one-half the number of chain ends. Here, it is worth considering the number of chain ends produced by each of the processes [17]. Neither propagation nor termination by combination produce any chain ends ($n=0$), while both initiation and termination by disproportionation produce one chain end ($n=1$), and transfer reactions actually create two chain ends ($n=2$). The steady-state approximation again allows the absolute number of each of these processes to be substituted by the overall rate of each:

$$dp_n = \frac{R_p}{\frac{1}{2}(R_i + R_{t,d} + R_{tr})} \qquad (1.36)$$

Among the distinct processes involved in the polymerization, termination by combination is noticeably absent in Equation 1.36, since combination contributes to neither the total number of polymerized monomer units nor the total number of chain ends in the final molecular weight distribution. Recalling the rate law for each of the processes in Equation 1.36 for a stationary polymerization and subsequently inverting the entire equation leads to a very useful relationship, which can be substantially simplified to give Equation 1.41.

$$R_i = 2fk_d[I] = 2(k_{t,d} + k_{t,c})[R^\bullet]^2 \qquad (1.37)$$

$$R_p = k_p[M][R^\bullet] \qquad (1.38)$$

$$R_{t,d} = 2k_{t,d}[R^\bullet]^2 \qquad (1.39)$$

$$R_{tr} = k_{tr}^M[M] + \sum_b k_{tr}^{T_b}[T_b][R^\bullet] \qquad (1.40)$$

$$\frac{1}{dp_n} = \frac{2k_{t,d} + k_{t,c}}{k_p^2[M]^2}R_p + \frac{k_{tr}^M}{k_p} + \sum_b \frac{k_{tr}^{T_b}}{k_p}\cdot\frac{[T_b]}{[M]} \qquad (1.41)$$

The summation of the last term in Equation 1.41 accounts for transfer to b different types of species, which typically include solvents, initiators, polymer chains, and any added CTA. Transfer to the monomer is separate from the summation because it cannot be avoided and, thus, must always be considered.

It is normal practice to provide a chain transfer constant (Eq. 1.21) for each of the different types of species that can accommodate transfer reactions:

$$C_M = \frac{k_{tr}^M}{k_p}; C_s = \frac{k_{tr}^S}{k_p}; C_I = \frac{k_{tr}^I}{k_p}; C_P = \frac{k_{tr}^P}{k_p}; C_T = \frac{k_{tr}^T}{k_p} \qquad (1.42)$$

If each of these transfer reaction replaces the summation in Equation 1.41, the following relationship to the inverse of the number-average degree of polymerization is obtained:

$$\frac{1}{dp_n} = \frac{2k_{t,d} + k_{t,c}}{k_p^2[M]^2}R_p + C_M + C_s\frac{[S]}{[M]} + C_I\frac{[I]}{[M]}$$
$$+ C_P\frac{[P]}{[M]} + C_T\frac{[T]}{[M]} \qquad (1.43)$$

If one considers an idealized case, where there is no transfer to the solvent (solvent-free polymerization), initiator, or polymer (e.g., in a low conversion regime), and there is no added transfer agent, Equation 1.41 can be further simplified:

$$\frac{1}{dp_n} = \frac{(1+\delta)k_t}{k_p^2[M]^2}R_p + C_M \qquad (1.44)$$

Equation 1.44 gives an important relationship between molecular weight and the transfer reaction to monomer. Even in the extreme case where termination becomes completely nonexistent (Eq. 1.45), the maximum attainable molecular weight is still limited by the transfer reaction to the monomer:

$$\lim_{k_t \to 0} \frac{1}{dp_n} = C_M \therefore dp_n^{max} = C_M^{-1} \quad (1.45)$$

For example, consider the polymerization of styrene performed at 100 °C. The transfer constant for styrene at this temperature is 2×10^{-4}; therefore, the maximum attainable degree of polymerization is 5000 even in the complete absence of any termination reactions. The same polymerization performed at 0 °C, at which the C_{sty} has a value of 1×10^{-5}, can lead to a degree of polymerization as high as 100,000.

The previous analysis allows determination of dp_n using the kinetic parameters in a steady-state polymerization; however, a complete characterization of the molecular weight distribution requires the first three moments of the chain length distribution (to provide M_n, M_w, and PDI). A statistical approach to analyze the inactive polymer chains can be used to calculate these quantities.

A generic polymer chain of length i is produced through $i - 1$ propagation reactions, after which the chain becomes inactive by termination (by disproportionation) or transfer. One can start by defining the probability of propagation, q, as shown in Equation 1.46:

$$q = \frac{R_p}{R_p + R_{tr} + R_t} \quad (1.46)$$

Next, the probability (or mole fraction χ) of forming a polymer chain of any given length can be derived. One simply calculates the probability of $i - 1$ propagation reactions, multiplied by the probability of any reaction that is not propagation:

$$\chi_{i,disp} = q^{i-1}(1-q) \quad (1.47)$$

Recalling the expressions for each of the moments of the chain length distribution (Eqs. 1.2–1.4), and substituting for χ from Equation 1.47, a series of easily solvable summations for each of these quantities is obtained:

$$\mu^{(0)} = \sum_{i=1}^{\infty} \chi_{i,disp} = \sum_{i=1}^{\infty} q^{i-1}(1-q) = -\frac{1-q}{q-1} = 1 \quad (1.48)$$

$$\mu^{(1)} = \sum_{i=1}^{\infty} i \cdot \chi_{i,disp} = \sum_{i=1}^{\infty} i \cdot q^{i-1}(1-q) = -\frac{1}{q-1} = (1-q)^{-1} \quad (1.49)$$

$$\mu^{(2)} = \sum_{i=1}^{\infty} i^2 \cdot \chi_{i,disp} = \sum_{i=1}^{\infty} i^2 \cdot q^{i-1}(1-q)$$
$$= \frac{q^2+q}{(q-1)^2 q} = (1+q)(1-q)^2 \quad (1.50)$$

From each of these moments, various important quantities such as the number-average and weight-average degrees of polymerization (dp_n and dp_w, respectively), and the PDI can be computed:

$$PDI = \frac{dp_w}{dp_n} = \frac{\mu^{(2)}\mu^{(0)}}{\mu^{(1)}\mu^{(1)}} = 1 + q \quad (1.51)$$

Using the results for the moments from this approach, the PDI is computed in Equation 1.51. Because q is the probability of propagation compared to chain inactivation events, the value for q must be very close to 1 for a polymer of any appreciable length to be produced. This finding shows that the PDI for a steady-state free radical polymerization terminated exclusively by disproportionation should be ~2.

Termination by combination complicates the situation slightly because an additional probability must be considered. In this case, chains of length n and m, respectively, with a combined chain length i, must first each be made and then combine to form the inactive polymer with length i. Because there are different combinations of chains with lengths n and m that can combine to form i, a summation must be done to calculate the mole fraction χ_i:

$$\chi_{i,comb} = \sum_{n=1}^{i-1} q^{n-1}(1-q)q^{m-1}(1-q) = (i-1) \cdot q^{i-2}(1-q)^2 \quad (1.52)$$

In the same way as it was derived for termination by disproportionation, χ is inserted into the expression for each of the moments of the chain length distribution. Again, these summations can be solved to give expressions for the first three moments:

$$\mu^{(0)} = \sum_{i=1}^{\infty} \chi_{i,comb} = \sum_{i=1}^{\infty} (i-1)(1-q)^2 q^{i-2} = 1 \quad (1.53)$$

$$\mu^{(1)} = \sum_{i=1}^{\infty} i \cdot \chi_{i,comb} = \sum_{i=1}^{\infty} i \cdot (i-1)(1-q)^2 q^{i-2}$$
$$= -\frac{2}{q-1} = 2(1-q)^{-1} \quad (1.54)$$

$$\mu^{(2)} = \sum_{i=1}^{\infty} i^2 \cdot \chi_{i,comb} = \sum_{i=1}^{\infty} i^2 \cdot (i-1)(q-1)^2 q^{i-2}$$
$$= \frac{2(q^3+2q^2)}{(q-1)^2 q^2}$$
$$= (2q+4)(1-q)^{-2} \quad (1.55)$$

The PDI can be computed in the same manner, which equals $1 + q/2$; and again, because q must be around 1, the polydispersity for a free radical polymerization in a stationary polymerization terminated exclusively by combination should equal 1.5. The polydispersity is lower in the case of termination purely by combination, due to the statistically random coupling of chains of different lengths:

$$\text{PDI} = \frac{dp_\text{w}}{dp_\text{n}} = \frac{\mu^{(2)}\mu^{(0)}}{\mu^{(1)}\mu^{(1)}} = \frac{(4+2q)(1-q)^{-2}}{4(1-q)^{-2}} = 1 + \frac{q}{2}$$

(1.56)

1.2.7 Exceptions and Special Cases

The previous sections address the kinetics for each of the processes involved in free radical polymerization, as well as the overall polymerization process. A steady-state approximation was used to determine the overall rate of polymerization and the chain length distribution. Practically, there are many exceptions to these approximations, including nonstationary polymerization and dead-end polymerization [50, 51], which are treated in more detail elsewhere.

There is also the case of reaction-controlled diffusion (briefly discussed in Section 1.2.4), closely associated with the Trommsdorff effect [55, 56], which leads to the loss of control even under isothermal conditions because the slow diffusion of radicals drastically decreases the rate of termination. This subsequently increases the concentration of radicals, as well as the rate of propagation relative to termination. Under these circumstances, polydispersity can increase significantly, easily reaching PDIs in excess of 10. In fact, the solutions found for polydispersity in a steady-state system in Section 1.2.7 generally underestimate the PDI values expected by a polymerization engineer due to various effects at high conversion and other deviations from steady-state conditions. It has also been recently shown that nanoconfinement of a free radical polymerization can actually lower the polydispersity [57–59].

Over the past two decades, new methodologies have been developed, which combine attributes of living polymerization and free radical polymerization, resulting in what is termed CRP [60]. It has become very attractive recently, due to its ability to polymerize a wide variety of monomers with low polydispersities and well-defined end-groups in a highly reproducible fashion. It encompasses a variety of techniques including but not limited to atom transfer radical polymerization (ATRP) [61, 62], reversible addition-fragmentation chain transfer (RAFT) [63], and nitroxide-mediated polymerization (NMP) [64, 65].

In a simplistic view, the control is achieved by using a reversible capping moiety which serves to render an actively growing polymer chain a nonreactive species (i.e., not a radical) for the majority of its time in the reaction mixture. This means that only a small fraction of the active polymer chains exist as macroradicals undergoing propagation reactions at any given time,

most of them being in a reversibly dormant state. This mechanism allows all the polymer chains to grow at approximately the same rate (i.e., much slower, taking hours or days instead of seconds), while drastically reducing the concentration of radicals and, thus, the associated side reactions.

1.3 CONDENSATION POLYMERIZATION

Condensation polymerization is defined as the polymerization where each addition of a monomer unit is accompanied by the elimination of a small molecule. It is used to synthesize some of the most important commodity polymers, including polyesters, polyamides, and polycarbonate.

Condensation polymerization also has a special place in polymer science history. The first truly synthetic polymer, Bakelite, was developed in 1907, as the condensation product of phenol and formaldehyde [66]. Meanwhile, Wallace Carothers pioneered polyester synthesis in the 1930s at Dupont and developed a series of mathematical equations to describe the kinetics, stoichiometry, and molecular weight distribution of condensation polymerizations.

Carothers categorized polymerizations into condensation and addition mechanisms [67], where a step-growth mechanism was synonymous with condensation polymerization. However, not all condensation polymerizations follow a step-growth mechanism. In particular, recent advancements have coerced condensation polymerizations to follow chainwise, and even "living" mechanisms. Nonetheless, the step-growth mechanism is still most common for condensation polymers, particularly among industrially relevant materials. The kinetic treatment will thus focus on the step-growth mechanism, with a separate section devoted to cases of living polycondensation.

1.3.1 Linear AB Step Polymerization

A wide variety of chemistries can be utilized to synthesize condensation polymers, typically producing polymers containing heteroatoms along the backbone. The truly distinctive feature of a stepwise mechanism is the reaction of functional groups from species of any size.

Flory advanced the understanding of step polymerization by postulating that such reactions were strictly random, meaning that reaction rates are independent of chain length [68, 69]. In this case, the problem becomes mathematically simple and probability can be used to compute the molecular weight distribution.

It is useful to start the kinetic analysis with an idealized case, which avoids complications that arise due to unequal stoichiometry, chain length-dependent reactivity, monofunctional impurities, cyclization, and reversible polymerization. The model addressed here is a linear AB step polymerization.

Any reaction in an AB step polymerization can be denoted as shown in Equation 1.57, where A represents one reactive group and B represents the complementary group:

$$(AB)_n + (AB)_m \rightarrow (AB)_{n+m} \qquad (1.57)$$

In the case of Nylon-11, a bioplastic derived from castor beans and one of the few industrially relevant AB-derived condensation polymers, A represents the carboxylic acid while B represents the amine in the monomer 1-aminoundecanoic acid:

$$\frac{d[A]}{dt} = \frac{d[B]}{dt} = -k[A][B] \qquad (1.58)$$

One starts by defining the rate constant for the polymerization in Equation 1.58. The choice of an AB system requires that the initial concentration of each monomer, $[A_0]$ and $[B_0]$ be equal at time zero, and the chemistry of amidation dictates that the rate of disappearance of each monomer also be equal.

As summarized in Table 1.1, it is recalled that in a stepwise mechanism, all species are treated as polymer, leading directly to Equation 1.59:

$$\frac{d[P]}{dt} = -k[P]^2 \qquad (1.59)$$

With the condition that [P] equals $[P_0]$ at time zero, Equation 1.59 has the following solution:

$$[P] = \frac{[P_0]}{1 + kt[P_0]} \qquad (1.60)$$

In this notation, P_i is a species with chain length i, meaning that monomer is denoted as P_1.

An expression for the rate of disappearance of the monomer species is written in Equation 1.61. Of some importance is the factor of 2, which is included because of the two indistinguishable reactions that lead to consumption of monomer (i.e., P_1 can be consumed either by the reaction of its amine with the carboxylic acid of P_i or by the reaction of its carboxylic acid with the amine of P_i):

$$\frac{d[P_1]}{dt} = -2k[P_1][P] \qquad (1.61)$$

To compute the entire molecular weight distribution, the rate of evolution of each species has to be known. Because a species with chain length i can be formed in $i - 1$ different ways, a summation must be used in the production term:

$$\frac{d[P_i]}{dt} = k\sum_{j=1}^{i-1}[P_j][P_{i-j}] - 2k[P_i][P] \qquad (1.62)$$

From here, it is evident that there is a set of infinite differential equations to be solved. The simplest way to confront this problem is to sequentially solve each differential equation and look for a pattern to emerge. This is possible because each successive solution depends on the previous solutions (i.e., larger species are derived from the combination of smaller species).

Substituting the expression for [P] from Equation 1.60 into the rate of disappearance of monomer $[P_1]$ gives the following:

$$\frac{d[P_1]}{dt} = -2k[P_1][P] = -2k[P_1][P_0] \cdot \frac{1}{1 + kt[P_0]} \qquad (1.63)$$

Note that the product term is unnecessary for monomeric species. The differential equation is easily separated and solved to give the solution for the concentration of monomer:

$$[P_1] = [P_0]\left(\frac{1}{1 + kt[P_0]}\right)^2 \qquad (1.64)$$

The next step is to write an expression for the evolution of dimer, which can only be produced by the reaction of two monomeric species with each other.

$[P_1]$ is substituted with the solution from Equation 1.64, while the value for [P] is still taken from Equation 1.60:

$$\frac{d[P_2]}{dt} = k[P_1]^2 - 2k[P_2][P] = k[P_0]^2\left(\frac{1}{1 + kt[P_0]}\right)^4$$
$$- 2k[P_2][P_0] \cdot \frac{1}{1 + kt[P_0]} \qquad (1.65)$$

The solution for this previous differential equation is more complex. With the condition that at time zero $[P_2] = 0$, the solution can be found by using the variation of constants method [70]:

$$[P_2] = [P_0]\left(\frac{1}{1 + kt[P_0]}\right)^2\left(\frac{kt[P_0]}{1 + kt[P_0]}\right) \qquad (1.66)$$

In the same manner, an expression is written for the evolution of trimer, produced by the reaction of dimer with monomer:

$$\frac{d[P_3]}{dt} = k[P_1][P_2] - 2k[P_3][P]$$
$$= k[P_0]^2\left(\frac{1}{1 + kt[P_0]}\right)^4\left(\frac{kt[P_0]}{1 + kt[P_0]}\right)$$
$$- 2k[P_3][P_0] \cdot \left(\frac{1}{1 + kt[P_0]}\right) \qquad (1.67)$$

Under the initial condition, $[P_3] = 0$, the following solution is obtained:

$$[P_3] = [P_0]\left(\frac{1}{1+kt[P_0]}\right)^2\left(\frac{kt[P_0]}{1+kt[P_0]}\right)^2 \quad (1.68)$$

Based on the aforementioned expressions, a general solution for the concentration of any given species $[P_i]$ can be postulated:

$$[P_i] = [P_0]\left(\frac{1}{1+kt[P_0]}\right)^2\left(\frac{kt[P_0]}{1+kt[P_0]}\right)^{i-1} \quad (1.69)$$

One can prove this by induction, starting with the assumption that this form is true for $[P_{i-1}]$ and inserting it into the kinetic equation for $[P_i]$:

$$\frac{d[P_i]}{dt} = (i-1)k[P_0]^2\left(\frac{1}{1+kt[P_0]}\right)^4\left(\frac{kt[P_0]}{1+kt[P_0]}\right)^{i-2}$$
$$- 2k[P_i][P_0]\left(\frac{1}{1+kt[P_0]}\right) \quad (1.70)$$

Since the solution to the homogenous equation is always the same, Equation 1.70 can be simplified:

$$[P_i] = [P_0]\left(\frac{1}{1+kt[P_0]}\right)^2\int_0^t(i-1)\left(\frac{kt[P_0]}{1+kt[P_0]}\right)^{i-2}$$
$$\left(\frac{1}{1+kt[P_0]}\right)^2 k[P_0]dt \quad (1.71)$$

The equation can then be integrated, giving the result postulated for the general form. Then, since this form was shown to be true for 1, 2, 3, and $i-1$, the validity of the general form is proven:

$$[P_i] = [P_0]\left(\frac{1}{1+kt[P_0]}\right)^2\left(\frac{kt[P_0]}{1+kt[P_0]}\right)^{i-1} \quad (1.72)$$

The general result for the concentration of P_i can be simplified by creating a simple expression for the conversion, p, of functional groups, derived from Equation 1.60:

$$p = \frac{[A]_0 - [A]}{[A]_0} = \frac{[B]_0 - [B]}{[B]_0} = \frac{[P]_0 - [P]}{[P]_0} = \frac{kt[P_0]}{1+kt[P_0]} \quad (1.73)$$

which can be used to derive a simplified expression for $[P_i]$:

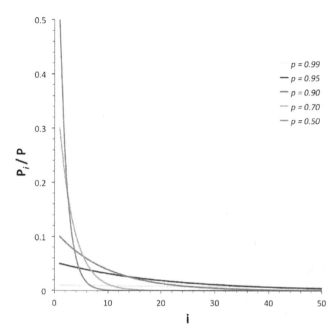

FIGURE 1.4 Geometric chain length distribution at different conversions. (*See insert for color representation of the figure.*)

$$[P_i] = [P_0](1-p)^2 \cdot p^{i-1} \quad (1.74)$$

The aforementioned expression is the geometric distribution or the Flory–Schulz distribution. The results can be illustrated by plotting the mole fraction of chain length for different values of conversion, p.

Figure 1.4 shows the chain length distribution for a geometric distribution for different values of p, while Figure 1.5 shows the corresponding molecular weight distribution (without taking into account the mass loss due to the condensate).

While the entire chain length distribution is shown in Figure 1.4 and Figure 1.5, polymer size is usually characterized by the moments of the distribution, as described in Section 1.1.2. From the results computed for the geometric chain distribution, one can solve for the moments in a straightforward way. By combining Equation 1.1 with 1.74, an expression for each of the first three moments can be written as follows:

$$\mu_0 = \sum_{i=1}^{\infty}[P_i] = [P_0]\sum_{i=1}^{\infty}(1-p)^2 \cdot p^{i-1} = [P_0](1-p)^2\sum_{j=0}^{\infty}p^j \quad (1.75)$$

$$\mu_1 = \sum_{i=1}^{\infty}i\cdot[P_i] = [P_0](1-p)^2\sum_{i=1}^{\infty}i\cdot p^{i-1}$$
$$= [P_0](1-p)^2\sum_{j=0}^{\infty}(j+1)\cdot p^j \quad (1.76)$$

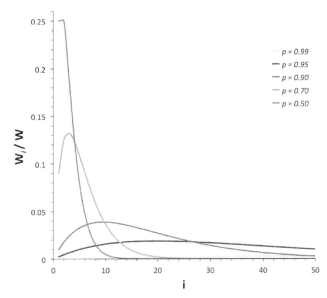

FIGURE 1.5 Geometric molecular weight distribution at different conversions. (*See insert for color representation of the figure.*)

$$\mu_2 = \sum_{i=1}^{\infty} i^2 \cdot [P_i] = [P_0](1-p)^2 \sum_{i=1}^{\infty} i^2 \cdot p^{i-1}$$

$$= [P_0](1-p)^2 \sum_{j=0}^{\infty} (j+1)^2 \cdot p^j \qquad (1.77)$$

While the conversion p can approach 1, it will never reach unity. Because p is always less than 1, each of the aforementioned summations converges to give the results for each of the moments as follows [71]:

$$\mu_0 = [P_0](1-p) \qquad (1.78)$$

$$\mu_1 = [P_0] \qquad (1.79)$$

$$\mu_2 = [P_0] \cdot \frac{(1+p)}{(1-p)} \qquad (1.80)$$

Next, the number-average and weight-average degrees of polymerization and the PDI can be computed:

$$dp_n = \frac{\mu_1}{\mu_0} = \frac{1}{(1-p)} \qquad (1.81)$$

$$dp_w = \frac{\mu_2}{\mu_1} = \frac{1+p}{1-p} \qquad (1.82)$$

$$\text{PDI} = \frac{\mu_2\mu_0}{\mu_1\mu_1} = 1+p \qquad (1.83)$$

Equation 1.81 is also known as the Carothers equation, which offers an expression for dp_n in terms of functional group conversion. Carothers equation clearly proves that high molecular weight can be achieved in a stepwise polymerization only by reaching very high conversion. Also, the polydispersity will approach 2 because conversion p must be close to 1 for a polymerization to attain any significant molecular weight.

1.3.2 Linear Step AA–BB Polymerization: Stoichiometric Imbalance

Despite the fact that the AB type of polymerization serves as a useful model for deriving the Carothers equation and gaining a basic understanding of step polymerization kinetics, most industrially relevant stepwise polymers are made using an AA–BB system. While this naturally simplifies monomer synthesis, it introduces a complicating factor into account, that of stoichiometry. In an AB system, perfect stoichiometry is assured. This does not hold for AA–BB systems, imbalances in stoichiometry leading to serious consequences for the molecular weight distribution, namely a severe reduction in molecular weight. While reaction engineers have many tools at their disposal for assuring the desired stoichiometry, it is still important to determine the results of unbalanced concentration of monomers:

$$r = \frac{[A]_0}{[B]_0} = \frac{p_B}{p_A} \leq 1 \qquad (1.84)$$

There are a number of ways to approach this problem. Assuming a system where B is the monomer in excess (described by Eq. 1.84), Flory's approach can be taken. If the chains are termed by their end-groups (i.e., AA-(AABB)$_n$-AA is an "odd-A" chain, AA-(AABB)$_n$-BB is an "even" chain, and BB-(AABB)$_n$-BB is an "odd-B" chain), the rate of evolution of each type of chain at length i can be determined, bearing in mind that at high conversion both "odd-A" and "even" chains will disappear. There are multiple statistical approaches to this problem, including those described by Case [72], Miller [73], and Lowry [74]. The results of these analyses are briefly presented in the following, other sources for a more rigorous mathematical treatment being available [70].

The number-average degree of polymerization is given by Equation 1.85 in terms of conversion and stoichiometric imbalance. In a situation where $r=1$ (i.e., perfect stoichiometry), this equation simplifies to Carothers equation (Eq. 1.81).

$$dp_n = \frac{1+r}{1+r-2rp} \qquad (1.85)$$

To address the question of how stoichiometric unbalance limits molecular weight, the effect at the limit where conversion p is equal 1 should be considered. The following equations give the expressions for the number-average and weight-average degrees of polymerization, and the PDI at full conversion:

$$dp_n = \frac{1+r}{1-r} \qquad (1.86)$$

$$dp_w = \frac{1+r}{1-r} + \frac{4r}{1-r^2} \qquad (1.87)$$

$$PDI = 1 + \frac{4r}{(1+r)^2} \qquad (1.88)$$

Carothers equation indicates that at full conversion, infinite molecular weight will be attained. However, in the case of 0.01% excess of monomer B ($r \cong 0.9999$), the number-average degree of polymerization will be 20,000. If that excess of monomer B rises to 1%, dp_n at full conversion will be only 201. This clearly proves the extreme limiting effect on the molecular weight of even a slight excess of one reagent. To avoid these problems, reaction engineers have designed a range of strategies for gaining the desired stoichiometry, including what amounts to a titration between carboxylic acids and amines in the synthesis of polyamides and the creation of a quasi-A$_2$ monomer from an AA–BB system during the synthesis of polyesters (see Section 1.3.4 for more details). One can see that the polydispersity is a monotonically decreasing function of the stoichiometric ratio r, where the PDI is equal to 2 in the case of perfect stoichiometry and is equal to 1 when $r = 1$ (i.e., only monomer is present because no reaction is possible). While it may seem that stoichiometric unbalance is entirely negative, it can be used intentionally with positive effects.

For example, consider a polyamide (e.g., nylon-6,6, see Section 1.3.4.3) made by starting with perfect stoichiometry and polymerized to a conversion of 99%. This polymer has a dp_n of 100, with a mixture of "odd-A," "even," and "odd-B" chains. Because the end-groups are still potentially reactive, if the polymer is subjected to heating, further amidation reactions are possible. This would change the molecular weight and potentially alter the mechanical properties of the polymer. Alternatively, a dp_n of 100 can be achieved by starting with a 2 mol% excess of B ($r \cong 0.98$) and reacting until nearly complete conversion is achieved. At full conversion, all the chains will be "odd-B" (i.e., all of the chain ends would be terminated by amines). In this scenario, additional heating will not alter the molecular weight distribution since no further amidation reactions can take place.

1.3.3 Effect of Monofunctional Monomer

The presence of monofunctional monomer has a similar effect as unequal stoichiometry on the molecular weight distribution in a stepwise polymerization. There are two scenarios where monofunctional monomer must be considered.

The first case is when the monofunctional monomer is an impurity, which will deleteriously limit the molecular weight; this is particularly problematic when high molecular polymer is desired. A monofunctional monomer can also be added to act as a chain stopper, thereby limiting the molecular weight and resulting in nonreactive chain ends, as discussed at the end of the previous section. Regardless of the intent, the effect on the molecular weight distribution is the same.

The stoichiometric ratio r' is defined in Equation 1.89 (a different variable was chosen to distinguish from the case of unequal stoichiometry):

$$r' = \frac{[P]_0}{[P]_0 + [P_{mono}]_0} \qquad (1.89)$$

The expressions for dp_n and dp_w are similar to those found in Section 1.3.2. The molecular weight is limited not only by the conversion p but also by the relative amount of monofunctional agent present in the system:

$$dp_n = \frac{1}{1 - r'p} \qquad (1.90)$$

$$dp_w = \frac{1 + r'p}{1 - r'p} \qquad (1.91)$$

1.3.4 Common Condensation Polymers Made by a Stepwise Mechanism

The previous sections describe the kinetics of a stepwise polymerization, which can be implemented using a wide array of different functional groups. This is unlike the case of the free radical polymerization, where the propagation step is always due to a radical adding across a carbon–carbon double bond.

Due to the wide variety of different chemistries employed to make condensation polymers by a stepwise process, a brief overview of some of the more common polymers made by this mechanism is given in the following subsections.

1.3.4.1 Polyesters Polyesters, polymers that contain an ester bond in the backbone of their repeating unit, are the most widely produced type of condensation polymer. In fact, PET (Scheme 1.3) is the third most highly produced commodity polymer in the world, trailing only polyethylene and polypropylene.

Polyethylene terephthalate, commonly referred as polyester, can be made by several slightly differing routes. In the terephthalic acid process, ethylene glycol is reacted with terephthalic acid at temperatures above 200 °C, which drives the reaction forward by removal of water. An alternative process utilizes ester exchange to reach high molecular weights.

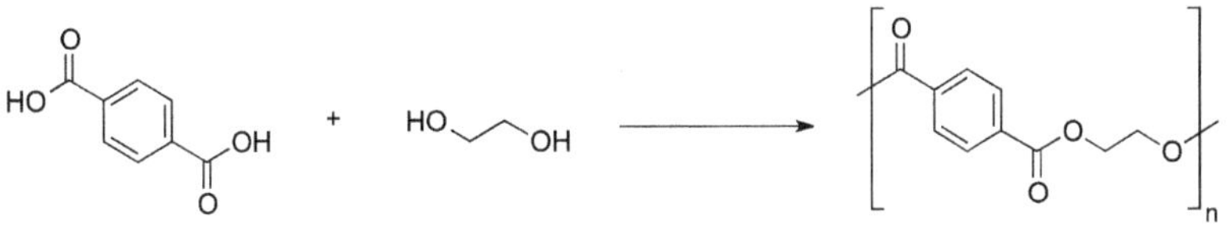

SCHEME 1.3   Poly(ethylene terephthalate) via the terephthalic acid approach.

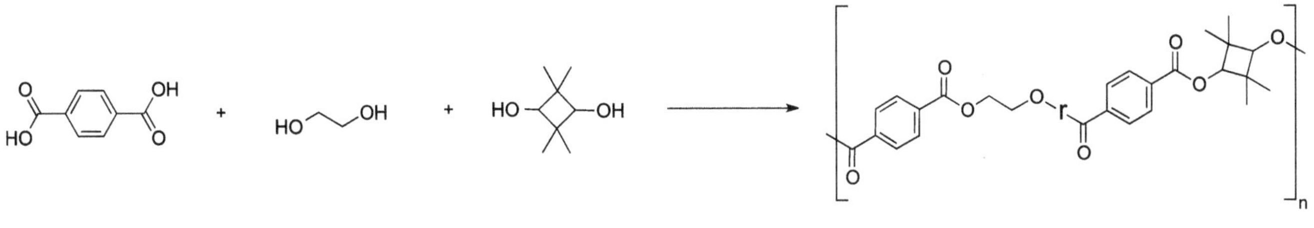

SCHEME 1.4   Copolymerization of terephthalic acid with ethylene glycol and TMCBDO to produce Tritan™ copolymer.

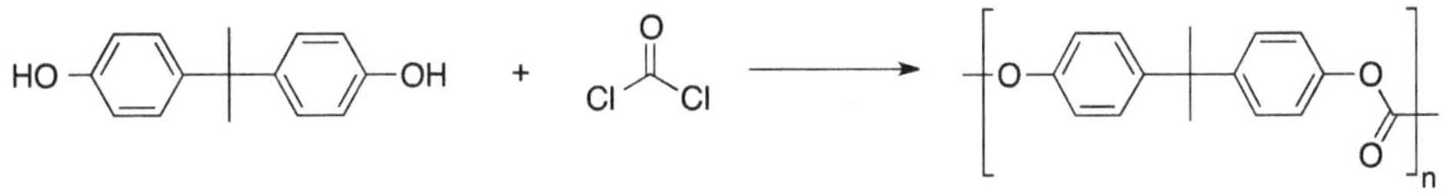

SCHEME 1.5   Reaction of bisphenol-A (BPA) with phosgene to make polycarbonate.

An initial esterification reaction occurs between an excess of ethylene glycol with dimethylterephthalate under basic catalysis at 150 °C. Removal of methanol by distillation drives the formation of bishydroxyethyl terephthalate, which can be considered an A_2 monomer. A secondary transesterification step performed at 280 °C drives polymer formation via ester exchange, which is pushed toward high molecular weight by removal of ethylene glycol via distillation.

Another polyester becoming increasingly important in the marketplace is Eastman's Tritan™ copolymer (Scheme 1.4), which has replaced polycarbonate in a variety of commercial products. Tritan™ is a modified PET copolymer, where a portion of the ethylene glycol is replaced by 2,2,4,4-tetramethyl-1,3-diol (TMCBDO).

The TMCBDO monomer imparts a higher glass transition temperature and improved mechanical properties, including resistance to crazing and efficient dissipation of applied stresses, while its diastereomeric impurity helps to prevent crystallinity and to keep Tritan amorphous, leading to improved clarity. While the real industrial feat may be the large-scale production of TMCBDO, which is made via a ketene intermediate, the interesting feature in the scope of this book is a step A_2–B_2, B_2' polymerization, where the two B_2 monomers have different reactivities (in fact, the *trans*- and *cis*-isomers of TMCBDO may also have different reactivities, but this has not been studied in detail to this point). As discussed in Section 1.1.2, the differing reactivity ratios could lead to gradient or blocky copolymers. However, in

the case of polyesterification, ester exchange reactions can serve to scramble the sequence and lead to a random distribution of monomer units even for unequal reactivities.

1.3.4.2 *Polycarbonate*

Polycarbonate (Scheme 1.5) is produced by the reaction of BPA with phosgene; it is the leaching of endocrine-disrupting BPA that has lead to its replacement in food and beverage containers. Nonetheless, polycarbonate is still used extensively as a building material, in data storage, and as bullet-resistant glass. In this reaction, the condensate is hydrochloric acid. Polycarbonate, despite its BPA-related problems, is a durable plastic that is flame retardant, heat resistant, and a good electrical insulator.

1.3.4.3 *Polyamides*

Polyamides are polymers containing amide bonds along the polymer backbone synthesized by the reaction of amine with carboxylic acid (or derivatives thereof, e.g., methyl ester, acyl halide). Proteins are polyamides made by a biosynthetic polycondensation, each having a specific sequence and monodisperse molecular weight distribution. Synthetic polyamides are not nearly as complex as their biological counterparts, but still have excellent properties. In particular, the hydrogen-bonding nature of the amide bond leads to high melting points and semicrystalline behavior, desirable traits for synthetic fibers.

The best-known class of polyamides is nylon. First discovered by Carothers in 1935, nylon-6,6 is produced by the condensation reaction between 1,6-hexanediamine and

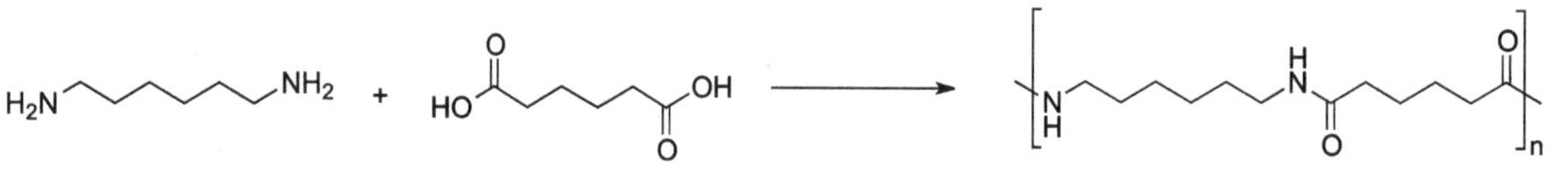

SCHEME 1.6   Reaction of 1,6-hexanediamine with adipic acid gives nylon-6,6.

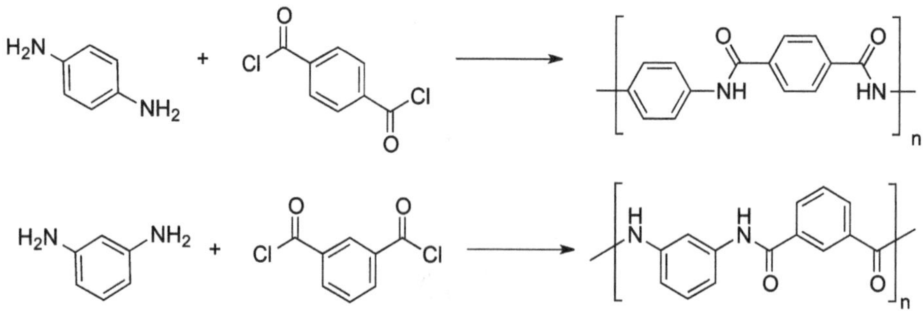

SCHEME 1.7   Polymerization of aryl amine with terephthaloyl chloride to give the *p*-aramid Kevlar and the *m*-aramid Nomex, respectively.

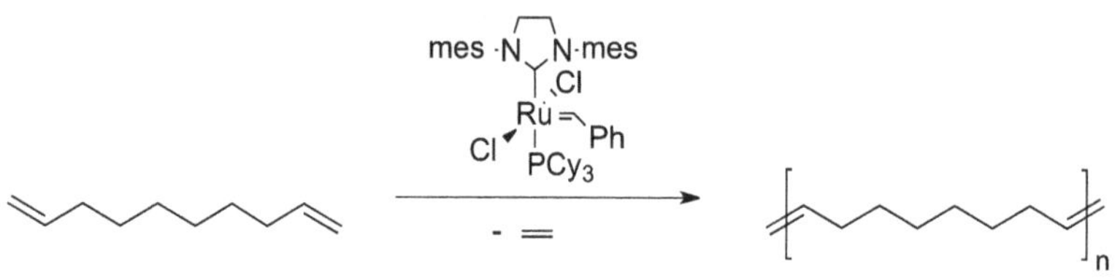

SCHEME 1.8   ADMET polymerization of 1,9-decadiene by Grubbs' second generation catalyst.

adipic acid (Scheme 1.6). It has a melting point of 265 °C and has been used as a fiber for a variety of applications, including in parachutes during World War II in the midst of worldwide silk shortages. A similar polymer, nylon-6, is made by the ring-opening polymerization of caprolactam, which is not a polycondensation reaction.

Polyamides can also be made by the reaction of amines with acyl halides, where the condensate is hydrochloric acid. This process is used to make aromatic polyamides, notably Kevlar and Nomex (Scheme 1.7). The reaction of *p*-phenylenediamine with terephthaloyl chloride results in the high performance *p*-aramid Kevlar. While Kevlar is expensive because processing requires the use of anhydrous sulfuric acid as solvent, its outstanding mechanical and thermal properties led to its use in demanding applications, including personal armor, bicycle tires, and racing sails.

When the corresponding *meta* monomers are used, the resulting polymer is Nomex. Nomex is more easily processed than Kevlar and since its fibers have excellent fire retardant properties, it is the material of choice for protective equipment for firefighters, fighter pilots, and racecar drivers.

1.3.4.4   *ADMET*   Whereas this Section might be more appropriately titled "polyolefins" to match more closely

with the previous subsections, the polymerization process itself is more notable than the products.

Acyclic diene metathesis (ADMET) [75] is the process by which a transition metal catalyst leads to a stepwise condensation polymerization of diene monomers, characterized by loss of gaseous ethylene and the production of linear polyolefins containing regular unsaturations along the polymer backbone (Scheme 1.8). In fact, many of the polymeric structures accessible by ADMET can be made by alternate mechanisms (e.g., 1,4-polybutadiene made by ADMET polymerization of 1,6-hexadiene is more commonly made by the anionic polymerization of 1,4-butadiene).

Nonetheless, ADMET is a versatile technique that allows the incorporation of a wide variety of functional groups into the resultant polymers. Scheme 1.9 shows the catalytic cycle of ADMET, controlled by the metathesis catalyst, which can be either ruthenium- [76, 77] or molybdenum-based [78, 79]. While the kinetics are controlled by the catalyst (there is no reaction in its absence), it still follows the kinetic picture described in Section 1.3.2. This is because the catalyst is removed from the chain end after each successful alkene metathesis reaction (i.e., coupling) and the olefin with which it subsequently reacts is statistically random.

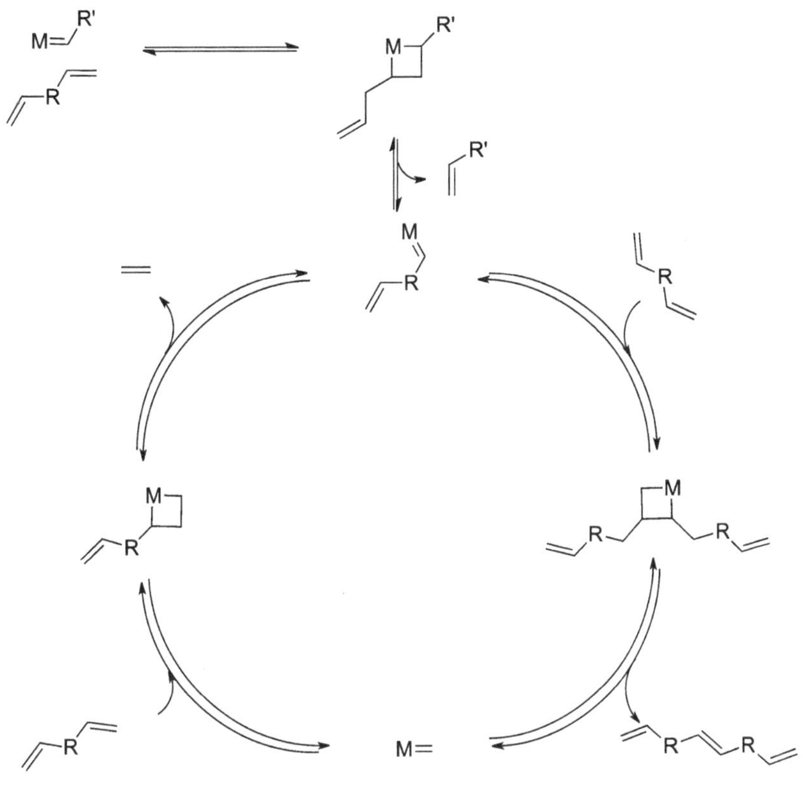

M = Ru, Mo

SCHEME 1.9   Generally accepted ADMET mechanism.

Acyclic diene metathesis polymerizations are often pushed to high molecular weight by solid-state reaction under high vacuum, while reaction under ethylene pressure causes depolymerization.

1.3.4.5   *Conjugated Polymers*

For the past 20 years, conjugated polymers have been made by polycondensation using Stille [80, 81] and Suzuki [82, 83] couplings. The Stille coupling reacts stannanes and aryl halides to form new carbon–carbon bonds [84], while the Suzuki coupling makes carbon–carbon bonds by coupling of boronic acids (or esters) and aryl halides [85, 86]. For Suzuki coupling, either an A_2–B_2 or AB system can be used. Even though using an AB monomer can help eliminate stoichiometric imbalance, A_2–B_2 systems are generally favored because it is simpler to synthesize the monomers. Stille couplings can run into problems with stoichiometry caused by the reduction of the Pd(II) catalyst to Pd(0) by the organotin monomer, and when homocoupling of the ditin monomer occurs. Because of these issues, oftentimes the catalyst and organotin monomer concentrations may be varied from equimolar to maintain proper stoichiometry.

A vast array of aromatic monomers has been polymerized by these techniques, including substituted benzenes, thio-phenes, fused thiophenes, pyrroles, pyrazines, ethylene and acetylene derivatives, and many more complex ring structures. More recently, techniques have been developed allowing chainwise and even "living" polymerization of many of the same basic monomer units using different mechanisms.

1.3.5   Living Polycondensation

The previous sections have focused on condensation polymerizations following a stepwise growth mechanism. However, a number of strategies have emerged which facilitate condensation polymerizations that would otherwise follow a stepwise growth mechanism to propagate via a chainwise or "living" mechanism. In fact, because condensation polymerization is defined only as a polymerization that releases a small molecule during each growth step, there are well-known condensation polymerizations that do not follow a stepwise mechanism under any circumstance.

The ring-opening polymerization of N-carboxyanhydride (NCA) monomers to give poly(peptides) proceeds via a chain-wise or "living" growth mechanism and has been studied in great detail over the past 15 years. This polymerization has been performed under a variety of conditions, including anionic [87],

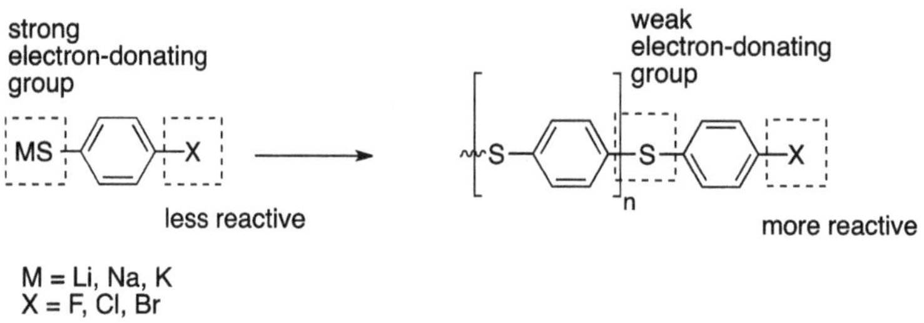

SCHEME 1.10 Polymeric aryl halide is more reactive than monomeric aryl halide.

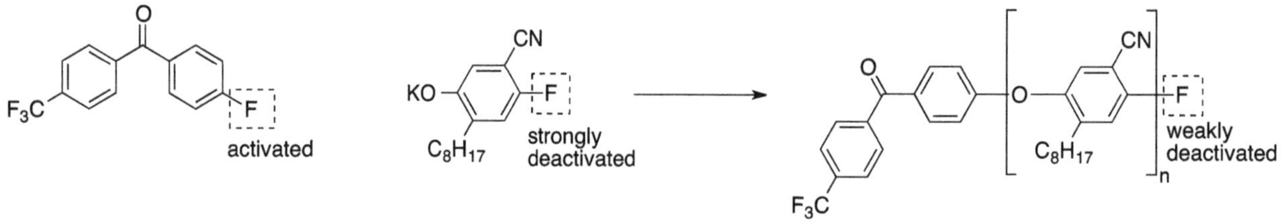

SCHEME 1.11 Potassium alkoxide strongly deactivates monomeric aryl fluoride relative to the polymeric aryl fluoride.

activated-monomer, and transition metal catalyzed [88], and in all cases the addition of one monomer unit is accompanied by the release of carbon dioxide. However, NCAs are not multifunctional monomers and cannot produce polymer via a stepwise mechanism and thus, reactions of this type will not be considered in this Section. The focus is on systems that can grow based on a stepwise mechanism, but on which strategies that alter the kinetic parameters were rather used to confer a chainwise or "living" behavior [89, 90].One of the most important assumptions made by Flory and Carothers was that of equal and random reactivity (i.e., chain length-independent) between any of the functional groups in the system. It is this assumption that leads to the growth kinetics and molecular weight distributions seen in stepwise polymerization.

One can start by considering Equation 1.62, which describes the rate of evolution for a polymer of chain length i. Equation 1.62 is rewritten in a way that separates the addition of a monomer unit, or P_1, from the addition of any other species, with k'' representing the rate for monomer addition and k' representing the rate constant for additions of species with $i \geq 2$:

$$\frac{d[P_i]}{dt} = k' \sum_{j=2}^{i-2} [P_j][P_{i-j}] + k''[P_1][P_{i-1}]$$
$$- 2k'[P_i] \cdot ([P] - [P_1]) - 2k''[P_i][P_1] \tag{1.92}$$

If Carothers's assumption that reactivity is equal and random reactivity holds, Equation 1.92 still equals the general form for step polymerization written in Equation 1.62. However, if a chemical system is designed such that $k'' \gg k'$, Equation 1.92 can be rewritten in a simpler form, which now resembles the rate of evolution for a chainwise system (see Eq. 1.33).

$$\frac{d[P_i]}{dt} = k''[P_1][P_{i-1}] - 2k''[P_i][P_1] \tag{1.93}$$

Efforts to influence condensation polymerizations to follow a "living" mechanism must favor the addition of monomer units to active chains over all other possible reactions. In other words, a chemical system must be designed such that k'' is much greater than k'. In some ways, compelling polycondensation to demonstrate "living" behavior is simpler than for free radical polymerizations because condensation polymerization is not affected by the various transfer and termination processes that plague free radical chemistry.

A successful method in creating conditions for living polycondensation takes the advantage of differing substituent effects to activate the polymer chain end relative to the monomer. With this approach, aromatic monomers have great use due to their propensity to be strongly activated or deactivated by substituents on the ring.

Early work performed by Lenz et al. in the 1960s demonstrated this approach by using electrophilic aromatic substitution to produce poly(phenylene sulfide) [91] (Scheme 1.10). In this case, an aryl halide is the electrophile, which is substituted by the metal thiophenoxide nucleophile. In the monomer, the metal sulfide is a strong electron-donating group, which deactivates the *para* position where electrophilic substitution must take place. Conversely, the polymer chain end is only weakly deactivated by the sulfide bond, rendering the polymeric aryl halide more reactive than the monomeric aryl halide. Unfortunately, Lenz was unable to characterize molecular weight distribution due to the insolubility of the resultant polymers.

Later work by Yokozawa et al. took advantage of the same principal to produce aromatic polyethers with PDIs < 1.1 [92]. In this case, the aryl fluoride was again strongly deactivated in the monomer by the electronic donating *p*-phenoxide (Scheme 1.11), while the chain end was only weakly deactivated by the ether bond *para* to the

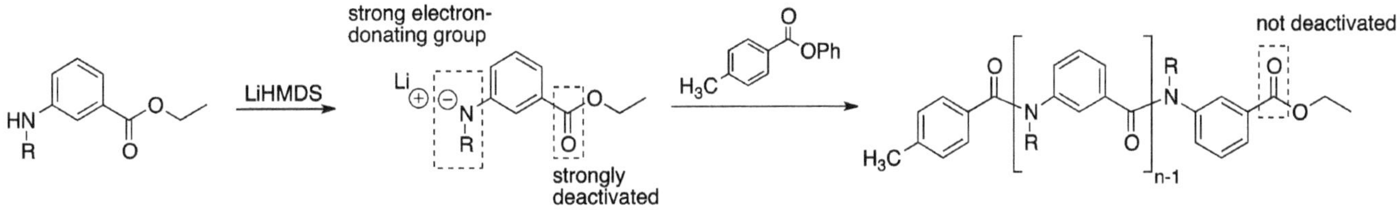

SCHEME 1.12   Inductively deactivated monomer in transamidation condensation polymerization.

chain-end aryl fluoride. Taking this concept a step further, the authors polymerized the deactivated monomer in the presence of 4-fluoro-4′-trifluoromethylbenzophenone, the aryl halide of which is activated. This molecule is much more reactive to electrophilic substitution than the monomer, it is effectively an initiator and leads to polymers with controlled molecular weight distributions and well-defined chain ends.

A similar strategy of monomer deactivation through an aromatic group has been used to successfully polymerize m-substituted monomers in a controlled fashion by using the inductive effect. An example is the transamidation of benzoate monomers to give poly(m-benzamides) [93]. The carbonyl of the monomer ester is strongly deactivated by the lithium amide in the m-position, discouraging its transamidation (Scheme 1.12). 4-Methylbenzoate was employed as an initiator, which is activated at the carbonyl by the p-methyl group. The resultant chain end is much more reactive to amidation than the monomer, which again results in a situation where $k'' >> k'$, leading to polymers with narrow molecular weight distributions.

Another method to activate the polymer relative to the monomer is to transfer a catalyst to the chain end. Catalyst-transfer living polycondensation has had an enormous impact in the field of conjugated polymers, providing facile routes to relatively monodisperse polythiophenes [94–97], polyphenylenes [98], polypyrroles [99], and polyfluorenes [100, 101] all made in a living fashion. Conjugated polymers have attracted great interest [102] due to their applications in organic optoelectronic devices including photovoltaics [103, 104], light-emitting diodes [105], organic field-effect transistors [106], and nonlinear optical devices. These applications highlight the importance of living polycondensation because conjugated polymers made by stepwise growth (e.g., Suzuki, Stille couplings) lack consistent molecular weight distributions and well-defined end-groups, leading to batch-to-batch variations that can affect materials performance.

One of the most highly researched conjugated polymers is poly(3-alkylthiophene) (underivatized polythiophene is insoluble). In a synthetic method developed by McCullough [94, 96] and further modified by Yokozawa [95, 97], Kumada catalyst-transfer polycondensation (KCTP), also commonly known as Grignard metathesis polymerization

(GRIM), is a nickel-catalyzed organometallic polycondensation that allows access to regioregular poly(3-alkylthiophene) and block copolymers thereof. While KCTP is affected by termination, as proved by the existence of maximum attainable molecular weights, in the same time, it exhibits a living behavior and consistently leads to polymers with PDIs ≤ 1.1. The initiating species are formed via transmetalation of two monomer molecules to produce the bis-organonickel compound. As shown in Scheme 1.13, the catalytic cycle during propagation involves transmetalation with a monomer unit, reductive elimination of the catalyst to add one unit to the propagating chain, and oxidative addition to reform the active Ni(0) species at the active chain end. The key step in this process is the ability of the nickel catalyst to reductively eliminate and subsequently reinsert into the terminal C–Br bond without diffusing into the reaction mixture (as compared to ADMET), thus, resulting in one polymer chain produced for each molecule of nickel catalyst (i.e., the molecular weight is inversely proportional to the catalyst loading).

Beyond its success in polymerizing thiophene derivatives, KCTP has been adapted not only to polymerize a variety of other monomers via a living chainwise growth mechanism, but also to produce an array of polymeric architectures. In particular, KCTP is capable of producing block copolymers by successive addition of monomer: a hallmark of living polymerization. Iovu et al. reported the synthesis of the first all-thiophene block copolymer by successive polymerization of 3-hexylthiophene and 3-dodedecylthiophene [94]. Subsequently, block copolymers comprising entirely of different conjugated polymers have been synthesized, including polymers containing pyrroles [99], phenylenes [99], fluorenes [101], thiophenes, and seleophenes [107]. Catalyst-transfer polycondensation has also been applied to a grafting-from strategy to synthesize brush copolymers [108]. as well as surface-initiated polymerization to produce mechanically robust polymer-coated objects [109, 110].

An attractive approach to controlled polycondensation developed by Yokoyama takes advantage of a biphasic system using phase-transfer catalysis (scheme 1.14) [111, 112]. The authors begin by dispersing the solid monomer, potassium-4-bromomethyl-2-n-octyloxybenzoate, in a nonsolvent. They also dissolved an initiator, 4-nitrobenzylbromide, into the liquid phase. By using 18-crown-6 as a phase transfer

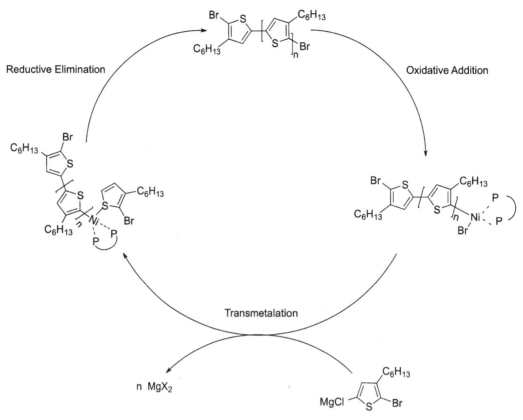

SCHEME 1.13   Catalytic cycle for Kumada catalyst-transfer polymerization of poly(3-hexylthiophene).

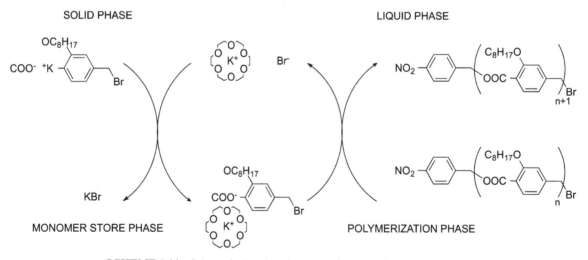

SCHEME 1.14   Schematic showing phase-transfer catalysis polycondensation.

catalyst, a small amount of the monomer is solubilized into the liquid phase, where it can react with the initiator (and later, the polymer chain end) to form a *p*-nitrobenzyl ester. A key factor in this approach is careful choice of the molar equivalents of the initiator and phase transfer agent, so that only a small monomer is allowed in solution at any time, sup-

pressing self-condensation. However, if the concentration of the crown ether is too low, only small amounts of monomer will enter the reactive liquid phase and homogenous growth of the polymer chains will not be possible. Therefore, accurately balancing solubility is critical. If the monomer is too soluble, the self-condensation reaction will not be sufficiently suppressed. On the other hand, the polymer chains have to be soluble in order to achieve significant molecular weight. Certainly, the fine balance of these parameters calls into question its versatility with other molecular systems. Nonetheless, it provides a rather simple way in which a stepwise mechanism can be converted to chainwise or even "living" polymerization.

1.4 CONCLUSIONS

Free radical and condensation polymerization reactions allow the synthesis of some of the world's most important commodity polymers, as well as cutting edge materials in research laboratories across the globe. Furthermore, recent advances in controlled polymerization techniques have offered the possibility to change the growth mechanism to a "living" polymerization in both cases. The kinetics and resulting molecular weight distributions for each polymerization have been discussed.

REFERENCES

[1] Kuhlmann R, Schnabel W. Flash-photolysis investigation on primary processes of sensitized polymerization of vinyl monomers. 2. Experiments with benzoin and benzoin derivatives. Polymer 1977;18:1163–1168.

[2] Buback M, Busch M, Kowollik C. Chain-length dependence of free-radical termination rate deduced from laser single-pulse experiments. Macromol Theory Simul 2000;9:442–452.

[3] North AM. *The Kinetics of Free Radical Polymerization.* Oxford: Pergamon Press; 1966.

[4] Odian G. *Principles of Polymerization.* 4th ed. Hoboken: John Wiley & Sons, Inc; 2004.

[5] Moad G, Rizzardo E, Solomon DH, Johns SR, Willing RI. Application of c-13 labeled initiators and c-13 NMR to the study of the kinetics and efficiency of initiation of styrene polymerization. Makromol Chem Rapid Commun 1984;5:793–798.

[6] Gruber HF. Photoinitiators for free-radical polymerization. Prog Polym Sci 1992;17:953–1044.

[7] Lehrle RS, Shortland A. A study of the purification of methylmethacrylate suggests that the thermal polymerization of this monomer is initiated by adventitious peroxides. Eur Polym J 1988;24:425–429.

[8] Olaj OF, Kauffmann HF, Breitenbach JW, Bieringer H. Diels–Alder intermediate as a chain-transfer agent in spontaneous styrene polymerization. 2. Evidence from comparison of chain-length distribution of spontaneously initiated and photoinitiated polymers. J Polym Sci Part C-Polym Lett 1977;15:229–233.

[9] Olaj OF, Kauffmann HF, Breitenbach JW. Spectroscopic measurements on spontaneously polymerizing styrene. 2. Estimation of reactivity of 2 Diels–Alder-isomers towards polymer radicals. Makromol Chem-Macromol Chem Phys 1977;178:2707–2717.

[10] Kauffmann HF, Olaj OF, Breitenbach JW. Spectroscopic measurements on spontaneously polymerizing styrene – Evidence for formation of 2 Diels–Alder isomers of different stability. Makromol Chem-Macromol Chem Phys 1976;177:939–945.

[11] Heuts JPA, Russell GT. The nature of the chain-length dependence of the propagation rate coefficient and its effect on the kinetics of free-radical polymerization. 1. Small-molecule studies. Eur Polym J 2006;42:3–20.

[12] Smith GB, Russell GT, Yin M, Heuts JPA. The effects of chain length dependent propagation and termination on the kinetics of free-radical polymerization at low chain lengths. Eur Polym J 2005;41:225–230.

[13] Deady M, Mau AWH, Moad G, Spurling TH. Evaluation of the kinetic-parameters for styrene polymerization and their chain-length dependence by kinetic simulation and pulsed-laser photolysis. Makromol Chem-Macromol Chem Phys 1993;194:1691–1705.

[14] Gridnev AA, Ittel SD. Dependence of free-radical propagation rate constants on the degree of polymerization. Macromolecules 1996;29:5864–5874.

[15] Heuts JPA, Sudarko RGG. First-principles prediction and interpretation of propagation and transfer rate coefficients. Macromol. Symposia 1996;111:147–157.

[16] Olaj OF, Vana P, Zoder M, Kornherr A, Zifferer G. Is the rate constant of chain propagation K_p in radical polymerization really chain-length independent? Macromol Rapid Commun 2000;21:913–920.

[17] Barner-Kowollik CVP, Davis TP. The kinetics of free-radical polymerization. In: Matyjaszewski K, Davis TP, editors. *Handbook of Radical Polymerization.* New York: John Wiley & Sons; 2002.

[18] Olaj OF, Schnoll-Bitai I. Solvent effects on the rate constant of chain propagation in free radical polymerization. Monatsh Chem 1999;130:731–740.

[19] Zammit MD, Davis TP, Willett GD, Odriscoll KF. The effect of solvent on the home-propagation rate coefficients of styrene and methyl methacrylate. J Polym Sci Part A-Polym Chem 1997;35:2311–2321.

[20] Coote ML, Davis TP, Klumperman B, Monteiro MJ. A mechanistic perspective on solvent effects in free-radical copolymerization. J Macromol Sci-Rev Macromol Chem Phys 1998;C38:567–593.

[21] Harrisson S, Mackenzie SR, Haddleton DM. Pulsed laser polymerization in an ionic liquid: strong solvent effects on propagation and termination of methyl methacrylate. Macromolecules 2003;36:5072–5075.

[22] Strehmel V, Laschewsky A, Wetzel H, Gornitz E. Free radical polymerization of n-butyl methacrylate in ionic liquids. Macromolecules 2006;39:923–930.

[23] Woecht I, Schmidt-Naake G, Beuermann S, Buback M, Garcia N. Propagation kinetics of free-radical polymerizations in ionic liquids. J Polym Sci Part A-Polym Chem 2008;46:1460–1469.

[24] Beuermann S, Buback M, Russell GT. Variation with pressure of the propagation rate coefficient in free-radical polymerization of methyl-methacrylate. Macromol Rapid Commun 1994;15:351–355.

[25] Buback M, Geers U, Kurz CH. Propagation rate coefficients in free-radical homopolymerizations of butyl methacrylate and dodecyl methacrylate. Macromol Chem Phys 1997;198: 3451–3464.

[26] Buback M, Kurz CH, Schmaltz C. Pressure dependence of propagation rate coefficients in free-radical homopolymerizations of methyl acrylate and dodecyl acrylate. Macromol Chem Phys 1998;199:1721–1727.

[27] Bovey FA, Kolthoff IM. Inhibition and retardation of vinyl polymerization. Chem Rev 1948;42:491–525.

[28] Tudos F, Foldesberezsnich T. Free-radical polymerization – inhibition and retardation. Prog Polym Sci 1989;14:717–761.

[29] Nikitin AN, Hutchinson RA. The effect of intramolecular transfer to polymer on stationary free radical polymerization of alkyl acrylates. Macromolecules 2005;38:1581–1590.

[30] O'Brien AK, Bowman CN. Impact of oxygen on photopolymerization kinetics and polymer structure. Macromolecules 2006;39:2501–2506.

[31] Horie K, Mita I, Kambe H. Fast reaction and micro-Brownian motion of flexible polymer molecules in solution. Polym J 1973;4:341–349.

[32] Ito K. Analysis of polymerization rates in radical polymerization with primary radical termination of styrene initiated by 2,2′-azobis(2,4-dimethylvaleronitrile). J Polym Sci Part A-Polym Chem 1972;10:931–936.

[33] Asua JM, Beuermann S, Buback M, Castignolles P, Charleux B, Gilbert RG, Hutchinson RA, Leiza JR, Nikitin AN, Vairon JP, van Herk AM. Critically evaluated rate coefficients for free-radical polymerization, 5. Propagation rate coefficient for butyl acrylate. Macromol Chem Phys 2004;205:2151–2160.

[34] Barner-Kowollik C, Buback M, Egorov M, Fukuda T, Goto A, Olaj OF, Russell GT, Vana P, Yamada B, Zetterlund PB. Critically evaluated termination rate coefficients for free-radical polymerization: experimental methods. Prog Polym Sci 2005;30:605–643.

[35] Allen PEMP, Patrick CR. *Kinetics and Mechanisms of Polymerization Reactions*. New York: Wiley; 1974.

[36] Benson SW, North AM. Kinetics of free radical polymerization under conditions of diffusion-controlled termination. J Am Chem Soc 1962;84:935–940.

[37] North AM, Reed GA. Diffusion-controlled termination during initial stages of free radical polymerization of methyl methacrylate. Trans Faraday Soc 1961;57:859–870.

[38] North AM, Reed GA. Diffusion-controlled polymerization of some alkyl methacrylates. J Polym Sci Part A-Gen Papers 1963;1:1311–1321.

[39] Achilias DS. A review of modeling of diffusion controlled polymerization reactions. Macromol Theory Simul 2007;16: 319–347.

[40] Buback M. Free-radical polymerization up to high conversion – a general kinetic treatment. Makromol Chem-Macromol Chem Phys 1990;191:1575–1587.

[41] Olaj OF, Zifferer G. Relative reaction probabilities in polymer–polymer reactions. 1. Investigations with Monte-Carlo model chains. Makromol Chem-Macromol Chem Phys 1988;189:1097–1116.

[42] Khokhlov AR. Influence of excluded volume effect on the rates of chemically controlled polymer–polymer reactions. Makromol Chem-Rapid Commun 1981;2: 633–636.

[43] Friedman B, Oshaughnessy B. Kinetics of intermolecular reactions in dilute polymer-solutions and unentangled melts. Macromolecules 1993;26:5726–5739.

[44] de Kock JBL, Van Herk AM, German AL. Bimolecular free-radical termination at low conversion. J Macromol Sci-Polym Rev 2001;C41:199–252.

[45] Olaj OF, Vana P. Chain-length dependent termination in pulsed-laser polymerization, 6. The evaluation of the rate coefficient of bimolecular termination K_T for the reference system methyl methacrylate in bulk at 25 degrees C. Macromol Rapid Commun 1998;19:533–538.

[46] Olaj OF, Vana P. Chain-length dependent termination in pulsed-laser polymerization, 5. The evaluation of the rate coefficient of bimolecular termination K_T for the reference system styrene in bulk at 25 degrees C. Macromol Rapid Commun 1998;19:433–439.

[47] Beuermann S, Buback M, Schmaltz C. Termination rate coefficients of butyl acrylate free-radical homopolymerization in supercritical CO_2 and in bulk. Ind Eng Chem Res 1999;38:3338–3344.

[48] Buback M, Kowollik C. Termination kinetics of methyl methacrylate free-radical polymerization studied by time-resolved pulsed laser experiments. Macromolecules 1998; 31:3211–3215.

[49] Hamielec AE, Hodgins JW, Tebbens K. Polymer reactors and molecular weight distribution, 2. Free radical polymerization in a batch reactor. Aiche J 1967;13:1087–1091.

[50] Tobolsky AV. Dead-end radical polymerization. J Am Chem Soc 1958;80:5927–5929.

[51] Tobolsky AV, Rogers CE, Brickman RD. Dead-end radical polymerization. 2. J Am Chem Soc 1960;82:1277–1280.

[52] Bamford CH, Barb WG, Jenkins AD, Onyan PF. *The Kinetics of Vinyl Polymerization by Radical Mechanisms*. Oxford: Pergamon Press; 1966.

[53] Williams RJJ. Statistics of free-radical polymerizations revisited using a fragment approach, 1. Bifunctional monomers. Macromolecules 1988;21:2568–2571.

[54] Dotson NA, Galvan R, Macosko CW. Structural development during nonlinear free-radical polymerizations. Macromolecules 1988;21:2560–2568.

[55] Norrish RGW, Smith RR. Catalysed polymerization of methyl methacrylate in the liquid phase. Nature 1942;150: 336–337.

[56] Trommsdorff E, Kohle H, Lagally P. Zur polymerisation des methacrylsauremethylesters. Makromol Chem-Macromol Chem Phys 1948;1:169–198.

[57] Begum F, Zhao H, Simon SL. Modeling methyl methacrylate free radical polymerization: reaction in hydrophilic nanopores. Polymer 2012;53:3238–3244.

[58] Begum F, Zhao, H, Simon, SL. Modeling methyl methacrylate free radical polymerization: reaction in hydrophobic nanopores. Polymer 2012;53:3261–3268.

[59] Iwasaki T, Yoshida J. Free radical polymerization in microreactors. Significant improvement in molecular weight distribution control. Macromolecules 2005;38:1159–1163.

[60] Braunecker WA, Matyjaszewski K. Controlled/living radical polymerization: features, developments, and perspectives. Prog Polym Sci 2007;32:93–146.

[61] Wang JS, Matyjaszewski K. Controlled living radical polymerization – halogen atom-transfer radical polymerization promoted by a Cu(I)Cu(II) redox process. Macromolecules 1995;28:7901–7910.

[62] Xia JH, Matyjaszewski K. Controlled/"living" radical polymerization. Atom transfer radical polymerization using multidentate amine ligands. Macromolecules 1997;30: 7697–7700.

[63] Chiefari J, Chong YK, Ercole F, Krstina J, Jeffery J, Le TPT, Mayadunne RTA, Meijs GF, Moad CL, Moad G, Rizzardo E, Thang SH. Living free-radical polymerization by reversible addition-fragmentation chain transfer: the raft process. Macromolecules 1998;31:5559–5562.

[64] Benoit D, Chaplinski V, Braslau R, Hawker CJ. Development of a universal alkoxyamine for "living" free radical polymerizations. J Am Chem Soc 1999;121:3904–3920.

[65] Hawker CJ, Bosman AW, Harth E. New polymer synthesis by nitroxide mediated living radical polymerizations. Chem Rev 2001;101:3661–3688.

[66] Baekeland LH. The synthesis, constitution, and uses of bakelite. Ind Eng Chem 1909;1:149–161.

[67] Carothers WH. Polymerization. Chem Rev 1931;8:353–426.

[68] Flory PJ. Molecular size distribution in linear condensation polymers. J Am Chem Soc 1936;58:1877–1885.

[69] Flory PJ. Fundamental principles of condensation polymerization. Chem Rev 1946;39:137–197.

[70] Dotson NA, Galvan R, Laurence RL, Tirrell M. *Polymerization Process Modeling.* New York: VCH Publishers; 1996.

[71] Taylor AE. *Advanced Calculus.* Boston: Ginn; 1955.

[72] Case LC. Molecular distributions in polycondensations involving unlike reactants, 2. Linear distributions. J Polym Sci 1958;29:455–495.

[73] Miller DR, Macosko CW. Average property relations for non-linear polymerization with unequal reactivity. Macromolecules 1978;11:656–662.

[74] Lowry GC. editor. *Markov Chains and Monte Carlo Calculations in Polymer Science.* New York: Dekker; 1970.

[75] Wagener KB, Boncella JM, Nel JG. Acyclic diene metathesis (admet) polymerization. Macromolecules 1991;24: 2649–2657.

[76] Grubbs RH, Chang S. Recent advances in olefin metathesis and its application in organic synthesis. Tetrahedron 1998; 54:4413–4450.

[77] Trnka TM, Grubbs RH. The development of L2x2ru=Chr olefin metathesis catalysts: an organometallic success story. Acc Chem Res 2001;34:18–29.

[78] Crowe WE, Goldberg DR. Acrylonitrile cross-metathesis – coaxing olefin metathesis reactivity from a reluctant substrate. J Am Chem Soc 1995;117:5162–5163.

[79] Schrock RR, Hoveyda AH. Molybdenum and tungsten imido alkylidene complexes as efficient olefin-metathesis catalysts. Angew Chem Int Ed 2003;42:4592–4633.

[80] Bao ZN, Chan WK, Yu LP. Exploration of the stille coupling reaction for the syntheses of functional polymers. J Am Chem Soc 1995;117:12426–12435.

[81] Carsten B, He F, Son HJ, Xu T, Yu L. Stille polycondensation for synthesis of functional materials. Chem Rev 2011;111:1493–1528.

[82] Sakamoto J, Rehahn M, Wegner G, Schlueter AD. Suzuki polycondensation: polyarylenes a la carte. Macromol Rapid Commun 2009;300:653–687.

[83] Schluter AD. The tenth anniversary of suzuki polycondensation (spc). J Polym Sci Part A-Polym Chem 2001;39: 1533–1556.

[84] Espinet P, Echavarren AM. The mechanisms of the stille reaction. Angew Chem Int Ed 2004;43:4704–4734.

[85] Kotha S, Lahiri K, Kashinath D. Recent applications of the Suzuki–Miyaura cross-coupling reaction in organic synthesis. Tetrahedron 2002;58:9633–9695.

[86] Suzuki A. Recent advances in the cross-coupling reactions of organoboron derivatives with organic electrophiles, 1995–1998. J Organomet Chem 1999;576:147–168.

[87] Aliferis T, Iatrou H, Hadjichristidis N. Living polypeptides. Biomacromolecules 2004;5:1653–1656.

[88] Deming TJ. Living polymerization of alpha-amino acid-n-carboxyanhydrides. J Polym Sci Part A-Polym Chem 2000;38:3011–3018.

[89] Yokoyama A, Yokozawa T. Converting step-growth to chain-growth condensation polymerization. Macromolecules 2007; 40:4093–4101.

[90] Yokozawa T, Yokoyama A. Chain-growth polycondensation: the living polymerization process in polycondensation. Prog Polym Sci 2007;32:147–172.

[91] Lenz RW, Smith HA, Handlovits CE. Phenylene sulfide polymers, 3. Synthesis of linear polyphenylene sulfide. J Polym Sci 1962;58:351–357.

[92] Suzuki Y, Hiraoka S, Yokoyama A, Yokozawa T. Chain-growth polycondensation of potassium 3-cyano-4-fluorophenolate derivatives for well-defined poly(arylene ether)s. Macromol Symp 2003;199:37–46.

[93] Sugi R, Yokoyama A, Furuyama T, Uchiyama M, Yokozawa T. Inductive effect-assisted chain-growth polycondensation. Synthetic development from para- to meta-substituted aromatic polyamides with low polydispersities. J Am Chem Soc 2005;127:10172–10173.

[94] Iovu MC, Sheina EE, Gil RR, McCullough RD. Experimental evidence for the quasi-"living" nature of the grignard metathesis method for the synthesis of regioregular poly(3-alkylthiophenes). Macromolecules 2005;38: 8649–8656.

[95] Miyakoshi R, Yokoyama A, Yokozawa T. Synthesis of poly (3-hexylthiophene) with a persity narrower polydispersity. Macromol Rapid Commun 2004;25:1663–1666.

[96] Sheina EE, Liu JS, Iovu MC, Laird DW, McCullough RD. Chain growth mechanism for regioregular nickel-initiated

cross-coupling polymerizations. Macromolecules 2004; 37:3526–3528.

[97] Yokoyama A, Miyakoshi R, Yokozawa T. Chain-growth polymerization for poly(3-hexylthiophene) with a defined molecular weight and a low polydispersity. Macromolecules 2004;37:1169–1171.

[98] Miyakoshi R, Shimono K, Yokoyama A, Yokozawa T. Catalyst-transfer polycondensation for the synthesis of poly(p-phenylene) with controlled molecular weight and low polydispersity. J Am Chem Soc 2006;128:16012–16013.

[99] Yokoyama A, Kato A, Miyakoshi R, Yokozawa T. Precision synthesis of poly (n-hexylpyrrole) and its diblock copolymer with poly(p-phenylene) via catalyst-transfer polycondensation. Macromolecules 2008;41:7271–7273.

[100] Huang L, Wu S, Qu Y, Geng Y, Wang F. Grignard metathesis chain-growth polymerization for polyfluorenes. Macromolecules 2008;41:8944–8947.

[101] Javier AE, Varshney SR, McCullough RD. Chain-growth synthesis of polyfluorenes with low polydispersities, block copolymers of fluorene, and end-capped polyfluorenes: toward new optoelectronic materials. Macromolecules 2010;43:3233–3237.

[102] Arias AC, MacKenzie JD, McCulloch I, Rivnay J, Salleo A. Materials and applications for large area electronics: solution-based approaches. Chem Rev 2010;110:3–24.

[103] Dennler G, Scharber MC, Brabec CJ. Polymer-fullerene bulk-heterojunction solar cells. Adv Mat 2009;21:1323–1338.

[104] McNeill CR, Greenham NC. Conjugated-polymer blends for optoelectronics. Adv Mat 2009;21:3840–3850.

[105] Grimsdale AC, Chan KL, Martin RE, Jokisz PG, Holmes AB. Synthesis of light-emitting conjugated polymers for applications in electroluminescent devices. Chem Rev 2009;109:897–1091.

[106] Allard S, Forster M, Souharce B, Thiem H, Scherf U. Organic semiconductors for solution-processable field-effect transistors (OFETs). Angew Chem Int Ed 2008;47:4070–4098.

[107] Hollinger J, Jahnke AA, Coombs N, Seferos DS. Controlling phase separation and optical properties in conjugated polymers through selenophene-thiophene copolymerization. J Am Chem Soc 2010;132:8546–8554.

[108] Khanduyeva N, Senkovskyy V, Beryozkina T, Horecha M, Stamm M, Uhrich C, Riede M, Leo K, Kiriy A. Surface engineering using kumada catalyst-transfer polycondensation (KCTP): preparation and structuring of poly(3-hexylthiophene)-based graft copolymer brushes. J Am Chem Soc 2009;131:153–161.

[109] Senkovskyy V, Khanduyeva N, Komber H, Oertel U, Stamm M, Kuckling D, Kiriy A. Conductive polymer brushes of regioregular head-to-tail poly(3-alkylthiophenes) via catalyst-transfer surface-initiated polycondensation. J Am Chem Soc 2007;129:6626–6632.

[110] Senkovskyy V, Tkachov R, Beryozkina T, Komber H, Oertel U, Horecha M, Bocharova V, Stamm M, Gevorgyan SA, Krebs FC, Kiriy A. "Hairy" poly(3-hexylthiophene) particles prepared via surface-initiated kumada catalyst-transfer polycondensation. J Am Chem Soc 2009;131: 16445–16453.

[111] Yokozawa T, Shimura H. Condensative Chain Polymerization. II. Prefential esterification of propagating end group in Pd-Catalyzed CO-Insertion polycondensation of 4-bromo-phenol derivatives. J Polym Sci Part A-Polym Chem 2000;37:2607–2618.

[112] Yokozawa T, Maeda D, Hiyama N, Hiraoka S. Chain-growth polycondensation in solid-liquid phase with ammonium salts for well-defined polyesters. Macromol Chem Phys 2001;202:2181–2186.

2

OVERVIEW OF CONTROLLED/LIVING POLYMERIZATION METHODS OF VINYL MONOMERS

ROBERT T. MATHERS, ANDREW J. D. MAGENAU, KRISTIN SCHRÖDER, AND KRZYSZTOF MATYJASZEWSKI

2.1 SCOPE

The purpose of this chapter is to provide an overview of controlled/living polymerization systems of vinyl monomers. The reader is referred to other reviews for discussions of ring-opening polymerizations [1–4]. After a discussion of the attributes of living polymerizations, an overview of anionic, cationic, coordination–insertion, and controlled radical polymerization (CRP) methods will follow. The discussion of CRP will focus on atom transfer radical polymerization (ATRP); however, other reviews covering reversible addition fragmentation chain transfer (RAFT) [5], organometallic-mediated radical polymerization (OMRP) [6, 7], and nitroxide-mediated polymerization (NMP) [8, 9] are available for the interested reader [10]. This chapter will conclude with a summary of progress in hybrids and complex polymeric architectures.

2.2 DEFINITION OF LIVING POLYMERIZATIONS

Living polymerizations are characterized by the absence of chain termination and chain transfer during polymerization [11–13]. In the first report of block copolymers, based on styrene and isoprene, Szwarc mentioned that "propagation should continue until all monomer is consumed" and adding more monomer or a different monomer should result in further propagation [14]. Although the ideal standard for controlled/living behavior includes quantitative and fast initiation, it was not part of the original definition proposed by Szwarc [11].

Generally, the degree to which a particular polymerization mechanism exhibits living behavior versus nonliving behavior depends on choice of monomer, solvent, temperature, initiator, and other experimental parameters which may include exclusion of water and/or oxygen. CRPs are usually less sensitive to impurities than ionic polymerizations such as those involving carbanions and carbenium ions. This distinction between living and nonliving polymerizations is not a step function, but rather a continuum whereby a polymerization system can gradually become more or less "living" based on the experimental and stoichiometric choices [15].

2.3 SIGNIFICANCE OF LIVING POLYMERIZATIONS

Living polymerizations afford a variety of options to control polymer size, dispersity, composition, and shape (architecture), as well as allowing specific and defined placement of useful chemical functionalities within these macromolecules. Common strategies for attaching chemical functionalities include the use of functional initiators and post-polymerization reactions. In addition, the absence of chain termination allows access to complex polymeric architectures, which may include block copolymers, multiblock polymers, star polymers, and bottlebrush copolymers.

Monitoring Polymerization Reactions: From Fundamentals to Applications, First Edition. Edited by Wayne F. Reed and Alina M. Alb.
© 2014 John Wiley & Sons, Inc. Published 2014 by John Wiley & Sons, Inc.

2.4   ATTRIBUTES OF A CONTROLLED/LIVING POLYMERIZATIONS

The degree of control or livingness during vinyl polymerizations is characterized by a number of attributes [16, 17]. The reality is that all polymerization systems need to be optimized to fully realize the potential of these attributes. Otherwise, polymerization systems may experience slow initiation, exchange problems, or irreversible termination reactions. Regardless of the polymerization mechanism, it is expected that living and controlled polymerizations will exhibit the following attributes:

1. Living polymerizations exhibit a linear first-order plot of monomer consumption (provided that initiation is fast) which indicates that the number of propagating chain ends is constant. Such a plot is obtained by considering the relationship between time (*x*-axis) and logarithmic monomer consumption. If the chain ends propagate at a constant rate in the absence of detectable amounts of chain termination, then a linear relationship results. Deviations from ideal linear behavior are observed as a result of termination events that decrease the slope, or slow initiation processes that increase the slope. However, this method is not sensitive to chain transfer.

2. Living polymerizations have a constant number of propagating chain ends as confirmed by a plot of number-average molecular weight versus conversion. In this regard, a constant number of chain ends necessitates a fast initiation period without chain transfer. In order to achieve these qualities, a judicious choice of experimental conditions (temperature, time, etc.) and component choices (monomer, solvent, initiator, etc.) is needed. If chain transfer occurs, then more polymer chains result and the molecular weight decreases below its theoretical value.

3. In the absence of termination and chain transfer, living polymerizations will produce polymers with molecular weight distributions (M_w/M_n) below 1.1. In order that all polymer chains propagate for approximately the same time, living polymerizations need quantitative initiation which is fast relative to propagation. As noted by Quirk and Lee, deliberate termination of 15% of the chain ends during a living polymerization still gave M_w/M_n values below 1.1 [16]. Although molecular weight distributions are not extremely sensitive to fractions of irreversible terminated polymer chains, M_w/M_n values are convenient for comparing the effect of experimental or stoichiometric modifications. However, by itself, M_w/M_n values do not represent rigorous criteria for establishing that a polymerization exhibits living behavior. Very often,

polymers with preserved end-functionality and predetermined molecular weight may have high dispersity, due to the slow exchange reactions between active and dormant species [18].

4. Living polymerizations produce polymers with controlled molecular weights. Since molecular weight greatly influences the physical properties of polymers, this is an important attribute of living polymerizations. This feature is enhanced by the predictable nature of monomer consumption, quantitative and fast initiation, and the absence of chain termination. As shown in Equation 2.1, the theoretical number-average molecular weight (M_n) for a particular conversion is determined by the grams of monomer and moles of initiator.

$$M_n = \left(\frac{\text{g monomer}}{\text{mol initiator}} \right) (\text{conversion}) \qquad (2.1)$$

5. A variety of postpolymerization reactions should be feasible. These include the ability to quantitatively functionalize the polymer chain end, add another monomer, or react the chain ends for the purpose of creating various architectures. In the case of block copolymers, it is important that initiation (crosspropagation) be fast relative to the subsequent propagation.

2.5   THERMODYNAMIC CONSIDERATIONS FOR CONTROLLED/LIVING POLYMERIZATIONS

A thermodynamic perspective of vinyl polymerizations can be elucidated from the Gibbs free energy equation. The prerequisite for polymerization is that the Gibbs free energy (ΔG) should be negative. Since ΔS in Equation 2.2 is often negative for chain growth processes due to the necessity of assembling vinyl monomers through σ-bonds, propagation has certain temperature limitations. When the polymerization temperature reaches the ceiling temperature (T_c) and above, ΔG equals zero and greater, and therefore propagation is no longer thermodynamically feasible because ΔG becomes positive. In Scheme 2.1, increasing the substituent size on alkenes dramatically increases steric hindrance in the resulting polymer and decreases T_c values. For example, the T_c value for α-methyl styrene ($T_c = 63\,°C$) is much lower than that for styrene ($T_c = 400\,°C$) and propylene ($T_c = 492\,°C$) [19]. Indeed, 1,1-diphenylethylene is not able to homopolymerize due to the excessive steric hindrance in the hypothetical resulting polymer.

$$\Delta G = \Delta H - T\Delta S \qquad (2.2)$$

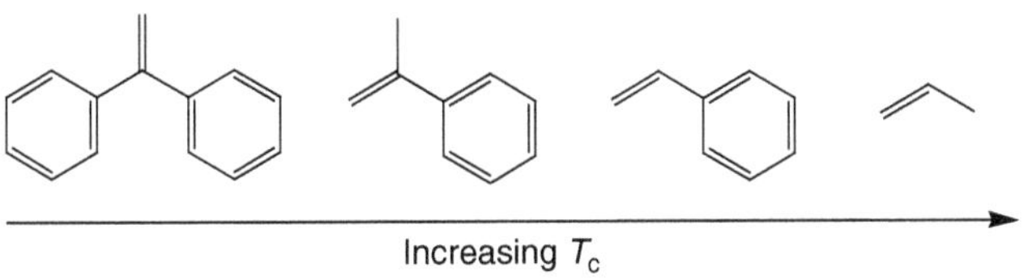

Increasing T_c

SCHEME 2.1 Influence of alkene substituents on the ceiling temperature (T_c).

2.6 TYPES OF LIVING POLYMERIZATIONS

2.6.1 Living Anionic Polymerizations

2.6.1.1 *Overview* In 1956, Szwarc was the first to report that sodium napthalenide in tetrahydrofuran (THF) at $-78\,°C$ would initiate the living anionic polymerization of styrene [11]. This seminal work resulted in the formation of carbanion chain ends, which would propagate with monomer in the absence of chain transfer or termination on a laboratory time frame. In addition, the living polymerization of styrene was a remarkable improvement over previous nonliving anionic systems that employed butylmagnesium bromide, phenylmagnesium bromide, triphenylmethylsodium, and sodium in liquid ammonia for polymerization of methacrylonitrile [20], as well as the copolymerization of methyl methacrylate and methacrylonitrile [21].

After Szwarc's 1956 publication, the success of living anionic polymerizations eventually led to the development of other living polymerization methods, which will be summarized in the following sections of this chapter.

Due to the high pK_a values of initiators and propagating chain ends, anionic polymerizations can undergo a variety of termination and chain transfer reactions. As a result, careful selection of solvent, initiator, and monomer is important for maintaining the living polymerization behavior. For example, acidic protons on monomers or the presence of alcohols will easily terminate carbanions.

2.6.1.2 *Monomers* Several general categories of vinyl monomers are suitable for anionic polymerizations. These include aromatic monomers (2-vinylpyridine, styrene), conjugated dienes (butadiene, isoprene), and alkyl methacrylates. In the case of vinyl monomers, adjacent substituents that stabilize an anion are most suitable for anionic polymerizations. Examples include substituents found in styrene, butadiene, isoprene, alkyl methacrylates, or cyano acrylate which stabilize propagating anions by electronic effects.

2.6.1.3 *Initiators* Anionic polymerizations utilize molecules with radical anions, Lewis bases, or nucleophiles such as carbanions, to initiate polymerization. Three important facets of initiation (Scheme 2.2a) are worth mentioning. First, in order to maximize initiator efficiency and obtain fast

initiation, the pK_a value of the initiator needs to be higher than the pK_a value of the propagating chain end. High initiator efficiency and a fast initiation period relative to the time required to propagate all of the monomer (Scheme 2.2b) are important aspects of living polymerizations. If the pK_a value for the initiator is much lower than the pK_a value for the propagating chain end, then irreversible initiation will not occur.

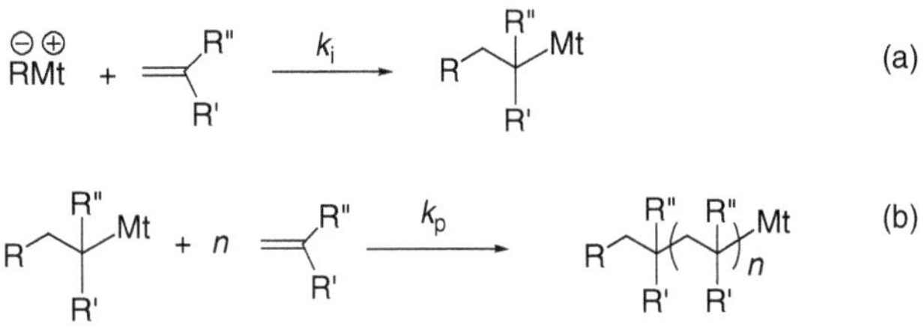

SCHEME 2.2 Anionic polymerization of vinyl monomers showing (a) initiation with RMt (Mt = Li, Na, K, MgBr) and (b) propagation of vinyl monomers.

Second, choosing an initiator with an appropriate pK_a value depends on the substituents attached to the vinyl monomer. Common examples of initiators include alkyllithiums (*n*-butyllithium, *sec*-butyllithium), Grignard reagents (RMgBr), or metal alkoxides (ROMt). Although initiators with high pK_a values, such as alkyllithiums ($pK_a \sim 45$), are quite useful for styrene, butadiene, and isoprene, the reactivity causes side reactions for some monomers. For example, in THF at $-78\,°C$, a large fraction of *n*-butyllithium will react with the carbonyl moiety in methyl methacrylate rather than the alkene [22]. In order to produce a more controlled initiation process, reacting *n*-butyllithium with 1,1-diphenylethylene creates an adduct (1,1-diphenylhexyllithium) which is more selective due the presence of the phenyl groups. Additionally, the pK_a value for 1,1-diphenylhexyllithium is similar to diphenylmethane ($pK_a = 32.2$) [23]. However, even though the pK_a value for 1,1-diphenylhexyllithium is substantially lower than that for alkyllithiums, this adduct is still able to efficiently initiate the polymerization of styrenes and dienes [24]. It should also be noted that the low pK_a values ($pK_a \sim 16$) for metal alkoxides (ROMt) are not sufficient for styrene, butadiene, and isoprene. Generally, metal alkoxides are only useful for more reactive monomers, such as cyano acrylates.

Third, the kinetics of initiation with Grignard reagents and alkyllithiums is further complicated by association of ion pairs. The association, as described by the Winstein spectrum (Scheme 2.3), is influenced by steric hindrance as well as temperature, solvent, and concentration. Normally, the position in this series (Scheme 2.3) varies from aggregated species in hydrocarbon solvents to contact ion pairs and solvent separated ion pairs in the presence of polar solvents. Decreasing the concentration and increasing solvent polarity shifts the equilibria toward the right. Experimentally, *sec*-butyllithium, which is tetrameric in hydrocarbon media, is more reactive at ambient temperature than hexameric *n*-butyllithium. Examples of hydrocarbon monomers which are well suited for *sec*-butyllithium include butadiene, isoprene, and styrene.

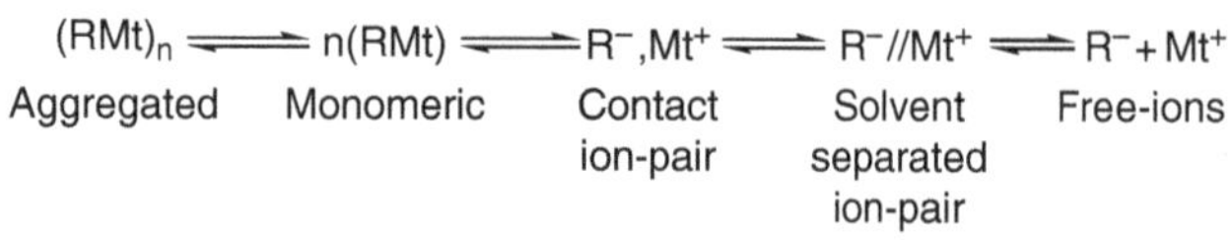

$$(RMt)_n \rightleftharpoons n(RMt) \rightleftharpoons R^-, Mt^+ \rightleftharpoons R^-//Mt^+ \rightleftharpoons R^- + Mt^+$$

| Aggregated | Monomeric | Contact ion-pair | Solvent separated ion-pair | Free-ions |

SCHEME 2.3 Winstein spectrum of carbanions.

2.6.1.4 *Chain-End Functionalization* The ability to functionalize living carbanion chain ends provides many options. Reaction of polystyryllithium (PSLi) with ethylene oxide and subsequent protonation in Scheme 2.4 yields a versatile strategy to obtain hydroxyl functionalized polystyrene [25].

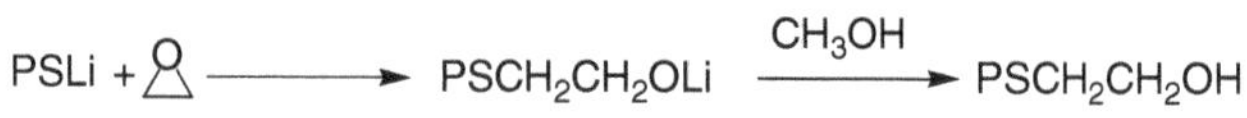

$$PSLi + \triangle\!\!O \longrightarrow PSCH_2CH_2OLi \xrightarrow{CH_3OH} PSCH_2CH_2OH$$

SCHEME 2.4 Functionalization of polystyryllithium (PSLi) with ethylene oxide.

A general functionalization method (Scheme 2.5) via hydrosilylation reactions provides options for nitriles [26], amines [27], and many other useful functionalities.

2.6.2 Living Cationic Polymerizations

2.6.2.1 *Overview* Cationic polymerizations proceed via a chain growth process which involves a carbenium ion (R_3C^+) at the chain end [19, 28]. Most examples of living cationic polymerizations utilize low temperatures ($-90\,°C$ to $-30\,°C$) and electrophilic solvents, such as dichloromethane. In contrast to anionic polymerizations that require Lewis bases as initiators, cationic polymerizations employ Brønsted acids or Lewis acids as initiators or coinitiators. For example, common Lewis acids include the halides of aluminum, boron, tin, and titanium.

2.6.2.2 *Monomers* Alkene monomers which contain an electron donating substituent that can stabilize a carbenium ion are well suited for cationic polymerizations. These include substituents based on nitrogen (*N*-vinyl carbazole),

oxygen (vinyl ethers), and carbon (4-methoxystyrene, α-methyl styrene, styrene, and isobutylene). As expected, in Scheme 2.6, the stabilization of the carbenium ion influences the monomer reactivity. For example, resonance stabilization of the phenyl ring in styrene provides similar stabilization as the hyperconjugation of two methyl groups in isobutylene.

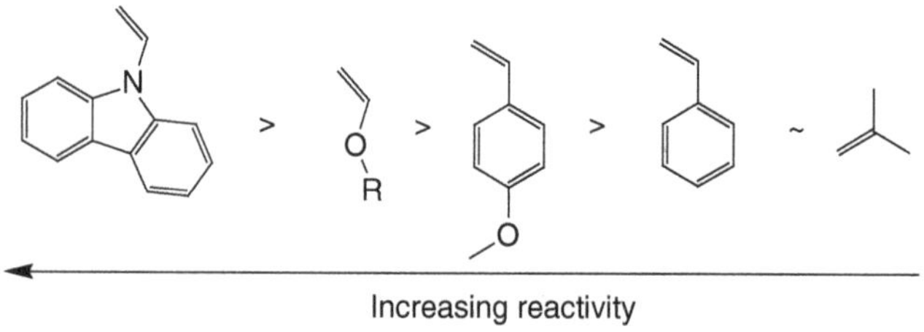

Increasing reactivity

SCHEME 2.6 Monomer reactivity for a series of common monomers which will polymerize with Lewis acids and Brønsted acids. Scheme adapted from Matyjaszewski K, *Cationic Polymerizations: Mechanisms, Synthesis, and Applications.* New York: Marcel Dekker; 1996. © 1996 Marcel Dekker, New York.

The first living cationic polymerization of isobutylvinyl ether was reported with HI/I_2 in 1984 [29]. Sawamoto and coworkers suggested the HI/vinyl ether adduct acts as an initiator while the I_2 acts as an activator [30]. In 1986, complexes of BCl_3 with esters, such as cumyl acetate, were reported to initiate the living polymerization of isobutylene [31]. Then, a living cationic polymerization of *N*-vinylcarbazole with HI in toluene ($-40\,°C$) and methylene chloride ($-78\,°C$) was reported in 1987 [30].

2.6.2.3 *Initiation* Cationic polymerizations can be initiated by either chemical [19] or physical methods, such as those involving the γ-radiation-induced polymerization [32]. For the purpose of conducting a living polymerization, chemical methods offer more control. Generally, Lewis acids coinitiators are utilized to convert an alkyl or aryl halides (Scheme 2.7) into a carbenium ion (Scheme 2.8), which then reacts with monomer. The strength of Lewis acid depends on atomic number (e.g., $GaX_3 > AlX_3 > BX_3$), oxidation state (e.g., $Sn^{IV}Cl_4 >> Sn^{II}Cl_2$), and electronegativity of the ligands (e.g., $TiCl_4 > TiCl_3OR$; $SnCl_4 > SnBr_4$) [19]. Overall, the reactivity of an initiator has a big influence on the kinetics of initiation. For example, if weak Lewis acids are paired with less reactive alkyl halides, then poor initiator efficiency will be expected from a slow initiation process.

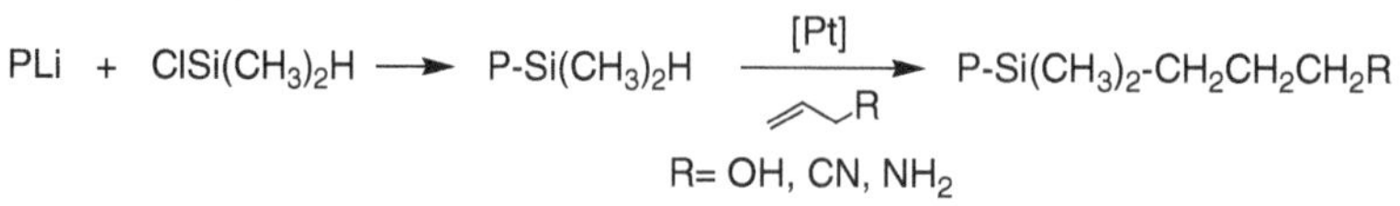

$$PLi + ClSi(CH_3)_2H \longrightarrow P\text{-}Si(CH_3)_2H \xrightarrow[\substack{\nwarrow R}]{[Pt]} P\text{-}Si(CH_3)_2\text{-}CH_2CH_2CH_2R$$

$$R = OH, CN, NH_2$$

SCHEME 2.5 General functionalization method for carbanion chain ends (P = polymer) using hydrosilylation chemistry.

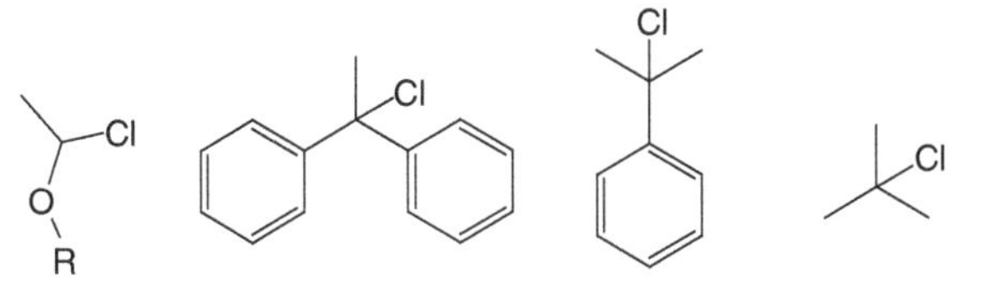

SCHEME 2.7 Examples of initiators for cationic polymerizations.

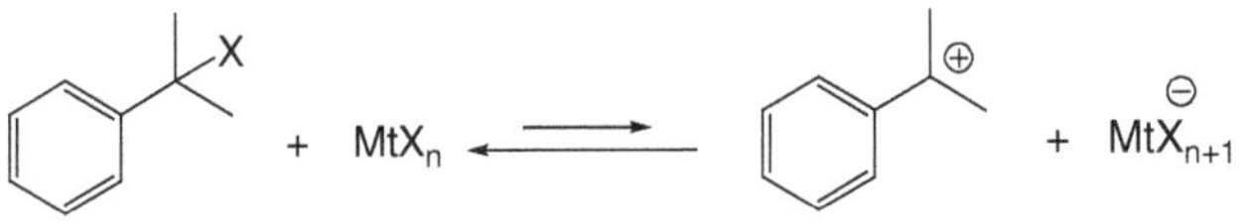

SCHEME 2.8 Reaction of initiator (cumyl halide) and coinitiator (MtX$_n$; X = Cl,Br) to form reversibly a carbenium ion.

2.6.2.4 *Propagation* After initiator and coinitiator react to form a carbenium ion (Scheme 2.8), the subsequent propagation of living chain ends with monomer (Scheme 2.9) should continue until all monomers are consumed. It has been reported that the propagation rate constants (k_p) for isobutylene polymerizations with TiCl$_4$, Me$_2$AlCl, and BCl$_3$ in hexanes/CH$_3$Cl are very high ($k_p \sim 10^9$ M^{-1} s^{-1}) and approach diffusion controlled limits [33, 34]. Since the carbocationic species in living and nonliving polymerization systems are similar, reducing the relative proportion of active sites through an equilibrium between active and dormant species minimizes termination and chain transfer [17].

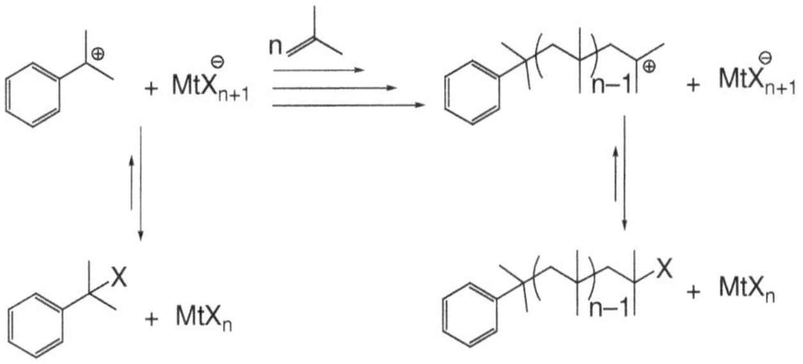

SCHEME 2.9 Propagation of monomer during isobutylene polymerization.

If the carbenium ion undergoes termination or chain transfer, the polymerization system deviates from living behavior. In Scheme 2.10, β-proton elimination is a common type of chain transfer that generates another carbenium ion while producing vinylidene (1,1-disubstituted) alkenes as

well as trisubstituted alkenes. Generally, temperature has a large impact on the degree to which β-proton elimination occurs. Increasing temperature favors chain transfer. Therefore, most controlled/living carbocationic polymerizations are carried out at low temperatures.

2.6.3 Coordination–Insertion Polymerizations

2.6.3.1 *Overview* A large number of catalysts based on vanadium [35–37], titanium [38–40], zirconium [41], hafnium [42], lanthanides (in particular neodymium, samarium, and ytterbium) [43], cobalt [44, 45], niobium [46], chromium [47], nickel [48], and palladium [49] provide well-defined polyolefins. Many of these systems are able to meet the requirements for living polymerizations by suppressing β-hydride and β-methyl eliminations, as well as chain transfer to cocatalysts, such as alkyl aluminums or methylaluminoxane (MAO). Since MAO is usually obtained as a liquid solution with residual trimethylaluminum, drying MAO to a white powder and removing residual trimethylaluminum can help minimize chain transfer to cocatlysts.

In many reports of living systems, suppression of chain transfer is often accomplished by low temperatures. For example, activation of dimethyl zirconocene [Cp$_2$ZrMe$_2$] with B(C$_6$F$_5$)$_3$ exhibits attributes of a living polymerization at −78 °C for propylene [50]. Although reducing the temperature helps obtain some characteristics of a living polymerization, monomer propagation is generally slowed and the resulting polymerization activity and molecular weight will be lower. Consequently, it is more desirable to find living catalyst systems which operate at or above ambient temperatures.

2.6.3.2 *Early-Metal Catalyst Systems* In 1979, a vanadium-based catalyst was the first polymerization system to exhibit living behavior below −65 °C [36, 37]. This V(acac)$_3$/ Et$_2$AlCl complex produced partially syndiotactic polypropylene ([r] = 0.81) with narrow molecular weight distributions (1.05–1.20) and a linear increase in molecular weight as a function of time. In addition to these two attributes of living polymerizations, further reports by Doi et al. established that vandanium catalysts could also produce chain-end functionalized polypropylene with iodine, aldehyde, and methacrylate groups [51, 52].

Fujita and coworkers at Mitsui reported the living polymerization of ethylene using a bis(phenoxyimine) titanium

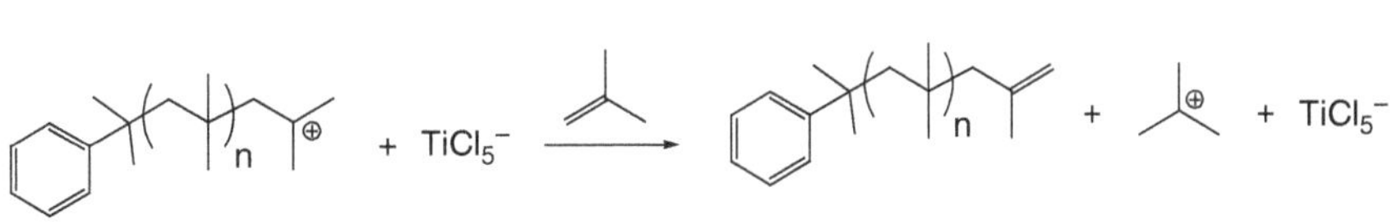

SCHEME 2.10 Chain transfer to monomer during isobutylene polymerization.

catalyst (Scheme 2.11, R=H, Ar=C$_6$H$_5$) in 1999 [39]. Coates and coworkers recognized that a combinatorial approach could identify interesting ligand modifications for extending this catalyst system (Scheme 2.11, R=^{t}Bu, Ar=C$_6$H$_5$) to propylene [40, 53]. Indeed, a 2,1-insertion was reported which allowed syndiotactic polypropylene due to chain-end control. Since the initial report by Fujita, this versatile class of bis-ligated titanium catalysts has also been extended to phenoxyketimine ligands [54] and indolide–imine ligands [55].

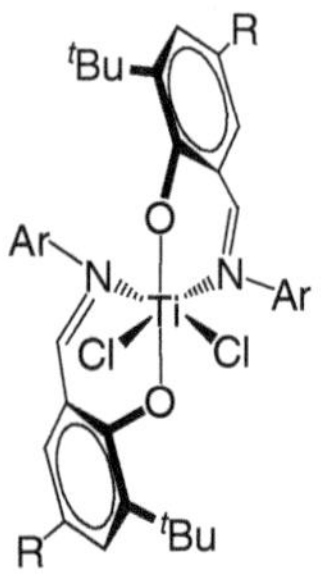

SCHEME 2.11 Titanium-based phenoxyimine catalysts as reported by Fujita in 1999 (R=H, Ar=C$_6$H$_5$) for living polymerization of ethylene, Coates in 2000 (R=^{t}Bu, Ar=C$_6$H$_5$) for syndiotactic polypropylene. Subsequently, fluorinated versions were reported by Coates (R=^{t}Bu, Ar=C$_6$F$_5$) and Fujita (R=H, Ar=C$_6$F$_5$).

In 2001, Coates (R=^{t}Bu, Ar=C$_6$F$_5$) [56] and Fujita (R=H, Ar=C$_6$F$_5$) [57] reported that the fluorination of phenoxyimine ligands (Scheme 2.11) gave catalysts which allowed the living polymerization of propylene. These catalysts were able to produce syndiotactic polypropylene with narrow molecular weight distributions ($M_n \leq 1.14$) and a linear increase in molecular weight versus conversion (based on polymer yield). The high degree of crystallinity obtained with the fluorinated phenoxyimine catalyst (Scheme 2.11, R=^{t}Bu) was accompanied by [*rrrr*] values as high as 96% and T_m values from 140 °C to 148 °C [56, 58].

As shown in Scheme 2.12, polymerization of 1,5-hexadiene proceeded to give an unusual insertion/isomerization mechanism that yielded vinyl-functionalized polyolefins [59]. Subsequently, the living nature of this polymerization system allowed formation of copolymers containing syndiotactic-polypropylene blocks or ethylene-*co*-propylene blocks. A general functionalization method was reported to prepare polyolefin architectures via cross metathesis [60]. Attaching alcohols, halides, acrylates, and fluorinated moieties onto the polymer chain was accomplished with a ruthenium-based metathesis catalyst.

Other researchers have reported living polymerizations with amine- or pyridine-containing ligands. Some select examples with these types of ligands with titanium, zirconium, and hafnium are worth mentioning due to their ability to polymerize α-olefins. For instance, Kol and coworkers have utilized various amine bis(phenolate) titanium complexes (Scheme 2.13) for living polymerization of 1-hexene [61] and 4-methyl-1-pentene [62].

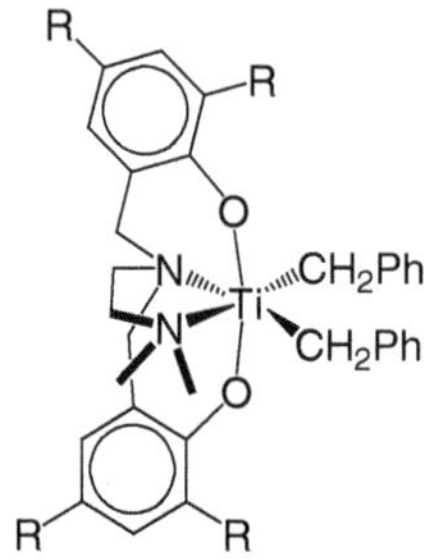

SCHEME 2.13 Amine-phenolate catalysts for living polymerization of 1-hexene (R = ^{t}Bu) and 4-methyl-1-pentene (R = Cl).

Activation of tridentate amine-diol titanium complexes with MAO also produced poly(1-hexene) with narrow molecular weight distributions [63]. Modification of C_1-symmetric hafnium catalysts (Scheme 2.14) [64] resulted in pyridylamido ligands (Scheme 2.15) for the polymerization of isotactic poly(α-olefins) with relatively narrow molecular weight distributions ($M_n \leq 1.20$) for molecular masses up to 152,000 g mole^{-1} [42].

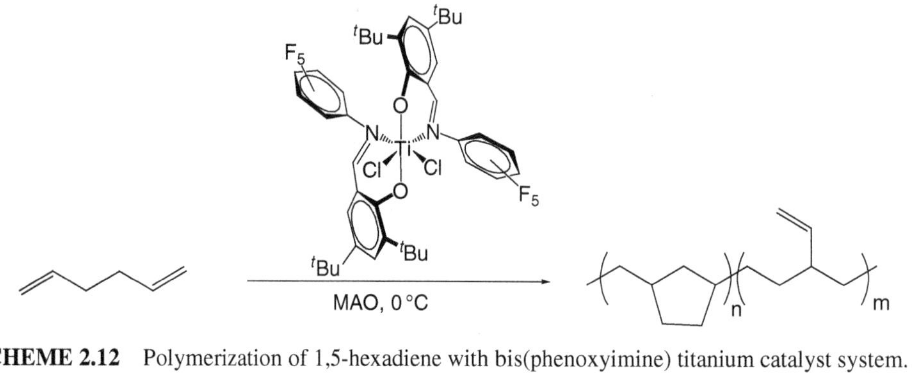

SCHEME 2.12 Polymerization of 1,5-hexadiene with bis(phenoxyimine) titanium catalyst system.

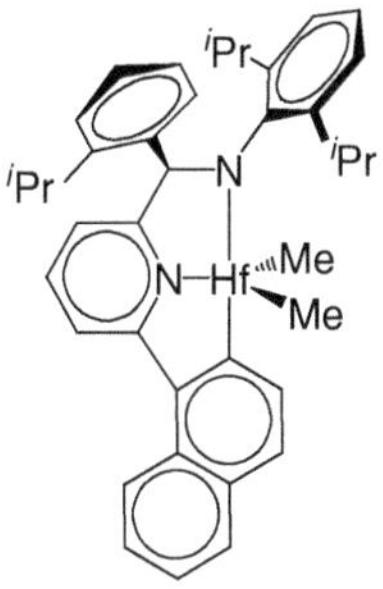

SCHEME 2.14 Pyridylamidohafnium complex for isotactic poly(α-olefins). From Boussie TR, Diamond GM, Goh C, Hall KA, LaPointe AM, Leclerc MK, Murphy V, Shoemaker JAW, Turner H, Rosen RK, Stevens JC, Alfano F, Busico V, Cipullo R, Talarico G. Nonconventional catalysts for isotactic propene polymerization in solution developed by using high-throughput-screening technologies. Angew Chem Int Ed 2006;45:3278–3283. © Wiley-VCH Verlag GmbH & Co. KGaA. Reproduced with permission.

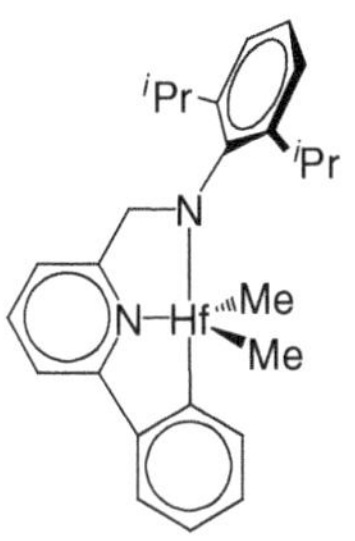

SCHEME 2.15 Modified pyridylamidohafnium complex for living polymerization of α-olefins after activation with $B(C_6F_5)_3$ to produce isotactic poly(α-olefins). Reprinted (adapted) with permission from Domski GJ, Lobkovsky EB, Coates GW. Polymerization of α-olefins with pyridylamidohafnium catalysts: living behavior and unexpected isoselectivity from a C_s-symmetric catalyst precursor. Macromolecules 2007;40:3510–3513. ©2007 American Chemical Society.

Sita and coworkers reported the living polymerization of 1-hexene and propylene with monocyclopentadienylzirconium amidinate catalysts (Scheme 2.16) [41]. Activation of the complex with 0.5 equivalent of $[PhNMe_2H][B(C_6F_5)_4]$ resulted in a mechanism that involves degenerative transfer of methyl groups between active cationic metal centers and dormant complexes combined with epimerization. The resulting living polymerization produces *atactic*-polypropylene. However, with 1 equivalent of $[PhNMe_2H][B(C_6F_5)_4]$, degenerative transfer is minimized and polypropylene has an isotactic microstructure ([*mmmm*]=0.71). Subsequently, this system was extended to a bimetallic zirconium complex [65] as well as a hafnium analogue [66] for coordinative chain-transfer polymerization (CCTP) with diethyl zinc [67].

The concept of degenerative transfer in coordination polymerization has been previously used for Fe-based systems [68, 69] and, commercially, to make block copolymers from simple olefins [70].

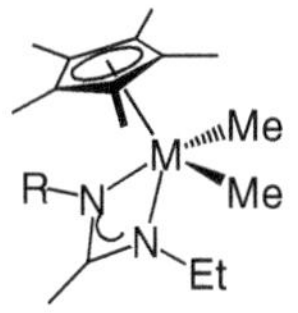

M = Zr, R = ^{t}Bu
M = Hf, R = Et

SCHEME 2.16 Monocyclopentadienylzirconium amidinate catalysts for living olefin polymerization.

2.6.3.3 *Late-Metal Catalyst Systems*

Compared to early metal systems, complexes based on late transition metals (group 8–11), for example, cobalt, nickel, and palladium, offer some interesting differences. First, since late metal are less oxophilic compared to titanium and zirconium, (co) monomers with certain types of functional groups are tolerated. For instance, palladium catalysts are able to polymerize norbornene monomers with pendant ester groups [71]. In some cases, polar monomers, such as acrylates, do not propagate, but provide options for chain-end functional or telechelic polymers [72]. Second, depending on catalyst structure, choice of monomer(s), and temperature, late metal catalysts may migrate along the polymer chain through a series of insertion/β-hydride elimination steps. In the case of Brookhart's α-diimine nickel complexes, this feature introduces branching during ethylene polymerizations and chain straightening during 1-octadecene polymerizations [73]. Other notable examples include Brookhart's cobalt complexes (Scheme 2.17), which polymerized ethylene (1 atm) at room temperature to yield end-functional polyethylene with narrow molecular weight distributions ($M_w/M_n=1.11$–1.16) [44, 45]. The catalyst was designed with an aryl spacer between the cobalt(III) and the functional group to prevent deactivation of the electrophilic metal center during chain migration.

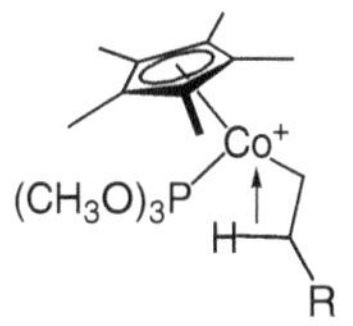

$(CH_3O)_3P$

R = CH_3, C_6H_5, C_6F_5,
Cl-C_6H_4, $(CH_2)_4Si(C_2H_5)_3$

SCHEME 2.17 Cobalt catalysts for living polymerization of ethylene.

Since Brookhart and coworkers initially polymerized ethylene, propylene, and 1-hexene with Ni(II) and Pd(II)-diimine catalysts (Scheme 2.18, R=R′=^{i}Pr) [74], numerous ligand modifications have resulted in an outstanding class of catalysts. Although this first report by Brookhart and coworkers was not a living polymerization system, subsequent modifications (Scheme 2.18, R=H, R′ = ′Bu) suppressed chain transfer during the polymerization of 1-hexene and 1-octadecene and allowed block copolymers [73]. Other researchers reported that Ni(II)-cyclophane diimine complexes would polymerize 1-hexene to give narrow molecular weight distributions ($M_w/M_n = 1.13$–1.22) [75].

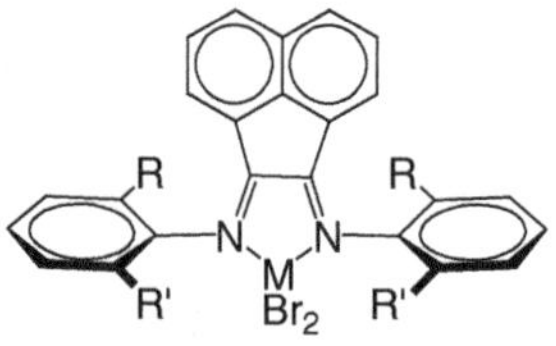

SCHEME 2.18 Brookhart's α-diimine catalyst in conjunction with MAO for nonliving (M = Ni and Pd, R = R′ = Pr) and living (M = Ni, R = H, R′ = ′Bu) polymerization of olefins.

2.7 CONTROLLED RADICAL POLYMERIZATIONS

2.7.1 Characteristics of Conventional Radical Methods

During a nonliving polymerization, such as conventional radical processes, four distinct mechanistic steps occur. As shown in Scheme 2.19 for monosubstituted vinyl monomers, these include initiation, propagation, irreversible termination by coupling or disproportionation, and chain transfer [76].

Initiation is most commonly accomplished by using thermally and photocatalytically unstable diazo or peroxide molecules. These thermally or photocatalytically activated initiators decompose to form radicals with a range of half-life values, which are a function of temperature [77].

Propagation is a repetitive process which attaches vinyl monomers to the growing polymer chain end through σ-bonds. A wide range of applicable monomers (Scheme 2.20) are suitable for conventional radical polymerizations, ranging from dienes to styrenics to (meth)acrylates.

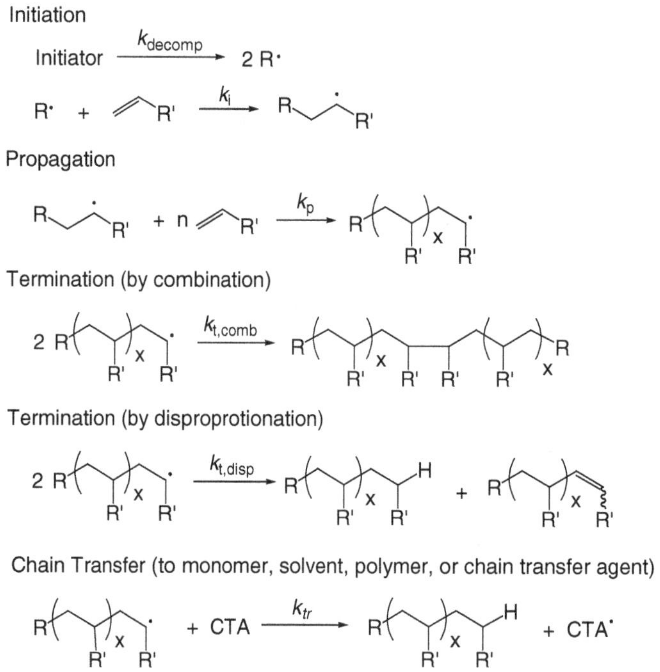

SCHEME 2.19 Mechanistic steps which occur during nonliving radical polymerizations.

R = H, CH$_3$

R = H, Cl, CN, OAc

R = N, CH, C-alkyl

R = H, alkyl
R′ = H, CH$_3$

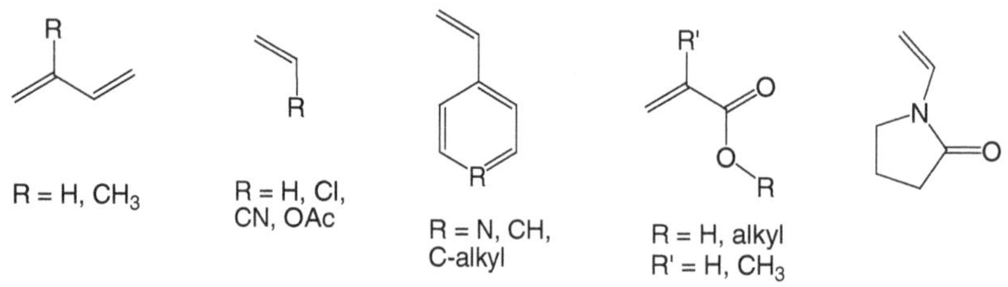

SCHEME 2.20 Suitable monomers for conventional radical polymerizations.

In addition to propagation with monomer, radicals also terminate by either combination with another radical to produce a σ-bond or by disproportionation via β-hydrogen abstraction. The propensity for disproportionation increases as the steric hindrance surrounding the radical increases. For example, a propagating methyl methacrylate radical would be more likely to undergo disproportionation compared to a propagating acrylate radical.

Chain transfer is a process which terminates a growing polymer chain while initiating a new propagating polymer chain at a rate which equals or exceeds propagation. Chain transfer may either be intentional with the addition of chain transfer agents (CTAs) to reduce the molecular weight or unintentional with chain transfer to solvent or monomer.

Each of the previously mentioned processes occur with defined rate constants of decomposition (k_d), propagation (k_p), termination (k_t), and chain transfer (k_{tr}).

2.7.2 Requirements for Controlled Radical Polymerizations

In the last two decades, tremendous improvements to conventional radical polymerizations have been made [10, 78, 79]. In general, converting nonliving systems into a controlled/living polymerization requires suppression of irreversible chain termination (Scheme 2.19) and chain transfer. Additionally, improving the initiation process, which is typically inefficient for conventional radical processes, is also necessary to produce a controlled polymerization.

2.7.3 Methods for Suppressing Termination

In order to reduce the contribution of termination processes (Scheme 2.19), controlled/living radical methods establish a fast equilibrium between dormant and active chain ends. Essentially, this equilibrium decreases the relative proportion of propagating radicals and allows them to reversibly deactivate rather than permanently terminate. The radical chain end concentration for CRPs (10^{-8}–10^{-9} M) is also much less than for ionic polymerizations (10^{-2}–10^{-3} M).

The two general methods for establishing a dynamic equilibrium are deactivation/activation (Scheme 2.21) and degenerative exchange (Scheme 2.22) processes. Methods based on deactivation/activation equilibria include NMP [9]

and ATRP [80]. The radical concentration for these methods is established based on the persistent radical effect (PRE) [81, 82]. In contrast, degenerative exchange processes, such as RAFT, have kinetics which are similar to conventional radical polymerizations [5].

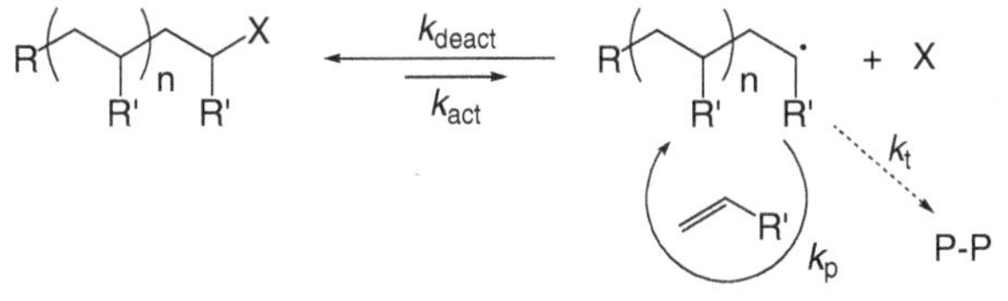

SCHEME 2.21 Propagation of vinyl monomers using a deactivation/activation equilibrium to suppress termination (k_t).

2.7.4 General Concepts of ATRP

ATRP is analogous to atom transfer radical addition reactions which are well known in the field of organic chemistry as Kharasch addition reactions [83]. These methods often utilize a transition metal complex based on copper, iron, ruthenium, and nickel to abstract a halogen and produce a carbon-based radical [84, 85]. Since the first reports in 1995 of living radical polymerizations based on copper(I) for styrene and methyl methacrylate [86] and ruthenium(II) for methyl methacrylate [87], this technique has become widely utilized in polymer science.

The versatility of ATRP stems from the wide range of conditions, initiators, catalysts (ligands, metals), monomers, solvents, and temperatures. An example of the versatility of this process is shown in Scheme 2.23, illustrating the wide range of monomers polymerizable by ATRP.

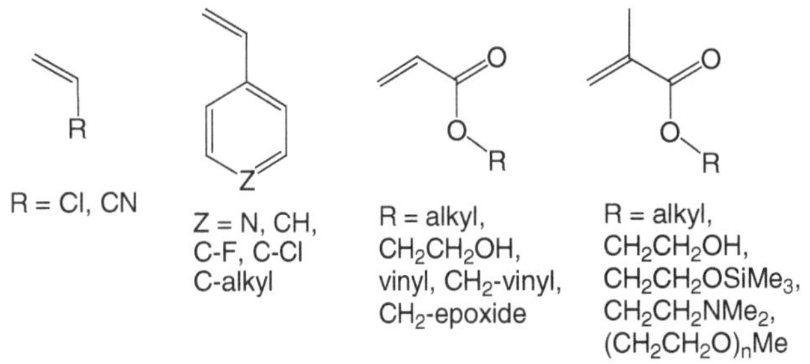

R = Cl, CN

Z = N, CH, C-F, C-Cl C-alkyl

R = alkyl, CH$_2$CH$_2$OH, vinyl, CH$_2$-vinyl, CH$_2$-epoxide

R = alkyl, CH$_2$CH$_2$OH, CH$_2$CH$_2$OSiMe$_3$, CH$_2$CH$_2$NMe$_2$, (CH$_2$CH$_2$O)$_n$Me

SCHEME 2.23 Selected examples of monomers which are suitable for ATRP.

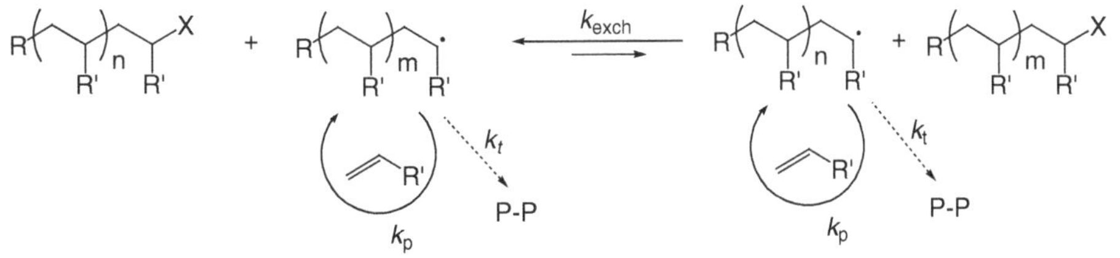

SCHEME 2.22 Polymerization of vinyl monomers using a degenerative exchange process to suppress termination during monomer propagation.

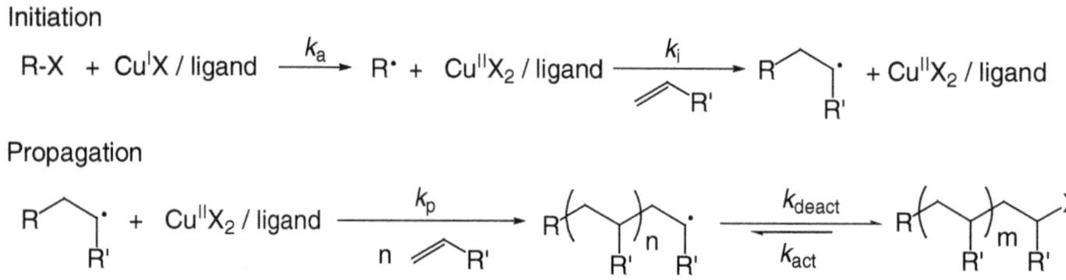

Initiation

$$R\text{-}X \ + \ Cu^{I}X \ / \ ligand \ \xrightarrow{\ k_a\ } \ R^{\bullet} \ + \ Cu^{II}X_2 \ / \ ligand \ \xrightarrow{\ k_i\ } \ R{\overset{\bullet}{\underset{R'}{\bigwedge}}} \ + \ Cu^{II}X_2 \ / \ ligand$$

Propagation

$$R{\overset{\bullet}{\underset{R'}{\bigwedge}}} \ + \ Cu^{II}X_2 \ / \ ligand \ \xrightarrow[n \ \diagup\!\!\diagup^{R'}]{\ k_p\ } \ R{\left(\bigwedge_{R'}\right)_n}{\overset{\bullet}{\underset{R'}{\bigwedge}}} \ \underset{k_{act}}{\overset{k_{deact}}{\rightleftharpoons}} \ R{\left(\bigwedge_{R'}\right)_m}{\underset{R'}{\bigwedge}}X$$

SCHEME 2.24   ATRP initiation and propagation leading to reversible deactivation of propagating radical.

2.7.4.1   *Mechanistic Aspects*   As shown in Scheme 2.24, a halogen atom is transferred from the alkyl halide initiator (R–X) to the metal center through homolytic cleavage (k_a) of the carbon–halogen bond to generate a carbon-based radical which subsequently adds to one or more vinyl monomers (k_i). During this inner sphere electron transfer (ISET) process, the metal atom is oxidized from Cu(I) to Cu(II) [88, 89]. Propagation (k_p) occurs until the radical is reversibly deactivated (k_{deact}) by Cu(II).

Although copper is commonly used, ATRP has been extended to a variety of transition metals including Ti, Mo, Re, Fe, Ru, Os, Rh, Co, Ni, and Pd [79, 80, 90]. Since ATRP minimizes termination through a fast equilibrium between dormant and active chain ends, the polymerization is able to proceed to high monomer conversion with predictable molecular weights based on stoichiometry.

2.7.4.2   *Performance of ATRP Catalysts*   Due to the wide variety of ligands, solvents, and experimental conditions that have been reported, comparison of various catalytic systems has been accomplished by describing the activation/deactivation equilibrium (K_{ATRP}) using Equation 2.3. The methodology for calculating K_{ATRP} involves radical trapping experiments [91], modified equations for the PRE [92], and electrochemical methods [93–95].

As shown in Scheme 2.25 and Scheme 2.26, typical values for K_{ATRP} range from 10^{-9} to 10^{-4} and strongly depend on ligand [96], initiator [97], solvent [93], and halogen atom [95]. Increasing the solvent polarity can influence K_{ATRP} values by one or two orders of magnitude [93, 98].

$$K_{ATRP} = \frac{k_{act}}{k_{deact}} \qquad (2.3)$$

2.7.4.3   *ATRP Methods*   A wide variety of ATRP methods have been reported in the last 15 years. These include normal ATRP starting with Cu(I)X and reverse ATRP which initially contains Cu(II)X$_2$ [80]. During these polymerizations, any termination reactions will produce Cu(II)X$_2$ and diminish the quantity of available Cu(I)X. As a result, typical catalyst loading are ~1 mol%.

Recently, advances have minimized the catalyst loading to ppm levels for iron [99] and copper [100] systems. Since

Cu(I)X is needed for activation, these methods utilize chemical additives [101] or electrochemical equipment (i.e., electrodes) [102] to recycle generated Cu(II)X$_2$ back to Cu(I)X. One example for a novel method is initiators for continuous activator regeneration (ICAR) which uses a radical sources [e.g., azobisisobutyronitrile (AIBN)] to convert the *in situ* formed Cu(II)X$_2$ back to Cu(I)X [101]. As a result, ICAR ATRP only requires 10–50 ppm catalyst.

Activators regenerated by electron transfer (ARGET) ATRP is another promising method, which reduces Cu(II)X$_2$ to Cu(I)X with chemical reducing agents such as ascorbic acid or Sn(ethylhexanoate)$_2$ [103].

2.8   ARCHITECTURAL POSSIBILITIES WITH CONTROLLED/LIVING POLYMERIZATION METHODS

Due to the advantages associated with controlled/living polymerizations, many strategies have emerged to synthesize polymer architectures with precisely defined branch points, controllable number of branches, and various functionalities. Although a comprehensive review of the numerous architectural possibilities is not possible given the scope of this chapter, a brief overview will be followed by highlights related to star polymers and graft/comb polymers.

In general, many methods fall into one of several categories which depend on what point the monomer is polymerized versus when the branch points are introduced. For example, branching can be introduced before, during, or after the monomer is polymerized. The core-first methodology illustrates the introduction of branching before the monomer is polymerized. In contrast, the arm-first approach involves reaction of a chain end with a core that contains an appropriate type and number of chemical functionalities. The introduction of branching during the polymerization of monomer has been exemplified with hyperbranched polymers [104, 105] using self-condensing vinyl polymerizations [106] and dendrimer-like polymers [107, 108].

The use of click chemistry [109] to prepare functional soft materials [110] has received a lot of attention in recent years. Click reactions are described as those organic transformations that are "selective and high yielding under

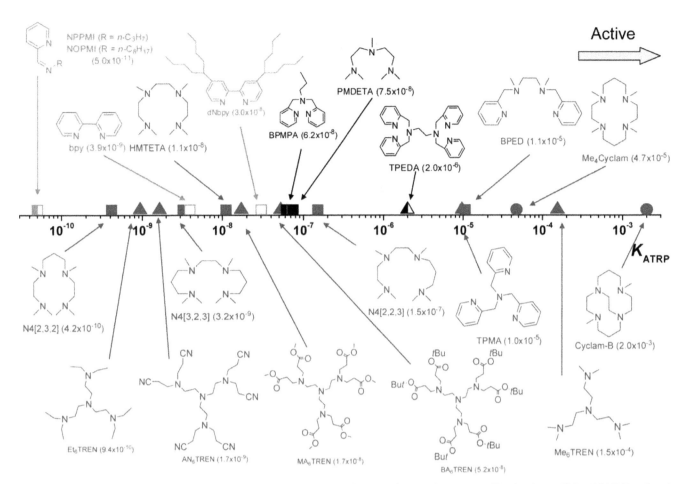

SCHEME 2.25 Influence of ligands on K_{ATRP} for polymerizations with ethyl 2-bromoisobutyrate (EtBriB) in MeCN at 22 °C. Reprinted (adapted) with permission from Tang W, Kwak Y, Braunecker WA, Tsarevsky NV, Coote ML, Matyjaszewski K. Understanding atom transfer radical polymerization: effect of ligand and initiator structures on the equilibrium constants. J Am Chem Soc 2008;130:10702–10713. © 2008 American Chemical Society.

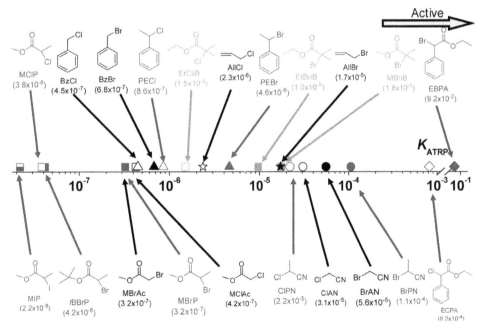

SCHEME 2.26 Influence of initiators on K_{ATRP} values for polymerizations using tris(2-pryidylmethyl)amine (TPMA) as ligand in conjunction with Cu(I)X. Reprinted (adapted) with permission from Tang W, Kwak Y, Braunecker WA, Tsarevsky NV, Coote ML, Matyjaszewski K. Understanding atom transfer radical polymerization: effect of ligand and initiator structures on the equilibrium constants. J Am Chem Soc 2008;130:10702–10713. © 2008 American Chemical Society.

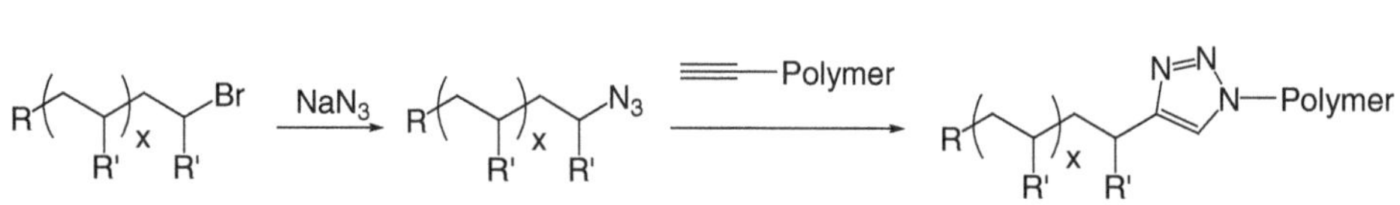

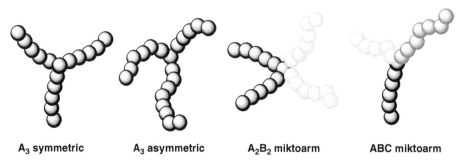

SCHEME 2.27   Reaction of bromo-terminated polymers made by ATRP with sodium azide for the purpose of click chemistry with alkyne-terminated polymers.

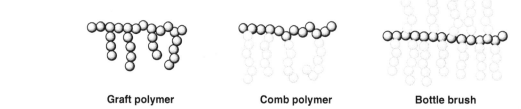

A₃ symmetric   **A₃ asymmetric**   **A₂B₂ miktoarm**   **ABC miktoarm**

SCHEME 2.28   Examples of star polymers obtained from various arm-first and core-first methodologies. (*See insert for color representation of the figure.*)

Graft polymer   **Comb polymer**   **Bottle brush**

SCHEME 2.29   Architectural possibilities involving long-chain branching. (*See insert for color representation of the figure.*)

mild conditions with little or no byproduct" [110]. Examples of click reactions include azide/alkyne and Diels–Alder cycloadditions, thiol-ene reactions, and certain types of Michael reactions [111]. In Scheme 2.27, the reaction of bromo-terminated polymers made by ATRP with sodium azide provides substrates for click chemistry which employ for arm-first methodology [112].

In the case of star polymers (Scheme 2.28), these include symmetric, asymmetric, and miktoarm stars, which are prepared by reacting active chain ends with a core using the arm-first method or via the core-first method. Miktoarm stars have been reported using anionic, cationic, and ATRP methodologies and typically have $AB_{n \geq 2}$, $A_{n \geq 2}$, $B_{n \geq 2}$, or $A_x B_y C_z$ compositions [12]. Since living anionic methods were available early on, many examples utilize the reaction of carbanion chain ends with chlorosilane compounds, 1,1-diphenylethylene derivatives, and 1,4-divinylbenzene crosslinking reactions [13, 113]. Typical stars made using chlorosilane chemistry have 3–6 arms while crosslinking reactions with 1,4-divinylbenzene provides options for a larger number of arms [10–30] ([13]). More recently, macroinitiators have been reacted with divinyl crosslinking agents, such as 1,4-divinylbenzene, ethylene glycol diacrylate, and

ethylene glycol dimethacrylate, to synthesize star polymers using CRP [114]. Many different CRP methods, such as normal ATRP [115] and ARGET ATRP [116], provide options for synthesizing star polymers.

Methods for introducing long-chain branching onto polymer chains have received much attention. Many examples include the synthesis of graft, comb, and bottlebrush polymers [117, 118] (Scheme 2.29). Milkovich was the first to report a strategy for introducing branching using the macromonomer concept by reacting living anionic chain ends with methacryloyl chloride [119]. Subsequently, this methacrylate-terminated polymer was utilized as a macromonomer. Since then, macromonomers have been made by cationic, ATRP, and coordination–insertion (living Ziegler–Natta) methods. These methods have also been extended to "grafting to," "grafting from," and "grafting through" strategies [110].

2.9   CONCLUSIONS

Living polymerization methods are an important part of macromolecular science and engineering. Since the initial reports of living anionic polymerizations, the field of living polymerization techniques has matured to include cationic, radical, and coordination–insertion methods. Additionally, these methods continue to advance toward the ideal standard for living polymerizations. In the case of controlled/living radical (CRP), significant improvements over conventional radical processes have been observed. These methods offer many of the benefits seen for living polymerization such as nearly quantitative functionalization, reasonably high molecular weight values, and formation of multiblock polymers.

ACKNOWLEDGMENTS

The support from NSF (DMR 09-69301 and CHE 10-26060) and Department of Energy (ER 45998) is gratefully acknowledged. Kristin Schröder acknowledges the Deutsche Forschungsgemeinschaft (DFG) for her postdoctoral scholarship.

REFERENCES

[1] Penczek S, Cypryk M, Duda A, Kubisa P, Somkowski, S. Living ring-opening polymerizations of heterocyclic monomers. Prog Polym Sci 2007;32:247–282.

[2] Bielawski CW, Grubbs RH. Living ring-opening metathesis polymerization. Prog Polym Sci 2007;32:1–29.

[3] Brunelle DJ, Aida T, editors. *Ring-Opening Polymerization: Mechanisms, Catalysis, Structure, Utility*. Munich: Hanser Publishers; 1993.

[4] Kamber NE, Jeong W, Waymouth RM, Pratt RC, Lohmeijer BGG, Hedrick JL. Organocatalytic ring-opening polymerization. Chem Rev 2007;107:5813–5840.

[5] Barner-Kowollik C, editor. *Handbook of RAFT Polymerization*. Weinheim: Wiley-VCH; 2008.

[6] Poli R. Relationship between one-electron transition-metal reactivity and radical polymerization processes. Angew Chem Int Ed 2006;45:5058–5070.

[7] Allan LEN, Perry MR, Shaver MP. Organometallic mediated radical polymerization. Prog Polym Sci 2012;37:127–156.

[8] Sciannamea V, Jérôme R, Detrembleur C. In-situ nitroxide-mediated radical polymerization (NMP) processes: their understanding and optimization. Chem Rev 2008;108:1104–1126.

[9] Hawker CJ, Bosman AW, Harth E. New polymer synthesis by nitroxide mediated living radical polymerizations. Chem Rev 2001;101:3661–3688.

[10] Matyjaszewski K, editor. *Controlled/Living Radical Polymerization: Progress in RAFT, DT, NMP & OMRP*. Volume 1024, ACS Symposium Series. Washington, DC: American Chemical Society; 2009.

[11] Szwarc M. 'Living' Polymers. Nature 1956;178:1168–1169.

[12] Müller AHE, Matyjaszewski K, editors. *Controlled and Living Polymerizations: From Mechanisms to Applications*. Weinheim: Wiley-VCH; 2009.

[13] Quirk RP, Hsieh HL. *Anionic Polymerization: Principles and Practical Applications*. New York: Marcel Dekker; 1996.

[14] Szwarc M, Levy M, Milkovich R. Polymerization initiated by electron transfer to monomer. A new method of formation of block polymers. J Am Chem Soc 1956;78:2656–2657.

[15] Matyjaszewski K. Ranking living systems. Macromolecules 1993;26:1787–1788.

[16] Quirk RP, Lee B. Experimental criteria for living polymerizations. Polym Int 1992;27:359–367.

[17] Matyjaszewski K, Sigwalt P. Unified approach to living and non-living cationic polymerization of alkenes. Polym Int 1994;35:1–26.

[18] Matyjaszewski K. Introduction to living polymerization. Living and/or controlled polymerization. J Phys Org Chem 1995;8:197–207.

[19] Matyjaszewski K, editor. *Cationic Polymerizations: Mechanisms, Synthesis, and Applications*. New York: Marcel Dekker; 1996.

[20] Beaman RG. Anionic chain polymerization. J Am Chem Soc 1948;70:3115–3118.

[21] Foster FC. The anionic copolymerization of methyl methacrylate-methacrylonitrile. J Am Chem Soc 1950;72:1370.

[22] Hatada K, Kitayama T, Okahata S, Yuki H. Studies on the mechanism of polymerization of methyl methacrylate in tetrahydrofuran with butyllithium at $-78\,^{\circ}$C by using perdeuterated monomer. Polymer J 1981;13:1045–1054.

[23] Bordwell FG. Equilibrium acidities in dimethyl sulfoxide solution. Acc Chem Res 1988;21:456–463.

[24] Quirk RP, Yoo T, Lee Y, Kim J, Lee B. Applications of 1,1-diphenylethylene chemistry in anionic synthesis of polymers with controlled structures. Adv Polym Sci 2000;153:67.

[25] Quirk RP, Mathers RT, Wesdemiotis C, Arnould MA. Investigation of ethylene oxide oligomerization during functionalization of poly(styryl)lithium using MALDI–TOF MS and NMR. Macromolecules 2002;35:2912–2918.

[26] Quirk RP, Janoski J, Chowdhury SR. Anionic synthesis of chain-end and in-chain, cyano-functionalized polystyrenes by hydrosilylation of allyl cyanide with silyl hydride-functionalized polystyrenes. Macromolecules 2009;42: 494–501.

[27] Quirk RP, Kim H, Polce MJ, Wesdemiotis C. Anionic synthesis of primary amine functionalized polystyrenes via hydrosilation of allylamines with silyl hydride functionalized polystyrenes. Macromolecules 2005;38:7895–7906.

[28] Kennedy JP, Ivan B. *Designed Polymers by Carbocationic Macromolecular Engineering*. Munich: Hanser; 1991.

[29] Miyamoto M, Sawamoto M, Higashimura T. Living polymerization of isobutyl vinyl ether with hydrogen iodide/iodine initiating system. Macromolecules 1984;17:265–268.

[30] Sawamoto M, Fujimori J, Higashimura T. Living cationic polymerization of N-vinylcarbazole initiated by hydrogen iodide. Macromolecules 1987;20:916–920.

[31] Faust R, Kennedy JP. Living carbocationic polymerization. Polym Bull 1986;15:317–323.

[32] Mah S, Yamamoto Y, Hayashi K. Radiation-induced cationic polymerization of α-methylstyrene enhanced by diphenyliodonium hexafluorophosphate. Macromolecules 1983;16:681–685.

[33] Sigwalt P, Moreau M. Carbocationic polymerization: mechanisms and kinetics of propagation reactions. Prog Polym Sci 2006;31:44–120.

[34] Sipos L, De P, Faust R. Effect of temperature, solvent polarity, and nature of Lewis acid on the rate constants in the carbocationic polymerization of isobutylene. Macromolecules 2003;36:8282–8290.

[35] Nomura K, Zhang S. FI catalysts for olefin polymerization-a comprehensive treatment. Chem Rev 2011;111:2342–2449.

[36] Doi Y, Ueki S, Keii T. Preparation of "living" polypropylenes by a soluble vanadium-based Ziegler catalyst. Makromol Chem Macromol Chem Phys 1979;180:1359–1361.

[37] Doi Y, Ueki S, Keii T. "Living" coordination polymerization of propene initiated by the soluble V(acac)$_3$-Al(C$_2$H$_5$)$_2$Cl system. Macromolecules 1979;12:814–819.

[38] Makio H, Terao H, Iwashita A, Fujita T. FI catalysts for olefin polymerization–a comprehensive treatment. Chem Rev 2011;111:2363–2449.

[39] Matsui S, Tohi Y, Mitani M, Saito J, Makio H, Tanaka H, Nitabaru M, Nakano T, Fujita T. New bis(salicylaldiminato) titanium complexes for ethylene polymerization. Chem Lett 1999;10:1065–1066.

[40] Tian J, Coates GW. Development of a diversity-based approach for the discovery of stereoselective polymerization catalysts: identification of a catalyst for the synthesis of syndiotactic polypropylene. Angew Chem Int Ed 2000;39:3626–3629.

[41] Harney MB, Zhang YH, Sita LR. Discrete, multiblock isotactic–atactic stereoblock polypropene microstructures of differing block architectures through programmable stereomodulated living Ziegler–Natta polymerization. Angew Chem Int Ed 2006;45:2400–2404.

[42] Domski GJ, Lobkovsky EB, Coates GW. Polymerization of α-olefins with pyridylamidohafnium catalysts: living behavior and unexpected isoselectivity from a C_s-symmetric catalyst precursor. Macromolecules 2007;40:3510–3513.

[43] Chen EY-X. Coordination polymerization of polar vinyl monomers by single-site metal catalysts. Chem Rev 2009;109:5157–5214.

[44] Brookhart M, Volpe AF, DeSimone JM, Lamanna WM. Cobalt(III)-catalyzed polymerizations of ethylene, oligomerizations of 1-hexene, and copolymerization of ethylene and 1-hexene. Polymer Prepr (ACS, Div Polym Chem) 1991;32:461.

[45] Brookhart M, DeSimone JM, Grant BE, Tanner MJ. Cobalt(III)-catalyzed living polymerization of ethylene: routes to end-capped polyethylene with a narrow molar mass distribution. Macromolecules 1995;28:5378–5380.

[46] Mashima K, Matsuo Y, Tani K. Unique complexation of 1,4-diaza-1,3-butadiene ligand on half-metallocene fragments of niobium and tantalum. Organometallics 1999;18:1471–1481.

[47] MacAdams LA, Buffone GP, Incarvito CD, Rheingold AL, Theopold KH. A chromium catalyst for the polymerization of ethylene as a homogeneous model for the phillips catalyst. J Am Chem Soc 2005;127:1082–1083.

[48] Boardman BM, Bazan GC. α-Iminocarboxamidato nickel complexes. Acc Chem Res 2009;42:1597–1606.

[49] Ittel SD, Johnson LK, Brookhart M. Late-metal catalysts for ethylene homo- and copolymerization. Chem Rev 2000;100:1169–1204.

[50] Fukui Y, Murata M, Soga K. Living polymerization of propylene and 1-hexene using bis-Cp type metallocene catalysts. Macromol Rapid Commun 1999;20:637–640.

[51] Doi Y, Murata M, Soga K. Reaction of carbon monoxide with living polypropylene prepared with a vanadium-based catalyst. Makromol Chem Rapid Commun 1984;5:811–814.

[52] Doi Y, Watanabe Y, Ueki S, Soga K. Synthesis of a propylene-tetrahydrofuran block copolymer via "living" coordination polymerization. Makromol Chem Rapid Commun 1983;4:533–537.

[53] Mason AF, Coates GW. Gel permeation chromatography as a combinatorial screening method: identification of highly active heteroligated phenoxyimine polymerization catalysts. J Am Chem Soc 2004;126:10798–10799.

[54] Reinartz S, Mason AF, Lobkovsky EB, Coates GW. Titanium catalysts with ancillary phenoxyketimine ligands for living ethylene polymerization. Organometallics 2003;22:2542–2544.

[55] Matsugi T, Kojoh S, Takagi Y, Inoue Y, Fujita T, Kashiwa N. New titanium complexes having two indolide-imine chelate ligands for living ethylene polymerization. Chem Lett 2001;(6):566–567.

[56] Tian J, Hustad PD, Coates GW. A new catalyst for highly syndiospecific living olefin polymerization: homopolymers and block copolymers from ethylene and propylene. J Am Chem Soc 2001;123:5134–5135.

[57] Saito J, Mitani M, Mohri J-i, Ishii S-i, Yoshida Y, Matsugi T, Kojoh S-i, Kashiwa N, Fujita T. Highly syndiospecific living polymerization of propylene using a titanium complex having two phenoxy-imine chelate ligands. Chem Lett 2001;(6):576–577.

[58] De Rosa C, Circelli T, Auriemma F, Mathers RT, Coates GW. Structure and physical properties of syndiotactic polypropylene from living polymerization with bis(phenoxyimine)-based titanium catalysts. Macromolecules 2004;37:9034–9047.

[59] Hustad PD, Coates GW. Functional propylene copolymers and block copolymers. J Am Chem Soc 2002;124:11578–11579.

[60] Mathers RT, Coates GW. Cross metathesis functionalization of polyolefins. Chem Commun 2004;422–423.

[61] Tshuva EY, Goldberg I, Kol M, Goldschmidt Z. Living polymerization of 1-hexene due to an extra donor arm on a novel amine bis(phenolate) titanium catalyst. Inorg Chem Commun 2000;3:611–614.

[62] Gendler S, Groysman S, Goldschmidt Z, Shuster M, Kol M. Polymerization of 4-methylpentene and vinylcyclohexane by amine bis(phenolate) titanium and zirconium complexes. J Polym Sci A Polym Chem 2006;44:1136–1146.

[63] Manivannan R, Sundararajan G. Latent bimodal polymerization of 1-Hexene by a titanium-based diastereomeric catalyst containing a *rac/meso*-aminodiol ligand. Macromolecules 2002;35:7883–7890.

[64] Boussie TR, Diamond GM, Goh C, Hall KA, LaPointe AM, Leclerc MK, Murphy V, Shoemaker JAW, Turner H, Rosen RK, Stevens JC, Alfano F, Busico V, Cipullo R, Talarico G. Nonconventional catalysts for isotactic propene polymerization in solution developed by using high-throughput-screening technologies. Angew Chem Int Ed 2006;45:3278–3283.

[65] Zhang W, Sita LR. Investigation of dynamic intra- and inter-molecular processes within a tether-length dependent series of group 4 bimetallic initiators for stereomodulated degenerative transfer living Ziegler–Natta propene polymerization. Adv Synth Catal 2008;350:439–447.

[66] Zhang W, Sita LR. Highly efficient, living coordinative chain-transfer polymerization of propene with ZnEt$_2$: practical production of ultrahigh to very low molecular weight amorphous atactic polypropenes of extremely narrow polydispersity. J Am Chem Soc 2008;130:442–443.

[67] Zhang W, Wei J, Sita LR. Living coordinative chain-transfer polymerization and copolymerization of ethene, α-olefins, and α,ω-nonconjugated dienes using dialkylzinc as "surrogate" chain-growth sites. Macromolecules 2008;41:7829–7833.

[68] Inoue Y, Matsugi T, Kashiwa N, Matyjaszewski K. Graft copolymers from linear polyethylene via atom transfer radical polymerization. Macromolecules 2004; 37:3651–3658.

[69] Britovsek GJP, Cohen SA, Gibson VC, Maddox PJ, Van Meurs M. Iron-catalyzed polyethylene chain growth on zinc: linear α-olefins with a Poisson distribution. Angew Chem Int Ed 2002;41:489–491.

[70] Hustad PD. Frontiers in olefin polymerization: reinventing the world's most common synthetic polymers. Science 2009;325:704–707.

[71] Breunig S, Risse W. Transition-metal-catalyzed vinyl addition polymerizations of norbornene derivatives with ester groups. Macromol Chem Phys 1992;193:2915–2927.

[72] Gottfried AC, Brookhart M. Living and block copolymerization of ethylene and α-olefins using palladium(II)-α-diimine catalysts. Macromolecules 2003;36:3085–3100.

[73] Killian CM, Tempel DJ, Johnson LK, Brookhart M. Living polymerization of α-olefins using Ni^{II}-α-diimine catalysts. Synthesis of new block polymers based on α-olefins. J Am Chem Soc 1996;118:11664–11665.

[74] Johnson LK, Killian CM, Brookhart M. New Pd(II)- and Ni(II)-based catalysts for polymerization of ethylene and α-olefins. J Am Chem Soc 1995;117:6414–6415.

[75] Camacho DH, Guan ZB. Living polymerization of α-olefins at elevated temperatures catalyzed by a highly active and robust cyclophane-based nickel catalyst. Macromolecules 2005;38:2544–2546.

[76] Odian G. *Principles of Polymerization*. 3rd ed. New York: John Wiley & Sons, Inc.; 2004.

[77] Moad G, Solomon DH. *The Chemistry of Radical Polymerization*. 2nd ed. Oxford: Elsevier; 2006.

[78] Matyjaszewski K, Davis TP, editors. *Handbook of Radical Polymerization*. Hoboken: John Wiley; 2002.

[79] Matyjaszewski K, editor. *Controlled/Living Radical Polymerization: Progress in ATRP*. Volume 1023, ACS Symposium Series. Washington, DC: American Chemical Society; 2009.

[80] Matyjaszewski K, Xia J. Atom transfer radical polymerization. Chem Rev 2001;101:2921–2990.

[81] Tang W, Fukuda T, Matyjaszewski K. Reevaluation of persistent radical effect in NMP. Macromolecules 2006;39:4332–4337.

[82] Fischer H. The persistent radical effect: a principle for selective radical reactions and living radical polymerizations. Chem Rev 2001;101:3581–3610.

[83] Gossage RA, van de Kuil LA, van Koten G. Diaminoarylnickel (II) "Pincer" complexes: mechanistic considerations in the Kharasch addition reaction, controlled polymerization, and dendrimeric transition metal catalysts. Acc Chem Res 1998;31:423–431.

[84] Clark AJ. Atom transfer radical cyclisation reactions mediated by copper complexes. Chem Soc Rev 2002;31:1–11.

[85] Iqbal J, Bhatia B, Nayyar NK. Transition metal-promoted free-radical reactions in organic synthesis: the formation of carbon-carbon bonds. Chem Rev 1994;94:519–564.

[86] Wang J-S, Matyjaszewski K. Controlled/"living" radical polymerization. atom transfer radical polymerization in the presence of transition-metal complexes. J Am Chem Soc 1995;117:5614–5615.

[87] Kato M, Kamigaito M, Sawamoto M. Polymerization of methyl methacrylate with the carbon tetrachloride/dichlorotris-(triphenylphosphine)ruthenium(II)/methylaluminum bis(2,6-di-tert-butylphenoxide) initiating system: possibility of living radical polymerization. Macromolecules 1995;28:1721–1723.

[88] Lin CY, Coote ML, Gennaro A, Matyjaszewski K. Ab initio evaluation of the thermodynamic and electrochemical properties of alkyl halides and radicals and their mechanistic implications for atom transfer radical polymerization. J Am Chem Soc 2008;130:12762–12774.

[89] Isse AA, Gennaro A, Lin CY, Hodgson JL, Coote ML, Guliashvili T. Mechanism of carbon-halogen bond reductive cleavage in activated alkyl halide initiators relevant to living radical polymerization: theoretical and experimental study. J Am Chem Soc 2011;133:6254–6264.

[90] di Lena F, Matyjaszewski K. Transition metal catalysts for controlled radical polymerization. Prog Polym Sci 2010;35:959–1021.

[91] Pintauer T, Zhou P, Matyjaszewski K. General method for determination of the activation, deactivation, and initiation rate constants in transition metal-catalyzed atom transfer radical processes. J Am Chem Soc 2002;124:8196–8197.

[92] Tang W, Tsarevsky NV, Matyjaszewski K. Determination of equilibrium constants for atom transfer radical polymerization. J Am Chem Soc 2006;128:1598–1604.

[93] Braunecker WA, Tsarevsky NV, Gennaro A, Matyjaszewski K. Thermodynamic components of the atom transfer radical polymerization equilibrium: quantifying solvent effects. Macromolecules 2009;42:6348–6360.

[94] Bell CA, Bernhardt PV, Monteiro MJ. A rapid electrochemical method for determining rate coefficients for copper-catalyzed polymerizations. J Am Chem Soc 2011;133:11944–11947.

[95] Pintauer T, Matyjaszewski K. Structural and mechanistic aspects of copper catalyzed atom transfer radical polymerization. Top Organomet Chem 2009;26:221–251.

[96] Tang W, Matyjaszewski K. Effect of ligand structure on activation rate constants in ATRP. Macromolecules 2006;39:4953–4959.

[97] Tang W, Kwak Y, Braunecker WA, Tsarevsky NV, Coote ML, Matyjaszewski K. Understanding atom transfer radical polymerization: effect of ligand and initiator structures on the equilibrium constants. J Am Chem Soc 2008;130:10702–10713.

[98] Bortolamei N, Isse AA, Magenau AJD, Gennaro A, Matyjaszewski K. Controlled aqueous atom transfer radical polymerization with electrochemical generation of the active catalyst. Angew Chem Int Ed 2011;50:11391–11394.

[99] Wang Y, Zhang Y, Parker B, Matyjaszewski K. ATRP of MMA with ppm levels of iron catalyst. Macromolecules 2011;44:4022–4025.

[100] Kwak Y, Magenau AJD, Matyjaszewski K. ARGET ATRP of methyl acrylate with inexpensive ligands and ppm concentrations of catalyst. Macromolecules 2011;44:811–819.

[101] Matyjaszewski K, Jakubowski W, Min K, Tang W, Huang J, Braunecker WA, Tsarevsky NV. Diminishing catalyst concentration in atom transfer radical polymerization with reducing agents. Proc Natl Acad Sci 2006;103:15309–15314.

[102] Magenau AJD, Strandwitz NC, Gennaro A, Matyjaszewski K. Electrochemically mediated atom transfer radical polymerization. Science 2011;332:81–84.

[103] Jakubowski W, Matyjaszewski K. Activators regenerated by electron transfer for atom-transfer radical polymerization of (meth)acrylates and related block copolymers. Angew Chem 2006;118:4594–4598.

[104] Matyjaszewski K, Gaynor SG, Kulfan A, Podwika M. Preparation of hyperbranched polyacrylates by atom transfer radical polymerization. 1. Acrylic AB* monomers in "living" radical polymerizations. Macromolecules 1997;30:5192–5194.

[105] Matyjaszewski K, Gaynor SG, Muller AHE. Preparation of hyperbranched polyacrylates by atom transfer radical polymerization. 2. kinetics and mechanism of chain growth for the self-condensing vinyl polymerization of 2-((2-Bromopropionyl)oxy)ethyl acrylate. Macromolecules 1997;30:7034–7041.

[106] Fréchet JMJ, Henmi M, Gitsov I, Aoshima S, Leduc MR, Grubbs RB. Self-condensing vinyl polymerization: an approach to dendritic materials. Science 1995;269:1080–1083.

[107] Matmour R, Gnanou Y. Combination of an anionic terminator multifunctional initiator and divergent carbanionic polymerization: application to the synthesis of dendrimer-like polymers and of asymmetric and miktoarm stars. J Am Chem Soc 2008;130:1350–1361.

[108] Konkolewicz D, Monteiro MJ, Perrier S. Dendritic and hyperbranched polymers from macromolecular units: elegant approaches to the synthesis of functional polymers. Macromolecules 2011;44:7067–7087.

[109] Kolb HC, Finn MG, Sharpless KB. Click chemistry: diverse chemical function from a few good reactions. Angew. Chem Int Ed 2001;40:2004–2021.

[110] Iha RK, Wooley KL, Nyström AM, Burke DJ, Kade MJ, Hawker C. Applications of orthogonal "click" chemistries in the synthesis of functional soft materials. Chem Rev 2009;109:5620–5686.

[111] Sumerlin BS, Vogt AP. Macromolecular engineering through click chemistry and other efficient transformations. Macromolecules 2010;43:1–13.

[112] Tsarevsky NV, Sumerlin BS, Matyjaszewski K. Step-growth "click" coupling of telechelic polymers prepared by atom transfer radical polymerization. Macromolecules 2005;38:3558–3561.

[113] Hadjichristidis N, Pitsikalis M, Pispas S, Iatrou H. Polymers with complex architecture by living anionic polymerization. Chem Rev 2001;101:3747–3792.

[114] Gao HF, Matyjaszewski K. Synthesis of functional polymers with controlled architecture by CRP of monomers in the presence of cross-linkers: from stars to gels. Prog Polym Sci 2009;34:317–350.

[115] Gao H, Matyjaszewski K. Synthesis of miktoarm star polymers via ATRP using the "in-out" method: determination of initiation efficiency of star macroinitiators. Macromolecules 2006;39:7216–7223.

[116] Burdynska J, Cho HY, Mueller L, Matyjaszewski K. Synthesis of star polymers using ARGET ATRP. Macromolecules 2010;43:9227–9229.

[117] Lee H.-i, Pietrasik J, Sheiko SS, Matyjaszewski K. Stimuli-responsive molecular brushes. Prog Polym Sci 2010;35:24–44.

[118] Sheiko SS, Sumerlin BS, Matyjaszewski K. Cylindrical molecular brushes: synthesis, characterization, and properties. Prog Polym Sci 2008;33:759–785.

[119] Holden G, Legge NR, Quirk RP, Schroeder HE, editors. *Thermoplastic Elastomers*. New York: Hanser Publishers; 1996.

3

STIMULI-RESPONSIVE POLYMERS VIA CONTROLLED RADICAL POLYMERIZATION

Joel D. Flores, Brooks A. Abel, DeeDee Smith, and Charles L. McCormick

3.1 INTRODUCTION

Polymers possessing inherent structural characteristics and functionality can be induced to respond to externally applied stimuli. While most commonly termed as "stimuli responsive," these macromolecules are also referred to as "smart," "intelligent," or "environmentally responsive" materials. Being either chemical or physical in nature, the responses may involve alteration in coil dimensions, secondary structures, or of molecular associations, as well as breakage or formation of chemical bonds (Fig. 3.1).

These physical and chemical events result from alterations of molecular interactions such as hydrogen bonding, hydrophobic association, and electrostatic interaction; occurrence of chemical reactions of the groups present along the polymer chain; and variations in the concentration. Externally applied stimuli may include changes in temperature, pH, and ionic strength. Application of mechanical force, interaction with chemical agents, and irradiation with light, electric, magnetic, or sonic energy can also trigger responses of polymers.

Research in the rapidly expanding field of stimuli-sensitive polymers includes the synthesis of responsive materials possessing novel functions and tunable properties, investigation of the thermodynamics and kinetics of the transitions involved, and elucidation of mechanisms operative at the molecular level [1]. The ability to precisely tailor the structures and properties of stimuli-responsive polymers has resulted in a plethora of advanced applications. For example, stimuli-responsive polymers are utilized in controlled and targeted delivery systems, (bio)sensors and diagnostics, separation science, electronics, industrial coatings, formulations, enhanced oil recovery, and water remediation, to name a few [1–3].

This chapter serves as a brief overview of the controlled radical polymerization methods, atom transfer radical polymerization (ATRP) and reversible addition–fragmentation chain transfer (RAFT) polymerization, and their utilization in the construction of solution-based stimuli-responsive polymers. The structure/property/behavior relationships of selected systems are discussed relative to changes in pH, temperature, and electrolyte concentration. Additionally, comprehensive reviews are cited for additional information.

3.2 SYNTHESIS OF STIMULI-RESPONSIVE POLYMERS

With their controllable, dynamic behavior, stimuli-responsive polymers can be considered mimics of naturally occurring polymers such as proteins, polysaccharides, and nucleic acids. Comparatively, synthetic polymers are quite primitive in terms of complexity in properties and functions; however, research in this area is very active, ranging from fundamental investigations to practical applications. Mimicking the properties and functions of biological materials requires precise control over structural attributes.

Design and synthesis of polymers with precise architecture, composition, molecular weight distribution, stereoregularity, block sequence, block length, and location of functional groups have become the ultimate goal in polymer synthesis. Progress toward this goal has been

Monitoring Polymerization Reactions: From Fundamentals to Applications, First Edition. Edited by Wayne F. Reed and Alina M. Alb.
© 2014 John Wiley & Sons, Inc. Published 2014 by John Wiley & Sons, Inc.

markedly accelerated with the advent of several controlled/living polymerization methodologies.

3.2.1 Controlled Polymerization Methods

Since the introduction of anionic polymerization by Szwarc [4–6] in the 1950s, a number of living polymerization methods have been identified. Living anionic and cationic ring opening polymerizations [7], group transfer polymerization (GTP) [8], living carbocationic polymerization [9, 10], ring-opening metathesis polymerization (ROMP) [11, 12], and the controlled/"living" radical polymerization (CRP) [13–15] techniques, such as stable free radical polymerization (SFRP) [16], nitroxide-mediated polymerization (NMP) [17], atom transfer radical polymerization (ATRP) [18, 19], and reversible addition–fragmentation chain transfer (RAFT) [20–25] polymerization have been developed and subsequently utilized in the syntheses of complex macromolecular structures (Fig. 3.2). Prior conventional techniques yielded largely noncontrolled, polydisperse systems.

An ideal living polymerization method [26–28] is characterized by a complete absence of chain termination and chain transfer events leading to conservation of active "living"

chain ends. Additionally, the rate of initiation must be substantially larger than the rate of propagation so that growth of the chains occurs simultaneously, leading to control over length and dispersity. In practice, termination reactions may be present in controlled/"living" (often referred to as "pseudo-living") polymerizations, but the formation of well-defined homopolymers as well as block copolymers can still be achieved. Generally, nearly all controlled and/or living polymerization methods possess a fast, dynamic equilibrium between dormant and active species either via reversible activation/deactivation process or degenerative chain transfer (Fig. 3.3).

This equilibrium minimizes, if not eliminates, the occurrence of termination, chain transfer, and other primary side reactions enabling the synthesis of polymers having well-defined structures. This is particularly critical in free radical–based polymerizations wherein growing chains can inherently terminate through disproportionation or bimolecular coupling reactions. For CRP methods, control is afforded by maintenance of sufficiently low and constant concentration of free radicals throughout the polymerization.

3.2.2 Advanced Polymer Architectures

A significant outcome from the discovery of controlled living polymerization methods is the ability to prepare advanced macromolecular architectures beyond linear homopolymers including copolymers, brush/pendant polymers, branched polymers, and star polymers (Fig. 3.2) [29–36].

Arguably, the most significant advancement to functional polymers is the ability to synthesize functional polymers and advanced architectures including block copolymers. A block

FIGURE 3.1 A responsive polymeric chain changes configuration upon application of a stimulus.

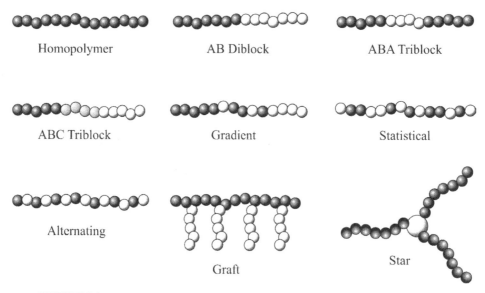

FIGURE 3.2 Polymer architectures via controlled/"living" polymerization techniques.

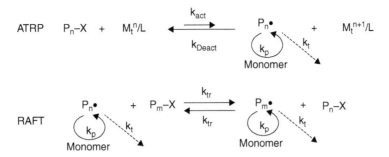

FIGURE 3.3 Examples of dynamic equilibria showing the reversible activation/deactivation (ATRP, X=halide) and degenerative chain transfer (RAFT, X=thiocarbonylthio) between active and dormant chains.

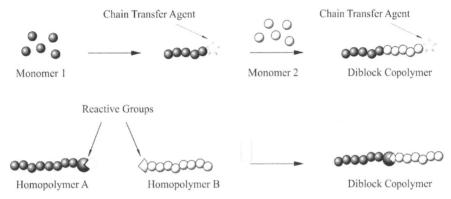

FIGURE 3.4 Routes for the synthesis of a diblock copolymer.

copolymer consisting of two or more monomers covalently attached in a linear chain may be synthesized via sequential copolymerization or by coupling separately prepared homopolymers (Fig. 3.4).

The former approach requires that the polymerizable groups and the functionalities of each monomer are amenable to the same polymerization technique. Furthermore, it should be emphasized that even for a particular polymerization technique and monomer-type combination, the order of monomer addition is critical for a successful block copolymerization.

The coupling of two or more homopolymers can, in principle, be employed to prepare copolymers when constituent blocks cannot be prepared via the same polymerization method or when there are issues in maintaining control or livingness of the polymerization process. Such an approach, however, presents some challenges. Coupling of polymers requires the presence of reactive functional groups at the chain ends. Various strategies such as the use of functional initiators and chain transfer agents, quenching the polymerization with functional capping agents, and post-polymerization modifications of the end groups have been developed [37–39]. The coupling reactions themselves must be highly efficient to obtain quantitative conversions.

Recently, the development of click chemistries [40] has demonstrated the utility of a few, well-established

chemical reactions for facile modification of polymers. The reader is directed to a recent review [41] for a comprehensive discussion of click chemistry and its various applications.

The increasing interest in block copolymers arises from unique solution and associative characteristics elicited by the intrinsic properties of each block (e.g., solubility, responsiveness, etc.). Block copolymers can assemble into an array of nanostructures. The polarities, dimensions, and molecular associations of each block determine the thermodynamics and kinetics of block copolymer assembly.

Furthermore, these block copolymers may be designed to contain stimuli-responsive components such that the assembly process may be triggered through the application of a stimulus. With unique properties and a host of potential applications, stimuli-responsive assemblies such as micelles, vesicles, bioconjugates, films, networks, and patterned surfaces have been prepared from block copolymers synthesized via controlled polymerization methods (Fig. 3.5) [1, 42].

In the sections that follow, amphiphilic block copolymers prepared via CRP techniques will be discussed. Representative examples of reports from the literature are briefly described to demonstrate solution properties and applications of stimuli-responsive copolymers.

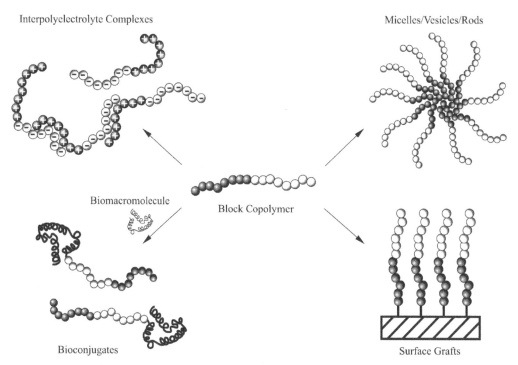

FIGURE 3.5 Ordered structures of stimuli-responsive block copolymers.

3.3 BLOCK COPOLYMERS IN SOLUTION

Amphiphilic block copolymers are polymers that can self-assemble into diverse structures in aqueous media above a critical micellization concentration (CMC) or critical aggregation concentration (CAC). Such block copolymers contain hydrophilic and hydrophobic segments. The amphiphilicity of these block copolymers may be tuned by incorporating stimuli-responsive blocks, which allow a dynamic micellization process (Fig. 3.6).

When dispersed in aqueous media, the hydrophobic component constitutes the core while the hydrophilic block corona serves to stabilize the micellar structure. In contrast to low molar mass surfactants, amphiphilic block copolymers have highly tunable composition and assembly characteristics. The constraints between the core-forming blocks, the interaction between chains in the corona, and the surface energy between the solvent and the core dictate the shape and size of the micellar structures [43]. An important property of amphiphilic block copolymers is the ability to enhance solubility of hydrophobic substances. As with low molar mass surfactants, the hydrophobic core of the micellar structures formed by amphiphilic block copolymers can serve as a compatible microenvironment for sequestration of water-insoluble compounds.

3.3.1 Temperature-Responsive Polymers

Arguably, the most studied stimuli-responsive polymers are systems that are sensitive to changes in temperature. Since the discovery of poly(*N*-isopropylacrylamide) (PNIPAm) in the 1950s [44], this class of responsive (co)polymers has experienced rapid growth, facilitated by the advent of CRP techniques that enable the fine-tuning of polymer properties.

Temperature-responsive (co)polymers exhibit a volume phase transition at a critical temperature. (Co)polymers exhibit a lower critical solution temperature (LCST) when they undergo phase transition (i.e., soluble to insoluble) upon heating. On the other hand, systems that become soluble upon heating have an upper critical solution temperature (UCST). Thermodynamically, the LCST and UCST behaviors of polymers are driven by entropic effects during the dissolution process involving the ordering/disordering of water molecules in the vicinity of the polymer chain and enthalpic effects originating from hydrogen bonding, hydrophobic association, and electrostatic interactions. Generally, these coil-to-globule transitions are manifested macroscopically as changes in copolymer solubility in a specific solvent system.

Comprehensive reviews on synthesis, properties, and applications of temperature-responsive (co)polymers include those of Winnik [43], McCormick [45], and others [46–49].

PNIPAm-based systems are the most extensively reported temperature-responsive polymers in literature. Other temperature sensitive systems incorporate analogues of NIPAm such as *N*-(*n*-propyl)acrylamide (nPAm), *N,N*-diethylacrylamide (DEA), and *N*-ethylmethylacrylamide (EMA) as well as derivatives of the amino acid L-proline; *N*-acryloyl-L-proline methyl ester (A-Pro-OMe) and *N*-acryloyl-4-trans-L-proline

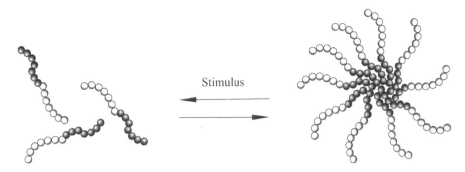

FIGURE 3.6 Reversible micellization in response to an external stimulus.

methyl ester (A-Hyp-OMe). Chemical structures of these monomers are shown in Fig. 3.7.

With an LCST just below the human body temperature, PNIPAm has been widely investigated for bio-related applications [50–52]. Tuning NIPAm-containing copolymer LCST close to the body temperature is essential for controlling efficacy of polymeric drug delivery vehicles and bioconjugates. The LCSTs of PNIPAm (co)polymers are dependent on both structure and composition. Copolymerization of NIPAm with hydrophilic monomers increases LCST, whereas incorporation of hydrophobic comonomers lowers the transition temperature. Furthermore, hydrophilic and hydrophobic terminal groups also affect PNIPAm LCST [53]. For example, incorporation of amino and hydroxyl groups to one end of PNIPAm chains remarkably increases LCST and slows the rate of phase transition. The opposite effects are observed when chain ends are hydrophobically modified.

There are numerous reports in the literature describing the synthesis of low polydispersity PNIPAm (co)polymers via CRP methods. For example, Binder et al. [54] described the synthesis of telechelic PNIPAm using functional NMP initiators modified via azide–alkyne cycloadditions. Efficient block copolymerization of NIPAm with styrene was achieved via sequential polymerization initiated by α-hydrogen alkoxyaminẽ [55]. A PNIPAm block was first prepared and then chain extended with styrene yielding amphiphilic diblock copolymers. Initial attempts to use ATRP to prepare PNIPAm resulted in polymers with broad polydisperisities [56], but this problem was circumvented by utilizing a more efficient ligand for the initiator complex [57].

Armes et al. [58] synthesized, via ATRP, biocompatible, thermoresponsive ABA triblock copolymers in which the outer A blocks are comprised of PNIPAm and the central B block is poly(2-methacryloyloxyethyl phosphorylcholine). This novel system is soluble in dilute aqueous solution but forms free-standing physical gels at 37 °C due to hydrophobic associations of the PNIPAm segments. The gelation process is reversible and shows potential applications in drug delivery and tissue engineering.

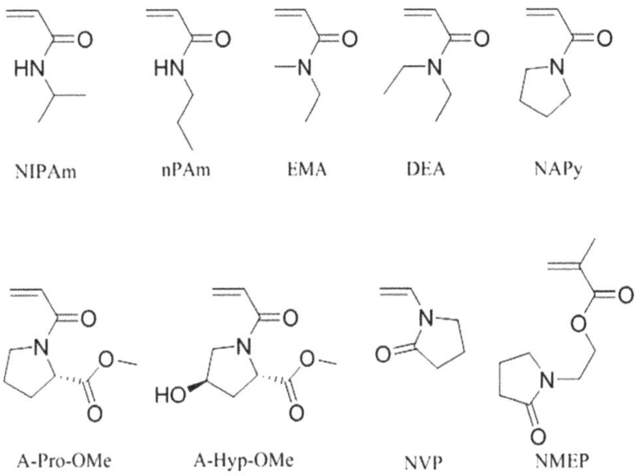

FIGURE 3.7 Typical monomers used for the synthesis of thermally responsive (co)polymers via controlled radical polymerization.

McCormick and coworkers polymerized NIPAm at room temperature via aqueous RAFT process [59]. Extending the study to prepare poly(N,N-dimethyl acrylamide(DMA)-b-NIPAm) diblock and poly(DMA-b-NIPAm-b-DMA) triblock copolymers, the group demonstrated that critical micelle temperature (CMT) was lower for diblock than for triblock copolymers, and decreased as the PNIPAm block length was increased [60]. The phase transition temperature of this polymer system may be adjusted from 10 °C to 36 °C by incorporating an additional pH-responsive block [61]. Thermoreversible hydrogels from RAFT-synthesized poly(NIPAm-b-DMA-b-NIPAm) triblock copolymers were also investigated as potential *in vitro* cell growth scaffolds [62].

Cao et al. [63] investigated the responsiveness of various N-alkyl substituted acrylamides. Multiblock copolymers consisting of nPAm, NIPAm and EMA (ethylmethylacrylamide) undergo multiple, stepwise phase transitions with each transition corresponding to the LCST of individual blocks. When the temperature is raised above the LCST of the first block, the copolymer self-assembles into micelles

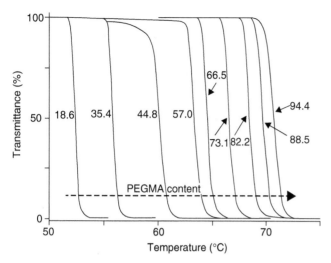

FIGURE 3.8   Phase transitions of P(DMAEMA-*stat*-PEGMA) copolymers at pH 7 with the indicated PEGMA contents (in weight %). Reprinted (adapted) with permission from Fournier D, Hoogenboom R, Thijs HML, Paulus RM, Schubert US. Tunable pH- and temperature-sensitive copolymer libraries by reversible addition fragmentation chain transfer copolymerizations of methacrylates. Macromolecules 2007;40:915–920. © 2007 American Chemical Society.

with collapsed blocks forming the core and the remaining water-soluble outer blocks stabilizing the structures in solution. Increasing the temperature leads to further collapse of the outer blocks.

Random copolymerization of dimethylaminoethyl methacrylate (DMAEMA) and poly(ethylene glycol)methyl ether methacrylate (PEGMA) using RAFT polymerization results in dually responsive (temperature and pH) copolymers [64]. When the amino groups of DMAEMA are protonated at pH 4, the copolymers do not exhibit an LCST. However, at pH 7 and 10, the observed LCST increases linearly with PEGMA content. Thus, the copolymer LCST (35–80 °C) can be tuned by adjusting solution pH and copolymer composition (Fig. 3.8).

Doubly-responsive molecular brushes consisting of statistical copolymers of di(ethylene glycol) methyl ether methacrylate (MEO$_2$MA) and either methacrylic acid (MAA) or DMAEMA have been synthesized by "grafting from" via ATRP [65]. The LCST of poly(MEO$_2$MA-*stat*-MAA) decreases with increasing molar fraction of MAA in the copolymer in deionized water but increases in buffer solution at pH 7. At pH 9, polymer hydrophobicity increases with ionization of the carboxylic acids to further raise the LCST. The LCST of poly(MEO$_2$MA-*stat*-DMAEMA) copolymers increase with increasing DMAEMA content at pH 4, 7, and 9. A bottlebrush terpolymer prepared from all three comonomers exhibits LCSTs at pH 4 and 7 but not at pH 9, which was attributed to the stronger ionization of MAA. The responsive nature

of the copolymer is enhanced by the densely graft structure of a brush copolymer.

3.3.2   pH-Responsive Polymers

Weak polyacids or weak polybases which contain ionizable groups (e.g., carboxylic acids or amines) are pH-responsive polymers. As the solution pH is changed, the degree of ionization changes polymer solubility, resulting in chain collapse and eventually aggregation. Carboxylic acid–containing polyacrylates and polymethacrylates are protonated at pH values below their pK_a, rendering the polymer hydrophobic and insoluble. However, these polymers become hydrophilic and water soluble when the carboxylic acid groups are ionized at higher pH values. On the other hand, weak polybases like poly(4-vinylpyridine) (P4VP) and poly(dimethyl ethyl amino methacylate) (PDEAMA) are protonated and cationically charged at pH values below their pK_a, but are neutral at higher pH.

Reversible protonation and deprotonation of the ionizable groups via adjustment of solution pH are utilized in the preparation of pH-controlled nanostructures. For details describing pH-induced phase transitions and applications of these copolymers, the reader is directed to a number of excellent reviews [1, 66, 67]. Examples of pH-responsive monomers that can be polymerized by CRP are shown in Figure 3.9.

The apparent pK_a values of weak polyacids and weak polybases may be tuned by proper design of (co)polymer structure and composition [66]. The phase transitions are sensitive to the balance of repulsive forces between charged entities and hydrophobic associations. When polymer chains are in the neutral form, electrostatic repulsive forces are diminished and hydrophobic interactions dominate, leading to aggregation of chains from aqueous environment. Other interactive forces such as hydrogen bonding may also trigger the collapse of polymers [68]. The pH range at which phase transition occurs can be modulated by either selecting monomers with the pK_a that matches the target pH or tuning the apparent pK_a of the polymer through the incorporation of hydrophobic moieties. For weak polyacids, addition of hydrophobic groups increases the phase transition pH as higher electrostatic repulsive forces are required to overcome the strong associations elicited by the hydrophobic groups. Opposite effects may be observed for weak polybases. Tuning hydrophobicity of pH-responsive copolymers can also be achieved by incorporating a temperature-sensitive block as illustrated in Section 3.3.1 [64, 65].

Carboxylic acid groups may deactivate ATRP initiators; however, carboxylic acid–containing monomers can be polymerized via ATRP by carefully adjusting the solution pH [69]. Matyjaszewski et al. reported the controlled copolymerization of DMAEMA with methyl methacrylate, methyl acrylate, and styrene yielding pH-responsive amphiphilic block copolymers [70]. Aqueous RAFT

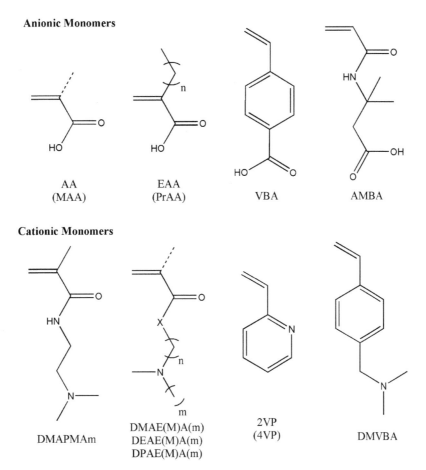

FIGURE 3.9 Typical monomers used for the synthesis of pH-responsive (co)polymers via controlled radical polymerization. X=O for acrylate (A) and NH for acrylamide (Am).

polymerization was utilized in the chain extension of poly(sodium 4-styrenesulfonate) with sodium 4-vinylbenzoic acid (VBA). The resulting diblock copolymer formed pH-responsive micelles and represents the first stimuli-responsive system reported via RAFT process. Other examples pH-responsive systems include copolymers of 3-acrylamido-3-methylbutanoate (AMBA) [71, 72], 6-acrylamidohexanoate [73], *N*-acryloylvaline [61], DMAEMA [64], *N*,*N*-dimethylvinylbenzylamine [74], and *N*-[3-(dimethylamino)propyl] methacrylamide [75].

Eisenberg et al. [76, 77] investigated self-assembled nanostructures from styrene and acrylic acid block copolymers. Crew-cut micelle-like aggregates of various morphologies were observed under near-equilibrium conditions. As the poly(acrylic acid) (PAA) content in block copolymer decreased, the morphology of the aggregates changed progressively from spheres to cylinders, to bilayers, and eventually to compound micelles. Webber et al. [78] described the reversible assembly of poly[(ethylene glycol)-*b*-(2-vinylpyridine)] diblock copolymers into micelles.

Schizophrenic assembly of dually responsive poly[(*N*,*N*-diethylaminoethyl methacrylate)-*b*-NIPAm] were reported

by Smith et al. [79]. The nanoassembly morphologies, dictated by the hydrophilic mass fraction, could be controlled by the polymer block lengths, solution pH, and temperature. Additional examples of pH-responsive systems are described under Section 3.4.

3.3.3 Salt-Responsive Polymers

Salt-responsive polymers, which contain charged groups along the polymer chains, may be classified as polyelectrolytes or polyzwitterions. Polyelectrolytes contain anionic or cationic charges along or pendent to the polymer backbone, whereas polyzwitterions have both anionic and cationic charges (Fig. 3.10) [80].

When both charges are on different repeat units, the polyzwitterions are termed polyampholytes; if both charges are on the same repeat unit, they are referred to as polybetaines. It should be noted that in addition to being salt responsive, polyelectrolytes, polyampholytes, and polybetaines may also be pH sensitive through incorporation of ionizable groups. Synthetic polyzwitterions have solution behavior and responsiveness comparable to proteins and nucleic acids [81].

Hence, preparation and fundamental investigation of this class of stimuli-responsive polymers are critical to the development of biomimetics. For additional information on the synthesis and applications of zwitterionic polymers the reader is referred to recent reviews [80, 82–84].

The repulsion between similarly charged groups on the chain provides polyelectrolytes with extended chain conformations and thus large hydrodynamic volumes in solution. However, when small molecule electrolytes (SMEs) are added, the ions screen repulsive forces between the charged groups on the chain, causing the polymer chain to adopt a more entropically favored conformation [80]. The extent of this behavior, termed the "polyelectrolyte effect," is contingent upon the identity of the charged groups on the polymer as well as the identity and concentration of the added SMEs [85, 86]. The intra- and intermolecular attractive forces between oppositely charged groups give polyzwitterions collapsed chain conformations, limiting their solubility in aqueous solutions. However, the ionic network may be disrupted by the addition of SMEs. This "anti-polyelectrolyte effect" results in dissolution of polyampholytes and polybetaines in aqueous media.

Although controlled synthesis of polyelectrolytes is relatively straightforward [25, 87], solubility problems, strong electrostatic interactions, and lack of tolerant polymerization methods have limited the preparation of well-defined polyzwitterions. Polybetaines with narrow distributions can be prepared by post-polymerization modification of precursor (co)polymers synthesized via GTP or NMP. Direct synthesis of carboxy-, phospho- and sulfobetaines via ATRP [88–92] or RAFT polymerization [92–96] has been reported (Fig. 3.11). The ATRP route usually employs protic organic solvents. The RAFT technique, on the other hand, allows polymerizations in homogeneous aqueous media. This is particularly useful for the synthesis of polyzwitterions which may require pH adjustment and higher ionic strengths.

Hydrophobically modified polyzwitterions may self-assemble in solution and undergo microphase separation in bulk. The balance of hydrophobic and hydrophilic interactions dictates the self-assembly and phase-separation processes [97]. The type of hydrophilic head group and geometry of the ionic groups determine the solution properties [98, 99]. Polyphosphobetaines with hydrophobic tails

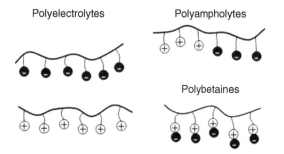

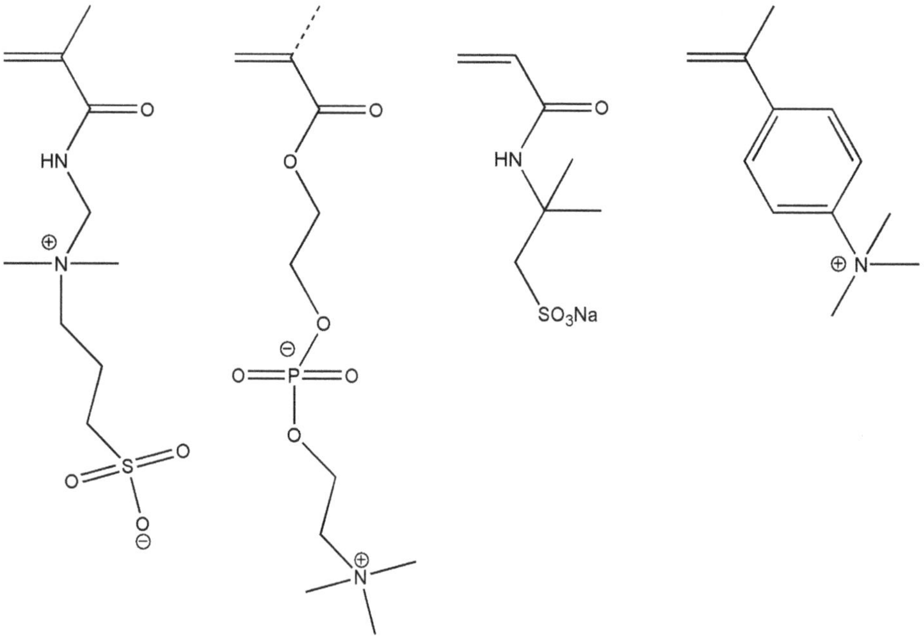

FIGURE 3.10 Generalized structures of salt-responsive polymers.

SPP
Zwitterionic Monomer

APC
Zwitterionic Monomer

AMPS
Permanently Anionic

VBTAC
Permanently Cationic

FIGURE 3.11 Typical monomers for the synthesis of salt-responsive (co)polymers.

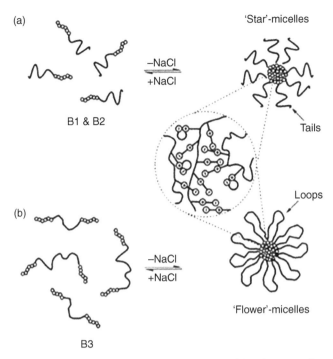

FIGURE 3.12 (a) Star-micelle formation with salt-responsive diblock copolymers B1 and B2 and (b) flower-micelle formation with salt-responsive triblock copolymer B3. Reprinted (adapted) with permission Donovan MS, Lowe AB, Sanford TA, McCormick CL. Sulfobetaine-containing diblock and triblock copolymers via reversible addition-fragmentation chain transfer polymerization in aqueous media. J Polym Sci A Polym Chem 2003;41:1262–1281. © 2003 American Chemical Society.

have drawn much attention as biomimics for phospholipid bilayers (i.e., liposomes) and as drug delivery vehicles [100]. Unlike small surfactants, fluoroalkyl end-capped polysulfobetaines exhibit unique solution properties [101, 102].

Diblock polyphosphobetaines based on phosphorylcholine monomers and NIPAm reversibly self-assemble in response to changes in solution temperature [91, 103, 104]. Donovan et al. reported the electrolyte-responsive assembly of diblock and triblock copolymers of sulfobetaine and DMA in water (Fig. 3.12) [94].

Selective post-polymerization modification of tertiary amine–containing polymers affords polysulfobetaines which are pH responsive [105, 106]. Biocompatible drug delivery vehicles have been made from the copolymers of phosphobetaine and pH-responsive alkyl methacrylates [88]. The well-controlled micelles exhibit pH-modulated drug uptake and release.

3.4 CROSS-LINKED MICELLES

The stimuli-induced assembly of block copolymers in solutions yields supramolecular structures, which may be utilized to capture, protect, and/or deliver active agents.

Block lengths and stimuli responsiveness of copolymers determine rigidity, stability, penetrability, size, and shell thickness of such associated structures. However, these assemblies can dissociate spontaneously into unimers when polymer concentrations fall below the CAC, resulting in burst or premature release of captured reagents and, thus, limiting practical applications. Therefore, a number of cross-linking methodologies have been developed to stabilize and control the assembly and disassembly of these structures [107].

For example, Li and coworkers [108] prepared a thermally responsive ABC triblock copolymer, poly(ethylene oxide)(PEO)-b-(DMA-stat-N-acryloyloxy succinimide)-b-PNIPAm. Upon formation of micelles above the LCST of PNIPAm, the activated ester groups in the middle block were reacted with ethylenediamine yielding covalently shell cross-linked micelles that would swell as the temperature was reduced below the LCST of PNIPAm. When cystamine was used instead of ethylenediamine, the cross-links became reversible as disulfide bridges were sensitive to redox active agents (Fig. 3.13) [109]. The release profile indicated that cross-linking of micelles slowed down the release of a model drug loaded into the core. In another strategy, a one-pot approach to reversibly cross-link micelles from a pH-responsive triblock copolymer was achieved utilizing a primary amine–imidoester reaction [110].

A cleavable, temperature-responsive polymeric cross-linker was utilized by Xu and coworkers [111] to stabilize micelles from PEO-b-PAPMA-b-poly((N,N-diisopropylamino)ethyl methacrylate) triblock copolymer. The PNIPAm cross-linker contained activated ester end groups that were reacted with the primary amines on the PAPMA middle block. The trithiocarbonate moiety located at the middle of PNIPAm cross-linker could then be degraded by aminolysis to break the cross-links. Temperature-responsive micelles and vesicles from diblock and triblock copolymers were shell cross-linked via interpolyelectrolyte complexation [108, 112]. The cross-links formed by the electrostatic interactions of oppositely charged polyelectrolytes could be disrupted by the addition of SME.

McCormick et al. reported reversible "self-locked" micelles from a dually responsive triblock copolymer [113]. In this system, protonation of the pH-responsive block led to the formation of micelles that were subsequently cross-linked through electrostatic interactions of the zwitterionic middle block. The micellization and cross-linking processes were reversible, triggered by changing solution pH and ionic strength.

Vesicles from thermally responsive PDEAEMA-b-PNIPAm block copolymers were cross-linked via the in situ reduction of gold salt [114]. The resulting gold-decorated structures remained intact when the temperature was lowered below the LCST of PNIPAm. The strategy allows simultaneous gold nanoparticle formation and cross-linking of the

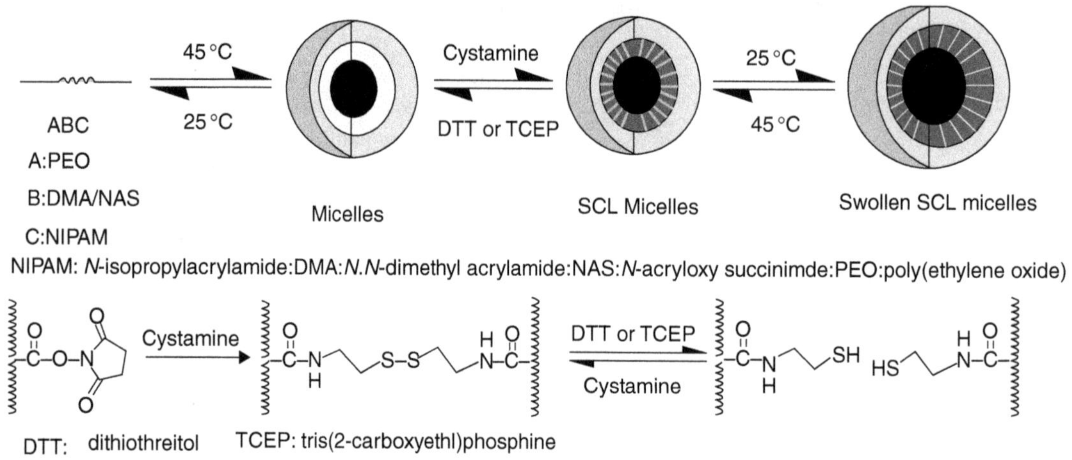

FIGURE 3.13   Formation of reversible shell-cross-linked micelles from PEO-*b*-P(DMA-*stat*-NAS)-*b*-PNIPAm triblock copolymers by reaction with cystamine. Reprinted (adapted) with permission from Xu XW, Smith AE, Kirkland SE, McCormick CL. Aqueous RAFT synthesis of pH-responsive triblock copolymer mPEO-PAPMA-PDPAEMA and formation of shell cross-linked micelles. Macromolecules 2008;41:8429–8435. © 2008 American Chemical Society.

assembled structures. In related work, Smith et al. [115] investigated the formation of various morphologies utilizing a similar polymer system. Parameters such as block length, solution pH, and ionic strength controlled the assembly of core-shell micelles, worm-like micelles, and vesicular structures. These assemblies were then cross-linked via the *in situ* gold nanoparticle formation [115].

3.5   STIMULI-RESPONSIVE COPOLYMERS IN BIOCONJUGATES

The preparation of smart polymer–protein conjugates is attractive since the responsive nature of the polymer can be conferred to the substrate to which it is attached. Such design provides a mechanism by which the activity or stability of the conjugated biomolecule may be controlled. In a number of reports [20, 116, 117], Sumerlin and coworkers established facile methodologies for conjugating stimuli-responsive polymers to proteins. In one study, a maleimide-containing RAFT chain transfer agent was conjugated to the protein and subsequently used in polymerization of NIPAm. The solubility and aggregation behavior of the bioconjugates were modulated by temperature. Grafting of PNIPAm from functionalized ATRP initiators containing biotin, lysozyme, and bovine serum albumin (BSA) resulted in bioconjugates in which biological function was maintained but could be rendered inactive at elevated temperatures [118–120].

Thermoresponsive synthetic glycopolymers were reported by Chen and coworkers via thiol–ene click addition [121]. The facile conjugation of glucose onto polymer scaffolds afforded self-assembled particles having specific affinity to lectins. The binding of saccharides to lectins may be exploited in the targeted delivery of drugs and other diagnostic agents. Similarly, with cancerous cells known to over-express folate-receptors at their surfaces, De and coworkers described the conjugation of folic acid to the terminal of thermoresponsive block polymers via azide–alkyne click cycloaddition [20]. Utilizing cationic triblock copolymers, York and coworkers demonstrated the facile conjugation of folic acid [122, 123]. These triblock copolymers were then utilized as nonviral vectors for the delivery of small interfering RNA (ribonucleic acid) in gene therapy. Additional examples on the utility of stimuli-responsive polymers in biorelated applications are described in several recent reviews [66, 67, 124, 125].

3.6   CONCLUSIONS

In this mini review, we have highlighted relevant examples of stimuli-responsive polymeric systems, their syntheses via CRP methods, and some selected applications. The field of stimuli-responsive polymers has experienced rapid growth over the last decade. Materials that are responsive to changes in the surrounding environment hold great potential for the construction of new functional materials. Recent discoveries in controlled living polymerizations and subsequent post-polymerization transformation allow polymer structures, compositions, and functionalities to be tailored, enabling the modulation of polymer properties. There is little doubt that research in this area will continue at an explosive pace as there is a need to improve existing technologies and develop functional materials to address future challenges.

ACKNOWLEDGMENTS

Support for this research from the MRSEC (DMR-0213883) and EPSCoR (EPS-0903787) programs of the National Science Foundation are gratefully acknowledged. Brooks A. Abel is a National Science Foundation Graduate fellow.

REFERENCES

[1] Hu J, Liu S. Responsive polymers for detection and sensing applications: current status and future developments. Macromolecules 2010;43:8315–8330.

[2] Magnusson JP, Saeed AO, Fernandez-Trillo F, Alexander C. Synthetic polymers for biopharmaceutical delivery. Polym Chem 2010;2:48–59.

[3] Liu F, Urban MW. Recent advances and challenges in designing stimuli-responsive polymers. Prog Polym Sci 2010;35:3–23.

[4] Szwarc M. 'Living' polymers. Nature 1956;176:1168–1169.

[5] Szwarc M, Levy M, Milkovich R. Polymerization initiated by electron transfer to monomer. A new method of formation of block copolymers. J Am Chem Soc 1956;78:2656–2657.

[6] Szwarc, M. Living polymers. Their discovery, characterization, and properties. J Polym Sci Part A Polym Chem 1998;36:ix–xv.

[7] Endo T. General mechanisms in ring-opening polymerization. In: Dubois P, Coulembier O, Raquez JM, editors. *Handbook of Ring-Opening Polymerization*. KGaA, Weinheim: Wiley-VCH Verlag GmbH & Co.; 2009. p 53.

[8] Webster OW. The discovery and commercialization of group transfer polymerization. J Polym Sci A Polym Chem 2000;38:2855–2860.

[9] Goethals EJ, Du Prez F. Carbocationic polymerizations. Prog Polym Sci 2007;32:220–246.

[10] Aoshima S, Kanaoka S. A renaissance in living cationic polymerization. Chem Rev 2009;109:5245–5287.

[11] Bielawski CW, Grubbs RH. Living ring-opening metathesis polymerization. Prog Polym Sci 2007;32:1–29.

[12] Bielawski CW, Grubbs RH. Living ring-opening metathesis polymerization. In: Müller AHE, Matyjaszewski K, editors. *Controlled Living Polymerization*. KGaA, Weinheim: Wiley-VCH Verlag GmbH & Co.; 2009. p 297.

[13] Moad G, Rizzardo E, Thang SH. Toward living radical polymerization. Acc Chem Res 2008;41:1133–1142.

[14] Matyjaszewski K, Davis TP, editors. *Handbook of Radical Polymerization*. New York: Wiley-Interscience; 2002. p 936.

[15] Braunecker WA, Matyjaszewski K. Controlled/living radical polymerization: features, developments, and perspectives. Prog Polym Sci 2007;32:93–146.

[16] Georges MK, Veregin RPN, Kazmaier PM, Hamer GK. Narrow molecular weight resins by a free-radical polymerization process. Macromolecules 1993;26:2987–2988.

[17] Hawker CJ, Bosman AW, Harth E. New polymer synthesis by nitroxide mediated living radical polymerizations. Chem Rev 2001;101:3661–3688.

[18] Matyjaszewski K, Xia JH. Atom transfer radical polymerization. Chem Rev 2001;101:2921–2990.

[19] Tsarevsky NV, Matyjaszewski K. "Green" atom transfer radical polymerization: from process design to preparation of well-defined environmentally friendly polymeric materials. Chem Rev 2007;107:2270–2299.

[20] De P, Gondi SR, Sumerlin BS. Folate-conjugated thermoresponsive block copolymers: highly efficient conjugation and solution self-assembly. Biomacromolecules 2008;9: 1064–1070.

[21] Moad G, Rizzardo E, Thang SH. Living radical polymerization by the RAFT process. Aust J Chem 2005;58:379–410.

[22] Moad G, Rizzardo E, Thang SH. Living radical polymerization by the RAFT process—A first update. Aust J Chem 2006;59:669–692.

[23] Moad G, Rizzardo E, Thang SH. Living radical polymerization by the RAFT process—A second update. Aust J Chem 2009;62:1402–1472.

[24] McCormick CL, Lowe AB. Aqueous RAFT polymerization: recent developments in synthesis of functional water-soluble (co)polymers with controlled structures. Acc Chem Res 2004;37:312–325.

[25] Smith AE, Xu XW, Mccormick CL. Stimuli-responsive amphiphilic (co)polymers via RAFT polymerization. Prog Polym Sci 2010;35:45–93.

[26] Hayes W, Rannard S. Controlled/'living' polymerization methods. In: Davis FJ, editor. *Polymer Chemistry: A Practical Approach*. Oxford: Oxford University Press; 2004. p 99.

[27] Smid J, Van Beylen M, Hogen-Esch TE. Perspectives on the contributions of Michael Szwarc to living polymerization. Prog Polym Sci 2006;31:1041–1067.

[28] Yagci Y, Tasdelen MA. Mechanistic transformations involving living and controlled/living polymerization methods. Prog Polym Sci 2006;31:1133–1170.

[29] Boyer C, Stenzel MH, Davis TP. Building nanostructures using RAFT polymerization. J Polym Sci A Polym Chem 2011;49:551–595.

[30] Cianga I, Cianga L, Yagci Y. New applications of living/controlled polymerization methods for synthesis of copolymers with designed architectures and properties. In: Dragan ES, editor. *New Trends in Nonionic (Co)polymer Hybrids*. New York: Nova Science Publishers; 2006. p 1–52.

[31] Gao H, Matyjaszewski K. Synthesis of functional polymers with controlled architecture by CRP of monomers in the presence of cross-linkers: from stars to gels. Prog Polym Sci 2009;34:317–350.

[32] Hadjichristidis N, Iatrou H, Pitsikalis M, Mays J. Macromolecular architectures by living and controlled/living polymerizations. Prog Polym Sci 2006;31: 1068–1132.

[33] Hadjichristidis N, Pitsikalis M, Iatrou H, Sakellariou G. Macromolecular architectures by living and controlled/living polymerizations. In: Müller AHE, Matyjaszewski K, editors. *Controlled Living Polymerization*. KGaA, Weinheim: Wiley-VCH Verlag GmbH & Co.; 2009. p 343.

[34] Matyjaszewski K, Tsarevsky NV. Nanostructured functional materials prepared by atom transfer radical polymerization. Nat Chem 2009;1:276–288.

[35] Nomura K. Precise synthesis of amphiphilic polymeric nano architectures utilized by metal-catalyzed living ring-opening metathesis polymerization (ROMP). In: Rizzo M, Bruno G, editors. *Surface Coatings*. New York: Nova Science Publishers; 2009. p 123–152.

[36] Ouchi M, Terashima T, Sawamoto M. Precision control of radical polymerization via transition metal catalysis: from dormant species to designed catalysts for precision functional polymers. Acc Chem Res 2008;41:1120–1132.

[37] Levy M. The impact of the concept of "Living Polymers" on material science. Polym Adv Technol 2007;18:681–684.

[38] Lo Verso F, Likos CN. End-functionalized polymers: versatile building blocks for soft materials. Polymer 2008;49: 1425–1434.

[39] Tasdelen MA, Kahveci MU, Yagci Y. Telechelic polymers by living and controlled/living polymerization methods. Prog Polym Sci 2011;36:455–567.

[40] Kolb HC, Finn MG, Sharpless KB. Click chemistry: diverse chemical function from a few good reactions. Angew Chem Int Ed 2001;40:2004–2021.

[41] Nandivada H, Jiang XW, Lahann J. Click chemistry: versatility and control in the hands of materials scientists. Adv Mat 2007;19:2197–2208.

[42] McCormick CL, Kirkland SE, York AW. Synthetic routes to stimuli-responsive micelles, vesicles, and surfaces via controlled/living radical polymerization. Polym Rev 2006;46: 421–443.

[43] Aseyev V, Tenhu H, Winnik FM. Non-ionic thermoresponsive polymers in water. Adv Polym Sci 2011;242:29–89.

[44] Heskins M, Guillet JE. Solution properties of PNIPAM. J Macromol Sci Chem 1968;2:1441–1455.

[45] McConaughy SD, Kirkland SE, Treat NJ, Stroud PA, McCormick CL. Tailoring the network properties of Ca^{2+} crosslinked aloe vera polysaccharide hydrogels for in situ release of therapeutic agents. Biomacromolecules 2008;9: 3277–3287.

[46] Aoshima S, Kanaoka S. Synthesis of stimuli-responsive polymers by living polymerization: poly(N-isopropylacrylamide) and poly(vinyl ether)s. Adv Polym Sci 2008;210 :169–208.

[47] Liu R, Fraylich M, Saunders BR. Thermoresponsive copolymers: from fundamental studies to applications. Colloid Polym Sci 2009;287:627–643.

[48] Schmaljohann D. Thermo- and pH-responsive polymers in drug delivery. Adv Drug Deliv Rev 2006;58:1655–1670.

[49] Xu J, Liu S. Polymeric nanocarriers possessing thermoresponsive coronas. Soft Matter 2008;4:1745–1749.

[50] Dong LC, Hoffman AS. Synthesis and application of thermally-reversible heterogels for drug delivery. J Control Release 1990;13:21–31.

[51] Winnik FM, Adronov A, Kitano H. Pyrene-labeled amphiphilic poly(N-isopropylacrylamides) prepared by using a lipophilic radical initiator: synthesis, solution properties in water, and interactions with liposomes. Canadian J Chem 1995;73:2030–2040.

[52] Chen G, Hoffman AS. Graft copolymers that exhibit temperature-induced phase transitions over a wide range of pH. Nature 1995;373:49–52.

[53] Chung JE, Yokoyama M, Aoyagi T, Sakurai Y, Okano T. Effect of molecular architecture of hydrophobically modified PNIPAM on the formation of thermoresponsive core-shell micellar drug carriers. J Control Release 1998;53:119–130.

[54] Binder WH, Gloger D, Weinstabl H, Allmaier G, Pittenauer E. Telechelic poly(N-isopropylacrylamides) via nitroxide-mediated controlled polymerization and "click" chemistry: livingness and "grafting-from" methodology. Macromolecules 2007;40:3097–3107.

[55] Harth E, Bosman A, Benoit D, Helms B, Frechet JMJ, Hawker CJ. A practical approach to the living polymerization of functionalized monomers: application to block copolymers and 3-dimensional macromolecular architectures. Macromol Symp 2001;174:85–92.

[56] Rademacher JT, Baum M, Pallack ME, Brittain WJ, Simonsick WJ. Atom transfer radical polymerization of N,N-dimethylacrylamide. Macromolecules 1999;33: 284–288.

[57] Masci G, Giacomelli L, Crescenzi V. Atom transfer radical polymerization of N-isopropylacrylamide. Macromol Rapid Commun 2004;25:559–564.

[58] Li C, Tang Y, Armes SP, Morris CJ, Rose SF, Lloyd AW, Lewis AL. Synthesis and characterization of biocompatible thermo-responsive gelators based on aba triblock copolymers. Biomacromolecules 2005;6:994–999.

[59] Convertine AJ, Ayres N, Scales CW, Lowe AB, McCormick CL. Facile, controlled, room-temperature RAFT polymerization of N-isopropylacrylamide. Biomacromolecules 2004;5:1177–1180.

[60] Convertine AJ, Lokitz BS, Vasileva Y, Myrick LJ, Scales CW, Lowe AB, McCormick CL. Direct synthesis of thermally responsive DMA/NIPAM diblock and DMA/NIPAM/DMA triblock copolymers via aqueous, room temperature RAFT polymerization. Macromolecules 2006;39:1724–1730.

[61] Lokitz BS, York AW, Stempka JE, Treat ND, Li Y, Jarrett WL, McCormick CL. Aqueous RAFT synthesis of micelle-forming amphiphilic block copolymers containing N-acryloylvaline. Dual mode, temperature/pH responsiveness, and locking of micelle structure through interpolyelectrolyte complexation. Macromolecules 2007;40:6473–6480.

[62] Kirkland SE, Hensarling RM, McConaughy SD, Guo Y, Jarrett WL, McCormick CL. Thermoreversible hydrogels from RAFT-synthesized BAB triblock copolymers: steps toward biomimetic matrices for tissue regeneration. Biomacromolecules 2007;9:481–486.

[63] Cao Y, Zhu XX, Luo J, Liu H. Effects of substitution groups on the RAFT polymerization of N-alkylacrylamides in the preparation of thermosensitive block copolymers. Macromolecules 2007;40:6481–6488.

[64] Fournier D, Hoogenboom R, Thijs HML, Paulus RM, Schubert US. Tunable pH- and temperature-sensitive copolymer libraries by reversible addition fragmentation chain

transfer copolymerizations of methacrylates. Macromolecules 2007;40:915–920.

[65] Yamamoto S-I, Pietrasik J, Matyjaszewski K. Temperature- and pH-responsive dense copolymer brushes prepared by ATRP. Macromolecules 2008; 41:7013–7020.

[66] Gil ES, Hudson SM. Stimuli-responsive polymers and their bioconjugates. Progr Polym Sci 2004;29:1173–1222.

[67] York AW, Kirkland SE, McCormick CL. Advances in the synthesis of amphiphilic block copolymers via RAFT polymerization: stimuli-responsive drug and gene delivery. Adv Drug Deliv Rev 2008;60:1018–1036.

[68] Aoki T, Kawashima M, Katono H, Sanui K, Ogata N, Okano T, Sakurai Y. Temperature-responsive interpenetrating polymer networks constructed with poly(acrylic acid) and poly(N,N-dimethylacrylamide). Macromolecules 1994;27:947–952.

[69] Ashford EJ, Naldi V, O'Dell R, Billingham NC, Armes SP. First example of the atom transfer radical polymerisation of an acidic monomer: direct synthesis of methacrylic acid copolymers in aqueous media. Chem Comm 1999;35:1285–1286.

[70] Zhang X, Matyjaszewski K. Synthesis of well-defined amphiphilic block copolymers with 2-(dimethylamino)ethyl methacrylate by controlled radical polymerization. Macromolecules 1999;32:1763–1766.

[71] Sumerlin BS, Donovan MS, Mitsukami Y, Lowe AB, McCormick CL. Water-soluble polymers. 84. Controlled polymerization in aqueous media of anionic acrylamido monomers via RAFT. Macromolecules 2001;34:6561–6564.

[72] Sumerlin BS, Lowe AB, Thomas DB, McCormick CL. Aqueous solution properties of pH-responsive AB diblock acrylamido copolymers synthesized via aqueous RAFT. Macromolecules 2003;36:5982–5987.

[73] Yusa S-I, Shimada Y, Mitsukami Y, Yamamoto T, Morishima Y. pH-responsive micellization of amphiphilic diblock copolymers synthesized via reversible addition fragmentation chain transfer polymerization. Macromolecules 2003;36:4208–4215.

[74] Mitsukami Y, Donovan MS, Lowe AB, McCormick CL. Water-soluble polymers. 81. Direct synthesis of hydrophilic styrenic-based homopolymers and block copolymers in aqueous solution via RAFT. Macromolecules 2001;34:2248–2256.

[75] Vasilieva YA, Thomas DB, Scales CW, McCormick CL. Direct controlled polymerization of a cationic methacrylamido monomer in aqueous media via the RAFT process. Macromolecules 2004;37:2728–2737.

[76] Zhang L, Eisenberg A. Multiple morphologies and characteristics of "crew-cut" micelle-like aggregates of polystyrene-b-poly(acrylic acid) diblock copolymers in aqueous solutions. J Am Chem Soc 1996;118:3168–3181.

[77] Zhang L, Yu K, Eisenberg A. Ion-induced morphological changes in "crew-cut" aggregates of amphiphilic block copolymers. Science 1996;272:1777–1779.

[78] Martin TJ, Prochazka K, Munk P, Webber SE. pH-dependent micellization of poly(2-vinylpyridine)-block-poly(ethylene oxide). Macromolecules 1996;29:6071–6073.

[79] Smith AE, Xu X, Kirkland-York SE, Savin DA, McCormick CL. "Schizophrenic" self-assembly of block copolymers synthesized via aqueous RAFT polymerization: from micelles to vesicles, paper number 143 in a series on water-soluble polymers. Macromolecules 2010;43:1210–1217.

[80] Lowe AB, McCormick CL. Synthesis and solution properties of zwitterionic polymers. Chem Rev 2002;102, 4177–4190.

[81] Chen S, Li L, Zhao C, Zheng J. Surface hydration: principles and applications toward low-fouling/nonfouling biomaterials. Polymer 2010;51:5283–5293.

[82] Armes SP, Robinson KL, Liu SY, Wang XS, Malet FLG, Furlong SA. Synthesis of new polymeric surfactants and dispersants via atom transfer radical polymerisation at ambient temperature. Spec Publ – R Soc Chem 2002;282:21–30.

[83] Buetuen V, Liu S, Weaver JVM, Bories-Azeau X, Cai Y, Armes SP. A brief review of 'schizophrenic' block copolymers. Reactive Funct Polym 2006;66:157–165.

[84] Srivastava N, Tiwari T. New trends in polymer electrolytes: a review. E-Polymers 2009 (paper 146):1–17.

[85] Holm C, Hofmann T, Joanny JF, Kremer K, Netz RR, Reineker P, Seidel C, Vilgis TA, Winkler RG. Polyelectrolyte theory. Adv Polym Sci 2004;166:67–112.

[86] Forster S, Abetz V, Muller AHE. Polyelectrolyte block copolymer micelles. Adv Polym Sci 2004;166:173–210.

[87] Lowe AB, McCormick CL. Reversible addition-fragmentation chain transfer (RAFT) radical polymerization and the synthesis of water-soluble (co)polymers under homogeneous conditions in organic and aqueous media. Prog Polym Sci 2007;32:283–351.

[88] Salvage JP, Rose SF, Phillips GJ, Hanlon GW, Lloyd AW, Ma IY, Armes SP, Billingham NC, Lewis AL. Novel biocompatible phosphorylcholine-based self-assembled nanoparticles for drug delivery. J Control Release 2005;104:259–270.

[89] Matsuura K, Ohno K, Kagaya S, Kitano H. Carboxybetaine polymer-protected gold nanoparticles: high dispersion stability and resistance against non-specific adsorption of proteins. Macromol Chem Phys 2007;208:862–873.

[90] Lobb EJ, Ma I, Billingham NC, Armes SP, Lewis AL. Facile synthesis of well-defined, biocompatible phosphorylcholine-based methacrylate copolymers via atom transfer radical polymerization at 20 degrees C. J Am Chem Soc 2001;123:7913–7914.

[91] Li CM, Madsen J, Armes SP, Lewis AL. A new class of biochemically degradable, stimulus-responsive triblock copolymer gelators. Angew Chem Int Ed 2006;45:3510–3513.

[92] Inoue Y, Watanabe J, Takai M, Yusa S, Ishihara K. Synthesis of sequence-controlled copolymers from extremely polar and apolar monomers by living radical polymerization and their phase-separated structures. J Polym Sci A Polym Chem 2005;43:6073–6083.

[93] Maeda Y, Mochiduki H, Ikeda I. Hydration changes during thermosensitive association of a block copolymer consisting of LCST and UCST blocks. Macromol Rapid Comm 2004;25:1330–1334.

[94] Donovan MS, Lowe AB, Sanford TA, McCormick CL. Sulfobetaine-containing diblock and triblock copolymers via reversible addition-fragmentation chain transfer polymerization in aqueous media. J Polym Sci A Polym Chem 2003;41:1262–1281.

[95] Stenzel MH, Barner-Kowollik C, Davis TP, Dalton HM. Amphiphilic block copolymers based on poly (2-acryloyloxyethyl phosphorylcholine) prepared via RAFT polymerisation as biocompatible nanocontainers. Macromol Biosci 2004;4:445–453.

[96] Arotcarena M, Heise B, Ishaya S, Laschewsky A. Switching the inside and the outside of aggregates of water-soluble block copolymers with double thermoresponsivity. J Am Chem Soc 2002;124:3787–3793.

[97] Kudaibergenov S, Jaeger W, Laschewsky A. Polymeric betaines: synthesis, characterization and application. Adv Polym Sci 2006;201:157–224.

[98] Johnson JA, Baskin JM, Bertozzi CR, Koberstein JT, Turro NJ. Copper-free click chemistry for the in situ crosslinking of photodegradable star polymers. Chem Comm 2008;44:3064–3066.

[99] Laschewsky A. Oligoethyleneoxide spacer groups in polymerizable surfactants. Coll Polym Sci 1991;269;785–794.

[100] Lee KY, Yuk SH. Polymeric protein delivery systems. Prog Polym Sci 2007;32:669–697.

[101] Sawada H, Umedo M, Kawase T, Baba M, Tomita T. Synthesis and properties of fluoroalkyl end-capped sulfobetaine polymers. J Appl Polym Sci 2004;92:1144–1153.

[102] Hadjichristidis N, Pispas S, Pitsikalis M. End-functionalized polymers with zwitterionic end-groups. Prog Polym Sci 1999;24:875–915.

[103] Miyazawa K, Winnik FM. Synthesis of phosphorylcholine-based hydrophobically modified polybetaines. Macromolecules 2002;35:2440–2444.

[104] Nedelcheva AN, Novakov CP, Miloshev SM, Berlinova IV. Electrostatic self-assembly of thermally responsive zwitterionic poly (N-isopropylacrylamide) and poly(ethylene oxide) modified with ionic groups. Polymer 2005;46: 2059–2067.

[105] Lowe AB, Billingham NC, Armes SP. Synthesis and properties of low-polydispersity poly(sulfopropylbetaine)s and their block copolymers. Macromolecules 1999;32 :2141–2148.

[106] Butun V, Bennett CE, Vamvakaki M, Lowe AB, Billingham NC, Armes SP. Selective betainisation of tertiary amine methacrylate block copolymers. J Mater Chem 1997;7:1693–1695.

[107] Read ES, Armes SP. Recent advances in shell cross-linked micelles. Chem Comm 2007;43:3021–3035.

[108] Lokitz BS, Convertine AJ, Ezell RG, Heidenreich A, Li Y, McCormick CL. Responsive nanoassemblies via interpolyelectrolyte complexation of amphiphilic block copolymer micelles. Macromolecules 2006;39:8594–8602.

[109] Li YT, Lokitz BS, Armes SP, McCormick CL. Synthesis of reversible shell cross-linked micelles for controlled release of bioactive agents. Macromolecules 2006;39: 2726–2728.

[110] Xu X, Smith AE, McCormick CL. Facile 'one-pot' preparation of reversible, disulfide-containing shell cross-linked micelles from a RAFT synthesized, pH-responsive triblock copolymer inwater at room temperature. Aust J Chem 2009;62:1520–1527.

[111] Xu XW, Smith AE, Kirkland SE, McCormick CL. Aqueous RAFT synthesis of pH-responsive triblock copolymer mPEO-PAPMA-PDPAEMA and formation of shell cross-linked micelles. Macromolecules 2008;41:8429–8435.

[112] Li Y, Lokitz BS, McCormick CL. Thermally responsive vesicles and their structural "locking" through polyelectrolyte complex formation. Angew Chem Int Ed 2006;45:5792–5795.

[113] Flores JD, Xu X, Treat NJ, McCormick CL. Reversible "self-locked" micelles from a zwitterion-containing triblock copolymer. Macromolecules 2009;42:4941–4945.

[114] Li Y, Smith AE, Lokitz BS, McCormick CL. In situ formation of gold-"decorated" vesicles from a RAFT-synthesized, thermally responsive block copolymer. Macromolecules 2007;40:8524–8526.

[115] Smith AE, Xu X, Abell TU, Kirkland SE, Hensarling RM, McCormick CL. Tuning nanostructure morphology and gold nanoparticle "locking" of multi-responsive amphiphilic diblock copolymers. Macromolecules 2009;42:2958–2964.

[116] Li M, De P, Gondi SR, Sumerlin BS. Responsive polymer-protein bioconjugates prepared by RAFT polymerization and copper-catalyzed azide-alkyne click chemistry. Macromol Rapid Commun 2008;29:1172–1176.

[117] Li M, De P, Li H, Sumerlin BS. Conjugation of RAFT-generated polymers to proteins by two consecutive thiol-ene reactions. Polym Chem 2010;1:854–859.

[118] Bontempo D, Maynard HD. Streptavidin as a macroinitiator for polymerization: in situ protein-polymer conjugate formation. J Am Chem Soc 2005;127:6508–6509.

[119] Heredia KL, Bontempo D, Ly T, Byers JT, Halstenberg S, Maynard HD. In situ preparation of protein "smart" polymer conjugates with retention of bioactivity. J Am Chem Soc 2005;127:16955–16960.

[120] Vázquez-Dorbatt V, Maynard HD. Biotinylated glycopolymers synthesized by atom transfer radical polymerization. Biomacromolecules 2006;7:2297–2302.

[121] Chen GJ, Amajjahe S, Stenzel MH. Synthesis of thiol-linked neoglycopolymers and thermo-responsive glycomicelles as potential drug carrier. Chem Commun 2009;45(10):1198–1200.

[122] York AW, Huang F, McCormick CL. Rational design of targeted cancer therapeutics through the multiconjugation of folate and cleavable siRNA to RAFT-synthesized (HPMA-s-APMA) copolymers. Biomacromolecules 2010;11:505–514.

[123] York AW, Zhang YL, Holley AC, Guo YL, Huang FQ, McCormick CL. Facile synthesis of multivalent folate-block copolymer conjugates via aqueous RAFT polymerization: targeted delivery of siRNA and subsequent gene suppression. Biomacromolecules 2009;10:936–943.

[124] de las Heras Alarcon C, Pennadam S, Alexander C. Simuli responsive polymers for biomedical applications. Chem Soc Rev 2005;34:276–285.

[125] Harada A, Kataoka K. Supramolecular assemblies of block copolymers in aqueous media as nanocontainers relevant to biological applications. Prog Polym Sci 2006;31:949–982.

4

REACTIONS IN HETEROGENEOUS MEDIA: EMULSION, MINIEMULSION, MICROEMULSION, SUSPENSION, AND DISPERSION POLYMERIZATION

RADMILA TOMOVSKA, JOSÉ C. DE LA CAL, AND JOSÉ M. ASUA

4.1 INTRODUCTION

The thermal control of bulk polymerization reactors is challenging because the high heat of reaction of the polymerization is accompanied by the low heat capacity and heat conductivity of the polymers and the high viscosity of the reaction mixture [1]. A better thermal control can be achieved by carrying out the polymerization in a phase dispersed in a basically inert continuous medium. This reduces the overall viscosity of the reaction medium allowing an efficient heat transfer. In addition, it lowers the rate of heat generation per unit volume.

A broad range of polymers are produced by polymerization in heterogeneous media, including polyolefins manufactured by slurry (high density polyethylene and isotactic polypropylene) and gas phase (linear low density polyethylene and high density polyethylene) polymerization; coatings and adhesives produced by emulsion and miniemulsion polymerization; flocculants obtained by inverse emulsion and microemulsion polymerization; poly(vinyl chloride) (PVC) and polystyrene produced by suspension polymerization; and toners synthesized by dispersion polymerization. As a whole, they represent more than 50% of the polymer produced worldwide [1].

In this chapter, emulsion, miniemulsion, microemulsion, suspension, and dispersion polymerization are discussed. Polyolefins are outside the scope of the chapter.

4.2 DESCRIPTION OF DIFFERENT POLYMERIZATION TECHNIQUES

Common features of the different polymerizations in dispersed media considered in this chapter are that the polymerization mostly proceeds through free radicals and the dispersed phase is stabilized by means of surface-active compounds (surfactants, suspension agents). The way in which the monomer dispersion is stabilized in the continuous medium largely defines the differences among these processes. Figure 4.1 summarizes the different options (with the exception of dispersion polymerization, discussed later).

In the case of the microemulsion polymerization, the use of appropriate surfactants allows achieving a very low interfacial tension and hence the monomer is dispersed in very small droplets. Often, a relatively high surfactant/monomer ratio (>10 wt%) is required. In emulsion polymerization, the monomer is dispersed in the continuous medium using a modest amount of surfactant (typically <2 wt% based on monomer). The resulting system comprises a dispersion of large monomer droplets (5–10 μm) and monomer-swollen micelles. Miniemulsion polymerization involves the use of a similar amount of surfactant as emulsion polymerization, but the monomer droplets are broken up to achieve a smaller size (80–500 nm) using an efficient homogenization device. The surfactant is adsorbed on the surface of the miniemulsion droplets; hence, there are no micelles in the system. This process is often used to prepare composite particles, where a preformed material is dissolved/dispersed in the monomer droplets.

Monitoring Polymerization Reactions: From Fundamentals to Applications, First Edition. Edited by Wayne F. Reed and Alina M. Alb.
© 2014 John Wiley & Sons, Inc. Published 2014 by John Wiley & Sons, Inc.

In the case of the suspension polymerization, the monomer is dispersed in large droplets stabilized by surface-active agents (inorganic and/or water-soluble polymers). No micelles are present in this system. Water is the most often used continuous phase. In this case, the monomers are at most sparingly soluble in water. For water-soluble monomers, an organic continuous medium is used while the dispersed phase is often the aqueous solution of the monomer. The monomer is either insoluble or sparingly soluble in the organic medium. These processes are referred to as inverse microemulsion/emulsion/miniemulsion/suspension polymerizations.

Upon addition of the initiator, the polymerization reaction proceeds, the characteristics of the products depending on the initial monomer dispersion. Thus, in the case of the microemulsion polymerization, small particles (20–60 nm) are formed; emulsion and miniemulsion polymerizations lead to polymer dispersions of similar size (most often 80–300 nm), but miniemulsion polymerization allows the production of composite particles not attainable otherwise. Suspension polymerization yields relatively large particles (50–1000 μm).

In dispersion polymerization (Figure 4.2), the reaction mixture is initially a homogeneous solution of monomer and stabilizer (or stabilizer precursor) in the continuous medium. Upon addition of the initiator, polymerization starts in the homogeneous phase and the polymer is formed, which is not soluble in the continuous medium, precipitates, and it is stabilized by the stabilizer. This technique yields monodisperse particles with sizes in the range of 1–5 μm.

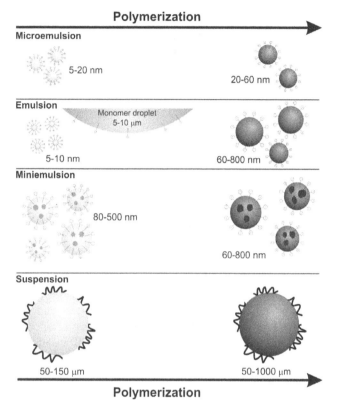

FIGURE 4.1   Schematic representation of microemulsion, emulsion, miniemulsion, and suspension polymerization.

4.3   EMULSION POLYMERIZATION

Emulsion polymerization is the leading technique to produce colloidal polymer dispersions. Carboxylated styrene–butadiene copolymers, acrylic and styrene–acrylic latexes, and vinyl acetate homopolymer and copolymers are the main polymer classes produced by this technique. These products are commercialized as dispersions and as dry products.

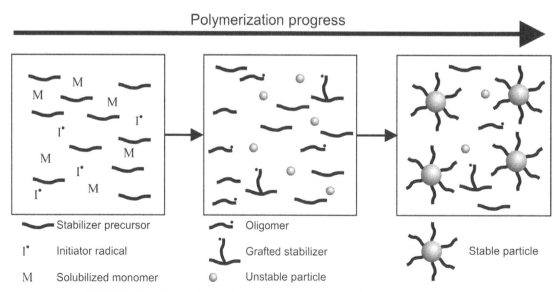

FIGURE 4.2   Schematics of dispersion polymerization using a stabilizer precursor.

The main markets for the dispersions are paints and coatings (26%), paper coating (23%), adhesives (22%), and carpet backing (11%) [2]. They are also used in such niche applications as diagnosis, drug delivery, and treatment [3]. The main dry products are styrene-butadiene rubber for tires, nitrile rubber, about 10% of the PVC production, acrylonitrile butadiene styrene (ABS), and redispersable powders for construction materials.

4.3.1 Description of the Process

Emulsion polymerization is mostly carried out in stirred tank reactors operated semicontinuously. The reason for using semicontinuous operation is that under similar reaction conditions, the polymerization rate is higher than in bulk (see Section), which makes the thermal control of the reactor difficult even with the relatively low overall viscosity of the reaction medium. Therefore, heat generation rate is controlled by feeding the monomers slowly. In addition, the semicontinuous operation allows better control of polymer characteristics. Continuous stirred tank reactors are used for the production of some high-tonnage emulsion polymers such as styrene-butadiene rubber.

In the semicontinuous process, the reactor is initially charged with a fraction of the formulation (monomers, emulsifiers, initiator, and water). The initial charge is polymerized in batch, and then the rest of the formulation is added over a certain period of time (typically 3–4 h). The monomers can be fed either as an aqueous preemulsion stabilized with emulsifier or as neat monomers. Monomers contain inhibitors to allow safe storage and are used without purification. The initiator is fed in a separate stream. The goal of the batch polymerization of the initial charge is to nucleate the desired number of polymer particles. Because particle nucleation is prone to suffer run-to-run irreproducibility, seeded semicontinuous emulsion polymerization is often used to overcome this problem. In this process, the initial charge contains previously synthesized latex (seed) and, eventually, a fraction of the formulation (monomers, emulsifiers, initiator, and water). Therefore, nucleation of new particles is minimized leading to better reproducibility.

Although batch emulsion polymerization is not frequently used, it will be discussed first because it is easier to understand as the fundamental processes occur in a sequential way, whereas in the semicontinuous and continuous operations the processes occur simultaneously.

4.3.1.1 *Batch Emulsion Polymerization* The monomers are dispersed in water in the presence of surfactants. The surfactants adsorb on the surface of the monomer droplets stabilizing them. In principle, any type of surfactant may be used, but in practice, anionic surfactants, nonionic surfactants, and mixtures thereof are mainly used. In most formulations, the amount of surfactant exceeds that needed amount in order to completely cover the monomer droplets and saturate the aqueous phase. The excess of surfactant forms micelles that are swollen with monomer.

Thermal initiators are used when the process is carried out at elevated temperatures (75–90 °C). Redox systems are used at lower reaction temperatures and if high rates of initiation are needed. Since most initiators are water soluble, the radicals are formed in the aqueous phase. These radicals are often too hydrophilic to directly enter into the organic phases. Consequently, they react with the monomer dissolved in the aqueous phase, forming oligoradicals that grow slowly because of the low concentration of monomer in the aqueous phase. Eventually, they become hydrophobic enough to be able to enter into the organic phases of the system. Because the total area of the micelles is about three orders of magnitude greater than that of the droplets, the entry of radicals into the micelles is more likely. The entering oligoradicals grow fast in the monomer-swollen micelle forming a polymer chain. The new species formed upon entry of a radical into a micelle is considered to be a polymer particle. The process of formation of polymer particles by the entry of radicals into micelles is called *heterogeneous nucleation* [4].

Polymer particles can also be formed when the oligoradicals grow in the aqueous phase beyond the length at which they are still soluble in water and precipitate. The precipitated polymer chain is stabilized by the emulsifier present in the aqueous phase and monomer diffuses into the new organic phase, which allows a fast growing of the polymer chain. The process of formation of polymer particles by precipitation of oligoradicals is called *homogeneous nucleation* [5].

Both homogeneous and heterogeneous nucleation may be operative in a given system. In general, homogeneous nucleation is predominant for monomers of relatively high water solubility (e.g., methyl methacrylate, 1.5 g/100 g of water, and vinyl acetate, 2.5 g/100 g of water) and heterogeneous nucleation is predominant for water-insoluble monomers (e.g., styrene, 0.045 g/100 g of water).

Regardless of the mechanism of particle nucleation (heterogeneous or homogeneous), the newly formed particles are very small, with an increased surface area upon particle growth. If the surfactant molecules do not diffuse rapidly enough to stabilize the surface of the fastly growing particles, they coagulate. This combined process is sometimes called *coagulative nucleation* [6]. The final number of particles is controlled by the ability of the surfactant to stabilize the particles formed by homogeneous and heterogeneous nucleation.

During nucleation, monomer droplets, monomer-swollen micelles, and monomer-swollen polymer particles coexist in the batch reactor. Polymer particles efficiently compete for radicals and, as their number increases, they become the main polymerization loci. The monomer consumed by free radical polymerization in the polymer particles is replaced by monomer that diffuses from the monomer droplets through the aqueous phase. Therefore, the size of the particles increases and that of the monomer droplets decreases. The number of micelles decreases because they become polymer particles upon entry of radicals, and also because they are destroyed to provide surfactant to stabilize both the polymer chains that precipitate in the aqueous phase and the increasing surface area of the growing polymer particles.

After a certain time, all the micelles disappear, the stage considered to be the end of nucleation; only limited formation of new particles may occur after this point since heterogeneous nucleation is not possible and there is no free surfactant available in the system to stabilize the particles formed by homogeneous nucleation. The stage of the batch emulsion polymerization in which particle nucleation occurs is called Interval I [4, 7]. At the end of Interval I, which typically occurs at a monomer conversion of about 5–10% (depending on the surfactant/monomer ratio), 10^{17}–10^{18} particles l^{-1} are formed. Unless coagulation occurs, the number of particles remains roughly constant during the rest of the batch process.

In Interval II, the system is composed of monomer droplets and polymer particles. The monomer consumed by polymerization in the polymer particles is replaced by monomer that diffuses from the monomer droplets through the aqueous phase. The mass transfer rate of monomers with water solubility equal or greater than that of styrene (0.045 g/100 g of water) is in most cases higher than the polymerization rate and, therefore, the monomer partitions between the different phases of the system according to thermodynamic equilibrium. In the presence of the monomer droplets, the concentration of the monomer in the polymer particles reaches a maximum value, kept roughly constant during Interval II. The transport of reactants (monomers, chain transfer agents) with water solubility lower than that of styrene from monomer droplets to polymer particles may be diffusionally limited.

Because of the polymerization and monomer transport, the polymer particles grow in size leading to the disappearance of the monomer droplets, which represents the end of Interval II. The monomer conversion at which Interval II ends depends on the maximum swelling of the polymer particles with the monomer. The higher the maximum swelling, the earlier the monomer droplets disappear. In general, the more water soluble the monomer is, the higher the maximum swelling and, hence, the lower the monomer conversion achieved at the end of Interval II. Thus, the transition from Interval II to Interval III occurs at about 40% conversion for styrene and at about 15% conversion for vinyl acetate. This means that most of the monomer polymerizes during Interval III. During this interval, the monomer concentration in the polymer particles decreases continuously.

4.3.1.2 Semicontinuous and Continuous Emulsion Polymerization
In semicontinuous reactors, monomers, surfactant, initiator, and water are continuously fed into the reactor. Monomer droplets form if the rate at which the monomer is fed into the reactor exceeds the polymerization rate. This is not a desirable situation because the presence of free monomer in the system lowers the capability for controlling the polymer characteristics [8].

In continuous stirred tank reactors, the whole formulation is continuously fed into the reactor and the product is continuously withdrawn. The composition of the outlet is the same as that of the reactor.

In semicontinuous and continuous systems, emulsion polymerization does not follow the sequence of events described earlier for a batch process. Nevertheless, the underlying processes are the same. Homogeneous nucleation requires that the oligoradicals in the aqueous phase grow beyond the critical length for precipitation. This is promoted by a low concentration of polymer particles (low solids content and/or large particle size), water-soluble monomers, and high generation rate of hydrophilic radicals. Heterogeneous nucleation requires the presence of micelles, namely, a high surfactant concentration. Therefore, nucleation of a new crop of particles can be achieved by adding extra amounts of surfactant at specific moments along the semicontinuous process. Intermittent nucleation occurs during emulsion polymerization reactions carried out in continuous stirred tank reactors [9, 10].

4.3.2 Polymerization Kinetics

Emulsion polymerization is a multiphase system; nonetheless, most of the polymerization occurs in the polymer particles. The concentration of radicals in the monomer droplets is very low because they are not efficiently capturing radicals from the aqueous phase. On the other hand, the concentrations of monomer and radicals in the aqueous phase are much lower than in the polymer particles. At first sight, this is surprising because radicals are formed in the aqueous phase. The reason is the so-called radical compartmentalization effect, which refers to the fact that the radicals distributed in different particles cannot terminate among them by bimolecular termination. This is the most distinctive kinetic feature of emulsion polymerization and has profound implications in both the polymerization rate and the polymer microstructure. In emulsion polymerization, the concentration of polymer particles is very high (5×10^{17} particles l^{-1} in a 50 wt% solids content latex with a particle size of 120 nm). Due to their small size, these particles contain only a small number of radicals per particle, in many cases of practical interest, less than one radical per particle. If only half of the particles contained one radical and the rest none, this results in an overall radical concentration of 4.2×10^{-7} mol l^{-1}, which is about one order of magnitude higher than the typical radical concentration in a continuous phase such as the aqueous phase. The radical compartmentalization effect decreases as the particle size increases, and above a certain particle size, the kinetics are essentially equal to those of bulk polymerization, namely, the particles behave as small bulk reactors.

The polymerization rate in emulsion polymerization is significantly higher than in bulk polymerization. In a latex, the overall concentration of radicals increases as the number of particles increases, for example, by decreasing the particle size at a given solids content. This provides further means of increasing polymerization rate (in addition to increasing temperature and initiator concentration).

Radical compartmentalization also results in longer lifetime of the radicals, which leads to higher molecular weight polymers. For the system described earlier, a polymer chain grows until a second radical enters into the polymer particle and terminates with the growing one. Therefore, the chain length is inversely proportional to the frequency of entry. For a given concentration of initiator, the frequency of radical entry decreases with the number of particles; therefore, the molecular weight increases. Consequently, in emulsion polymerization it is possible to simultaneously increase the polymerization rate and the molecular weight by simply increasing the number of particles, feature not possible in any other polymerization technique in which the radicals are not compartmentalized (bulk, solution, suspension).

4.3.2.1 Polymerization Rate

For a homopolymerization, the polymerization rate per unit volume of the reactor, R_p, is given by

$$R_p = k_p[M]_p \frac{\bar{n} N_p}{N_A V} \quad (\text{mol}\,l^{-1}s^{-1}) \qquad (4.1)$$

where k_p is the propagation rate coefficient ($l\,mol^{-1}s^{-1}$), $[M]_p$ is the concentration of monomer in the polymer particles ($mol\,l^{-1}$), $\bar{n}$ is the average number of radicals per particle, N_A is the Avogadro's number, N_p is the number of polymer particles in the reactor, and V is the volume of the reactor (l).

In the case of copolymerization, the copolymer-averaged rate coefficient for propagation should be used instead of k_p and $[M]_p$ represents the total monomer concentration in the polymer particles [11].

4.3.2.2 Average Number of Radicals per Particle

The average number of radicals per particle, $\bar{n}$, can be easily calculated from the population balance of particles with n radicals considering that for most practical cases, the pseudo-steady-state assumption can be applied to the radicals in the polymer particles and in the aqueous phase [12]:

$$\bar{n} = \frac{2k_a[P_{tot}]_w}{k_d + (k_d^2 + 4k_a[P_{tot}]_w c \Psi)^{0.5}} \qquad (4.2)$$

$$\Psi = \frac{2(2k_a[P_{tot}]_w + k_d)}{2k_a[P_{tot}]_w + k_d + c} \qquad (4.3)$$

where k_a is the entry rate coefficient ($l\,mol^{-1}s^{-1}$), which should be estimated for each system, $[P_{tot}]_w$ is the concentration of radicals in the aqueous phase ($mol\,l^{-1}$), k_d (s^{-1}) is the rate coefficient for radical desorption from particles, and c is the pseudo-first-order rate coefficient for termination in the polymer particles given by

$$c = \frac{k_t}{2v_p N_A} \quad (s^{-1}) \qquad (4.4)$$

where k_t is the termination rate constant ($l\,mol^{-1}\,s^{-1}$) and v_p is the volume of the monomer-swollen polymer particle.

The concentration of radicals in the aqueous phase, $[P_{tot}]_w$, can be calculated by means of the material balance for the radicals in the aqueous phase under quasi-steady-state conditions (see Chapter 1):

$$0 = 2fk_I[I]_w + k_d\bar{n} \frac{N_p}{N_A V_w} - k_{tw}[P_{tot}]_w^2 - k_a[P_{tot}]_w \frac{N_p}{N_A V_w} \qquad (4.5)$$

where radical formation from a water-soluble thermal initiator is considered, f is the efficiency factor of the initiator radicals, k_I is the rate coefficient for initiator decomposition (s^{-1}), $[I]_w$ is the concentration of the thermal initiator in the aqueous phase ($mol\,l^{-1}$), and k_{tw} is the termination rate coefficient in the aqueous phase ($l\,mol^{-1}s^{-1}$).

The solution of the system of algebraic Equations 4.2–4.5 includes the three limiting cases of the pioneering work of Smith and Ewart [7], summarized in Table 4.1. Case 1 ($\bar{n} << 0.5$) corresponds to a system in which the radical desorption rate is much faster than the rate of radical entry. Case 2 ($\bar{n} = 0.5$) corresponds to a system in which the rate of radical desorption is zero and instantaneous termination occurs when a radical enters a polymer particle already containing one radical. In Case 3, the concentration of radicals in the polymer particle approaches that of bulk polymerization ($\bar{n} >> 0.5$). For Case 2, the polymerization rate is proportional to the number of particles. For Cases 1 and 3, the polymerization rate is independent of the number of polymer particles if radical termination in the aqueous phase is negligible and it increases with N_p when the termination in the aqueous phase is significant.

4.3.2.3 Monomer Partitioning

At any moment during the polymerization, the monomers partition among the

TABLE 4.1   Smith–Ewart Limiting Cases

Smith–Ewart Limiting Case	Experimental Conditions	$\bar{n}$ Range	Equation to Calculate $\bar{n}$	$\bar{M}_n^{inst}$	$\bar{M}_w^{inst}$
Case I	1. Small particles (<100 nm) 2. Relatively water-soluble monomers or relatively water-soluble CTAs 3. Low rate of generation of radicals from the initiator 4. Large number of particles	$\bar{n} \ll 0.5$	$\bar{n} \approx \dfrac{k_a[P_{tot}]_w}{2k_a[P_{tot}]_w + k_d}$	$\bar{M}_n^{inst} \approx \dfrac{k_p[M]_p}{2fk_I[I]_w N_A} \dfrac{N_p w_m}{V_w}$	$\bar{M}_w^{inst} \approx 2\bar{M}_n^{inst}$
Case II	1. No chain transfer to small molecules (i.e., monomers and CTAs), or these small molecules are highly water insoluble 2. Fast bimolecular termination rate 3. The polymer particles are relatively small (typically dp < 200 nm)	$\bar{n} = 0.5$	$\bar{n} = 0.5$	$\bar{M}_n^{inst} \approx \dfrac{k_p}{k_{tr}^{mon}} w_m$	$\bar{M}_w^{inst} \approx 2\bar{M}_n^{inst}$
Case III	1. Large particles (dp > 200 nm) 2. High initiator concentrations or redox initiators 3. Slow termination rates (gel effect)	$\bar{n} \gg 0.5$	$\bar{n} = \left(\dfrac{k_a[P_{tot}]_w}{2c}\right)^{0.5}$	$\bar{M}_n^{inst} \approx \dfrac{k_p[M]_p N_p}{(2c_d + c_c)\mu_0 N_A} w_m$	Termination by disproportionation: $\bar{M}_w^{inst} \approx 2\bar{M}_n^{inst}$ Termination by combination: $\bar{M}_w^{inst} \approx 1.5\bar{M}_n^{inst}$

$c = \dfrac{k_t}{2v_p N_A}$, $\mu_0 = \dfrac{\bar{n} N_p}{N_A}$; w_m = molecular weight of the repeat unit in the polymer

different phases. If the mass transfer of the monomers is faster than the rate of polymerization, monomer partitioning is ruled by thermodynamics. For a multimonomer system, the calculation of the concentrations of the monomers in the different phases involves the simultaneous solution of the thermodynamic equilibrium equations and the material balances. The equilibrium equations can be given in terms of the Morton equations [13, 14], which account for the effect of the interfacial tensions and the Flory–Huggins interaction parameters. However, the use of these equations in multimonomer systems is limited by the availability of the parameters of the model. In these cases, the use of partitioning coefficients is advised [15].

Equilibrium Equations:

$$K_i^j = \frac{\phi_i^j}{\phi_i^w} \qquad j = \text{polymer particles, droplets} \qquad (4.6)$$

Material Balances:

$$\phi_{pol}^p + \sum_i \phi_i^p = 1 \qquad (4.7)$$

$$\phi_{water}^w + \sum_i \phi_i^w = 1 \qquad (4.8)$$

$$\sum_i \phi_i^d = 1 \qquad (4.9)$$

$$V_p \phi_i^p + V_d \phi_i^d + V_w \phi_i^w = V_i \qquad (4.10)$$

$$V_w \phi_{water}^w = V_{water} \qquad (4.11)$$

$$V_p \phi_{pol}^p = V_{pol} \qquad (4.12)$$

where K_i^j is the partition coefficient of monomer i between the phase j and the aqueous phase; ϕ_i^j is the volume fraction of monomer i in phase j; V_p, V_d, and V_w are the volumes of monomer-swollen particles, monomer droplets, and aqueous phase, respectively; and V_i, V_{pol}, and V_{water} are the volumes of monomer i, polymer, and water, respectively.

Efficient methods for solving Equations 4.6–4.12 are available [16, 17].

4.3.2.4 *Particle Nucleation* Particle nucleation may occur through both entry of radicals into micelles (heterogeneous nucleation) and precipitation of oligoradicals in the aqueous phase (homogeneous nucleation).

For heterogeneous nucleation, the dependence of the number of particles on surfactant and initiator concentrations is given by [7, 18, 19]

$$\left(\frac{N_p}{V_w}\right) \div \left(\frac{2fk_I[I]_w N_A \phi_{pol}^p}{r_v}\right)^{1-z} \left(\frac{a_s S_T}{V_w}\right)^z \quad 0.6 \le z \le 1$$

(4.13)

where r_v is the volumetric growth rate of one polymer particle ($1 s^{-1}$), S_T is the total amount of surfactant in the reactor (mol), and a_s is the area of the polymer particles covered by 1 mol of surfactant under saturation conditions ($m^2 mol^{-1}$).

In Equation 4.13, z approaches unity as the water solubility of the monomer increases (e.g., $z \approx 0.6$ for styrene and $z \approx 0.8$ for MMA).

The rate of formation of particles by homogeneous nucleation is the rate at which the oligoradicals growing in the aqueous phase exceed the maximum soluble length. The critical length depends on the composition of the oligoradical, and hence radicals formed from the initiator (i.e., containing inorganic moieties) and from the desorbed radicals should be distinguished. The oligomers formed from the initiator should have a certain length (δ_z) so that being hydrophobic enough, they could enter into the polymer particles; oligomers formed from the desorbed radicals may directly enter into the particles.

The rate of formation of particles by homogeneous nucleation is the rate of propagation of oligoradicals of critical length:

$$R_{nuc} = k_p[M]_w \frac{(P_{Ijcrit} + P_{Micrit})}{V} N_A \quad (particles\, 1^{-1} s^{-1})$$

(4.14)

where $[M]_w$ is the concentration of monomer in the aqueous phase (mol 1^{-1}) and P_{Ijcrit} and P_{Micrit} are the oligoradicals of critical length (with $j_{crit} > i_{crit}$) formed from the initiator and from desorbed radicals (mol), respectively.

P_{Ijcrit} and P_{Micrit} are calculated from the balances of radicals of both types in the aqueous phase assuming that pseudo-steady-state conditions are applied:

$$P_{Ijcrit} = \alpha_1^{jcrit-\delta_z+1} \alpha_2^{\delta_z-1} \frac{2fk_I I}{k_p[M]_w} \quad (mol)$$

(4.15)

$$P_{Micrit} = \frac{k_d \bar{n} N_p}{k_p[M]_w N_A} \alpha_1^{icrit} \quad (mol)$$

(4.16)

where α_1 is the probability of propagation of radicals able to enter into the polymer particles (generated from desorbed radicals and from the initiator with lengths equal or greater than δ_z) and α_2 is the probability of propagation of radicals generated from the initiator of length shorter than δ_z.

It is assumed that the radicals generated from the initiator with a length shorter than δ_z are too hydrophilic to enter into the organic phase. These probabilities are given by

$$\alpha_1 = \frac{k_p[M]_w}{k_p[M]_w + k_{tw}[P_{tot}]_w + k_a \frac{N_p}{N_A V_w} + \delta_{(mic)} k_{am} \frac{N_m}{N_A V_w}}$$

(4.17)

$$\alpha_2 = \frac{k_p[M]_w}{k_p[M]_w + k_{tw}[P_{tot}]_w}$$

(4.18)

where $\delta_{(mic)} = 0$ if only homogeneous nucleation applied and $\delta_{(mic)} = 1$ if the two nucleation mechanisms occur.

In systems including rather water-soluble monomers and surfactant concentrations high enough so that micelles are present, particle nucleation may be formed by both heterogeneous and homogeneous mechanisms. When the particles formed by heterogeneous and/or homogeneous nucleation are not well stabilized by the emulsifier, particle coagulation occurs. Both detailed [20, 21] and simplified [22] modeling for this case are given elsewhere. It is worth pointing out that particle stability does not require full surface coverage. Actually, most commercial latexes are only partially covered by emulsifier.

4.3.3 Molecular Weights

In emulsion polymerization, the molecular weights strongly depend on the average number of radicals per particle. Detailed mathematical models for the calculation of linear [23] and nonlinear [24–34] polymers for any value of $\bar{n}$ are available. A detailed discussion of this issue is outside the scope of this chapter. Instead, particular solutions for the limiting cases of Smith–Ewart [7] are presented in Table 4.1 where for Case 3, it was considered that the main chain growth termination event was bimolecular termination.

Linear polymers are formed from monofunctional monomers with a polymerization scheme that does not include chain transfer to polymer (e.g., styrene and methyl methacrylate). In Smith–Ewart Cases 1 and 2, the probability of having particles with more than one radical is almost negligible, allowing the system to be considered formed by particles with no radicals and particles with one radical. This is called a zero–one system. In such systems, the inactive polymer chains are formed in particles containing one radical by chain transfer to monomer and by instantaneous termination upon entry of one radical. In Smith–Ewart Case 3, the number of radicals per particle is large and the reaction kinetics approach the bulk polymerization case.

Nonlinear polymers are formed by (i) transfer of a radical from a polymeric radical to another polymer chain (intermolecular chain transfer to polymer), (ii) polymerization with terminal double bonds (resulting from disproportionation, transfer to monomer, and chain scission), and (iii) polymerization of multifunctional monomers. Nonlinear polymers are also formed by intramolecular chain transfer (backbiting), but this process does not affect molecular weights. Processes (i)–(iii) reactivate dead polymer chains that increase in length through subsequent propagation until they terminate again. Each polymer chain may suffer several activation–deactivation cycles. As a consequence, the molecular weight of the polymer increases and polymer networks are formed.

The mathematical modeling of these polymers are based on either population balances [24–31] or Monte Carlo methods [32–35]. The output of these models includes a detailed description of the polymer architecture (sol MWD, gel fraction, cross-linking points, long chain branching). The models are rather complex and will not be discussed here. Nevertheless, for cases of practical interest, the average molecular weights can be calculated with a modest computational effort [36].

4.3.4 Particle Morphology

Particles with special morphologies, that is, particles containing different phases, present certain advantages in many applications such as impact modifiers for polymer matrices [37], low minimum film forming temperature nonblocking coatings [38], and opacifiers [39]. These latexes are manufactured by seeded semicontinuous polymerization using a seed with a composition different from that of the polymer produced by polymerization of the monomer.

Figure 4.3 shows several types of morphologies of the particles produced. The incompatibility of the seed polymer with the newly formed polymer results in phase separation and cluster formation. The size of the clusters increases by both polymerization and cluster aggregation. In addition, the clusters tend to reach an equilibrium morphology, which minimizes the interfacial energy of the system and depends on the polymer–polymer and polymer–water interfacial tensions [40–44].

The final morphology heavily depends on the kinetics of cluster migration [45, 46]. The driving force for the migration is the minimization of the interfacial energy and the resistance increases with the internal viscosity of the particles. Metastable morphologies can be achieved under starved conditions (high resistance for cluster migration due to the high internal viscosity of the particles) and promoting grafting reactions (low polymer–polymer interfacial tensions).

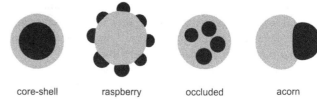

core-shell   raspberry   occluded   acorn

FIGURE 4.3 Examples of particle morphologies.

Equilibrium morphologies may be attained if the internal viscosity of the particle is low, the polymers are very incompatible (high interfacial tensions), and during long processes.

4.3.5 Inverse Emulsion Polymerization

In this process, an aqueous solution of a water-soluble monomer is dispersed in an organic continuous phase using a nonionic surfactant of low hydrophilic-lipophilic balance (HLB = 4–6). The polymerization is initiated using oil-soluble initiators and the mechanisms involved are similar to those occurring in emulsion polymerization [47, 48].

A major application is the synthesis of high molecular weight water-soluble polymers (e.g., polymers and copolymers of acrylamide, acrylic acid, and its salts) for flocculants and tertiary oil recovery. Other uses are the synthesis of polyaniline/CdSe quantum dots composites [49], hybrid polyaniline/carbon nanotube nanocomposites [50], polyaniline-montmorillonite nanocomposites [51], or in reversible addition–fragmentation chain-transfer-controlled radical polymerization (RAFT) [52].

4.4 MINIEMULSION POLYMERIZATION

Figure 4.1 shows a simplified view of miniemulsion polymerization in which the miniemulsion droplets are polymerized yielding polymer particles. When a monomer (or a mixture of monomers) able to diffuse through the aqueous phase is polymerized, the final product is similar to that obtained during emulsion polymerization reactions. In this case, miniemulsion polymerization does not provide any advantage over emulsion polymerization, which is technologically simpler. However, emulsion polymerization is of limited use when a highly water-insoluble monomer is polymerized because its diffusion from the monomer droplets to the polymer particles is hindered. The need of material diffusion through the aqueous phase precludes the use of emulsion polymerization for the production of polymer–polymer and polymer–inorganic hybrids.

Miniemulsion polymerization is advantageous for the production of complex colloidal polymer dispersions since, due to droplet nucleation, no mass transfer through the aqueous phase is needed. Therefore, it allows the synthesis of a broad range of materials not obtainable by the means of other polymerization processes in dispersed media. Some examples of these polymer materials are (i) dispersed polymers with well-defined microstructure obtained by controlled radical polymerization [53–56], (ii) waterborne linear polyethylene produced by catalytic polymerization [57–59], (iii) easy incorporation of hydrophobic monomers [60], (iv) polymer–polymer hybrid latexes [61–63], and (v) and polymer–inorganic waterborne dispersions [64–66]. Reviews of miniemulsion polymerization are available [67–71].

A successful miniemulsion polymerization requires (i) the formation of small droplets; (ii) their stabilization against coagulation and diffusional degradation (Ostwald ripening); (iii) the nucleation of most of the droplets, avoiding other nucleation mechanisms and coagulation (with particles and droplets); (iv) achieving high monomer conversion; and (v) controlling polymer architecture and particle morphology.

4.4.1 Miniemulsification

The industrial implementation of the miniemulsion polymerization is being hindered by the miniemulsification process. At first sight, this may look surprising because liquid–liquid dispersions have been extensively studied [72–75]. However, most of these studies involved droplets in the supra-micron range, the volume fraction of the dispersed phase was in most cases below 20%, and in general, the viscosity of the dispersed phase was low. For most practical applications, high solids content (about 50 wt%) composite miniemulsions with droplet size below 100 nm and relatively high viscosities of the dispersed phase are required. The formation of these miniemulsions is challenging.

Miniemulsions can in principle be produced by using rotor-stators, sonicators, static mixers and high pressure homogenizers. It has been reported that the rotor-stator was not effective enough to reduce the droplet size of the dispersed phase to the required size [76]. Sonication was effective to prepare small volumes of miniemulsions with small droplet sizes, but it is not well adapted for scale-up. Static mixers are efficient for moderate solids content (< 43 wt%) and low viscosity dispersed phase [77, 78]. However, their performance in high solids content systems (≥ 50 wt%) and their use for preparing miniemulsions with high viscosity organic phase remains to be demonstrated. High pressure homogenization allowed preparing small size, high solids miniemulsions containing a highly viscous dispersed phase [76, 79]. High pressure homogenization is a promising choice for industrial scale as large capacity (21,0001 h^{-1} at 400 bar) equipment is available.

The process through which the miniemulsion droplets are formed in this equipment is illustrated in Figure 4.4. The coarse emulsion is pumped through the narrow gap of the valve. At the entrance of the gap, the droplets suffer elongational flow but the time that the droplets spend at the entrance of the gap is too short to break the droplets. The flow through

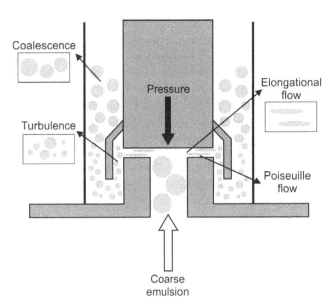

FIGURE 4.4 Mechanisms for miniemulsion formation in a high-pressure homogenizer.

the gap is basically a Poiseuille flow that does not affect the droplets. Maximum turbulence intensity occurs at the exit of the valve, which causes the breakup of the elongated droplets. The newly formed droplets are not well covered by the surfactant, and hence they may coagulate. Therefore, droplet size is the result of two consecutive mechanisms: droplet breakup and coagulation. The final droplet size is determined by the mechanism giving the largest droplet size. Droplet breakup is the size-controlling mechanism at low pressures in the valve and high emulsifier concentrations, when the amount of emulsifier is enough to stabilize the droplets formed. Droplet coagulation determines the size of the droplets at high pressures and relatively low emulsifier concentrations, not enough to stabilize the large surface area created [79].

4.4.2 Droplet Stabilization

Once the miniemulsion is formed, the droplets should remain stable until they are polymerized. Droplet stability is reduced by both coagulation and degradation by monomer diffusion. The stability against coagulation is enhanced by using adequate surfactants; all the knowledge developed for the stability of colloids applies here [80–83]. Degradation by monomer diffusion is characteristic of colloidal dispersions of compounds that have water solubility in some degree and present a significant interfacial tension with the continuous phase. Because of their small size, the contribution of the surface energy to the chemical potential of the monomer in the droplet is substantial, increasing as the droplet size increases. Therefore, the chemical potential of the monomer in the smaller droplets is higher than in the larger ones and the monomer diffuses from the smaller to the larger droplets. Consequently, in a system composed only by monomer droplets, the smaller droplets eventually disappear leading to the degradation of the dispersion by the formation of very large droplets. This process is called Ostwald ripening [84, 85].

A way to avoid the degradation is to include in the droplets a water-insoluble compound that cannot diffuse through the aqueous phase. In this case, due to the differences in surface energy, monomer diffuses from the smaller droplets to the larger ones and the concentration of monomer in the smaller droplets decreases, whereas it increases in the larger droplets. This counteracts the effect of the surface energy on the chemical potential and smaller and larger droplets are at equilibrium (although their composition is different, the smaller droplets being richer in the water-insoluble compound). Thermodynamically, the smaller the molecular weight of the water-insoluble compound, the smaller the differences in composition between small and large droplets [86, 87]. Therefore, miniemulsions are conveniently prepared by using small molecular weight water-insoluble compounds called costabilizers. Hexadecane is an efficient costabilizer but it does not polymerize and increases the content of volatile organic compounds (VOCs) in the latex. Water-insoluble monomers such as lauryl acrylate are preferred. Polymers can also be used to stabilize the miniemulsions against Ostwald ripening but in relatively high concentrations because they are less efficient. High molecular weight water-insoluble compounds are often called hydrophobes.

4.4.3 Droplet Nucleation

Droplet nucleation is the distinctive feature of miniemulsion polymerization [88] that allows incorporating hydrophobic compounds into polymer particles, because mass transfer through the aqueous phase is avoided. As mentioned earlier, this has led to the development of a wide range of useful polymer materials accessible only through this technique. The successful synthesis of many of these materials requires an efficient nucleation of the monomer droplets. This can be illustrated by considering the synthesis of hybrid dispersions. In this process, a preformed material is dissolved/dispersed in a monomer mixture containing a costabilizer (low molecular weight highly water-insoluble compound) and the mixture is dispersed in an aqueous solution of surfactants to obtain a miniemulsion. The challenge is to transform most of the droplets into composite particles maintaining the same polymer/material ratio in all particles.

Figure 4.5 illustrates this challenge for a batch miniemulsion polymerization initiated by a water-soluble thermal initiator. The radicals produced in the aqueous phase by decomposition of the initiator are hydrophilic and cannot enter into the polymer particles. Therefore, they react with

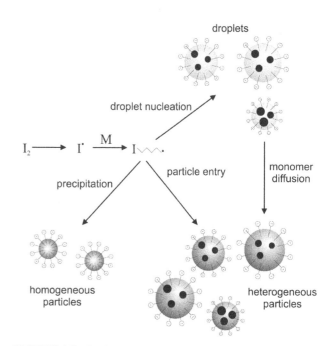

FIGURE 4.5 Particle nucleation in miniemulsion polymerization.

the monomer dissolved in the aqueous phase. Once the oligoradicals reach a certain length, they become surface active and may enter into miniemulsion droplets and polymer particles or they may continue growing in the aqueous phase and precipitate when they reach a critical length above which they are not soluble in water. Precipitation leads to the formation of new polymer particles through homogeneous nucleation, which, as discussed previously, leads to polymer particles containing only the polymer formed by polymerization of the monomer, which is in this case undesirable. The condition for avoiding homogeneous nucleation is that the oligoradicals are captured by the droplets and polymer particles before they reach the critical length for precipitation. In practice, this requires a large number of particles and droplets per unit volume, namely, high solids content and small droplet and particle sizes. These requirements are more stringent under conditions that promote homogeneous nucleation: use of monomers with high water solubility and high propagation rate constants, and high generation rate of water soluble radicals. In addition, an efficient droplet nucleation requires a fast nucleation of all the droplets in order to avoid monomer mass transfer from droplets to growing particles, which would lead to the production of particles with different compositions. Fast nucleation implies a high radical generation rate, which may enhance homogeneous nucleation. Therefore, oil-soluble initiators are often used. Furthermore, the miniemulsion should be colloidally stable and particle–particle coagulation and particle–droplet coagulation should be minimized.

Industrial production of waterborne polymer dispersions is mainly carried out by conventional emulsion polymerization in 15–60 m³ reactors operating in a semicontinuous mode, and likely this will be the choice for the industrial implementation of miniemulsion polymerization.

Droplet nucleation in semicontinuous processes with miniemulsion feeding is complex since new droplets are continuously entering into the reactor. These processes can be summarized as follows [89]: (i) polymerization of the initial charge proceeds as in batch miniemulsion polymerization; (ii) during the first stages of the semicontinuous process, a large fraction of the droplets is nucleated because the number of particles in the system is still moderate and, hence, the droplets efficiently compete for radicals; and (iii) as the polymerization proceeds, the number of particles in the system increases and it becomes less likely that the droplets entering into the reactor capture radicals. Therefore, the monomer in the droplets diffuses to the polymer particles and eventually the droplets disappear by coagulation with the polymer particles or by nucleation. In the latter case, they yield relatively small particles richer in the water-insoluble components because they have lost a substantial part of the monomer. The requirements for colloidal stability are stronger than in the case of batch system.

4.4.4 Polymerization Kinetics

In miniemulsion homopolymerization, once the polymer particles are nucleated, the processes controlling the particle growth are the same as in emulsion polymerization. The presence of a costabilizer affects the monomer partitioning, which may affect copolymer composition, molecular weight, and polymer architecture [90–92]. Limiting conversion has been reported in the production of some polymer–polymer hybrid latexes, such as alkyd–acrylic [61] and acrylic–poly(dimethylsiloxane) (PDMS) divinyl terminated [93]. These preformed polymers contained double bonds that can yield stable radicals acting as radical sinks. Nevertheless, complete monomer conversion was achieved using both high reaction temperatures for thermal initiators and redox initiators fed semicontinuosly [93, 94].

4.4.5 Control of Polymer Architecture and Particle Morphology

Miniemulsion polymerization is a unique way to synthesize complex multiphase polymer dispersions that have the potential to synergistically combine the properties of their components. Because of the good compatibility among the phases enhances synergy, grafting is sought. In the case of polymer–polymer systems, this is achieved by using preformed polymer containing functional groups (double bonds, isocyanate, epoxy, etc.) that can react with the radicals and the monomers used in the polymerization. As a consequence, both polymer architecture and particle morphology are affected. Grafting adds complexity to the polymer architecture.

Thus, in alkyd–acrylic systems, the final product contains free acrylic polymer, grafted acrylic–alkyd polymer, and free alkyd resin [95]. Polyurethane–acrylic hybrids present a polymer network formed by acrylic chains linked by both polyurethane chains and pure acrylic connections [76].

As in the case of emulsion polymerization, particle morphology is ruled by the interplay between thermodynamics and kinetics. Equilibrium morphologies are reached when the internal viscosity of the polymer particle is low. Thus, due to the plasticizing effect of the alkyd resin, equilibrium morphologies are usually reached for alkyd/acrylic systems [96]. The equilibrium morphology is affected by the presence of graft copolymer that reduces the interfacial tension between the polymer phases in the particle. Methods to calculate the equilibrium morphology of multiphase polymer particles are available [43].

4.4.6 Inverse Miniemulsion Polymerization

Inverse miniemulsions are produced by dispersing aqueous solutions of polar monomers in organic solvents using nonionic surfactants, cosurfactant (insoluble in the organic phase), and an efficient homogenization device [97]. Polymerization of miniemulsions has been used to produce nanocomposite particles containing hydrophilic polymers [98, 99].

4.5 MICROEMULSION POLYMERIZATION

Microemulsions are thermodynamically stable systems. Oil-in-water (O/W) microemulsions are mixtures of monomer(s), water, surfactant, and, in some cases, cosurfactant. The cosurfactant is a surface-active compound that, in combination with the surfactant, reduces the interfacial tension between the monomer and the aqueous phase to very low values, ensuring the thermodynamic stability of the microemulsion. Alcohols are often used as cosurfactants. The low interfacial tension results in a frequent fluctuation in size and shape of the microemulsion droplets. In water-in-oil (W/O) microemulsions, a mixture of water-soluble monomers and water are dispersed in an organic solvent with the help of a surfactant. The use of a cosurfactant is not needed often because the monomers are surface active. The amount of surfactant required in microemulsion polymerization (≥10 wt%) is substantially higher than that used in emulsion polymerization. The droplet (swollen micelle) size of the both O/W and W/O microemulsions is in the range of 5–20 nm in diameter. Since these small droplets only weakly scatter light, the microemulsions are transparent. Bicontinuous microemulsions are sometimes formed using blends of nonionic surfactants [100]. Microemulsion polymerization has been reviewed [101].

Polymerization of O/W microemulsions is referred to as microemulsion polymerization and that of W/O microemulsions as inverse microemulsion polymerization. Both of them proceed in a similar manner. The radicals are produced in the continuous phase and react with the monomer dissolved in the continuous medium. Once they become surface active, they are able to enter into the monomer-swollen micelles, which become polymer particles [102, 103].

For high surfactant concentrations, the number of monomer-swollen micelles is much larger than the number of the polymer particles for the entire process and hence nucleation of new particles may continuously occur during the polymerization. Homogeneous nucleation is negligible. The monomer needed to form the polymer diffuses from the micelles to the particles. At the end of the process, the system is composed by polymer particles and surfactant micelles. The final particle size (20–60 nm) is larger than the size of the initial micelles (5–20 nm). Polymerization of bicontinuous microemulsions leads to larger particle sizes (50–100 nm) than that of the globular microemulsions.

Another consequence of the existence of micelles during the whole process is that the concentration of monomer in the polymer particles continuously decreases during the process due to its partitioning between polymer particles and micelles. The combination of an increasing number of particles and a decreasing monomer concentration leads to a maximum in the evolution of the polymerization rate during the process.

Because of the huge number of micelles, most of particles receive only one radical during the process, resulting in the formation of a single polymer chain of a very high molecular weight. Under these circumstances, the only termination event is the chain transfer to monomer, which determines the maximum molecular weight:

$$\bar{M}_n = \frac{k_p}{k_{tr}^{mon}} w_m \qquad (4.19)$$

where k_{tr}^{mon} is the rate coefficient for chain transfer to monomer and w_m is the molecular weight of the monomer.

Molecular weights higher than those predicted by Equation 4.19 have been reported [104]. However, when an alcohol is used as cosurfactant, much lower molecular weights are obtained, due to the chain transfer reactions.

The main drawbacks of microemulsion polymerization are that high concentrations of surfactant are used and low solids contents are obtained. At lower surfactant concentrations, larger particle sizes are obtained. In addition, during the polymerization, the number of micelles may be similar and even lower than the number of the particles; in this case, the polymer particles efficiently compete for radicals resulting in several polymer chains per particle and, consequently, in lower molecular weights.

The use of semicontinuous microemulsion polymerization has allowed to partially overcome the main drawbacks of microemulsion polymerization, leading to small particle size (about 20–50 nm) and relatively high solids contents (up to 40 wt%), obtained with relatively low surfactant concentrations (8 wt%) [105, 106].

Inverse microemulsion polymerization is used for the production of water-soluble polymers of high molecular weight, with applications in enhanced oil recovery, as flocculants in water treatments, thickeners for coatings, and retention aids in papermaking.

Nanoporous materials have been produced by the polymerization of bicontinuous microemulsions containing polymerizable surfactants in addition to water and monomer [107]. Polymerizable surfactants, also called surfmers, are amphiphilic substances containing a polymerizable double bond [108]. A similar technology has been used to synthesize ion-conductive membranes [109], luminescence films [110], and polymer–inorganic nanocomposites [111].

4.6 SUSPENSION POLYMERIZATION

In suspension polymerization, the monomer is dispersed in water forming relatively large droplets (50–150 μm) stabilized by suspension agents, which are surface-active inorganic particles and/or water-soluble polymers. The process is carried out in stirred tank reactors with a capacity up to 150 m³. Polymerization occurs in the monomer droplets using a free radical initiator soluble in the monomer. During polymerization, the droplets transform from a low viscosity liquid to a sticky polymer–monomer mixture of progressively increasing internal viscosity, and finally become solid particles. Two types of suspension polymerization may be distinguished: *suspension bead polymerization*, which leads to nonporous spherical beads formed when the polymer is soluble in the monomer, and *suspension powder polymerization*, which occurs when the polymer is not soluble in the monomer and precipitates during polymerization, resulting in opaque porous polymer particles [112, 113].

Polystyrene, expandable polystyrene (containing a volatile C4–C6 hydrocarbon), high impact polystyrene, styrene copolymers with acrylonitrile (SAN) and ABS, and poly(methyl methacrylate) are the main products of the suspension bead polymerization. PVC is the main polymeric material produced by suspension powder polymerization.

A key point in suspension polymerization is the control of the particle size distribution, which depends on the interplay between breakage and coalescence of droplets and particles. The initial droplet size distribution depends on the type and concentration of the suspension agent, the flow patterns in the reactor, the phase ratio, and on the physicochemical characteristics of the system and its evolution during the process. For the initial monomer suspension and at low monomer conversions, the viscosity of the monomer–polymer phase is low and breakage occurs easily. Therefore, the droplet/particle size is controlled by breakage. As the polymerization reaction proceeds, the viscosity of the monomer–polymer phase increases and breakage becomes more difficult. In this period, coalescence leads to an increase in particle size because the high internal viscosity may completely restrict breakage. Above a certain monomer conversion, the particles are sufficiently hard and no coalescence occurs. Therefore, the particle size distribution remains constant from this point called the identification point. For polystyrene, the identification point is about 70% conversions [114] and for PVC, about 35–40% [115].

4.6.1 Suspension Agents

Water-soluble polymers and inorganic particles are used as suspension agents. The former are frequently referred to as protective colloids and consist of hydrophilic and hydrophobic segments. Their efficiency depends on the appropriate balance between these segments; molecular weight is of considerably less significance. The protective colloids adsorb on the surface of the droplets/particles reducing the interfacial tension and, hence, reducing the energy required for breakage. In addition, the formation of a protective film at the interface reduces the coalescence. Therefore, an increase in the suspension agent concentration results in a smaller particle size. However, an excess of protective colloid may enhance the formation of much smaller emulsion particles by homogeneous nucleation. This problem is more acute as the solubility of the monomers and initiators in the continuous phase increases. The use of water-soluble inhibitors (e.g., NH_4SCN or a copper salt) may suppress the formation of emulsion particles. Both natural (e.g., cellulose derivatives) and synthetic polymers (styrene–maleic anhydride copolymer, poly(vinyl pyrrolidone), poly(vinyl alcohol) [PVA]) are used as protective colloids. Finally, divided inorganic solids (Pickering stabilizers) may also be used as suspension agents. Common choices include barium sulfate, talc, aluminum hydroxide, hydroxyapatite, tricalcium phosphate, calcium oxalate, magnesium carbonate, and calcium carbonate. Their main advantage is that they can be easily removed after the polymerization by using a dilute acid, thereby minimizing polymer contamination.

4.6.2 Particle Size Distribution

The final particle size distribution depends on the droplet/particle breakage and coalescence occurring during the whole process. Droplet breakage occurs when the hydrodynamic stress exceeds the interfacial and viscoelastic energies [116]. Droplet coalescence depends on the energy dissipation rate, the physical properties of the systems (viscosity and density of the continuous phase, and interfacial tension), volume fraction of the dispersed phase, and the efficiency of the suspension agent [117]. Because some critical properties vary during polymerization, the existing correlations for the droplet size of liquid–liquid dispersions are of limited use for the prediction of the particle size in suspension polymerization and mathematical models accounting for the dynamic evolution of the particle size distribution [118] should be used. Empirical equations for the average particle size in specific systems are available [119–122], but their use is limited to the system for which they were developed.

The particle size shows a U-shaped curve with respect the energy dissipation per unit mass of fluid [123]. At low values of the energy dissipation, the particle size decreases with the energy dissipation because of the more intensive breakage, but at higher values, achieved by increasing the agitation rate and/or the diameter of the stirrer, particle size increases with energy dissipation because of the enhanced coalescence. The increase of the volume fraction of the dispersed phase leads to an increase of the particle size due to the higher coalescence frequency. The particle size also shows a U-shaped curve with respect the degree of hydrolysis of the PVA [123]. By decreasing the degree of hydrolysis, the particle size decreases until about 73% hydrolysis, likely due to a better stabilization of the droplets and particles achieved with a more hydrophilic PVA. However, above this point the particle size increases, perhaps because the adsorption equilibrium of the PVA shifts toward the aqueous phase.

4.6.3 Polymerization Kinetics

The size of the suspension droplets/particles is so large that there is no compartmentalization effect and the kinetics are almost identical to those occurring in the equivalent bulk process. However, the presence of protective colloids that may suffer chain transfer to polymer leading to graft copolymers causes some differences with bulk polymerization.

4.6.4 Particle Morphology

Bead suspension polymerization yields homogeneous particles unless the formation of a specific morphology is sought by polymerizing a second monomer on a previously formed suspension particle [124]. In the suspension powder polymerization, used to produce about 90% of the PVC, a porous particle is produced because the polymer is not soluble in the monomer [112].

During the process, the first polymer chains produced within the suspension droplets precipitate forming unstable microdomains (10 nm). Aggregation of these microdomains leads to the formation of primary particles. The primary particles are swollen with monomer (~27 wt%) and grow by polymerization and by coagulation with other primary particles. Massive aggregation of these particles leads to the formation of a three-dimensional network within the monomer droplet. The primary particles continue to grow and because the reaction temperature is higher than the glass transition temperature of the polymer–monomer mixture, both internal (among the microdomains) and external (among the primary particles) fusions occur. Finally, a porous particle is obtained. The porosity of the particle should be controlled to achieve a fast residual monomer removal rate and a large plasticizer uptake capacity, favored by a high porosity, and a mechanically resistant particle, which requires a lower porosity. The control of particle porosity is achieved by using secondary suspension agents, which are more hydrophobic than the primary suspension agents and, hence, soluble in the monomer phase. They adsorb on the primary particles, stabilizing them, and, thus, controlling the formation of the three-dimensional network.

4.6.5 Microsuspension Polymerization

Microsuspension polymerization is a process used in the PVC industry to produce resins for plastisols [125]. In this process, which resembles miniemulsion polymerization, a mixture of monomer and an oil-soluble initiator are dispersed in an aqueous solution of surfactants using intensive shear. The monomer droplets are polymerized yielding particles usually <2 μm, which are normally isolated by spray drying as they cannot be separated by centrifuging or filtering. These particles are solid and nonporous. The polymer particles are larger than the monomer droplets (0.1–2 μm) because the combined effect of the Ostwald ripening (as no costabilizer is used in the formulation) and droplet/particle coalescence.

4.6.6 Semisuspension Polymerization

In this case, a preformed polymer dissolved in monomer is dispersed in water to form a suspension. The monomer is polymerized using an oil-soluble initiator. It is claimed that this process allows obtaining morphologies not attainable by conventional suspension polymerization [126].

4.6.7 Inverse Suspension Polymerization

Inverse suspension polymerization of water-soluble monomers is used to produce superwater-absorbent polymers [127].

4.7 DISPERSION POLYMERIZATION

Dispersion polymerization is the method of choice to obtain micron-size (1–15 μm) monodisperse particles in a single batch process [128]. This size range is between that produced by emulsion polymerization (0.06–0.8 μm) and that of the suspension polymerization (50–1000 μm).

Dispersion polymerization is basically a precipitation polymerization. The reaction medium is a good solvent for the monomers, initiator, and stabilizer (or stabilizer precursor), but it does not dissolve the polymer. Polymerization starts as a homogeneous phase and the polymer formed precipitates forming unstable nuclei. The nuclei are stabilized by the stabilizer present in the system, which can be a block copolymer in which one block has an affinity for the surface of the precipitated polymer, or it can be formed during the polymerization by grafting to a soluble stabilizer precursor. Alternatively, a polymerizable macromonomer may be used as stabilizer precursor. The stabilizer provides steric stabilization. Nucleation ends when the number of stable polymer particles increases to a point in which all new nuclei are captured by the existing stable particles.

For stabilization by grafting to stabilizer precursors, the relationship between the particle size and the operation variables is [129, 130]

$$R = x^{1/3} \left(\frac{3M_0}{\rho N_A} \right)^{2/3} \left(\frac{w_s}{C_s M_{s0} a_{scrit}} \right)^{1/2} \quad (4.20)$$
$$\left(\frac{0.386 k_2}{4\pi k_t} \right)^{1/2} \left(\frac{k_t}{2k_I f[I]_0} \right)^{1/12}$$

where x is the monomer conversion, M_0 is the initial amount of monomer (g l^{-1}), ρ is the density of the polymer (g cm^{-3}), N_A is the Avogadro's number, w_s is the molecular weight of the stabilizer (g mol^{-1}), C_s is the normalized chain transfer constant to stabilizer precursor, M_{s0} is the initial amount of stabilizer (g l^{-1}), a_{scrit} is the area of the polymer particles covered by one molecule of stabilizer under saturation conditions (cm^2), k_t is the termination rate constant (l mol^{-1} s^{-1}), k_2 is the coalescence rate coefficient (l mol^{-1} s^{-1}), f is the initiator efficiency, $[I]_0$ is the initiator concentration (mol l^{-1}), and k_I is the initiator decomposition rate constant (s^{-1}).

For stabilization with macromonomers, the dependence of the particle size with respect the operation variables can be obtained using Equation 4.20 in which $1/C_s$ is substituted by r_1, where r_1 is the reactivity ratio of the monomer in copolymerization with the macromonomer [131].

Polymerization occurs in both the continuous phase and the particles. Because of the compartmentalization of the radicals among the polymer particles, the main polymerization locus changes from the continuous phase at the beginning of the process to the dispersed phase after the nucleation period. The polymerization in the particles is mainly initiated by radicals entering from the continuous phase with some contribution from the initiator located in the particles [132].

Dispersion polymerization in organic hydrocarbon media was first developed by Osmond and Wagstaff [133] as a way to reduce the viscosity of the solvent-based coatings. Almog et al. [134] showed that monodisperse polymer particles could be obtained by means of dispersion polymerization in polar solvents, which has been used to produce toners, instrument calibration standards, column packing materials for chromatography, spacers for liquid crystal displays, and biomedical and biochemical applications [128]. Dispersion polymerization in supercritical carbon dioxide is an environmental-friendly alternative to the use of organic solvents [135]. Core-shell particles [136], polymer–inorganic hybrids [137], and highly cross-linked particles [138] have been synthesized by dispersion polymerization. Anionic polymerization [139], controlled radical polymerization [140], and addition polymerization [141] have been implemented in dispersion. Electrically conducting polymers (polypyrrole and polyaniline) have been produced by oxidative dispersion polymerization [142].

REFERENCES

[1] Asua JM, editor. *Polymer Reaction Engineering*. Oxford: Blackwell Publishing; 2007.

[2] Schmidt-Thümmes J, Schwarzenbach E, Lee DI. Applications in the paper industry. In: Urban D, Takamura K, editors. *Polymer Dispersions and Their Industrial Applications*. Weinheim: Wiley-VCH Verlag GmbH; 2002. p 75–101.

[3] Delair T. Applications biomédicales des latex synthétiques. In: Daniel JC, Pichot C, editors. *Les Latex Synthétiques. Élaboration, Propriétés, Applications*. Paris: Lavoisier; 2006. p 699–718.

[4] Harkins WD. A general theory of the mechanism of emulsion polymerization. J Am Chem Soc 1947;69:1428–1444.

[5] Priest WJ. Particle growth in the aqueous polymerization of vinyl acetate. J Phys Chem 1952;56:1077–1082.

[6] Fitch RM, Tsai CH. Particle formation in polymer colloids. III. Prediction of the number of particles by homogeneous nucleation theory. In: Fitch RM, editor. *Polymer Colloids*. New York: Plenum Press; 1971. p 73–102.

[7] Smith WV, Ewart RH. Kinetics of emulsion polymerization. J Chem Phys 1948;16:592–599.

[8] Echeverria A, Leiza JR, de la Cal JC, Asua JM. Molecular weight distribution control in emulsion polymerization. AIChE J 1998;44:1667–1679.

[9] Greene R, González RA, Poehlein GW. Continuous emulsion polymerization-steady state and transient experiments with vinyl acetate and methyl methacrylate. ACS Symp Ser 1976;24:341–358.

[10] Kiparissides C, MacGregor JF, Hamielec AE. Continuous emulsion polymerization of vinyl acetate. Part: II Parameter estimation and simulation studies. Can J Chem Eng 1980;58:48–55.

[11] Hutchinson RA, Penlidis A, Asua JM, *Polymer Reaction Engineering*. Oxford: Blackwell Publishing; 2007. p 140.

[12] Li BG, Brooks BW. Prediction of the average number of radicals per particle for emulsion polymerization. J Polym Sci A Polym Chem 1993;31:2397–2402.

[13] Morton M, Kaizerman S, Altier WJ. Swelling of latex particles. Colloid Interface Sci 1954;9:300–312.

[14] Ugelstad J, Mork PC, Mfutakamba HR, Soleimany E, Nordhuus I, Nustad K, Schmid R, Berge A, Ellingsen T, Aune O. Thermodynamics of swelling of polymer, oligomer and polymer–oligomer particles. Preparation and application of monodisperse polymer particles. In: Poehlein GW, Otewill RH, Goodwin JW, editors. *NATO ASI Series, Series E: Applied Sciences*, Volume 67, Science and Technology of Polymer Colloids, v. 1; 1983. p 51–99.

[15] Gugliotta LM, Arzamendi G, Asua JM. Choice of monomer partition model in mathematical modeling of emulsion copolymerization systems. J Appl Polym Sci 1995;55:1017–1039.

[16] Omi S, Kushibiki K, Negishi M, Iso M. A generalized computer modeling of semi-batch n-component emulsion copolymerization system and its applications. Zairyo Gijutsu 1985;3:426–441.

[17] Armitage PD, de la Cal JC, Asua JM. Improved methods for solving monomer partitioning in emulsion copolymer systems. J Appl Polym Sci 1994;51;1985–1990.

[18] Nomura M, Harada M, Eguchi W, Nagata S. Kinetics and mechanism of the emulsion polymerization of vinyl acetate. ACS Symp Ser 1976;24:102–121.

[19] Hansen FK, Ugelstad J. The effect of desorption in micellar particle nucleation in emulsion polymerization. Makromol Chem 1979;180:2423–2434.

[20] Richards JR, Congalidis JR, Gilbert RG. Mathematical modeling of emulsion copolymerization reactors. J Appl Polym Sci 1989;37:2727–2756.

[21] Araújo PHH, de la Cal JC, Asua JM, Pinto JC. Modeling particle size distribution (PSD) in emulsion copolymerization reactions in a continuous loop reactor. Macromol Theor Simul 2001;10:769–779.

[22] Urretabizkaia A, Arzamendi G, Asua JM. Modeling semicontinuous emulsion terpolymerization. Chem Eng Sci 1992;47:2579–2584.

[23] Lichti G, Gilbert RG, Napper DH. Theoretical predictions of the particle size and molecular weight distributions in emulsion polymerization. In: Piirma I, editor. *Emulsion Polymerization*. New York: Academic Press; 1982. p 93–144.

[24] Tobita H, Hamielec AE. Modeling emulsion copolymerization: crosslinking kinetics. Makromol Chem Macromol Symp 1990;35–36:193–212.

[25] Charmot D, Guillot J. Kinetic modelling of network formation in styrene-butadiene emulsion copolymers: a comparative study with the generalized form of Flory's theory of gelation. Polymer 1990;31:352–360.

[26] Tobita H. Molecular weight distribution in free-radical cross-linking copolymerization. Macromolecules 1993;26:836–841.

[27] Teymour F, Campbell JD. Analysis of the dynamics of gelation in polymerization reactors using the "numerical fractionation" technique. Macromolecules 1994;27:2460–2469.

[28] Arzamendi G, Asua JM. Modeling gelation and sol molecular weight distribution in emulsion polymerization. Macromolecules 1995;28:7479–7490.

[29] Butté A, Storti G, Morbidelli M. Evaluation of the chain length distribution in free-radical polymerization. 2. Emulsion polymerization. Macromol Theor Simul 2002;11:37–52.

[30] Zubitur M, Ben Amor S, Bauer C, Amram B, Agnely M, Leiza JR, Asua JM. Multimonomer emulsion copolymerization in presence of inhibitors. Chem Eng J 2004;98:183–198.

[31] Ghielmi A, Fiorentino S, Storti G, Morbidelli M. Molecular weight distribution of crosslinked polymers produced in emulsion. J Polym Sci A Polym Chem 1998;36:1127–1156.

[32] Tobita H, Yamamoto K. Network formation in emulsion crosslinking copolymerization. Macromolecules 1994;27: 3389–3396.

[33] Tobita H, Uemura Y. Microgel formation in emulsion copolymerization. 1. Polymerization without seed latex. J Polym Sci B Polym Phys 1996;34:1403–1413.

[34] Tobita H. Molecular-weight distribution in nonlinear emulsion polymerization. J Polym Sci B Polym Phys 1997;35: 1515–1532.

[35] Arzamendi G, Leiza JR. Molecular weight distribution (soluble and insoluble fraction) in emulsion polymerization of acrylate monomers by Monte Carlo simulations. Ind Eng Chem Res 2008;47:5934–5947.

[36] Barandiaran MJ, de la Cal JC, Asua JM. Emulsion polymerization. In: Asua JM, editor. *Polymer Reaction Engineering*. Oxford: Blackwell Publishing; 2007. p 248–254.

[37] Lovell PA, Pierre D. Rubber-toughened plastics. In: Lovell PA, El-Aasser MS, editors. *Emulsion Polymerization and Emulsion Polymers*. Chichester: John Wiley & Sons; 1997. p 657–695.

[38] Schuler B, Baumstark R, Kirsch S, Pfau A, Sandor M, Zosel A. Structure and properties of multiphase particles and their impact on the performance of architectural coatings. Prog Org Coat 2000;40:139–150.

[39] Mc Donald CJ, Devon MJ. Hollow latex particles: synthesis and applications. Adv Colloid Interface Sci 2002;99:181–213.

[40] Sundberg DC, Durant YG. Latex particle morphology, fundamental aspects: a review. Polym React Eng 2003;11: 379–432.

[41] Chen YC, Dimonie V, El-Aasser MS. Effect of interfacial phenomena on the development of particles morphology in a polymer latex system. Macromolecules 1991;24:3779–3787.

[42] González-Ortiz LJ, Asua JM. Development of particle morphology in emulsion polymerization. 1. Cluster dynamics. Macromolecules 1995;28:3135–3145.

[43] Reyes Y, Asua JM. Modeling multiphase latex particle equilibrium morphology. J Polym Sci A Polym Chem 2010; 48:2579–2583.

[44] Reyes Y, Paulis M, Leiza JR. Modeling the equilibrium morphology of nanodroplets in the presence of nanofillers. J Colloid Interface Sci 2010;352:359–365.

[45] González-Ortiz LJ, Asua JM. Development of particle morphology in emulsion polymerization. 2. Cluster dynamics in reacting systems. Macromolecules 1996;29:383–389.

[46] González-Ortiz LJ, Asua JM. Development of particle morphology in emulsion polymerization. 3. Cluster nucleation and dynamics in polymerizing systems. Macromolecules 1996;29:4520–4527.

[47] Hernandez-Barajas J, Hunkeler DJ. Inverse-emulsion polymerization of acrylamide using block copolymeric surfactants: mechanism, kinetics and modelling. Polymer 1997; 38:437–447.

[48] Capek I, Fialova L, Berek D. On the kinetics of inverse emulsion polymerization of acrylamide. Des Monomers Polym 2008;11:123–137.

[49] Haldorai Y, Nguyen VH, Shim J-J. Synthesis of polyaniline/Q-CdSe composite via ultrasonically assisted dynamic inverse emulsion polymerization. Colloid Polym Sci 2011; 289:849–854.

[50] Suckeveriene RY, Zelikman E, Mechrez G, Tzur A, Frisman I, Cohen Y, Narkis M. Synthesis of hybrid polyaniline/carbon nanotube nanocomposites by dynamic interfacial inverse emulsion polymerization under sonication. J Appl Polym Sci 2011;120:676–682.

[51] Sun F, Pan Y, Wang J, Wang Z, Hu C, Dong Q. Synthesis of conducting polyaniline-montmorillonite nanocomposites via inverse emulsion polymerization in supercritical carbon dioxide. Polym Composite 2011;31:163–172.

[52] Ouyang L, Wang L, Schork FJ. Synthesis and nucleation mechanism of inverse emulsion polymerization of acrylamide by RAFT polymerization: a comparative study. Polymer 2011;52:63–67.

[53] Qiu J, Charleux B, Matyjaszewski K. Controlled/living radical polymerization in aqueous media: homogeneous and heterogeneous systems Prog Polym Sci 2011;26:2083–2134.

[54] Li M, Matyjaszewski K. Reverse atom transfer radical polymerization in miniemulsion. Macromolecules 2003;36:6028–6035.

[55] Cunningham MF. Controlled/living radical polymerization in aqueous dispersed systems. Prog Polym Sci 2008;33:365–398.

[56] Braunecker WA, Matyjaszewski K. Controlled/living radical polymerization: features, developments, and perspectives. Prog Polym Sci 2007;32:93–146.

[57] Tomov A, Broyer JP, Spitz R. Emulsion polimerization of ethylene in water medium catalysed by organotransition metal complexes. Macromol Symp 2000;150:53–58.

[58] Skupov KM, Marella PR, Hobbs JL, McIntosh LH, Goodall BL, Claverie JP. Catalytic copolymerization of ethylene and norbornene in emulsion. Macromolecules 2006;39:4279–4281.

[59] Yu SM, Berkefeld A, Goettker-Schnetmann I, Mueller G, Mecking S. Synthesis of aqueous polyethylene dispersions with electron-deficient neutral nickel(II) catalysts with enolatoimine ligands. Macromolecules 2007;40:421–428.

[60] Chern CS, Chen T. Miniemulsion polymerization of styrene using alkyl methacrylates as the reactive cosurfactant. Colloid Polym Sci 1997;275:546–554.

[61] Wang ST, Schork FJ, Poehlein GW, Gooch JW. Emulsion and miniemulsion copolymerization of acrylic monomers in the presence of alkyd resin. J Appl Polym Sci 1996;60: 2069–2076.

[62] Kim JW, Kim JY, Suh KD. Preparation of epoxy acrylate emulsion using mixed surfactants and its polymerization. Polym Bull 1996;36:141–148.

[63] Goikoetxea M, Minari RJ, Beristain I, Paulis M, Barandiaran MJ, Asua JM. Polymerization kinetics and microstructure of

waterborne acrylic/alkyd nanocomposites synthesized by miniemulsion. J Polym Sci A Polym Chem 2007;47: 4871–4885.

[64] Erdem B, Sudol ED, Dimonie V, El-Aasser MS. Encapsulation of inorganic particles via miniemulsion polymerization. III. Characterization of encapsulation. J Polym Sci A Polym Chem 2000;38:4431–4440.

[65] Bechthold N, Tiarks F, Willert M, Landfester K, Antonietti M. Miniemulsion polymerization: applications and new materials. Macromol Symp 2000;151:549–555.

[66] Diaconu G, Mičušík M, Bonnefond A, Paulis M, Leiza JR. Macroinitiator and macromonomer modified montmorillonite for the synthesis of acrylic/MMT nanocomposite latexes. Macromolecules 2009;42:3316–3325.

[67] Capek I, Chern CS. Radical polymerization in direct mini-emulsion systems. Adv Polym Sci 2001;155: 101–165.

[68] Antonietti M, Landfester K. Polyreactions in miniemulsions. Prog Polym Sci 2002;27:689–757.

[69] Asua JM. Miniemulsion polymerization. Prog Polym Sci 2002;27:1283–1346.

[70] Schork FJ, Luo Y, Smulders W, Russum JP, Butte A, Fontenot K. Miniemulsion polymerization. Adv Polym Sci 2005;175: 129–255.

[71] Landfester K. Synthesis of colloidal particles in miniemulsions. Ann Rev Mater Res 2006;36:1351–1374.

[72] Shinnar R, Church JM. Statistical theories of turbulence in predicting particle size in agitated dispersions. J Ind Eng Chem 1960;52:253–256.

[73] Chatzi EG, Kiparissides C. Steady-state drop-size distributions in high holdup fraction dispersion systems. AIChE J 1995;41:1640–1652.

[74] Calabrese RV, Changm TPK, Dang PT. Drop breakup in turbulent stirred-tank contactors. Part I: Effect of dispersed-phase viscosity. AIChE J 1986;32:657–666.

[75] Schultz S, Wagner G, Urban K, Ulrich J. High-pressure homogenization as a process for emulsion formation. Chem Eng Technol 2004;27:361–368.

[76] Lopez A, Degrandi E, Creton C, Canetta E, Keddie JL, Asua JM. Waterborne polyurethane-acrylic hybrid nanoparticles by miniemulsion polymerization: applications in pressure-sensitive adhesives. Langmuir 2011;27: 3878–3888.

[77] Ouzineb K, Lord C, Lesauze C, Graillat C, Tanguy PA, McKenna T. Homogenisation devices for the production of miniemulsions. Chem Eng Sci 2006;61:2994–3000.

[78] El-Jaby U, Farzi G, Bourgeat-Lami E, Cunningham M, McKenna TFL. Emulsification for latex production using static mixers. Macromol Symp 2009;281:77–84.

[79] Manea M, Chemtob A, Paulis M, de la Cal JC, Barandiaran MJ, Asua JM. Miniemulsification in high-pressure homogenizers. AIChE J 2008;54:289–297.

[80] Derjaguin BV, Landau LD. Theory of the stability of strongly charged lyophobic sols and of the adhesion of strongly charged particles in solution of electrolytes. Acta Physicochim URSS 1941;14:633–662.

[81] Verwey EJW, Overbeek JTG. *Theory of the Stability of Lyophobic Colloids. The Interaction of Particles having an Electric Double Layer.* Amsterdam: Elsevier; 1948. p 413–414.

[82] Goodwin JW. *Colloids and Interfaces with Surfactants and Polymers. An Introduction.* West Sussex: John Wiley & Sons; 2004.

[83] Fitch RM. *Polymer Colloids: A Comprehensive Introduction.* London: Academic Press; 1997.

[84] Higuchi WI, Misra J. Physical degradation of emulsions via the molecular diffusion route and the possible prevention thereof. J Pharm Sci 1962;51:459–466.

[85] Davies SS, Smith A. The influence of the disperse phase on the stability of oil-in-water emulsions. In: Smith AL, editor. *Theory and Practice of Emulsion Technology.* New York: Academic Press; 1976.

[86] Ugelstad I. Swelling capacity of aqueous dispersions of oligomer and polymer substances and mixtures thereof. Makromol Chem 1978;179:815–817.

[87] Ugelstad I, Kaggerud KH, Hansen FK, Berge A. Absorption of low molecular weight compounds in aqueous dispersions of polymer-oligomer particles. 2. A two step swelling process of polymer particles giving an enormous increase in absorption capacity. Makromol Chem 1979;180:737–744.

[88] Ugelstad J, El-Aasser MS; Vanderhoff JW. Emulsion polymerization: initiation of polymerization in monomer droplets. J Polym Sci Polym Lett 1973;11:503–513.

[89] Rodriguez R, Barandiaran MJ, Asua JM. Particle nucleation in high solids miniemulsion polymerization. Macromolecules 2007;40:5735–5742.

[90] Delgado J, EI-Aasser MS, Silebi CA, Vanderhoff, JW. Miniemulsion copolymerization of vinyl acetate and butyl acrylate. IV. Kinetics of the copolymerization. J Polym Sci A Polym Chem 1990;28:777–794.

[91] Wu XQ, Schork FJ. Batch and semibatch mini/macroemulsion copolymerization of vinyl acetate and comonomers. Ind Eng Chem Res 2000;39:2855–2865.

[92] González I, Paulis M, de la Cal JC, Asua JM. Miniemulsion polymerization: effect of the segregation degree on polymer architecture. Macromol React Eng 2007;1:635–642.

[93] Rodriguez R, Barandiaran MJ, Asua JM. Polymerization strategies to overcome limiting monomer conversion in silicone-acrylic miniemulsion polymerization. Polymer 2008;49:691–696.

[94] Minari RJ, Goikoetxea M, Beristain, I, Paulis M, Barandiaran MJ, Asua JM. Post-polymerization of waterborne alkyd/ acrylics. Effect on polymer architecture and particle morphology. Polymer 2009;50:5892–5900.

[95] Minari RJ, Goikoetxea M, Beristain I, Paulis M, Barandiaran MJ, Asua JM. Molecular characterization of alkyd/acrylic latexes prepared by miniemulsion polymerization. J Appl Polym Sci 2009;114:3143–3151.

[96] Goikoetxea M, Minari RJ, Beristain I, Paulis M, Barandiaran MJ, Asua JM. A new strategy to improve alkyd/acrylic compatibilization in waterborne hybrid dispersions. Polymer 2010;51:5313–5317.

[97] Capek I. On inverse miniemulsion polymerization of conventional water-soluble monomers. Adv Colloid Interface Sci 2010;156:35–61.

[98] Zu ZZ, Wang CC, Yang WL, Deng YH, Fu SK. Encapsulation of nanosized magnetic iron oxide by polyacrylamide via inverse miniemulsion polymerization. J Magn Magn Mater 2004;277:136–143.

[99] Cao Z, Wang Z, Herrmann C, Ziener U, Landfester K. Narrowly size-distributed cobalt salt containing poly(2-hydroxyethyl methacrylate) particles by inverse miniemulsion. Langmuir 2010;26:7054–7061.

[100] Candau F, Pabon M, Anquetilq J-Y. Polymerizable microemulsions: some criteria to achieve an optimal formulation. Colloids Surf A Physicochem Eng Asp 1999;153:47–59.

[101] Chow PY, Gan LM. Microemulsion polymerizations and reactions. Adv Polym Sci 2005;175:257–298.

[102] Guo JS, Sudol ED, Vanderhoff JW, El-Aasser MS. Particle nucleation and monomer partitioning in styrene O/W microemulsion polymerization. J Polym Sci A Polym Chem 1992;30:691–702.

[103] Guo JS, Sudol ED, Vanderhoff JW, El-Aasser MS. Modeling of the styrene microemulsion polymerization. J Polym Sci A Polym Chem 1992;30:703–712.

[104] Co CC, Cotts P, Burauer S, de Vries R, Kaler EW. Microemulsion polymerization. 3. Molecular weight and particle size distributions. Macromolecules 2001;34:3245–3254.

[105] He G, Pan Q, Rempel GL. Synthesis of poly(methyl methacrylate) nanosize particles by differential microemulsion polymerization. Macromol Rapid Commun 2003;24:585–588.

[106] Smeets NMB, Moraes RP, Jeffery A, McKenna TFL. A new method for the preparation of concentrated translucent polymer nanolatexes from emulsion polymerization. Langmuir 2011;27:575–581.

[107] Liu J, Teo WK, Chew CH, Gan LM. Nanofiltration membranes prepared by direct microemulsion copolymerization using poly(ethylene oxide) macromonomer as a polymerizable surfactant. J Appl Polym Sci 2000;77:2785–2794.

[108] Escudero Sanz FJ, Lahitte JF, Remigy JC. Membrane synthesis by microemulsion polymerisation stabilised by commercial non-ionic surfactants. Desalination 2006;199:127–129.

[109] Gan LM, Chow PY, Liu Z, Hana M, Quek CH. The zwitterion effect in proton exchange membranes as synthesised by polymerisation of bicontinuous microemulsions. Chem Commun 2005;2005(35):4459–4461.

[110] Moy HY, Chow PY, Yu WL, Wong KMC, Yam VWW, Gan LM. Ruthenium (II) complexes in polymerised bicontinuous microemulsions. Chem Commun 2002;9:982–983.

[111] Palkovits R, Althues H, Rumplecker A, Tesche B, Dreier A, Holle U, Fink G, Cheng CH, Shantz DF, Kaskel S. Polymerization of w/o microemulsions for the preparation of transparent SiO$_2$/PMMA nanocomposites. Langmuir 2005;21:6048–6053.

[112] Kotoulas C, Kiparissides C, Asua JM. *Polymer Reaction Engineering*. Oxford: Blackwell Publishing; 2007.

[113] Yuan HG, Kalfas G, Ray WH. Suspension polymerization. J Macromol Sci C Polym Rev 1991;31:215–299.

[114] Kiparissides C, Achilias DS, Chatzi E. Dynamic simulation of primary particle-size distribution in vinyl chloride polymerization. J Appl Polym Sci 1994;54:1423–1438.

[115] Cebollada AF, Schmidt MJ, Farber JN, Capiati NJ, Valles EM. Suspension polymerization of vinyl chloride. I. Influence of viscosity of suspension medium on resin properties. J Appl Polym Sci 1989;37:145–166.

[116] Alvarez J, Alvarez J, Hernández M. A population balance approach for the description of particle size distribution in suspension polymerization reactors. Chem Eng Sci 1994;49:99–113.

[117] Coulaloglou CA, Tavlarides LL. Description of interaction processes in agitated liquid–liquid dispersions. Chem Eng Sci 1977;32:1289–1297.

[118] Kotoulas C, Kiparissides C. A generalized population balance model for the prediction of particle size distribution in suspension polymerization reactors. Chem Eng Sci 2006;61:332–346.

[119] Hopff H, Lüssi H, Gerspacher P. Zur kenntnis der perlpolymerisation. 1. Mitt. grundlagen. Makromol Chem 1964;78:24–36.

[120] Hopff H, Lüssi H, Gerspacher P. Zur kenntnis der perlpolymerisation. 2. Mitt. praktische anwendung der dimensionsanalyse auf das system MMA – mowiol 70/88. Makromol Chem 1964;78:37–46.

[121] Abu-Ayanaa YM, Mohsena RM. Study of some variables affecting particle size and particle size distribution of suspension polymerization of methyl methacrylate. Polym Plast Technol Eng 2005;44:1503–1522.

[122] Farahzadi H, Shahrokhi M. Dynamic evolution of droplet/particle size distribution in suspension polymerization of styrene. Iran J Chem Eng 2010;7:49–60.

[123] Smallwood PV. Vinyl chloride polymers, polymerization. In: Mark H, editor. *Encyclopedia of Polymer Science and Engineering*. Volume 17. New York: John Wiley & Sons; 1985.

[124] Gonçalves OH, Asua JM, Araujo PHH, Machado RAF. Synthesis of PS/PMMA core–shell structured particles by seeded suspension polymerization. Macromolecules 2008;4:6960–6964.

[125] Issacs H, Schwartz R, Garti N, Lerner F. Microsuspension polymerization of vinyl chloride. Tenside Det 1987;24:220–226.

[126] Mahabadi HK, Wright D. Semi-suspension polymerization process. Macromol Symp 1996;111:133–146.

[127] Wang G, Li M, Chen X. Inverse suspension polymerization of sodium acrylate. J Appl Polym Sci 1997;65:789–794.

[128] Kawaguchi S, Ito K. Dispersion polymerization. Adv Polym Sci 2005;175:299–328.

[129] Ito K, Kawaguchi S. Poly(macromonomers): homo-and copolymerization. Adv Polym Sci 1999;142:129–178.

[130] Paine AJ. Dispersion polymerization of styrene in polar solvents. 7. A simple mechanistic model to predict particle size. Macromolecules 1990;23:3109–3117.

[131] Kawaguchi S, Winnik MA, Ito K. Dispersion copolymerization of n-butyl methacrylate with poly(ethylene oxide) macromonomers in methanol–water. comparison of experiment with theory. Macromolecules 1995;28:1159–1166.

[132] Sáenz JM, Asua JM. Kinetics of the dispersion copolymerization of styrene and butyl acrylate. Macromolecules 1998;31:5215–5222.

[133] Osmond, DWJ, Wagstaff, I. Properties of polymer dispersions prepared in organic liquids. In: Barrett KEJ, editor. *Dispersion Polymerization in Organic Media*. Chichester: John Wiley & Sons; 1975. p 243–271.

[134] Almog Y, Reich S, Levy M. Monodisperse polymeric spheres in the micron size range by a single step process. Braz Polym J 1982;14:131–136.

[135] Shiho H. Dispersion polymerization of acrylonitrile in supercritical carbon dioxide. Macromolecules 33:1565–1569.

[136] Kimoto M, Yamamoto K, Hioki A, Inoue Y. Preparation of core–shell particles by dispersion polymerization with a poly(ethylene oxide) macroazoinitiator. J Appl Polym Sci 2008;110:1469–1476.

[137] Jun JB, Hong JK, Park JG, Suh KD. Preparation of monodisperse crosslinked organic–inorganic hybrid copolymer particles by dispersion polymerization. Macromol Chem Phys 2003;204:2281–2289.

[138] Lee KC, Wi HA. Highly crosslinked micron-sized, monodispersed polystyrene particles by batch dispersion polymerization. Part I: Batch, delayed addition, and seeded batch processes. J Appl Polym Sci 2010;115:297–307.

[139] Muranaka M, Kitamura Y, Yoshizawa H. Preparation of biodegradable microspheres by anionic dispersion polymerization with PLA copolymeric dispersion stabilizer. Colloid Polym Sci 2007;285:1441–1448.

[140] Saikia PJ, Lee JM, Lee BH, Choe S. Reversible addition fragmentation chain transfer mediated dispersion polymerization of styrene. Macromol Symp 2007;248:249–258.

[141] Ramanathan LS, Shukla PG, Sivaram S. Synthesis and characterization of polyurethane microspheres. Pure Appl Chem 1998;70:1295–1299.

[142] Sreeja R, Najidha S, Quamara JK, Predeep P. Development of transparent flexible conducting thin films by in-situ dispersion polymerization of pyrrole (DBSA) in pre-vulcanized nr latex development of transparent flexible conducting thin films by in-situ dispersion polymerization of pyrrole (DBSA) in pre-vulcanized NR latex. Mater Manuf Process 2007;22:384–387.

SECTION 2

POLYMERIZATION CHARACTERIZATION AND MONITORING METHODS

5

POLYMER CHARACTERISTICS

Wayne F. Reed

5.1 INTRODUCTION

Polymer characteristics include molecular weight and composition distribution, conformations, spatial dimensions, hydrodynamic properties such as intrinsic viscosity, and many others. These characteristics, in turn, affect macroscopic properties such as crystallinity, tensile and compressive strength, rheology, processability, etc.

Polymer characterization is a well-developed field in and of itself, and involves many methods, some of which are discussed in detail in subsequent chapters. One of the main challenges of online polymerization monitoring has been to translate these characterization techniques from the off-line analytical laboratory to the reactor itself. This chapter focuses chiefly on the properties of polymer molecules themselves, with a very small amount of introductory concepts concerning viscoelastic and rheological behavior in concentrated polymer solutions and melts, and on solid-state properties.

5.2 POLYMER CONFORMATIONS AND DIMENSIONS

5.2.1 Rotational Isomeric Model

Polymers consist of many monomeric subunits held together by covalent bonds.[1] In general, there is some rotational freedom between monomers in a polymer chain, and this leads to the possibility of many different configurations for any given polymer.

[1] There are also noncovalent polymers, such as those assembled through hydrogen bonds.

Figure 5.1 shows a fixed covalent bond angle, θ, between monomers, and a free $360°$ azimuth rotation angle, ϕ.

In reality, free rotations often meet with steric or electrostatic hindrances that yield a finite number of rotational energy minima. The most common situation is of rotational isomers, or rotamers, around a single bond for tetrahedral bonding. According to the angular rotational energy scheme of Figure 5.2, there is an energy minimum each $120°$, for a total of three rotamers [1]. Even with this restriction there are still an enormous number of configurations for a polymer consisting of N monomers, namely 3^{N-2}. For a small polymer, for example, polyacrylamide, with 100 monomers (a mere $7100 \, \text{g} \, \text{mol}^{-1}$ polymer) there are 5.7×10^{46} possible configurations. Each configuration would require 9 binary bits of information to specify. A computer capable of outputting 1 trillion bits per second would require 1.6×10^{28} years to represent this number of configurations, far longer than the estimated age of the observable universe of 1.3×10^9 years.

Given the enormous number of configurations for even short chains, polymers are best characterized by averages, such as are determined through the principles of probability theory, thermodynamics, and statistical mechanics.

5.2.2 Polymer Conformations

The simplest possible polymer chain has free rotations of its noninteracting monomers. The natural result of this is the so-called random coil, whose end-to-end probabilities obey the diffusion (or random walk) equation. When intermonomeric forces in the polymer chain come into play, then polymers can adapt other conformations besides the random coil. Such forces can be those due to steric interaction, electrostatic effects, including charge–charge,

Monitoring Polymerization Reactions: From Fundamentals to Applications, First Edition. Edited by Wayne F. Reed and Alina M. Alb.
© 2014 John Wiley & Sons, Inc. Published 2014 by John Wiley & Sons, Inc.

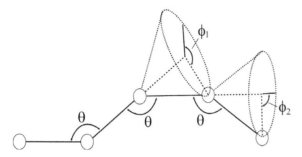

FIGURE 5.1 Model of a polymer with fixed bond angle θ between monomers and rotational angle ϕ.

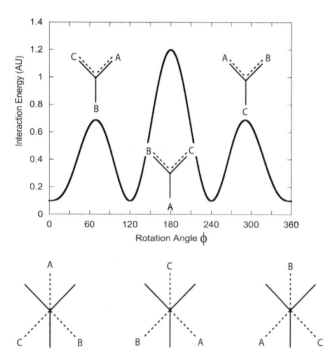

FIGURE 5.2 Rotational isomeric state model. Interaction energy between adjacent monomers as a function of the bond rotation angle ϕ, showing the eclipsed rotational states leading to energy maxima. The antieclipsed states, corresponding to energy minima, are shown.

charge–dipole, and dipole–dipole, hydrophobic interactions, and hydrogen bonding. The particular amino acid composition of a given protein, for example, provides multiple sources of these various forces, allowing the protein to fold into a globule, form secondary structures, such as α-helices and β-sheets, and stretch into fiber conformations. In fact, random coils, globules, and rods are the major conformational classes of polymers. After these, one can differentiate a wide variety of other "architectures," such as random branching, hierarchical branching, cross-linking, dendrimers, stars, and so on, as illustrated in Figure 5.3.

The characteristic spatial dimension of polymers in the different conformational classes go roughly as a function of the polymer mass M: for spheroidal particles, $D \propto M^{1/3}$;

Random Coil

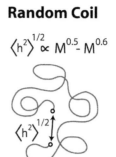

Globule

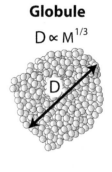

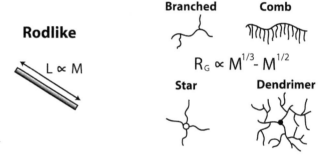

FIGURE 5.3 Families of polymer conformations, including the scaling law between the characteristic linear dimension and the polymer mass for each family.

for ideal random coils, the root mean square end-to-end length $<h^2>^{1/2} \propto M^{1/2}$; for random coils with excluded volume, $<h^2>^{1/2} \propto M^{3/5}$; and for rodlike polymers, length $L \propto M$. Branched type architectures usually have characteristic dimensions in the range from $M^{1/3}$ to $M^{3/5}$.

For a freely rotating polymer chain (i.e., without a fixed bond angle or rotameric states), the solution to the 3-D diffusion equation for the end-to-end length probability distribution $P(N, h)$ is

$$P(N,h) = \left(\frac{3}{2N\pi b^2} \right)^{3/2} \exp\left(\frac{-3h^2}{2Nb^2} \right) \quad (5.1)$$

where N is the total number of monomers in the chain ($N \gg 1$, the so-called long chain approximation), b is the monomer-to-monomer bond length, and h is the chain's end-to-end length. $P(N, h)$ is the probability per unit volume that a given h lies between h and $h+dh$.

The mean square end-to-end length is immediately found from this distribution to be

$$\left\langle h^2 \right\rangle = Nb^2 \quad (5.2)$$

A very interesting feature of this distribution function is that it describes the origin of elasticity for rubber and other elastic polymers. Namely, the amount of work done on a polymer in changing h is equal to the change in free energy, which at constant T is

$$dW = dG = dU - TdS \qquad (5.3)$$

where dU is the change in internal energy and dS the change in entropy.

For an ideal chain there is no interaction among monomers, so that $dU = 0$. Hence $dG = -TdS$, and using the statistical interpretation of entropy in terms of probability

$$S = k_B \ln(P) = k_B \left[\ln\left(\left(\frac{3}{2N\pi b^2} \right)^{3/2} \right) - \frac{3h^2}{2Nb^2} \right] \qquad (5.4)$$

Assigning the work done in changing h to the potential yields a linear restoring force,

$$F = -\frac{dG}{dh} = -\frac{3k_B T}{2Nb^2} h \qquad (5.5)$$

Hence, it is the decrease in entropy as a chain is stretched that yields the elastic restoring force. Typically, a polymer chain can be stretched to many times its most entropically relaxed coil state. This is in strong contrast to the origin of elasticity in "hard" materials, such as metals, where it is the minute interatomic displacement in steep potential wells that provides the restoring force, and the entropic portion of G plays no role.

5.2.3 Polymer Dimensions for Random Coils

Consider the random coil of Figure 5.4, where the covalent bond angle is θ and the three rotameric states have potential energies $U(\phi)$, of the type seen in Figure 5.2. The thermodynamic average over $\cos \phi$ is

$$<\cos\phi> = \frac{\int \cos(\phi)\exp[-U(\phi)/k_B T]d\cos(\phi)}{\int \exp[-U(\phi)/k_B T]d\cos(\phi)} \qquad (5.6)$$

The mean square end-to-end length, $<h^2>$, is given, again in the long chain limit, by

$$\left\langle h^2 \right\rangle = Nb^2 \left(\frac{1+\cos\theta}{1-\cos\theta} \right)\left(\frac{1+<\cos\phi>}{1-<\cos\phi>} \right) \qquad (5.7)$$

The usual tetrahedral bond angle of $109.5°$ hence tends to stretch the polymer out, as do rotameric states for which $<\cos\phi> > 0$.

The latter equation shows that $<h^2>$ continues to be proportional to Nb^2 despite the "stiffening" due to the bond and rotational angles, when these are greater than $90°$.

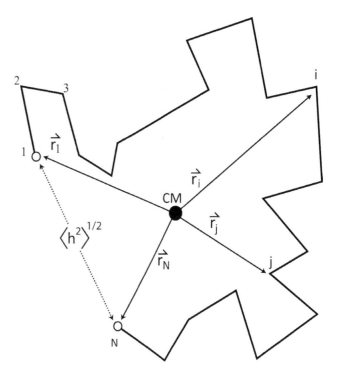

FIGURE 5.4 Monomer numbering and coordinate position scheme with respect to a polymer's center of mass (CM), used to compute $<S^2>$ in Equation 5.11.

$$\left\langle h^2 \right\rangle = \alpha Nb^2 \qquad (5.8)$$

It is convenient to introduce a statistical segment length l_k, such that

$$\left\langle h^2 \right\rangle = N_k l_k^2 = Ll_k \qquad (5.9)$$

where N_k is the number of statistical segment lengths; l_k is then a measure of a polymer's stiffness.

For example, typical synthetic polymers, such as polystyrene and polyacrylamide, have l_k on the order of 1 nm, whereas a semirigid polyelectrolyte such as hyaluronic acid has l_k on the order of 10 nm in high-added electrolyte solutions [2], and DNA is on the order of 100 nm [3]. The polymer contour length, L, is the length of a polymer following its contour, similar to the length traveled along a winding road, and is equivalent to the end-to-end length of the polymer when fully stretched. By this definition, $L = Nb = l_k N_k$. Substituting this latter in the aforementioned equations yields

$$l_k = \alpha b \qquad (5.10)$$

5.2.4 Radius of Gyration

While theories for polymer conformations and dimensions are most easily formulated in terms of $<h^2>$, what is normally measured experimentally by light and other scattering techniques are the mean square radius of gyration, $<S^2>$, which is the mass-averaged mean square distance of all monomers from the polymer's center of mass,

$$< S^2 >= \frac{\sum_{i=1}^{N} m_i r_i^2}{M} \tag{5.11}$$

where m_i is the mass of monomer i, r_i is its distance from the polymer's center of mass, and M is the mass of the entire polymer.

For ideal random coils, the relationship between $<h^2>_0$ and $<S^2>_0$ is simply

$$< S^2 >_0 = \frac{< h^2 >_0}{6} \tag{5.12}$$

The subscript "0" in $<S^2>_0$ refers to the fact that this is the mean square radius of gyration for an ideal coil, without any long-range intramolecular interactions.

5.2.5 Persistence Length of Semiflexible Polymers

The discussion in the earlier sections shows how polymer dimensions can increase or decrease when restrictions such as bond and rotational angles are taken into account.

There are many other factors that can lead to the stretching or shrinking of polymer chains, including short-range and long-range effects. Short-range effects are usually considered among neighboring polymers, whereas "long-range forces" are those resulting from monomers distant from a given monomer bending back through space into the vicinity of the given monomer and interacting with it. These latter effects are usually subsumed into the theory of intramolecular excluded volume, briefly discussed in the following section.

An elegant statistical mechanical approach to the dimensions of flexible polymers was developed by Landau and Lifschitz [4], who considered a long, flexible thread of contour length L and positive bending energy β_b (J×m in SI units). The lowest energy state is when the thread is straight, but thermal fluctuations allow it to bend and "wobble" in a complex way that tends to shrink $<h^2>^{1/2}$ below the full contour length L. By considering the thermodynamically averaged bending angle over infinitesimal lengths along the chain, and integrating over the whole length L, the following expression was found for $<h^2>_0$

$$< h^2 >_0 = 2L_p^2 \left[\frac{L}{L_p} - 1 + \exp\left(\frac{-L}{L_p} \right) \right] \tag{5.13}$$

which introduces the persistence length, L_p.

From the analysis, L_p is physically equal to the ratio of bending energy β_b to thermal energy, that is,

$$L_p = \frac{\beta_b}{k_B T} \tag{5.14}$$

L_p provides a scale length for how a polymer bends through space after proceeding a length l along the chain, via

$$< \cos \theta >= \exp\left(\frac{-l}{L_p} \right) \tag{5.15}$$

where $<\cos \theta>$ is the thermodynamically averaged cosine of the angle formed by the initial spatial direction of a polymer segment and the tangent to the polymer a contour distance l away from the initial point.

In other words, L_p measures how far the polymer chain "persists" in a direction as it gradually loses track of its starting direction. In this interpretation, L_p is the contour distance along the chain from a starting point at which $<\cos \theta>= e^{-1}$, that is, the angle between starting tangent and the tangent at l is 68.8°.

Equation 5.13 for $<h^2>_0$ provides a bridge between the rod and the coil limits of the polymer. In the rod limit $L_p >> L$, so that

$$< h^2 >_0 = L^2, \text{rod limit} \tag{5.16}$$

and the coil limit when $L_p << L$

$$< h^2 >_0 = 2LL_p, \text{coil limit} \tag{5.17}$$

Comparing the coil limit to that for the random coil by Equation 5.9 provides the connection between the statistical segment length and the persistence length

$$l_k = 2L_p \tag{5.18}$$

Landau and Lifschitz [4] provided the further vector analysis needed to relate $<h^2>$ to the more experimentally useful $<S^2>$. For a polymer with no long-range intramolecular interactions,

$$< S^2 >_0 = \frac{LL_p}{3} - L_p^2 + \frac{2L_p^3}{L} - 2\left(\frac{L_p^4}{L^2} \right) \left[1 - \exp\left(\frac{-L}{L_p} \right) \right] \tag{5.19}$$

The coil limit of this expression is frequently useful

$$< S^2 >_0 = \frac{LL_p}{3}, \text{coil limit} \tag{5.20}$$

5.2.6 Excluded Volume

All the aforementioned expressions for polymer dimensions $<h^2>_0$ and $<S^2>_0$ are for ideal polymers. An ideal polymer is like an ideal gas in that, while the ideal gas molecules are noninteracting point-like particles, the

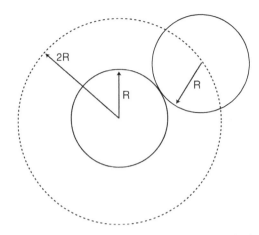

FIGURE 5.5 The excluded volume between two hard spheres, as computed in Equation 5.22.

monomers in the polymer chain have no net interaction with each other, although they are constrained to each other by chemical bonds.

In a polymer solution, the condition in which interparticle attractive and repulsive forces cancel each other out, yielding an ideal polymer solution, is termed the "theta condition," discussed further in this chapter.

Deviations from ideality can be conceptualized in terms of the excluded volume (EV) concept. If the net interparticle force is repulsive then the excluded volume is termed positive, and termed negative for net attractive interparticle force. The forces leading to deviations from ideality are the same as those listed earlier. A particularly simple situation is that of two interacting hard spheres. If each sphere has a radius R, then Figure 5.5 shows how the total excluded volume (EV) between them is that of a sphere of radius $2R$, so that

$$EV = \frac{32\pi R^3}{3} \qquad (5.21)$$

In general, the excluded volume can be computed from knowledge of the interparticle potential $U(r_{12}, \theta_{12}, \phi_{12})$ by

$$EV = \int_0^{2\pi} \int_0^{\pi} \int_0^{\infty} \left\{ 1 - \exp\left[\frac{-U(r_{12}, \theta_{12}, \phi_{12})}{k_B T} \right] \right\} r_{12}^2 \, dr_{12}$$
$$\sin\theta_{12} d\theta_{12} \, d\phi_{12} \qquad (5.22)$$

where r_{12} is the interparticle center-to-center distance, and the spherical altitude and azimuth angles θ_{12} and ϕ_2 respectively, allow for noncentral forces, such as dipolar and magnetic forces, as well as attractive hydrophobic and repulsive electrostatic "patches" in more complex structures, such as globular proteins.

For central forces, equivalent sphere models are often robust enough for assessing interparticle effects. Steric repulsion lends itself naturally to hard sphere approximation, whereas charge–charge effects in polyelectrolytes let the "soft" spherically symmetric electric field be represented by equivalent spheres [5–8].

The theory of intramolecular excluded volume for random coil polymer chains has been elaborated to high levels of complexity [9–11]. In a major simplification to an inherently very difficult problem, Flory proposed a scaling argument for estimating the excluded volume effect by minimizing the free energy, $G = U - TS$, of N monomers confined to a volume represented by R^3. U is proportional to the number of two body interactions of the N monomers per volume, or $U \propto N^2/R^3$. As seen earlier, $S = k_B \ln(P) \propto h^2/N$, and taking h as the characteristic length scale containing the N monomers, R, yields

$$G \propto \frac{N^2}{R^3} - T\frac{R^2}{N} \qquad (5.23)$$

Minimizing G with respect to R gives the scaling relation

$$R \propto N^{3/5} \qquad (5.24)$$

The more complex theories as well as numerical approaches give an exponent around 0.583, which is only slightly different from the simple Flory result.

In several theories, the effects of excluded volume are combined into a static expansion factor α_s relating the total, experimentally measured mean square radius of gyration $<S^2>$ to the value with no long-range intramolecular forces accounted for, $<S^2>_0$

$$<S^2> = \alpha_s^2 <S^2>_0 \qquad (5.25)$$

The importance of excluded volume in thermodynamics of polymer solutions is discussed later in terms of virial coefficients.

A useful, although nonrigorous, approach to assessing the possibility of branching, cross-linking, or collapse of a polymer is to use the experimental value of $<S^2>$, which will inseparably contain any excluded volume effects, and equate this in the coil limit to an expression involving the "apparent persistence length," L_p', which itself embodies both the intrinsic and excluded volume effects in $<S^2>$ [12, 13]:

$$<S^2> = \frac{LL_p'}{3} \qquad (5.26)$$

Knowing the mass of the polymer allows the computation of L_p', which can then be compared to other values of L_p or L_p' for similar polymers. For example, L_p' for a typical linear polystyrene in THF is around 0.7 nm. If experimental determinations of L_p' are made using Equation

5.26 and they are found to be smaller than the values for known linear molecules of the same type, then this is evidence of branching, cross-linking, aggregation, or collapse.

5.2.7 Branching

Polymer branching is ubiquitous and profoundly modifies polymer properties. It is often sought as an important feature of polymers, sometimes appears as an inevitable result or side reaction, and has biological precedent especially among polysaccharides and branched fatty acids. The topic of branching occurs throughout this book, for example, in Chapters 2, 4, 9, 10, 18, and 20. Chapter 9 describes some size exclusion chromatography (SEC) methods for branching detection and analysis. Other SEC methods for branching analysis have been presented [14–17].

There are many types of branching such as regular, random, star, comb, brush, dendrimer, etc., some of which are depicted in Figure 5.3. It is beyond the scope of this book to enter into the many mechanisms and models related to branching. There are many sources that treat this [18–20].

From the practical polymer characterization point of view, the issues are usually how to detect when branching is present and how to quantify it. While there are some relatively straightforward means for detection branching in a polymer sample, a meaningful analysis normally requires a model, which itself requires some knowledge or conjecture of the mechanism and type of branching involved for each particular case.

A widespread method proposed for quantifying branching is due to Zimm and Stockmayer [21]. This defines the branching ratio BR for a given polymer mass M in terms of the experimentally mean squared radius of gyration of a branched polymer $<S^2>_{branched}$ to that of a linear analog of the same polymer.

$$BR \equiv \frac{<S^2>_{branched}}{<S^2>_{linear}} \qquad (5.27)$$

BR can often be measured by light scattering if $<S^2>$ is large enough (typically $<S^2>^{1/2} > 10$ nm is achievable for standard multiangle light scattering (MALS) instruments). Such measurements can be made in batch, or, increasingly by multidetector SEC.

An alternate branching ratio, sometimes termed BR$'$, can be determined from the ratio of intrinsic viscosities for a given M

$$BR' \equiv \frac{[\eta]_{branched}}{[\eta]_{linear}} \qquad (5.28)$$

BR$'$ is often easier to measure experimentally, since viscosity measurements do not have the spatial dimension lower limit of MALS. It is also possible to define a branching ratio using the ratio of hydrodynamic radii for branched and unbranched polymers $R_{H,branched}/R_{H,unbranched}$, determined from dynamic light scattering (DLS).

Normally, BR and BR$'$ are less than unity since the branched structures are denser than the corresponding linear polymers. To proceed to an analytical interpretation of BR or BR$'$ requires a model, such as star or random branching.

A star polymer with n branches, for example, has

$$BR = \frac{3n - 2}{n^2} \qquad (5.29)$$

5.3 MOLECULAR WEIGHT DISTRIBUTIONS (MWDs) AND AVERAGES

Polymer populations in which all chains have identical masses are termed "monodisperse." Proteins and DNA of the same type are necessarily monodisperse, as they are coded for precise monomer sequences. In general, however, polymer populations contain a wide distribution of masses. Such populations are termed "polydisperse." While most synthetically produced polymers are polydisperse, many biopolymers are also produced in polydisperse populations; for example, a variety of polysaccharides, glycoproteins, etc.

Some methods of producing polymers can lead to distinct subpopulations, widely separated in mass. When there is a distribution with only a single maximum, the population is termed "unimodal." When two or more identifiable maxima exist within a polymer population, it is termed "multimodal"; that is, "bimodal" for two subpopulations, "trimodal" for three, and so on.

As in any situation where probability distributions are involved, useful averages are defined that act as hallmarks for related properties. Some of the important distributions commonly found in polymer populations are outlined in the following section, as derived from the binomial distribution (BD). Complementary approaches to these and other distributions can be found in Chapters 1 and 16 of this book. In particular, Chapter 16 uses the method of moments and z-transform techniques to treat MWD.

5.3.1 Common MWDs

5.3.1.1 *Binomial Distribution (BD)* The BD describes a wide variety of situations in which the probability of the outcome of a certain event, q, remains fixed for all trials; for example, the probability of rolling a certain number of a die.

If N independent events occur, each with the same probability q for success, the probability for failure in each event is $(1-q)$. The probability that j events out of N are successful is given by the BD.

$$W(N,j,q) = \frac{N!}{(N-j)!\,j!} q^j (1-q)^{N-j} \qquad (5.30)$$

The factorial terms take account of the number of sequences in which N events could occur to give a total of j successes and $N-j$ failures. The average number of events that occur is computed in the usual way

$$<j(N,q)> = \frac{\sum_{j=0}^{N} jW(N,j,q)}{\sum_{j=0}^{N} W(N,j,q)} = qN \qquad (5.31)$$

where a summation rule has been used to solve for the sums shown.

For example, a coin is tossed 10 times ($N=10$). The probability of obtaining two heads ($j=2$) during this trial is, according to Equation 5.31, is $45/1024 = 4.4\%$. The probability of obtaining five heads is $252/1024 = 24.6\%$. The average number of heads for many individual trials of ten tosses each is five.

5.3.1.2 Geometric Distribution (GD)

This distribution is a special case of the BD and is ubiquitous in polymer science. It arises when only one sequence out of the $N!/[(N-j)!\,j!]$ possible sequences is relevant for the success of obtaining j successes.

If ideal free-radical polymerization chains are formed very quickly with respect to the total time of monomer conversion, then any given propagating chain sees an environment (concentration of radicals, monomers, solvent, etc.) that changes little during its formation. It is of interest to know after N free-radical events, what the probability is that there were $N-1$ consecutive events ($j=N-1$) that chained monomers to the free radical, and a final "chain killing" event that terminated the chain by a radical–radical encounter. Although there are $N!/[(N-1)!\,1!]=N$ different sequences of events that give $N-1$ monomers and one terminated radical in a chain, the only physically possible sequence is when the terminated radical is at the end of the chain.

The probability q of adding a monomer, instead of terminating via a collision with a free radical, is

$$q = \frac{\text{Rate of propagation}}{\text{Rate of propagation} + \text{Rate of termination}} = \frac{k_p M}{k_p M + k_t R} \qquad (5.32)$$

The probability that an R_t radical is terminated on the next step is $(1-q)$. The probability that it adds a monomer is q. The probability that this two unit chain terminates on the

next step is $(1-q)q$. In general, the probability that a chain of length i is reached is

$$P_i = (1-q)q^{i-1} \qquad (5.33)$$

It is seen that Equation 5.33 is the BD in the case where $j=N-1$ and the combinatorial term is omitted, for reasons just discussed. If n_i is the number of chains of chain length i, where n_{total} is the total number of chains, then n_i will simply be equal to

$$n_i = n_{\text{total}} P_i \qquad (5.34)$$

since P_i is a normalized distribution. The GD also describes step growth polymerization, where q is replaced by p, the degree of polymerization.

Other relevant distributions derived from the BD are the Poisson and Gaussian distributions.

5.3.1.3 Poisson Distribution (PD)

This is the BD in the special case where $N >> i$ and p is small. In this case,

$$W(j) = \frac{m^j e^{-m}}{j!} \qquad (5.35)$$

where m is the average number of successful events, $m=Np$ (which is small).

The distribution is completely determined by the single parameter m. It is, however, not a continuous function, and so still requires summation techniques, rather than integration, to determine average values from it.

The width, or standard deviation, of $W(i)$ for the PD is sqrt(m). An important property of the PD is that the ratio of the width to the average value is $1/\text{sqrt}(m)$. Hence, this ratio decreases as m increases. Pollsters often quote their sampling error simply by using $1/\text{sqrt}(m)$, where m is the number of people sampled.

A familiar example of the Poisson distribution in physics is in radioactive decay of long-lived isotopes. Here, there are a very large number of radioactive nuclei, N, but a short number of "successful events" i (i.e., radioactive decays) during a typical measurement interval, and the probability p, of any given nucleus decaying during this time is small. $M=Np$ is the average count rate for a measurement, but as Equation 5.35 shows, there is a significant probability of finding values other than m.

The Poisson distribution is applicable to chain growth reactions where initiation of propagating species is very quick, termination is a negligible mechanism, and the time it takes for the chain to grow is comparable to the time it takes for complete conversion of monomer. Such polymerization reactions are often termed "living polymerization," because of the fact that if there is no termination mechanism the polymer is always "alive." This means that if after all monomer has been chained into polymer molecules, more monomer is added to the solution, the chains will grow further.

In this case, there is a population of a fixed number of slowly growing chains. The number of initiated monomers at the outset of the reaction is equal to the concentration of initiator I_0; for example, lithium alkyls. This number does not change throughout the reaction. Over any short interval, there is an average chain size $<i>$, and the number of monomers in a chain is small compared to the total number of monomers in the reactor. The chain length distribution $W(i)$ is

$$W(i) = e^{-\tau} \frac{\tau^{i-1}}{(i-1)!} \qquad (5.36)$$

where τ is related to both monomer conversion and the ratio of converted polymer to initial initiator concentration I_0 via

$$\tau = \frac{M_0 - M}{I_0} \qquad (5.37)$$

Obviously, τ increases as the reaction proceeds and monomers are added to growing chains. What is remarkable, however, is that since the width to average of the chain length distribution is $1/\sqrt{\tau}$, the polydispersity in living polymerization actually decreases as the reaction proceeds. Living polymerization reactions are of great practical importance in producing low polydispersity polymers.

5.3.1.4 *Gaussian Distribution (GaD)* This is obtained from the BD in the case where $q = 0.5$, and $N >> j$, and the average $Nq = m >> 1$. This is a continuous function, completely defined by two parameters: the mean, $<x>$ (indicated by m in the foregoing distributions), and the width (or standard deviation), σ. Here j has been changed to x for the GaD to emphasize its continuous nature. The form of the GaD is

$$W(x) = \frac{\exp\left[\frac{-(x - <x>)^2}{2\sigma^2}\right]}{\sigma\sqrt{2\pi}} \qquad (5.38)$$

where $W(x)$ is a normalized distribution.

A form for chain length distributions that is found to arise experimentally in many situations is the log-normal distribution

$$W(\ln x) = \frac{\exp\left[\frac{-(\ln x - <\ln x>)^2}{2\sigma^2}\right]}{\sigma\sqrt{2\pi}} \qquad (5.39)$$

where $W(\ln x)d(\ln x)$ is the probability of there being chains in the interval $(\ln x, \ln x + d\ln x)$.

5.3.2   Computation of Molecular Weight Averages

Computing molecular weight averages is covered in Chapters 1 and 16. The following summary is provided for continuity in this chapter. Given a probability distribution function $W(i)$, the average of any variable $Q(i)$ is computed according to

$$\langle Q(i) \rangle = \frac{\sum_{i=i_{min}}^{i_{max}} Q(i)W(i)}{\sum_{i=i_{min}}^{i_{max}} W(i)} \qquad (5.40)$$

where i_{min} is the minimum, physically meaningful value of i, often equal to 0 or 1, and i_{max} is the maximum value, which can often be taken as infinite.

If the distribution function is a continuous function $W(x)$ (e.g., the Gaussian distribution), then

$$\langle Q(x) \rangle = \frac{\int_{x_{min}}^{x_{max}} Q(x)W(x)\,dx}{\int_{x_{min}}^{x_{max}} W(x)\,dx} \qquad (5.41)$$

The discussion in the earlier section was in terms of the chain length probability distribution, W_i, or the chain length number distribution, $n_i = n_{total}W_i$. Oftentimes, it is of direct interest to work with the polymer mass distribution $n(M_i)$, which is the number of polymers of mass M_i present in a population

$$N(M_i) = n_{total}W(M_i) \qquad (5.42)$$

In probability theory, it is customary to define the *moments* of a distribution. The jth moment μ_j of a distribution W_i is defined as

$$\mu_j \equiv \langle i^j \rangle = \frac{\sum_i i^j W(i)}{\sum_i W(i)} \qquad (5.43)$$

The zeroth moment of a distribution is 1, the first moment is $<i>$, the second moment is $<i^2>$, etc. The higher moments of a distribution hence compute successively higher averages of the distributions of the independent variable; for example, in classical statistical thermodynamics the mean square velocity $<v^2>$ is the second moment of the Maxwell–Boltzmann speed distribution for an ideal gas, and is directly related to average kinetic energy $<KE> = m<v^2>/2$, and hence to temperature [$<KE> = 3k_BT/2$ for a monatomic gas].

For a one-parameter distribution, it suffices to know any moment (higher than the zeroth) to completely define the distribution. The geometric and Poisson distributions are examples of one-parameter distributions. For a two-parameter distribution, such as the Gaussian, two moments (higher than the zeroth) completely define the distribution, and so on.

The moments of a polymer's mass distribution are used to define the important types of averages corresponding to a given quantity Q. The first moment corresponds to taking a simple number average over the distribution. If Q is the mass itself then the number average mass, $<M>_n$ (usually abbreviated as M_n), is given for discrete and continuous distributions respectively as

$$M_n \equiv \langle M \rangle_n = \frac{\sum_i M_i n(M_i)}{\sum_i (M_i)} \qquad (5.44a)$$

$$M_n \equiv \langle M \rangle_n = \frac{\int_{M_{min}}^{M_{max}} M n(M) dM}{\int_{M_{min}}^{M_{max}} n(M) dM} \qquad (5.44b)$$

where the second equation applies if $n(M)$ is a continuous distribution. In that case, $n(M)dM$ represents the number of chains in the mass interval from M to $M+dM$. Notice that $<M>_n$ is the ratio of the first to zeroth moments of the distribution,

$$\langle M \rangle_n = \frac{\mu_1}{\mu_0} \qquad (5.45)$$

The weight average (more properly termed mass average, but polymer tradition uses the former term) of any quantity Q of a polymer population is defined as

$$\langle Q(i) \rangle_w = \frac{\sum_i Q(i) M_i n(M_i)}{\sum_i M_i n(M_i)} \qquad (5.46a)$$

$$\langle Q(x) \rangle_w = \frac{\int_{M_{min}}^{M_{max}} Q(M) M n(M) dM}{\int_{M_{min}}^{M_{max}} M n(M) dM} \qquad (5.46b)$$

The weight average of a quantity hence uses the total mass of M_i (which is $M_i n(M_i)$ to weight the average rather than the number of masses. The weight average mass, $<M>_w$, normally abbreviated M_w, is hence given, for discrete and continuous distributions, respectively, by

$$M_w \equiv \langle M_w \rangle = \frac{\sum_i M_i^2 n(M_i)}{\sum_i M_i n(M_i)} \qquad (5.47a)$$

$$M_w \equiv \langle M \rangle_w = \frac{\int_{M_{min}}^{M_{max}} M^2 n(M) dM}{\int_{M_{min}}^{M_{max}} M n(M) dM} \qquad (5.47b)$$

$<M>_w$ is the ratio of second to first moments of the $n(M)$ distribution

$$\langle M \rangle_w = \frac{\mu_2}{\mu_1} \qquad (5.48)$$

A third average that is frequently important in polymer science is the z-average. It is defined, for a quantity Q, as

$$\langle Q \rangle_z = \frac{\sum_i Q(i) M_i^2 n(M_i)}{\sum_i M_i^2 n(M_i)} \qquad (5.49a)$$

$$\langle Q \rangle_z = \frac{\int_{M_{min}}^{M_{max}} Q(M) M^2 n(M) dM}{\int_{M_{min}}^{M_{max}} M^2 n(M) dM} \qquad (5.49b)$$

The z-average mass, $<M>_z$, is the ratio of the third to second moments of $n(M)$

$$\langle M \rangle_z = \frac{\mu_3}{\mu_2} \qquad (5.50)$$

It is customary in polymer science to deal with polydispersity indices. The two most common are M_w/M_n and M_z/M_w. Ideal polymer reactions have well-defined ratios, which are often approached in real polymerization reactions. M_w/M_n is so frequently used in this context that it often simply referred to as the "PDI," or polydispersity index.

It is also important to note that the concentration, $C(M)$, where $C(M)dM$ is the concentration of polymer in the interval M to $M+dM$, is related to $n(M)$ (where $n(M)dM$ is the number density of particles in M to $M+dM$) by

$$C(M) = M n(M) \qquad (5.51)$$

with the normalization condition that

$$c_{total} = \int_0^\infty C(M) dM \qquad (5.52)$$

where c_{total} is the total concentration of polymer in a population. The quantity $C(M)$ is used experimentally so often that M_n, M_w, and M_z are given explicitly here

$$M_n = \frac{\int C(M) dM}{\int \frac{C(M)}{M} dM} \qquad (5.53a)$$

$$M_w = \frac{\int M C(M) dM}{\int C(M) dM} \qquad (5.53b)$$

$$M_z = \frac{\int M^2 C(M) dM}{\int M C(M) dM} \qquad (5.53c)$$

As examples of these averages, M_n and M_w are given for the GD by

$$M_n = \frac{m}{1-q} \qquad (5.54a)$$

$$M_w = m \frac{1+q}{1-q} \qquad (5.54b)$$

where q is the probability that a propagating chain will add a monomer rather than terminate and m is the monomer mass. For the PD,

$$M_n = m(1+\tau) \qquad (5.55a)$$

$$M_w = m \frac{1+3\tau+\tau^2}{1+\tau} \qquad (5.55b)$$

Several techniques exist for determining different averages of M. These include membrane osmometry for direct measurement of M_n, and light scattering for M_w. Other techniques determine averages of other quantities, which, via a model, can be related to one or more of the above averages. Determination of intrinsic viscosity, for example, generally yields a weight average intrinsic viscosity, which can be related to a moment of M, if the relation between viscosity and M is known. Likewise, DLS can yield a z-average diffusion coefficient directly, which can be related to M.

Currently, the only practical way of accurately determining an MWD is via a separation technique. The most widely used methods include SEC, field flow fractionation (FFF), capillary hydrodynamic fractionation, and gel electrophoresis. Since knowledge of $C(M)$ furnishes the most complete description of a polymer distribution, there is intensive development and application of these techniques currently in progress. SEC is covered in great detail in Chapter 9.

5.4 POLYMER HYDRODYNAMIC CHARACTERISTICS

5.4.1 Diffusion

Due to thermal agitation, particles in a liquid will move about randomly in a process described by the diffusion equation. This can be applied to either the concentration of particles at a point $\vec{r}$ at a time t, $c(\vec{r}, t)$ or, for a single particle, to the probability of finding it at a displacement $\vec{r}$ from its initial position at time t, $P(\vec{r}, t)$.

Either function can be used in the diffusion equation; for example, for $P(\vec{r}, t)$

$$\frac{\partial P}{\partial t} = D\nabla^2 P \qquad (5.56)$$

For spherical symmetry the general solution is

$$P(\vec{r}, t) = P(r, t) = \frac{1}{(4\pi Dt)^{3/2}} \exp\left(\frac{-r^2}{4Dt}\right) \qquad (5.57)$$

For a particle at $r=0$ at $t=0$, the root mean square displacement from the origin $<r^2>^{1/2}$ after time t can be found from this to be

$$<r^2>^{1/2} = \sqrt{6Dt} \qquad (5.58)$$

For a spherical particle, D is related to the hydrodynamic diameter R_H, the solution viscosity η, and the temperature via the well-known Stokes–Einstein equation

$$D = \frac{k_B T}{6\pi\eta R_H} \qquad (5.59)$$

where $f=6\pi\eta R_H$ is the Stokes friction factor for a spherical particle.

The friction factor arises whenever motion of a particle in a viscous medium is subjected to any type of applied force, F_{ap}. In these cases, the particle reaches a terminal velocity v_t, which in low velocity systems such as macromolecules under normal conditions is proportional to the friction factor; that is, $f=\alpha v_t$. At terminal velocity, F_{ap} is equal and opposite to f.

The applicability of D via Equation 5.59 has been amply considered for the case of macromolecules, and in the case of monodisperse polymers, R_H represents an "equivalent hydrodynamic diameter."

When particles begin to interact hydrodynamically, the mutual diffusion coefficient D_m deviates from the single particle or self-diffusion coefficient, D_0 [22]. In such cases, a useful first-order expansion involving the particle concentration and interaction parameter, K_{int}, is [23, 24]

$$D_m = D_0 \left[1 + K_{int}c + O(c^2)\right] \qquad (5.60)$$

5.4.1.1 *Determining D from DLS* As discussed in depth in Section 8.2, DLS provides a widespread means of determining D.

Here, it is interesting to see how the diffusion probability links to the light scattering autocorrelation function. Denoting $F_P(\vec{q}, t)$ the spatial Fourier transform of $P(r, t)$, that is

$$F_{\rm p}(\vec{q},t) = \frac{1}{(2\pi)^3} \int P(\vec{r},t) e^{i\vec{q}\cdot\vec{r}} d^3r \qquad (5.61)$$

The spatial Fourier transform of the aforementioned diffusion equation is

$$\frac{\partial F_{\rm p}(\vec{q},t)}{\partial t} = -Dq^2 F_{\rm p}(\vec{q},t) \qquad (5.62)$$

hence,

$$F_{\rm p}(\vec{q},t) = F_{\rm p}(\vec{q},0)\exp(-q^2 Dt) \qquad (5.63)$$

It can be shown in a rather lengthy derivation [25] that $F_{\rm p}(\vec{q},t)$ is proportional to the scattered electric field heterodyne autocorrelation function, which, in turn is directly related to the scattered electric field homodyne autocorrelation function. It is this latter that is most frequently measured in DLS, where the scattered intensity is directly autocorrelated, rather than the scattered electric field.

The electric field autocorrelation function can be obtained in a heterodyne system, in which the scattered light is mixed with unscattered light from the laser source, thus obtaining a beat frequency. The characteristic exponential decay rate for the heterodyne correlation function is $q^2D/2$; that is, one-half the decay rate of the homodyne autocorrelation function. Sometimes there will be a partial heterodyne character to the autocorrelation function if unwanted stray light from the incident laser mixes with the scattered light, termed "accidental heterodyning" [16].

Thus, in either homodyne or heterodyne mode, the DLS autocorrelation function can yield D at any given q from an exponential fit of the form in Equation 5.63. For polydisperse systems, DLS provides an average over the diffusion coefficient distribution of the particles. A number of approaches have been developed for analyzing polydisperse systems, including the robust cumulant method (see Section 8.2.2), histograms, and the inverse Laplace transform method (see Section 8.2.3). To turn the coefficient distribution into the MWD requires a known relationship between D and M, often of the form $D = AM^{\tilde{\gamma}}$.

The widely used cumulant method yields the z-averaged diffusion coefficient, D_z. This means that the "hydrodynamic radius" thus obtained by DLS in the cumulant analysis is actually the reciprocal of the z-averaged reciprocal, $R_{\rm H}$

$$R_{\rm H,DLS} = \frac{1}{\left\langle \dfrac{1}{R_H} \right\rangle_z} = \frac{k_{\rm B}T}{6\pi\eta D_z} \qquad (5.64)$$

5.4.1.2 *Depolarized Scattering and Rotational Diffusion*

Most flexible and semiflexible polymers are isotropic and hence do not depolarize incident polarized light. For particles with shape anisotropy there will be a tensor polarizability and the scattered light will have depolarized components. Depolarized scattering principles and nomenclature are covered in Sections 8.1, 8.2.4, and 8.4.

Besides translational diffusion, anisotropic particles can also undergo rotational diffusion. A rodlike particle has a rotational diffusion coefficient, Θ, which can be measured from the autocorrelation function for depolarized scattered light; for example, by placing an analyzing polarizer at right angles to the incident electric field polarization in front of a detector in the scattering plane (which is perpendicular to the incident polarization direction).

The complete correlation function $I_{\rm VH}(\vec{q},t)$ is the product of the rotational and translational portions,

$$I_{\rm VH}(\vec{q},t) = A\beta^2 \exp\left[-\left(6\Theta + q^2 D\right)t\right] \qquad (5.65)$$

where β is the optical anisotropy defined for a cylindrically symmetric anisotropic particle in terms of its polarizability components parallel and perpendicular to its symmetry axis, $\alpha_\parallel$ and $\alpha_\perp$, respectively, as

$$\beta = \left(\alpha_\parallel - \alpha_\perp\right) \qquad (5.66)$$

Equation 5.65 shows that measuring the correlation function over several values of q (i.e., at different angles or incident wavelengths) allows extrapolation to $q=0$, yielding 6Θ as the intercept and D as the slope.

5.4.2 Viscosity

A fluid is said to be *viscous* when there is spatial inhomogeneity in the fluid's velocity field (adapted with permission from Reference 26). Such inhomogeneity can arise both from differential momentum transport of fluid and the presence of particles within the fluid that alter the velocity field, which contribute additional dissipative *internal friction* to the fluid. The amount of viscosity that a macromolecule contributes to a fluid is easy to measure and can be related to conformational and other properties of the macromolecule, making viscosity measurements a valuable characterization tool.

Spatial variations in the velocity field are specified locally by the tensorial gradient terms, $\partial v_i/\partial x_j$, where $\mathbf{v}_i$ is the velocity component in the i-direction of a three-component space and x_j is the jth directional component of the space. The gradient terms have dimensions of s^{-1}, and are referred to as shear rates, $\dot{\gamma}$.

Stokes and Einstein [27] made the initial, arduous computations of the frictional factors of spherical particles in a fluid, and found the additional viscosity they contribute to be proportional to the fluid's own viscosity and a factor, which is characteristic of the particle's own geometry and mass distribution, the *intrinsic viscosity*, $[\eta]$. The total viscosity of a fluid, whose pure viscosity is η_s and which contains particles of concentration c and having an intrinsic viscosity $[\eta]$, is given by

$$\eta = \eta_s \left\{ 1 + [\eta]c + k_p[\eta]^2 c^2 + O(c^3) \right\} \qquad (5.67)$$

where k_p is a constant related to the hydrodynamic interactions between polymer chains, usually around 0.4 for neutral, coil polymers [28], and $O(c^3)$ represents terms of order c^3 and higher.

It is customary to define the reduced viscosity, η_r, as

$$\eta_r \equiv \frac{\eta - \eta_s}{\eta_s c} = [\eta] + k_p[\eta]^2 c + O(c^2) \qquad (5.68)$$

where $[\eta]$ is the extrapolation to zero concentration and zero shear rate of η_r.

Determination of $[\eta]$ hence requires that η_s of the sample solvent and the total viscosity of the fluid containing the macromolecules η be measured. In some analytical techniques, such as SEC and ACOMP, the concentration is usually low enough that to a good approximation $[\eta] \cong \eta_r$, where η_r is measured by combining viscometer and concentration detector data. Importantly, $[\eta]$ is a direct measure of the ratio of a polymer's hydrodynamic volume V_H to its molar mass M.

For example, $[\eta]$ for an ideal random coil in Θ-solvent conditions is [29]

$$[\eta] = \frac{\Phi_v}{M} \left(\sqrt{6} < S^2 >_0^{1/2} \right)^3 \qquad (5.69)$$

where $\Phi_v = 2.56 \times 10^{23}$ and $<S^2>_0$ is the unperturbed (Θ-condition) mean square radius of gyration; the value of Φ_v changes with chain architecture and the perturbing effects of hydrodynamic interactions [28, 30].

There is often a scaling law between $[\eta]$ and M of the form

$$[\eta] = GM^{\gamma} \qquad (5.70)$$

where γ is sometimes termed the Mark–Houwink exponent and G is a scaling constant.

$<S^2>$ often also has a relationship to M of the form

$$< S^2 >^{1/2} = AM^{\beta} \qquad (5.71)$$

and viscosity exponents are expected to bear a relationship close to

$$\gamma = 3\beta - 1 \qquad (5.72)$$

For example, a random coil in Θ-conditions has $\beta = \gamma = 0.5$, whereas a random coil with excluded volume has $\beta = 0.6$ and $\gamma = 0.8$. Knowledge of G and γ also allows computation of the viscosity averaged mass M_η, which lies between M_n and M_w for $\gamma < 1$. Viscosity measurements can also be useful for assessing branching, since branched polymers will have smaller values of $[\eta]$; than their unbranched analogs of the same molar mass.

It is interesting to note that in the years of early polymer science viscosity was thought to increase linearly with polymer mass (Staudinger's law), since each additional monomer would add another identical friction factor to the chain. Debye found that for ideal coil polymers with many monomers, the hydrodynamic interaction among them becomes strong, a sort of hydrodynamic shielding develops, and the viscosity becomes proportional to the hydrodynamic volume divided by the mass, as assumed earlier [31]. This leads to the scaling law for an ideal coil in the so-called nondraining limit of $[\eta] \propto M^{1/2}$. The nondraining limit alludes to the hydrodynamic shielding among segments, such that there is little solvent flow through the polymer coil as it moves relative to the solvent.

Several geometries can be used to create velocity gradients in fluids for the computation of η. The Navier–Stokes equation provides the basis for finding the relationship between η, the geometry and applied forces. One of the most common arrangements is the capillary viscometer, for which the Poiseuille solution to the Navier–Stokes equation, which has a parabolic flow profile, is used:

$$\eta = \frac{\pi R^4 \Delta P}{8LQ} \qquad (5.73)$$

where Q is the flow rate of solution through the capillary (in $\text{cm}^3 \, \text{s}^{-1}$) of radius R across whose length L there is a pressure drop ΔP.

If z is taken as the direction of fluid flow in the capillary and r is the distance from the center of the capillary, then the shear rate is

$$\dot{\gamma}(r) = \frac{\partial v_z}{\partial r} = -\frac{4Qr}{\pi R^4} \qquad (5.74)$$

The average shear rate in the capillary is found by integration over the capillary cross section to be

$$\left| \dot{\gamma}_{ave} \right| = \frac{8Q}{3\pi R^3} \qquad (5.75)$$

A typical value is $860 \, \text{s}^{-1}$, given $Q = 1 \, \text{ml min}^{-1}$, and $R = 0.0254 \, \text{cm}$.

TABLE 5.1 Various Static and Dynamic Characteristics of Polystyrene in THF at 30 °C[a]

M (g mol^{-1})	A_2 (cm^3-mol g^{-2})	$<S^2>^{1/2}$ (nm)	R_H (nm)	D (cm^2 s^{-1})	Time (s) to Diffuse 100 μm	$[\eta]$ (cm^3 g^{-1})
10^4	1.39E-03	3.63	2.59	1.96E-06	8.52	9.43
10^5	7.21E-04	13.19	9.40	5.39E-07	30.94	48.36
10^6	3.73E-04	47.88	34.14	1.48E-07	112.32	248.03
10^7	1.93E-04	173.84	123.96	4.09E-08	407.81	1272.03

[a] $\eta = 0.00438$ Poise for THF at 30 °C.

Single capillary viscometers are remarkably easy and inexpensive to construct using a modern differential pressure transducer with a double T-network and capillary of desired L and R. Capillary bridge viscometers, based on the Wheatstone bridge principle of null measurement, are inherently more sensitive than single capillary viscometers, and can largely eliminate temperature and pump pressure fluctuations. The latter are most useful for SEC but are normally not suitable for long-term flow measurements, however, since a hold-up volume is used in one or more bridge arms, which leads to zeroing out of the differential signal with time. Single and bridge capillary viscometer results have been exhaustively compared [32].

5.4.3 Case Study of Polystyrene Characteristics in Tetrahydrofuran (THF)

The earlier sections have detailed the static and dynamic properties of polymers. To obtain an idea of the magnitudes of these quantities, data from various authors are summarized here for polystyrene in THF, taken as a very common case of a linear polymer in a good solvent.

Using low polydispersiy molecular weight standards for polystyrene over the wide range of 10^4–10^7 g mol^{-1}, the following relationships to molecular weight M were found in THF at 30 °C [33–35]: $<S^2>^{1/2}$ (nm) $= 0.0209 M^{0.56}$, A_2 (cm^3-mol g^{-2}) $= 0.0194 M^{-0.286}$, $[\eta]$ (cm^3 g^{-1}) $= 0.01363 M^{0.71}$, D (cm^2 s^{-1}) $= 3.4 \times 10^{-4} M^{-0.56}$. Values for these characteristics are shown for a range of polystyrene masses in Table 5.1.

Table 5.1 shows $<S^2>^{1/2}/R_H = 1.40$ for PS in THF. This is close to the theoretical value for an ideal random coil in the nonfree draining limit.

5.5 ELECTRICALLY CHARGED POLYMERS AND COLLOIDS

5.5.1 Poisson–Boltzmann Equation and the Debye Screening Length

Charged polymers, or polyelectrolytes, are the basis of life. Polynucleic acids—DNA and RNA—proteins, and many polysaccharides bear electrical charge. In some cases, the charges may be all of the same sign, for example, DNA is polyanionic due to the phosphate groups in its double helix backbone, whereas proteins contain ionizable groups of both charge signs, and are termed "polyampholytes." Hence, the net charge of proteins varies with pH, generally being negative at high pH, passing through an isoelectric point at which the positive and negative charges balance to zero net charge, and becoming net positive at low pH. Some biopolymers have zwitterionic groups, which contain one positive and one negative moiety, and hence have dipolar character. An important class of such zwitterions is the phosphatidylcholines.

Similarly, many man-made polymers and colloids have electrical charge, especially those used in aqueous systems, such as water purification, pharmaceuticals and biotechnology, oil recovery, personal care, metallurgy, papermaking, food and agricultural applications, etc.

One long-standing approach to characterizing liquids containing electrically charged particles is to combine the Poisson equation for electrostatic potential $V(r, \theta, \phi)$ with the Boltzmann distribution for mobile charges in the electrostatic potential. For a simple electrolyte of the form $A_{Z_+}B_{Z_-}$, where Z_+ is the number of positive charges on the cationic group and Z_- the number of negative charges on the anionic group, the Poisson–Boltzmann equation (PBE) results:

$$\nabla^2 V = -\frac{\rho_0}{\varepsilon}\left[\exp\left(\frac{-Z_+ eV}{k_B T}\right) - \exp\left(\frac{Z_- eV}{k_B T}\right)\right] \quad (5.76)$$

where ρ_0 is the bulk concentration of electrolyte and ε is the liquid's permittivity, $\varepsilon = \varepsilon_0 D$, ε_0 is the permittivity of free space and D is the liquid's dielectric constant, and SI (MKSA) units are used.

This second order nonlinear partial differential equation is difficult to solve in a general form analytically, but it can be solved to arbitrarily high accuracy in any geometry and boundary conditions of interest by numerical computation, including commercially available packages for such calculations.

For symmetric electrolytes, $Z_+ = Z_-$, (e.g., NaCl), the right-hand side of the PBE can be linearized with cancellation of both constant and second-order terms. If a spherically symmetric potential is involved, $V = V(r)$, the linearized PBE can be expressed as

$$\frac{d^2(rV)}{dr^2} = \kappa^2(rV) \tag{5.77}$$

where κ is given by Equation 5.80.

Linearization is appropriate when

$$\frac{ZeV}{k_B T} \ll 1 \tag{5.78}$$

Two particularly important features result from this linearized PBE. The first is the form of the solution, which is a "screened potential," such as also found in plasma physics [36], where A is a constant potential, related to boundary conditions.

$$V(r) = A\frac{e^{-\kappa r}}{r} \tag{5.79}$$

The second feature is the screening parameter κ which appears in the linearized Equation 5.77 and in the solution, Equation 5.79. Its reciprocal $1/\kappa$ is the characteristic range of the screened potential and is sometimes termed the Debye–Hückel screening length. It is composed of the following constants:

$$\frac{1}{\kappa} = \left(\frac{k_B T \varepsilon}{2Z^2 e \rho_0}\right)^{1/2} \tag{5.80}$$

For aqueous solutions at 25 °C

$$\frac{1}{\kappa} = \frac{1}{Z}\frac{0.305\,\text{nm}}{(\rho_0(M))^{1/2}} \tag{5.81}$$

where the following SI constants are used: $k_B = 1.38 \times 10^{-23}$ J K^{-1}, $\varepsilon_0 = 8.85 \times 10^{-12}$ C^2 N^{-1}–m^2, $\varepsilon = 80\ \varepsilon_0$ for water, $e = 1.6 \times 10^{-19}$ C. ρ_0 is the molar concentration of the electrolyte (mol l^{-1}). For water with 150 mM univalent salt at 25 °C, $1/\kappa = 0.8$ nm, and at 0.001 M $1/\kappa = 9.7$ nm. For distilled, neutral water, the H$^+$/OH$^-$ balance of 10^{-7} mol l^{-1} gives the limiting value of $1/\kappa = 1\ \mu$m. Dissolved CO_2 in water can add carbonic acid that raises this "native" level of electrolyte and can typically decrease the screening length to about 0.25 μm.

The aforementioned leads to the definition of ionic strength as a useful auxiliary concept when treating charged particles in solution. Ionic strength, I, is defined as

$$I = \frac{1}{2}\sum_{i=1}^{N}[A_i]z_i^2 \tag{5.82}$$

where $[A_i]$ is the molar concentration of species I and the sum is over all N electrically charged species in solution.

The Debye screening length, κ, is inversely proportional to the square root of I:

$$\frac{1}{\kappa}(m) = \left(\frac{\varepsilon k_B T}{2eI}\right)^{1/2} \tag{5.83}$$

Strong differences in the solution to V for symmetric versus asymmetric salts and for monovalent versus higher valence salts begin to grow as $zeV/k_B T$ grows greater than its linearization constraint [37].

5.5.2 Electrostatic Persistence Length (EPL) and Electrostatic Excluded Volume (EEV)

In the Gedanken experiment where a flexible polymer in a solution that is initially neutral becomes electrically charged with charges of a single sign, one would imagine that the polymer would stretch out as the charges along the chain repel each other. That the chain would not stretch out completely follows from the fact that the entropic penalty in the free energy for complete stretching is very high, since there is only a single state (all trans) corresponding to complete stretching among 3^{N-2} states. Hence, the challenging physical problem is to compute how far the chain stretches out in response to the electrical charge on the polymer, as well as the ionic strength of the supporting medium, that will minimize free energy, $G = U - TS$, with respect to minimizing the repulsive electrostatic energy term U, by stretching out, while maximizing the entropic terms $-TS$, by resisting the stretching. One approach to formulating this problem is by introducing the electrostatic persistence length (EPL), L_e.

One can further imagine that as the neutral polymer becomes charged, the intermolecular repulsive energy will increase, and hence the excluded volume will also increase, due both to the increasing strength of the electric field in the vicinity of the charged polymer, and to the fact that as it stretches out in space it will also occupy and hence exclude more volume. This problem has been formulated in terms of the electrostatic excluded volume (EEV).

Odijk [38] and Fixman [39], separately, derived an expression for EPL as a perturbation to the inherent persistence length $L_{p,0}$, to obtain a total persistence length L_p

$$L_p = L_{p,0} + L_e \tag{5.84}$$

where $L_{p,0}$ is the persistence length of the polymer when it is electrically uncharged.

Their equivalent expression for L_e is

$$L_e = \frac{\xi^2 \kappa^{-2}}{12\lambda_B}\left[3 - \frac{8}{y} + e^{-y}\left(y + 5 + \frac{8}{y}\right)\right] \quad (5.85)$$

where ξ is the number of elementary charges per Bjerrum length, κ is the Debye–Hückel screening parameter, and $y = \kappa L$.

With knowledge of $L_{p,0}$ obtained at very high ionic strength, L_p can hence be computed using $L_{p,0}$ and L_e from Equation 5.85, so that $<S^2>_0$ can be computed by Equation 5.19. The measured value $<S^2>$ is related to $<S^2>_0$ via the static expansion factor α_s, introduced earlier in Equation 5.25.

There are several theories that relate α_s to a commonly used perturbation parameter z, given by

$$z = \left(\frac{3}{2\pi L_k^2}\right)(EV)N_k^{1/2} \quad (5.86)$$

where L_k and N_k are the (Kuhn) statistical segment length and number, respectively, and $L_k = 2L_p$ in the coil limit. Here (EV), introduced in Equation 5.22, is the excluded volume between two charged rodlike segments, for which Fixman and Skolnick [40] arrived at the expression

$$EV = 8L_p^2\kappa^{-1}\int_0^{\pi/2}\sin^2\theta\int_0^{w/\sin\theta} x^{-1}(1-e^{-x})\,dx\,d\theta \quad (5.87)$$

where

$$w = 2\pi\xi^2\kappa^{-1}e^{-\kappa d} \quad (5.88)$$

where d is the rod diameter and ξ is given by Equation 5.91.

Of the several expressions relating z to α_s, one that has found considerable utility when $N_k > 2$ is the Gupta–Forsman expression [41].

$$\alpha_s^5 - \alpha_s^3 \approx \frac{134}{105}\left(1 - 0.885N_k^{-0.462}\right)z \quad (5.89)$$

These outlined expressions have proven to be a robust approach to predicting polyelectrolyte dimensions for varying linear charge density and ionic strength conditions [8, 12, 42].

5.5.3 Counterion Condensation

The aforementioned Bjerrum length, l_B, is the distance between charges of valence Z along a chain at which the electrostatic repulsive energy between these charges is equal to the thermal energy k_BT, that is,

$$l_B = \frac{Z^2 e^2}{4\pi\varepsilon k_B T} \quad (5.90)$$

For example, for water at $T = 300\,K$ and monovalent charge groups ($Z = 1$), $l_B = 0.718\,nm$. In a simple straight-line chain of charges, it was shown that if charges were situated closer than l_B, then the repulsive electrostatic energy would exceed k_BT and counterions would condense onto the charges until the spacing between charges was greater than or equal to l_B. While this is an oversimplified version of counterion condensation theory, and leads to a sharp condensation threshold, which has been shown experimentally to be smooth rather than sharp (e.g., Chapter 13), the essential notion has been amply demonstrated experimentally by conductivity and also online reaction monitoring [43–45].

A dimensionless linear charge density, ξ, is normally used as a convenient parameter for counterion condensation and polyelectrolyte theories. It is defined as

$$\varepsilon = \frac{\text{Number of elementary charges}}{\text{Bjerrum length}} \quad (5.91)$$

The rule of thumb on counterion condensation is that $\xi \leq 1$; that is, that there will be equal to or less than one elementary charge per Bjerrum length.

Many elaborations on counterion condensation theory and its consequences exist [46–49].

5.5.4 Electrophoretic Mobility and "Zeta" Potential

Charged polymers and colloids exhibit a wide variety of electrokinetic behavior with terms such as "electrophoresis," "electro-osmosis," "streaming potential," "sedimentation potential," and others. These have been thoroughly reviewed [50].

Among the most important characteristics associated with electrokinetic phenomena are the electrophoretic mobility and "zeta potential" (or ζ-potential). Electrophoretic mobility, μ_e, is the steady-state drift velocity, v_d, a charged particle obtains per unit of applied uniform electric field E

$$\mu_e = \frac{v_d}{E} \quad (5.92)$$

The drift velocity results when the viscous drag force on the particle balances the electric force.

The ζ-potential is the electric potential at the surface of hydrodynamic slip of a charged particle moving through a medium [51]. Smoluchowski's treatment provides a simple link between μ_e and ζ for the case where the Debye length is much smaller than the particle's characteristic size, a

$$\mu_e = \frac{\varepsilon\zeta}{\eta}, \quad a\kappa \gg 1 \quad (5.93)$$

where η is the solvent viscosity.

For nanoparticles or very low ionic strength situations, when the Debye length is greater than the particle size, there is a different numerical factor

$$\mu_e = \frac{2\varepsilon\zeta}{3\eta}, \quad a\kappa < 1 \tag{5.94}$$

ζ-potential is a particularly convenient parameter for gauging the stability of colloidal solutions. Generally speaking, the larger the value of ζ the greater the electrostatic interparticle repulsion among particles and hence the greater the stability. As a benchmark, the electrostatic potential energy of an elementary charge equivalent to $k_B T$ at room temperature is 25.7 mV. In practice, ζ values greater than this lead to good colloidal stability, whereas systems usually become less stable as values decrease below this.

While a number of electrophoretic and related techniques have been developed, one of the preeminent means of ζ-potential determination is via electrophoretic light scattering. One means of achieving this is by direct measurement of the Doppler shift of light scattered from particles moving with v_d in an electric field. Another means that is up to three orders of magnitude more sensitive, and hence applicable to very low electrophoretic mobilities of charged particles in nonpolar solvents is the phase analysis procedure [52, 53]. This method analyzes the time rate of change of the phase of the light scattered from charged particles moving in the applied electric field, rather than measuring the frequency shift itself. This allows for much smaller periods of application of the applied field, which can also have a much higher frequency than in the standard Doppler velocimetry approach.

5.6 THERMODYNAMICS OF POLYMER SOLUTIONS

When a polymer is dissolved in a good solvent, the free energy is lowered, and when dissolved in a poor solvent, the free energy increases. The way in which polymer molecules interact with the solvent, with each other, and with themselves (e.g., excluded volume effects, self-organization of copolymer chains, etc.) lends itself to a thermodynamic description. In the latter, the usual thermodynamic parameters of temperature, concentration, pressure, and interparticle potentials come into play.

Concentration regimes. It is convenient to consider three polymer concentration regimes, although there is no sharp boundary among them. In dilute solutions, polymer molecules have enough free spatial volume to accommodate them without overlapping on average with other polymer molecules. As concentration increases, the so-called overlap concentration c^* is eventually reached, where the spatial volume available for each molecule in the solution is approximately equal to the average spatial volume occupied by the molecule. Since $[\eta]$ measures hydrodynamic volume per mass of a polymer molecule, its reciprocal is the mass/hydrodynamic volume occupied by the polymer and hence is a good, rough, easily measured gauge of c^*

$$c^* \approx \frac{1}{[\eta]} \tag{5.95}$$

As concentration increases above c^*, the congestion of polymer molecules increases until the concentrated regime, c^{**}, is reached. In this regime, segments of individual chains are so intermingled with segments of other chains that the chains become ideal (no excluded volume) in the polymer melt limit [54].

5.6.1 Chemical Potential and Related Concepts

Central concepts in treating polymer solution thermodynamics are the Gibbs free energy G (an extensive variable)

$$G = U + PV - TS \tag{5.96}$$

and the chemical potential μ_i, which is the partial molar free energy for solution component i (components can include the solvent, polymer, counterions, salts, other polymers, etc.) at constant P, T, and concentrations of the other components j.

$$\mu_i = \left.\frac{\partial G}{\partial n_i}\right|_{T,P,n_j} \tag{5.97}$$

The Gibbs–Duhem relation follows immediately from these quantities under constant T and P

$$\sum_1^N n_i d\mu_i = 0 \tag{5.98}$$

that is, the changes in chemical potential among N species are related to each other by this summation rule. One consequence of this rule is that for a species i partitioned among different phases a, b, c, ... in a multicomponent system (e.g., surfactant molecules present individually and in micellar form in a soap solution), the chemical potential is the same for all of them; that is,

$$\mu_{i,a} = \mu_{i,b} = \mu_{i,c} \tag{5.99}$$

μ_i for a component in an ideal solution is given in terms of the substance's chemical potential in a pure state μ_i^0 and the associate entropy of mixing term $RT\ln(x_i)$

$$\mu_i = \mu_i^0 + RT \ln(x_i) \qquad (5.100)$$

where x_i is the mole fraction of component i in the solution.

In nonideal solutions, there are interactions among components that lead to a deviation from the earlier mentioned ideal case. For example, as polymer concentration in a good solution increases, the osmotic pressure increase has a positive second derivative with concentration, whereas light scattering has a negative second derivative. An empirical form of handling this is via the activity coefficient, γ; $\mu_i = \mu_i^0 + RT \ln(\gamma x_i)$. A more analytical approach often used treats the deviations from ideality in terms of the virial coefficients A_2, A_3, A_4, and higher terms. Measurement of the virial coefficients is experimentally feasible. Using them, the chemical potential for the solvent in terms of solute (polymer) concentration, c_p, is

$$\mu_s = \mu_s^0 - RTv_s^0 \left(\frac{c_p}{M} + A_2 c_p^{\,2} + A_3 c_p^{\,3} + A_4 c_p^{\,4} + \cdots \right)$$
$$(5.101)$$

where v_s^0 is the partial molar specific volume of the solvent.

The notion of osmotic pressure Π can be extracted from this. A simple way of understanding how Π arises is to consider solvent in a container of uniform temperature divided in two portions α and β by membrane permeable to solvent but impermeable to polymer The polymer resides in only side β of the container partition. By the Gibbs–Duhem relation, the solvent chemical potential strives to be the same on both sides, but the polymer chemical potential does not, since it cannot cross the membrane. This requires that $\mu_{s,\alpha} = \mu_{s,\beta}$, but since $x_{s,\alpha} = 1$ in the α side, so that $\mu_{s,\alpha} = \mu_s^0$, and $x_{s,\beta} < 1$ on the β side, this equality can only be obtained if there is an additional term to $\mu_{s,\beta}$. Since T is equal on both sides there must be an energy term on the β side involving pressure. A brief analysis yields

$$\mu_{s,\beta} = \mu_s^0 + RT \ln\left(\gamma x_s \right) + v_s^0 \Pi = \mu_s^0 \qquad (5.102)$$

Equating this to the right-hand term of Equation 5.101 yields

$$\Pi = RT \left(\frac{c_p}{M} + A_2 c_p^{\,2} + A_3 c_p^{\,3} + A_4 c_p^{\,4} + \cdots \right) \qquad (5.103)$$

For an ideal solution, $A_2 = A_3 = A_4 = A_N = 0$, so that the osmotic pressure reduces to the so-called van't Hoff law, which closely resembles the ideal gas equation

$$\Pi = RT \frac{c_p}{M} \qquad (5.104)$$

The action of Π is to produce a pressure similar to a hydrostatic or gas pressure; for example, at high enough polymer content in side β it is possible to burst the separating membrane due to high Π if the volume of side β is not allowed to expand. The equivalence of hydrostatic and osmotic pressure is seen in a U-tube osmometer which is divided in half by the membrane and osmotic pressure in side β, which is left open to volume expansion in this device, causes the solvent height difference Δh in side β to rise to a level where the hydrostatic pressure $\rho g \Delta h = \Pi$ (where ρ is the liquid density).

A_2 is directly related to the excluded volume EV discussed earlier, via

$$A_2 = \frac{N_A}{2M^2} \text{EV} \qquad (5.105)$$

Computation of A_3 and higher becomes increasingly difficult to compute theoretically even for hard spheres [55], and the complications become far greater for interacting polymer chains. Whereas A_2 involves pairwise interparticle potentials, A_3 involves three body interactions in different combinations, and A_4 involves four body interactions. Boltzmann derived the following for hard spheres

$$A_3 = \frac{5MA_2^{\,2}}{8} \qquad (5.106)$$

and

$$A_4 = 0.28695 M^2 A_2^{\,3} \qquad (5.107)$$

5.6.2 Relation of Light Scattering by Polymer Solutions to Thermodynamic Fluctuations

While light scattering is covered in detail in Chapter 8, it is interesting to make the link between osmotic pressure and light scattering in this discussion. A frequent starting point for analyzing light scattering data is to consider that the scattered intensity is given by

$$I_s(q) = S(q)P(q) \qquad (5.108)$$

where q is the usual scattering vector, whose magnitude is

$$q = \frac{4\pi n}{\lambda} \sin\left(\frac{\theta}{2} \right) \qquad (5.109)$$

and $P(q)$ is the form factor of the scatterer, and $S(q)$ is the interparticle structure factor.

There are a number of approaches to computing $I_s(q)$. The fluctuation theory [56] at $q=0$ is particularly useful and is outlined in the following.

By the fluctuation theory, scattering intensity I_s from a volume element V of scatterers at distance r from the scattering volume can be described in terms of the mean square fluctuations from average polarizability α_0 (in units of cm^3) in a liquid

$$\alpha = \alpha_0 + \delta\alpha \qquad (5.110)$$

by (in CGS units)

$$\frac{I_s}{I_0} = \frac{16\pi^4 \sin^2\phi}{\lambda^4 r^2} < \delta\alpha^2 > \qquad (5.111)$$

for vertically polarized incident light of intensity I_0; where $<\alpha^2> = \alpha_0^2 + 2\alpha_0 < \delta\alpha > + < \delta\alpha^2 >$ is the time average mean square polarizability of the element V, ϕ is the altitude angle measured from the direction of E-field polarization of the incident light.

In most practical polymer characterization work, measurements are made in the scattering plane, $\phi = \pi/2$, where the scattering intensity is maximum. The inverse proportionality to r^2 reflects the conservation of energy of a wave propagating outward from its dipolar source.

Equation 5.111 shows that the scattering is inversely proportional to λ^4, which Lord Rayleigh first used to explain why a cloudless sky is blue; that is, the gas molecules in the upper atmosphere preferably scatter short wavelength blue light compared to longer wavelength red light. Likewise, at dawn and dusk as one looks in the vicinity of the rising or setting sun, the light reaching the eye has traversed a long swath of atmosphere, so that most of the blue light has been scattered away, leaving a reddish hue.

Only $< \delta\alpha^2 >$, the mean square fluctuation term, contributes to scattering since the α_0^2 term corresponds to the homogeneous state of the liquid without fluctuations, and is hence similar to a "crystal term" for which there is no scattering other than in the forward direction, $\theta = 0$. Likewise the term involving $< \delta\alpha > = 0$ by definition of the average.

Fluctuations $\delta\alpha$ can be caused by fluctuations in pressure, temperature, and concentration. Considering polymer or colloid solutions at constant T and P, and invoking a form of the Clausius–Mossotti equation (again in CGS units) [56]

$$\frac{4\pi\alpha}{V} = n^2 - n_0^2 \qquad (5.112)$$

where α is the total polarizability of element V, n is the total liquid index of refraction, and n_0 is the pure solvent index of refraction, and considering that the scatterer adds a small increment Δn to n_0, $n = n_0 + \Delta n$, the I_s in the scattering plane ($\phi = \pi/2$) from element V at $q = 0$ is given by

$$\left.\frac{I_s}{I_0}\right|_{q=0} = \frac{4\pi^2 n_0^2 (\partial n/\partial c)^2 V^2 \sin^2\phi}{\lambda^4 r^2} < \delta c^2 > \qquad (5.113)$$

where c is the concentration of polymer (g cm^{-3}) and $<\delta c^2>$ is the mean square concentration fluctuation in the volume V at constant pressure and temperature.

Hence, the scattering problem involves finding expressions for $<\delta c^2>$, which can be found by taking a Boltzmann weighted average over all fluctuations, where the fluctuations are treated as a continuum of concentration states, each with a free energy $G(\delta c)$

$$< \delta c^2 > = \frac{\int_{-\infty}^{\infty} (\delta c)^2 \exp\left[\frac{-G(\delta c)}{k_B T}\right] d(\delta c)}{\int_{-\infty}^{\infty} \exp\left[\frac{-G(\delta c)}{k_B T}\right] d(\delta c)} \qquad (5.114)$$

Expanding $G(\lambda c)$ in a power series to second order gives

$$G(\delta c) \cong G(c_0) + \left.\frac{\partial^2 G}{\partial c^2}\right|_{c_0} \frac{(\delta c)^2}{2} \qquad (5.115)$$

since $\left.\frac{\partial G}{\partial c}\right|_{c_0} = 0$. Equation 5.115 in 5.114 yields

$$< \delta c^2 > = \frac{k_B T}{\left.\frac{\partial^2 G}{\partial c^2}\right|_{c_0}} \qquad (5.116)$$

Expressing G in terms of the partial chemical potentials of the solvent μ_1 and polymer μ_2, and invoking the Gibbs–Duhem relation, Equation 5.98 yields

$$< \delta c^2 > = \frac{-k_B T v_1}{V\left(\frac{\partial \mu_1}{\partial c}\right)} c \qquad (5.117)$$

where v_1 is the molal volume of solvent.

From the usual virial expansion treatment of chemical potential

$$\mu_1 = G_1^0 - RT v_1 c\left(\frac{1}{M} + A_2 c + A_3 c^2 + \cdots\right) \qquad (5.118)$$

where $R = k_B N_A$ is the universal gas constant. Hence,

$$\left.\frac{I_s}{I_0}\right|_{q=0} = \frac{KVc}{r^2\left(\frac{1}{M} + A_2 c + A_3 c^2 + \cdots\right)} \sin^2\phi \qquad (5.119)$$

where, for vertically polarized incident light,

$$K = \frac{4\pi^2 n_0^2 \left(\frac{\partial n}{\partial c}\right)^2}{N_A \lambda^4} \quad (5.120)$$

The Rayleigh scattering ratio $R(q)$, first introduced in Section 8.1, is obtained from Equation 5.119 by multiplying by r^2 and dividing by V. At $q = 0$ this gives

$$R(q=0) = \frac{r^2}{V}\frac{I_s}{I_0}\bigg|_{q=0} = \frac{Kc}{\left(\frac{1}{M} + A_2 c + A_3 c^2 + \cdots\right)}\sin^2\phi \quad (5.121)$$

$R(q=0)$ has the units of cm^{-1} and can be interpreted as the fraction of incident light scattered per steradian of solid angle per centimeter of path length traveled in the scattering medium.

Considering Equation 5.103, $R(q=0)$ can also be seen to be proportional to osmotic compressibility $\partial c/\partial \pi$. Hence, highly osmotically compressible solutions will exhibit high levels of light scattering, whereas osmotically incompressible solutions will exhibit little scattering. A good example of highly osmotically compressible solutions are those containing large neutral polymers, or large polyelectrolytes at high ionic strength; that is, low values of A_2. An example of an osmotically incompressible solution is one containing polyelectrolytes (large or small) with little or no added supporting electrolyte (no added ionic strength). The polyelectrolytes are unshielded and avoid each other leading to very little concentration fluctuation; that is, A_2 is very large and very little light is scattered above the solvent level.

When the scatterers are polydisperse, M is replaced by weight average molar mass M_w, and the virial coefficients by more complex averages, which continue to be represented for convenience by A_2, A_3, etc.

5.6.3 Flory–Huggins Mean-Field Theory

The essence of the Flory–Huggins model is to treat a polymer solution as a 3-D lattice composed of n_{total} lattice sites, equally spaced in all dimensions. Each lattice site contains either a solvent or monomer of a polymer chain, so that $n_{total} = n_s + Nn_p$, where n_s and n_p are the number of solvent and polymer molecules, respectively, and N is the degree of polymerization.

The theory yields the Gibbs free energy, G, and related quantities in terms of (i) the entropy of mixing and (ii) the enthalpy of mixing. The theory is formulated in terms of the volume fraction, ϕ, of monomers in the lattice (and $1 - \phi$ solvent molecules in the lattice) and the Flory–Huggins χ-parameter, which characterizes the energy of polymer–

polymer (u_{pp}), solvent–solvent (u_{ss}), and polymer–solvent (u_{ps}) interactions [57, 58].

$$\frac{\Delta G_{mix}}{n_{total}k_B T} = \frac{\phi}{N}\ln\phi + (1-\phi)\ln(1-\phi) + \chi\varphi(1-\phi) \quad (5.122)$$

The first two terms on the right represent the entropy of mixing. The first of these is that due to the mixing of the polymer and the second to the mixing of the solvent. For long chain polymers, $N > 100$, the first of these terms becomes negligible so that only the mixing entropy of the solvent remains important.

The enthalpy of mixing is given in terms of the Flory–Huggins, χ, which is computed as

$$\chi = \frac{\left[u_{ps} - \left(u_{ss} + u_{pp}\right)\right]}{k_B T} \quad (5.123)$$

where u_{ps}, u_{ss}, and u_{pp} refer to the polymer–solvent, solvent–solvent, and polymer–polymer potentials, respectively.

It is seen that χ is inversely proportional to T but is also affected by other interaction factors, which can themselves also be T dependent; that is, χ for real substances can have non-monotonic dependence on T, and can produce complex phase diagrams, mentioned in Section 5.6.4. The chemical potential of the polymer can be determined from the partial derivative of free energy with respect to n_p at constant n_s and n_t to yield

$$\frac{\Delta\mu_p}{k_B T} = \ln\phi - (N-1)(1-\phi) + \chi N(1-\phi)^2 \quad (5.124)$$

where the relation

$$\frac{\partial\phi}{\partial n_p}\bigg|_{T,n_s} = \frac{N(1-\phi)^2}{n_s} \quad (5.125)$$

is useful in the derivation. Similar reasoning leads to the chemical potential of the solvent

$$\frac{\Delta\mu_s}{k_B T} = \ln(1-\phi) + \left(1 - \frac{1}{N}\right)\phi + \chi\phi^2 \quad (5.126)$$

Using the relationship between solvent chemical potential and the specific volume of solvent (i.e., of a lattice site), $\Pi v_s = -\Delta\mu_s$ yields

$$\frac{\Pi v_s}{k_B T} = \frac{\phi}{N} - \ln(1-\phi) - \phi - \chi\phi^2 \quad (5.127)$$

For dilute solutions, expansion of $\ln(1-\phi)$ in a power series yields

$$\frac{\Pi v_s}{k_B T} = \frac{\phi}{N} + \left(\frac{1}{2} - \chi\right)\phi^2 + O\left(\phi^3\right) \qquad (5.128)$$

Comparing the second term to the virial coefficient expansion of osmotic pressure yields the following connection between A_2 and χ,

$$A_2 = N_A v_s \left(\frac{N}{M}\right)^2 \left(\frac{1}{2} - \chi\right) \qquad (5.129)$$

As noted, under Θ-conditions the polymer solution becomes ideal, repulsive and attractive forces cancel each other out, and the virial coefficients become 0. Under Θ-conditions, $\chi = 1/2$. In a good solvent, $A_2 > 0$ so $\chi < 1/2$. In a poor solvent, $A_2 < 0$ so $\chi > 1/2$.

χ (and hence also A_2) is useful for determining coexistence and phase separation behavior of polymer solutions, discussed next.

5.6.4 Phase Behavior

The miscibility of multicomponent systems depends on how the total free energy of mixing of the solution changes as the composition and temperature of the solution change. Using the free energy of mixing from the lattice model, Equation 5.122 allows phase behavior to be computed as follows. The chemical potential for the polymer chains is the derivative of this free energy with respect to n_p at constant T and V. When there is a separation of the system into two phases, for example, high concentration of solvent with some polymer in phase 1 and high concentration of polymer with some solvent 2 in phase 2, the chemical potential of polymer is the same in both phases, and likewise for the solvent. This requires that the tangent be the same in both phases; that is, the common tangent condition is

$$\left.\frac{\partial\left(\dfrac{\Delta G_{mix}}{n_{total}k_B T}\right)}{\partial\phi}\right|_1 = \left.\frac{\partial\left(\dfrac{\Delta G_{mix}}{n_{total}k_B T}\right)}{\partial\phi}\right|_2 \qquad (5.130)$$

Tracing out the values at which ϕ occurs for each value of T (and hence χ) defines the binodal or curve between the phases, sometimes also termed the "coexistence" or "phase stability" curve.

An interesting aspect of phase stability is that there can be concentrations at which the solution is metastable against phase separation. This arises when the second derivative of $\Delta G_{mix}/n_{total}k_B T$ is 0. Then, the system is stable against phase

separation for small changes in ϕ in either direction; that is, the free energy is lower at the inflection point of $\Delta G_{mix}/n_{total}k_B T$ for the system staying unseparated at that point than what the total $\Delta G_{mix}/n_{total}k_B T$ would be if separated into two phases, one an amount $+\Delta\phi$ from the inflection point and the other $-\Delta\phi$ from the inflection point. Large perturbations, such as shaking or nucleation sites, however, will lead to destabilization and phase separation, similar to the case of a supercooled liquid remaining liquid until some perturbation (e.g., mechanical shock) sends it into phase separation. The curve that defines the metastable region is called the spinodal line. Between the bimodal and spinodal lines resides the metastable region. A graphical example is shown to illustrate these concepts.

Figure 5.6 shows a hypothetical $\Delta G_{mix}/n_{total}k_B T$ versus the volume fraction of polymer ϕ for different values of T and corresponding χ, where χ was taken to vary, for example only, as $\chi = -0.65 + 490/T$ and $N = 2$ (a dimer). One example of how the equal tangents are found graphically is shown by a dashed line in the figure and how they translate to the binodal points on the bimodal curve in the phase diagram overset in the figure is shown by the dotted lines. Also shown for the same curve are the inflection points which translate up to the spinodal curve in the phase diagram, also shown with dotted lines. The portions of the diagram where the solution is phase separated, single phase, and metastable are also indicated.

In the hypothetical case of Figure 5.6, χ decreases with temperature from 0.98 at 283 K to 0.60 at 360 K; that is, the solvent becomes better at higher temperatures. The temperature at which the bimodal and spinodal curves meet is termed the upper critical solution temperature (UCST). Below the UCST, the system separates into two phases.

There are also systems for which χ increases with temperature, for example, in a form $\chi = A - B/T$. In this case the bimodal and spinodal lines open upward, visualized by rotating the top curve in Figure 5.6, 180° around the x-axis. In this case, the system will have a lower critical solution temperature (LCST); that is, the phase separation occurs when the system is heated; for example, polyethylene in hexane under 5 bar pressure [59].

There are also cases where χ is not a monotonic function of temperature and may have a maximum and/or minimum value versus T. In such cases, it is possible to describe phase shapes that do not resemble parabolas, such as in Figure 5.6. Poly(oxyethylene) in water, for example, has a χ versus T with a peak, and this leads to an oval shape in the T versus ϕ plane with the two phase separation within the oval [58]. An example of χ versus T with a minimum includes low mass poly(styrene) in acetone, which has an LCST in an upward parabola above a downward parabola with a UCST. It is not a contradiction that the LCST is higher than the UCST; if the system goes from demixed to single phase upon heating then a UCST is crossed, and continued heating the single phase in

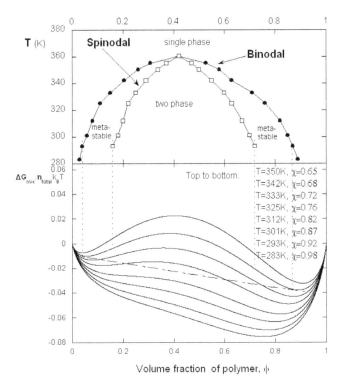

FIGURE 5.6 Top: The phase diagram for a hypothetical binary system of a solvent and second molecule with $N=2$. Bottom: The free energy versus volume fraction ϕ, at various temperatures, from which the phase diagram is constructed. Dashed lines show examples of how the phase diagram relates to the free energy functions.

this latter case eventually leads to demixing again at the LCST, which is higher than the UCST.

Another example of a system with a minimum in χ versus T is high molar mass poly(styrene) in acetone, for which an "hour glass" figure forms in the T versus ϕ plane, inside of which the system is phase separated at any temperature.

5.7 RHEOLOGY

The earlier discussion of polymer diffusion and viscosity involves time-independent quantities that occur under no externally applied shear conditions (diffusion) and vanishingly small, time-independent shear (intrinsic viscosity). Polymers in melts and in solution constitute complex fluids with intricate time-dependent responses to stresses and flows that reflect molecular motions on many different length and timescales.

The two extremes in rheology are (i) the pure, inviscid liquid that is incapable of supporting any shear stress and immediately flows in response to such stress and (ii) a solid that neither flows nor irreversibly deforms in response to a shear stress. Most real materials exist between these extremes. Bose–Einstein condensation in helium is a rare

exception, where all atoms are phase locked into the ground state, similar to superconductivity, and have no viscosity at all while in this superfluid state.

The Deborah number, De, is a dimensionless parameter which is the ratio of the time of relaxation of a material in response to a shear stress to the time of experiment or observation. $De=0$ for the extreme of inviscid matter (e.g., superfluid helium) and $De \rightarrow \infty$ for a pure solid. De was taken not only as an index defining the continuum from pure liquid to pure solid but also as a jocular shield for the sometimes misplaced use of "theology" where, in fact, "rheology" is concerned [59]. "All things flow before God" is the paraphrase of the prophetess Deborah that indicates all things have $De=0$ when the timescale of observation is infinite.

5.7.1 Newtonian and Non-Newtonian Polymer Liquids

Polymer motions, such as bond rotations among different configurations, reptations through space, intermolecular overlap interactions, etc., lead to viscoelastic responses in polymer solutions that can span many orders of magnitude in time, or in its conjugate variable, the frequency at which stress is applied. Liquids whose viscosity does not change with frequency of stress are termed "Newtonian" fluids. Most polymeric fluids deviate from Newtonian behavior under certain conditions, and are termed "non-Newtonian."

Shear thinning (thixotropy) is one of the most common manifestations of non-Newtonian behavior in polymeric liquids [61]. Increased shear can lead to partial alignment of polymers or colloid particles with the flow, thus decreasing viscosity. Examples include latex paint, blood, and syrups. Shear thickening is the opposite phenomenon (antithixotropy) whereby the material becomes more viscous or stiffer with increasing shear, often due to shear-induced organization, such as partial crystallization. Quicksand and aqueous solutions of cornstarch are examples of shear-thickening materials.

In practical terms, polymer rheology and especially its non-Newtonian aspects are of critical importance in processing polymers. The relationship between applied stress and polymer flow determines the processability of polymers in applications such as extrusion, fiber spinning, and film blowing.

5.7.2 Dynamic Mechanical Analysis and Other Viscoelastic Analyses

An important means of assessing the rheological behavior of polymer melts and concentrated solutions is the use of dynamic mechanical analysis. This consists in testing the mechanical response, or strain, of a polymeric material to a time-dependent stress, or, conversely, measuring the stresses generated by time-dependent strains. For simplicity, a sinusoidal driving function is normally used. In the case of controlled shear stress, $\gamma(t)$

$$\gamma(t) = \gamma_0 \sin(\omega t) \ \text{(controlled input)}$$

$$\frac{\sigma(t)}{\gamma_0} = G'(\omega, T)\sin(\omega t) + G''(\omega, t)\cos(\omega t)$$

$$\text{(response function)} \qquad (5.131)$$

The coefficient of the in-phase portion, often termed the dynamic storage modulus, $G'(\omega, T)$, measures the purely elastic response of the material, and is normally very sensitive to both frequency, ω, and temperature. The out-of-phase portion, termed the dynamic loss modulus G'', measures the energy dissipative response of the material. The phase shift between the input frequency and the response is usually denoted by δ and is the ratio of the moduli:

$$\tan \delta = \frac{G''}{G'} \qquad (5.132)$$

A purely viscous material with no elasticity will have $\delta = \pi/2$, whereas a loss-free elastic material will have $\delta = 0$. A typical polymer will show a complex interplay between how G' and G'' behave versus ω. It is not unusual for G' and G'' to cross each other once or more as ω increases.

Because polymer rheology is a deeply complex field [62–64], there are many other methods to assess different types of viscoelastic and mechanical behavior of polymers in concentrated solutions, melts, and rubbery states.

The Mooney viscosity is the torque generated on a spindle rotating at constant angular velocity, immersed in a polymeric material between heated dies. It is one of the most widely used industrial measures of bulk viscoelastic properties of polymeric materials, especially elastomers and rubbers. Oftentimes, the Mooney viscosity is of greater practical significance than the dynamic mechanical determination of G' and G''. The notion of Mooney viscosity is well grounded in the principles of continuum mechanics but involves the empirically determined Mooney viscosity and related parameters.

The melt-flow index is an extremely simple test, and essentially consists of measuring the amount of polymer melt flowing through a tube of specific dimensions under a specific pressure over a specified amount of time. Experimentally, the apparatus is a small extruder. The index obtained has a sort of inverse relationship to viscosity at any given temperature, and is indirectly related to polymer molecular weight.

5.7.3   Solid-State Properties

Polymers in the solid state exhibit extraordinarily diverse properties. Polymers can be made exceptionally hard, and scratch-, impact-, and high temperature–resistant, while they can also be made highly ductile and flexible. Additionally, their electrical and optical properties can be tailored in many unique ways.

The details of mechanical behavior are rooted in a number of factors; molecular weight distribution, polymer tacticity, degree of crystallinity, inclusions, defects, incorporation of specific modifiers, polymer blending, processing conditions, etc.

Many of the methods used for characterizing solid materials are applicable to polymers in the solid state; x-ray diffraction and crystallography, thermal analyses such as differential scanning calorimetry, mechanical testing for ductility, brittleness, compressive, tensile, and torsional stress–strain relationships, impact, fatigue, creep, etc.

The glass transition is of particular importance for polymers that are in a glassy state when solid. This state is typical of other solid-state amorphous materials; that is, no long-range crystal order, although there can be regions and domains of crystallinity. Upon heating a polymer in the glassy state, a glass transition temperature, T_g, is found where the polymer loses its solidity and becomes rubbery. A number of other thermodynamic parameters also change at T_g. The subject has been extensively reviewed [65].

A new generation of polymer electronics, including photovoltaics, transistors, and flexible electronic materials, is being developed based on conducting and doped polymers [66–69]. Polymers have unique optical properties where they are used in the vast field of polymer optics, to fabricate devices such as lenses, polarizers, fiber optics and other waveguides, liquid crystals, novel photonic materials, nanooptical components, etc. [70, 71]. Some polymers, including most biopolymers, have marked optical activity and can be analyzed and used in areas of polarimetry, and linear and circular dichroism and birefringence. Polymer optics also extends to nonlinear optical applications, such as second harmonic generation [72, 73].

5.8   CHARACTERISTICS OF POLYMERIZATION REACTIONS

The foregoing overview of polymer characteristics leads up to a consideration of which of these are important candidates for online monitoring. There is a long history of attempts at online monitoring using GPC, in order to obtain MWD at discrete intervals during polymerization reactions.

With multidetector GPC, it is possible not only to obtain absolute MWD combining static light scattering with concentration detection, but also to have the radius of gyration and intrinsic viscosity distributions.

Many other methods have been used for monitoring polymer characteristics during reactions. A nonexhaustive list of these includes measuring the increasing amperage flowing through a stirring motor in a reactor needed to maintain constant rpm as the reacting medium becomes more viscous, or directly measuring the torque on the stirrer, adapting melt-flow determinations to discrete online measurements, use of *in situ* viscoelastic probes in reactors, ultrasound, and turbidity.

In situ infrared and Raman scattering are dealt with extensively in Chapter 6. Monitoring by *in situ* calorimetry, conductivity, pH, densimetry, and rheological measurements are covered extensively in Chapter 7. Specific applications of these and other methods to monitoring are presented in Chapters 18–21.

Because the reaction medium is normally quite concentrated, however, rheological and other measurements often only indirectly measure molecular mass and other single chain properties, because the interactions between polymer chains often dominate signals from undiluted reactor contents. Automatic continuous online monitoring of polymerization reactions (ACOMP) provides a solution for this. ACOMP is covered in Chapters 11–13.

5.8.1 Monomer Conversion

Monitoring the conversion of monomers is in itself not a direct measure of polymer characteristics. Rather, it furnishes kinetic information on the progress of the reaction that is vital to knowing how long to run a synthetic process step before proceeding to the next step or terminating a reaction. Additionally, information on conversion is often necessary in order to obtain polymer concentration, which is needed for light scattering determination of molecular mass, and for viscometric determination of reduced or intrinsic viscosity. In the case of copolymerization, being able to follow the conversion of comonomers allows computation of the instantaneous average polymer composition and of the average composition drift during the reaction.

The actual kinetics in large-scale reactions may deviate from the well-characterized results presented in detail in Chapters 1, 2, 16, and 17. These deviations can result from spatial inhomogeneity and fluctuations of temperature and reagents in the reactor, reactor fouling, contaminated reactor and/or reagents, varying effects of exothermicity, Trommsdorf effect, reagent feed rates, etc. From this point of view, it is sometimes important to measure levels of initiators, catalysts, and other supporting reagents.

Another aspect of conversion monitoring that is often of critical importance in manufacturing is the level of residual monomer, since these levels are often tightly regulated by government and other agencies. This often requires that monomer levels in final products be under 500 ppm, or as low as even 10 ppm.

5.8.2 Trends in Molecular Weight and Intrinsic Viscosity during Polymer Reactions

Polymer characteristics that one wishes to monitor during polymerization reactions were covered in this and other chapters. Hence, it is frequently of interest to monitor average molecular weights, or entire MWD, $[\eta]$ averages and distributions, degree of branching, particle sizing, etc. The challenge is how to actually make multiple, even continuous measurements of these quantities *during the polymerization reaction*. These characteristics are often difficult to measure in their own right even on single, off-line samples. Polymer characterization of single, off-line samples is in itself a highly developed and complex field, which is continuously advancing.

In general, the MWD and its averages will evolve during polymerization. The MWD can broaden, as in many free-radical polymerization reactions, narrow, as in reactions with strong living character, or stay constant, such as in controlled semibatch reactions that target a constant MWD. M_w, frequently the most useful average, can decrease during polymerization, as is common in free-radical polymerization when the initiator concentration is essentially constant and radical transfer reactions are not dominant, increase, such as in living reactions, and also in reactions where cross-linking occurs, and in "dead-end" reactions, or stay the same, as in some radical transfer reactions and semibatch reactions targeted to constant MWD.

Table 5.1 shows a schematic of M_w evolution for different types of polymerization reactions.

While $[\eta]$ is related to M the relationship can change if the polymer architecture changes. Hence, if branching, cross-linking, or aggregation occur, M will normally increase considerably, whereas $[\eta]$ may show little or no change [74]. Conversely, in the case of copolymeric polyelectrolytes with both charged and uncharged monomers, $[\eta]$ can vary strongly at low ionic strength when composition drifts toward higher charged monomer incorporation, even if M changes little.

Derivitization reactions can often produce the same effect just mentioned, where $[\eta]$ increases while M remains nearly constant. Derivitizations of neutral polymers, for example, polyacrylamide, neutral polysaccharides, etc., to produce charged polymers by base carboxylation, quaternization, sulfation, etc., will display this effect when the process is monitored [75].

5.8.3 Particulates and Other Phenomena during Polymer Synthesis

A very widespread issue in polymer production is the existence of nano-, micro-, and macroscopic particulates. In some cases, these particulates are integral to the process, such as bacteria in bioreactors, cell fragments in bioextraction, or latex particles in emulsion polymerization. In most cases, however, the particulates are unwanted and can be detrimental both to the product and the reactors. Such particulates include physical aggregates, branched and cross-linked microgels, microcrystals, impurity particles, and macroscopic coagula. This topic is covered in depth in Section 14.3.

In the biotechnology industry focused on producing monoclonal antibodies and therapeutic proteins, the problem of protein aggregation is ubiquitous. This is dealt with in detail in Chapter 20 and in Section 14.1.

5.9 SUMMARY

Important characteristics that describe static mass, conformations, and dimensions of polymer molecules have been surveyed. This has been followed by hydrodynamic properties such as diffusion and viscosity. A separate section has been used to describe the salient aspects of charged polymers and colloids in solution. From there, the collective properties of polymers were briefly introduced in terms of their solution thermodynamics, the relationship of these to the scattering of light, and to phase behavior and transitions. Concentrated polymer solutions and melts become extraordinarily complex, with time response behavior depending on polymer architecture and interactions, and this has been briefly discussed in the area of rheology. In the solid-state limit of rheology, polymers take on myriad applications in materials engineering applications, in electronics, optics, and other areas.

Since polymerization monitoring is the focus of this book, these characteristics set the stage for the quantities that are useful to measure during polymerization reactions. Because the latter involve characteristics including and beyond the polymers themselves, there is a strong focus on monomer kinetics and composition distributions in reaction monitoring.

REFERENCES

[1] Gedde UW. *Polymer Physics*. Dordrecht, the Netherlands: Kluwer Academic Publishers; 1995. p 19–24.

[2] Ghosh S, Li X, Reed CE, Reed WF. Persistence length and diffusion behavior of high molecular weight hyaluranate. Biopolymers 1991;30:1101–1112.

[3] Wenner JR, Williams MC, Rouzina I, Bloomfield VA. Salt dependence of the elasticity and overstretching transition of single DNA molecules. Biophys J 2002;82:3160–3169.

[4] Landau LD, Lifschitz EM. In: *Statistical Physics*. Oxford: Pergamon Press; 1980. Ch. 12.

[5] Doty P, Steiner RF. Macro-ions. I. Light-scattering theory and experiments with bovine serum albumin. J Chem Phys 1952; 20:85–94.

[6] Li X, Reed WF. Polyelectrolyte properties of proteoglycans. J Chem Phys 1991;94:4568.

[7] Benmouna M, Weill G, Benoit H, Akcasu Z. Scattering from charged macromolecules. I. Static structure factor. J Phys (Les Ulis France) 1982;43:1679–1685.

[8] Sorci GA, Reed WF. Electrostatically enhanced second and third virial coefficients, viscosity and interparticle correlations for linear polyelectrolytes. Macromolecules 2002;35: 5218–5227.

[9] Yamakawa H. *Modern Theory of Polymer Solutions*. London: Harper and Row; 1971.

[10] McMillan WG, Jr, Mayer JE. The statistical thermodynamics of multicomponent systems. J Chem Phys 1945;13:276–305.

[11] Hansen J-P, McDonald IR. *Theory of Simple Liquids*. 2nd ed. London: Academic Press; 1990.

[12] Ghosh S, Li X, Reed CE, Reed WF. Apparent persistence length and diffusion behavior of high molecular weight hyaluronate. Biopolymers 1990;30:1101–1112.

[13] Reed WF, Ghosh S, Medjahdi G, François J. Dependence of polyelectrolyte apparent persistence lengths, viscosity, and diffusion on ionic strength and linear charge density. Macromolecules 1991;24:6189–6198.

[14] Cotts PM, Guan Z, McCord E, McLain S. Novel branching topology in polyethylenes as revealed by light scattering and C NMR. Macromolecules 2000;33:6945–6952.

[15] Podzimek S, Vlcek T. Characterization of branched polymers by SEC coupled with a multiangle light scattering detector. II. Data processing and interpretation. J Appl Polym Sci 2001;82:454–460.

[16] Gaborieau M, Castignolles P. Size-exclusion chromatography (SEC) of branched polymers and polysaccharides. Anal Bioanal Chem 2011;399:1413–1423.

[17] Kratochvil P. Characterization of branched polymers. Macromol Symp 2000;152:279–287.

[18] Burchard W. Solution properties of branched macromolecules. Adv Polym Sci 1999;143:113–194.

[19] Flory PJ. Molecular size distribution in three dimensional polymers 1. Gelation. J Am Chem Soc 1941;63: 3083–3100.

[20] Stockmayer WH. Theory of molecular size distribution and gel formation in branched-chain polymers. J Chem Phys 1943;11:45–55.

[21] Zimm BH, Stockmayer WH. The dimensions of chain molecules containing branches and rings. J Chem Phys 1949;17: 1301–1313.

[22] Dhont JKG. *An Introduction to Dynamics of Colloids*. Amsterdam: Elsevier; 1996.

[23] Dorshow R, Nicoli DF. The effect of hydrodynamics on the diffusivity of charged macromolecules: application to BSA. J Chem Phys 1981;75:5853–5856.

[24] Carter JM, Phillies GDJ. Second-order concentration correction to the mutual diffusion coefficient of a suspension of hard Brownian spheres. J Phys Chem 1985;89: 5118–5124.

[25] Berne BJ, Pecora R. In: *Dynamic Light Scattering*. Krieger Publishing Co; 1976. Chs. 4 & 5.

[26] Reed WF. Fundamentals of static light scattering and viscometry in SEC and related methods. In: Striegel AM, editor. *Multiple Detection Size-Exclusion Chromatography*. ACS Symposium Series 893. Washington, DC: American Chemical Society; 2004.

[27] Landau LD, Lifshitz EM. *Fluid Mechanics*. London: Pergamon Press; 1959.

[28] Huggins ML. The viscosity of dilute solutions of long-chain molecules. IV. Dependence on concentration. J Am Chem Soc 1942;64:2716–2718.

[29] Flory PJ, Fox TG. Treatment of intrinsic viscosities. J Am Chem Soc 1951;73:1904.

[30] Kirkwood JG, Riseman J. The intrinsic viscosities and diffusion constants of flexible macromolecules in solution. J Chem Phys 1948;16:565.

[31] Debye P. The intrinsic viscosity of polymer solutions. J Chem Phys 1946;14:636–639.

[32] Norwood DP, Reed WF. Comparison of single capillary and bridge viscometers as size exclusion chromatography detectors. Int J Polym Anal Char 1997;4:99–132.

[33] Venkataswamy K, Jamieson AM, Petschek RG. Static and dynamic properties of polystyrene in good solvents: ethylbenzene and tetrahydrofuran. Macromolecules 1986;19: 124–133.

[34] Schulz GV, Baumann H. Thermodynamic behavior, expansion coefficient, and viscosity number of polystyrene in tetrahydrofuran. Makromol Chem 1968;114:122–138.

[35] Stacy CJ, Kraus G. Second virial coefficients of homopolymers and copolymers of butadiene and styrene in tetrahydrofuran. J Polym Sci: Polym Phys Ed 1979;17:2007–2012.

[36] Block LP. A double layer review. Astrophys Space Sci 1978;55:59–83.

[37] Sorci GA, Reed WF. Effect of ion type and valence on polyelectrolyte conformations and interactions. Macromolecules 2004;37:554–565.

[38] Odijk T. Polyelectrolytes near the rod limit. J Polym Sci 1977;15:477–483.

[39] Skolnick J, Fixman M. Electrostatic persistence length of a wormlike polyelectrolyte. Macromolecules 1977;10: 9444–9448.

[40] Fixman M, Skolnick J. Polyelectrolyte excluded volume paradox. Macromolecules 1978;11:863–867.

[41] Gupta SK, Forsman WC. General treatment of random-flight statistics with and without excluded volume. Macromolecules 1972;5:779–785.

[42] Reed WF, Ghosh S, Medjahdi G, Francois J. Dependence of polyelectrolyte apparent persistence lengths, viscosity, and diffusion on ionic strength and linear charge density. Macromolecules 1991;24:6189–6198.

[43] Wilson RW, Bloomfield VA. Counter-ion-induced condensation of deoxyribonucleic-acid—light-scattering study. Biochemistry 1979;18:2192–2196.

[44] Hinderberger D, Spiess HW, Jeschke G. Dynamics, site binding, and distribution of counterions in polyelectrolyte solutions studied by electron paramagnetic resonance spectroscopy. J Phys Chem B 2004;108:3698–3704.

[45] Kreft T, Reed WF. Experimental observation of crossover from noncondensed to counterion condensed regimes during free radical polyelectrolyte copolymerization under high-composition drift condition. J Phys Chem B 2009;113: 8303–8309.

[46] Manning GS. Limiting laws and counterion condensation in polyelectrolyte solutions II. Self diffusion of the small ions. J Chem Phys 1969;51:934–938.

[47] Oosawa F. *Polyelectrolytes*. New York: Marcel Dekker; 1971.

[48] Deserno M, Holm C, May S. Fraction of condensed counterions around a charged rod: comparison of Poisson–Boltzmann theory and computer simulations. Macromolecules 2000;33:199–206.

[49] Chu JC, Mak CH. Inter and intrachain attractions in solutions of flexible polyelectrolytes at nonzero concentration. J Phys Chem 1999;110:2669–2679.

[50] Delgado AV, Gonzalez-Caballero F, Hunter RJ, Koopal LK, Lyklema J. Measurement and interpretation of electrokinetic phenomena (IUPAC technical report). Pure Appl Chem 2005;77:1753–1850.

[51] Lyklema J. *Fundamentals of Interface and Colloid Science*. Volume 2. London: Elsevier Ltd; 1995.

[52] Miller JF, Schaetzel K, Vincent BJ. The determination of very small electrophoretic mobilities in polar and nonpolar colloidal dispersions using phase analysis light scattering. J Colloid Interface Sci 1991;143:532–554.

[53] Tscharnuter WW. Mobility measurements by phase analysis. Appl Opt 2001;40:3995–4003.

[54] de Gennes PG. Polymer melts. In: *Scaling Concepts in Polymer Physics*. Ithaca: Cornell University Press; 1979.

[55] Hansen J-P, McDonald IR. In:*Theory of Simple Liquids*. 2nd ed. London: Academic Press; 1986. Ch. 4.

[56] Tanford C. In: *Physical Chemistry of Macromolecules*. New York: John Wiley & Sons; 1963. Ch. 4.

[57] Flory PJ. In: *Principles of Polymer Chemistry*. Ithaca: Cornell University Press; 1953.

[58] Teraoka I. *Polymer Solutions*. New York: Wiley Interscience; 2002. Ch. 2.

[59] Elias HG. In: *An Introduction to Polymer Science*. Weinheim: VCH; 1997. Ch. 6.

[60] Reiner M. The Deborah number. Physics Today 1964;17:62–65.

[61] Barnes HA. Thixotropy, a review. J Non-Newtonian Fluid Mech 1997;70:1–33.

[62] Doi M, Edwards SF. *In: The Theory of Polymer Dynamics*. New York: Oxford University Press; 1986.

[63] Rubinstein M, Colby RH. *In: Polymer Physics*. New York: Oxford University Press; 2003.

[64] Klempner D, Sperling LH, Utracki LA, editors. *Intepenetrating Polymer Networks*. Advances in Chemistry Series 239. Washington, DC: American Chemical Society; 1994.

[65] Eisenberg A. The glassy state and the glass transition. In: *Physical Properties of Polymers: ACS Professional Reference Book*. Washington, DC: American Chemical Society; 1993.

[66] Venkataraman D, Russell TP. Polymer electronics: power from polymers. J Polym Sci B: Polym Phys 2012;50:1013–1028.

[67] Bruetting W, editor. *Physics of Organic Semiconductors*. Weinheim: Wiley-VCH; 2005.

[68] Hu W, Nakashima H, Wang E, Furukawa K, Li H, Luo Y, Shuai Z, Kashimura Y, Liu Y, Torimitsu K. Advancing conjugated polymers into nanometer-scale devices. Pure Appl Chem 2006;78:1803–1822.

[69] Leising G, Stadlober B, Haas U, Haase A, Palfinger C, Gold H, Jakopic G. Nanoimprinted devices for integrated organic electronics. Microelectron Eng 2006;83:831–838.

[70] Picken SJ. Liquid crystal polymer optics. Polym Prepr (American Chemical Society) 1999;40:1167–1168.

[71] Powell S, Fisher D. Polymer optics gain increased precision. Laser Focus World 2007;43:111–117.

[72] Dai DR, Hubbard MA, Park J, Marks TJ, Wang J, Wong GK. Rational design and construction of polymers with large second-order optical nonlinearities. Synthetic strategies for enhanced chromophore number densities and frequency doubling temporal stabilities. Mol Cryst Liq Cryst 1990;189:93–106.

[73] Kajzar F, Chollet P-A. Poled polymers and their application in second harmonic generation and electrooptic modulation devices. Adv Nonlinear Optics 1997;4:1–46.

[74] Farinato RS, Calbick J, Sorci GA, Florenzano FH, Reed WF. Online monitoring of the final divergent growth phase in the stepgrowth polymerization of polyamines. Macromolecules 2005;38:1148–1158.

[75] Paril A, Alb AM, Reed WF. Online monitoring of the evolution of polyelectrolyte characteristics during postpolymerization modification processes. Macromolecules 2007;40: 4409–4413.

6

INFRARED (MIR, NIR), RAMAN, AND OTHER SPECTROSCOPIC METHODS

ALEXANDRE F. SANTOS, FABRICIO M. SILVA, MARCELO K. LENZI, AND JOSÉ C. PINTO

6.1 INTRODUCTION

Polymerization industries strive to increase production rates of high-quality products with the lowest costs, employing the most flexible and safest processes. This is motivated by the fierce commercial competition in the field and by the increase of customer expectations and demands for nonpolluting processes and plants [1]. These challenges can be faced more efficiently with the help of process modeling, control and optimization techniques. Many of these techniques have been developed and are available specifically for polymerization processes [2–4]. However, since a universal technique that can be used in all cases is not available, each particular case must be analyzed carefully for selection of the most suitable set of process control techniques. Some issues must be taken into account during the analysis, including the complexity of the polymerization mechanism, the mathematical model to be used, the reactor behavior, and the control objectives pursued, among others [5, 6].

In order to achieve the production of a polymer resin with specified end-use properties, the availability of efficient control techniques may not be enough. The gap between the polymerization process and the control technique must be bridged with accurate and robust instrumentation for online monitoring so that the controller can be fed with the current state of the process. Figure 6.1 illustrates the control configurations normally used to control the process. The lack of such instruments, which must be able to measure and monitor the quality of the polymer resin, has been long recognized as the most important problem in the field of polymerization reactor control [7]. Several studies provide good surveys on techniques, sensors, and instruments normally used for monitoring of polymerization reactors [7–12].

The complex nature of typical polymerization systems is one of the main obstacles for the development of adequate monitoring instruments [2–13]. Despite the significant efforts made for the development of reliable and robust monitoring techniques, most polymer properties cannot be measured online and have to be evaluated with very laborious techniques [8–11]. Moreover, it is difficult to correlate molecular properties (such as molecular weight averages, particle size averages, copolymer composition, among others) and final end-use properties (such as brightness, thermal and mechanical resistance, stiffness, etc.), whose quantification is much more complex and usually determined for final products only through standardized test piece characterizations [14, 15]. Additionally, sample preparation may require the execution of many distinct unit operations in series, such as extraction, dilution, dissolution, drying, evaporation, precipitation, among others, introducing undesirable long time delays. For online applications, sampling devices are often too sophisticated and complicated, requiring careful design and installation procedures, which very frequently cannot be performed in real industrial sites.

Finally, available monitoring instruments are usually very expensive, so that the installation and maintenance costs can be prohibitive for most industrial facilities. It has also to be taken into account that the lab environment is completely different from the industrial environment, because of the dimensions of reactor vessels, feed lines, physical distances, and so on. For all the reasons presented earlier, it is no surprise there is a big gap between the control applications

Monitoring Polymerization Reactions: From Fundamentals to Applications, First Edition. Edited by Wayne F. Reed and Alina M. Alb.

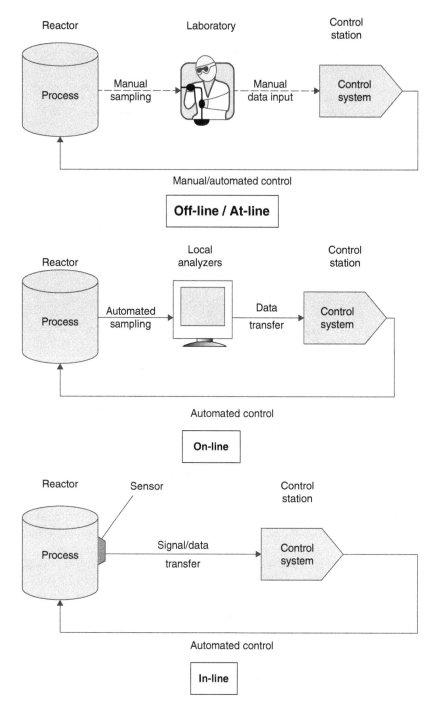

FIGURE 6.1 Process analytical techniques. From Hergeth WD. On-line monitoring of polymerization and cure. Sens Update 1999;5: 191–242. © Wiley-VCH Verlag GmbH & Co. KGaA. Reproduced with permission.

developed in the research lab and the actual control applications implemented industrially [2–4, 8–13].

Recently, monitoring techniques have been greatly improved by the combination of spectroscopic methods and fiber optics technology, which allow for the *in-situ* and in-line acquisition of process data, allowing for reduction of time delays normally involved with sample preparation.

Among them, techniques based on near infrared spectroscopy (NIRS) have certainly become extremely important ones for practical reasons discussed in the following sections. Useful applications have also been developed for middle infrared (MIR), visible (VIS), and ultraviolet (UV) spectroscopic techniques. However, usually spectroscopic methods cannot be used as *ad hoc* monitoring techniques, so that some sort of calibration must be performed and validated with actual process data. For this reason, the use of chemometric methods for instrument calibration is very popular in this field. Then, the in-line implementation of calibration models allows for the development of *soft-sensors*, which can be used to infer properties of the final polymer resins [16].

A review of spectroscopy as a tool for polymerization process monitoring and control is given in this chapter. Initially, fundamental principles of spectroscopy are reviewed, including a brief discussion about the nature and properties of the electromagnetic wave and its interaction with matter, followed by a more detailed discussion about NIR radiation. In the second part of this chapter, calibration techniques used most frequently for the development and implementation of control applications are briefly discussed. Several applications of spectroscopy to other fields and to other control of polymerization reactors are also included in order to illustrate the versatility of spectroscopic methods for the development of monitoring and control strategies of chemical processes. The final part of the chapter presents relevant applications of the spectroscopic methods for *in-situ* and in-line monitoring and control of solution, suspension, and emulsion polymerization processes.

6.2 FUNDAMENTALS

6.2.1 Interaction of Electromagnetic Radiation with Matter

Wave can be defined as a perturbation of space or of a material medium, which modifies the local properties of the medium and transports energy while it propagates through the space. If perturbations are periodical, they can lead to formation of periodical waves and oscillatory spatio-temporal patterns. If perturbations are caused by the periodical displacement of electrical charges, then the magnetic and electric fields are periodically modified in the space, giving birth to electromagnetic waves, which are generically called as electromagnetic radiation or simply radiation. A propagating electromagnetic wave is usually described in terms of its characteristic oscillation frequency and energy content, and can interact with the environment in different ways, as well described in standard textbooks [17–30].

Besides being reflected and transmitted, radiation can also be absorbed and scattered by matter. As chemical bonds are not rigid connections, the atoms that constitute the molecules can also present different vibration modes around the chemical bonds. Electrons that constitute the chemical bond can be excited and yield different energy level spectra [25].

When a molecule absorbs a photon, the energy content of the molecule increases and an energy transition occurs [25]. The energy transition can occur only when the photon

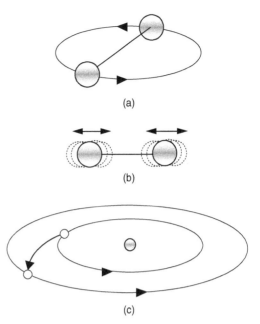

FIGURE 6.2 Representation of energy transitions in the molecular and atomic levels: (a) molecular rotation; (b) molecular vibration; (c) electronic transition.

absorbed can provide energy for an allowed energy transition. The fundamental energy transitions of interest are related to the rotation of the molecule (or groups of atoms), to the vibration of bonded atoms and to the excitation of electrons. Excitation of protons and neutrons require too much energy and are disregarded here. Figure 6.2 illustrates the energy transitions.

It can be shown [26, 27] that the allowed energy transitions of the molecule are

$$\Delta E_r = k\frac{h}{2\pi\Gamma} = hf \tag{6.1}$$

$$\Delta E_v = k\frac{h}{2\pi}\sqrt{\frac{C}{\mu}} = hf \tag{6.2}$$

$$\Delta E_e = hc\frac{\mu}{m_e}R_\infty Z^2\left(\frac{1}{k_2^2} - \frac{1}{k_1^2}\right) = hf \tag{6.3}$$

where ΔE is the allowed energy transition; k, k_1, and k_2 ($k_2 > k_1$) are natural numbers; Γ is the moment of inertia of the molecule; C is the strength of the chemical bond (see Table 6.1 for typical C values); μ is the reduced mass of the oscillator; m_e is the mass of the electron; R_∞ is the Rydberg constant ($R_\infty = 109737\ cm^{-1}$); Z is the charge of the nucleus around which the electron moves; and the subscripts r, v, and e stand for the rotational, vibrational, and electronic energy transitions, respectively.

TABLE 6.1 Wave Numbers of Fundamental Absorption Peaks and Force Constants for Some Diatomic Molecules[a]

Molecule	Fundamental Absorption Band—υ(cm^{-1})	Force Constant, C (dyne cm^{-1})
HF	2907	9.7×10^5
HCl	2886	4.8×10^5
HBr	2559	4.1×10^5
HI	2230	3.2×10^5
CO	2143	18.4×10^5

[a]From Reference [27].

For simple two-body oscillators,

$$\mu = \frac{m_1 m_2}{m_1 + m_2} \tag{6.4}$$

$$\Gamma = \mu d^2 \tag{6.5}$$

where m is the atomic mass and d is the average distance between the centers of mass of the two atoms.

Equation 6.1 indicates that the higher the moment of inertia of the molecule, the lower the energy barrier required for rotational energy transitions. Typical values of ΔE_r are in the range 10^{-4} eV $< \Delta E_r < 10^{-3}$ eV, which correspond to the microwave and far infrared (IR) regions of the electromagnetic spectrum [27]. Therefore, if the molecule presents a permanent electric dipole, it can either emit or absorb photons in the microwave/far IR regions for rotational energy transitions at well-defined frequencies (wavelengths). Thus, if the frequency of incident electromagnetic wave is varied in this range, absorption peaks will be observed at certain characteristic frequencies. However, absorption spectra in this range are usually weak and broad because of significant thermal noise. For instance, at ambient temperature the translational kinetic energy of typical molecules is of order 10^{-2}–10^{-1} eV, which means that very small changes of kinetic energy can promote rotational energy transitions [27].

Equation 6.2 indicates that the energy barrier for vibrational energy transitions depends on the strength of chemical bonds and on the reduced mass of the oscillator. The higher the strength of the chemical bond and the lower the mass of the oscillator, the higher the energy barrier required for energy transition. Typical values of C are presented in Table 6.1. If these values are inserted into Equation 6.2, typical vibrational energy transitions fall in the characteristic IR region, between 800 and 5000 nm. Given the much higher amounts of energy required for vibrational energy transitions, thermal noise is not significant in this spectral region at usual ambient temperatures. For this reason, IR spectral analysis constitutes a powerful technique for the identification and characterization of chemical functions in organic molecules and of chemical constituents in multicomponent mixtures [25, 27].

Based on Equation 6.3, the energy barriers required for electronic energy transitions are typically within the range of 1–10 eV [25]. Therefore, electronic energy transitions fall within the VIS and UV regions of the spectrum. For this reason, absorbance spectra in the VIS and UV regions of the spectrum are useful for interpretation of the electronic configuration of atoms and molecules. As electronic transitions are very sensitive to the atomic configuration and do not significantly depend on the much weaker intermolecular interactions, the analysis of VIS and UV spectra may be very useful when interatomic and intermolecular interactions must be neglected. For instance, a typical application is the detection of certain metal components in complex organic media, such as petroleum [29].

Equations 6.1–6.3 show that energy transitions are allowed for certain multiples of a fundamental frequency. In this case, energy transitions occur between nonadjacent energy levels and are usually named *overtones*. Similarly, multiple energy transitions can occur simultaneously in complex molecules, where different fundamental rotational and vibrational modes are possible. These complex energy transitions are usually called *combinations* [25, 27]. Figure 6.3 illustrates these concepts for the simple water molecule. As overtones and combinations involve larger

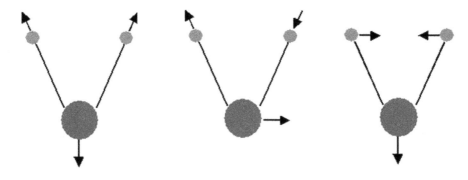

FIGURE 6.3 Some fundamental vibration modes for water.

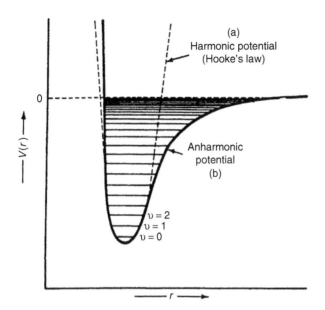

FIGURE 6.4 Anharmonic behavior of a chemical vibration.

amounts of energy, these energy transitions are usually less probable than the fundamental ones, leading to weaker absorption peaks.

Equations 6.1–6.3 assume that the chemical oscillators behave ideally, which is not true. Actually, the chemical bond (electron clouds) limits the approach of the different nuclei, when it is compressed, and eventually breaks, when it is stretched. Figure 6.4 illustrates the actual anharmonic behavior of the chemical bonds. According to the figure, nonlinear shifts can be observed for overtones, which tend to accumulate at a certain maximum energy transition barrier. For this reason, overtones tend to concentrate within certain ranges of the radiation spectrum [27].

Besides absorption, a light beam can also be scattered when it passes through a material medium [17, 18, 27]. When an electromagnetic wave impinges on a single molecule, a dipole moment is induced by the electromagnetic field. The induced dipole then vibrates with the frequency of the incident light wave, which is the source of the scattered light. When the frequency and phase of the scattered light are the same as those of the incident light, the scattering is said to be coherent or elastic; otherwise, scattering is said to be quasi-elastic. When a monochromatic and polarized light beam is considered, the well-known Rayleigh formula shows that:

$$I_S = \frac{8}{3}\pi\left(\frac{2\pi}{\lambda}\right)\alpha^2 \qquad (6.6)$$

where I_S is the fraction of scattered light, γ is the wavelength of the incident wave, and α is the polarizability of the molecule.

Simple mathematical solutions can be derived for dilute chemical solutions [31]. Light scattering is treated in detail in Chapters 5 and 8. Here, it is sufficient to observe that the fraction of scattered light decreases with the inverse fourth power of the wavelength of the incident wave and increases with the square of the polarizability of the medium.

It is important to note that the frequencies of incident and scattered light may sometimes be different. If the incident radiation is intense and monochromatic with frequency v, the scattered light contains a strong component of frequency v (Rayleigh scattering) and weak radiation signals of frequency $v \pm \Delta v$ (Raman scattering). If the incident frequency is changed, the strong radiation signal changes accordingly, but the frequency shift, Δv, of the weak radiation remains the same. This characteristic spectrum of frequency shift values is usually called the Raman spectrum [32, 33]. The Raman spectrum is generated because the scattering molecules may have their energy levels changed during the scattering process. The symmetric negative and positive frequency shifts correspond to loss of photon energy to a molecular bond (Stokes shift) and absorption of energy from a molecular bond (anti-Stokes shift), respectively. Molecules are polarized during the scattering process and the Raman spectra can be obtained and used to characterize nonpolar molecules that do not produce rotational and vibrational absorption spectra, making Raman spectroscopy complementary to IR spectroscopy. However, the Raman signal is usually very weak and difficult to interpret in concentrated solutions due to multiple scattering and absorption phenomena. It is also to be separated from the much stronger Rayleigh scattering signal, mentioned earlier. Nonetheless, Raman techniques have been recently developed for monitoring chemical systems and have been reviewed in the literature [33, 34].

6.2.2 Near Infrared Radiation

Most spectroscopic techniques developed for the in-line monitoring and control of polymerization reactions are based on NIRS, including Raman applications. In 1800, Sir William F. Herschel performed experiments using a thermometer and a glass prism to identify which radiation in the VIS spectrum was responsible for the heat of sunlight [35]. He found the heating effect increased in the dark portion beyond the red end of the spectrum, showing that there was light radiation beyond the VIS and Herschel called it infrared (IR—"below red").

Different radiation regimes present in the electromagnetic spectrum are shown in Figure 6.5. The near infrared (NIR) spectral region comprises the interval between 800 and 2500 nm, while the usual mid infrared (MIR) spectral region covers the interval between 2500 and 25,000 nm [30]. The main characteristics of the NIR spectral region are the following: [30]

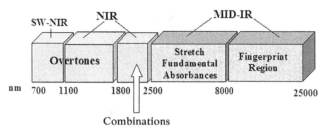

FIGURE 6.5 Fundamental vibrational phenomena in the NIR and MIR regions.

- The absorption bands observed in the MIR spectrum can be assigned to fundamental vibrations, whereas absorption bands observed in the NIR spectrum are due to overtones and combinations (see Fig. 6.10 and Fig. 6.12).
- Absorption in the NIR region is much less intense than in the MIR region as absorption decreases by a factor of 10–100 in moving from the fundamental transition to the first overtone.
- Combinations and overtones of the –OH, –NH, and –CH absorption bands present appreciable intensity in the NIR region, which means that NIR spectra may contain significant amount of information about the organization of most organic solvents and solutions.
- The smaller absorptivities and larger scattering coefficients at the shorter NIR wavelengths lead to larger scattering/absorption ratios and allow for the development of effective diffuse–reflection analytical techniques.

It is interesting to observe that the reduced intensity of NIR absorptions requires the use of larger optical pathlengths. However, what looks like a disadvantage actually constitutes a significant advantage over MIR and UV spectroscopy for in-line process analysis, as probes can be designed to avoid the formation of stagnant zones in the sampling window of *in-situ* process probes. Similarly, the larger scattering/absorption ratios, usually seen as a disadvantage for spectroscopy analysis, constitute a major advantage for analysis of the morphological characteristics of heterogeneous systems, as in usual polymerization reaction media, and development of effective Raman spectroscopy. Finally, the superposition of overtones and combinations of –XH functionalities in the NIR region allows for the analysis of intermolecular and intramolecular interactions in complex organic media.

Due to the reasons discussed earlier, the NIR spectrum may contain information on the following set of process parameters: composition of complex chemical solutions, molecular configuration and conformation, crystallinity of solid (polymer) samples, anisotropy (from polarization measurements), size and shape of heterogeneous domains,

solution properties (viscosity, degree of molecular association), and so on. Therefore, the rapid growth of the number of practical applications based on the NIRS is not surprising [30, 36]. However, the development of useful NIR applications rely heavily on the use of advanced mathematical and statistical modeling tools, as most times NIR absorption bands cannot be assigned to specific molecular structures in a straightforward manner. Thus, NIRS requires the simultaneous development of both hardware and software technologies.

6.2.3   Measurement Technologies

Although IR and UV spectroscopic methods had been established on sound technical and fundamental grounds since the nineteenth century, NIR and Raman technologies developed very slowly until the 1970s for several reasons. First, efficient NIR detectors became available only after the development of PbS detectors in 1950s. Second, spectroscopists considered the NIR region and the Raman spectra confusing, with many weak and overlapping peaks of numerous overtone and combination bands, making peak assignment very difficult. However, the development of tungsten–halogen filament lamps, of fiber optics technology, and, especially, of computer resources and advanced mathematical tools for the efficient handling of massive data sets led to a fast growth of the NIR and Raman technologies. Efficient lamps and detectors allowed for the development of efficient hardware technologies and significant reduction of noise to signal ratios, while development of mathematical tools allowed for interpretation of very large and complex data sets, through the implementation of modeling and calibration procedures with the help of process computers.

It is important to emphasize that the development of fiber optics technology is a fundamental cornerstone that allowed for the development of real in-line and *in-situ* monitoring spectroscopic techniques, as the sampling device can be placed at very harmful environments, while the spectrometer still sits in a process control room. Without the support of fiber optics technology, samples have to be prepared and placed inside the illuminated chambers (as performed in the lab since the nineteenth century) or pumped through sampling windows (as performed in advanced systems intended for process and product development, such as automatic continuous online monitoring of polymerization reactions (ACOMP) [37–41] in order for spectral data to be obtained.

It is particularly important in this case, the fact that fused silica (or quartz) presents excellent transmittance properties in the NIR region (with much poorer transmittance properties in the MIR, VIS, and UV regions) and is insensitive against water. As a matter of fact, this explains why NIR-based technologies developed much faster in real industrial applications, when compared to MIR, VIS, and UV spectroscopic methods. Efficient transmission of fundamental IR, VIS, and UV radiations through long distances with the help of fiber optics still constitutes an important technological challenge [42].

Different authors reviewed the field of spectroscopy and described available commercial technology and instrumentation [10, 27, 30, 33, 43–45]. The commercial area is not reviewed here because of the fast evolution of the available technology and continuous modification of the commercial scenario. Besides, information provided by spectrometer manufacturers and suppliers must be carefully checked by the analyst, as expressions such as "in-line," "in situ," "accuracy," and "scan speed" are often presented with different meanings and within distinct contexts.

Depending on how the sample is illuminated by the radiation source, commercial instruments can be classified as dispersive or Fourier Transform (FT) spectrometers [25, 27]. Dispersive instruments separate the individual frequencies of energy emitted by the radiation source with the help of a prism or a grating. The detector then evaluates the amount of energy absorbed at each particular wavelength used to illuminate the sample. Fourier Transform instruments were developed in order to overcome some limitations encountered with dispersive spectrometers and particularly to increase the scan speed. In FT spectrometers, absorptions are measured simultaneously for all wavelengths in the interval of interest, so that the original radiation does not have to be separated into its individual components. In order to allow for proper spectral interpretation, a simple optical device called interferometer is used. The technique consists in dividing the original light beam into two distinct optical beams. The first beam illuminates a fixed flat mirror while the second beam illuminates a flat mirror that can be displaced in the space. When the two reflected beams are combined, the beams interfere with each other and produce a signal (interferogram) that contains different amounts of information for each radiation frequency that comes from the source. By collecting absorption data for different mirror displacements, absorptions for the individual wavelengths can be recovered with the help of mathematical deconvolution techniques, such as the Fourier transformation.

In principle, FT spectrometers have several advantages over classical dispersive instruments: faster scanning speeds, higher detector sensitivity (because of the higher intensity of sample illumination), simpler mechanical design, internal wavelength calibration, constant resolution, and negligible stray light. Therefore, FT spectrometers are normally regarded as more accurate and reproducible than dispersive instruments. In spite of that, most instruments available nowadays are of the dispersive type, although new FT–NIR spectrometers have been introduced in the market every year since the mid-1990s. This is because the performance of dispersive instruments is less sensitive to mechanical vibrations and perturbations than FT spectrometers, which makes dispersive instruments more robust for applications in harsh industrial environments. According to manufacturers of FT

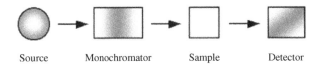

Near-Infrared Transmittance (NIT)

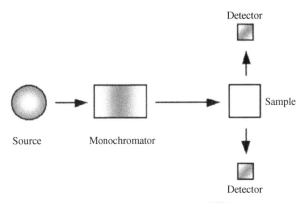

Near-Infrared Reflectance (NIR)

FIGURE 6.6 Representation of transmittance or reflectance spectroscopy.

spectrometers, however, modern FT instruments are as robust as standard dispersive spectrometers.

Depending on how response data are collected, measuring techniques can be normally classified as transmittance or reflectance (or diffuse–reflectance) [10, 23, 25, 27], as illustrated in Figure 6.6. In transmittance measurements, the emitted radiation crosses a sampling window and the transmitted fraction is collected and sent to the light detector. Transmission spectra contain indirect information about absorbed, scattered, and reflected portions of the original illuminating radiation. In reflectance measurements, the emitted radiation is partially reflected at the interface probe/sample and sent to the light detector. Reflection spectra also contain indirect information about absorbed, scattered, and reflected portions of the original illuminating radiation. Transmission measurements are normally indicated for analysis of partially transparent samples, constituted by homogeneous systems or dispersions of large particles in a continuous matrix. Reflection measurements are normally indicated for analysis of opaque samples and for dispersions of small particles in a continuous matrix. However, it is very difficult to present an unambiguous set of rules of thumb for proper selection of the sampling mode, as manufacturers of spectrometers provide very different probe designs for sample analysis (see Fig. 6.7 for typical probe configurations). For instance, transmittance and reflectance signals can be combined in transflectance probes, as illustrated in Figure 6.7. Therefore, preliminary analysis of the performance of the particular spectroscopic technique of interest is strongly recommended whenever an application is sought for spectroscopy.

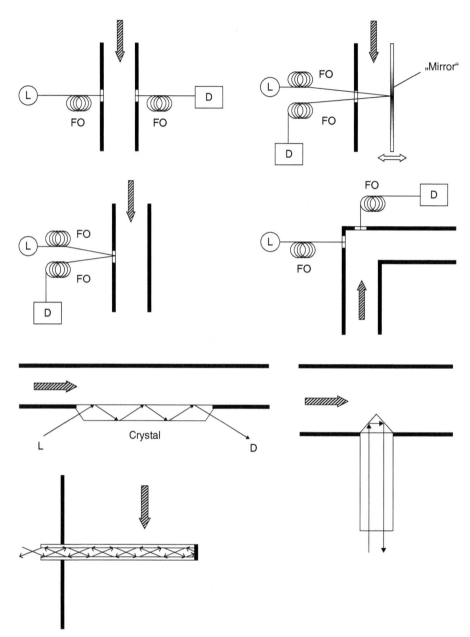

FIGURE 6.7 Typical probes used for NIR spectroscopy. From Hergeth WD. On-line monitoring of polymerization and cure. Sens Update 1999;5:191–242. ©Wiley-VCH Verlag GmbH & Co. KGaA. Reproduced with permission.

6.3 MODEL BUILDING AND CALIBRATION

Given the complex nature of NIR and Raman spectra, it is very difficult to provide *ad-hoc* quantitative interpretation of spectral data. For this reason, careful building and validation of calibration models are of fundamental importance prior to the development of a useful application of modern spectroscopic technology. Therefore, model building and calibration constitute a very important issue for those interested in monitoring and controlling chemical processes with the help of spectroscopic methods.

Calibration models developed for spectroscopic applications are generally represented as linear functions of the measured responses obtained at certain wavelengths. Commonly, multivariate linear regression models in the form

$$y = \boldsymbol{x}^T \boldsymbol{b} + e \qquad (6.7)$$

are used, where y is a process response that must be evaluated from the available spectral data, $\boldsymbol{x}$, and e is a residual, which contains both experimental and calibration errors. If a calibration data set is available and contains different experimental points $(y_i, \boldsymbol{x}_i)$, $i = 1, 2, \ldots, n$, where n is the size of the calibration data set, then the regression coefficients $\boldsymbol{b}$ can be estimated with the help of standard regression techniques, such as multiple linear regression (MLR), principal component regression (PCR), and partial least squares (PLS).

Multiple linear regression is normally employed when the user defines the particular set of wavelengths that must be used for quantitative model calibration. For this reason, MLR is employed when the most significant spectral modifications are concentrated on certain narrow regions of the spectra. This usually happens when a single process response (for instance, monomer conversion in solution homopolymerizations) can be measured as a function of a single process perturbation (for instance, batch time, if it is assumed that initial process conditions are kept constant). Multiple linear regression calibration techniques are used more frequently in the MIR, VIS, and UV spectral regions, because of the easier assignment of observed peaks to molecular structure and medium composition.

Principal component regression and PLS are normally employed when the user does not (or cannot) define the particular set of wavelengths that should be used for model calibration. In this case, the regression technique must provide both the vector of regression coefficients $\boldsymbol{b}$ and the most suitable set of wavelengths to be used for model calibration. Therefore, PCR and PLS calibration models are based on the information available within the whole spectral region defined by the user. These models are commonly employed when process responses (for instance, individual monomer conversions in solution copolymerizations) are functions of multiple process perturbations (e.g., batch time and initial compositions of chemical constituents).

The main challenge concerning multivariate calibration is the construction of an accurate model that relates the n spectra recorded at p wavelengths to the values (for instance, concentration of m constituents in n samples) obtained by independent or reference methods [46]. The calibration procedure starts with the construction of the data matrix $X_{n \times p}$, defined as:

$$X = \begin{bmatrix} x_1^{\mathrm{T}} \\ \vdots \\ x_n^{\mathrm{T}} \end{bmatrix} \qquad (6.8)$$

which contains the spectral responses collected by the NIR instrument for the n calibration samples and a data matrix $Y_{n \times m}$, defined as:

$$Y = \begin{bmatrix} y_1^{\mathrm{T}} \\ \vdots \\ y_n^{\mathrm{T}} \end{bmatrix} \qquad (6.9)$$

which contains the process responses to be evaluated from the NIR spectra.

Equation 6.9 assumes that multiple process responses (m dependent variables) are analyzed simultaneously, although most calibration models are built for single process responses even when monitoring of multiple process variables is sought. In these cases, it is much more common to develop different calibration models for each particular process response than to build a single calibration model for the whole set of process variables. Usually, Y contains a set of properties obtained through off-line analyses with the help of standard laboratory techniques, such as gravimetry, titration, chromatography, viscosimetry, conductivimetry, nuclear magnetic resonance (NMR), and light scattering, among others [47].

Surveying the many different techniques used for the development of calibration models and the many distinct applications of calibration models is not sought here. Detailed reviews in the area are available elsewhere [46–49]. However, it is important to mention here that an overview of the recent literature indicates that calibration models are almost always based on MLR, PCR, and PLS techniques, regardless of the particular field of application. (Illustrative examples of actual applications are presented in the following sections.)

Attempts to use alternative nonlinear calibration models have been reported in few studies. Among them, the usual approach is the splitting of the calibration region into different domains where distinct linear calibration models are built [49–54]. Neural network (NN) calibration models have also been used [55–57], but the obtained results were not proven to be better than the results obtained with the more traditional linear techniques.

The near infrared spectra may be subject to perturbations caused by a very large number of factors, such as temperature fluctuations, scattering in turbid samples, alignment of gratings, mirrors and lamps, mechanical vibrations due to stirring, among others. For this reason, spectral data are usually preprocessed before the development of calibration models in order to reduce the possible influence of extraneous factors on model performance. In fact, preprocessing of spectral data can be regarded as one of the most important steps during model building and therefore must be carefully taken into consideration by the analyst.

Numerous techniques have been developed to improve the robustness and transportability of spectral calibration models [58–68]. Among them, the computation of first and second derivatives of spectral data with respect to the wavelength and the application of multiplicative scatter corrections (MSCs) certainly are the most popular ones. Derivatives of spectral data allow for efficient removal of baseline shifts and for magnification of small changes of the spectral signal. However, derivation of spectral data could increase the effects of noisy measurements. MSC was developed to reduce the effect of scattered light on diffusion and transmission spectra, but can also be used to remove varying background spectra with nonscattering origins [68, 69].

Some very interesting practical issues related to building of calibration models have been discussed in the literature [70, 71]. First, the usual belief that spectral models can never perform better than the primary reference technique used for calibration is not correct. Actually, spectral models usually perform much better than the original reference technique, as model calibration filters experimental errors committed during the preparation and characterization of the reference samples. Therefore, spectral models can be used to "correct" the reference values obtained off-line [71]. Another interesting point is that linear spectral models present nonlinear properties and do account for nonlinear effects [70]. This is because the spectral data respond nonlinearly to perturbations of the operation conditions. For instance, the spectra of chemical mixtures are not weighted sums of the spectra of the individual constituents, but nonlinear transformations of the individual spectra due to occurrence of chemical and physical association effects. Thus, the relationship between the calibration model and the properties of the system is nonlinear, due to the nonlinear transformation of the spectral signal. This probably explains why the use of nonlinear models for spectral calibration has not led to significant improvement of calibration models.

6.3.1 Calibration Techniques

The fundamentals of some of the most common techniques used for model building are discussed in the following sections. Detailed description of the techniques can be found in the literature [46, 49].

6.3.1.1 *Multiplicative Scatter Correction (MSC)* The main assumption used to perform the MSC is that the different spectra collected for a certain sample contain a broad and changing background from differential scattering at each wavelength. Thus, different spectra may contain different information for the same analyzed sample.

Assuming that spectral changes occur at random around the mean spectrum and are multiplicative, as predicted by scattering theories, one can write

$$x_i = \alpha_i \bar{x} + \beta_i \mathbf{1} + e_i \qquad (6.10)$$

where $\bar{x}$ is the mean spectrum, α_i and β_i are model parameters, which can be calculated with the help of standard linear least squares procedures, and $\mathbf{1}$ is a vector of $\mathbf{1}$s. Therefore, the corrected spectrum can be represented as

$$x_i^{\mathrm{MSC}} = \frac{x_i - \beta_i \mathbf{1}}{\alpha_i} \qquad (6.11)$$

where the superscript MSC stands for the MSC corrected spectrum.

This very simple preprocessing technique may allow for very significant improvement of calibration models and quantitative spectral interpretation [68, 69].

6.3.1.2 *Multiple Linear Regression (MLR)* The MLR procedure consists in minimizing the observed differences between experimental data and model predictions, as defined by the quadratic function

$$S = \sum_{i=1}^{n} \sum_{j=1}^{m} \left(y_{ij}^{e} - y_{ij}^{c} \right)^2 \qquad (6.12)$$

where the superscripts e and c stand for the experimental and calculated values, respectively.

Assuming that Equation 6.7 can be used for model calibration, then the usual least squares procedure leads to the following solution [72]

$$B = (X^{\mathrm{T}}X)^{-1} X^{\mathrm{T}}Y \qquad (6.13)$$

where B is the matrix of model parameters, defined as

$$Y = XB = X \begin{bmatrix} b_1 & \ldots & b_m \end{bmatrix} \qquad (6.14)$$

Equation 6.13 shows that a solution can only be obtained if spectral data are linearly independent; otherwise, the matrix $(X^{\mathrm{T}}X)$ cannot be inverted. Inversion problems can also arise if the matrix $(X^{\mathrm{T}}X)$ is ill conditioned. In such cases, special numerical techniques should be used to provide the inverse of $(X^{\mathrm{T}}X)$. For this reason, MLR is a method indicated for well-posed problems. This method may fail if process outputs respond nonlinearly to spectral changes or if the characteristic wavelengths in X are not selected properly [46, 49].

6.3.1.3 *Principal Component Regression (PCR) and Partial Least Squares (PLS)* For PCR and PLS calibration model building, it is assumed that spectral data can be decomposed as

$$x_i \approx \sum_{j=1}^{c} t_{ij} p_j \qquad (6.15)$$

where c is the number of independent components p_j used for spectral decomposition and t_{ij} are the projection coefficients of x_i on to p_j. Equation 6.15 can be written as

$$X = \begin{bmatrix} x_1^{\mathrm{T}} \\ \vdots \\ x_n^{\mathrm{T}} \end{bmatrix} \approx \begin{bmatrix} t_{11} & \cdots & t_{1c} \\ \vdots & \ddots & \vdots \\ t_{n1} & \cdots & t_{nc} \end{bmatrix} \begin{bmatrix} p_1^{\mathrm{T}} \\ \vdots \\ p_c^{\mathrm{T}} \end{bmatrix} = \begin{bmatrix} t_1 & \cdots & t_c \end{bmatrix} \begin{bmatrix} p_1^{\mathrm{T}} \\ \vdots \\ p_c^{\mathrm{T}} \end{bmatrix} = TP^{\mathrm{T}}$$

$$(6.16)$$

where p_i and t_i are normally called as the ith loading vector and the ith score, respectively.

The loading vectors, also called principal components, can be regarded as "fundamental spectra," while the scores can be interpreted as "the number of fundamental spectra" needed to reconstruct the actual spectrum. Thus, the same "fundamental spectra" are used to represent all spectra in the set, while the number of "fundamental spectra" varies from spectrum to spectrum (and thus from sample to sample).

The maximum number of loadings equals the number of spectra in the data set ($c \leq n$). This leads to the obvious representation $x_i = p_i$. However, most of the spectral variation can usually be described in terms of few loadings, which means that the dimension of spectral variations is usually much smaller than the dimension of x_i (the total number of wavelengths analyzed and recorded in the spectrum).

Equation 6.16 can also be written as [73]

$$X = U \wedge V^T \tag{6.17}$$

where the columns of U contain the eigenvectors of $(X^T X)$, the columns of V contain the eigenvectors of (XX^T), and the diagonal matrix Λ^2 contains the eigenvalues of either matrix.

If Equations 6.16 and 6.17 are compared to each other, one may conclude that the columns of P contain the eigenvalues of (XX^T), while the columns of T contain the scaled eigenvectors of $(X^T X)$. If the eigenvalues (and respective eigenvectors) of $(X^T X)$ are ordered in a list of decreasing values, then very good representation of X is possible by retaining the leading eigenvalues (and respective eigenvectors) of the list only. Therefore, the traditional principal component analysis (PCA) consists in computing the leading eigenvalues (and respective eigenvectors) of $(X^T X)$ and (XX^T) in order to obtain a proper representation of X in a space of lower dimension, where calibration model building must be performed.

The number of selected eigenvalues (and eigenvectors) depends on the required precision and on the noise level. The higher the precision required for representation of X, the larger the number of eigenvalues needed in Equation 6.16. The larger the noise level, the smaller the number of eigenvalues recommended for representation of X, as nonsignificant effects must be discarded.

If Equation 6.7 is written as

$$Y = (TP^T)^T B + E \tag{6.18}$$

which means that X is replaced by its projection (TP^T), then the minimization of Equation 6.12 through the usual least squares procedure also leads to the regression coefficients in Equation 6.13, where X must be replaced by its projection (TP^T).

This is the well-known PCR procedure. It is important to observe that model predictions can only be obtained if the experimental spectrum x_i is projected onto P first. It is also important to observe that the user does not have to select particular wavelengths to build PCR calibration models. The relative importance of each particular region of the spectrum for the quantitative interpretation of the process responses is automatically recorded in P. This contributes with significant increase of the robustness and predictive capacity of calibration models.

PLS is closely related to PCR. During the decomposition of X, PCR is not influenced by the observed process responses Y. However, the first factors obtained through PCA, which present the maximum spectral variation, are not necessarily the most useful for model calibration. For instance, if assume that a single factor is used for model building, then, using Equation 6.16 to decompose the original independent spectral data

$$X = t_1 p_1^T + E_1 \tag{6.19}$$

and

$$y = t_1 b + e_1 \tag{6.20}$$

However, from Equation 6.19

$$t_1 = \frac{Xp_1}{p_1^T p_1} \tag{6.21}$$

Assuming that the size of p_1 is equal to 1, then Equation 6.20 can be written as

$$y = bXp_1 + e_1 \tag{6.22}$$

so that the least squares solution becomes

$$(X^T X)(bp_1) = X^T y \tag{6.23}$$

However, the matrix $(X^T X)$ cannot be inverted, as the spectral dimension (number of columns of X) is usually much larger than the sample dimension (number of lines of X). Equation 6.23 shows that the direction p_1 that leads to the best description of the process responses depends simultaneously on X and y. Equation 6.23 also shows that a numerical solution has to be sought, as a closed analytical solution is not available.

During the calibration step, the PLS technique assumes that the spectral data set X can be decomposed in the form of Equation 6.16. PLS factors are then computed with the help of iterative numerical procedures, such as the popular nonlinear iterative partial least squares (NIPALS) algorithm, as described in standard texts [46, 74, 75]. The PLS factors can be regarded as rotations of the PCA factors computed in order to maximize the correlation with the process responses, as shown in Equation 6.23. Thus, PLS achieves comparable calibration accuracy with fewer factors in the calibration model, when compared to models obtained with PCR.

6.3.1.4 *Nonlinear Calibration Approaches* Spectral data can respond nonlinearly to process perturbations due to deviations of the Lambert–Beer law, to the nonlinear characteristics of light detectors or to interactions among analytes. Sources of nonlinear behavior and techniques for the detection of important nonlinear effects in spectral responses have been discussed in the literature [25, 76]. In order to cope with the nonlinear features of spectral data sets, different approaches have been applied to build calibration models. These calibration approaches have almost always been based on NN models and locally weighted regression (LWR) models.

NN models are nonlinear mathematical structures built by summing up iteratively nonlinear transformations of linear combinations of certain input variables. NN models can assume many different configurations. In the simplest case, usually called as the feed-forward NN structure, three different layers are employed (Fig. 6.8): the input layer, the hidden layer, and the output layer. The input layer is fed by values of a number of input variables, generally the spectral data measured at certain wavelengths. The output layer provides the desired process response. The backpropagation procedure is normally used to estimate the NN model parameters [77]. The nonlinear transformation generally used at each particular node of the NN model is a sigmoidal activation function, defined as

$$f(x_1, x_2, \ldots, x_n) = \frac{1}{1 + e^{(w_1 x_1 + \cdots + w_n x_n + b)}} \quad (6.24)$$

where w_i and b are the model parameters to be determined.

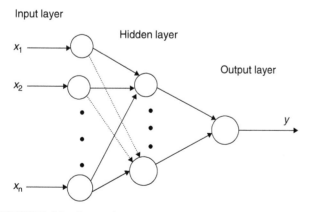

FIGURE 6.8   A neural network structure for a multiple input and single output problems.

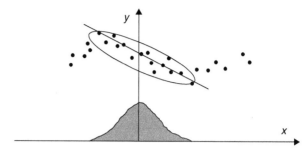

FIGURE 6.9   Schematic representation of the LWR procedure.

Although NN models are capable of describing very complex nonlinear behaviors, as shown extensively in the literature, the use of NN models for spectral calibrations has not led to any significant improvement of model responses [55–57]. Nevertheless, numerous successful attempts have been made to use NNs for the interpretation of spectral data [78–80].

LWR is a memory-based method that explicitly retains the original calibration data set and uses it whenever a prediction is needed. Predictions are obtained by performing model regression around a particular point of interest, using training data that are "local" to that point [81]. LWR can easily cope with nonlinear effects when these effects can be locally approximated with a simple model [51, 54, 69].

Figure 6.9 shows a schematic diagram of the LWR method.

6.4   MONITORING AND CONTROL OF POLYMERIZATION REACTIONS

Spectroscopic techniques have been employed extensively for monitoring and control of processes in different fields. Since a detailed review of the applications of spectroscopic techniques in distinct areas is certainly beyond the objectives of the chapter, the interested reader should refer to textbooks and surveys for additional details [10, 27, 30, 33, 43, 44]. It is also important to emphasize that most publications available in the field of polymer and polymerization reactions make use of spectrometers for off-line characterization of polymer properties. Typical applications include identification of polymer materials [82], evaluation of copolymer and polymer blend compositions [83, 84], evaluation of monomer and polymer compositions during polymerizations [85], determination of additive content in polymer samples [86, 87], and estimation of end-use properties of polymer materials. End-use properties analyzed include the degree of crystallinity of polymer samples [88], the degree of orientation of polymer films [85], the hydroxyl number of polyols [89], the melt flow index of polymer pellets [90], and the intrinsic viscosity of polymer powders [91], the morphology of polymer pieces and of heterogeneous polymer materials [92, 93], ageing properties of elastomers [94], and composition of copolymers [95, 96] and polymer blends [97].

Relatively few applications have been reported for the online monitoring and control of polymerization processes based on spectroscopic techniques. Within this relatively small group, most studies have focused on the final processing of the polymer resin [98–101], not on the polymer production stage. In addition, the literature describing actual implementations of control schemes based on spectroscopic data is scarce. Despite that, the number of studies that report actual *in-situ* and online monitoring and control of reaction vessels is growing rapidly, with applications in solution, emulsion, and suspension polymerizations, for both step growth and chain reactions, and including the monitoring and control of compositions, molecular weight averages and average particle sizes. These results are reviewed in the following sections.

It is important to emphasize, though, that successful applications and real experimental results have been reported in the open literature only for small lab-scale reactors. Therefore, the detailed analysis of benefits and drawbacks of spectroscopy-based technology at plant site is not possible yet. One of the major problems regards the fouling of probes during real and long operation periods. This problem has not been discussed in the open literature, because fouling is not a real problem in most laboratory applications, as reaction batches are relatively short and probes can be cleaned between successive experiments. It is also well known that externals can promote the formation of polymer agglomerates in heterogeneous reactions. This has not been analyzed either, because agglomerates are usually formed during long reaction periods. Finally, an important issue regards the stability of spectrometers at plant site and sensitivity of spectral data to fluctuations of process parameters. Frequent changes of internal spectrometer components, such as probes, lamps and detectors, can disturb the performance of calibration models (and therefore of monitoring and control techniques) very significantly, and may lead to rejection of the technique in an industrial environment. More important, the fluctuation of operation conditions in large reaction vessels, possibly due to modifications of the feed characteristics, to the unavoidable existence of mechanical vibrations, and to the internal heat and mass flow properties of the reacting mixture, can disturb the collection of spectral data and the performance of calibration models. This problem has not been addressed either, as the laboratory environment is usually much more stable and much less aggressive than the industrial site. For the reasons cited earlier, despite the promising features of spectroscopy-based technology for monitoring and control applications in the polymerization field, it is certain that a lot of work has yet to be done at plant site prior to drawing of a definite picture of the technology.

6.4.1 Solution and Bulk Polymerizations

Solution and bulk polymerization processes are widely used for production of polymer resins. In typical solution and bulk processes the initiator (or catalyst), the comonomers, and the final polymer are soluble in the reaction medium, which may contain a solvent (solution process) or not (bulk process). Thus, reaction proceeds in a homogeneous reacting mixture. The main advantages of solution processes are related to the low viscosity and homogeneous properties of the reaction medium, while the main disadvantages are related to the lower reaction rates, productivities, and molecular weight averages of the final product, when compared to bulk and emulsion processes. Solution and bulk reactions can be performed in batch, semibatch, and continuous operation modes, in many distinct reactor configurations. Reactions can follow step or addition mechanisms [102].

The main process control challenges in solution and bulk polymerizations are the control of molecular weight averages [103], molecular weight distribution (MWD) [104], monomer conversion [105], and copolymer composition and copolymer composition distribution (CCD) [106, 107]. The development of mathematical models for solution and bulk processes is relatively easy, as it is not necessary to take into account mass transfer between different phases. The main challenge in this field is the proper description of the glass and gel effects, which is usually performed with the help of empirical models [108]. Therefore, reliable process models are usually available for solution and bulk processes.

The use of the spectroscopic technology for monitoring and control of monomer conversion and copolymer composition is straightforward, as spectral data are normally sensitive to changes of composition. However, monitoring and control of molecular weight, MWD, and CCD without the help of a process model is a controversial subject. In principle, molecular associations that can be modified by the size and internal structure of polymer chains can lead to modifications of spectral data; therefore, calibration models can probably be developed for the first moments of MWD and CCD, allowing for implementation of monitoring and control schemes of molecular weight averages, MWD, and CCD. However, there is no study available in the literature showing unequivocally that chain size and structure can be monitored with the use of spectroscopy without a process model. It is important to observe that the independent evaluation of the leading moments of MWD and CCD should be based only on the collected spectral data, and not on the history of the polymerization. Development of calibration models that use initial feed compositions and feed flowrate and temperature profiles as inputs for prediction of the first moments of the MWD and CCD should be regarded as attempts to replace the detailed process model for an empirical predictive structure.

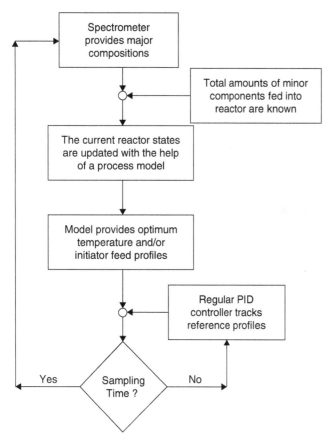

FIGURE 6.10 Proposed scheme for control of monomer conversion and Mw in free-radical solution polymerizations.

Therefore, the reported monitoring and control schemes can be represented generically as depicted in Figure 6.10.

6.4.1.1 *Process Monitoring* It seems that Long et al. [109] were the first to report the successful use of NIR spectral data for in-line and *in-situ* evaluation of monomer conversion in a polymerization process. The authors studied styrene and isoprene polymerizations in solutions of tetra-hydrofuran and cyclohexane, using anionic catalysts. Spectral data were obtained with a dispersive instrument, using standard transflectance probes. Monomer conversion of the vinyl protons in the monomer to methylene protons in the polymer was easily monitored with the help of MLR calibration techniques when the solids content was in the range of 10–20% in weight. However, NIR spectral data were used for off-line kinetic studies and off-line estimation of kinetic constants for homopolymerizations and copolymerizations. The use of NIRS was justified because of the high sampling frequencies required to monitor the fast reactions, as batch times were always smaller than 20 min. Based on preliminary data, the authors suggested that NIRS could be used to detect the sequence distributions for tapered block copolymers, the geometric isomer content, and reactivity ratios for free-radical copolymerization, but

it seems that additional conclusive results have never been presented.

Almost simultaneously, Aldridge et al. [110] used NIRS to monitor monomer conversion *in situ* and in real time in bulk methyl methacrylate (MMA) polymerizations performed inside a mold. The authors used a dispersive instrument to observe the spectral changes at 868 and 890 nm, related to the disappearance of the third overtone of the CH stretch of the vinyl group and the concomitant appearance of the CH of the methylene group. Spectral changes were monitored with the help of the second-derivative spectra of the samples and posterior subtraction of the initial monomer spectrum. Simple linear calibrations were performed and shown to be in good agreement with conversion data obtained with independent procedures.

De Thomas et al. [111] studied the production of polyure-thanes and showed that NIRS can be used successfully to monitor the course of the reaction in real time. Spectral data were obtained with a dispersive instrument, using standard transflectance probes. An MLR model was derived for the quantitative determination of isocyanate concentrations during the urethane polymerization reaction. Model predictions were used to build statistical process control charts and to detect trends along the polymerization reaction. The authors suggested that the integration of NIRS with process control routines could lead to improvements of product quality and consistency, while minimizing reaction time. However, model predictions were not used as feedback information for any sort of correction of the process trajectory. Similar studies were performed by Dallin [112] for prediction of the acid number during the production of polyesters.

Barrès [113] used NIRS to monitor the chemical modification (catalyzed esterification) of a molten ethylene–vinyl alcohol copolymer by octanoic acid in a twin screw extruder. Extrusion samples were characterized off-line, for calibration purposes, with the use of ^{1}H-NMR and residual free acid titration. Multivariate PLS calibration was performed successfully to predict the analyte concentrations.

Although the use of dispersive NIR instruments has been reported more frequently in the literature, FT–NIRS has also been used to monitor solution polymerization reactions. The free-radical styrene polymerization in super-critical carbon dioxide was studied by Beuermann et al. [114] using FT–NIR spectrometry. The free-radical bulk copolymerization of styrene and divinylbenzene in thin pirex tubes was studied by Zetterlund et al. [115] using FT–NIR spectrometry for online data acquisition. Wang et al. [116] studied the isothermal curing of epoxy resins in thin films using FT–NIRS and reviewed the use of NIRS for kinetic studies of epoxy curing processes. Obtained spectral data were used for building of kinetic models and estimation of model parameters off-line. Lin and Stansbury [117] used real-time FT–NIRS to monitor monomer and water concentrations simultaneously during cationic vinyl ether photopolymerizations. The area of the NIR peak was used to monitor the system composition, as calibrated by gravimetric analysis. This approach allowed for analysis of the moisture influence on the polymerization reactions. Darcos et al. [118] also used FT–NIRS to monitor monomer conversion during copper-mediated living radical polymerizations, with the help of an inert fiber–optic probe. In all cases, FT–NIR results correlated excellently with results obtained by ^{1}H-NMR and were found to be more accurate because it provided many more data points and avoided sampling during the polymerization reaction.

Raman-NIR spectrometers have also been used frequently to monitor solution polymerization reactions. Haigh et al. [119] reported the construction and use of reaction cells for the in-line evaluation of composition during solution polymerizations using FT-Raman spectroscopy in the NIR region. Copolymerization reactions of styrene and vinyl imidazole in 1-4-dioxane were studied and results were used for kinetic model building. However, the reaction cells and techniques could be inadequate for actual control applications, because it is necessary to circulate the polymer solution through a thin glass tube illuminated by the light source. Moreover, calibration MLR model predictions and independent NMR composition data were shown to be in disagreement, which indicates that the developed FT-Raman technique was not completely ready for use in real reaction systems at that time. Clarkson et al. [120] and Damoun et al. [121] have used similar techniques and reaction cells to study the homopolymerizations of styrene and acrylates and the homopolymerization of MMA. Urlaub et al. [122] also studied the bulk polymerization reaction of cyanacrylates in thin films in-line using NIR-Raman spectrometers. Similar comments can be made about the limitations of the experimental apparatus used in all cases for actual monitoring and control of polymerization reactions.

The development of Raman techniques that do not require the circulation of the polymer solution through illuminated sampling cells increased the general interest in the use of the spectroscopic techniques for the purposes of process monitoring and control. De Clercq et al. [123] used FT-Raman spectroscopy in the NIR region to study the solution polymerization of norbornene in dichloromethane, using a single-site heterogeneous metathesis catalyst. Spectra were acquired by placing the probe at a short distance from the reactor during the reaction. The laser light was focused into the reaction mixture through the glass reactor wall and an optimum distance was determined by adjusting the probe distance in order to maximize the intensity of the Raman signal.

Although used less frequently, NIR-based reflection spectroscopy has also been reported for monitoring of polymerization reactions. Scherzer et al. [124, 125] used NIR reflection spectroscopy to monitor the conversion of double bonds in acrylate coatings after irradiation with UV light or electron beams. Quantitative analysis of the spectroscopic data was performed with linear calibration models, using off-line FT–IR and high-precision liquid chromatography data as references.

Monitoring of solution polymerization reactions has also been performed with the help of spectral data collected in the IR region. As a matter of fact, the improvement of the fiber optics transmission technology in the IR region has led to the steady increase of the number of applications in the IR region. Puskas et al. [126] used Attenuated Total Reflection (ATR) probes and FT–MIR spectrometers for *in-situ* monitoring of living carbocationic polymerizations of isobutylene and styrene in solutions of methyl chloride and methyl cyclohexane. The MLR calibration models were developed for monomer conversions and shown to be in excellent agreement with off-line data obtained through conventional gravimetry. Results were used off-line for kinetic model building and estimation of kinetic constants. The obtained kinetic constants were used to predict the molecular weight averages with the help of detailed mechanistic models, which were shown to be in good agreement with experimental data obtained independently.

After Puskas et al. [126], Yu et al. [127] used Raman FT–IR spectroscopy to monitor the polymerization of acrylic acid (AA) in aqueous solution in sampling cells. Pasquale and Long [128] reported the use of real-time IR spectroscopy to monitor monomer conversion in the solution controlled free radical polymerization of styrene initiated by benzoyl peroxide in the presence of TEMPO. A kinetic model was built and apparent rate constants were calculated off-line with the spectral data. Roper et al. [129] studied the effect of the chemical structure on the reactivity of alkenes used in thiol-alkene photopolymerizations with real-time transmission IR spectroscopy. Photoreactions were conducted in sampling cells through the sandwiching of the samples between two sodium chloride salt plates at a thickness of approximately $20\,\mu m$. IR absorption spectra were obtained under continuous UV irradiation. The characteristic IR absorbance bands were used to monitor monomer conversions. Messman and Storey [130] studied the kinetics of ring opening polymerizations for *rac*-lactide measured with *in-situ* FT–IR spectroscopy, using ATR probes. Simple linear calibration models were built with independent size exclusion chromatograph data for monitoring purposes. The method was used to study the effects of different experimental variables, such as the initiator and catalyst concentrations, on the rate of polymerization. The carbocationic solution copolymerization of isobutylene and styrene was also investigated by Puskas et al. [131] using real-time FT–IR monitoring. Sahre et al. [132] used ATR–FT–IR spectroscopy to monitor the reaction of isocyanates and amines, which leads to hyperbranched poly(urea-urethane)s. It was possible to verify that the individual reaction steps of this complex polyreaction could be monitored efficiently with the proposed technique. Mabilleau et al. [133] studied the solution polymerization of 2-(hydroxyethyl)methacrylate using a redox initiation system. The kinetics of the reaction was investigated in sampling cells with Raman microscopy in the IR region. As observed experimentally, calibration could be performed with simple linear models. Tracht et al. [134] used transmission IR spectroscopy to monitor the catalytic activity in different solution cationic polymerizations.

Studies regarding the monitoring of solution polymerization reactions in the UV–VIS spectral region are scarce, because of the radiation transmission problems. Santhosh et al. [135] investigated the electrochemical copolymerization of diphenylamine (DPA) with ortho-toluidine (OT) in sulphuric acid medium by cyclic voltammetry. *In-situ* UV–VIS spectroelectrochemical studies on copolymerization were carried out in sampling cells using indium tin oxide (ITO)-coated glass plate as working electrode for different feed ratios of DPA and OT. The molar compositions of the copolymers were determined based on the UV–VIS spectra and used for the determination of reactivity ratios.

In some applications, a combination of different spectral technologies was employed to provide improved performance of the monitoring system. For example, Florenzano et al. [136] reported the use of *in-situ* NIRS coupled with the ACOMP for the simultaneous monitoring of weight average molecular mass M_w, intrinsic viscosity, and monomer conversion, during methyl MMA polymerization. An advantage of *in-situ* NIR is that it furnishes immediate information on the conversion in the reactor, whereas ACOMP relies on continuous withdrawal and dilution of a small stream of reactor fluid. Simultaneous monomer conversion data from *in-situ* NIR and from ACOMP (derived from both refractive index and UV absorption) were compared and found in good agreement.

It seems that Puskas et al. [126] were the first that coupled the spectral data and a process model to predict the average molecular weight of the final polymer material. However, predictions of molecular weight data were performed off-line with the help of the kinetic models. Cherfi and Févotte [137] and Cherfi et al. [138] used dispersive NIRS to monitor monomer conversion and weight-average molecular weight of the final polymer material produced during MMA polymerizations in solutions of toluene. Immersion transmission probes were used for sampling and calibration models were obtained with PLS techniques. NIR spectra used for calibration were recorded during batch and semibatch reactions. Off-line gravimetry and Gel Permeation Chromatography (GPC) were used as reference methods to provide the conversion and the average molecular weight data sets required for calibration. Model predictions were then validated for different operation conditions in both batch and semibatch modes. Standard errors for prediction of monomer conversion and weight-average molecular weight were equal to 2.6% (absolute) and 5.7% (relative), respectively. These two studies extended previous results reported by Olinga

et al. [139], who had monitored the MMA homopolymerization in toluene in a more limited set of operation conditions using FT–NIR transmission spectroscopy.

It is very important to mention that the structure of calibration models used to predict the weight-average molecular weight of the final polymers were not reported by Cherfi et al. [138]. Based on the authors' analysis, it seems that the initial feed compositions and feed profiles were used for model calibration. Therefore, it seems that the calibration model developed for M_w was actually replacing the detailed process model required for prediction of molecular weight averages, as discussed previously. This means that the model developed for M_w would be unable to respond to process perturbations not included in the input data set and that independent evaluation of M_w based solely on the NIR spectra was not possible.

Lousberg et al. [140] used an FT–NIR spectrometer to monitor the bulk styrene polymerization, using a flow cell for sampling. Because of the intrinsic limitations of the flow cell, the maximum analyzed monomer conversion was equal to 35%. Partial least squares was used to build the calibration model, leading to average calibration errors of 0.32% (absolute). The model performance was insensitive to variations of M_w, indicating that the NIR spectra could not be used for prediction of molecular weight averages.

6.4.1.2 Process Control   The first attempt to use NIRS to control a solution polymerization process was presented by Fontoura et al. [141]. The authors studied the free-radical styrene polymerization in toluene and aimed to control monomer conversions and weight-average molecular weight during semibatch reactions. First an MLR calibration model was built for monomer conversion, using online spectra and off-line conversion data collected during a couple of polymerization tests. Off-line conversion values were obtained through standard gravimetrical analysis. Kalman filter techniques were then implemented to allow for the estimation of the molecular weight averages, using an accurate mechanistic process model as reference. The initial feed conditions, the feed rate profiles, and the temperature profiles were used with the spectral data as inputs of the Kalman filter. Standard GPC analysis was used as reference technique for calibration of the Kalman filter. A model-based predictive control strategy was devised and used to control the monomer conversion and the molecular weight averages, through manipulation of monomer feed flow rates, initiator feed flow rates, and/or reaction temperature profiles. According to this procedure, monomer concentrations obtained with the NIR model were used for recomputation of optimum feed rate and temperature profiles at each sampling time, which were then used as references for the controller. Very good results were achieved for both monomer conversion and M_w, indicating that NIRS can be combined efficiently with process models for actual and complex control applications.

Nogueira et al. [142] applied NIRS to monitor and control the synthesis of high molecular weight polyurethanes in solutions of dimethyl formamide. The polyurethane synthesis follows step-growth reaction mechanisms and is usually performed in two steps. A polymer of low molecular weight is produced during the first reaction step and used as reagent during the second reaction step, when a chain length extensor is added to the system. The polymer properties depend on the average molecular weight of the final polymer material and on the average size of the oligomer blocks prepared during the first reaction stage. Typically, reaction recipes are prepared in the lab (or though evolutionary procedures) and implemented at plant site without any sort of feedback for correction of process drifts and/or perturbations.

It is well known that conversion of chemical groups and chain sizes are correlated strongly in condensation reactions [102]. It is also well known that the final average size of polymer chains obtained through condensation reactions are extremely sensitive to small variations of feed composition [102]. Therefore, controlling the composition and average molecular weight of the final polymer may not be trivial. Nogueira et al. [142] proposed the simultaneous use of an NIR spectrometer and a shear metering device for simultaneous control of polymer composition and molecular weight, as shown in Figure 6.11. In this case, the NIR spectrometer would be used to monitor and control the composition of the

reaction medium during the first reaction stage. Compositions can be followed and controlled closely with the help of calibration models, through manipulation of the initial feed compositions. When the desired composition (monomer conversion) is reached, the NIR spectrometer can provide the correct amount of extensor required to meet the M_w specification. The control activity is then transferred to the shear instrument, which is used to monitor and control the evolution of the molecular weights, through manipulation of feed rates of extensor and/or reticulation agents.

Othman et al. [143, 144] used NIRS to control a polymerization reactor and produce polymer solutions with well-defined monomer conversions and molecular weight averages. Calibration models for monitoring of both the average polymer molecular weight and the concentration of monomer were developed with the PLS calibration technique. With the help of a process model, a nonlinear input–output linearizing geometric controller was developed and applied to control the polymer molecular weight by manipulating the inlet flow rate of the monomer. The control strategy was validated online during the solution polymerization of AA in an industrial pilot scale reactor.

Raman spectroscopy was also used for the control of monomer compositions in solution polymerizations. Van den Brink et al. [145] studied the solution polymerization of n-butyl acrylate in dioxane and monitored the monomer concentration by online Raman spectroscopy. It was shown that it is possible to control the monomer to solvent ratio with the help of a feedback monomer addition strategy, while the spectroscopic data are analyzed in real time. The proposed control strategy was validated by disturbing a semicontinuous polymerization process on purpose and verifying that the process could be operated without abrupt changes in the monomer to solvent ratio and the monomer flow rate.

6.4.2 Suspension Polymerization

Suspension polymerization processes are widely used for the production of polymer beads. In typical suspension processes, an organic phase constituted by initiator (or catalyst), comonomers, and the final polymer are suspended in an aqueous phase, which contains additives and residual monomer. Thus, reaction proceeds in a heterogeneous reacting mixture. The main advantages of suspension processes are easy purification of the polymer material, the low viscosity of the reaction medium, and the reduction of the effective heat of reaction, as water absorbs significant amounts of the heat released by the reaction. The main disadvantages are the lower productivities (when compared to bulk processes) and the sticking characteristics of the suspended polymer, which explains why continuous suspension processes have not been used commercially so far. Reactions can follow step or addition mechanisms [102].

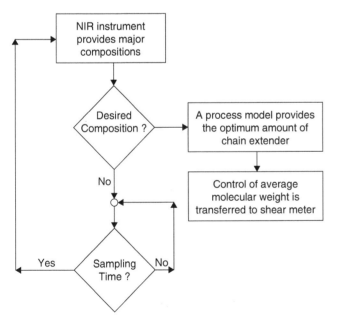

FIGURE 6.11 Proposed scheme for control of monomer conversion and M_w in solution polycondensations. From Nogueira ES, Borges CP, Pinto JC. In-line monitoring and control of conversion and weight-average molecular weight of polyurethanes in solution step-growth polymerization based on near infrared spectroscopy and torquemetry. Macromol Mat Eng 2005;290: 272–282. ©Wiley-VCH Verlag GmbH & Co. KGaA. Reproduced with permission.

Besides the challenges in control described previously for solution processes, the control of the average particle sizes and particle size distributions (PSDs) of final polymer particles is also of fundamental importance, as compounding, processing, and bulk-handling properties are affected by these variables. The development of mathematical models based on mass and energy balances for suspension processes is relatively easy, as the suspended organic droplets may be regarded as small bulk reactors. Therefore, reliable process models are usually available for suspension processes. However, the proper description of PSD of polymer particles may be extremely complex, as rates of droplet breakage and droplet coalescence vary tremendously along the reaction course. When compared to other polymerization processes, the literature regarding the control of suspension processes is scarce. Most control procedures developed in the field regard the off-line computation of optimum temperature profiles for reaching specified values of monomer conversion and molecular weight averages at the end of the batch [146]. Yuan et al. [147] and Kiparissides [5] present detailed descriptions of typical suspension processes and of their characteristic control challenges.

6.4.2.1 *Process Monitoring and Control*   The use of the spectroscopic technology for monitoring and control of suspension processes may not be straightforward, as spectra can be very sensitive to changes of composition (due to absorption) and particle sizes (due to scattering). Santos et al. [55, 148] showed that changes of the agitation speed and of the chemical composition can lead to marked changes of the spectral signal in the NIR region. Particularly, Santos et al. [55] showed that NIR spectral responses to changes of the agitation speed present hysterisis when suspending agents are added to the suspension medium, which is a clear indication of independent evaluation of the suspended particle sizes. Therefore, depending on the characteristics of the analyzed chemical system and on the particular set of selected operation conditions, good calibration models can be developed for both chemical compositions and average particle sizes or cannot be developed at all for any property of interest. For this reason, it cannot be said that the acquisition of spectra will be useful for a particular control application in a particular suspension process without evaluation of actual spectral data and attempts to develop a calibration model. This probably explains why the number of spectroscopic applications in this field is so small, when compared to the field of solution polymerizations.

Santos et al. [55, 148] studied the styrene suspension polymerization using NIRS. A dispersive NIR instrument equipped with a transflectance probe was used to monitor the reaction course and very good PLS calibration models were developed for the final average particle sizes. Calibration models were then used as references for the implementation of an in-line procedure for control of average

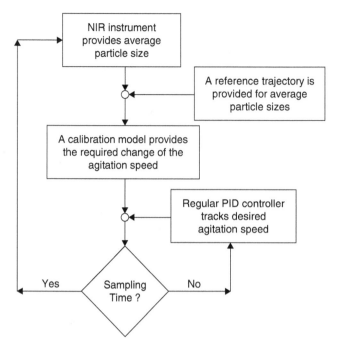

FIGURE 6.12   Proposed scheme for control of average particle sizes in free-radical suspension polymerizations. This material is reproduced with permission of John Wiley & Sons, Inc. from Santos AF, Lima EL, Pinto JC. Control and design of average particle size in styrene suspension polymerization using NIRS. J Appl Polym Sci 2000;77:453–462. © 2000 John Wiley & Sons, Inc.

particle sizes, as illustrated in Figure 6.12. In short, the agitation speed is manipulated in order to make the average particle size, as predicted by the PLS model, to be equal to the desired size value.

Afterward, Santos et al. [149] tried to extend the approach described previously to control MMA suspension polymerization reactions, but observed that the evolution of particle sizes is too fast in this case, requiring modification of the control scheme. In the case of MMA polymerization, the feedback control was only possible during the first moments of the reaction, when the spectral data were sensitive to modification of the agitation speed and suspending agent concentrations.

It is very important to observe that Santos et al. [55, 148] were unable to develop good calibration models for monomer conversion, as variations of the spectral signal were largely dominated by scattering phenomena. However, Silva et al. [150] and Pereira et al. [151] showed that NIRS could be used to control the polymer composition in copolymerizations of AA and vinyl acetate (VA) performed in suspension reactors. In this case, AA is partitioned between the aqueous and the organic phases, and the scattering signal does not overshadow the absorption AA signal obtained in the aqueous phase. As a consequence, very good PLS calibration models could be developed for the AA aqueous concentration, using NIR spectra obtained in a dispersive

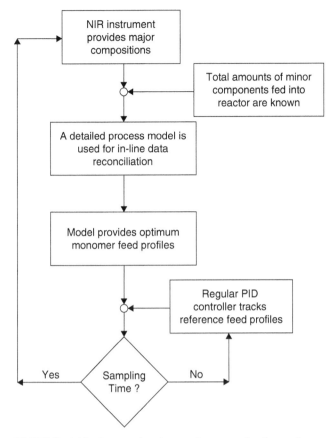

FIGURE 6.13 Proposed scheme for control of copolymer composition in semibatch suspension AA/VA copolymerizations. Reprinted (adapted) with permission from Silva FM, Lima EL, Pinto JC. Control of copolymer composition in suspension copolymerization reactions. Ind Eng Chem Res 2004;43:7312–7323. © 2004 American Chemical Society.

instrument equipped with a transflectance probe. Based on the AA concentration, a mechanistic process model was used as reference for computation of the concentrations of the main chemical species in the aqueous and organic phases and for the online optimization and control of monomer feed and temperature profiles, as described in Figure 6.13.

Faria et al. [152–156] used dispersive NIR technology to monitor and control morphological properties of poly(vinyl chloride) (PVC) resins. It was shown that NIR spectra collected in-line and in real time are sensitive to changes of morphological characteristics of PVC resins, such as bulk density, cold plasticizer absorption, and average particle diameter. As a consequence, advanced control strategies could be devised and implemented for real-time control of morphological properties, through manipulation of concentrations of suspending agents and of agitation speeds.

Vickers et al. [157] were the first to report the use of Raman spectroscopy to monitor suspension polymerization reactors. They used a fiber optic probe to monitor the reaction course *in situ*. Good calibration models were difficult to

be developed to monitor monomer conversions because of the complex nature of the reacting system. For this reason, calibration models were built with samples obtained during the reaction and withdrawn from the reactor vessel.

Santos et al. [158] reported a study on suspension polymerizations using Raman spectroscopy in the NIR region. The authors observed that the Raman spectra were affected by the PSD and showed that it was possible to monitor the evolution of monomer conversion during batch reactions using an external Raman probe. The Raman spectra were able to indicate abnormal process behavior during suspension polymerization reactions.

6.4.3 Emulsion Polymerization

Emulsion polymerization is a widely used industrial process for the production of synthetic polymer colloids or latexes of several different types of polymers, used in a variety of applications, such as synthetic rubber, coatings, paints, adhesives, binders, and so on. A typical emulsion polymerization recipe includes water, monomer(s), surfactant, water-soluble initiator, and additives (buffers, modifiers, and chain-transfer agents, among others) leading to a heterogeneous reaction mixture, composed of submicron solid polymer particles dispersed in an aqueous medium. Commercial recipes can include monomer contents varying between 30% and 55% in mass. Detailed description of emulsion polymerization systems can be found elsewhere [13, 159–162].

The control challenges in emulsion polymerization processes are similar to the ones discussed previously for suspension processes. Different from suspension processes, however, intense research activity has been conducted in the field, aiming to control emulsion polymerization reactions and the final latex properties. It is important to emphasize, though, that the lack of proper instrumentation for online measurement of latex properties has limited the development of control schemes basically to the control of latex composition. In these cases, control schemes depend heavily on the availability of complex process models for in-line evaluation of unmeasured properties, such as the MSD and the PSD. Comparatively, the online control of PSDs has received little attention, and few studies are devoted to this subject.

6.4.3.1 *Process Monitoring*   Regarding the use of spectroscopic data in emulsion polymerizations for monitoring and control purposes, few studies are found in the open literature. This is probably due to the complexity of emulsion polymerization systems, which involve different phases (aqueous phase, monomer droplets, and polymer particles), several compounds (water, monomer, polymer, initiator, stabilizer, and buffers) providing spectra of difficult interpretation. Moreover, polymer particles and monomer can scatter light to such an extent that the use of spectra for quantitative purposes may require the application of special mathematical

techniques to isolate the spectral features of each compound, as described previously for suspension processes.

The NIR-Raman spectroscopy was used by Wang et al. [163] to study the kinetics of styrene polymerizations in glass reaction flasks. Wang et al. [164], Özpozan et al. [165], Al-Khanbashi et al. [166], Urlaub et al. [122], Bauer et al. [167], Van Den Brink et al. [168], McCaffery and Durant [169], and Elizalde et al. [170] performed similar studies forVA, styrene/butyl acrylate, MMA, cyanacrylate, styrene/butadiene, styrene and butyl acrylate/MMA polymerizations in emulsion and miniemulsion reactions. These studies showed that NIR-Raman spectral data obtained in-line during emulsion polymerizations could be used for kinetic model building and kinetic analysis.

Using FT–MIR spectrometers equipped with ATR probes, Chatzi et al. [171], Kammona et al. [172], Hua and Dubé [173], and Roberge and Dubé [174] obtained similar results for 2-ethylhexyl acrylate/styrene and VA/butyl acrylate/MMA emulsion homo- and copolymerizations. Particularly, Hua and Dubé [173] present a review about the use of FT–IR–ATR spectroscopy for kinetic studies in polymerization systems. In all cases, MLR or PLS calibration models were used for interpretation of spectral data.

The first attempt to use NIRS to characterize a polymer emulsion latex in a broad range of operation conditions for use in production environments was presented by Gossen et al. [175]. Gossen et al. [175] combined UV and NIRS in order to predict the composition and particle diameter of styrene/MMA copolymer latices. A dispersive instrument equipped with a transflectance probe was used to collect spectral data. Different from the UV spectroscopy analysis, it was not necessary to dilute the polymer latex to collect useful NIR spectra, which is a major advantage of the NIR technique. In general, PLS calibration models based on the NIR spectra could yield good predictions of concentrations of the major constituents of the latex, but poor predictions of average particle sizes. This led the authors to suggest that NIR spectral data might depend on the shape of the PSD. It is important to emphasize that Gossen et al. [175] analyzed synthetic samples prepared through the modification of samples produced in batch reactions.

The prediction of average particle sizes of polymer particles suspended in emulsions from NIR spectra is a very controversial issue. The wavelength of NIR radiation is usually much larger than typical emulsion particles (usually below 100 nm), which means that the reaction medium may be regarded as homogeneous for most practical reasons. Therefore, the poor prediction of particle size reported by Gossen et al. [175] is not surprising. Vieira et al. [176] also attempted and failed to predict particle size averages based on NIR spectra of polymer emulsions with particle sizes within the usual range found in emulsion polymerizations. However, Reis et al. [177] and Chicoma et al. [178] reported the successful calibration and prediction of polymer parti-

cles in different emulsion polymerization systems, using dispersive instruments and standard transmission NIR technology.

Wu et al. [179, 180] investigated the use of NIRS to monitor the concentrations of water, styrene, and polystyrene in-line during emulsion polymerization reactions, using a dispersive instrument equipped with a reflectance probe. The authors built PLS calibration models for these variables based on the second derivative of the spectra obtained in the wavelength range of 700–1100 nm. A mechanistic model was implemented to provide proper filtering of the NIR calibration model predictions. Reliable predictions were obtained in the solids content range of 0–24%, even when process disturbances were introduced, such as the discontinuous modification of monomer content, reactor temperature, and agitation speed.

Comparative analyses of the performances of different spectroscopic techniques, when applied for monitoring of emulsion polymerization reactors, were made in several studies. Elizalde et al. [181] compared the performances of calorimetry and Raman spectroscopy for monitoring of solids contents and monomer conversions. They concluded that the overall performances were similar, although Raman spectroscopy led to more precise results for reactions performed in starved conditions. Reis et al. [182] compared the performances of Raman and standard transmission NIR spectroscopic methods for monitoring of emulsion polymerizations and concluded that both techniques had similar performances when applied to estimation of monomer conversions, average particle sizes, and solids contents.

Elizalde et al. [183] discussed issues related to building and maintenance of calibration models. Particularly, they showed that the model performance was subject to frequent variations, because of unavoidable maintenance and repairing of the instruments. They showed that an effective strategy for updating of calibration models is the combination of the original PLS model with a new PLS model that accounts for the model mismatch of the old model. Reis et al. [184] also presented strategies for building of reliable calibration models to monitor emulsion polymerization reactions.

6.4.3.2 Process Control   Vieira et al. [176, 185, 186] studied copolymerizations of butyl acrylate and MMA in emulsion reactors and reported for the first time the use of the NIR spectral signal to perform the feedback control of an emulsion polymerization reactor. Vieira et al. [176, 185, 186] used a dispersive instrument equipped with a transflectance probe to collect the NIR spectra and showed that robust and reliable PLS calibration models could be developed for the prediction of the major constituents of the emulsion. A detailed process model was used as reference to provide estimates of unmeasured process variables, such as average particle sizes and the polymer average molecular weight. The NIR model predictions were corrected in-line with the

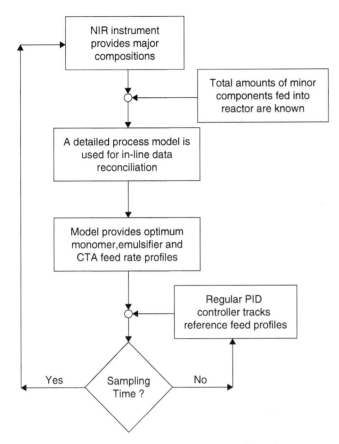

FIGURE 6.14 Proposed scheme for control of copolymer composition and average molecular weights in semibatch methyl methacrylate/butyl acrylate emulsion copolymerizations. Reprinted (adapted) with permission from Vieira RAM, Sayer C, Lima EL, Pinto JC. Closed-loop composition and molecular weight control of a copolymer latex using near-infrared spectroscopy. Ind Eng Chem Res 2002;41:2915–2930. © 2002 American Chemical Society.

use of reference model and data reconciliation procedure. The reference model and the current process states were used to compute optimum feed flow rates of the monomers and chain transfer agents during semibatch reactions in order to achieve specified copolymer compositions and average molecular weights. The general control scheme is presented in Figure 6.14.

Vieira et al. [176] also showed that the NIR spectra of polymer latices were sensitive to the presence of monomer droplets in the reaction medium. This is in accordance with the results previously reported by Santos et al. [55] for polymer suspensions, as monomer droplets in emulsion polymerizations have sizes typically in the same range as observed in suspension polymerizations. This could be used to avoid the formation of monomer droplets during starved polymerizations and to detect abnormal process operation conditions, considerably improving the process operation.

Santos et al. [187] used NIRS to monitor monomer conversion and copolymer composition during styrene/α-methyl styrene emulsion copolymerization reactions. An empirical PLS regression model was derived to correlate NIR spectra obtained in the wavelength range of 800–1900 nm with the overall monomer conversion, using a dispersive instrument equipped with a transflectance probe. In this case, however, detailed characterization of copolymer composition can be too expensive, as NMR analysis of copolymer samples obtained along the various polymerization tests must be performed. As the reactivity ratios of styrene and α-methyl styrene are known, an alternative procedure was implemented and tested. Pseudo-experimental polymer composition data were generated with a kinetic model and used as reference values for PLS calibration models. Surprisingly, the obtained NIR calibration model was able to perform very accurate in-line predictions of polymer composition from NIR spectra, in agreement with actual polymer composition data measured by NMR. This procedure can be very useful for development of practical applications, as reactivity ratios are known with reasonable accuracy for a large number of experimental systems and detailed analysis of copolymer compositions is usually expensive both in terms of costs and in terms of time. A similar model-based technique can be used to enlarge the calibration data set collected for calibration purposes.

Josephy and Breyer [188] filed a patent regarding the use of an NIR spectrometer to monitor the quality of latices of halogen-containing vinyl polymers during emulsion polymerizations. The authors claimed the invention of a monitoring device that includes a measuring probe placed inside the reactor vessel, connected to the NIRS through optical fibers. The authors claim that the light scattered by the polymer particles can be related to the mean particle diameter, the total solids content, the overall monomer conversion, the PSD, and the total particle surface area.

Van Den Brink et al. [189] and Ito et al. [190] reported the successful use of NIR-Raman spectroscopy for monitoring emulsion copolymerizations of VeoVa 9 (vinyl neononanoate)/butyl acrylate and of styrene/butadiene/MMA/acrylonitrile. Particularly, Ito et al. [190] reported the successful development of MLR calibration models for the average particle sizes of the polymer emulsions. Van Den Brink et al. [189] performed the online control of the copolymerization by using the Raman spectra to calculate the monomer concentrations and the copolymer composition in real time with MLR calibration models. The obtained results showed that the noise in the prediction values of monomer composition increased steadily with conversion, leading to poor prediction of monomer compositions and recommending the use of data reconciliation procedures, as performed by Vieira et al. [186]. In spite of that, a closed-loop control strategy based on the in-line optimization of the feed flow rate of butyl acrylate was implemented, using a process model as reference and allowing for the production of latices with controlled composition.

6.4.4 Other Polymerization Systems

Very little has been reported about the use of spectroscopic methods for monitoring and control of other polymerization systems. Lenzi et al. [191] reported that the NIR spectra collected in a dispersive instrument with a transflectance probe may contain very useful information about the structure of core-shell polystyrene beads produced through simultaneous semibatch emulsion/suspension polymerizations. Lenzi et al. [192] developed a polymerization technique that combines recipes of typical emulsion and suspension polymerizations to produce core-shell polymer beads. More interesting, the appearance of the core-shell structure always led to qualitatively different NIR spectra that could not have been obtained with polymer suspensions, polymer emulsions, or mixtures of polymer suspensions and emulsions. As described by Lenzi et al. [191], different spectral peaks could be detected in the wavelength region constrained between 1700 and 1900 nm when the core-shell structure developed.

Domingo et al. [193] presented an interesting study in which transmission and Raman-IR spectra were used to monitor the polymerization of silicon-based repellents and consolidants in stones *in situ* and in real time. Jiang et al. [194] used online FT Raman spectroscopy to monitor dispersion anionic polymerizations of styrene and 1,3-butadiene. Sapk et al. [195] and Souza et al. [196] used IR-Raman spectroscopy and transmission NIRS to monitor the evolution of aniline polymerizations. Spectral data were used to detect the formation of water soluble products in the first case and for building of a kinetic model, in the second case.

6.5 CONCLUSIONS

A review of the fundamentals and applications of spectroscopic methods for monitoring and control of polymerization reactions was made in this chapter. It was shown that spectroscopic methods are widely used in polymerization processes for monitoring and control of monomer and polymer composition, average particles size, and average molecular weight. Most applications make use of radiation in the NIR region, because of the current stage of radiation transmission technology. Additionally, in the vast majority of the applications, monitoring of monomer composition is made in lab-scale reactors, which leaves room for the development of monitoring and control procedures in real industrial reactors.

REFERENCES

[1] Ogunnaike BA. A contemporary industrial perspective on process control theory and practice. Annu Rev Control 1996;20:1–8.

[2] Embiruçu M, Lima EL, Pinto JC. A survey of advanced control of polymerization reactors. Polym Eng Sci 1996;36:433–447.

[3] Seborg DE. A perspective on advanced strategies for process control (revisited). European Control Conference. Karlsruhe, Germany; 1999. p 1–32.

[4] Vieira RAM, Embiruçu M, Sayer C, Lima EL, Pinto JC. Control strategies for complex chemical processes: applications in polymerization processes. Comput Chem Eng 2003;27:1307–1327.

[5] Kiparissides C. Polymerization reactor modeling: a review of recent developments and future directions. Chem Eng Sci 1996;51:1637–1659.

[6] Giudici R. Polymerization reaction engineering: a personal overview of the state-of-art. Lat Am Appl Res 2000;30:351–356.

[7] Ray WH, Villa CM. Nonlinear dynamics found in polymerization processes – a review – a detailed model analysis. Chem Eng Sci 2000;55:275–290.

[8] Chien DCH, Penlidis A. On-line sensors for polymerization reactors. J Macromol Sci – Rev Macromol Chem Phys 1990;C30:01–42.

[9] Kammona O, Chatzi EG, Kiparissides C. Recent developments in hardware sensors for the on-line monitoring of polymerization reactions. J Macromol Sci, Part C: Polym Rev 1999;39:57–134.

[10] Hergeth WD. On-line monitoring of polymerization and cure. Sens Update 1999;5:191–242.

[11] Fonseca GE, Dubé MA, Penlidis A. A critical overview of sensors for monitoring polymerizations. Macromol React Eng 2009;3:327–373.

[12] Frauendorf E, Hergeth WD. Industrial polymerization monitoring. Macromol Symp 2011;30:1–5.

[13] Dimitratos J, Eliçabe G, Georgakis C. Control of emulsion polymerization reactors. AIChE J 1994;40:1993–2021.

[14] Latado A, Embiruçu M, Mattos Neto AG, Pinto JC. Modeling of end-use properties of poly(propylene/ethylene) resins. Polym Test 2001;20:419–439.

[15] Nele M, Latado A, Pinto JC. Correlating polymer parameters to the entire molecular weight distribution: application to melt index. Macromol Mat Eng 2006;291:272–278.

[16] Araujo PHH, Sayer C, De La Cal JC, Asua, Lima EL, Pinto JC. Utilization of neural networks as soft sensors to monitor emulsion polymerization reactions (average particle diameter and conversion). Lat Am Appl Res 2001;31:525–531.

[17] Smith GS. *An Introduction to Classical Electromagnetic Radiation*. Cambridge: Cambridge University Press; 1997.

[18] Green JA. *Electromagnetic Radiation: Fundamentals and Applications*. 2nd ed. Wichita: Greenwood Research; 2002.

[19] Inam US, Inan A. *Electromagnetic Waves*. Englewood Cliffs: Prentice-Hall; 2000.

[20] D'Agostino S. *A History of the Ideas of Theoretical Physics*. New York: Kluwer Academic Publishers; 2002.

[21] Pedrotti FL, PedrottI LM, Pedrotti LS. *Introduction to Optics*. 3rd ed. New Jersey: Pearson Prentice Hall; 2007.

[22] Benett CA. *Principles of Physical Optics*. Hoboken: John Wiley & Sons; 2008.

[23] Urban MW. *Attenuated Total Reflectance Spectroscopy of Polymers: Theory and Practice*. Polymer Surfaces and

Interface Series. Washington, DC: American Chemical Society; 1996.

[24] Dahm DJ, Dahm KD. *Interpreting Diffuse Reflectance and Transmittance: A Theoretical Introduction to Absorption Spectroscopy of Scattering Materials*. Chichester: NIR Publications; 2007.

[25] Hollas JM. *Modern Spectroscopy*. 4th ed. Chichester: John Wiley & Sons; 2004.

[26] Eisberg R, Resnick R. *Quantum Physics of Atoms, Molecules, Solids, Nuclei and Particles*. New York: John Wiley & Sons; 1974.

[27] Barton II, FE. Theory and principles of near infrared spectroscopy. Spectrosc Eur 2002;14:12–18.

[28] Sonnessa AJ. *Introduction to Molecular Spectroscopy*. New York: Reinhold Publishing Corporation; 1966.

[29] Nadkarni RA. *Modern Instrumental Methods of Elemental Analysis of Petroleum Products and Lubricants*. Philadelphia: American Society for Testing and Materials; 1991.

[30] Burns DA, Ciurczak EW. *Handbook of Near-Infrared Analysis*. 2nd ed. New York: Marcel Dekker; 2001.

[31] Lindner P, Zemb T. Neutron, *X-Ray and Light Scattering: Introduction to an Investigative Tool for Colloidal and Polymeric Systems*. Amsterdam: North-Holland; 1991.

[32] Ferraro JR, Nakamoto K, Brown CW. *Introductory Raman Spectroscopy*. 2nd ed. San Diego: Academic Press; 2003.

[33] Smith E, Dent G. *Modern Raman Spectroscopy: A Practical Approach*. Chichester: John Wiley & Sons; 2005.

[34] Lyon AL, Keating CD, Fox AP, Baker BE, He L, Nicewarner SR, Mulvaney SP, Natan MJ. Raman spectroscopy. Anal Chem 1998;70:341–361.

[35] Herschel WF. Experiments on the refrangibility of the invisible rays of the sun. Philos Trans R Soc 1800;90:284–292.

[36] Santos AF, Silva FM, Lenzi MK, Pinto JC. Monitoring and control of polymerization reactors using NIR spectroscopy. Polym Plast Technol Eng 2005;44:1–61.

[37] Alb AM. Automatic Continuous Online Monitoring of Polymerization Reactors (ACOMP): Progress in Characterization of Polymers and Polymerization Reactions. PhD Thesis, Tulane University, New Orleans.

[38] Reed WF. Automatic Continuous Online Monitoring of Polymerization reactors (ACOMP). Polym News 2004;29: 271–279.

[39] Alb AM, Reed WF. Recent advances in Automatic Continuous Online Monitoring of Polymerization reactors (ACOMP). Macromol Symp 2008;271:15–25.

[40] Alb AM, Drenski MF, Reed WF. Automatic Continuous Online Monitoring of Polymerization reactors (ACOMP). Polym Int 2008;57:390–396.

[41] Alb AM, Reed WF. Fundamental measurements in online polymerization reaction monitoring and control with a focus on ACOMP. Macromol React Eng 2010;4: 470–485.

[42] Mitschke F. *Fiber Optics: Physics and Technology*. Berlin: Springer; 2010.

[43] Davies AMC. NIR instrument companies: the story so far. NIR News 1999;10:14–15.

[44] Workman JJ, Burns DA. Commercial NIR instrumentation. In: Burns DA, Ciurczak EW, editors. *Handbook of Near-Infrared Analysis*. 2nd ed. New York: Marcel Dekker; 2001.

[45] Assi S, Watt R, Moffat T. Comparison of laboratory and hand-held Raman instruments for the identification of counterfeit medicines. Spectroscopy 2011;(June Suppl):36–47.

[46] Otto M. *Chemometrics – Statistics and Computer Application in Analytical Chemistry*. Weinheim: Wiley-VCH; 1999.

[47] Beebe KR, Kowalki BR. An introduction to multivariate calibration and analysis. Anal Chem 1987;59:1007A–1017A.

[48] Davies AMC, Giangiacomo R. Near Infrared Spectroscopy: Proceedings of the 9th International Conference. Charlton: NIR Publications; 2000.

[49] Haaland DM, Easterling RG, Vopicka DA. Multivariate least-squares methods applied to the quantitative spectral analysis of multicomponent samples. Appl Spectrosc 1985;39:73–84.

[50] Geladi P, MacDougall D, Martens H. Linearization and scatter-correction for near-infrared reflectance spectra of meat. Appl Spectrosc 1985;39:491–500.

[51] Naes T, Isaksson T. Locally weighted regression of diffuse near infrared transmittance spectroscopy. Appl Spectrosc 1992;46:34–43.

[52] Oman SD, Naes T, Zube A. Detecting and adjusting for non-linearities in calibration of near-infrared data using principal components. J Chemom 1993;7:195–212.

[53] Almoy T, Haugland E. Calibration methods for NIRS instruments: a theoretical evaluation and comparisons by data splitting and simulations. Appl Spectrosc 1994;48:327–332.

[54] Ge Z, Cavinato AG, Callis JB. Noninvasive spectroscopy for monitoring cell density in a fermentation process. Anal Chem 1994;66:1354–1362.

[55] Santos AF, Lima EL, Pinto JC. In-line evaluation of average particle size in styrene suspension polymerizations using near-infrared spectroscopy. J Appl Polym Sci 1998;70:1737–1745.

[56] Alam MK, Stanton S, Hebner GA. Plastics analysis at two national laboratories. Part A. Resin identification using near-infrared spectroscopy and neural networks. In: Burns DA, Ciurczak EW, editors. *Handbook of Near-Infrared Analysis*. 2nd ed. New York: Marcel Dekker; 2001.

[57] Janik LJ, Forrester ST, Rawson A. The prediction of soil chemical and physical properties from mid-infrared spectroscopy and combined Partial Least-Squares regression and Neural Network (PLS-NN) analysis. Chemom Intell Lab Syst 2009;97:179–188.

[58] De Noord OE. The influence of data preprocessing on the robustness and parsimony of multivariate calibration models. Chemom Intell Lab Syst 1994;23:65–70.

[59] De Noord OE. Tutorial: multivariate calibration standardization. Chemom Intell Lab Syst 1994;25:85–97.

[60] Bouveresse E, Massart DL. Standardisation of near-infrared spectrometric instruments: a review. Vib Spectrosc 1996;11: 3–15.

[61] Sun J. Statistical analysis of NIR data: data pretreatment. J Chemom 1997;11: 525–532.

[62] Gil AJ, Romera R. On robust Partial Least Squares (PLS) methods. J Chemom 1998;12:365–378.

[63] Swierenga H, De Weijer AP, Van Wijk RJ, Buydens LMC. Strategy for constructing robust multivariate calibration models. Chemom Intell Lab Syst 1999;49:1–17.

[64] Swierenga H, Wülfert F, De Noord OE, De Weijer AP, Smilde AK, Buydens LMC. Development of robust calibration models in near infra-red spectrometric applications. Anal Chim Acta 2000;411:121–135.

[65] Hansen PW. Pre-processing method minimizing the need for reference analyses. J Chemom 2001;15:123–131.

[66] Janné K, Pettersen J, Lindberg N-O, Lundstedt T. Hierarchical Principal Component Analysis (PCA) and Projection to Latent Structure (PLS) technique on spectroscopic data as a data pre-treatment for calibration. J Chemom 2001;15:203–213.

[67] Macho S, Rius A, Callao MP, Larrechi MS. Monitoring ethylene content in heterophasic copolymers by near-infrared spectroscopy; standardisation of the calibration model. Anal Chim Acta 2001;445:213–220.

[68] Boysworth MK, Booksh KS. Aspects of multivariate calibration applied to near-infrared spectroscopy. In: Burns DA, Ciurczak EW, editors. *Handbook of Near-Infrared Analysis.* 2nd ed. New York: Marcel Dekker; 2001.

[69] Naes T, Isaksson T, Kowalski B. Locally weighted regression and scatter correction for near-infrared reflectance data. Anal Chem 1990;62:664–673.

[70] Di Foggio R. Examination of some misconceptions about near-infrared analysis. Appl Spectrosc 1995;49:67–75.

[71] Coates DB. Is near infrared spectroscopy only as good as the laboratory reference values? An empirical approach. Spectrosc Eur 2002;14:24–26.

[72] Kleinbaum DG, Kupper LL, Muller KE, Nizam A. *Applied Regression Analysis and Other Multivariable Methods.* New York: Duxbury Press; 1998.

[73] Jolliffe IT. *Principal Component Analysis.* New York: Springer-Verlag; 1986.

[74] Geladi P, Kowalski BR. Partial least-squares regression: a tutorial. Anal. Chim Acta 1986;185:1–17.

[75] Bjorsvik H-R, Martens H. Data analysis: calibration of NIR instruments by PLS regression. In: Burns DA, Ciurczak EW, editors. *Handbook of Near-Infrared Analysis.* 2nd ed. New York: Marcel Dekker; 1986.

[76] Centner V, De Noord OE, Massart DL. Detection of nonlinearity in multivariate calibration. Anal Chim Acta 1998;376:153–168.

[77] Pollard IF, Broussard MR, Garrison DB, San KY. Process identification using neural networks. Comput Chem Eng 1992;16:253–270.

[78] Bertrand E, Blanco M, Maspoch S, Ortiz MC, Sánchez MS, Sarabia LA. Handling intrinsic non-linearity in near-infrared reflectance spectroscopy. Chemom Intell Lab Syst 1999;49:215–224.

[79] Blanco M, Coello J, Iturriaga H, Maspoch S, Pagés J. Calibration in non-linear near infrared reflectance spectroscopy: a comparison of several methods. Anal Chim Acta 1999;384:207–214.

[80] Rantanen J, Räsänen E, Antikainen O, Mannermaa J-P, Yliruusi J. In-line moisture measurement during granulation with a four-wavelength near-infrared sensor: an evaluation of process-related variables and a development of non-linear calibration model. Chemom Intell Lab Syst 2001;56:51–58.

[81] Cleveland WS, Devlin SJ, Grosse E. Regression by local fitting; methods, properties, and computational algorithms. J Econ 1988;37:87–114.

[82] Van Den Broek W, Derks EPPA Van De Ven EW, Wienke D, Geladi P, Buydens LMC. Plastic identification by remote sensing spectroscopic NIR imaging using Kernal Partial Least Squares (KPLS). Chemom Intell Lab Syst 1996;35:187–197.

[83] Giammarise A. Near infrared analysis of the styrene content of copolymers with aliphatic acrylates and methacrylates. Anal Lett 1969;2:117–121.

[84] Miller CE, Edelman PG, Ratner BD, Eichinger BE. Near-infrared spectroscopy analyses of poly(ether urethane urea) block copolymers. Part I: Bulk composition. Appl Spectrosc 1990;44:576–580.

[85] Siesler HW. Near infrared spectroscopy for polymer reaction and process control: synthesis, processing and recycling. NIR News 1995;6:3–6.

[86] Weyer L. Near-infrared spectroscopy of organic-substances. Appl Spectrosc Rev 1985;21:1–43.

[87] Hall JW, Grzybowski DE, Monfre SL. Analysis of polymer pellets obtained from two extruders using near infrared spectroscopy. J Near Infrared Spectrosc 1993;1:55–62.

[88] Miller CE, Eichinger BE. Determination of crystallinity and morphology of fibrous and bulk polyethylene terephthalate by near-infrared diffuse reflectance spectroscopy. Appl Spectrosc 1990;44:496–504.

[89] Jones C, Brown JA. Polyether polyol monitoring using near infrared process photometers. J Adv Instrum 1983;38:429–438.

[90] Hansen MG, Vedula S. In-line fiber-optic near-infrared spectroscopy: monitoring of rheological properties in an extrusion process. Part I. J Appl Polym Sci 1998;68:859–872.

[91] Zhu C, Hieftje GM. Near-infrared analysis of chemical constituents and physical characteristics of polymers. Appl Spectrosc 1992;46:69–75.

[92] Miller CE, Edelman PG, Ratner BD, Eichinger BE. Near-infrared spectroscopy analyses of poly(ether urethane urea) block copolymers. Part II: Phase separation. Appl Spectrosc 1990;44:581–586.

[93] Miller CE, Eichinger BE. Analysis of rigid polyurethane foams by near-infrared diffuse reflectance. Appl Spectrosc 1990;44:887–894.

[94] Lachenal G, Stevenson I, Celette N. Near-infrared transmittance spectroscopy for radiochemical ageing of EPDM. Analyst 2001;126:2201–2206.

[95] Furukawa T, Watari M, Siesler HW, Ozaki Y. Discrimination of various poly(propylene) copolymers and prediction of their ethylene content by near-infrared and Raman spectroscopy in combination with chemometric methods. J Appl Polym Sci 2003;87:616–625.

[96] Tuchbreiter A, Kappler B, Stockmann R, Mülhaupt R, Honerkamp J. Near infrared reflection spectroscopy: a versatile tool for rapid characterization of olefin copolymers and high-throughput experiment. Macromol Mat Eng 2003;288:29–34.

[97] Shield SR, Ghebremeskel GN. Use of mid- and near-infrared techniques as tools for characterizing blends of copolymers of styrene–butadiene and acrylonitrile–butadiene. J Appl Polym Sci 2003;88:1653–1658.

[98] Hansen MG, Khettry A. In-line monitoring of molten polymers: near infrared spectroscopy, robust probes, and rapid data analysis. Polym Eng Sci 1994;34:1758–1766.

[99] Khettry A, Hansen MG. Real-time analysis of ethylene vinyl acetate random copolymers using near infrared spectroscopy during extrusion. Polym Eng Sci 1996;36:1232–1243.

[100] Rohe T, Becker W, Kölle S, Eisenreich N, Eyerer P. Near infrared (NIR) spectroscopy for in-line monitoring of polymer extrusion processes. Talanta 1999;50:283–290.

[101] Nagata T, Ohshima M, Tanigaki M. In-line monitoring of polyethylene density using near infrared (NIR) spectroscopy. Polym Eng Sci 2000;40:1107–1113.

[102] Odian G. *Principles of Polymerization*. 3rd ed. New York: John Wiley & Sons; 1991.

[103] Ellis MF, Taylor TW, Jensen KF. Online molecular-weight distribution estimation and control in batch polymerization. AIChE J 1994;40:445–462.

[104] Crowley TJ, Choi KY. Experimental studies on optimal molecular weight distribution control in a batch-free radical polymerization process. Chem Eng Sci 1998;53:2769–2790.

[105] Vega MP, Lima EL, Pinto JC. Modeling and control of tubular solution polymerization reactors. Comput Chem Eng 1997;21:S1049–1054.

[106] Zaldívar C, Iglesias G, Del Sol O, Pinto JC. On the preparation of acrylic acid/vinyl acetate copolymers with constant composition. 4. Modeling batch and continuous free-radical AA/VA copolymerization. Polymer 1997;38:5823–5833.

[107] Zaldívar C, Del Sol O, Iglesias GD. On the preparation of acrylic acid/vinyl acetate copolymers with constant composition. 1. Copolymerization reactivity Ratios. Polymer 1997;39:245–246.

[108] Pinto JC, Ray WH. The dynamic behavior of continuous solution polymerization reactors—VIII. A full bifurcation analysis of a lab-scale copolymerization reactor. Chem Eng Sci 1995;50:1041–1056.

[109] Long TE, Liu HY, Schell BA, Teegarden DM, Uerz DS. Determination of solution polymerization kinetics by near-infrared spectroscopy. 1. Living anionic-polymerization process. Macromolecules 1993;26:6237–6242.

[110] Aldridge PK, Kelly JJ, Callis JB. Noninvasive monitoring of bulk polymerization using short-wavelength near-infrared spectroscopy. Anal Chem 1993;85:3581–3585.

[111] De Thomas FA, Hall JW, Monfre ESL. Real-time monitoring of polyurethane production using near-infrared spectroscopy. Talanta 1994;41(3):425–431.

[112] Dallin P. NIR analysis in polymer reactions. Proc Control Qual 1997;9(4):167–172.

[113] Barrès C. In-line near infrared monitoring of esterification of a molten ethylene–vinyl alcohol copolymer in a twin screw extruder. Polym Eng Sci 2006;46:1613–1624.

[114] Beuermann S, Buback M, Isemer C, Wahl A. Homogeneous free-radical polymerization of styrene in supercritical CO_2. Macromol Rapid Commun 1999;20:26–32.

[115] Zetterlund PB, Yamazoe H, Yamada B. Propagation and termination kinetics in high conversion free radical copolymerization of styrene/divinylbenzene investigated by electron spin resonance and Fourier-transform near-infrared spectroscopy. Polymer 2002;43:7027–7035.

[116] Wang Q, Storm BK, Houmoller LP. Study of the isothermal curing of an epoxy prepreg by near-infrared spectroscopy. J Appl Polym Sci 2003;87:2295–2305.

[117] Lin Y, Stansbury JW. Near-infrared spectroscopy investigation of water effects on the cationic photopolymerization of vinyl ether systems. J Polym Sci: Part A Polym Chem 2004;42:1985–1998.

[118] Darcos V, Monge S, Haddleton DM. In situ Fourier transform near infrared spectroscopy monitoring of copper mediated living radical polymerization. J Polym Sci Part A: Polym Chem 2004;42:4933–4940.

[119] Haigh J, Brookes A, Hendra PJ, Strawn A, Nicholas C, Purbrick M. The design and construction of a cell for the in-situ monitoring of copolymerisation reactions using FT-Raman spectroscopy. Spectrochim Acta Part A 1997;53:9–19.

[120] Clarkson J, Mason SM, Williams KPJ. Bulk radical homopolymerization studies of commercial acrylate monomers using near-infrared Fourier-transform Raman-spectroscopy. Spectrochim Acta Part A: Mol Biomol Spectrosc 1991;47:9–10, 1345–1351.

[121] Damoun S, Papin R, Ripault G, Rousseau M, Rabadeaux JC. Radical polymerization of methyl-methacrylate in solution monitored and studied by Raman-spectroscopy. J Raman Spectrosc 1992;23:385–389.

[122] Urlaub E, Popp J, Roman VE, Kiefer W, Lankers M, Rössling G. Raman spectroscopic monitoring of the polymerization of cyanacrylate. Chem Phys Lett 1998;298:177–182.

[123] De Clercq B, Smellinckx T, Hugelier C, Maes N, Verpoort F. Monitoring of the polymerization of norbornene by in-line fiber-optic near IR-FT Raman. Appl Spectrosc 2001;55:1564–1567.

[124] Scherzer T, Mehnert R, Lucht H. On-line monitoring of the acrylate conversion in UV photopolymerization by near-infrared reflection spectroscopy. Macromol Symp 2004;205:151–162.

[125] Scherzer T, Muller S, Mehnert R, Volland A, Lucht H. In-line monitoring of the conversion in photopolymerized acrylate coatings on polymer foils using NIR spectroscopy. Polymer 2005;46:7072–7081.

[126] Puskas JE, Lanzendörfer MG, Pattern WE. Mid-IR real-time monitoring of the carbocationic polymerization of isobutylene and styrene. Polym Bull 1998;40:55–61.

[127] Yu J, Liu H-Z, Chen JY. FT-Raman spectroscopy for monitoring the polymerization of acrylic acid in aqueous solution. Chin J Polym Sci 1999;17:603–606.

[128] Pasquale AJ, Long TE. Real-time monitoring of the stable free radical polymerization of styrene via in-situ mid-infrared spectroscopy. Macromolecules 1999;32:7954–7957.

[129] Roper TM, Guymon CA, Jonsson ES, Hoyle CE. Influence of the alkene structure on the mechanism and kinetics of thiol–alkene photopolymerizations with real-time infrared spectroscopy. J Polym Sci Part A: Polym Chem 2004;42:6283–6298.

[130] Messman JM, Storey RF. Real-time monitoring of the ring-opening polymerization of rac-lactide with in situ attenuated total reflectance/Fourier transform infrared spectroscopy with conduit and diamond-composite sensor technology. J Polym Sci Part A: Polym Chem 2004;42:6238–6247.

[131] Puskas JE, Chan P, McAuley KB, Kaszas G, Shaikh S. Real-time FTIR monitoring of the carbocationic copolymerization of isobutylene with styrene. Macromol Symp 2006;240:18–22.

[132] Sahre K, Elrehim MHA, Eichhorn K-J, Voit B. Monitoring of the synthesis of hyperbranched poly(urea-urethane)s by real-time Attenuated Total Reflection (ATR)-FT-IR Spectroscopy. Macromol Mat Eng 2006;291:470–476.

[133] Mabilleau G, Cincu C, Basle MF, Chappard D. Polymerization of 2-(hydroxyethyl)methacrylate by two different initiator/accelerator systems: a Raman spectroscopic monitoring. J Raman Spectrosc 2008;39:767–771.

[134] Tracht U, Leiberich R, Wiesner U, Paul H-I. Activity monitoring for a polymerization catalyst system. Macromol Symp 2011;302:208–215.

[135] Santhosh P, Gopalan A, Vasudevan T, Wen T-C. Studies on monitoring the composition of the copolymer by cyclic voltammetry and in situ spectroelectrochemical analysis. Eur Polym J 2005;41:97–105.

[136] Florenzano FH, Enohnyaket P, Fleming V, Reed WF. Coupling of near infrared spectroscopy to automatic continuous online monitoring of polymerization reactions. Eur Polym J 2005;41:535–545.

[137] Cherfi A, Fevotte G. On-line conversion monitoring of the solution polymerization of methyl methacrylate using near-infrared spectroscopy. Macromol Chem Phys 2002;203:1188–1193.

[138] Cherfi A, Fevotte G, Novat C. Robust on-line measurement of conversion and molecular weight using NIR spectroscopy during solution polymerization. J Appl Polym Sci 2002;85:2510–2520.

[139] Olinga A, Winzen R, Rehage H, Siesler HW. Methyl methacrylate on-line polymerisation monitoring by light-fibre Fourier transform near infrared transmission spectroscopy and Fourier transform mid infrared/attenuated total reflection spectroscopy. J Near Infrared Spectrosc 2001;9:19–24.

[140] Lousberg HHA, Boelens HFM, Le Comte EP, Hoefsloot HCJ, Smilde AK. On-line determination of the conversion in a styrene bulk polymerization batch reactor using near-infrared spectroscopy. J Appl Polym Sci 2002;84:90–98.

[141] Fontoura JMR, Santos AF, Silva FM, Lenzi MK, Lima EL, Pinto JC. Monitoring and control of styrene solution polymerization using NIR spectroscopy. J Appl Polym Sci 2003;90:1273–1289.

[142] Nogueira ES, Borges CP, Pinto JC. In-line monitoring and control of conversion and weight-average molecular weight of polyurethanes in solution step-growth polymerization based on near infrared spectroscopy and torquemetry. Macromol Mat Eng 2005;290:272–282.

[143] Othman NS, Fevotte G, Peycelon D, Egraz J-B, Suau J-M. Control of polymer molecular weight using near infrared spectroscopy. AIChE J 2004;50:654–664.

[144] Othman NS, Peycelon D, Fevotte G. Monitoring and control of free-radical polymerizations using near-infrared spectroscopy. Ind Eng Chem Res 2004;43:7383–7391.

[145] Van Den Brink M, Van Herk AM, German AL. On-line monitoring and control of the solution polymerization of n-butyl acrylate in dioxane by Raman spectroscopy. Process Control Qual 1999;11:265–275.

[146] Cavalcanti MJR, Pinto JC. Modeling and optimization of suspension SAN polymerization reactors. J Appl Polym Sci 1997;65:1683–1701.

[147] Yuan HG, Kalfas G, Ray WR. Suspension polymerization – a review. J Macromol Sci – Rev Macromol Chem Phys 1991;C31:215–299.

[148] Santos AF, Lima EL, Pinto JC. Control and design of average particle size in styrene suspension polymerization using NIRS. J Appl Polym Sci 2000;77:453–462.

[149] Santos JGF, Way DV, Melo PA, Nele M, Pinto JC. Analysis of near infrared spectra during Methyl Methacrylate (MMA) suspension polymerizations. Macromol Symp 2011;299–300:57–65.

[150] Silva FM, Lima EL, Pinto JC. Control of copolymer composition in suspension copolymerization reactions. Ind Eng Chem Res 2004;43:7312–7323.

[151] Pereira HL, Machado F, Lima EL, Pinto JC. In-line monitoring of vinyl acetate/acrylic acid batch copolymerizations through near infrared spectroscopy. Macromol Symp 2011;299–300:1–9.

[152] Faria JR, Machado F, Lima EL, Pinto JC. Monitoramento In-Situ e em Tempo Real de Variáveis Morfológicas do PVC Usando Espectroscopia NIR. Polímeros: Ciência e Tecnologia 2009;19:95–104.

[153] Faria J. R.; Machado F, Lima EL, Pinto JC. Monitoring of vinyl chloride suspension polymerization using NIRS. 1. Prediction of morphological properties. Comput-Aided Chem Eng 2009;27:327–332.

[154] Faria J. R.; Machado F, Lima EL, Pinto JC. Monitoring of vinyl chloride suspension polymerization using NIRS. 2. Proposition of a scheme to control morphological properties of PVC. Comput-Aided Chem Eng 2009;27:1329–1334.

[155] Faria JR, Machado F, Lima EL, Pinto JC. In-line monitoring of vinyl chloride suspension polymerization with near infrared spectroscopy. 1. Analysis of morphological properties. Macromol React Eng 2010;4:11–24.

[156] Faria JR, Machado F, Lima EL, Pinto JC. In-line monitoring of vinyl chloride suspension polymerization with near

infrared spectroscopy. 2. Design of an advanced control strategy. Macromol React Eng 2010;4:486–498.

[157] Vickers TJ, Lombardi DR, Sun B, Wang H, Mann CK. Calculating inaccessible Raman reference spectra for use in monitoring a suspension polymerization. Appl Spectrosc 1997;51:1251–1253.

[158] Santos JC, Reis MM, Machado RAF, Bolzan A, Sayer C, Giudici R, Araújo PHH. Online monitoring of suspension polymerization reactions using Raman spectroscopy. Ind Eng Chem Res 2004;43:7282–7289.

[159] Min KW, Ray WH. On the mathematical modeling of emulsion polymerization reactors. J Macromol Sci – Rev Macromol Chem 1974;C11:177–255.

[160] Gilbert RG, Napper DH. The direct determination of kinetic-parameters in emulsion polymerization systems. J Macromol Sci – Rev Macromol Chem 1983;C23:127–186.

[161] Penlidis A, MacGregor JF, Hamielec AE. A theoretical and experimental investigation of the batch emulsion polymerization of vinyl acetate. Polym Proc Eng 1985;3:185–218.

[162] Gilbert RG. *Emulsion Polymerization: A Mechanistic Approach*. London, UK: Academic, Harcourt and Brace; 1995.

[163] Wang C, Vickers TJ, Schlenoff JB, Mann CK. In-situ monitoring of emulsion polymerization using fiberoptic Raman-spectroscopy. Appl Spectrosc 1992;46:1729–1731.

[164] Wang C, Vickers TJ, Mann CK. Use of water as an internal standard in the direct monitoring of emulsion polymerization by fiberoptic Raman-spectroscopy. Appl Spectrosc 1993;47:928–932.

[165] Özpozan T, Schrader B, Keller S. Monitoring of the polymerization of vinylacetate by near IR FT Raman spectroscopy. Spectrochim Acta Part A: Mol Biomol Spectrosc 1997;53:1–7.

[166] Al-KhanbashI A, Dhamdhere M, Hansen M. Application of in-line fiber-optic Raman spectroscopy to monitoring emulsion polymerization reactions. Appl Spectrosc Rev 1998;33:115–131.

[167] Bauer C, Amram B, Agnely M, Charmot D, Sawatzki J, Dupuy N, Huvenne J-P. On-line monitoring of a latex emulsion polymerization by fiber-optic FT-Raman spectroscopy. Part I: Calibration. Appl Spectrosc 2000;54:528–535.

[168] Van Den Brink M, Pepers M, Van Herk AM, German AL. Emulsion (Co) polymerization of styrene and butyl acrylate monitored by on-line Raman spectroscopy. Macromol Symp 2000;150:121–126.

[169] McCaffery TR, Durant YG. Application of low-resolution Raman spectroscopy to online monitoring of miniemulsion polymerization. J Appl Polym Sci 2002;86:1507–1515.

[170] Elizalde O, Leiza JR, Asua JM. On-line monitoring of all-acrylic emulsion polymerization reactors by Raman spectroscopy. Macromol Symp 2004;206:135–148.

[171] Chatzi EG, Kammona O, Kiparissides C. Use of a midrange infrared optical-fiber probe for the on-line monitoring of 2-ethylhexyl acrylate/styrene emulsion copolymerization. J Appl Polym Sci 1997;63:799–809.

[172] Kammona O, Chatzi EG, Kiparissides C. On-line monitoring of emulsion copolymerization using a MIR probe in combination with factor analysis. Dechema Monogr 1998;134:365–373.

[173] Hua H, Dubé MA. In-line monitoring of emulsion homo- and copolymerizations using ATR-FTIR spectrometry. Polym React Eng 2002;10:21–40.

[174] Roberge S, Dube MA. Inline monitoring of styrene/butyl acrylate miniemulsion polymerization with attenuated total reflectance/Fourier transform infrared spectroscopy. J Appl Polym Sci 2007;103:46–52.

[175] Gossen PD, Macgregor JF, Pelton RH. Composition and particle diameter for styrene/methyl methacrylate copolymer latex using UV and NIR spectroscopy. Appl Spectrosc 1993;47:1852–1870.

[176] Vieira RAM, Sayer C, Lima EL, Pinto JC. Detection of monomer droplets in a polymer latex by near-infrared spectroscopy. Polymer 2001;42:8901–8906.

[177] Reis MM, Araújo PHH, Sayer C, Giudici R. In situ near-infrared spectroscopy for simultaneous monitoring of multiple process variables in emulsion copolymerization. Ind Eng Chem Res 2004;43:7243–7250.

[178] Chicoma DL, Sayer C, Giudici R. In-line monitoring of particle size during emulsion polymerization under different operational conditions using NIR spectroscopy. Macromol React Eng 2011;5:150–162.

[179] Wu CC, Danielsen JDS, Callis JB, Eaton MT, Ricker NL. Remote in-line monitoring of emulsion polymerization of styrene by short-wavelength near-infrared spectroscopy. 1. Performance during normal runs. Process Control Qual 1996;8:1–23.

[180] Wu CC, Danielsen JDS, Callis JB, Eaton MT, Ricker NL. Remote in-line monitoring of emulsion polymerization of styrene by short-wavelength near-infrared spectroscopy. 2. Performance in the face of process upsets. Process Control Qual 1996;8:25–40.

[181] Elizalde O, Azpeitia M, Reis MM, Asua JM, Leiza JR. Monitoring emulsion polymerization reactors: calorimetry versus Raman spectroscopy. Ind Eng Chem Res 44, 7200–7207.

[182] Reis MM, Araújo PHH, Sayer C, Giudici R. Comparing near infrared and Raman spectroscopy for on-line monitoring of emulsion copolymerization reactions. Macromol Symp 2004;206:165–178.

[183] Elizalde O, Asua JM, Leiza JR. Monitoring of emulsion polymerization reactors by Raman spectroscopy: calibration model maintenance. Appl Spectrosc 2005;59:1280–1285.

[184] Reis MM, Araújo PHH, Sayer C, Giudici R. Development of calibration models for estimation of monomer concentration by Raman spectroscopy during emulsion polymerization: facing the medium heterogeneity. J Appl Polym Sci 2004;93:1136–1150.

[185] Vieira RAM, Sayer C, Lima EL, Pinto JC. In-line and in-situ monitoring of semi-batch emulsion copolymerizations using near-infrared spectroscopy. J Appl Polym Sci 2002;84:2670–2682.

[186] Vieira RAM, Sayer C, Lima EL, Pinto JC. Closed-loop composition and molecular weight control of a copolymer latex

using near-infrared spectroscopy. Ind Eng Chem Res 2002;41:2915–2930.

[187] Santos AF, Nogueira RF, Lima EL, Pinto JC. Monitoramento em Linha da Copolimerização de Estireno/Alfa-Metilestireno em Emulsão por Espectrofotometria de Infravermelho Próximo. Anais do 6° CBPOL – Congresso Brasileiro de Polímeros, Gramado – RS.

[188] Josephy C, Breyer P. Device for monitoring latex properties during emulsion polymerization. US Patent 6,335,527. 2002.

[189] Van Den Brink M, Pepers M, Van Herk AM, German AL. On-line monitoring and composition control of the emulsion copolymerization of VeoVa 9 and butyl acrylate by Raman spectroscopy. Polym React Eng 2001;9:101–133.

[190] Ito K, Kato T, Ona T. Non-destructive method for the quantification of the average particle diameter of latex as water-based emulsions by near-infrared Fourier transform Raman spectroscopy. J Raman Spectrosc 2002;33:466–470.

[191] Lenzi MK, Lima EL, Pinto JC. Detecting core–shell structure formation using near infrared spectroscopy. J Near Infrared Spectrosc 2006;14:179–187.

[192] Lenzi MK, Silva FM, Lima EL, Pinto JC. Semi-batch styrene suspension polymerization processes. J Appl Polym Sci 2003;89:3021–3038.

[193] Domingo C, Buergo MA, Sánchez-Cortés S, Fort R, Garcia-Ramos JV, Gomez-Heras M. Possibilities of monitoring the polymerization process of silicon-based water repellents and consolidants in stones through infrared and Raman spectroscopy. Prog Org Coat 2008;63:5–12.

[194] Jiang J-H, Ozaki Y, Kleimann M, Siesler HW. Resolution of two-way data from on-line Fourier-Transform Raman spectroscopic monitoring of the anionic dispersion polymerization of styrene and 1,3-butadiene by Parallel Vector Analysis (PVA) and Window Factor Analysis (WFA). Chemom Intell Lab Syst 2004;70:83–92.

[195] Sapk M, Akbulut U, Batchelder DN. Monitoring of electro-initiated polymerization of aniline by Raman microprobe spectroscopy. Polymer 1998;40:21–26.

[196] Souza FG, Anzai TK, Rodrigues MVA, Melo PA, Nele M, Pinto JC. In situ determination of aniline polymerization kinetics through near infrared spectroscopy. J Appl Polym Sci 2009;112:157–162.

7

CALORIMETRY, CONDUCTIVITY, DENSIMETRY, AND RHEOLOGICAL MEASUREMENTS

Jose R. Leiza and Timothy McKenna

7.1 INTRODUCTION

In this chapter, four techniques for monitoring polymerization reactors are introduced: calorimetry, conductimetry, densimetry, and rheological measurements. The methods are presented together as they are all indirect methods; in other words, they are used to measure some state (e.g., temperature) or physical property (e.g., viscosity) in the reactor, which is then reinterpreted using reasonable kinetic and process models to infer important information such as the polymerization rate or the molecular weight in real time (or with minimal delay).

The major advantages of all methods presented here are rapidity and cost. In the case of calorimetry, all one needs are the basic (accurate) measurements of flow rates in and out of the reactor, and representative temperatures to estimate the reaction rate. Rheological measurements can be obtained by measuring the torque on the agitator shaft. Conductimetry and densimetry require dedicated hardware. In fact, only densimetry requires additional equipment with any substantial budget. All provide information for interpretation and control very quickly. Of course, there are limitations to each method, discussed throughout the chapter.

7.2 REACTION CALORIMETRY

7.2.1 Introduction to Reaction Calorimetry in Polymerization Reactors

Reaction calorimetry is an appropriate technique to monitor polymerization reactions because of the exothermic nature of the conversion of monomer(s) to (co)polymer (enthalpies of polymerization of typical monomers are in the range of 50–200 kJ mol^{-1}). Reaction calorimetry is noninvasive, rapid, and robust, and it only requires accurate temperature measurements of the reaction mixture and the heating/cooling liquid of the jacket, which are relatively easy to carry out.

Figure 7.1 shows schematically a chemical reactor with the heat (sources and sinks) contributions that play a role in the overall energy balance of the reactor.

The basis of reaction calorimetry is the computation of the heat of reaction, Q_r, from the energy balances of the reactor [1–4] (reactor content, heating/cooling jacket, and, in some cases of the reactor lid [5]). The energy balances of the reactor content and heating/cooling jacket assuming that both are perfectly mixed are given by

$$\frac{d}{dt}(C_{p,tot}T_R) = Q_r + Q_{inflow} - Q_f + Q_{stir} + Q_{cal} - Q_{loss} - Q_{reflux} \tag{7.1}$$

$$\frac{d}{dt}(C_{p,J}T_J) = Q_{J,in} - Q_{J,out} + Q_f - Q_{J,loss} \tag{7.2}$$

where $C_{p,tot}$ is the heat capacity of the filled reactor (including reactor content, reactor inserts, and reactor), $C_{p,J}$ is the heat capacity of the jacket, Q_{inflow} is the sensible heat due to the feeding stream to the reactor, Q_f is the heat flow from the reactor to the jacket, Q_{stir} is the heat input due to stirring the (viscous) reaction mixture, Q_{cal} is the heat input due to calibration or other means heaters immersed in the reaction mixture, Q_{loss} is the heat loss to the surroundings, Q_{reflux} is the

Monitoring Polymerization Reactions: From Fundamentals to Applications, First Edition. Edited by Wayne F. Reed and Alina M. Alb.
© 2014 John Wiley & Sons, Inc. Published 2014 by John Wiley & Sons, Inc.

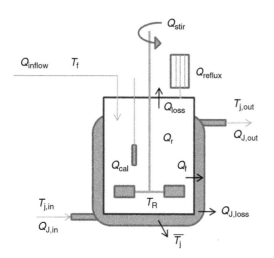

FIGURE 7.1 Schematic of a semibatch stirred tank reactor with indication of the heat contributions.

heat loss by condensation, $Q_{J,in}$ and $Q_{J,out}$ are the inlet and outlet sensible heat due to the stream of fluid entering and exiting the jacket, and $Q_{J,loss}$ is the heat loss of the jacket to the surrounding.

The heat of reaction, Q_r is given by

$$Q_r = V_R \sum_i r_i (-\Delta H_{r,i}) \qquad (7.3)$$

where V_R is the volume of the reaction mixture, r_i is the rate of reaction (in this case polymerization rate), and $\Delta H_{r,i}$ is the enthalpy of reaction i.

The heat contributions in Equations 7.1 and 7.2 are defined as follows:

$$Q_{inflow} = \sum_i \dot{m}_{F,i} c_{p,i} (T_{f,i} - T_R) \qquad (7.4)$$

$$Q_f = UA(T_R - T_J) \qquad (7.5)$$

$$Q_{stir} = P_0 N_s^3 d_{stir}^5 \rho_{mix} \qquad (7.6)$$

$$Q_{cal} = VI \qquad (7.7)$$

$$Q_{loss} = U_e A_e (T_R - T_e) \qquad (7.8)$$

$$Q_{reflux} = \dot{m}_c c_{p,c} (T_{c,in} - T_{c,out}) \qquad (7.9)$$

$$Q_{J,in} - Q_{J,out} = \dot{m}_J c_{p,J} (T_{j,in} - T_{j,out}) \qquad (7.10)$$

$$Q_{J,loss} = U_{j,e} A_{j,e} (T_J - T_e) \qquad (7.11)$$

where $\dot{m}_{F,i}$, $c_{p,i}$, and $T_{f,i}$ are the mass flow rate, heat capacity, and temperature of stream i, U is the overall heat transfer

coefficient, A is the reactor to jacket heat transfer area, P_0 is the power number, N_s is the stirrer speed, d_{stir} is stirrer diameter, ρ_{mix} is the density of the reaction mixture, V and I are the voltage and current of the electrical heater, U_e and A_e are the heat transfer coefficient and heat transfer area of the reactor to the surroundings, T_e is the temperature of the surrounding area, $\dot{m}_c$ and $c_{p,c}$ are the mass flow rate of the fluid in the reflux condenser, $T_{c,in}$ and $T_{c,out}$ are the inlet and outlet temperatures of the fluid in the reflux condenser, $\dot{m}_J$ and $c_{p,J}$ are the mass flow rate and heat capacity of the heating/cooling fluid in the jacket, $T_{j,in}$ and $T_{j,out}$ are the inlet and outlet temperatures of the fluid in the jacket, and $U_{j,e}$ and $A_{j,e}$ are the heat transfer coefficient and transfer area from the jacket to the surroundings.

It is important to note that the heat contribution by the stirrer (Eq. 7.6) is an approximation since one needs to account for motor efficiency and not all power will actually be transformed into heat.

Several operation modes of reaction calorimetry for laboratory equipment are possible:

7.2.1.1 Adiabatic calorimetry
An adiabatic reaction calorimeter is characterized by thermal insulation of the reaction mixture from the surroundings. Consequently, the heat released by reaction is stored within the reaction mixture. Thus, the temperature gradient of the reaction is proportional to conversion. Equation 7.1 reduces to

$$\frac{d}{dt}(C_{p,tot} T_R) = Q_r$$

7.2.1.2 Isothermal calorimetry
In isothermal calorimetry, the reactor temperature is kept constant by jacket temperature control. In practice, it is very difficult to keep the reactor strictly isothermal because of the nonzero heat transfer resistance of the reactor wall and deviations and delays in the temperature control. Therefore, the accumulation term should be considered in Equation 7.1:

$$\frac{d}{dt}(C_{p,tot} T_R) = Q_r - UA(T_R - T_J)$$

7.2.1.3 Isoperobolic calorimetry
In this method, the jacket temperature is maintained constant by using a sufficiently high jacket flow rate or a bath and the reactor temperature is not controlled and changes during reaction. The heat balance is the same as for the isothermal case.

Two methods are used to compute the heat of reaction, Q_r, depending if only Equation 7.1 is solved, namely, the heat balance of the reactor content, or either Equations 7.1 and 7.2 are simultaneously solved [6]. The first method is known as *heat flow (flux) calorimetry* and the second as *heat balance calorimetry*.

7.2.2 Heat Flow Calorimetry

In this method, Q_r is calculated by inverting Equation 7.1 that mainly consists of measurable temperatures (T_r and T_j) and flow rates as well as known physical properties as shown in Equation 7.12 in which for simplicity only the most important heat contributions have been considered.

$$Q_r = C_{p,tot} \frac{dT_R}{dt} + UA(T_R - T_J) - \sum_i \dot{m}_{F,i} c_{p,i}(T_{f,i} - T_R) + Q_{loss}$$

(7.12)

An accurate estimation of Q_r requires the heat flow from the reactor to the jacket, Q_f, to be correctly estimated. The heat flow is a function of the overall heat transfer coefficient, the heat transfer area, and the temperature difference between the reactor and jacket. Typically, the temperature difference between the reactor and jacket is large enough to make heat flow calorimetry very sensitive. The heat transfer area is constant (or only slightly affected due to differences in densities of monomer and polymer) in batch processes, but significantly changes during the course of a semibatch polymerization. If feeding flow rates are known, the heat transfer area can also be computed. However, the overall heat transfer coefficient, U, is not constant during a polymerization reaction and might change considerably because it depends on several parameters that include the viscosity and density of the reaction mixture, the stirring rate (hydrodynamics of the reactor), and the heat transfer through the reactor/jacket interface (influence of film formation or fouling in the reactor walls). Therefore, an accurate estimation of the evolution of the overall heat transfer coefficient is a must for reliable computation of the heat of reaction. Additional complications arise in the case of semibatch reactors where the overall heat transfer surface area A changes. In that case, it is important to have an estimate of the product UA.

Several approaches have been proposed in the literature to estimate the overall heat transfer coefficient [7–12]. The most common approach implemented in commercial lab scale calorimeters (like the RC1 calorimeter from Mettler-Toledo), uses a two point calibration (using a calibration heater of a known power) [13]. The calibration is carried out at the beginning and at the end of the polymerization (in absence of reaction) and then the value of U is interpolated. There are two main drawbacks with this approach: The first is that the heat of reaction cannot be obtained online because the expected change of U during the polymerization reaction (typically a decrease) cannot be calculated until the end of the reaction, and the second drawback is that the off-line calculated Q_r depends on the interpolation of the U values calculated at the beginning and at the end of the experiment. Although commercial equipment allows for different interpolations methods (linear, proportional to conversion, etc.), significant errors can be made in the computation of the

evolution of heat of reaction. The larger the difference between the U values at the beginning and the end, the larger the deviation on the evolution of the heat of reaction and the enthalpies of polymerizations calculated from it.

Different alternatives have been presented to circumvent this issue in heat flow calorimetry. A priori off-line determination of the dependence of UA [9], adaptive calorimetry using an additional off-line measurement [12] and cascade state estimation observers [14] proven to work, will be discussed in the following section. Obviously, another alternative is to use *heat balance calorimetry* and to solve the energy balances given by Equations 7.1 and 7.2 simultaneously to compute the evolution of the heat of reaction, Q_r, and the overall heat transfer coefficient, UA. This approach will be addressed in Section 7.2.3.

7.2.2.1 Estimation of the Heat Transfer Coefficient, *UA*

As discussed in section 7.2.2, the overall heat transfer coefficient changes during a polymerization reaction. Typically, U decreases with polymerization time (with increasing conversion) because the viscosity of the reaction medium increases, as well as the polymer concentration (solids content). Online measurements of the viscosity of the reaction medium would allow estimating the evolution of U, but this measurement is not readily available in polymerization reactors although it can be inferred from torque measurements of the agitation system (see Section 7.5). Alternatively, empirical equations that relate the overall heat transfer coefficient with the conversion (or solids content) of the polymerization have been used [9, 15].

The overall heat transfer coefficient can be estimated online by using an additional process measurement (e.g., gravimetric conversion or solids content) together with state (parameter) estimation techniques to update the value of the overall heat transfer coefficient. This approach referred as adaptive calorimetry has been mainly exploited by Fevotte and coworkers [12] to monitor emulsion (co) polymerization reactions. They used a dependence of U with conversion

$$U(X(t)) = U_{init} + P_1 X(t-1) + P_2 X^2(t-1) \quad (7.13)$$

where $X(t-1)$ is the estimated overall monomer conversion at sampling time $t-1$, and P_1 and P_2 are the parameters to be estimated.

These parameters were estimated online using a parameter estimation algorithm that minimizes the least squares of the off-line-measured gravimetric conversion and the estimated conversion calculated based on a third-order state-space representation that included the energy balance of the reactor (Eq. 7.1).

BenAmor et al. [14] developed a high-gain cascade state observer based on the energy balance of Equation 7.1 to monitor the heat of reaction online and the heat

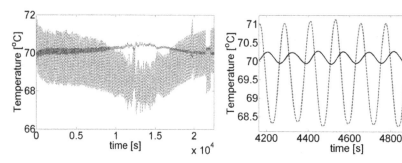

FIGURE 7.2   Oscillations in the reactor (solid) and jacket (dashed) temperatures acquired during a semibatch reaction (left) and zoomed region (right).

transfer coefficient without the need of any additional measurement and with no prior knowledge of the dependence of the overall heat transfer coefficient with conversion (or solids content). The only requirement was the knowledge of the initial UA. Then, the cascade state observer allowed the sequential estimation of Q_r and UA based on reactor and jacket temperature measurements. The technique was applied in a commercial lab-scale reactor (3 l) and pilot-plant reactor (250 l) to monitor semibatch emulsion copolymerization reactions. The main drawback of the proposed technique is that the state observers used are only observable when $T_r \neq T_j$. The authors show that during the polymerization, for instance, during the nucleation stage, the jacket and reactor temperatures cross and hence violate this condition. However, the observer behaves well during the starved feeding period of the polymerization.

7.2.2.2 *Temperature Oscillation Calorimetry*   A more elegant way to estimate online the overall heat transfer coefficient without any additional measurement was developed by Carloff [11] by the technique known as *temperature oscillation calorimetry, TOC*. In this approach, the unknown product UA is computed from the analysis of the sine-shaped oscillations, which are superposed on either the reactor temperature or jacket temperature. The objective is to decouple the slow dynamic of the chemical heat production from the fast dynamic variable heat transfer during the reaction. The oscillations can be achieved either by adding a calibration heater to the system or by adding a sine signal to the set point of either T_R or T_J. Figure 7.2 shows the evolution of the reactor and jacket temperatures in a reaction calorimeter where a sine wave temperature modulation was superimposed on the reactor jacket temperature.

For the evaluation of the resulting oscillations, Carloff [11] used an algorithm requiring the user to choose a parameter β which was not trivial. An improved evaluation algorithm was introduced by Tietze et al. [10] who proposed to compute UA from

$$UA = \frac{C_{P,R}\omega}{\tan\left(\arccos\left[\dfrac{\delta T_R}{\delta T_J}\right]\right)} = \frac{C_{P,R}\omega}{\sqrt{1-\left(\dfrac{\delta T_R}{\delta T_J}\right)^2}}\left(\frac{\delta T_R}{\delta T_J}\right)$$

(7.14)

where δT_R and δT_J are the amplitudes of the corresponding temperature oscillations (Fig. 7.2).

This formula was derived describing the sine oscillation by complex numbers and by separating the real and the imaginary parts. Alternatively, it can be derived from classical system theory [16]. The simplified amplitudes δT_R and δT_J are computed from

$$\bar{T}\left(-\frac{\pi}{\omega}\right) = \frac{\omega}{2\pi}\int_{-\frac{\pi}{\omega}}^{0} T(t)\,dt$$

$$\delta T\left(-2\frac{\pi}{\omega}\right) = \sqrt{\int_{-3\frac{\pi}{\omega}}^{-\frac{\pi}{\omega}}\left(T(t)-\overline{T(t)}\right)^2}$$

(7.15)

If this method is applied online, it leads to a delay in the estimate as data from the past 1.5 periods $(-3\pi/\omega)$ is used and the values of δT_R and δT_J have to be appointed to the time $-2\pi/\omega$ (one cycle time ago).

Figure 7.3 shows the evolution of the oscillating reactor and jacket temperatures and the estimated product of UA obtained during the solution polymerization of MMA in ethyl acetate [11].

De Luca and Scali [17] compared the evaluation algorithms of Carloff [11] and Tietze [10] and two further algorithms and concluded that the algorithm of Equation 7.14 was the most robust in case of disturbances, although all the algorithms were mathematically equivalent. Lüdke [18] extended the approach of Equation 7.14 by computing the amplitudes from a Fourier transform instead of subsequent integrations. This yields less-noisy results and decreases the delay.

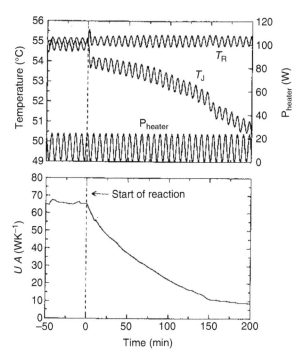

FIGURE 7.3 Evolution of the oscillating reactor and jacket temperatures, electric heater voltage and calculated overall heat transfer coefficient. With permission of VCH Verlagsgesellchaft from Carloff R, Pross A, Reichert K. Temperature oscillation calorimetry in stirred-tank reactors with variable heat-transfer. Chem Eng Technol 1994;17:406–413.

Mauntz [5] developed new approaches for TOC to cope with nonideal situations in the sinusoidal oscillations superimposed either to T_R or T_J. He demonstrated that under these circumstances, algorithms that do not assume a specific shape of the temperature trajectories performed better than the original algorithm developed by Tietze et al. [10].

7.2.3 Heat Balance Calorimetry

Heat balance calorimetry is based on an energy balance on the cooling jacket using Equation 7.2 in order to identify the term Q_f that can then be used in Equation 7.1 to calculate Q_r. The energy balance on the cooling jacket in Equation 7.2 can be simplified in many cases using a pseudo-steady-state approximation and substituting Equation 7.10 for the difference in heat flow in and out of the jacket:

$$Q_f \approx Q_{J,in} - Q_{J,out} - Q_{J,loss} \approx \dot{m}_J c_{p,J} (T_{j,in} - T_{j,out}) \quad (7.16)$$

If one lumps the heat losses through the jacket with heat losses elsewhere, the overall energy balance for Q_r in Equation 7.12 can be rewritten:

$$Q_r = C_{p,tot} \frac{dT_R}{dt} + \dot{m}_J c_{p,J} (T_{j,in} - T_{j,out}) \\ - \sum_i \dot{m}_{F,i} c_{p,i} (T_{f,i} - T_R) + Q_{loss} \quad (7.17)$$

The advantage of using heat balance calorimetry is that Q_f can be measured without knowing a value for the overall heat transfer coefficient or the heat transfer surface area. The major disadvantage is that in order for this approach to be practical, the difference between the reactor inlet and outlet temperatures needs to be measurably different from zero. The only way to achieve this in laboratory and small pilot scale reactors is to use low flow rates in the cooling jacket, which can cause temperature control to be sluggish, and often inadequate. On the other hand, in the case of industrial scale reactors, heat balance calorimetry can be quite useful as the path taken by the cooling fluid in the jacket is long enough that there is an appreciable difference between the inlet and outlet temperatures. In this case, Equation 7.17 can be used to measure the amount of energy generated in the reactor [19].

7.2.4 Reaction Calorimetry in Industrial Reactors

Contrary to lab reactors in which heat balance calorimetry cannot be applied, because the large flow rates of the heating/cooling fluids makes $T_{j,in} \sim T_{j,out}$, in industrial scale reactors, both heat flow and heat balance calorimetry can be applied. However, as U is often not known in this case or significantly changes during the course of the polymerization, heat balance calorimetry is the preferred option.

Gesthuisen et al. [16] showed by simulation that although TOC was claimed [10] to be applicable in reactors as large as $1 m^3$, the actual constraints of the technique (e.g., the high oscillation frequency required as the time constant of the reactor increases) make the technique only suitable in reactors with high jacket mass flow rates and small reactors ($<5 l$).

Furthermore, with certain reasonable approximations, heat balance calorimetry can be used with a minimum of effort in industrial reactors. The heat loss term in the energy balance, Q_{loss} is typically proportional to the surface area of the reactor (Eq. 7.11). Since the specific surface area per unit volume in large reactors is very small, the loss term, even if nonzero, is small with respect to the heat generation term. This means that as a first approximation, if the reactor is operating at a steady state or under isothermal conditions, one can simplify the energy balance to

$$Q_r \approx \dot{m}_J c_{p,J} (T_{j,in} - T_{j,out}) - \sum_i \dot{m}_{F,i} c_{p,i} (T_{f,i} - T_R) \quad (7.18)$$

Thus, if measurements of the inlet and outlet jacket temperatures, the reactor and feed temperature, as well as the flow rate of coolant and feed are known, one can obtain a rapid estimate of Q_r. Even if it becomes necessary to include a value for Q_{loss}, the method proposed by Othman et al. [20] can be used to estimate this parameter during the preparation of the reactor.

Therefore, for large-scale reactors in which the inlet and outlet temperatures of the heating/cooling fluid are significantly different, the heat-balance approach is the most appropriate to compute online the heat of reaction and the heat transfer coefficient.

7.2.5   Monitoring Polymerization Reactions by Means of Reaction Calorimetry

Any of the calorimetric approaches discussed in the earlier sections will provide quasi-instantaneous and continuous information on the heat generated during the polymerization reaction, Q_r. This information is by itself very useful because one can quickly learn about inhibition periods or monomer accumulation [21], autoacceleration processes [22], or other changes occurring during the polymerization [23]. Reaction calorimetry can also provide useful information for process design such as the necessary cooling power, the adiabatic temperature rise, and the heat transfer for scale-up.

However, when monitoring and controlling polymerization reactors, one would like to get not only global reaction kinetics, but also information related to the state variables of the reactor; namely, the concentration of monomer(s), polymer (conversion), concentration of radicals, molar masses, and so on. Obviously, this information cannot be directly obtained from the heat of reaction, but for some cases, the use of simple polymerization models in combination with the heat of reaction allows the state variables to be monitored noninvasively in real time and at much lower cost than when using dedicated instruments (near-infrared spectroscopy, mid-range infrared spectroscopy, Fourier transformed infrared spectroscopy, and/or Raman spectrometers).

7.2.5.1   *Polymerization Rate*   The overall polymerization rate is the only information that can be directly obtained from the online heat of reaction. In a homopolymerization reaction, it can be computed directly, $R_p(t) = Q_r(t)/V_r(-\Delta H_r)$ $(mol\,l^{-1}\,s^{-1})$. For (co)polymerization reactions, strictly speaking, the overall polymerization rate cannot be computed from this equation because the enthalpy of polymerization of each monomer is typically different. However, using an average $(\overline{\Delta H_r})$ will suffice in many cases and a rough estimation can also be obtained from the equation. If a more accurate value is needed, information about the reactivity ratios must be used as it would be discussed in Section 7.2.5.3.

7.2.5.2   *Overall Monomer Conversion*   Overall monomer conversion in a polymerization reaction is defined as the fraction of monomer converted to polymer. In homopolymerization reactions, the overall calorimetric monomer conversion can be directly computed from the heat of reaction as follows:

$$X_c(t) = \frac{\int_0^t Q_r(t)\,dt}{\int_0^{t_f} Q_r(t)\,dt} \tag{7.19}$$

where t_f is the final polymerization time.

The calorimetric conversion calculated by Equation 7.19 might slightly differ from the gravimetric/molar monomer conversion if these do not reach 100% conversion, which is typically the case. Alternatively, the overall calorimetric conversion is readily calculated by a similar expression using the nominal amount of monomer employed in the formulation:

$$X_c(t) = \frac{\int_0^t Q_r(t)\,dt}{\sum_i M_{0,i}(-\Delta H_{r,i})} \tag{7.20}$$

where $M_{0,i}$ is the total amount of monomer i in the formulation (mol) and $(-\Delta H_{r,i})$ the enthalpy of polymerization of monomer i.

Implicit in this equation is the assumption that cross-propagation reactions produce the same heat as the corresponding homopropagation reactions; namely that $\Delta H_{r,ii} = \Delta H_{r,ij} = \Delta H_{r,i}$ [24].

7.2.5.3   *Individual Monomer Concentration*   Although, strictly speaking, the polymerization rate of each monomer and hence free monomer concentration and the polymer composition are not observable from the overall heat of the reaction, Q_r, it is possible to infer them using an open-loop observer based on a model of the process [23, 25]. The overall polymerization heat, Q_r, is related to the rates of reaction of the individual monomers by Equation 7.3 assuming that the heats of cross-propagation are equal to those of homopolymerization [24]:

$$Q_r = R_{pA}(-\Delta H_{r,A}) + R_{pB}(-\Delta H_{r,B}) \tag{7.21}$$

where R_{pA} and R_{pB} are the polymerization rates of monomers A and B $(mol\,s^{-1})$ and $(-\Delta H_{r,A})$ and $(-\Delta H_{r,B})$ enthalpies of polymerization of the monomers A and B $(J\,mol^{-1})$.

For instance, the polymerization rate, R_{pi}, in emulsion polymerization is given by

$$R_{pi} = \left[\sum_j k_{pji}P_j\right][M_i]_p \frac{\overline{n}N_p}{N_A} \qquad i,j = A,B \tag{7.22}$$

where k_{pji} is the propagation rate constant of a radical having ultimate unit of type j with monomer i, $[M_i]_p$ is the concentration of monomer i in the polymer particles, P_j is the time averaged probability of finding a free radical with ultimate unit type j in the polymer particles, $\bar{n}$ is the average number of radicals per particle, N_p is the total number of polymer particles, and N_A is Avogadro's constant.

The ratios of the individual polymerization rates can be calculated as follows (Mayo–Lewis equation):

$$f = \frac{R_{pA}}{R_{pB}} = \frac{r_A + [M_B]_p / [M_A]_p}{r_B \left([M_B]_p / [M_A]_p \right)^2 + [M_B]_p / [M_A]_p}$$

(7.23)

where r_A and r_B are the reactivity ratios of monomer A and B, respectively.

The monomer partitioning between the different phases is accounted for by using the partition coefficients and by solving the material balances for the different phases:

$$K_i^j = \frac{\phi_i^j}{\phi_i^w}, j = \text{polymer particles, droplets} \quad (7.24)$$

Material balances:

$$\phi_{pol}^p + \sum_i \phi_i^p = 1$$

$$\phi_{water}^w + \sum_i \phi_i^w = 1$$

$$\sum_i \phi_i^d = 1$$

$$V_p \phi_i^p + V_d \phi_i^d + V_w \phi_i^w = V_i$$

$$V_w \phi_{water}^w = V_{water}$$

$$V_p \phi_{pol}^p = V_{pol}$$

(7.25)

where K_i^j is the partition coefficient of monomer i between the phase j and the aqueous phase, ϕ_i^j is the volume fraction of monomer i in phase j; V_p, V_d, and V_w are the volumes of polymer, monomer droplets and water; V_i, V_{pol}, and V_{water} are the volumes of monomer i, polymer, and aqueous phase.

The material balances in a semibatch polymerization reactor for a copolymer system are as follows:

$$\frac{dN_i}{dt} = F_{i,in} - R_{pi} V$$

(7.26)

where N_i is the amount of monomer i in the reactor at time t (mol), $F_{i,in}$ is the flow rate of monomer i (mol s^{-1}), and R_{pi} is the polymerization rate of monomer i (mol s^{-1}).

Combining Equations 7.21, 7.23, and 7.26, the material balances can be written as a function of the heat released by polymerization and the reactivity ratios [23, 25]:

$$\frac{dN_A}{dt} = F_{A,in} - \frac{Q_r}{(-\Delta H_A) + (-\Delta H_B) / f}$$

(7.27)

$$\frac{dN_B}{dt} = F_{B,in} - \frac{Q_r}{(-\Delta H_A) f + (-\Delta H_B)}$$

(7.28)

The open-loop observer formed by Equations 7.23, 7.27, and 7.28 has been successfully used to estimate the unreacted amounts of monomers in different monomer systems under different conditions [25–29]. Once the unreacted amounts of the monomers (N_A and N_B) are estimated during the course of the polymerization, the conversion and copolymer compositions can be readily estimated for batch and semibatch reactors. The accuracy of the estimation depends on the reactivity ratios and enthalpies of polymerization for each monomer that can be obtained experimentally or are available in the literature.

Figure 7.4 shows the comparison of instantaneous and overall conversion and free monomer during the seeded semibatch emulsion copolymerizations of n-butyl acrylate/ methyl methacrylate (BA/MMA) calculated online from calorimetric measurements and off-line by gravimetry [29]. These results show that calorimetry predicts overall conversion well but that the prediction of instantaneous conversion and free monomer is less accurate, especially when the polymerization is carried out under starved conditions and/or the monomer concentrations are very low in the reactor, as for the MMA in Figure 7.4d. Under these polymerization conditions, other more accurate techniques (but less robust and more expensive) such as Raman spectroscopy might be better suited for monitoring free monomer concentration [29].

Note that for homogenous polymerizations, the same open-loop observer can be applied in a more straightforward manner because monomer partitioning need not be considered.

7.2.5.4 Instantaneous Number-Average Molar Masses

In free-radical polymerization when the main chain terminating event is by chain transfer to chain transfer agents (CTA), the molar mass of the (co)polymer instantaneously produced is only a function of the polymerization temperature and the ratio of the monomer concentration over the CTA concentration. Thus, the instantaneous number-average molar mass can be computed as

$$M_{n,i} = \frac{k_p [M]}{k_{tr,CTA} [CTA]} w_M = \frac{1}{c_{tr,CTA}} \frac{[M]}{[CTA]} w_M$$

(7.29)

where $k_{tr,CTA}$ is the chain transfer to CTA rate coefficient, [CTA] is the concentration of CTA, and w_M is the molar mass of the repeating unit.

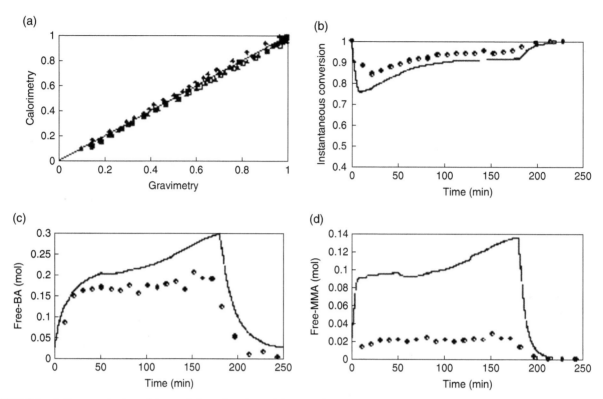

FIGURE 7.4   Online monitoring of the seeded semibatch emulsion copolymerization of BA/MMA = 50/50: calorimetry versus gravimetry in (a) overall conversion, (b) instantaneous conversion, and (c) and (d) free monomers BA and MMA. Line is calorimetry and dots gravimetry. Reprinted (adapted) with permission from Elizalde O, Azpeitia M, Reis MM, Asua JM, Leiza JR. Monitoring emulsion polymerization reactors: calorimetry versus Raman spectroscopy. Ind Eng Chem Res 2005;44:7200–7207. © 2005 American Chemical Society.

Therefore, if the ratio of the monomer to CTA concentration can be computed online, the instantaneous number-average molar masses can be inferred.

Vicente et al. [30] used the heat of reaction and the open-loop observers developed in Section 7.2.5.3 to determine the concentration of monomer and CTA and hence to infer the instantaneous number-average molar masses during emulsion homo- and copolymerization reactions. In addition, the authors used the inferred values for online control of the molar mass distributions of copolymers with predefined distributions. They demonstrated that polymer latexes with unimodal MMD with the minimum achievable polydispersity index in free-radical polymerization (PI = 2) and bimodal distributions could be easily produced in linear polymer systems [15, 30].

7.2.5.5   *Fouling* Industrial reactors are more likely to suffer film formation and fouling on the reactor walls because of the nonstop daily operation, higher solids content, and the less rigorous cleaning procedures as compared with small size lab reactors.

Film formation and fouling on the reactor walls reduce the heat transfer coefficient of the reactor and hence makes

the control of the reactor temperature and the safety of the plant more challenging. The poor control of the temperature might also lead to off-spec polymer production with consequent economic impact.

Reaction calorimetry, as discussed in Sections 7.2.2 and 7.2.3, allows the heat transfer coefficient of the reactor to be estimated. The information of the heat transfer coefficient calculated on a daily basis can be very useful to detect the level of fouling in the reactor walls and to anticipate and therefore carry out actions that minimize its deleterious effects on the performance of the reactors [31].

Furthermore, state estimation techniques such as the "boot-strapping" method developed by BenAmor et al. [14] allow one to account for fouling in real time. The method assumes that the overall heat transfer coefficient (multiplied by the surface) changes very little in the course of the time required to measure the reactor temperature. Thus, the reactor temperature is used to calculate first a heat generation rate over the course of 1–2 s, assuming constant heat transfer conditions, then a heat transfer coefficient assuming constant rate for a very short time. If fouling occurs over a reasonable timescale, it can be accounted for by the software sensor in-line.

7.3 CONDUCTIVITY

The electrical conductivity of a fluid is a quantitative measure of its ability to carry an electrical current, and therefore depends to a large extent on the concentration of ionic species. Given that the conductivity of pure water is extremely low (limited to 0.0548 $\mu S\,cm^{-1}$ at 25 °C by the H_2O dissociation constant into H^+ and OH^- when no added ions are present), this technique will be sensitive to changes in ionic concentration. So, while it is not impossible to be used for the online monitoring of solution or melt phase processes, it is better suited for use in emulsion and miniemulsion polymerization reactions where ionic surfactants and initiators are commonly employed.

The use of conductivity as an online technique is an extension of the well-known analytical method used to measure the critical micelle concentration (CMC) of different anionic surfactants. To measure the CMC, a surfactant such as sodium dodecyl sulfate (SDS) is slowly added to deionized water, and the conductivity recorded. As shown in Figure 7.5, the conductivity increases until the CMC is reached. From this point on, dodecyl sulfate molecules start to form micelles. Due to the larger aggregate size and lower mobility of the micelles, the rate of change of the conductivity signal changes above the CMC. A plot of the conductivity signal for increasing surfactant concentrations at constant temperature ideally produces two straight lines at the CMC. Note that the conductivity will change with temperature.

If this concept is taken a step further, in the case of emulsion polymerization reactions, the conductivity cell can respond to changes of ionic surfactant concentrations, and allowing for the determination of surfactant dynamics in the media, which are related to particle nucleation and/or coagulation phenomena. As particles are created and grow, surfactant molecules present in the reactor are adsorbed from the aqueous phase onto the polymer particle surface. If ionic stabilization is employed, the surfactant molecules lose mobility after adsorption, thereby causing the conductivity to drop. On the other hand, coagulation leads to a decrease in the specific surface of a latex, and causes the conductivity to increase.

As an example, the group of Schork monitored the conductivity of miniemulsions before and during polymerization and interpreted a lack of change in the conductivity as an indication that the surface area of the organic phase did not change and that the number of droplets at the beginning of the process equaled the number of particles at the end [32, 33]. This type of qualitative in-line monitoring is very useful in the sense that it allows to obtain real-time insight into the evolution of the surface area of a latex, and therefore the number (and possibly size) of the particles in the reactor in ways that no other technique can; certainly not at such a low cost.

Other qualitative applications of this type include a study of the differentiation between the intervals of emulsion polymerization [34] and mechanisms of nucleation [35, 36], the swelling of emulsion polymers [37], the partitioning of surfactants [38], and the monitoring of the state of emulsion polymerizations [39]. It has also been demonstrated that particle coagulation could be detected by conductivity measurements [40].

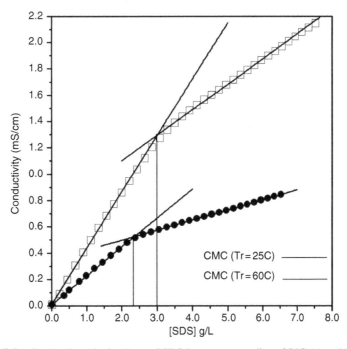

FIGURE 7.5 Conductimetric titrations of SDS in aqueous media at 25 °C (•) and 60 °C (□).

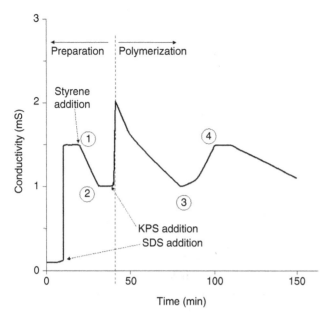

FIGURE 7.6 Evolution of the conductivity during the emulsion polymerisation of styrene.

It is therefore logical that conductivity measurements can be used to monitor the dynamics of particle nucleation, growth, and coalescence in a reactive situation.

Consider the experimental curve in Figure 7.6 that shows the conductivity as a function of time during the preparation and subsequent polymerization of a styrene emulsion at 60 °C using potassium persulfate (KPS) as the free radical initiator and SDS as the anionic surfactant [41]. Initially, the conductivity is nearly zero in deionized water, and increases when SDS is introduced (here, at two times the CMC of SDS). The conductivity then decreases when monomer is added because a portion of the surfactant is absorbed on the surface of the monomer droplets. With the aid of a calibration curve, the difference in conductivity between points 1 and 2 in Figure 7.6 indicates precisely how much surfactant is absorbed on the surface of the organic phase. When the initiator is added, there is a sharp increase due to the negatively charged sulfate radicals that are formed. The decrease in the signal from the peak of conductivity following KPS addition to point 3 corresponds to the creation of new particle surface that needs to be stabilized. The increase from point 3 to point 4 corresponds to the coalescence of understabilized droplets as the system reorganizes itself [42].

Since the exact nature of all factors that contribute to the conductivity signal in a complex milieu like a latex is not known [41], it is difficult to develop a purely mechanistic approach to linking latex properties. However, it is possible to develop a semiempirical model to link measurements of the conductivity shown in Figure 7.6 to changes in the surface area of the particle phase in the reactor as a function of

the latex composition and the reactor temperature [42–44]. Assuming that the conductivity signal is the sum of contributions from surfactant molecules present in the latex as free molecules, as micelles, and as adsorbed species, the following expression can be written [43]:

$$\sigma = \sigma_0 + (\xi_0)\frac{m_e^{aq}}{V^{aq}} + (\xi_1)\frac{m_e^{mic}}{V^{aq}} + (\xi_2)\frac{1}{m_p}\frac{m_e^{ads}}{V^{aq}} \quad (7.30)$$

The number of particles is related to the conductivity signal through the last term in Equation 7.30, which gives the contribution of surfactant absorbed on the particle surface per mass of particle. Note that in this simplified model, there is no explicit contribution from components other than an ionic surfactant. In a batch reactor under conditions where the initiator decomposes slowly, the contribution of components like persulfate initiators to the conductivity signal is included in the constant baseline contribution (σ_0); if this is not realistic, additional terms would need to be added. Model parameters were estimated by fitting the model responses to experimental conductivity data obtained for different polymerization runs. The final set of model parameters (Eq. 7.31) was so large that the effects of temperature and ionic strength had to be taken into account in the conductivity signal.

$$\begin{bmatrix} \sigma_0 \\ \xi_0 \\ \xi_1 \\ \xi_2 \end{bmatrix} = \begin{bmatrix} \hat{\sigma}_0 & \sigma_0^T & \sigma_0^E & \sigma_0^{ET} & \sigma_0^{EE} & \sigma_0^{EET} \\ \hat{\xi}_0 & \xi_0^T & \xi_0^E & \xi_0^{ET} & \xi_0^{EE} & 0 \\ \hat{\xi}_1 & \xi_1^T & \xi_1^E & \xi_1^{ET} & \xi_1^{EE} & \xi_1^{EET} \\ \hat{\xi}_2 & \xi_2^T & \xi_2^E & \xi_2^{ET} & \xi_2^{EE} & 0 \end{bmatrix} \cdot \begin{bmatrix} 1 \\ (T_0 - T_r) \\ [AIS] \\ (T_0 - T_r)[AIS] \\ [AIS]^2 \\ (T_0 - T_r)[AIS]^2 \end{bmatrix}$$

$$(7.31)$$

where [AIS] is the concentration of anionic surfactant, T_0 and T_r are reference and reactor temperatures, and the different elements of the first matrix are empirical parameters expressing the dependence of parameters in Equation 7.31 on the temperature and surfactant concentrations.

Since this model is empirical, it is necessary to estimate model parameters for different formulations. Nevertheless, as shown in Figure 7.7, this modeling approach can be used to predict the number of polymer particles in the latex without need of online measurements of particle size.

In conclusion, online conductivity is a useful tool for the online monitoring emulsion and miniemulsion polymerizations that are stabilized by anionic surfactants. If applied in a

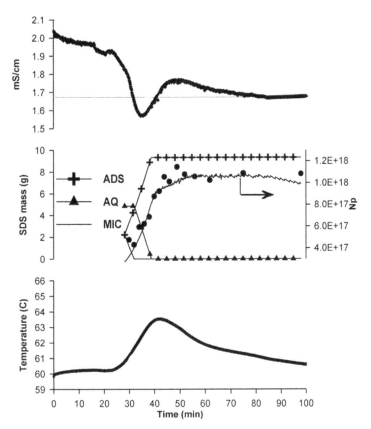

FIGURE 7.7 Conductivity measurements (top), model predictions, and off-line measurements of the number of particles per liter of emulsion, and surfactant partitioning between micelles (MIC), aqueous phase (AQ), and adsorbed on the surface (ADS) (middle) and reactor temperature for the emulsion polymerization of styrene using SDS as the surfactant and KPS as the free radical initiator. This material is reproduced with permission of John Wiley & Sons, Inc. from Santos AF, Pinto JC, McKenna TFL. On-line monitoring of the evolution of number of particles in emulsion polymerization by conductivity measurements. Part II. Model validation. J Appl Polym Sci 2004;91:941–952. © 2004 John Wiley & Sons, Inc.

qualitative manner, one can interpret the conductivity signal in terms of the evolution of the number and/or size of the particles and droplets in the reactor, and detect either particle nucleation or coagulation events.

7.4 DENSIMETRY

Another useful technique for measuring the progress of polymerization reactions is densimetry. This technique relies on density differences between the various components in the reactor in order to estimate the conversion. Under typical reaction conditions, the density of monomers is on the order of 0.8–0.9 g cm^{-3} and that of polymers generally superior to 1 g cm^{-3}. If a precise estimate of the density of the reactor contents can be obtained, it is possible to use a mass balance to calculate the amount of polymer produced as a function of time, and therefore to find the instantaneous conversion [46–52].

The U-tube densimeter is the most commonly used apparatus for online measurements [51]. When used in this manner, the U-tube of the densimeter is connected to a pump which continually recirculates the fluid to be measured through the cell at a constant temperature. The operating principle of the densimeter is based on the measurement of the period of oscillation (T) of a U-tube filled with the solution to be analyzed. The frequency measurements are converted into density measurements by means of the following empirical correlation:

$$\rho_{\text{measured}} = A(T^2 - B) \qquad (7.32)$$

where A and B are temperature-dependent parameters that must be determined via simple calibration of the apparatus.

The densimeter should contain a thermocouple accurate to within at least $\pm 0.1\,^{\circ}\text{C}$.

The measured density of the reactor contents (ρ_{measured}) is related to the densities of the individual components and the conversion by the following general expression (ignoring the contribution of the initiator):

$$\rho_{measured} = \frac{Total\ mass}{Total\ volume}$$

$$= \frac{N_s^{tot} M_s + \sum N_i^{tot} M_i}{\dfrac{N_s^{tot} M_s}{\rho_s} + \dfrac{\sum x_i N_i^{tot} M_i}{\rho_p} + \sum \left(\dfrac{(1-x_i) N_i^{tot} M_i}{\rho_i} \right)}$$

(7.33)

where ρ_s, ρ_p, and ρ_i are the densities of the solvent, copolymer, and monomer i, respectively; M_s and M_i are the molar mass of solvent and monomer i, respectively; and N_s^{tot} and N_i^{tot} are the total number of moles of solvent and monomer added to the reactor up to the time the density measurement is taken.

As written, Equation 7.33 is valid for solution polymerization reactions with one solvent. In principle, it should be straightforward to extend this concept to emulsion (or other dispersed phase reactions) by including mass and volume terms for the emulsifier, and any other species added to the reactor.

In the special case of a homopolymerization, or a multi-component polymerization with no composition drift (e.g., at azeotrope), the polymer density does not depend on its composition and therefore Equation 7.33 provides a direct, measurable estimate of monomer conversions.

However, in most multicomponent situations, the polymer density does depend on its overall composition and it is therefore necessary to account for this by using either an empirical expression, or, as is done here, assuming that the density of the copolymer is equal to the sum of the densities of the component homopolymers, ρ_{pi}, weighted by their individual mass fractions in the following manner:

$$\rho_p = \sum x_i N_i^{tot} M_i \left(\sum \frac{x_i N_i^{tot} M_i}{\rho_i} \right)^{-1}$$

(7.34)

It is therefore required to know the individual monomer conversions in order to use densimetry in the case of multiple monomer systems [51].

Despite the apparent simplicity of the technique, there are several challenges to be overcome before it can be used reliably online. For instance, it is imperative that bubbles be avoided in the transfer lines and the U-tube [50]. In the case of polymerizations involving more than one monomer, it is also necessary to have a separate online measurement for each monomer in order to obtain the value of the individual conversions in Equation 7.33. Furthermore, precise density data are required, often as high as four significant figures in order to obtain reliable estimates of the conversion in simple solution polymerizations [52]. Such data are not always

available, and, when they are, require very accurate temperature control in the sampling tubes. Further complications arise in the event of emulsion polymerizations since flocculation of the latex can be a problem. The presence of monomer droplets can also pose certain problems related to stability. Finally, as the viscosity of the reactor contents increases, it can be difficult to take meaningful samples in real time.

7.5 RHEOLOGICAL SENSORS

7.5.1 Introduction

Rheology is the science of deformation and flow. Viscosity, η, is a measure of the stress (τ) required to make a fluid to deform (flow) at a desired strain rate ($\dot{\gamma} = dv_x / dy$).

$$\tau_{yx} = \eta \dot{\gamma}_{yx}$$

(7.35)

A unique feature of polymers is that even a small concentration (c) of a high molar mass polymer considerably enhances the viscosity of a solution as compared with the viscosity of the solvent, η_0. The capacity of the polymer to increase the solution viscosity is described in terms of its intrinsic viscosity, $[\eta]$:

$$[\eta] = \lim_{c \to 0} \frac{\eta - \eta_0}{c \eta_0} = \lim_{c \to 0} \frac{\eta_s}{c}$$

(7.36)

where η_s is the specific viscosity.

The dependence of the intrinsic viscosity on the molar mass of the polymer is given by the Mark–Houwink relationship

$$[\eta] = K M^\alpha$$

(7.37)

where K and α are known as the M–H coefficients obtained by fitting experimental data for a given polymer–solvent system.

In a dilute polymer solution, the dependence of the viscosity on the concentration is almost linear, that is, the intrinsic viscosity is independent of the concentration. However, as the concentration of polymer in solution increases (above the overlap concentration, c^*, when polymer coils get entangled with each other), the dependence on the concentration is not linear. At high molar masses and concentrations, the dependence of viscosity on concentration and molar mass can be expressed by the following equation [53]:

$$\eta \sim c^{4-5} M^{3.4}$$

(7.38)

During a homogenous (bulk or solution) polymerization reaction, both the concentration of polymer and molar mass change and hence viscosity should also change. Therefore, online monitoring of the viscosity could give access to process information and it could also be used to infer molar mass.

In polymerizations in disperse phase (emulsion, suspension, and dispersion polymerizations), the viscosity of the dispersed systems depends on the volume fraction of the dispersed polymer, ϕ, particle shape and size, particle size distribution, interparticle interaction, and shear rate [54]. For monodisperse particles at low ϕ, viscosity is proportional to the volume fraction of the disperse phase as given by the Einstein equation:

$$\eta = \eta_0(1 + 2.5\phi) \tag{7.39}$$

For concentrated dispersions where interparticle interactions are dominant, several theoretically based, semiempirical and empirical equations have been reported in the literature [55–57]. As for homogeneous polymerization systems, the viscosity of disperse phase polymerization changes; hence, indirect information of particle number and size can be obtained by monitoring the viscosity. Furthermore, since the viscosity of the polymerization media affects the heat transfer coefficient, online monitored viscosity will be very useful for understanding process changes and safety during the polymerization.

7.5.2 Monitoring Viscosity in Polymerization Reactions

Rotational (torque measurements) or capillary process viscometers, as well as ultrasound and tube oscillations [58] can be used to measure viscosity online during a polymerization reaction. The first two techniques have been widely applied in the monitoring of both chain growth polymerizations and step-growth polymerizations.

In capillary viscometers, one measures the pressure drop when a liquid is flowing with a constant laminar flow in a capillary. Typically, the constant flow is achieved by a pump and the pressure drop is measured by a pressure transducer. The pressure drop is then directly proportional to the viscosity according to the Hagen–Poiseuille law (Eq. 7.40) with r and L being the radius and length of the capillary, respectively, and Q the volumetric flow rate.

$$\eta = \frac{\pi r^4 \Delta P}{8QL} \tag{7.40}$$

Reed and coworkers [59, 60] were the first to develop a robust polymerization setup to online monitor reduced viscosity in free-radical and living polymerization. The methodology developed by the authors is based on the continuous extraction and dilution of a small stream from the reactor on which measurements are being made while it flows though a multidetector platform. Among the detectors used, a single capillary connected to a pressure transducer measures the pressure drop and thus, by combining with concentration data computed from spectroscopic detection, allows the reduced viscosity to be measured online during the polymerization. The same approach has also been used successfully by Alb et al. [61] in the case of heterogeneous phase polymerization.

Torque measurements have been used for online monitoring of the viscosity of polymerization reactions. The advantage of the torque measurement as compared with that of the capillary is that no treatment (dilution and flow) of the reaction medium is needed. Several examples of monitoring chain-growth polymerization reactions [62, 63] and step-growth polymerization reactions (specially curing reactions) can be found in the literature [64].

7.6 CONCLUSIONS

Four different types of indirect sensors for the monitoring of polymerization reactors have been considered, each with different applications and advantages.

Calorimetry can be used to estimate the rate of heat generation, and, if enough is known about the kinetics of the reaction of interest, it can also be used to follow the evolution of the rate of reaction and composition of a multiple component system in real time. While the applications discussed here focused mostly on free-radical polymerization, calorimetry can be used for essentially any exothermic or endothermic reaction (polymerization or otherwise), and on reactors containing any number of phases. It is an inexpensive technique that relies for the most part on measurements that are commonly taken in most reaction settings. Given its flexibility, low cost, and simplicity, calorimetry is widely spread both in research laboratories and industrial settings. In the latter case, heat balance calorimetry is an excellent option if accurate information is available for the flow rate through the cooling jacket. All said, calorimetry is probably the most used (and certainly the most generic) of the methods presented in this chapter.

Conductivity measurements are more specific in their application; being of particular use in the case of aqueous dispersions of particles, stabilized by some type of ionic stabilizer. The information obtained with this technique is the conductivity of the continuous phase. In the case of emulsion polymerizations stabilized with ionic surfactants, this is related to the concentration of free surfactant, which, when combined with absorption isotherms, for example, or an empirical model, can be used to follow the evolution of the surface area of a latex. This is a promising method, but given its complexity and the need to develop more robust means of linking the conductivity to properties of interest, it has not found widespread use in commercial production at the current time.

Densimetry is another technique with very limited applications in the area of polymer production. The fact that accurate, reliable measurements must be obtained, coupled with operational difficulties, such as fouling, means that it is not easy to apply densimetry with confidence in many cases. However, for single-phase systems, where the relationship between the density and the composition of the reaction medium is understood, it can be useful.

Finally, rheological measurements are also found in many contexts. They can certainly provide a rapid estimate of the state of the reactor contents. If only from a safety standpoint, this is useful in itself. Also, with an appropriate relationship between the viscosity and a property of interest, it can be used to furnish reliable information about the order of magnitude of the molar mass, the solid content in the reactor, and eventually to detect problems such as flocculation in an emulsion polymerization reaction.

REFERENCES

[1] Moritz HU. Polymerization calorimetry—a powerful tool for reactor control. In: Reichert KH, Geiseler W, editors. *Polymer Reaction Engineering*. Weinheim: VCH Verlag; 1989. p 248–266.

[2] McKenna T, Othman S, Fevotte G, Santos A, Hammouri H. An integrated approach to polymer reaction engineering: a review of calorimetry and state estimation. Polym React Eng 2000;8:1–38.

[3] Schmidt C, Reichert K. Reaction calorimeter a contribution to safe operation of exothermic polymerizations. Chem Eng Sci 1988;43:2133–2137.

[4] Karlsen L, Villadsen J. Isothermal reaction calorimeters. 1. A literature-review. Chem Eng Sci 1987;42:1153–1164.

[5] Mauntz W. A contribution to observation and time-optimal control of emulsion co-polymerization reactions [Ph.D.]. Dortmund (Germany): Technische Universitat Dortmund; 2010.

[6] Regenass W. Calorimetric monitoring of industrial-chemical processes. Thermochim Acta 1985;95:351–368.

[7] Bonvin D, Devalliere P, Rippin D. Application of estimation techniques to batch reactors. 1. Modeling thermal effects. Comp Chem Eng 1989;13:1–9.

[8] Karlsen L, Villadsen J. Isothermal reaction calorimeters. 2. Data treatment. Chem Eng Sci 1987;42:1165–1173.

[9] deBuruaga IS, Arotcarena M, Armitage PD, Gugliotta LM, Leiza JR, Asua JM. On-line calorimetric control of emulsion polymerization reactors. Chem Eng Sci 1996;51: 2781–2786.

[10] Tietze A, Ludke I, Reichert K. Temperature oscillation calorimetry in stirred tank reactors. Chem Eng Sci 1996;51: 3131–3137.

[11] Carloff R, Pross A, Reichert K. Temperature oscillation calorimetry in stirred-tank reactors with variable heat-transfer. Chem Eng Technol 1994;17:406–413.

[12] Fevotte G, Barudio I, Guillot J. An adaptive inferential measurement strategy for on-line monitoring of conversion in polymerization processes. Thermochim Acta 1996;289: 223–242.

[13] Grob B, Riesen R, Vogel K. Reaction calorimetry for the development of chemical-reactions. Thermochim Acta 1987;114:83–90.

[14] BenAmor S, Colombie D, McKenna T. Online reaction calorimetry. Applications to the monitoring of emulsion polymerization without samples or models of the heat-transfer coefficient. Ind Eng Chem Res 2002;41:4233–4241.

[15] Vicente M, Leiza JR, Asua JM. Maximizing production and polymer quality (MWD and composition) in emulsion polymerization reactors with limited capacity of heat removal. Chem Eng Sci 2003;58:215–222.

[16] Gesthuisen R, Kramer S, Niggemann G, Leiza JR, Asua JM. Determining the best reaction calorimetry technique: theoretical development. Comp Chem Eng 2005;29:349–365.

[17] De Luca P, Scali C. Temperature oscillation calorimetry: robustness analysis of different algorithms for the evaluation of the heat transfer coefficient. Chem Eng Sci 2002;57: 2077–2087.

[18] Lüdke IS. Untersuchurgen zur Methode der Temperaturchwingungs-kalorimetrie in einem Ruhrkesselreaktor. Berlin: Technischen Universitat Berlin; 1999.

[19] Hvalaa N, Aller F, Mitevaa T, Kukanjab D. Modelling, simulation and control of an industrial, semi-batch, emulsion-polymerization reactor. Comp Chem Eng 2011;35: 2066–2080.

[20] Othman N, Santos A, Févotte G, McKenna TF. Monitoring of emulsion polymerisations: a study of reaction kinetics in the presence of secondary nucleation. Can J Chem Eng 2002; 80:88–104.

[21] deBuruaga IS, Echevarria A, Armitage PD, de la Cal JC, Leiza JR, Asua JM. On-line control of a semibatch emulsion polymerization reactor based on calorimetry. AIChE J 1997;43:1069–1081.

[22] De la Rosa L, Sudol E, El-Aasser M, Klein A. Emulsion polymerization of styrene using reaction calorimeter. I. Above and below critical micelle concentration. J Polym Sci Part A: Polym Chem 1999;37:4054–4065.

[23] Gugliotta LM, Leiza JR, Arotcarena M, Armitage PD, Asua JM. Copolymer composition control in unseeded emulsion polymerization using calorimetric data. Ind Eng Chem Res 1995;34:3899–3906.

[24] Urretabizkaia A, Sudol ED, Elaasser MS, Asua JM. Calorimetric monitoring of emulsion copolymerization reactions. J Polym Sci Part A: Polym Chem 1993;31: 2907–2913.

[25] de Buruaga IS, Leiza JR, Asua JM. Model-based control of emulsion terpolymers based on calorimetric measurements. Polym React Eng 2000;8:39–75.

[26] de Buruaga IS, Armitage PD, Leiza JR, Asua JM. Nonlinear control for maximum production rate of latexes of well-defined polymer composition. Ind Eng Chem Res 1997;36: 4243–4254.

[27] Gugliotta LM, Arotcarena M, Leiza JR, Asua JM. Estimation of conversion and copolymer composition in semicontinuous emulsion polymerization using calorimetric data. Polymer 1995;36:2019–2023.

[28] Vicente M, Leiza JR, Asua JM. Simultaneous control of copolymer composition and MWD in emulsion copolymerization. AIChE J 2001;47:1594–1606.

[29] Elizalde O, Azpeitia M, Reis MM, Asua JM, Leiza JR. Monitoring emulsion polymerization reactors: calorimetry versus Raman spectroscopy. Ind Eng Chem Res 2005;44: 7200–7207.

[30] Vicente M, BenAmor S, Gugliotta LM, Leiza JR, Asua JM. Control of molecular weight distribution in emulsion polymerization using on-line reaction calorimetry. Ind Eng Chem Res 2001;40:218–227.

[31] Wu R. Dynamic thermal analyzer for monitoring batch processes. Chem Eng Prog 1985;81:57–61.

[32] Reimers JL, Schork FJ. Predominant droplet nucleation in emulsion polymerization. J Appl Polym Sci 1996;60: 25–262.

[33] Tsavalas JG, Gooch JW, Schork FJ. Water-based crosslinkable coatings via miniemulsion polymerization of acrylic monomers in the presence of unsaturated polyester resin. J Appl Polym Sci 2000;75:916–927.

[34] Wang I, Schork FJ. Miniemulsion polymerization of vinyl acetate with nonionic surfactant. J Appl Polym Sci 1994;54:2157–2164.

[35] Kuhn I, Tauer K. Nucleation in emulsion polymerization: a new experimental study. Macromolecules 1995;28:8122–8128.

[36] Janssen RQF, Van Well WJM, Van Herk AM, German AL. Determination of the maximum swellability of polymer by monomer using conductivity measurements. J Colloid Interf Sci 1995;175:461–469.

[37] Boutti S, Graillat C, McKenna TF. A look at surfactant partitioning in polymeric latexes using conductivity measurements. Eur Polym J 2004;40:2671–2677.

[38] Zhao F. *Online conductivity and stability in the emulsion polymerization of n-butyl methacrylate* [Ph.D.]. Bethlehem (PA): Lehigh University; 2011.

[39] Janssen RQF. *Polymer encapsulation of titanium dioxide: efficiency, stability and compatibility* [Ph.D.]. Eindhoven (the Netherlands): Eindhoven University of Technology; 1994.

[40] Santos AF, Pinto JC, McKenna TFL. In-situ monitoring of emulsion polymerisation of MMA and of styrene using conductimetry and calorimetry. Macromol Symp 2005;226: 157–166.

[41] Fortuny M, Graillat C, Araujo P, Pinto JC, McKenna TF. Modelling the nucleation stage in batch emulsion polymerization. AIChE J 2005;51:2521–2533.

[42] Noel LJF, Janssen RQF, van Well WJM, van Herk AM, German AL. Determination of the maximum swellability of polymer by monomer using conductivity measurements. J Coll Inter Sci 1995;175:461–469.

[43] Santos AF, Pinto JC, McKenna TFL.On-line monitoring of the evolution of number of particles in emulsion polymerization by conductivity measurements. Part I. Model formulation. J Appl Polym Sci 2003;90:1213–1226.

[44] Santos AF, Pinto JC, McKenna TFL. On-line monitoring of the evolution of number of particles in emulsion polymerization by conductivity measurements. Part II. Model validation. J Appl Polym Sci 2004;91:941–952.

[45] Graillat C, Santos AF, Pinto JC, McKenna TFL. On-line monitoring of emulsion polymerisation using conductivity measurements. Macromol Symp 2004;206:433–442.

[46] Schork FJ, Ray WH. On-line monitoring of emulsion polymerisation reactor dynamics. ACS Symp Series 1981;165: 505–510.

[47] Chien DCH, Penlidis A. On-line sensors for polymerization reactors. J Macromol Sci: Rev Macromol Chem Phys 1990; C30:1–27.

[48] Ponnuswamy S, Shah SL, Kiparissides C. On-line monitoring of polymer quality in a batch polymerisation reactor. J Appl Polym Sci 1986;32:3239–3246.

[49] Canegallo S, Storti G, Morbidelli M, Carrà S. Densimetry for on-line conversion monitoring in emulsion homo- and copolymerization. J Appl Polym Sci 1993;47:961–979.

[50] Barudio I, Févotte G, McKenna TF. Density data for copolymer systems: butyl acrylate/vinyl acetate homo- and copolymerisation in ethyl acetate. Eur Polym J 1999;35: 775–780.

[51] McKenna TF, Févotte G, Graillat C, Guillot J. Joint use of calorimetry, densimetry and mathematical modelling for multiple component polymerisations. Trans I Chem E 1996;74A:340–348.

[52] Kammona O, Chatzi EG, Kiparissides C. Recent developments in hardwaresensors for the on-line monitoring of polymerization reactions. J Macromol Sci: Rev Macromol Chem Phys 1999;C39:57–134.

[53] Agarwal UD. Polymer properties through structure. In: Meyer T, Keurentjes J, editors. *Handbook of Polymer Reaction Engineering*. Weinheim: Wiley-VCH; 2005. p 679–720.

[54] Guyot A, Chu F, Schneider M, Graillat C, McKenna T. High solid content latexes. Prog Polym Sci 2002;27:1573–1615.

[55] Arevalillo A, do Amaral M, Asua J. Rheology of concentrated polymeric dispersions. Ind Eng Chem Res 2006;45: 3280–3286.

[56] Pishvaei M, Graillat C, McKenna T, Cassagnau P. Experimental investigation and phenomenological modeling of the viscosity-shear rate of bimodal high solid content latex. J Rheol 2007;51:51–69.

[57] Takamura K, van de Ven T. Comparisons of modified effective medium theory with experimental data on shear thinning of concentrated latex dispersions. J Rheol 2010;54:1–26.

[58] Dietrich T, Freitag A, Schlecht U. New micro viscosity sensor—A novel analytical tool for online monitoring of polymerization reactions in a micro reaction plant. Chem Eng J 2010;160:823–826.

[59] Catalgil-Giz H, Giz A, Alb A, Reed W. Absolute online monitoring of a stepwise polymerization reaction: polyurethane synthesis. J Appl Polym Sci 2001;82:2070–2077.

[60] Florenzano F, Strelitzki R, Reed W. Absolute, on-line monitoring of molar mass during polymerization reactions. Macromolecules 1998;31:7226–7238.

[61] Alb AM, Reed W. Fundamental measurements in online polymerization reaction monitoring and control with a focus on ACOMP. Macromol React Eng 2010;4:470–485.

[62] Ponnuswamy S, Shah S, Kiparissides C. Online monitoring of polymer quality in a batch polymerization reactor. J Appl Polym Sci 1986;32:3239–3253.

[63] Neef C, Ferraris J. MEH-PPV: improved synthetic procedure and molecular weight control. Macromolecules 2000;33: 2311–2314.

[64] Castro J, Lopez Serrano F, Camargo R, Macosko C, Tirrell M. Onset of phase-separation in segmented urethane polymerization. J Appl Polym Sci 1981;26:2067–2076.

8

LIGHT SCATTERING

Guy C. Berry

8.1 INTRODUCTION

Electromagnetic scattering (light, x-ray, and neutron) has long been used to characterize a wide variety of material properties, including especially thermodynamic, dynamic, and structural features. The scattering arises from fluctuations in the contrast factor appropriate to the wavelength of the scattered radiation, for example, the refractive index in the case of light scattering, the principal subject of this chapter. The subject here is limited to a relatively narrow subset of these studies, focusing on the characterization of polymeric solutes in dilute solutions, or particles in dilute dispersions. The properties of interest are those that may be obtained via dynamic scattering and from the static, or total intensity, scattering. The dynamic, or quasi-elastic, scattering involves an analysis of the fluctuations of the scattering arising on the timescale of the solute (or particle) motion in the dilute solution, via an autocorrelation analysis of the fluctuations in the photon count statistics of the scattered light. Static scattering refers to the time-averaged intensity of light measured under conditions designed to average over the fluctuations in the intensity that are used with advantage in dynamic light scattering; static scattering is also referred to by the terms elastic static or absolute intensity scattering.

Both dynamic and static scattering measurements are made as functions of the scattering angle θ, with a detector in the plane containing the incident and scattered beam; that angle determines the modulus $q = (4\pi/\lambda) \sin(\theta/2)$ of the scattering vector (the wavenumber), where λ is the wavelength of the incident light in the scattering medium, given by $\lambda = \lambda_0/n$, with λ_0 being the wavelength *in vacuo* and n the refractive index of the medium. Often, the incident beam is vertically polarized in the plane defined by the incident and

scattered beams, and the vertically polarized component of the scattered beam is isolated, and that will be assumed to be the case unless otherwise specified, as with depolarized scattering. As would be anticipated, a number of monographs and review chapters are available on the subjects of interest here, some of which are included in the bibliography as papers of particular relevance, review chapters or books on static scattering [1–31], dynamic scattering [13, 25–35], and static scattering from large particles [36–47], and review chapters or books on polymer solution properties [48–54].

Dynamic scattering makes use of the (unnormalized) photon count autocorrelation function $G^{(2)}(\tau; q)$ computed from the photon count statistics for a given q [13, 14, 32, 34, 35]:

$$G^{(2)}(\tau;q) = <\tilde{n}(t;q)\ \tilde{n}(\tau+t;q)>_t \qquad (8.1)$$

where $\tilde{n}(t; q)$ is the number of photons detected at angle θ over the time interval t to $t + \Delta t$, and the average is over the time t of the data acquisition, which must exceed the longest correlation times in the photon count statistics, see the following text; the effect of the interval Δt on the result is discussed in detail elsewhere [13]. In the limit with τ longer than all such correlation times, $G^{(2)}(\tau; q)$ tends to the asymptotic limit $G^{(2)}(\infty; q)$, with $[G^{(2)}(\infty; q)]^{1/2}$ proportional to the time-averaged intensity $I(q)$ of the scattered light.

Analysis of the static scattering involves the Rayleigh ratio $\boldsymbol{R}(q)$, given by $r^2 I(q)/VI_{INC}$, where r is the distance between the scattering centers and the detector, V the scattering volume, and I_{INC} the intensity of the incident light [5, 13, 32]. With analog detectors, the measured signal is proportional to $I(q)$. Notation to indicate the dependence of $G^{(2)}(\tau; q)$ and $\boldsymbol{R}(q)$ on solute concentration c is suppressed here, for convenience but, for example,

Monitoring Polymerization Reactions: From Fundamentals to Applications, First Edition. Edited by Wayne F. Reed and Alina M. Alb.
© 2014 John Wiley & Sons, Inc. Published 2014 by John Wiley & Sons, Inc.

$R(q) = R(q,c)$, $G^{(2)}(\tau; q) = G^{(2)}(\tau; q,c)$, etc., and this notation will be used where needed for clarity.

The dependence of both static and dynamic scattering on q includes effects from the interference among the scattered rays as determined at the detector. As depicted in Figure 8.1, this interference arises from the difference in path lengths reaching the observer from light scattered from different scattering centers in the sample. With static scattering from dilute solutions (or dispersions), the interference is principally among rays scattered from a particular molecule (or particle), with the interference of rays from different particles a much weaker effect, that is, points j and k in Figure 8.1 would predominantly refer to points on the same molecule (or particle). These effects vanish as θ tends to zero since in that case all paths have the same length, but (usually) cause $R(q)$ to decrease with increasing q. As elaborated in the following, the resulting tangent of $R(q,c)/c$ versus q^2 for small q extrapolated to infinite dilution is used to determine the root-mean-square radius of gyration R_G, requiring $(qR_G)^2 > 0.05$ for reasonable accuracy in the estimate of R_G, placing a lower bound on the value of R_G that may be determined of about $R_G > \lambda/20$. By contrast, the fluctuations that determine the dynamic scattering arise principally from the fluctuating interference of the scattered rays from different molecules (or particles) as the position of these move relative to each other, that is, points j and k in Figure 8.1 on different molecules (or particles), with interference effects among rays scattered from a single molecule (or particle) playing a secondary role. In particular, the analysis of $G^{(2)}(\tau; q)$ extrapolated to infinite dilution gives information on the translational diffusion constant D_T, or equivalently the hydrodynamic radius, R_H, with the lower bound on R_H not set by q, but rather by the upper bound on the intensity fluctuation frequency that

may be determined and separated from fluctuations in the intensity of the scattering from that due to the solvent. This will typically allow R_H to be determined to much smaller values than is possible with R_G. Representative expressions for R_G and R_H are given in Table 8.1 for later reference.

In the scattering to be discussed here, unless otherwise stated, it will be assumed that the Rayleigh–Gans–Debye (RGD) approximation is valid, such that R_G/λ is small enough, and the optical contrast weak enough that it may be assumed that the electric field giving rise to the dipole radiation of the scattered light is that of the incident radiation propagating in the medium [11, 24, 36, 37]. This approximation, which permits a considerable simplification in the interpretation of the scattering, will generally be valid even for very high molecular weight threadlike polymers in dilute solution since the density of the polymer in a domain of order R_G in radius will be small, of the order of 1%, but appreciable error can be realized with large particles differing in refractive index from the solvent, giving rise, for example, to the Mie scattering regime for spheres, with effects described in the following.

Since both polarized and depolarized scattering are included in the following, it is convenient to establish a nomenclature to describe the polarization state of the scattered intensity relative to the incident beam. The scattered intensity and corresponding Rayleigh ratio will be designated by $I_{Si}(q)$ and $R_{Si}(q)$, respectively, where the subscripts S and i designate the polarization of the scattered and incident light, respectively, with respect to the scattering plane. More precisely, we will be interested in the excess Rayleigh ratio, equal to the Rayleigh ratio for the solution less that for the solvent, but notation to this effect is suppressed for convenience except to note that the excess scattering is an explicit function of the concentration c of the solute. The components $R_{Hv}(q,c)$ and $R_{Vv}(q,c)$ comprise contributions designated $R_{iso}(q,c)$, $R_{aniso}(q,c)$, and $R_{cross}(q,c)$ [10, 24]:

$$R_{Hv}(q,c) = R_{aniso}(q,c) \tag{8.2}$$

$$R_{Vv}(q,c) = R_{iso}(q,c) + \left(\frac{4}{3}\right)R_{aniso}(q,c) + R_{cross}(q,c) \tag{8.3}$$

where $R_{aniso}(q,c)$ and $R_{cross}(q,c)$ vanish for an isotropic solute with $R_{Hv}(q,c) = 0$, so that $R_{Vv}(q,c) = R_{iso}(q,c)$; $R_{cross}(0,c) = 0$ in any case.

The subscript iso will be suppressed for convenience when considering this behavior, with $R_{Vv}(q,c)$ denoted simply as $R(q,c)$. If unpolarized incident light is used, as was often the case prior to the now nearly universal use of plane polarized light generated by lasers as the source for the incident light, the scattered light will comprise $R_{Vv}(q,c) + R_{Vh}(q,c)$ if a vertical polarization analyzer is used, or these plus $R_{Hv}(q,c) + R_{Hh}(q,c)$ if no analyzer is used, with $R_{Hv}(q,c) = R_{Vh}(q,c)$ and $R_{Hh}(q,c) = \cos^2(\theta)R_{Vv}(q,c)$ in the RGD regime.

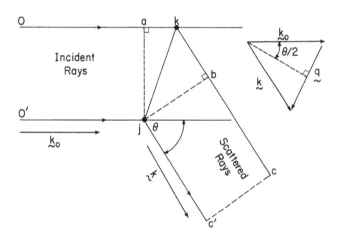

FIGURE 8.1 The wave vector q defined as the difference $k_0 - k$ between the incident wave vector k_0 and the scattered wave k. The diagram depicts the difference in the paths oakbc and o'j'c' for the rays scattered by point scattering centers j and k separated by r_{jk}. The path difference ak+kb, equal to $q \cdot r_{jk}$ gives rise to interference effects important in light scattering analysis.

TABLE 8.1 Relations for R_H, R_G^2, and the Intramolecular Interference Function $P(q,0)$ for Several Model Structures[a] (Rayleigh–Gans–Debye Regime)

Model	Length Scales	R_H	R_G^2	$P(q,0)$[b]
Random-flight linear coil ("infinitely thin")	L=contour length $\hat{a}$=persistence length	$(2/3)(\hat{a}L/3)^{1/2}$	$\hat{a}L/3$	$(2/u^2)[u-1+\exp(-u)]$
Persistent (wormlike) linear chain ("infinitely thin")	L=contour length $\hat{a}$=persistence length	$(2/3)(\hat{a}L/3)^{1/2}Z(\hat{a}/L, d/L)c$	$(\hat{a}L/3)S(\hat{a}/L)c$	See Equation 8.50
Rod ($L \gg d$)	L=length R=radius	$L/\{2\ln(3L/4R)\}$	$L^2/12$	$(2/x^2)[x\mathrm{Si}(x)-1+\cos(x)]d$
Disk ("infinitely thin")	R=radius	R/π	$R^2/2$	$(2y^2)[1-J_1(2y)/y]d$
Cylinder	L=length R=radius		$L^2/12+R^2/2$	See Table 8.5 in [11]
Sphere	R=radius	R	$3R^2/5$	$(9/y^6)[\sin(y)-y\cos(y)]^2$
Spherical shell	R=radius (outer) Δ = shell thickness	R	$(3R^2/5)\left(\dfrac{1-[1-(\Delta/R)]^5}{1-[1-(\Delta/R)]^3}\right)$	See Equation 8.61, with $\tilde{w}_A=1$, $R_B=R$, and $R_A=R_B-\Delta$
Spherical shell ("infinitely thin")	R=radius (outer)	R	R^2	$[\sin(y)/y]^2$
Spheroid	$2R_1$=unique axis length $2R_2$=transverse axis length $p=R_2/R_1$	See Table 8.1 in [13]	$R_1^2\left(\dfrac{2+p^2}{5}\right)$	See Table 8.5 in [11]

[a]Unless given in the following, the original citations for the expressions presented in the preceding for $P(q,0)$, and many more may be found in Reference [11].
[b]$u=R_G^2q^2$; $y=Rq$; $x=Lq$.
[c]$Z(\hat{a}/L, d/L) \approx \{1+(16\hat{a}/27L)\ln^2(3L/2d)\}^{-1/2}$; a Padé approximates to the exact expression, see [49]. $S(\hat{a}/L)=1-3(\hat{a}/L)+6(\hat{a}/L)^2-6(\hat{a}/L)^3[1-\exp(-L/\hat{a})] \approx (1+4\hat{a}/L)^{-1}$.
[d]$\mathrm{Si}(\dots)$ and $J_1(\dots)$ are the sine integral and the Bessel function of the first order and kind, respectively.

The following is presented in three principal sections, one for the interpretation of dynamic and two for static scattering, with most of the treatment concerned with scattering in the RGD regime, complemented by some mention of scattering in the Mie regime for large spheres.

8.2 DYNAMIC LIGHT SCATTERING

8.2.1 The Autocorrelation Function

The rise of dynamic scattering to a position of prominence in the characterization of dilute solutions of polymers and dispersions of particles occurred with the development of lasers to provide temporally and spatially coherent light with an intensity suitable for scattering measurements on dilute systems, and equipment for computer-aided data acquisition and analysis. The optical requirements for dynamic light scattering, described elsewhere [13], are now present in many commercially available light scattering photometers. The experimentally measured photon-count autocorrelation $G^{(2)}(\tau; q,c)$ is usually determined in what is called a homodyne mode, meaning that the photon count arises only from the solution under study, and does not include a component from, say, the scattering from an external source, as is done in the so-called heterodyne mode described in the following [13, 32].

The normalized photon-count autocorrelation function, $g^{(2)}(\tau; q,c)$ is calculated by normalization of $G^{(2)}(\tau; q,c)$ by $G^{(2)}(\infty; q,c)$. This normalized function is related to the function $g^{(1)}(\tau; q,c)$ for which (for most solutes) [13, 14, 32–35]:

$$g^{(2)}(\tau; q, c) = 1 + f_C\{g^{(1)}(\tau; q, c)\}^2 \qquad (8.4)$$

Here, $f_C=g^{(2)}(0; q,c) - 1$, a measure of the coherence in the scattered light as controlled by the detector optics, is expected to be independent of q and c. Further, for an optically isotropic solute, it is usually possible to represent $g^{(1)}(\tau; q,c)$ by the expression [13, 14, 34, 35]:

$$g^{(1)}(\tau; q, c) = \sum r_\mu(q,c)\exp[-\tau\gamma_\mu(q,c)] \qquad (8.5)$$

where the sum is over all of the components present in the sample, with $\gamma_\mu(q,c)$ the weight factor for component μ with relaxation rate $\gamma_\mu(q,c)$, $\Sigma r_\mu(q,c)=1$; for a monodisperse solute $g^{(1)}(\tau; q,c)=\exp[-\tau\gamma_\mu(q,c)]$.

For a homopolymer with a distribution of molecular weights, $r_\mu(0,c)$ reduces to $w_\mu M_\mu/M_w$ in the limit of infinite dilution and zero scattering angle; for $q>0$, the intramolecular interference function $P(q,c)$ discussed in the section on static scattering must be included for large enough size to make $P(q,c)<1$, that is, for homopolymers, $r_\mu(0,c)\approx w_\mu M_\mu P_\mu(q,c)/M_w P_{LS}(q,c)$, introducing a dependence of the weighting factor on q, where the subscript LS designates an average imposed by the light scattering experiment. More complex expressions may be required for very high-molecular-weight polymer chains, asymmetric particles, optically anisotropic solute (see the following text) or certain other systems [13, 32]. In general, these effects will not invalidate the use of Equation 8.5 if $qR_G<1$ [13].

A revision of Equation 8.4 is necessary if the source of constant intensity is mixed with the scattered light, as may be done, for example, in investigation of the mobility of a charged solute in an electric field by the effects on $G^{(2)}(\tau; q,c)$ in a heterodyne mode. In such a case, $g^{(2)}(\tau; q,c)$ is given by:

$$g^{(2)}(\tau; q, c) = 1 + 2r_S(q)[1-r_S(q)]f_C^{1/2}\left|g^{(1)}(\tau; q, c)\right|$$
$$+[1-r_S(q)]^2 f_C\{g^{(1)}(\tau; q, c)\}^2 \qquad (8.6)$$

where $r_S(q)$ is the fraction of the light contributed by the stationary source.

Thus, if $r_S(q)$ approaches unity, the dynamic response no longer depends on $\{g^{(1)}(\tau; q,c)\}^2$, but rather on $|g^{(1)}(\tau; q,c)|$. In electrophoretic scattering, in which an electric field is imposed on the sample of a charged solute (or particle) with $r_S(q)\approx 1$, the response is modulated by a cosine factor that contains information on the electrophoretic mobility [13, 32].

Another situation in which Equation 8.6 may be necessary may occur if the sample contains a very slow moving component that scatters light very strongly in comparison with all other components in the sample, providing an essentially stationary source, with an effective $r_S(q)\approx 1$, especially for small q. This circumstance may arise, for example, if the solution of an otherwise well-dissolved solute also contains large aggregates, with appreciably smaller diffusion constant than that of the dissolved solute. Although this may be difficult to deduce from the shape of the observed $g^{(2)}(\tau; q,c)$, an unusually small value for the extrapolated value of $g^{(2)}(0; q,c)$ can be an indication of the presence of a quasi-stationary component.

8.2.2   Cumulant Analysis

A cumulant expression is often used to analyze the behavior for small τ [13, 14, 34, 35]:

$$\ln[g^{(2)}(\tau; q, c)-1]^{1/2} = \ln(f_C)^{1/2} - K^{(1)}(q,c)\tau$$
$$+\frac{K^{(2)}(q,c)\tau^2}{2!} + \cdots \qquad (8.7)$$

where $K^{(n)}(q,c)$ is the nth cumulant, with:

$$K^{(1)}(q,c) = \sum r_\mu(q,c)\gamma_\mu(q,c) \qquad (8.8)$$

$$K^{(2)}(q,c) = \sum r_\mu(q,c)\gamma_\mu^2(q,c) - \{K^{(1)}(q,c)\}^2 \qquad (8.9)$$

Inspection shows that $K^{(2)}(q,c)=0$ for a monodisperse solute; $\ln[g^{(2)}(\tau; q,c) - 1]^{1/2}$ versus τ should be accurately linear if Equation 8.7 applies with a monodisperse solute, with deviations from linearity anticipated for samples with molecular weight or structural dispersity. The programs provided with some commercially available dynamic light scattering photometers refer to $K^{(2)}/(K^{(1)})^2$ as a "polydispersity index," but the exact significance of that term will depend on the structure under study. For example, for rodlike chains, $K^{(2)}/(K^{(1)})^2=[(M_w/M_n) - 1]/2$ [13]. Since in general $\gamma_\mu(q,c)\propto \tau_\mu q^2$ in effecting a cumulant analysis, it is useful to consider $\ln[g^{(2)}(\tau; q,c) - 1]^{1/2}$ as a function of τq^2 with data obtained at several scattering angles; deviation from superposition will reflect any appreciable dependence of $r_\mu(q,c)$ on q.

For polarized scattering from optically isotropic solute, the first cumulant in the limit of small q is used to compute the mutual diffusion coefficient $D_M(c)$:

$$K^{(1)}(q, c) = -\frac{1}{2}\lim_{\tau=0}\frac{\partial \ln[g^{(2)}(\tau; q, c)-1]}{\partial \tau} \qquad (8.10a)$$

$$\lim_{q=0} K^{(1)}(q,c) = D_M(c)q^2 \qquad (8.10b)$$

It is useful to express $D_M(c)$ in terms of a diffusion scaling length $a_{LS}(c)$ to remove the effects of the viscosity η_s for comparison of data in different solvents and at different temperatures:

$$a_{LS}(c) = \frac{kT}{6\pi\eta D_M(c)} \qquad (8.11)$$

Then, in the limit of infinite dilution, for which $D_M(c)$ is equal to the translational diffusion constant $D_{T;LS}$, $a_{LS}(c)$ extrapolates to the light scattering averaged hydrodynamic radius $R_{H;LS}=kT/6\pi\eta_s D_{T;LS}$. Examples of $R_{H;LS}$ for several solute (or particle) structures heterodisperse in molecular weight are given in Table 8.2.

The concentration dependence of $a_{LS}(c)$ (or $D_M(c)$) is normally less than that of $Kc/\mathbf{R}(q,c)$ described in the section on static scattering. For example, for dilute solutions of polymeric solutes:

$$a_{LS}(c) = R_{H;LS}\{1 - k_D[\eta]c + \cdots\} \qquad (8.12)$$

with $k_D=(2 - k_2)(A_2M/[\eta]) - k_1$, where $[\eta]$ is the intrinsic viscosity, A_2 is the second virial coefficient (elaborated

TABLE 8.2 Light Scattering Average Mean-Square Radius of Gyration and Hydrodynamic Radius

	$R^2_{G;LS}$	$R_{H;LS}$
Exact relation[a]	$(1/M_w)\Sigma w_\mu M_\mu R^2_{G;\mu^2}$	$M_w / \Sigma w_\mu M_\mu R^{-1}_{H;\mu}$
Approximate for[b] $R_H \approx R_G \propto M^{\varepsilon/2}$	$(R^2_G / M^\varepsilon)\mathrm{M}^{\varepsilon+1}_{(\varepsilon+1)} / M_w$	$(R_H / M^{\varepsilon/2})M_w / M^{1-\varepsilon/2}_{(1-\varepsilon/2)}$
Random-flight coil[c]; $\varepsilon = 1$	$(R^2_G / M)M_w$	$(R_H / M^{1/2})M_w / M^{1/2}_{(1/2)}$
		$\approx (R_H / M^{1/2})M^{1/2}_w (M_w / M_n)^{0.10}$
Rodlike chain; $\varepsilon = 2$	$(R^2_G / M^2)M_z M_{z+1}$	$(R_H/M)M_w$
Sphere[c]; $\varepsilon = 2/3$	$(R^2_G / M^{2/3})M^{5/3}_{(5/3)} / M_w$	$(R_H / M^{1/3})M_w / M^{2/3}_{(2/3)}$
	$\approx (R^2_G / M^{2/3})M^{2/3}_z (M_w / M_z)^{0.10}$	$\approx (R_H / M^{1/3})M^{1/3}_w (M_w / M_n)^{0.10}$

[a] For optically isotropic solute, only molecular-weight diversity, with sums over all components.

[b] $M_{(\alpha)} = \left(\sum w_\mu M^\alpha_\mu\right)^\alpha$; for example., $M_{(\alpha)}$ is M_n, M_w, $(M_w M_z)^{1/2}$, and $(M_w M_z M_z + 1)^{1/3}$ for $\alpha = -1$, 1, 2, and 3, respectively.

[c] Approximations for a solute with a Schulz–Zimm (two-parameter exponential) distribution for M; $M(\alpha)=M_w\Gamma(1+h+\alpha)/\Gamma(1+h)]1/\alpha/(1+h)$, see [53].

in the section on static scattering), and both k_1 and k_2 have been estimated for various solute geometries, for example, for linear flexible chain polymers $k_1 \approx 1.2$ and $k_2 = B(4\pi R^3_H / 3M[\eta])$, with $B \approx 1$ to 2.23 [13, 28, 49, 54]. It should be noted that k_D may be negative, even if $A_2 > 0$.

8.2.3 Inverse Laplace Transform Analysis

The cumulant analysis does not exhaust the information available in the observed $g^{(1)}(\tau; q,c)$; further information on the distribution of the scattering components may be determined via an inversion Laplace transform (ILT) of Equation 8.5 to give $r_\mu(q,c)$ as a function of $\gamma_\mu(q,c)$. Unfortunately, this inversion is an ill-posed problem, in which case a number of such sets of $r_\mu(q,c)$ as a function of $\gamma_\mu(q,c)$ will produce equivalent estimates of $g^{(1)}(\tau; q,c)$ within the experimental limits of the determination of that function [13, 14, 34, 35, 55, 56]. However, many of these sets will exhibit physically untenable features, for example, negative values of $r_\mu(q,c)$, unreasonably rapid oscillations in $r_\mu(q,c)$ as a function of $\gamma_\mu(q,c)$ or sets that do not sum to the observed $K^{(1)}$ according to Equation 8.8. This can be minimized by placing suitable constraints on the analysis (e.g., $r_\mu(q,c) \geq 0$), and by recognizing the limitations on the resolution possible in the inversion presented by the experimentally limited range of τ available and the precision in the determination of $g^{(1)}(\tau;q,c)$. Although reasonably useful algorithms have been developed and are included in the software of some commercially available equipment, the inversion is sufficiently delicate that care should be exercised in the use of such an analysis.

An example of results obtained with an ILT analysis on a homopolymer sample with molecular weight dispersity is given in Figure 8.2. In that analysis, in which the algorithm was nonnegatively constrained to exclude any $r_\mu(q,c)<0$, the ratio $\gamma_\mu / \gamma_{\mu-1}$ of n successive relaxation rates was constant, starting with the smallest and largest values of the γ_μ-set used

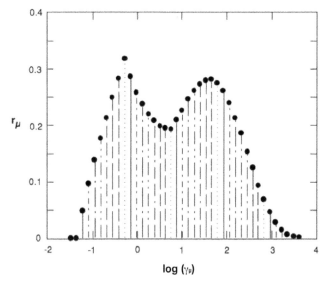

FIGURE 8.2 Example of results on an inverse Laplace transform of data on $g^{(2)}(q,c)$ exhibiting nonexponential behavior. As discussed in the text, eight sets of analyses, each with five terms, but different shortest relaxation rates were applied. The sets are distinguished by the differing lines (solid, dashed, etc.). The data are from [57], as reported in [30]. Copyright 1999, John Wiley & Sons Ltd.

being consistent with the largest and smallest τ, respectively, and the value of n being the largest value that provided stable results in the analysis, reproduced the observed $g^{(2)}(\tau; q,c)$ and an estimate of $K^{(1)}(q,c)$ calculated using Equation 8.8 and the data on $r_\mu(q,c)$ as a function of $\gamma_\mu(q,c)$ from the ILT analysis that agreed with that obtained from the initial tangent per Equation 8.7. In the example shown, $n=5$, producing a result with five terms to represent the experimental data on $g^{(2)}(\tau; q,c)$; for the data shown, a larger n, which would have produced more points in the distribution function led instead to an unstable ILT analysis. Although not necessary to improve the representation of $g[2](\tau; q,c)$, the process was repeated eight times to provide a more useful visual representation of the distribution of $r_\mu(q,c)$ as a function of $\gamma_\mu(q,c)$.

8.2.4   Depolarized Scattering [57]

Finally, although not often used, it may be noted that if the scattering elements are optically anisotropic, the depolarized scattering for rodlike chains may provide additional information. Thus, for the autocorrelation function $G^{(2)}_{Hv}(\tau; q)$ obtained for the scattering in the horizontal plane with vertically polarized incident light may be used to obtain $g^{(2)}_{Hv}(\tau; q)$, for which

$$g^{(2)}_{Hv}(\tau;q) - 1 = [g^{(2)}_{Hv}(0;q) - 1]$$
$$\left\{ \sum r_{Hv;\mu} \exp[-\tau(q^2 D_{T;\mu} + 6 D_{R;\mu})] \right\}^2 \quad (8.13)$$

where $D_{T;\mu}$ and $D_{R;\mu}$ are the translational and rotational diffusion constants of component μ, respectively, $\Sigma r_{Hv;\mu} = 1$, and $r_{Hv;\mu}$ may depend on q.

It may be noted that the term involving $D_{R;\mu}$ does not depend on q, and will become the dominant contribution to $g[2]_{Hv}(\tau; q) - 1$ as q tends to zero.

8.3   STATIC LIGHT SCATTERING; ISOTROPIC SYSTEMS

8.3.1   General Relation for the Rayleigh Ratio

Certain general relations used with the static scattering in the RGD limit from isotropic solutions in the absence of depolarized scattering ($R_{Hv}(q,c) = 0$) are gathered in this section. In these limits, for a monodispersed solute, the Rayleigh ratio may be expressed in the form [4, 5, 10, 11, 13, 24, 49, 58]

$$R(q,c) = KcMP(q,c)\{1 - cM\tilde{B}(c)P(q,c)Q(q,c)\} \quad (8.14)$$

where $K = K'n_{medium}^2\psi^2$ with $K' = 4\pi^2/N_A\lambda_0^4$, ψ is a contrast factor elaborated below, M is the molecular weight, $\tilde{B}(c)$ accounts for q-independent intermolecular interactions and $P(q,c)$ and $Q(q,c)$ are q-dependent intramolecular and intermolecular structure factors, respectively; both $P(0,c)$ and $Q(0,c)$ are unity, and $P(q,c) < 0$ for $q > 0$. (Note: The nomenclature for $Q(q,c)$ here differs from that in [13].) Although Equation 8.14 must be used to represent the effects of dispersity in solute size, composition, etc., alternative forms obtained by rearrangement of its terms find use in most studies of scattering from dilute systems (Eq. 8.15) [5, 13, 49]:

$$R(q,c) = KcMS(q,c) \quad (8.15a)$$

$$S(q,c) = P(q,c)F(q,c) \quad (8.15b)$$

where $F(q,c)$ may be expressed in either of two equivalent forms:

$$F(q,c) = 1 - cB(c)P(q,c)Q(q,c) \quad (8.16a)$$

$$F(q,c)^{-1} = 1 + c\Gamma(c)P(q,c)H(q,c) \quad (8.16b)$$

with $B(c) = M\tilde{B}(c)$, and where $H(q,c)$ and $\Gamma(c)$ may be expressed in terms of $B(c)$ and $Q(q,c)$ [1, 10, 24, 28, 49]:

$$\Gamma(c) = \frac{B(c)}{[1 - cB(c)]} \quad (8.17)$$

$$H(q,c) = \frac{[1 - cB(c)]Q(q,c)}{F(q,c)} \quad (8.18)$$

As elaborated in the following, $F(0,c)$ will generally not be equal to unity except at infinite dilution. (Note: In alternate representations of Equation 8.15, some authors use the notation $S(q,c)$ for a different function.) For the scattering in the limit of infinite dilution, $P(q,0)$ may be expressed in a series in terms of $R_G^2 q^2$, with R_G being the root-mean-square radius of gyration averaged over all orientations of the solute [10, 11, 49]:

$$P^{-1}(q,0) = 1 + \frac{R_G^2 q^2}{3} + \cdots \quad (8.19)$$

For rigid particles $P(q,c) = P(q,0)$, and as elaborated in the following, this is a reasonable approximation for flexible chain polymers in dilute solutions.

The expression for $F(q,c)$ given in Equation 8.18 is particularly useful for dilute solutions and dispersions since $H(q,c) \approx 1$ in most cases for such systems (charged solute or particles in a salt-free aqueous solution being an exception), motivating the analysis of experimental data in the form for monodisperse solute (or particles):

$$\frac{KcM}{R(q,c)} = \frac{1}{S(q,c)} = \frac{1}{P(q,c)} + c\Gamma(c)H(q,c) \quad (8.20)$$

Since the interest in this chapter focuses on dilute solutions, in most cases, one may expand $\Gamma(c)$ in a Maclaurin series in c [1, 48, 49, 59]:

$$c\Gamma(c) = 2A_2Mc + 3A_3Mc^2 + \cdots \quad (8.21)$$

where A_2 and A_3 are the second and third virial coefficients, respectively, to obtain the result that finds use with monodisperse dilute solutions of isotropic solute (or particles):

$$\frac{Kc}{R(q,c)} = \frac{1}{MP(q,0)} + [2A_2Mc + 3A_3Mc^2 + \cdots]H(q,c) \quad (8.22)$$

For scattering from a dilute polymer solution or (particle) comprising a solvent and C components differing in some way (e.g., molecular weight, refractive index, etc.), with $\psi_{j,v}$ and $m_{j,v}$, the refractive index increment and the molecular weight, respectively, of the jth element on the vth component, and n_v is the number of elements in the vth component, with a component, comprising structures identical in composition, molecular weight, and (statistical) structure, present at weight fraction wv [5, 10, 24], and assuming that $H(q,c) \approx 1$ for a dilute system:

$$\frac{Kc}{R(q,c)} = \frac{1}{M_{LS}P_{LS}(q,0)} + [2A_2Mc + 3A_3Mc^2 + \cdots] \quad (8.23)$$

$$M_{LS} = \psi^{-2} \sum_v^C w_v M_v^{-1} \left\{ \sum_j^{n_v} \psi_{j,v} m_{j,v} \right\}^2 \quad (8.24)$$

$$M_{LS} P_{LS}(q,0) = \psi^{-2} \sum_v^C w_v M_v^{-1} \sum_j^{n_v} \sum_k^{n_v} \psi_{j,v} \psi_{k,v} m_{j,v} m_{k,v}$$
$$\left\langle \sin(q|r_{jk}|_v) / q|r_{jk}|_v \right\rangle \quad (8.25)$$

$$R_{G;LS}^2 = 3 \lim_{q \to 0} \frac{\partial P_{LS}^{-1}(q,0)}{\partial q^2}$$
$$= \frac{1}{2M_{LS}\psi^2} \sum_v^C w_v M_v^{-1} \sum_j^{n_v} \sum_k^{n_v} \psi_{j,v} \psi_{k,v} m_{j,v} m_{k,v} \left\langle r_{jk}^2 \right\rangle_v \quad (8.26)$$

$$\psi = \left(\frac{\partial n}{\partial c} \right)_w = \sum_v^C w_v \psi_v \quad (8.27)$$

Further, for a system for which all of the components exhibit only molecular-weight dispersity:

$$\tilde{B}_{2,LS}(c) = M_w^{-2} \sum_v^C \sum_\mu^C w_v M_v w_\mu M_\mu \tilde{B}_{v\mu}(c) \approx 2A_{2,LS} + \cdots \quad (8.28)$$

$$A_{2,LS} = M_w^{-2} \sum_v^C \sum_\mu^C w_v M_v w_\mu M_\mu A_{2,v\mu} \quad (8.29)$$

where $A_{2,v\mu}$ is the value for a monodispersed chain with $M_\mu = M_v$ if $\mu = v$, but is a function of the two molecular weights if $\mu \neq v$ [5].

By comparison, $A_{2,\Pi}$ determined from the osmotic pressure for a sample dispersed in M is given by [5]:

$$A_{2,\Pi} = \sum_v^C \sum_\mu^C w_v w_\mu A_{2,v\mu} \quad (8.30)$$

Theoretical evaluations of $A_{2,v\mu}$ for $v \neq \mu$ have proved elusive, but comparisons of $A_{2,LS}$ and $A_{2,\Pi}$ have been made on the basis of approximations for this function, and the use of the Schulz–Zimm distribution of M, comparing the functions

$\Omega_{LS} = A_{2,LS} / kM_w^{-1}$ and $\Omega_\Pi = A_{2,\Pi} / kM_n^{-1}$, where $k = A_2/M^{-\gamma}$ is a constant characteristic of the solute/solvent pair [24, 53].

8.3.2 Scattering at Zero Angle and Infinite Dilution

Since $P(q,c)$, $Q(q,c)$, and $H(q,c)$ all reduce to unity for $q = 0$, Equations 8.14 and 8.19 simplify. Further, as noted in the preceding, since $\Gamma(c)$ may be expanded in a Maclaurin series in c [1, 48, 49, 59],

$$c\Gamma(c) = 2A_2Mc + 3A_3Mc^2 + \cdots \quad (8.31)$$

where A_2 and A_3 are the second and third virial coefficients, respectively.

These relations form the basis for the determination of M and A_2 (and sometimes A_3 as well) from scattering data on dilute solutions. With the use of this expansion:

$$\frac{KcM}{R(0,c)} = 1 + 2A_2Mc + 3A_3Mc^2 + \cdots \quad (8.32)$$

The term in c^2 can produce considerable curvature with increasing c that may be reduced by examining $[KcM/R(0,c)]^{1/2}$, for which [13, 60]:

$$\left(\frac{KcM}{R(0,c)} \right)^{1/2} = 1 + 2A_2Mc + \left[\frac{(3\gamma_3 - 1)}{2} \right](A_2M)^2 c^2 + \cdots \quad (8.33)$$

where $\gamma_3 = A_3M/(A_2M)^2$; the curvature will be reduced if $\gamma_3 \approx 1/3$.

For example, for spheres interacting through a hard-core potential, $\gamma_3 = 5/8$ [5, 61], reducing the curvature arising from the term in c^2. A similar effect is realized in the use of the square-root plot for polymer chains in a good solvent, but this strategy is not helpful in a system for which $A_2 = 0$, occurring at the Flory theta temperature Θ such that the integral over the pair-wise potential among solute polymer chain elements vanishes [48, 49, 54, 59]; Θ is analogous to the Boyle temperature for a gas, for which the second virial coefficient is zero, but the third is a small positive number owing to nonpair-wise interactions among three molecules. Similarly, a small $A_3 > 0$ has been reported for polymer solutions under Flory Θ conditions [28].

Similarly, for a sample with component disparity,

$$\frac{Kc}{R(q,c)} = \frac{1}{M_{LS}} + [2A_{2,LS}c + 3A_{3,LS}c^2 + \cdots] \quad (8.34)$$

For a single solute comprising scattering elements all with the same refractive index so that $\psi_v = \psi$ for all v, but heterodisperse in molecular weight, as for a homologous series of a polymer, and remembering that $M_v = \sum_j^{n_v} m_{j,v}$, inspection of the expression for M_{LS} in the RGD limit gives the well-known result:

$$M_{LS} = \sum_v^C w_v M_v = M_w \qquad (8.35)$$

with M_w the weight average molecular weight, reducing further to $M_{LS} = M$ for a monodisperse solute.

In some cases, it is desirable to employ a mixed solvent, for example, to obtain a proper solution with copolymers, or to suppress electrostatic interactions with a polyelectrolyte. In this case, preferential distribution of the solvent components in the vicinity of the polymer can complicate the analysis, but use of the refractive index increment $(\partial n/\partial c_v)_\Pi$ determined at constant temperature, pressure and at osmotic equilibrium of the solution and the solute-free solvent mixture (e.g., by dialysis) gives the simple result [5, 15]:

$$M_{LS} = \left\{ (\partial n/\partial c)_\Pi (\partial n/\partial c)_w^{-1} \right\}^2 M_w \qquad (8.36)$$

The ratio $(\partial n/\partial c)_\Pi/(\partial n/\partial c)_w$ provides a measure of the preferential solvation of the solute by a component of the solvent, with deviations from unity most prevalent when the preferentially solvating component is present at low concentration in the mixed solvent [5, 15, 62, 63].

The scattering from polymers containing two or more chemically diverse repeating units is often of interest owing to the advantage such structures may confer to useful mechanical, electronic, optical, etc., properties. As will be seen, $M_{LS} = M_w$ for random and alternating copolymers or regular block copolymers, with uniform structures among the molecules. It should be noted that M_w for the entire copolymer is obtained, even though one of the comonomers may be isorefractive with the solvent, and therefore not contribute to the observed scattering. For a system with dispersity in both composition and chain length, the aforementioned general expression for M_{LS} may be written in the form [5, 13, 64, 65]:

$$M_{LS} = \psi^{-2} \sum_v^C w_v M_v^{-1} \left\{ \psi_v M_v + \sum_j^{n_v} \Delta \psi_{j,v} m_{j,v} \right\}^2 \qquad (8.37a)$$

$$\begin{aligned} \Delta m_{j,v} &= m_{j,v} - m_v \\ \Delta \psi_{j,v} &= \psi_{j,v} - \psi_v \end{aligned} \qquad (8.37b)$$

where $m_v = M_v/n_v$ and $\Sigma_j \Delta m_{j,v} = \Sigma_j \Delta \psi_{j,v} = 0$.

If either $\Delta \psi_{j,v}$ or $\Delta m_{j,v}$ are 0 for all v, as for a homologous series of chains heterodisperse in chain length, or a blend of such chains, this simplifies to:

$$M_{LS} = \psi^{-2} \sum_v^C \Psi_v^2 w_v M_v \qquad (8.38)$$

For a copolymer such that ψ_v takes on one of two values, ψ_A or ψ_B, for the scattering elements with molecular weights m_A and m_B, respectively, $\psi = w_A \psi_A + (1 - w_A) \psi_B$ with w_A the weight fraction of component A. In this case, the relation for M_{LS} may be expressed in the form [65]:

$$M_{LS} = (1 + 2\mu_1 Y + \mu_2 Y^2) M_w \qquad (8.39a)$$

$$\mu_k = \sum_v^C w_v M_v \Delta w_v^k \qquad (8.39b)$$

where $\Delta w_v = w_{Av} - w_A = w_B - w_{Bv}$ and $Y = (\psi_A - \psi_B)/\psi$, with w_{Av} and $w_{Bv} = 1 - w_{Av}$ the weight fractions of A and B in component v, respectively.

In principle, Y may be adjusted by the choice of solvent, offering the opportunity to determine M_{LS} as a function of Y, and hence evaluate the parameters μ_1 and μ_2, along with M_w with M_{LS} from as few as three solvents with different refractive indices. If the composition is independent of molecular weight, μ_1 vanishes, and μ_2 depends on the composition, falling in the range $0 \leq \mu_2 \leq w_A w_B$. Further examples have been discussed in detail [65], along with an alternative form of the relation for M_{LS}.

Although the RGD approximation will almost always be adequate for nonabsorbing, or even moderately absorbing threadlike molecules, such as flexible or semiflexible coils, rodlike or helical chains, etc., owing to the sparse density of scattering elements in the "intramolecular" domain, that approximation will fail for particles if they are large enough, depending on the refractive index of the solvent in which they are dispersed. This failure reflects changes in the phase of the incident beam as it propagates through the particle. The scattering beyond the RGD regime for a number of particle shapes has been treated [10, 36–38, 42–46], but the example here is limited to the use of the Mie theory, which describes the scattering from nonabsorbing spherical solutes [36]. Two parameters are critical in defining the crossover from the RGD regime to that for which the Mie theory is required: $\alpha = 2\pi R/\lambda$, with R being the sphere radius and the ratio $\tilde{n} = n_{solute}/n_{medium}$ of the refractive indices of solute and medium holding the solute. Evaluation of M_{LS} requires the introduction of two functions: $h_{sph}(\tilde{n}) = 3(\tilde{n}+1)/[2(\tilde{n}^2+2)]$, used as a multiplier to modify ψ, and $m_{sph}(\tilde{n},\alpha)$ appearing in the ratio between M_{LS} and the true molecular weight.

Thus, for nonabsorbing spheres dispersed in size, but all with the same $\tilde{n}$ [10],

$$M_{LS} = \sum_v^C w_v M_v [m_{sph}(\tilde{n}\alpha_v)]^2 \qquad (8.40)$$

An analytical expression for $m_{sph}(\tilde{n},\alpha)$ was used to compute $M_{LS}/M = [m_{sph}(\tilde{n},\alpha)]^2$ for monodispersed spheres for a range of $\tilde{n} - 1$, for several values of α to obtain the results shown

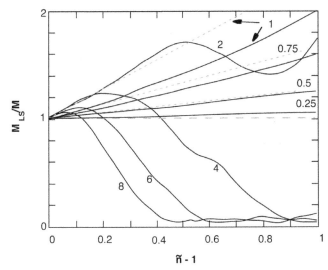

FIGURE 8.3 The ratio M_{LS}/M ($= [m_{sph}(\tilde{n},\alpha)]^2$) of the light scattering averaged molecular weight M_{LS} for monodisperse spheres of radius R to the molecular weight M as a function of the relative refractive index $\tilde{n}$ for the indicated values of the size parameter $\alpha = 2\pi R/\lambda$. The dashed lines give the limiting behavior for small $\tilde{n} - 1$. With permission from Berry GC. Light scattering, classical: size and size distribution classification. In: Meyers RA, editor. *Encyclopedia of Analytical Chemistry*. New York: John Wiley & Sons; 2000. p 5413–5448. © 2000, John Wiley & Sons Ltd.

in Figure 8.3 [10]. For $\tilde{n}\alpha \leq 1$, $m_{sph}(\tilde{n},\alpha)$ may be expanded in the series [36]:

$$m_{sph}(\tilde{n}\ \alpha_{v}) = 1 + j_{sph}(\tilde{n})\alpha^2 |\tilde{n}-1| + \cdots \qquad (8.41)$$

$$j_{sph}(\tilde{n}) = \frac{(\tilde{n}+1)(\tilde{n}^4 + 27\tilde{n}^2 + 38)}{[15(\tilde{n}^2 + 2)(2\tilde{n}^2 + 3)]} \qquad (8.42)$$

As may be seen in Figure 8.3, the range for which the use of this approximation gives reasonable estimates of M_{LS}/M for monodispersed spheres is limited to small $|\tilde{n} - 1|$ and $\alpha < 0.75$, but in that regime M_{LS} may be expressed in the form (with $2\pi R/\lambda M^{1/3}$ a constant for a sphere of radius R and molecular weight M) [36]:

$$M_{LS} = M_{w}\left\{1 + 2\left(\frac{2\pi R}{\lambda M^{1/3}}\right)^2 M_z^{2/3} j(\tilde{n})|\tilde{n}-1| + \cdots \right\} \geq M_{w} \qquad (8.43)$$

8.3.3 Scattering at Arbitrary Angle and Infinite Dilution

In this case, the scattering in the RGD regime for isotropic systems, for which $R_{Hv}(q,c)=0$, is represented by Equations 8.23–8.25:

$$\left[\frac{R(q,c)}{c}\right]^0 = KM_{LS}P_{LS}(q,0) \qquad (8.44)$$

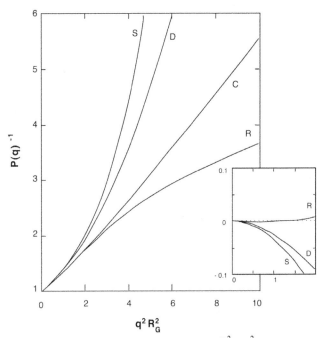

FIGURE 8.4 Examples of $P(q,0)^{-1}$ versus $R_{G;LS}^2 q^2$ for random-flight linear chains (C), rodlike chains (R), disks (D), and spheres (S); expressions for $P(q)$ for these cases are given in Table 8.1. The inset shows the ratio of the logarithm of $P(q,0)$ divided by $P_{RF}(q,0)$ versus $R_G^2 q^2$ for the coil with the same R_G^2 for these cases, showing a close agreement of $P(q,0)$ for coils and rodlike chains up to $R_{G;LS}^2 q^2 \approx 2$. With permission from Berry GC, Cotts PM. Static and dynamic light scattering. In: Pethrick RA, Dawkins JV, editors. *Modern Techniques for Polymer Characterisation*. London: John Wiley & Sons; 1999. p 81–108. © 1999, John Wiley & Sons Ltd .

Expressions for M_{LS} are given in the preceding, and formulae for $P(q,0)$ for monodisperse solutes (or particles) and the corresponding values of R_G^2 calculated with Equations 8.25 and 8.26 are given in Table 8.1 for a number of structures of interest, along with plots of $P^{-1}(q,c)$ versus $R_G^2 q^2$ in Figure 8.4; many more expressions for $P(q,0)$ and R_G^2 may be found in [11, 22].

The expression for $P_{RF}(q,0)$ for the monodisperse random-flight chain model, with $R_G^2 = \hat{a}M/3M_L$ (where $M_L = M/L$), is widely used for the conformation of flexible chain polymers. Expansion of this expression in the limit of small $u = (qR_G)^2$ gives [13, 66]:

$$P_{RF}^{-1}(q,0) = 1 + \frac{u}{3} + \frac{u^2}{36} + \cdots \qquad (8.45)$$

$$P_{RF}^{-1/2}(q,0) = 1 + \frac{u}{6} + \frac{u^3}{1080} + \cdots \qquad (8.46)$$

The lack of a term in u^2 for the expansion in $P_{RF}^{-1/2}(q,0)$ shows the resultant substantial reduction of the curvature makes this form useful in the evaluation of R_G^2 from initial tangent of $P(q,0)^{-1/2}$ versus q^2 for monodisperse flexible chain polymers.

Although the expression in Table 8.1 for $P(q,0)$ for the random-flight chain is specifically valid for linear chains, an approximation useful for branched random-flight chains is obtained if the expression $R_G^2 = \hat{a}L/3$ appearing in that $P(q,0)$ for the linear chain is replaced by the $R_G^2 = g\hat{a}L/3$, where the parameter $g = \left(R_G^2 \right)_{br} / \left(R_G^2 \right)_{lin}$ is known for a wide range of branched structures [25, 67]. This simple approximation can yield an estimate of $P(q,0)$ for the branched chain that is useful within a few percent for a range of branched structures for $(R_G^2)_{br} q^2 < 2$ [10, 67].

Light scattering photometers using laser sources for the incident light have been adapted to serve as detectors on the effluent from chromatographic analyzers, principally via size exclusion chromatography (SEC) [30, 68, 69]. The effluent is well suited for light scattering analysis since it is usually free of extraneous particles (unless the column packing sheds such debris). In addition, the narrowing of the size distribution at a given eluent sample simplifies the analysis. Commercial photometers are available to determine the scattering over a range of angles to give the analysis maximum use of the scattering analysis [69], but in an echo of the dissymmetry method used in early light scattering practice based on measurements at 45, 90, and 135° [3], a commercial unit is also available for measurement at two fixed angles [68]. Unlike the original dissymmetry method, in the current version, the two angles used, 15 and 90°, are not symmetric but that does not alter the strategy to determine $R_{G;LS}^2 / \lambda^2$ and $[R(0,c)]/[R(q_{\theta=90},c)]$ as functions of the dissymmetry parameter $Z_{15/90} = [R(q_{\theta=15},c)]/[R(q_{\theta=90},c)]$, given by the ratio of the intensities at these two angles, corrected for any difference in the scattering volumes at the two angles; with the advent of ubiquitous computers, such calculations of $Z_{15/90}$ for comparison with the experimental data no longer requires the use of printed tables as in earlier practice, but may be done automatically for the structure of interest (coil, rod, sphere, etc.) in computers associated with the light scattering equipment [18, 70]. The limitation to only one measurement at low angle could lead to some problems if the sample contains aggregates, as the scattering from a few aggregates could be emphasized in the scattering at 15°, but that may be less likely in a solution emerging from an SEC column than for solutions without such filtering. In addition, since the eluent at a given elution volume has a narrow size distribution, the effects of dispersity on $P(q,0)$ may be neglected (unless the sample is a mixture of structures, e.g., linear and branched chains, etc.). As shown in Figure 8.4, $P(q,0)$ tends to be similar for several structures for small q, as the limiting linear behavior for small q is approached, facilitating the use of the two-angle method for $R_G^2 q^2$ up to about 1 [70]. Furthermore, as mentioned in the

preceding, $P(q,0)$ for the linear random-flight chain retains its approximate usefulness for a range of branched chain, and even rodlike, polymers for this range of $R_G^2 q^2$ [10, 67], provided the actual value of R_G^2 is used, or vice versa, determination of R_G^2 / λ^2 from the dissymmetry analysis using $P(q,0)$ for the linear random-flight chain will yield a reasonable approximation to R_G^2 / λ^2 for a range of branched polymer analytes.

In the limit of very large u, $P_{LS;RF}(q,0)$ for the (monodisperse) random-flight chain model takes the form [5, 64]:

$$\lim_{R_G q \gg 1} P_{RF}^{-1}(q,0) = \left(\frac{1}{2} \right) \left[1 + \frac{\hat{a}Lq^2}{6} + O\left[(R_G^2 q^2)^{-1} \right] \right] \quad (8.47)$$

By comparison, the asymptotic behavior at large q is very different for the rodlike chain model. Thus, for monodispersed rodlike chains, the asymptotic behavior at large $R_G^2 q^2$ is given by [5]:

$$\lim_{R_G q \gg 1} P_{ROD}^{-1}(q,0) = \frac{2}{\pi} + \frac{Lq}{\pi} + O\left[(R_G^2 q^2)^{-1} \right] \quad (8.48)$$

Accordingly, in the limit of large $R_G^2 q^2$, it may be seen that $\partial[c/R(q,c)]^0/\partial q$ is proportional to L/M, and does not provide information on the molecular size.

Inspection of Equations 8.47 and 8.48 reveals that these two models give very different asymptotic behavior, with $q^2 P(q,0)$ tending to a constant given by $2/(\hat{a}L/3)$ for the random-flight chain, but becoming linear in q for the rodlike chain. In fact, the random-flight model, and the resultant asymptotic expressions given by Equation 8.47 cannot be valid for $q > \hat{a}^{-1}$, as then the scattering must crossover to the behavior of local short "rodlike" elements along the chain, of length $q^{-1} < \hat{a}$, such that $P^{-1}(q,0)$ becomes similar to the behavior expected for a rodlike chain in the limit $q > \hat{a}^{-1}$, indicating the need to improve the expression for $P_{RF}(q,0)$ for the monodisperse random-flight chain model for more realistic behavior in that limit. Since this would require a fairly large $\hat{a}$ for the range of q accessible in light scattering, this points to improved models for the wormlike chain model; the improvement would be more relevant for flexible chain polymers with smaller $\hat{a}$ for the range of q accessible in neutron or small angle x-ray scattering.

Considerations based on the wormlike chain model suggest three regimes describing the crossover from the low q to very high q model, with $P^{-1}(q,0)$ linear in $R_G^2 q^2$ for low $R_G^2 q^2$, passing through a regime described by the asymptotic behavior for the random-flight chain at intermediate q before adopting the asymptotic behavior for a rodlike chain for $q > \hat{a}^{-1}$ [71]. This behavior forms the basis of analyses based on the behavior of $(\hat{a}L/3)q^2 P(q,0)$ versus $\hat{a}q$: Region I for $R_G^2 q^2 \ll 1$, for which $(\hat{a}L/3)q^2P(q,0)$ increases monotonically with increasing q; Region II for $1/R_G^2 < q^2 < \hat{a}^{-2}$, for which $(\hat{a}L/3)q^2P(q,0)$ tends to plateau with the value of 2 given by Equation 8.47 (for a monodisperse solute); and Region III for still larger q, for which $(\hat{a}L/3)q^2P(q,0)$ increases linearly with q. Wormlike chain models afford a means to represent $P(q,0)$ for these three regimes. In the first approximation, the intersection of lines from Region II to Region III occurs at a crossover q^* given by the value of q for which the asymptotic forms given by the random-flight and rodlike models coincide [71]:

$$\hat{a}q^* \approx (6/\pi)S(\hat{a}/L)^{-1} \approx (6/\pi)(1 + 4\hat{a}/L) \quad (8.49)$$

The calculation of $P(q,0)$ for the wormlike chain model has proved to be elusive, and to date the best model is presented in a complicated numerical format [58, 72]. Some results calculated using that model are displayed in Figure 8.5 for a range of the parameter $L/\hat{a}$, showing $(\hat{a}L/3)\,q^2P(q,0)$ versus $\hat{a}q$, displaying the three regimes described in the preceding [24]. Another format found in the literature, with $(Lq\pi)$ $P(q,0)$ versus $\hat{a}q$ is also shown. Both of these formats have found use in the literature in an attempt to determine $\hat{a}$ from experimental data. The complex behavior exhibited in Figure 8.5 may be represented by a fairly simple Padé relation for $L/\hat{a} > 5$ [24]:

$$P(q,0) \approx \left(P_{\text{W-RF}}^m(q,0) + \left(\frac{1 - \exp\left[-(\hat{a}q)^2\right]}{1 + Lq^2/\pi} \right)^m \right)^{1/m} \quad (8.50)$$

where $m \approx 3$, and with $P_{\text{W-RF}}(q,0)$ calculated using R_G^2 for the wormlike chain model with the relevant value of $\hat{a}/L$ in the expression for $P_{\text{RF}}(q,0)$ given in Table 8.1 (i.e., in place of the value $\hat{a}L/3$ for the random-flight model); the second term in the brackets is devised to go to zero as q tends to zero, and to give the correct asymptotic behavior for a rodlike chain for larger q.

The preceding has implicitly assumed "thin" molecular structures, and requires some modification for the analysis of the scattering from particles, such as wormlike micelles that have been of considerable interest in recent years. Despite the relatively large values of L and $\hat{a}$ of these, their values of $\tilde{n}$ close to unity allow use of the RGD approximation, making investigation of these possible by light scattering with a simple change to account for scattering in Region III: the scattering is assumed to arise from the radius R_c of a the circular cross section of the structure, with the relevant intramolecular interference function given by [11, 73]:

$$P_{\text{III}}(q,0) \approx \left(\frac{\pi}{Lq} \right) P_{\text{section}}(q,0) \quad (8.51a)$$

$$P_{\text{section}}(q,0) \approx \left(\frac{2J_1(R_c q)}{R_c q} \right)^2 \approx \exp\left[\frac{-(R_c q)^2}{4} \right] \quad (8.51b)$$

where $J_1(\ldots)$ is the first-order Bessel function of the first kind; inspection shows that the exponential (Guinier) approximation is within 10% of the Bessel function relation for $R_c q < 2$, with the deviation increasing rapidly for larger $R_c q$.

Expressions for $P_{\text{section}}(q,0)$ for other cross-section geometries are available, including hollow structures [11, 73, 74].

Evaluations of $P_{\text{LS}}(q,0)$ and $R_{\text{G:LS}}^2$ via Equations 8.25 and 8.26 for samples with component diversity for some of the models represented in Table 8.1 are elaborated in the following, starting with components with only molecular-weight diversity, for which simplified versions of Equations 8.28 and 8.29 apply, followed by components with chemical structure

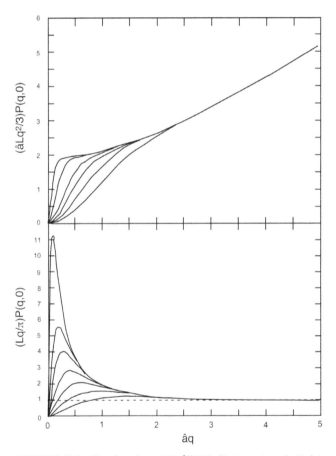

FIGURE 8.5 The functions $(\hat{a}Lq^2/3)P(q,0)$ (upper) and (Lq/π) $P(q,0)$ (lower) versus $\hat{a}q$ for the Kratky–Porod wormlike chain model [58, 72], for chains of contour length L and persistence length $\hat{a}$. For convenience of comparison, the values of $L/\hat{a}$ used are the same as among those in an alternative bilogarithmic representation $(L/2\hat{a})P(q,0)$ versus $2\hat{a}q$ presented in the literature [19]: 640, 160, 80, 40, 20, 10, 5 for the curves from top to bottom in the lower panel, and all of these except 160 in the upper panel for the curves from left to right [24]. Copyright 2008, Springer Science+Business, LLC.

diversity. In the first case, the $\psi_{j,v}$ and $m_{j,v}$ are identical for all factors, and may be factored from the summations, giving:

$$M_w P_{LS}(q,0) = \sum_v^C w_v M_v \frac{1}{n_v^2} \sum_j^{n_v} \sum_k^{n_v} \left\langle \sin\left(q\left|r_{jk}\right|_v\right) / q\left|r_{jk}\right|_v \right\rangle \quad (8.52)$$

$$R_{G;LS}^2 = \frac{1}{2M_w} \sum_v^C w_v M_v \sum_j^{n_v} \sum_k^{n_v} r_{jkv}^2 = \frac{1}{M_w} \sum_v^C w_v M_v R_{G;v}^2 \quad (8.53)$$

For a number of the models given in Table 8.1, $R_{G;v}^2 \propto M^\varepsilon$, for example, ε is equal to 2/3, 1, and 2 for spheres, random-flight chains, and rods, respectively, facilitating calculation of $R_{G;LS}^2$ and R_H with the results shown in Table 8.2. The expression for $R_{G;LS}^2$ for the wormlike model does not follow the power-law behavior, but may still be examined for dispersity in M described by the Schulz–Zimm distribution function, for which $(1+h)/h = M_w/M_n$, $(2+h)/(1+h) = M_z/M_w$, etc.) [53]:

$$R_{G;LS}^2 = \left(\frac{L_z \hat{a}}{3}\right) S_{LS}\left(\frac{\hat{a}}{L_z}\right) \quad (8.54a)$$

$$S_{LS}(Z) = 1 - 3Z + 6Z^2 \frac{2+h}{1+h}$$
$$- 6Z^3 \frac{(2+h)^2}{h(1+h)^2}\{1 - [1 + Z/(2+h)]^{-h}\} \quad (8.54b)$$

$$\approx \left(1 + \frac{4\hat{a}}{L_{z+1}}\right)^{-1} \quad (8.54c)$$

with $R_{G;LS}^2$ reducing to $L_z \hat{a}/3$ for coils at small $\hat{a}/L$ and to $L_z L_{z+1}/12$ for rods at large $\hat{a}/L$, as expected.

Evaluation of $P_{LS}(q,0)$ for a flexible coil disperse in M gives [1, 53]:

$$P_{LS;RF}(q,0) = \left(\frac{2}{rM_w q^2}\right) \left\{ \begin{array}{l} 1 - \left(\dfrac{1}{rM_n q^2}\right) \times \\[2mm] [1 - M_n \sum_v^C w_v M_n^{-1} \exp(-rM_n q^2)] \end{array} \right\} \quad (8.55)$$

with $r = R_G^2/M = \hat{a}/3M_L$ a polymer-specific parameter, independent of chain length in the random-flight model. For heterodispersity in M characterized by the Schulz–Zimm distribution function, the final summation in $P_{LS;RF}(q,0)$ may be completed to give:

$$P_{LS;RF}(q,0) = (2/rM_w q^2)\{1 - (1/rM_n q^2)[1 - (1 + rM_n q^2/h)^{-h}]\} \quad (8.56)$$

where $M_w/M_n = (1+h)/h$, $M_z/M_w = (2+h)/(1+h)$, etc., such that h tends to infinity for a monodisperse sample.

For the special case with $h=1$, so that $M_w/M_n = 2$, $M_z/M_w = 3/2$, etc., this reduces to an exact result linear in q^2 (for all q) [1, 53]:

$$P_{LS;RF}^{-1}(q,0) = 1 + \frac{(rM_z q^2)}{3} \quad (8.57)$$

Clearly, in this case, use of the "square-root" analysis based on Equation 8.46 for a monodisperse random-flight chain would not be appropriate. In the limit of large $R_{G;LS}^2 q^2$, the expression for $P_{LS;RF}(q,0)$ is given by:

$$\lim_{rM_z q^2 \gg 1} P_{LS;RF}^{-1}(q,0) = \left(\frac{M_w}{2M_n}\right)\{1 + rM_n q^2 + O[(R_G^2 q^2)^{-1}]\} \quad (8.58)$$

In principle, information on the molecular weight distribution could be obtained via Equation 8.58 if experimental limitations permitted scattering at large enough $R_{G;LS}^2 q^2$ to reach this asymptotic regime; in practice, that is seldom the case with light scattering, but the relevant regime could be reached by a combination of light scattering with the scattering from electromagnetic radiation at smaller λ, for example, neutron scattering. Expressions for $P_{LS}(q,0)$ for other dispersed solute structures with the Schluz–Zimm distribution in M are available, including rodlike chains [5, 60].

The full expressions for $P_{LS}(q,0)$ and $R_{G;LS}^2$ via Equations 8.25 and 8.26 must be applied to account for chemical diversity among the scattering elements, with m_{iv} and ψ_{iv} differing among the scattering elements, even for the case for which all chains have the same molecular weight and composition, allowing only for variation in the sequence of the scattering elements among the chains in a copolymer (or in a particle), for example, a block or alternating copolymer, stratified particles, etc. Here, the case is considered for an optically isotropic solute in the RGD approximation, comprising two optically different types of scattering elements A and B, for a solute (or particles) with no molecular-weight dispersity, allowing only for variability in the distribution. The components of the two different scattering elements within the structure are characterized by weight fractions w_A and $w_B = 1 - w_A$, molar weights m_A and m_B, and contrast factors ψ_A and ψ_B, respectively, with $\psi \neq 0$, such that the expression for $P_{LS}(q,0)$ and $R_{G;LS}$ given by Equations 8.25 and 8.26 may be expressed, respectively, as [24, 65]:

$$P_{LS}(q,0) = \tilde{w}_A P_A(q,0) + (1 - \tilde{w}_A)P_B(q,0) + \tilde{w}_A(1 - \tilde{w}_A)P_{AB}(q,0) \quad (8.59)$$

$$R_{G;LS}^2 = \tilde{w}_A R_{G;A}^2 + (1 - \tilde{w}_A)R_{G;B}^2 + \tilde{w}_A(1 - \tilde{w}_A)\Delta_{AB}^2 \quad (8.60)$$

with $\tilde{w}_A = w_A \psi_A / \psi$ and $P_A(q,0)$ and $P_B(q,0)$ the interference functions for the structures occupied by the type A and B elements, respectively, and $P_{AB}(q,0)$ is a cross-term dependent on the distribution of components A and B in the copolymer (particle).

Expansion of $P_{AB}(q,0)$ in q^2 gives a leading term $\Delta_{AB}{}^2 q^2/3$, with $\Delta_{AB}{}^2$ defined in the corresponding expression for $R_{G;LS}^2$, showing that as expected $P_{AB}(0,0)$ must be zero to give $P_{LS}(0,0)=1$ for arbitrary $\tilde{w}_A$. Since $\tilde{w}_A$ may be positive, negative, or zero, the initial tangent $\partial P_{LS}^{-1}(q,0)/q^2$ may also adopt positive, negative, or zero values, leading to the same behavior for $R_{G;LS}^2$, $R_{G;A}^2$ and $R_{G;B}^2$, the mean-square radii of gyration of the portions of the solute molecules (particles) comprising only type A and type B scattering elements, respectively, w_A the weight fraction of type A, and Δ_{AB}^2 the mean-square separation of the center of mass for these portions. The possible dependence of the parameters $R_{G;LS}^2$, $R_{G;A}^2$, and $R_{G;B}^2$ on solvent is neglected herein; this would be expected to be less of an issue with particles than for flexible chain polymers owing to the likely dependence of R_G^2 on the solvent used in solution for the latter; this effect is suppressed, if not quite eliminated, in neutron scattering where the contrast may be varied by changes in the deuterium content of the solvent, with modest effect on R_G^2 [65]. The function Δ_{AB}^2 is known for a solute dispersed in molecular weight and for certain graft copolymers [65]. In one important example of block copolymers comprising n_{blk} blocks each of the A and B blocks, $\Delta_{AB}^2 = 2\left(R_{G;A}^2 + R_{G;B}^2\right)/n_{blk}$, showing that Δ_{AB}^2 tends to zero with increasing n_{blk}, as expected since $R_{G;LS}^2 = R_{G;A}^2 = R_{G;B}^2$ for an alternating copolymer with high degree of polymerization.

Interest in stratified spheres has been motivated by research on coated nanoparticles. For such a system, with a sphere of radius R_A of scattering elements of type A, coated by a shell with radius R_B of scattering elements of type B, $P_A(q,0)$ and $P_B(q,0)$ are the intramolecular interference functions for spheres of radius R_A and R_B, respectively, $P_{LS;RF}^{-1}(q,0)$ and $R_{G;LS}^2$ via Equations 8.25 and 8.26 are given by [11]:

$$P_{LS}(q,0) = \left(\begin{array}{c} \tilde{w}_A P_{sphere}^{1/2}(q,0) \\ +(1-\tilde{w}_A)\dfrac{R_B^3 P_{sphere}^{1/2}(q,0) - R_A^3 P_{sphere}^{1/2}(q,0)}{R_B^3 - R_A^3} \end{array} \right)^2 \quad (8.61)$$

$$R_{G;LS}^2 = \left(\frac{3}{5}\right)\left[\tilde{w}_A R_A^2 + (1-\tilde{w}_A)\frac{(R_B^5 - R_A^5)}{(R_B^3 - R_A^3)} \right] \quad (8.62)$$

Here, $P_{sphere;A}(q,0)$ and $P_{sphere;B}(q,0)$ are the interference functions for spheres of radius R_A and R_B, respectively. If $\psi_A = 0$, this result reduces to $P_{LS}(q,0)$ for a shell of thickness $\Delta = R_B - R_A$ filled with the solvent, per the entry in Table 8.1. If $\psi = 0$, then $[R(0,c)/c]^0 = 0$, but for $q>0$, $[R(q,c)/c]^0$ may be computed with Equation 8.59 with $\tilde{w}_A$ being replaced by

$w_A \psi_A$ to obtain $[R(q,c)/c]^0$ displaying a series of maxima and minima for $q>0$, with a superficial similarity to $R(q,c)/c$ observed with charged spheres owing to electrostatic interactions among the spheres. The model leading to Equation 8.60 for a dense shell coating a dense core has been extended to a dense sphere, ellipsoid, or cylinder coated by linear flexible chain polymers attached to the surface of the core, motivated by work on such structures to develop nanostructures for a range of applications [75]. The result is fairly complicated, but perhaps of practical interest nonetheless since the model may approximate nanoparticles important in some applications.

As would be anticipated, scattering for particles too large to obey the approximations for the RGD regime introduce appreciable change in the behavior described in the preceding text. The effects on $P(q,0)$ and R_G^2 for the conditions described in Figure 8.3 are given elsewhere [24]; as would be anticipated, a considerable literature exists on the scattering beyond the RGD regime [7, 33, 36–38, 42–44, 46, 47, 76–80].

8.3.4 Scattering at Arbitrary Angle and Dilute Concentration

For a monodisperse solute with optically isotropic, identical solute, the full expression for $R(q,c)$ given by Equation 8.19, must be used. In principle, though $P(q,c)$ may depend on c for a flexible chain polymer, such an effect is expected to be weak in a dilute solution, though not in a more concentrated solution [1, 49], and such a dependence is, of course, nil for a rigid particle. Given that circumstance, an equivalent, but more convenient form for dilute solutions is given by [28, 49]:

$$\frac{KcM}{R(c,q)} = \frac{1}{P(q,0)} + c\Gamma(c)\tilde{H}(q,c) \quad (8.63)$$

where the dependence of $P(q,c)$ and $H(q,c)$ on c in Equation 8.19 has been subsumed in $\tilde{H}(q,c)$, such that for dilute solutions [28, 49],

$$c\Gamma(c)\tilde{H}(q,c) = 2\hat{\psi}_2 W_2 \hat{c}$$
$$+ \left\{ 3\hat{\psi}_3 W_3(q) + 4\begin{bmatrix} P(q,0)W_2(q)^2 \\ -W_3(q) \end{bmatrix} \hat{\psi}_2^2 \right\} \hat{c}^2 + \cdots \quad (8.64)$$

with $\Gamma(c)$ given by Equation 8.31 in the preceding section.

The functions $W_2(q)$ and $W_3(q)$ have been calculated for the random-flight chain model [49], and as given in the preceding, $\hat{\psi}_\mu = A_\mu M(M/N_A R_G^3)^{\mu-1}$ and $\hat{c} = N_A R_G^3/M$. For example, for the so-called single-contact approximation for dilute solutions of random-flight polymers, $W_2(q)=1$, and $\tilde{H}(q,c) \approx 1$ to order $\hat{c}$ [1, 58]. In this case, plots of $Kc/R(q,c)$

versus q^2 at constant c are expected to be parallel, with intercepts $Kc/R(0,c) = M^{-1}[1 + c\Gamma(c)]$ and an initial tangent for small q given by $\partial[Kc/R(q,c)]/\partial q^2 = R_G^2/M$. In addition, plots of $Kc/R(q,c)$ versus c at constant q are also expected to be parallel, with intercepts $[Kc/R(q,c)]^0 = 1/[MP(q,0)]$ and initial tangents for small c given by $\partial[Kc/R(q,c)]/\partial c = 2A_2$.

This dual parallelism gave rise to the double-extrapolation method of the "Zimm plot," [1] in which $Kc/R(q,c)$ is plotted versus $q^2 + kc$, with k an arbitrary constant adjusted to give a satisfactory separation of the data at constant q and c, to take advantage of the anticipated parallelism in evaluating the experimental data for artifacts or error. It may be noted that the square-root plots of $P^{-1/2}(q,0)$ versus q^2 mentioned in the preceding as an aid to the analysis of high molecular weight flexible chain polymers with a narrow molecular weight distribution would not lead to plots of parallel plots of $[Kc/R(q,c)]^{1/2}$ versus q^2 at fixed c, nor would the suggested plots of $[Kc/R(q,c)]^{1/2}$ versus c at fixed q be parallel, making the double-extrapolation method of dubious use, even though each square-root plot by itself may improve the determination of the relevant intercept and initial tangent.

Although $P(q,c)$ does not depend on c for rigid spheres interacting through a hard-core potential, the intermolecular interference function $H(q,c)$ does depend on q, with $H(q,c) \approx P^{1/2}(2q,0)/P(q,0)$ for small c in dilute solutions [2]; a more complex behavior on both q and c is expected for more concentrated solutions [24, 81]. The dependence of $H(q,c)$ on q is normally a small effect in dilute dispersions of spheres interacting through a hard-core potential, but can have an appreciable impact in the scattering from charged spheres dispersed in an aqueous medium without any supporting electrolyte to suppress electrostatic interactions among the spheres. It is convenient to start with the representation for $R(q,c)/KMc$ for a homogeneous solute with optically isotropic identical scattering elements and a monodisperse solute:

$$\frac{R(c,q)}{KMc} = \frac{P(q,c)}{1 + c\Gamma(c)P(q,c)H(q,c)} \quad (8.65)$$

For large κ^{-1}, corresponding to a low concentration of an electrolyte in the solution with the charged solute, the substantial increase in $\Gamma(c)$ can lead to an approximate form of this expression provided $H(q,c) > 0$:

$$\frac{R(c,q)}{KM} = \frac{1}{\Gamma(c)H(q,c)} \quad (8.66)$$

In this form, $H(q,c)$ given in the preceding gives a maximum in $H^{-1}(q,c)$ for uncharged spheres for $(qR)_{max} \approx 2.25$ (in fact, since $H(q,c) = 0$ for this $(qR)_{max}$ for the expression given in the preceding, a singularity is

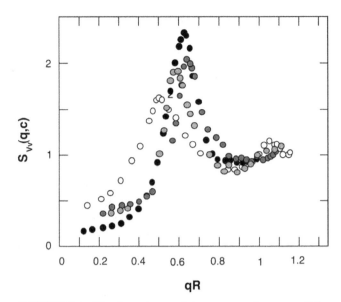

FIGURE 8.6   The dependence of the structure factor on qR for polystyrene spheres ($R = 45$ nm) immersed in deionized water adapted from figures in [83], with the number concentration n/particles·mm^{-3} = 2.53, 5.06, 7.59, and 10.12 for the circles with increasing depth of the shading [10]. Copyright 2000, John Wiley & Sons Ltd.

predicted for monodisperse spheres). The data in Figure 8.6 for charged polystyrene spheres in deionized water reveal a substantial maximum in $R(q,c)$ as expected, but the maximum occurs at both smaller qR than suggested by the relation for uncharged spheres, and is broader than the peak given by that relation. Similar to the observation earlier that $A_{2,LS} \propto M_L^2 k^{-1}$ for charged rodlike chains [82], the low ionic strength of these dispersions suggests that the relevant dimension for the maximum in $1/[\Gamma(c)H(q,c)]$ may occur for a solute dimension larger than the sphere diameter R, perhaps proportional to κ^{-1}, or equivalently, for a smaller value of q. In addition, the relatively soft nature of the potential could correspond to a broadened peak compared to that predicted for monodispersed uncharged spheres [83].

8.4   STATIC LIGHT SCATTERING: ANISOTROPIC SYSTEMS

8.4.1   General Relation for the Rayleigh Ratio

The depolarized scattering in the RGD regime from a solute with anisotropic scattering elements may be expressed as a function of the ensemble-averaged mean-square optical anisotropy per molecule $\langle \gamma_v^2 \rangle$ [5, 10, 24, 39, 84]:

$$[R_{aniso}(0,c)]^0 = \left(\frac{1}{15}\right)K'(2\pi N_A)^2 \sum_v^C w_v M_v^{-1} \gamma_v^2 \quad (8.67a)$$

$$\gamma_v^2 = \left\{ \begin{array}{l} \left(\dfrac{3}{2}\right)\left[\left(\tilde{\alpha}_{11} - \tilde{\alpha}_{22}\right)^2 + \left(\tilde{\alpha}_{11} - \tilde{\alpha}_{33}\right)^2 + \left(\tilde{\alpha}_{22} - \tilde{\alpha}_{33}\right)^2\right] \\ +3\left[\tilde{\alpha}_{12}^2 + \tilde{\alpha}_{13}^2 + \tilde{\alpha}_{23}^2\right] \end{array} \right\}_v$$

(8.67b)

with the $\tilde{\alpha}_{ij}$ the components of the polarizability tensor.

For isotropic scatterers, $\tilde{\alpha}_{ij} = \left(\tilde{\alpha}_{iso}/3\right)/\delta_{ij}$ and $\langle\gamma_v^2\rangle = 0$, giving $[R_{aniso}(0,c)/c]^0 = 0$. As elaborated in the following, this is almost always a good approximation for flexible chain polymers with a reasonably high molecular weight. Detailed evaluation of $\langle\gamma^2\rangle$ has been given for some polymers in the frame of the rotational isomeric state model [50]. A useful simplification arises for polymers with identical cylindrically symmetric polarizabilities for all chain elements, where $\tilde{\alpha}_{\parallel}$ and $\tilde{\alpha}_{\perp}$ are the principal polarizabilities of the scattering elements. In that case (for $(\partial n/\partial c)_w \neq 0$), Equation 8.67 may be put in the form [10, 24, 84]:

$$\left[\frac{R_{aniso}q,c}{c}\right]^0 = K' n_{medium}^2 \left(\frac{\partial n}{\partial c}\right)_w^2 M_{LS;Hv}$$

(8.68a)

$$M_{LS;Hv} = \left(\frac{3}{5}\right) M_w \delta_{LS}^2$$

(8.68b)

$$\delta_{LS}^2 = M_v^{-1} \sum_C^v w_v M_v \delta_v^2$$

(8.68c)

$$\delta_v^2 = \left(\frac{3}{2}\right)\frac{\delta_0^2}{n_v} \sum_j^{n_v} \sum_k^{n_v} \left[\langle\cos^2\beta_{jk}\rangle - 1\right]$$

(8.68d)

with β_{jk} being the angle between the major axes associated with scattering elements j and k on a chain with n_v elements, each with molecular weight m_0, $\langle...\rangle$ denotes an ensemble average and δ_0 is given by:

$$\delta_0 = \frac{\tilde{\alpha}_{\parallel} - \tilde{\alpha}_{\perp}}{\tilde{\alpha}_{\parallel} + 2\tilde{\alpha}_{\perp}}$$

(8.69)

8.4.2 Scattering at Zero Angle and Infinite Dilution

As discussed in section , $R_{Hv}(q,c)$ and $R_{Vv}(q,c)$ are expressed in terms of $R_{iso}(q,c)$, $R_{iso}(q,c)$, and $R_{cross}(q,c)$ by Equations 8.2 and 8.3, respectively, with $R_{cross}(0,c)=0$. The vertical and horizontal components of light scattered with vertically polarized light from a dilute solution extrapolated to zero scattering angle, $R_{Vv}(0,c)$ and $R_{Hv}(0,c)$, respectively, for a monodisperse solute with identical anisotropic scattering elements are given by [13, 28, 84–86]:

$$\frac{Kc}{R_{Vv}(0,c)} = \frac{1}{M(1+4\delta^2/5)}\left\{1 + 2\frac{(1-\delta^2/10)}{(1+4\delta^2/5)}A_2Mc + ...\right\}$$

(8.70)

$$\frac{Kc}{R_{Hv}(0,c)} = \frac{5}{3M\delta^2}\{1 - A_2Mc/4 + ...\}$$

(8.71)

with the latter limited to rodlike chains.

It may be noted that the dependence of $Kc/R_{Hv}(0,c)$ for rodlike chains is both smaller in magnitude than that of $Kc/R_{Vv}(0,c)$, and opposite in sign. The weak dependence of $Kc/R_{Vv}(0,c)$ on c indicates that to a reasonable approximation, $Kc/R_{Vv}(0,c) \approx [Kc/R_{Vv}(0,c)]^0$, a result found in scattering reported for a rodlike chain, albeit for an example with $M_w/M_n \approx 1.5$ [82].

The parameters $M_{LS;Vv}$ and $M_{LS;Hv}$ corresponding, respectively, to the observable $(R_{Vv}(0,c)/Kc)^0$ and $(R_{Hv}(0,c)K/c)^0$ are given by:

$$\left(\frac{R_{Vv}(q,c)}{Kc}\right)^0 = M_{LS;Vv} = \left[1 + \left(\frac{4}{5}\right)\delta_{LS}^2 M_w\right]$$

(8.72)

$$\left(\frac{R_{Hv}(q,c)}{Kc}\right)^0 = M_{LS;Hv} = \left(\frac{3}{5}\right)\delta_{LS}^2 M_w$$

(8.73)

Two limiting cases for chain molecules are provided by a rodlike chain, for which $\beta_{jk}=0$ for all j and k, so that $\delta_v = \delta_0$, and a random coil with uncorrelated orientations among the scattering elements, so that only terms with $j=k$ contribute to the sum, giving $\delta_v = \delta_0/n_v = m_0\delta_0/M_v$. For these cases:

$$\text{rod: } M_{LS;Hv} = \left(\frac{3}{5}\right)\delta_0^2 M_w$$

(8.74)

$$\text{coil: } M_{LS;Hv} = \left(\frac{3}{5}\right)\delta_0^2 m_0$$

(8.75)

It may be seen that $M_{LS;aniso}$ does not provide information on M_w for a random coil chain owing to the lack of orientational correlation among the scattering elements, and that $M_{LS;aniso}$ approaches a maximum value of $(3/5)M_w$ for a rodlike chain if $\delta_0=1$, and may be much smaller for lower values of δ_0. The wormlike chain model provides a crossover between these two limits. In this model, a chain is characterized by it persistence length $\hat{a}$, contour length L, and mass per unit length $M_L=M/L$, approaching rodlike behavior for increasing $\hat{a}/L$, and coil behavior in the opposite limit. For this model [13, 84, 87],

$$\frac{\delta_v^2}{\delta_0^2} = \left(\frac{2Z_v}{3}\right)\left\{1 - \left(\frac{Z_v}{3}\right)\left[1 - \exp\left(\frac{-3}{Z_v}\right)\right]\right\}$$

(8.76)

with $Z_v = \hat{a}/L_v$.

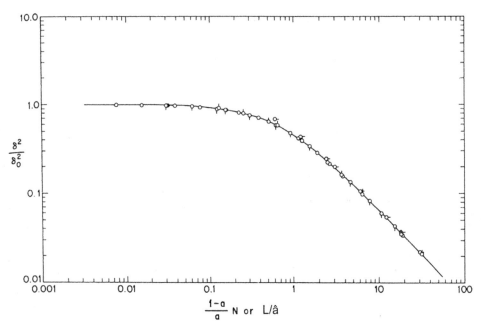

FIGURE 8.7 The δ/δ_0^2 versus $L/\hat{a}$ for the wormlike chain (line) and for a chain with free rotation about N bonds with valance angles $(\pi - \alpha_{val})$ versus $[(1-a)/a]N$, where $a = \cos(\alpha_{val})$, for $\alpha_{val} = 5°$, 10°, 20°, 40° for pips absent, right, down, and up, respectively. With permission from Berry GC. Properties of an optically anisotropic heterocyclic ladder polymer (BBL) in solution. J Polym Sci: Polym Symp 1978;65:143–172. © 1978 Wiley Blackwell.

As may be seen in Figure 8.7, δ_v^2/δ_0^2 approaches the rod-like limit of unity for $L_v/\hat{a} \ll 1$ and the coil limit $2\hat{a}/3L_v$ for $L_v/\hat{a} \gg 1$, with transition developing for $L_v/\hat{a} \approx 1$. A numerically similar result is obtained for the model with free rotation about N bonds with valance angles $(\pi - \alpha_{val})$. As seen in Figure 8.7, the results for the model, with $a = \cos(\alpha_{val})$, are essentially equivalent to the results for the wormlike chain model [88].

8.4.3 Scattering at Arbitrary Angle and Infinite Dilution [84]

In this case, the scattering in the RGD regime for identical anisotropic scattering elements at infinite dilution may be expressed in the form as given in the preceding, with anisotropic scattering elements, expressions $[\boldsymbol{R}_{Vv}(q,c)/c]^0$ and $[\boldsymbol{R}_{Hv}(q,c)/c]^0$ are given by:

$$\left[\frac{\boldsymbol{R}_{Vv}(q,c)}{c}\right]^0 = K'n_{medium}^2\left(\frac{\partial n}{\partial c}\right)_w^2 M_{LS;Vv}P_{LS;Vv}(q,0) \quad (8.77)$$

$$\left[\frac{\boldsymbol{R}_{Hv}(q,c)}{c}\right]^0 = K'n_{medium}^2\left(\frac{\partial n}{\partial c}\right)_w^2 M_{LS;Hv}P_{LS;Hv}(q,0) \quad (8.78)$$

The intramolecular interference functions are given by [5, 13, 39, 84, 87]:

$$\left[1+\left(\frac{4}{5}\right)\delta^2\right]P_{LS;Vv}(q,0) = p_1(X)+\left(\frac{4}{5}\right)\delta^2 p_3(X)$$
$$+(\delta-2\delta^2)m_1(X)+\left(\frac{9}{8}\right)\delta^2 m_2(X)+\delta^2 m_3(X) \quad (8.79)$$

$$P_{LS;Hv}(q,0) = p_3(X)+\left(\frac{5}{16}\right)\sin^2\left(\frac{\theta}{2}\right)m_2(X) \quad (8.80)$$

where $X = (12\,u)1/2$, with $u = R_G^2 q^2$, and the p_μ and m_μ given by:

$$p_1(X) = (2X^2)\{X\mathrm{Si}(X)+\cos X - 1\} \approx 1 - (1/3)u+\cdots \quad (8.81a)$$

$$p_2(X) = (6X^3)\{X - \sin X\} \approx 1 - (3/5)u+\cdots \quad (8.81b)$$

$$p_3(X) = (10X^5)\{X^3 + 3X\cos X - 3\sin X\} \approx 1 - (3/7)u+\cdots \quad (8.81c)$$

$$m_1 = p_1 - p_2 \approx (4/15)u+\cdots \quad (8.82a)$$

$$m_2 = 3p_1 - p_2 - 2p_3 \approx (16/35)u+\cdots \quad (8.82b)$$

$$m_3 = p_3 - p_2 \approx (6/35)u+\cdots \quad (8.82c)$$

It may be noted that $P_{LS;Hv}(q,0)$ depends explicitly on the scattering angle θ, or alternatively on $q/(4\pi/\lambda) = \sin(\theta/2)$;

further, the functions m_μ all vanish as q goes to zero and $P_{LS;Vv}(q,0)$ reduces to $P(q,0)$ for a monodisperse optically isotropic rodlike chain if $\delta = 0$ (with $\boldsymbol{R}_{Hv}(q,c)=0$).

Owing to the dependence of δ on the ratio $\hat{a}/L$ of the persistence length $\hat{a}$ to the chain contour length L for wormlike chains, δ decreases rapidly from its intrinsic value δ_0 with decreasing $\hat{a}/L$, for example, as discussed in the preceding. For a wormlike chain comprising a helical structure the twisting of the orientation along the chain will further decrease δ_0. Thus, the effects of any anisotropy on $[\boldsymbol{R}_{Vv}(q,c)/c]^0$ may often be too small for accurate evaluation for helical chains. Consequently, since δ^2 appears directly as a factor in $[\boldsymbol{R}_{Hv}(q,c)/c]^0$, the depolarized scattering affords an advantage in the evaluation of the anisotropy of a wormlike chain, with the attendant possible evaluation of $\hat{a}/L$ and hence $\hat{a}$ using L from data on $[\boldsymbol{R}_{Vv}(q,c)/c]^0$ [24].

The expressions for $P_{LS;Vv}(q,0)$ and $P_{LS;Hv}(q,0)$ may be used to obtain the corresponding values of $\left[R_{G;LS}^2\right]_{Vv}$ and $\left[R_{G;LS}^2\right]_{Hv\,Hv}$, respectively, to give [13, 39, 84, 87, 88]:

$$\left[\boldsymbol{R}_{G;LS}^2\right]_{Vv} = \frac{\sum_\nu^C w_\nu M_\nu \left[1 - \left(\frac{4}{5}\right)f_{1;\nu}\delta_\nu + \left(\frac{4}{7}\right)\left(f_{1;\nu}\delta_\nu\right)^2\right]\boldsymbol{R}_{G;\nu}^2}{\sum_\nu^C w_\nu M_\nu \left[1 + \left(\frac{4}{5}\right)\delta_\nu^2\right]}$$

(8.83)

$$\left[\boldsymbol{R}_{G;LS}^2\right]_{Hv} = \frac{\sum_\nu^C w_\nu M_\nu \left[\left(f_{1;\nu}\delta_\nu\right)^2\right]\boldsymbol{R}_{G;\nu}^2}{\sum_\nu^C w_\nu M_\nu \left[\delta_\nu^2\right]}$$

(8.84)

with δ being the anisotropy introduced in the preceding.

The functions $f_{i;\nu}$ are known as a function of the ratio $\hat{a}/L$ of the persistence length $\hat{a}$ and contour length [84]. In the limit of $\hat{a}/L \gg 1$, all of the $f_{i;\nu}$ are unity, with both δ_ν and the $f_{i;\nu}$ decreasing with increasing $\hat{a}/L_\nu$. Because of the feature, it is a reasonable approximation to evaluate $\left[R_{G;LS}^2\right]_{V_\nu}$ and $\left[R_{G;LS}^2\right]_{H_\nu}$ with these relations, using the δ_ν corresponding to the appropriate $\hat{a}/L_\nu$, but with all of the $f_{i;\nu}$ set equal to unity.

8.5 NOMENCLATURE

Definition of principal symbols used throughout the text:

A_2, A_3	The second virial coefficient and third virial coefficient, respectively; see Equation 8.21
$\tilde{B}(c)$	The thermodynamic interaction function; see Equation 8.16
$B(c)$	A thermodynamic interaction function, equal to $M\tilde{B}(c) = [F(0,c)-1]/c$, see Equation 8.16
$F(q,c)$	The intermolecular structure factor $\boldsymbol{R}(q,c)/K_c$ $MP(q,c)$; see Equation 8.15
$H(q,c)$	The function $[F(q,c)^{-1} - 1]/c\Gamma(c)P(q,c)$; see Equation 8.16
$\boldsymbol{K}$	An optical constant relating intensities to the Rayleigh ratio $K=K'n^2_{medium}$.
K'	A factor, equal to $4\pi^2/N_A\lambda_0^4$, appearing in $\boldsymbol{K}$
L	Chain contour length
M	Molecular weight
M_L	The mass per unit length, M/L
M_w	The weight average molecular weight
N_A	Avogadro's constant
$P(q,c)$	The intramolecular structure factor; see Equation 8.14
$Q(c)$	The function $[F(q,c) - 1]/cB(c)P(q,c)$; see Equation 8.14
R_G^2	Mean-square radius of gyration
$\boldsymbol{R}(q,c)$	The excess Rayleigh ratio at wavenumber q, for solute concentration c (wt/vol)
$S(q,c)$	The total structure factor $\boldsymbol{R}_{iso}(q,c)/KcM$; see Equation 8.15
$\hat{a}$	Persistence length for semiflexible chains
c, c_μ	The solute concentration (wt/vol); concentration of solute component μ
h	A parameter in the Schulz–Zimm molecular weight distribution function (e.g., $1+h^{-1}=M_w/M_n$), see Table 8.1
m_ν	The molecular weight of the νth scattering element
n_{medium}	Refractive index of the medium
$(\partial n/\partial c)_w$	The refractive index increment
$\tilde{n}(t; q)$	The number of photons count at angle θ over the time interval t to $t+\Delta t$
q	The modulus $q=(4\pi/\lambda)\sin(\theta/2)$ wave vector $\boldsymbol{q}$
w, w_μ	The solute weight fraction; weight fraction of solute component μ
θ	The scattering angle
α	The parameter $2\pi R/\lambda$ for a sphere of radius R
δ^2	Mean-square molecular optical anisotropy; see Equation 8.68
δ_0	Optical anisotropy of a scattering element with molecular weight m_0
λ	The wavelength of light in the scattering medium; λ_0 the same *in vacuo*
ρ	The density (wt/vol)
ψ, ψ_μ	The contrast factor for optically isotropic media; the same for component μ
$\Gamma(c)$	The thermodynamic interaction function $\{F(0,c)^{-1} - 1\}/c$; see Equation 8.16

Definitions of the principal subscripts used throughout the text:

LS   A subscript indicating the average of a function or parameter obtained in light scattering (e.g., M_{LS}, $R^2_{G;LS}$, etc.)

μ A solute component in a mixture (e.g., c_μ)

iso   The isotropic component of a function (e.g., $R_{iso}(q,c)$)

aniso   The anisotropic component of a function (e.g., $R_{aniso}(q,c)$)

Hv   A property determined using the horizontally plane polarized component of the light scattered using vertically plane polarized incident light (e.g., $R_{Hv}(q,c)$, $P_{Hv}(q,c)$, etc.)

Vv   A property determined using the horizontally plane polarized component of the light scattered using vertically plane polarized incident light (e.g., $R_{Vv}(q,c)$, $P_{Vv}(q,c)$, etc.)

REFERENCES

[1] Zimm BH. The scattering of light and the radial distribution function of high polymer solutions. J Chem Phys 1948;16:1093–1099.

[2] Guinier A, Fournet, G. *Small-Angle Scattering of X-Rays*. New York: John Wiley & Sons, Inc.; 1955.

[3] Stacey KA. *Light-Scattering in Physical Chemistry*. New York: Academic Press; 1956.

[4] McIntyre D, Gornick F, editors. *Light Scattering from Dilute Polymer Solutions*. International Science Review Series 3. New York: Gordon and Breach; 1964.

[5] Casassa EF, Berry GC. Light scattering from solutions of macromolecules. In: Slade Jr, PE, editor. *Polymer Molecular Weights, Pt 1*. New York: Marcel Dekker Inc.; 1975. p 161–286.

[6] Glatter O, Kratky O, editors. *Small Angle X-Ray Scattering*. New York: Academic Press; 1982.

[7] Pangonis WJ, Heller W, Jacobson, A. *Tables of Light Scattering Functions for Spherical Particles*. Detroit: Wayne State University Press; 1957.

[8] Kerker M, editor. *Selected Papers on Light Scattering*. SPIE Milestone Series 951. Bellingham: SPIE International Society for Optical Engineering; 1988.

[9] Stockmayer WH. Reminiscences of "light scattering in multicomponent systems". J Polym Sci Part B: Polym Phys 1999;37:642–643.

[10] Berry GC. Light scattering, classical: size and size distribution classification. In: Meyers RA, editor. *Encyclopedia of Analytical Chemistry*. New York: John Wiley & Sons; 2000. p 5413–5448.

[11] Kerker, M. *The Scattering of Light, and Other Electromagnetic Radiation*. New York: Academic Press; 1969.

[12] Huglin MB, editor. *Light Scattering from Polymer Solutions*. London: Academic Press; 1972.

[13] Berry GC. Light scattering. In: Mark HF, Overberger CG, Bikales NM, Menges FM, editors. *Encyclopedia of Polymer Science and Engineering*. Volume 8. New York: John Wiley & Sons; 1987. p 721–794.

[14] Schmitz KS. *An Introduction to Dynamic Light Scattering by Macromolecules*. Boston: Academic Press; 1990.

[15] Eisenberg H. *Biological Macromolecules and Polyelectrolytes in Solution*. London: Oxford University Press; 1976.

[16] Sadler DM. Neutron scattering from solid polymers. In: Allen G, Bevington JC, editors, *Comprehensive Polymer Science*. New York: Pergamon Press; 1988. p 731–763.

[17] Casassa EF. Particle scattering factors in Rayleigh scattering. In: Immergut EH, Branderup J, editors. *Polymer Handbook*. 3rd ed. New York: John Wiley & Sons; 1989. p VII/485–491.

[18] Mourey TH, Coll H. Size-exclusion chromatography with two-angle laser light-scattering (SEC-TALLS) of high molecular weight and branched polymers. J Appl Polym Sci 1995;56:65–72.

[19] Pedersen JS, Schurtenberger P. Scattering functions of semiflexible polymers with and without excluded volume effects. Macromolecules 1996;29:7602–7612.

[20] Pedersen JS, Gerstenberg MC. Scattering form factor of block copolymer micelles. Macromolecules 1996;29:1363–1365.

[21] Stockmayer WH. Light scattering in multi-component systems. J Chem Phys 1950;18: 58–61.

[22] Pedersen JS. Structure factors effects in small-angle scattering from block copolymer micelles and star polymers. J Chem Phys 2001;114:2839–2846.

[23] Pedersen JS. Modelling of small-angle scattering data from colloids and polymer systems. In: Lindner P, Zemb T, editors. *Neutrons, X-Rays Light*. New York: Elsevier; 2002. p 391–420.

[24] Berry GC. Total intensity light scattering from solutions macromolecules. In: Borsali R, Pecora R, editors. *Soft-Matter Characterization*. New York: Springer; 2008. p 41–131.

[25] Burchard W. Static and dynamic light scattering from branched polymers and biopolymers. Adv Polym Sci 1983;48:1–124.

[26] Kirkwood JG, Goldberg RJ. Light scattering arising from composition fluctuations in multi-component systems. J Chem Phys 1950;18:54–57.

[27] Burchard W. Static and dynamic light scattering approaches to structure determination in biopolymers. In: Harding SE, Sattelle DB, Bloomfield VA, editors. *Laser Light Scattering in Biochemistry*. Cambridge, UK: Royal Society Chemistry; 1992. p 3–22.

[28] Berry GC. Static and dynamic light scattering on moderately concentrated solutions: isotropic solutions of flexible and rodlike chains and nematic solutions of rodlike chains. Adv Polym Sci 1994;114:233–290.

[29] Brown W, editor. *Light Scattering: Principles and Development*. New York: Oxford University Press; 1996.

[30] Berry GC, Cotts PM. Static and dynamic light scattering. In: Pethrick RA, Dawkins JV, editors. *Modern Techniques for*

Polymer Characterisation. London: John Wiley & Sons; 1999. p 81–108.

[31] Brown W, Mortensen K, editors. *Scattering in Polymeric and Colloidal Systems.* Newark: Gordon and Breach Publishers; 2000.

[32] Berne BJ, Pecora, R. *Dynamic Light Scattering.* New York: John Wiley & Sons; 1976.

[33] Dahneke BE, editor. *Measurement of Suspended Particles by Quasi-Elastic Light Scattering.* New York: Wiley-Interscience; 1983.

[34] Brown W, editor. *Dynamic Light Scattering.* Oxford: Clarendon Press; 1993.

[35] Chu B. *Laser Light Scattering.* 2nd ed. Boston: Academic Press; 1991.

[36] van de Hulst HC. *Light Scattering by Small Particle.* New York: Wiley; 1957.

[37] Bohren CF, Huffman DR. *Absorption and Scattering of Light by Small Particles.* New York: John Wiley & Sons; 1983.

[38] Weiner BB. Particle and droplet sizing using Fraunhofer diffraction. In: Barth HG, editor. *Modern Methods of Particle Size Analysis.* New York: John Wiley & Sons; 1984. p 135–172.

[39] Horn P. Light scattering in solutions of anisotropic macromolecules. *Annalen der Physik* 1955;10:386–434.

[40] Hirleman ED. Optimal scaling of the inverse Fraunhofer diffraction particle sizing problem: The linear system produced by quadrature. In: Gouesbet G, Gréhan G, editors. *Optical Particle Sizing: Theory and Practice.* New York: Plenum Press; 1988. p 135–146.

[41] Hirleman ED. Modeling of multiple scattering effects in Fraunhofer diffraction particle analysis. In: Gouesbet G, Gréhan G, editors. *Optical Particle Sizing: Theory and Practice.* New York: Plenum Press; 1988. p 159–175.

[42] Gouesbet G, Gréhan G, editors. *Optical Particle Sizing: Theory and Practice.* New York: Plenum Press; 1988.

[43] Glatter O, Hofer, M. Particle sizing of polydisperse samples by Mie-scattering. In: Gouesbet G, Gréhan G, editors. *Optical Particle Sizing: Theory and Practice.* New York: Plenum Press; 1988. p 121–133.

[44] Bertero M, Boccacci P, De Mol C, Pike ER. Particle-size distributions from Fraunhofer diffraction. In: Gouesbet G, Gréhan G, editors. *Optical Particle Sizing: Theory and Practice.* New York: Plenum Press; 1988. p 99–105.

[45] Brown DJ, Weatherby EJ, Alexander, K. Shape, concentration and anomalous diffraction effects in sizing solids in liquids. In: Gouesbet G, Gréhan G, editors. *Optical Particle Sizing: Theory and Practice.* New York: Plenum Press; 1988. p 351–362.

[46] Barber PW, Hill SC. *Light Scattering by Particles: Computational Methods.* Singapore: World Scientific; 1990.

[47] Sorensen CM, Scattering and absorption of light by particles and aggregates. In: Birdi KS, editor. *Handbook of Surface and Colloid Chemistry.* Boca Raton: CRC Press; 1997. p 533–558.

[48] Flory PJ. *Principles of Polymer Chemistry.* Ithaca: Cornell University Press; 1953.

[49] Yamakawa H. *Modern Theory of Polymer Solutions.* New York: Harper and Row; 1971.

[50] Flory P. *Statistical Mechanics of Chain Molecules.* New York: Wiley-Interscience; 1979.

[51] de Gennes P-G. *Scaling Concepts in Polymer Physics.* Ithaca: Cornell University Press; 1979.

[52] Kurata M. *Thermodynamics of Polymer Solutions*; translated by Hiroshi Fujita. New York: Harwood Academic Publishers; 1982.

[53] Berry GC. Molecular weight distribution. In: Bever MB, editor. *Encyclopedia of Materials Science and Engineering.* Oxford: Pergamon Press; 1986. p 3759–3768.

[54] Casassa EF, Berry GC. Polymer solutions. In: Allen G, Bevington JC, editors, *Comprehensive Polymer Science.* New York: Pergamon Press; 1988. p 71–120.

[55] Kim SH, Ramsay DJ, Patterson GD, Selser JC. Static and dynamic light scattering of poly(α-methylstyrene) in toluene in the dilute region. *J Polym Sci Part B: Polym Phys* 1990;28:2023–2056.

[56] Ostrowsky N, Sornette D, Parker R, Pike ER. Exponential sampling method for light scattering polydispersity analysis. *Optica Acta* 1981;28:1059–1070.

[57] Yue S, Berry GC. Moderately concentrated solutions of polystyrene. Part 5. Static and dynamic light scattering in bis(2-ethylhexyl) phthalate. *Macromol Symp* 1995;98:1005–1027.

[58] Yamakawa H. *Helical Wormlike Chains in Polymer Solutions.* New York: Springer-Verlag; 1997.

[59] Fujita H. *Polymer Solutions.* Amsterdam: Elsevier; 1990.

[60] Berry GC. Thermodynamic and conformational properties of polystyrene. I. Light-scattering studies on dilute solutions of linear polystyrenes. *J Chem Phys* 1966;44:4550–4564.

[61] Stockmayer WH, Casassa EF. The third virial coefficient in polymer solutions. *J Chem Phys* 1952;20:1560–1566.

[62] Benoit HC, Strazielle C. Interpretation of preferential adsorption using random phase approximation theory. *Collect Czech Chem Commun* 1995;60:1641–1652.

[63] Casassa EF. Interpretation of Rayleigh scattering by polymers in mixed solvents. *Makromol Chem* 1971;150:251–254.

[64] Benoit H. Light scattering by dilute solutions of high polymers. In: Kerker M, editor. *Electromagnetic Scattering.* Elmworth: Pergamon Press; 1963. p 285–301.

[65] Benoit H, Froelich D. Application of light scattering to copolymers. In: Huglin MB, editor. *Light Scattering from Polymer Solutions.* New York: Academic Press; 1972. p 467–501.

[66] Schurtenberger P. Contrast and contrast variation in neutron, X-ray and light scattering. In: Lindner P, Zemb T, editors. *Neutrons, X-Rays Light.* New York: Elsevier; 2002. p 145–170.

[67] Casassa EF, Berry GC. Angular distribution of intensity of Rayleigh scattering from comblike branched molecules. *J Polym Sci: Part A-2* 1966;4:881–897.

[68] Frank R, Frank L, Ford NC. Molecular characterization using a unified refractive index-light-scattering intensity detector. In: Provder T, Urban MW, Barth HG, editors. *Chromatographic Characterization of Polymers.* Advanced Chemical Series 247. Washington, DC: Oxford University Press; 1995. p 109–121.

[69] Wyatt PJ. Multiangle light scattering: the basic tool for macromolecular characterization. Instrum Sci Technol 1997;25:1–18.

[70] Terao K, Mays JW. On-line measurement of molecular weight and radius of gyration of polystyrene in a good solvent and in a theta solvent measured with a two-angle light scattering detector. Eur Polym J 2004;40:1623–1627.

[71] Peterlin, A. Light scattering by non-Gaussian macromolecular coils. In: Kerker M, editor, *Electromagnetic Scattering*. Elmworth: Pergamon Press; 1963. p 357–375.

[72] Yoshizaki T, Yamakawa H. Scattering functions of wormlike and helical wormlike chains. Macromolecules 1980;13:1518–1525.

[73] Porod G. The dependence of small-angle x-ray scattering on the shape and size of colloidal particles in dilute systems. IV. Acta Phys Austriaca 1948;2:255–292.

[74] Mays H, Mortensen K, Brown W. Microemulsions studied by scattering techniques. In: Brown W, Mortensen K, editors. *Scattering in Polymeric and Colloidal Systems*. Amsterdam: Gordon & Breach; 2000. p 249–325.

[75] Terashima T, Motokawa R, Koizumi S, Sawamoto M, Kamigaito M, Ando T, Hashimoto T. In situ and time-resolved small-angle neutron scattering observation of star polymer formation via arm-linking reaction in ruthenium-catalyzed living radical polymerization. Macromolecules 2010;43:8218–8232.

[76] Asano S, Sato M. Light scattering by randomly oriented spheroidal particles. Appl Optics 1980;19:962–974.

[77] Hodkinson JR. Particle sizing by means of the forward scattering lobe. Appl Optics 1996;5:839–844.

[78] Wiscombe WJ. Improved Mie scattering algorithms. Appl Optics 1980;19:1505–1509.

[79] Strawbridge KB, Hallett FR. Polydisperse Mie theory applied to hollow latex spheres: an integrated light-scattering study. Can J Phys 1992;70:401–406.

[80] Hahn DK, Aragon SR. Mie scattering from anisotropic thick spherical shells. J Chem Phys 1994;101:8409–8417.

[81] Kinning DJ, Thomas EL. Hard-sphere interactions between spherical domains in diblock copolymers. Macromolecules 1984;17:1712–1718.

[82] Sullivan VJ, Berry GC. Light scattering studies on dilute solutions of semiflexible polyelectrolytes. Int J Polym Anal Ch 1995;2:55–69.

[83] Grüner F, Lehmann W. On the long time diffusion of interacting Brownian particles. In: Degiorgio V, Corti M, Giglio M, editors. *Light Scattering in Liquids and Macromolecular Solutions*. New York: Plenum Press; 1980. p 51–69.

[84] Berry GC. Properties of an optically anisotropic heterocyclic ladder polymer (BBL) in solution. J Polym Sci: Polym Symp 1978;65:143–172.

[85] Benoit H. Determination of the dimensions of anisotropic macromolecules by means of light scattering. Makromol Chem 1956;18/19:397–405.

[86] Benoit H, Stockmayer WH. Influence of the interaction of assembled particles on the scattering of light. J de Physique et le Radium 1956;17:21–26.

[87] Nagai K. Theory of light scattering by an isotropic system composed of anisotropic units with application to the Porod-Kratky chain. Polym J 1972;3:67–83.

[88] Benoit H. Depolarization of light scattered by chainlike molecules. Comptes Rendus 1953;236:687–689.

9

GPC/SEC AS A KEY TOOL FOR ASSESSMENT OF POLYMER QUALITY AND DETERMINATION OF MACROMOLECULAR PROPERTIES

Daniela Held and Peter Kilz

9.1 INTRODUCTION

Size exclusion chromatography (SEC), gel permeation chromatography (GPC), and gel filtration chromatography (GFC) are all synonyms for the most successfully applied separation technique for the characterization of macromolecules, (bio)polymers and proteins. Although the term SEC is used in general for separations in aqueous solutions and GPC for separations in organic eluents, the same theoretical background, setup, advantages, and limitations are valid. However, with respect to the separation mechanism SEC is the most descriptive name [1].

Size exclusion chromatography has multiple advantages. First of all, it is a fractionating technique providing access to distribution information as well as to property averages, for example, all molar mass averages, in a single measurement. Since in most cases nondestructive detectors are used, it is possible to recollect the sample and/or to investigate the fractions with other analytical methods. Basic laboratory equipment can be used and it is easy to run the samples. Furthermore, a SEC system can be fully adapted to the characterization needs. Different types of columns are available, allowing to meet goals as high sample throughput/fast analysis, high information depth/high resolution, saving solvent/green chemistry or high sample loading/small-scale clean-up (Table 9.1). The use of hyphenated techniques [2] and the addition of specialty detectors allow to increase the information content and to measure different results and distributions simultaneously with just one injection. This short summary explains why SEC is an indispensable tool when working with macromolecules.

9.2 BASIC CONCEPTS OF SEC

Many components of SEC systems are well known and ubiquitous in an analytical lab. Size exclusion chromatography systems require at least an isocratic pump, a high-pressure injection system (manual or automated), one or more SEC column(s), one or more detector(s), and data evaluation units with software to acquire, calibrate, and analyze data. Other components, such as degassers or column compartments are often optional and depend on the particular application and on the lab environment and conditions. Although SEC and high-performance liquid chromatography (HPLC) seem to be very similar in instrumentation requirements, the theoretical background and separation mechanism of both techniques are entirely different. The major differences in the practice of SEC are outlined in Table 9.2.

9.2.1 Separation Mechanism and Modes of Chromatography

In nearly all cases, a macromolecular solution consists of chains with different chain lengths (molecular weights), and, in the case of copolymers, of chains with different comonomer composition. Each chain has a volume in solution, hydrodynamic volume, which defines the size of the chain in

Monitoring Polymerization Reactions: From Fundamentals to Applications, First Edition. Edited by Wayne F. Reed and Alina M. Alb.
© 2014 John Wiley & Sons, Inc. Published 2014 by John Wiley & Sons, Inc.

TABLE 9.1 SEC Analysis Goals with Column and Method Recommendations

	Column Requirements	Detector Requirements
Fast analysis time/high sample throughput	HighSpeed column(s) (dimensions 20 × 50 mm) Shorter column(s) (reduced resolution)	
Green chemistry	Micro columns (smaller inner diameter)	Low dead volume, small detector cells
High resolution	Column bank with 2–4 analytical columns in combination	
True molar masses	High-resolution columns/column banks	Molar mass–sensitive detector (light scattering, viscometry, MS) + concentration detector (RI, UV/DAD)
Structure determination	High-resolution columns/column banks	Molar mass–sensitive detector (light scattering, viscometry, MS) + concentration detector (RI, UV/DAD)
Sample fractionation	Preparative columns (large inner diameter)	

TABLE 9.2 Important Differences between High-Performance Liquid Chromatography (HPLC) and Size Exclusion Chromatography (SEC)

	HPLC	SEC
Sample preparation	Fast dissolution, sample degradation unlikely	Slow dissolution, avoid shear, μ-wave, etc.
Chromatogram appearance	Many narrow peaks	Broad peak (one) (plus trash peaks in RI)
Results	(a) Qualitative analysis (b) Quantitative analysis	(a) Molar mass averages (b) Molar mass distribution
Information derived from	(a) Peak sequence (b) Peak area	(a) Absolute peak position (b) Peak shape
Calibration	Detector response > amount	Retention > molar mass
Detection	Single detector (UV DAD) (MS)	Multiple detectors (UV, RI, LS, viscometry, FTIR, MS, etc.)

a particular solvent. If such a polydisperse mixture is injected into a column with pores and a defined surface polarity/chemistry, a chromatographic separation will occur.

The basic principle of chromatography separation can be described by simple thermodynamic principles applying the distribution coefficient K:

$$K = \frac{a_s}{a_m} = \exp\left(\frac{-\Delta G}{RT}\right) = \exp\left(\frac{(-\Delta H + T\Delta S)}{RT}\right) \quad (9.1)$$

where a is activity (concentration) of the molecule in the stationary phase (indexed s) and in the mobile phase (indexed m) and ΔG is free energy change between the species in the stationary phase and in the mobile phase, respectively.

In SEC separation, the enthalpic contribution, ΔH, to the free energy term is negligible if there is no energetic interaction between analyte and solvent:

$$K_{SEC} = \exp\left(\frac{\Delta S}{R}\right), \quad 0 < K_{SEC} < 1, \ \Delta H \approx 0 \quad (9.2)$$

where ΔS is the entropy loss when a molecule enters the pore of the stationary phase.

In the case of nonsteric exclusion of the molecule from parts of the stationary phase, the retention can be described by the enthalpic term alone:

$$K_{HPLC} = \exp\left(\frac{-\Delta H}{R}\right), \quad K_{HPLC} \geq 1, \ T\Delta S \ll 0 \quad (9.3)$$

where ΔH is the enthalpy change when a molecule is adsorbed by the stationary phase.

Equations 9.2 and 9.3 describe the two ideal extremes of chromatography (SEC and HPLC), when there is no change in entropy or enthalpy, respectively.

Another mode of chromatographic behavior was predicted and verified [3] when enthalpic and entropic contributions to the distribution coefficient balance out, that is, when the change in free energy disappears ($\Delta G = 0$). This mode is called liquid chromatography at the critical adsorption point (LCCC). The polymeric nature of the sample (i.e., the repeating units) does not contribute to the retention of the species. Only structural "defects" like end-groups, comonomers, or branching points contribute to the separation.

Figure 9.1 illustrates the different modes of chromatography and shows the retention volume dependence of the molar mass.

To understand the different chromatographic modes, let us consider the separation of poly(methyl methacrylate) (PMMA) on a nonmodified silica column as an example. The separation can take place in the pores of the columns or depending on the column surface. In tetrahydrofuran (THF), a medium polar eluent, the PMMA elutes in SEC mode

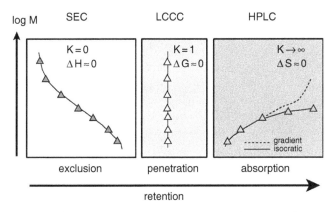

FIGURE 9.1 Major features of chromatographic modes of separation of macromolecules with different molar masses.

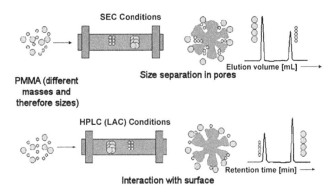

FIGURE 9.2 Comparison of SEC and HPLC separation of PMMA on a silica column. The choice of solvent defines the separation mode.

because the dipoles of the methyl methacrylate (MMA) repeating units are masked by the dipoles of THF. Molecules with a hydrodynamic volume larger than the pores will not be retained at all and elute first. After that, molecules with a hydrodynamic volume small enough to penetrate the larger pores will elute. At the end, molecules with a hydrodynamic volume small enough to fit in all of the pores will elute. In this case, the retention is often dominated by end-groups and/or the chemistry of the molecule. Using the nonpolar solvent toluene as eluent on the same column, the separation is governed by adsorption in HPLC mode. The dipoles of the carbonyl group in the PMMA will interact with the dipoles on the surface of the stationary phase. In this case, the lower molar masses (smaller hydrodynamic volumes) will elute before the higher molar masses.

Figure 9.2 shows the difference between the two separation modes for this example.

9.2.2 Calculation of Molar Mass Averages

Size exclusion chromatography is the most commonly used method to measure molar mass distributions (MMDs also termed "molecular weight distributions," MWD, in this book

and elsewhere) and molar mass averages. However, SEC separates based on the hydrodynamic volume and the molar mass information is only available when a correlation between molar mass and elution volume has been established by a calibration (Section 9.2.3) or absolute detection (Section 9.4.2).

The calculation of the molecular weight averages uses the slice method [1–4]. The eluted peak is separated into equidistant volume slices and the elution volume is translated into the molar mass.

The molecular weight averages are defined and calculated by:

- number average molecular weight:

$$\bar{M}_n = \frac{\mu_0}{\mu_{-1}} = \frac{\sum_i h_i}{\sum_i \frac{h_i}{M_i}} \tag{9.4}$$

- weight average molecular weight:

$$\bar{M}_w = \frac{\mu_1}{\mu_0} = \frac{\sum_i h_i M_i}{\sum_i h_i} \tag{9.5}$$

- z-average molecular weight:

$$\bar{M}_z = \frac{\mu_2}{\mu_1} = \frac{\sum_i h_i M_i^2}{\sum_i h_i M_i} \tag{9.6}$$

where h_i is the weight and M_i is the molar mass of the fraction i and μ is the μth moment of the distribution.

The width of the MMD can be described by the polydispersity, D, also called polydispersity index, PDI:

$$PDI = \frac{\bar{M}_w}{\bar{M}_n} \tag{9.7}$$

However, molar mass averages represent reduced information only and do not describe a polydisperse sample comprehensively. The macroscopic properties of macromolecules can better be derived from their MMD, $w(M)$. Two samples can have the same molar mass averages but very different MMDs and therefore macroscopic properties. It is possible to derive the molar mass averages from the MMD but not vice versa. The MMD can be calculated from the signal heights, $h(V)$.

The differential distribution, $w(M)$, of the molar mass, M, is defined as:

$$w(M) = \frac{dm}{dM} \tag{9.8}$$

where dm/dM is the mass fraction of polymer in the dM interval.

By simple transformations, $w(M)$ can be expressed by quantities measured by SEC directly:

$$w(M) = \frac{h(V)}{M(V)\sigma(V)}; \quad \sigma(V) = \frac{d\lg M}{dV} \qquad (9.9)$$

where $h(V)$ is the detector signal, $\sigma(V)$ is the slope of the calibration curve, and $M(V)$ is the molar mass.

The correction with the slope of the calibration curve is necessary, because the data recording is linear in time while the molecular mass does not increase linearly. This means that the number of polymer chains with the same concentration on the high molecular weight part of the chromatogram is much smaller than on the low-molecular weight part. Only with strictly linear calibration curves, for a very limited number of set-ups, the correction is not needed.

In many HPLC-based analysis software packages with SEC module, the correction with the slope of the calibration curve is not done. This leads to wrong MMDs and partially wrong results, for example, amount below/above a defined molar mass border. The error caused by the missing correction will increase with sample polydispersity and when the data recording frequency is reduced. The qualitatively introduced correction by the slope of the calibration curve can associated with the first derivative of the calibration curve.

The cumulative distribution, $I(M)$, will be used as a normalization condition resulting from:

$$I(M) = \int_0^\infty w(M)\, dM \qquad (9.10)$$

The molecular weight averages can be calculated from the moments, i, of the MMD, as described earlier:

$$\mu_i = \int_0^\infty M^i w(M)\, dM \qquad (9.11)$$

where μ_i is the ith moment of the MMD

A list of additional GPC/SEC results yields the following summary:

Value Description:

$[\eta]$—intrinsic viscosity either measured using an online viscometer or calculated using the Mark–Houwink equation
V_p—elution volume at peak maximum of the elugram
M_p—molecular weight at peak maximum of the elugram
A—peak area
$w\%$—mass fraction within defined molecular weight limits
$I(M)$—molecular weight at a given mass fraction

9.2.3 Calibration

The primary information obtained from SEC is not the molar mass, but the apparent concentration at an elution volume. Only with the matching SEC calibration and the concentration profile from the concentration detector can the molar mass average and the MMD be obtained. Size exclusion chromatography is therefore a relative method if no absolute detection is employed [1]. The SEC calibration is based on assigning a molar mass to an elution volume (calibration of x-axis). This is in contrast to HPLC, where the detector response (signal intensity and peak area) is calibrated and assigned to a concentration (calibration of y-axis).

The relation of molar mass to elution volume can be determined using one of the following methods or setups:

- SEC calibration with molar mass reference materials, for example, with
  - narrow MMD
  - broad MMD
  - use of an online viscometer and a concentration detector and universal calibration of the SEC system
  - use of a light scattering and a concentration detector
  - use of a mass spectrometer and a concentration detector

Size exclusion chromatography calibration with reference materials with narrow MMD is by far the most commonly used and most accurate method for calibration [5]. The standards come with certificates showing at least the molar mass averages. They are used for conventional calibration or for universal calibration with or without an online viscometer. The calibration curve is created by measuring the elution volumes of the reference materials and by plotting them versus the logarithm of the molar masses (in general the molar mass at the peak maximum). Then a fit function, which describes the shape of the calibration curve, has to be chosen. Unfortunately, there is no general fit function that can be used for all columns/column banks, but users have to select a proper fit function based on multiple criteria.

Most calibration curves, often even the ones for linear or mixed-bed columns, have a sigmoidal shape. This is in agreement with the fundamental separation characteristics and in contrast to other calibrations in chromatography, where linear calibration curves for the peak area plotted versus the concentration are obtained.

A SEC calibration curve can be divided into three distinct regions, as shown in Figure 9.3. Region I is the exclusion limit of the column/column bank, where no separation takes place. The pores are too small and all molecules, independent of their size, elute at the same volume. Region II is the optimum separation range. Molecules are separated according to their hydrodynamic volume in solution. Large molecules with high molar masses and volumes elute first; molecules with lower molar masses and hydrodynamic volumes elute later, at higher elution volumes. Region III is called total penetration volume, the region where the separation is influenced by interaction with the column material and by the size of the molecule alike.

In most cases, SEC fit functions are based on polynomial functions with a degree of 3 (cubic) or higher [4–7]. There are also special dedicated fit functions (PSS calibration functions) available that are based on polynomial functions, but optimized to avoid typical pitfalls.

Three decision criteria can help to decide if the proper function has been selected. These are the regression coefficient, R^2, the deviation of the calibration point from the fitted value (e.g., average deviation), and the slope of the calibration curve.

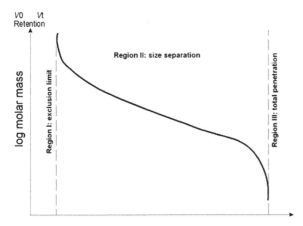

FIGURE 9.3 Generalized shape of SEC calibration curves covering the complete separation range (V_0 represents complete exclusion from pores; V_t represents total penetration into pores).

Table 9.3 illustrates this decision-making process. It shows the regression coefficients and the average deviation for all data points for identical calibration data fitted with different functions. It is obvious that the regression coefficient alone is not a proper parameter to select the best fit function. Large average deviations are observed even for a regression coefficient very close to 1. If the SEC software provides the regression coefficient as the only selection criteria, a value of $R^2 > 0.999$ should be achieved.

In addition, the table shows that when polynomial functions with a higher degree are selected, the regression coefficient and the average deviation become smaller.

However, it is not physically meaningful to use the function with the highest degree, even though this will always generate the lowest average deviation. More important than small deviations is that the shape of the calibration curve be in general agreement with the separation mechanism. A good measure for a physically meaningful fit is the first derivative of the calibration curve, the slope of the calibration curve.

Figure 9.4 shows an ideal first derivative for a calibration curve. The slope is constant for the optimum separation range and changes only close to the exclusion limit and the

TABLE 9.3 Influence of the Calibration Fit Function on the Regression Coefficient and the Average Deviation

Calibration Model	R^2	Average Deviation (%)
Linear	0.9925	30.2
Polynomial 3 (cubic)	0.9986	10.4
Polynomial 5	0.9995	7.35
Polynomial 7[a]	0.9999	3.57
PSS Polynomial 7	0.9998	4.92

[a]First derivative is discontinuous; this function should not be used.

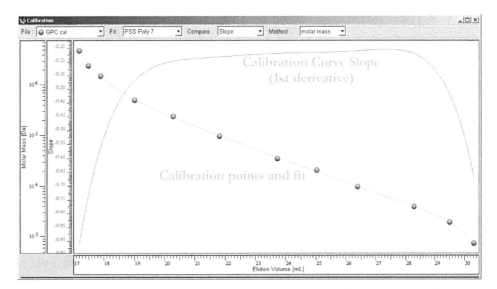

FIGURE 9.4 Example of a good SEC calibration with small deviations and continuous first derivative which covers the complete column pore volume.

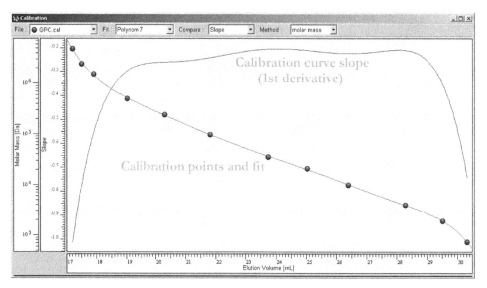

FIGURE 9.5 Poor SEC calibration fit (same data points as in Fig. 9.4) with small deviations but discontinuous first derivative which will lead to artifacts (e.g., shoulders) in the molar mass distribution.

total penetration volume. If a higher polynomial fit function of seventh degree is chosen (Fig. 9.5) local maxima and minima, lacking any physical significance, will appear. This function fit should therefore be avoided since it can produce artifacts in the MMD (e.g., shoulders) that are not related to the sample characteristics.

Therefore, the optimum function fit is the one with the lowest deviations that still has a constant slope without maxima or minima.

However, one of the major limitations in SEC is that the separation is based on the hydrodynamic volume. This does not only depend on the molar mass of the molecule but also on its chemical nature and topology. Hence, a calibration curve created from reference materials is strictly only valid for samples with the same chemistry and topology. For other samples, apparent molar masses will be obtained. It is still possible to compare the samples, but it is not possible to measure accurate molar masses.

Several solutions are available to overcome this dilemma:

a. First of all, there are many different types of reference materials available to create the matching calibration curve for many samples. In addition, universal calibration with Mark–Houwink coefficients and broad calibration methods are available.

b. Secondly, online viscometers can be used to measure a universal calibration curve [6]. Here, the logarithm of the hydrodynamic volume, the molar mass multiplied by the intrinsic viscosity, is plotted against the elution volume. Universal calibration curves are valid for all types of polymers and copolymers independent on the topology. For creating a universal curve and choosing a fit function the same rules apply as for a conventional

calibration curve. The only difference is that the intrinsic viscosity measured using the viscosity detector is additionally used.

c. Thirdly, molar mass sensitive detectors as online light scattering detectors and MS detectors allow measuring the molar mass in every elution volume directly.

The solutions (b) and (c) will be discussed in more detail in Section 9.4.

9.3 SEC ASPECTS OF SEPARATION

Size exclusion chromatography separations require interaction-free diffusion of the sample molecules into and out of the pores of the stationary phase. In general, this goal is easier to achieve in organic eluents than in aqueous solutions. In aqueous mobile phases, more parameters (e.g., type of salt, salt concentration, pH, addition of organic modifier, and concentration of cosolvent) have to be adjusted correctly. In addition, due to the presence of charged functional groups, hydrophobic, and/or hydrophilic regions in the molecule, water soluble macromolecules have more possibilities to interfere with the stationary phase.

A proper SEC experiment has to be balanced with respect to polarities. In order to obtain a true and pure SEC separation, the polarity of stationary phase (column material), the polarity of eluent, and the polarity of sample have to be matched. This is visualized by the "magic triangle" (Fig. 9.6). Dominance of size separation is only maintained in the center of the triangle (bright area), where the overall system is balanced. Otherwise, specific interactions will occur, which will overlay with the normal SEC elution behavior.

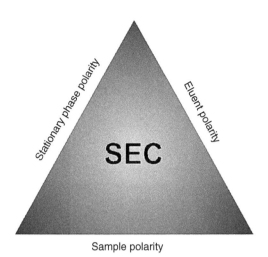

FIGURE 9.6 Balancing the polarities of the phase system in SEC applications for interaction-free separations.

9.3.1 SEC Column Selection

The major parameter for column selection is the intended application [7, 8]. A balance of mobile phase polarity in comparison with the polarity of the stationary phase and sample polarity is important for pure SEC separations. In general, users will select their columns according to the mobile phase they need to use. Stationary phase materials can be either silica or polymeric based. Table 9.4 shows an overview of stationary phases with different polarities typically used in SEC.

After selecting the best column material, other column parameters have to be taken into account to meet the analysis goals. Here, the column dimensions, the particle size, and the porosity are of importance.

9.3.1.1 *Column dimensions*
Different column dimensions are available. Most common are analytical SEC columns with an inner diameter between 0.7 and 0.8 cm and a length of 25–30 cm. Columns of 60 cm are still available but rarely used. Preparative columns with the same length but a larger diameter (up to several centimeters) allow to fractionate larger sample amounts. (Semi)micro-columns with

approximately the same length and a smaller inner diameter are used with dedicated instrumentation for saving solvent and if only small sample amounts are available.

9.3.1.2 *Particle size*
The smaller the particles are, the better is the resolution. However, high molar masses and high viscous solvents require larger particle sizes. So, there is an optimum particle size depending on the application. The influence of the particle size is described in more detail in Section 9.3.2.

9.3.1.3 *Porosity*
Another important parameter for column selection is the proper choice of sorbent porosity. The molar mass range of the samples to be investigated determines the column porosity. The larger the pores, the higher molar mass samples can be characterized. Unfortunately, there is no general nomenclature, which will allow easy selection of column pore sizes. Each manufacturer has its own system for pore-size designation. The easiest method to find out which columns will be useful for a selected task uses the calibration curve which every manufacturer shows in their literature. In general, SEC columns can be either single porosity columns with narrow pore-size distribution or linear (also called mixed-bed) or multipore columns with a very broad pore-size distribution. SEC separation capacity is limited by the available pore volume and depends on solvent type, column dimensions, and the slope of the calibration curve. The highest selectivity for a separation is determined by the lowest slope of the calibration curve.

In the case of the single porosity columns, the separation capacity is concentrated in a narrow molar mass range. This yields a calibration curve with a flat or shallow slope in this region (cf. Fig. 9.3). Therefore, single porosity columns have a limited molar mass separation range, but a high resolution in that range. In contrast, columns with a broad pore-size distribution provide a larger separation range and the calibration curve has a steeper slope and therefore less resolution. Linear or mixed-bed columns are often used either in QC for fast-screening experiments, or to identify the molar mass range of a sample so that it can be investigated on a matching single porosity column bank with higher precision.

TABLE 9.4 SEC Stationary Phase Materials

Polarity	Polymer Packing			Inorganic Packing		
	Non	Medium	Polar	Non	Medium	Polar
Chemistry	Styrene–divinylbenzene	Acrylic, polyester	OH-acrylic	SiO_2–C_{18}	SiO_2-diol	SiO_2
Solvents	THF, TCM, toluene	DMF, NMP, DMAc, DMSO	Water (pH 1.5–13)		Aqueous (pH < 9)	THF
Samples	PS, PMMA	PU, starch,	PEG/PEO,		Proteins,	PS, PB,
	PVC, PC, PE,	cellulose,	pullulan,		peptides,	PIB
	resins, PP, etc.	polyimide, etc.	dextran,		PLA,	
			PAA, etc.		polyester,	
					POM, etc.	

9.3.2 SEC Method Optimization

A very simple approach can be applied to increase the resolution and/or the separation range. Instead of just using one column, multiple columns are combined to a column combination or a column bank. Two to four columns (plus a pre- or guard column) are typical in SEC. A column combination or column bank provides more available pore volume for more efficient separations. If two columns with the same pore sizes (single porosity or linear/mixed bed/ multipore) are combined, the calibration curve becomes flatter and the resolution increases by a factor of 1.4 whereas the

separation time increases by a factor of 2 (Fig. 9.7). The molar mass range increases if columns with different porosities are combined.

Figure 9.8 and Figure 9.9 show a comparison of the same sample mixture analyzed on two different column banks. In Figure 9.8, the columns are optimized for the characterization of low molar masses while the column bank in Figure 9.9 is optimized for the separation of medium molar masses. This example illustrates also the influence of the slope of the calibration curve on the resolution, as well as the difficulty for interlaboratory comparison of chromatograms: the look of chromatograms or the raw data depends of the columns

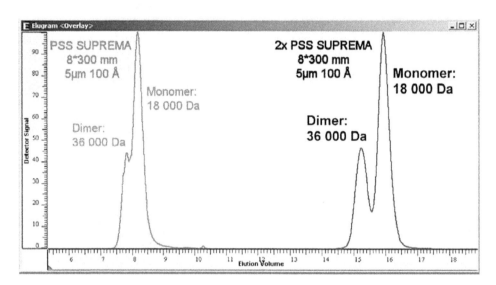

FIGURE 9.7   Comparison of SEC resolution enhancement by increasing the column length by a factor of 2.

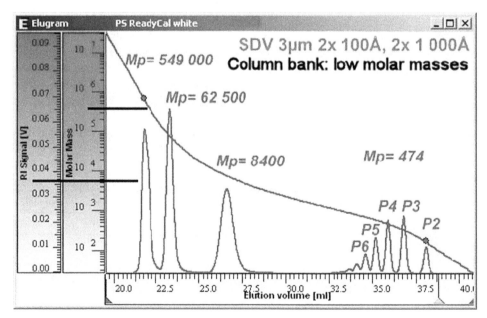

FIGURE 9.8   Separation of a poly(styrene) standards cocktail on a SEC column bank optimized for oligomer separation by combining narrow pore-size columns; the flat calibration curve indicates best resolution at low molar mass (column details shown in figure).

used. Therefore, it is always recommended to compare the MMD instead of the chromatograms or raw data.

Disadvantages of column banks are that price, pressure, analysis time, and eluent consumption increase. An increased pressure might result in the need to reduce the flow rate and/ or to increase the temperature to have better chromatographic conditions, especially for high molar mass macromolecules. In addition, there is the potential danger of porosity mismatch for all column types, linear/mixed bed or single porosity alike.

Porosity mismatch often shows itself in peak shoulders which might be misinterpreted as better resolution, but are artifacts of a column bank due to non-matching porosities. This phenomenon can also be observed if nonmatching

porosities are mixed in one linear/mixed bed column to provide a wide linear separation range.

Method optimization with respect to a better resolution includes also adjusting all parameters that improve the mass transfer. The following are the parameters to be chosen.

9.3.2.1 *Particle size* Plate height and column permeability decrease with the particle diameter. Smaller particle size columns provide, therefore, a better resolution. This concept, which led to the development of Ultra High Performance Liquid Chromatography (UHPLC), can also be adapted to SEC taking some peculiarities into account. Figure 9.10 shows a comparison of a protein mixture

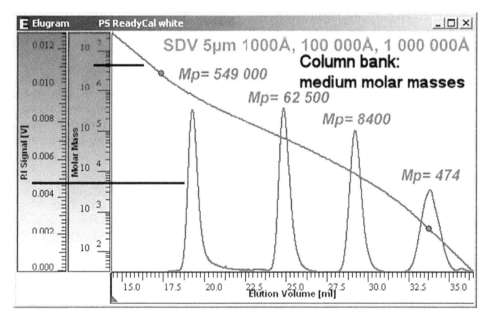

FIGURE 9.9   Separation of the same poly(styrene) standards cocktail as in Figure 9.8 on a SEC column bank optimized for medium molar masses. The lower resolution in the low molar mass region results in a single peak instead of multiple peaks for each single oligomer.

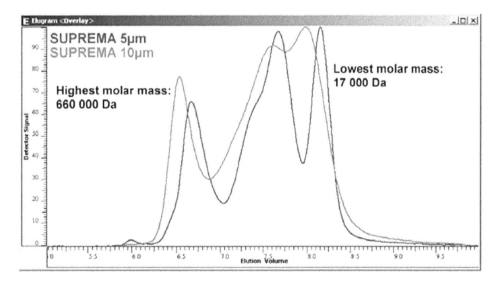

FIGURE 9.10   Influence of particle size of packing material on the resolution of a protein mixture under otherwise identical conditions.

measured on the same column material with different particle sizes. The mass transfer for the 5 μm material is much better, resulting in an increased resolution. Therefore, if the molar mass and rigidness of the macromolecules permit and no shear degradation occurs, the higher prices for small particle columns are a good investment in higher resolution.

The general rule of thumb is that oligomers in low-viscous solvents and proteins allow particle sizes around 3 μm, for medium molar masses 5–10 μm particles are used, and for high molar masses and high viscous solvents 10–20 μm particles sizes are applied.

9.3.2.2 *Flow rate* A flow rate of 1 ml min^{-1} is often used for analytical SEC columns with an inner diameter between 7 and 8 mm as the flow rate with the best compromise between resolution and analysis time. Especially for higher molar masses, a decrease of the flow rate leads to higher resolution. Columns with larger inner diameter are best operated with higher flow rates while columns with smaller inner diameter are used with lower flow rates (see Section 9.3.3).

9.3.2.3 *Temperature* Generally, temperature increase also leads to a better resolution due to the enhanced mass transfer. However, this is not applicable for all macromolecules. For example, poly(ethylene glycol) (PEG) in aqueous solution shows a better resolution at lower temperatures. High temperature SEC systems, completely heated, are needed for macromolecules only soluble at elevated temperatures, for example, polyethylene (PE) or polypropylene (PP).

Table 9.5 summarizes various ways to optimize SEC separations by proper selection and combination of SEC columns.

TABLE 9.5 Optimization of SEC Separation

Task	Optimized by
Better peak separation	Addition of similar columns; use of 3:m columns, if $M < 100$ kg mol^{-1}; use of 2D chromatography (cf. Chapter 5)
Better separation of high molar mass fractions	Addition of column(s) with large pore width/large porosity
Better separation of small molecules, additives, solvents, eluent, etc.	Addition of column(s) with small pore width/small porosity
Avoiding exclusion peaks	Addition of column with large pore width/large porosity
Not all compounds detected	Change phase system (column, eluent) use universal detector (RI)
Faster analyses	Use of HighSpeed columns
Faster calibrations	Use of premixed calibration cocktails
Fast screening of unknown samples	Use of (less) linear/mixed-bed columns
Better reproducibility	Use of internal standard correction (flow marker

9.3.3 High-Speed SEC Analyses

Recent trends have been focused on increasing the analytical throughput in order to increase productivity. QC and combinatorial chemistry demand the optimization of high-throughput methods. Increased analytical throughput can also save time and resources in production-related fields. In combinatorial research, high-throughput analytical techniques are a bare necessity, because of the huge numbers of sample to be analyzed.

There have been several approaches to overcome the traditionally slow SEC separations in column banks [9]. Most of them are column related; others replace separation with simplified sample preparation. Cloning existing methods and instrumentation is also used, but has disadvantages as it is the most costly approach. Benefits and limitations of each method are summarized in Table 9.6.

9.3.3.1 *Cloning of SEC Systems* The number of processed samples can be increased proportionally by increasing the number of identical systems. The time and analytical requirements for each sample are not changed but the number of samples per hour can be increased. Cloning SEC instruments and methods is straightforward and can be done in most environments, since no change in analytical methods is necessary.

The approach, however, is clearly limited by the availability of important resources such as lab space, operators, instrumentation, and software licenses. Cloning systems can become very costly and time and effort for instrument maintenance and operation increases proportionally.

Parallelization of analytical processes has so far not been very successful. In such setups, only the column is setup in parallel while solvent delivery, injection, detection, and data processing are multiplexed. These systems will no longer be as simple in operation and maintenance as the cloned systems.

9.3.3.2 *Small Particle Technology* Reducing particle size of the SEC column packing reduces the time requirements in SEC due to the increased separation efficiency. Hence, columns can have smaller dimensions while maintaining the resolution. This approach has been used for many years and the development of UHPLC has created a new interest in separation techniques in general. Nowadays, column bank lengths have dropped from several meters to 60–90 cm, with current SEC column particle size of 5 μm and below, compared to about 100 μm in the early 1960s. During the same period, time requirements dropped from about 6 h to less than 1 h.

Unfortunately, this approach is limited, because of the high shear rates in columns packed with small particles, due to potential polymer degradation. However, materials with reduced particle size have been successfully applied in the

TABLE 9.6 Overview of Strategies for Fast SEC Separations

Method	Benefits	Limitations	Beneficial for
Instrument cloning	No method change, easy to implement, no additional training	High investment cost, high maintenance, higher operating cost, more people, more space, limited throughput gain	Sample increase of up to 3×
HighSpeed column	No method change, uses existing equipment, 1:1 application transfer, no additional training, minimizes investment (column only), SEC separations in 1 min time gain ca. 10× No additional shear, high efficiency, runs with conventional software	No eluent savings	QC/QA, increased throughput (10×), use with existing methods
Flow injection analysis	Uses existing equipment, saves eluent	No separation, limited time gain, not applicable for copolymers/blends, requires molar mass, sensitive detectors, only primary information (conc., M_w, IV), needs method change, needs special software	Samples difficult to separate Utilize existing instruments
Overlaid injections	No method change, uses existing equipment, no additional training, low cost	Needs overlaid injection, ready software	QC/QA known samples
Small columns	Uses existing equipment, minimizes investment, saves eluent, runs with current software	Limited time savings, needs method adaption, optimization of injection volumes, detection systems, shear degradation, low efficiency, needs training, limited throughput increase	Low resolution applications, low time-saving requirements, single detector applications

case of proteins. Also, new solutions are available for aqueous systems [9].

9.3.3.3 *HighSpeed SEC Columns*

The pore volume of the column packing has been shown to be one of the major factors influencing peak resolution in SEC. The reduction of the column length could, in theory, reduce the time requirements of the separation substantially. However, several limitations predicted by chromatographic theory have to be considered. A study of the influence of column dimensions on fast SEC separations has been reported [9]. It has been found difficult to optimize and transfer existing methods; in many cases new equipment has to be purchased.

True HighSpeed separations with good resolution require dedicated columns allowing fast flow rates and easy access to the pores by the solute. In addition, they provide high separation volumes. In optimized system column dimensions, length and inner diameter have to be matched.

Figure 9.11 shows the overlay of a separation of poly(styrene) standards on an analytical and a high-speed column. Virtually, no loss of resolution is observed while the analysis time is reduced by a factor of about ten to less than 2 min.

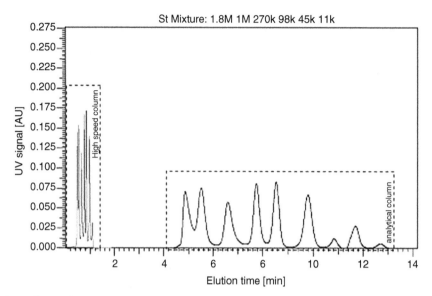

FIGURE 9.11 Comparison of analytical (black trace) and HighSpeed (gray trace) SEC separation showing similar resolution and eluent consumption at 10× reduced analysis time with HighSpeed SEC column.

9.3.3.4 *Overlaid Injections* This method is used to cut down the analysis time by a factor of 2 without changing the equipment or columns. It utilizes the fact that about 50% of the SEC elution time is needed to transport the solutes through the interstitial volume of the columns. It is not possible that sample elutes in this time. This allows injecting another sample before the current one is totally eluted. The optimum injection interval t_{min} can be calculated from the separation properties of the instrument:

$$t_{min} = \frac{V_t - V_0}{F} \qquad (9.12)$$

where V_t is the total penetration volume of column(s), V_0 is the total exclusion volume of column(s), and F is the volume flow rate.

The required parameters are easily determined from a molar mass calibration curve.

Nowadays, this method can be combined with proper software to automate data acquisition and data processing. It is easy to use and requires no additional investment and method modifications.

9.3.3.5 *Flow Injection Analysis* A completely different approach to cut down on analysis time avoids separation and injects samples directly into the detector cell(s). This method has been adapted from environmental analysis, where it is called flow injection analysis (FIA).

For fast polymer characterization, HPLC equipment is used for sample handling but it requires high-cost, molar mass–sensitive detectors (such as light scattering and/or viscometry; see Section 9.4.2) to obtain bulk property values from each detector (M_w and/or intrinsic viscosity,

respectively). A concentration detector will only measure the polymer content in a sample, which can also be determined in a simpler and cheaper way with various other well-established methods. The FIA method applied to polymers requires expensive and well-maintained equipment and does not save a lot of time and solvent, in addition to the fact that no distribution information is available.

9.4 SEC ASPECTS OF DETECTION

At least one SEC detector is required to detect the eluting sample. The MMD is then derived from the slice information of the measured concentration, the molar mass (from a calibration curve or measured directly using additional detectors as light scattering detectors or mass spectrometers), and the slope of the calibration curve.

The requirements for detectors in SEC are the same as that for detectors in other methods: first of all, it must be able to detect the sample in the desired application; the ability to detect a broad range of samples is a definite plus. Other parameters, for example, sensitivity/detection limit, linearity, baseline drift, noise, cell volume, and ease-of-use have to be taken into account. Since the detector design is discussed in many HPLC publications [10, 11], this text will focus on the applicability and use of detectors in polymer analysis.

In contrast to HPLC, the combined parallel or in series use of detectors based on different principles is one of the major advantages in modern SEC experiments. It allows the access to more detailed sample information, sometimes to absolute molar masses, sometimes to other types of distributions present in complex polymer samples

TABLE 9.7 Summary of SEC Applications with Different Detector Combinations

Detector Combination	Applicable for
UV/VIS with UV/VIS or any other detector combination for UV/RI/IR/ELSD	Copolymer characterization: copolymer composition distribution, copolymer molar mass
UV/VIS with RI	As above plus Heparin analysis according to pharmaeuropa end-group analysis
RALLS/LALLS/MALLS with RI/UV/ELSD	Homopolymer characterization: absolute molar masses and molar mass distribution MALLS only: radius of gyration averages and size distribution, polymer structure, branching
Viscometer with RI/UV/ELSD	Specific and intrinsic viscosity, molar masses and molar mass distribution based on Benoit's universal calibration hydrodynamic radius, branching, polymer structure, and size distribution
RALLS with viscometer and RI/UV/ELSD (Triple detection)	Homopolymer characterization: absolute molar masses and molar mass distribution Specific and intrinsic viscosity, molar masses and molar mass distribution based on Benoit's universal calibration hydrodynamic radius, branching, polymer structure, and size distribution
MALLS with viscometer and RI/UV/ELSD	Homopolymer characterization: absolute molar masses and molar mass distribution radius of gyration (averages and distribution), polymer structure, branching Specific and intrinsic viscosity, molar masses and molar mass distribution based on Benoit's universal calibration hydrodynamic radius, branching, polymer structure, size distribution
UV/RI with FTRI/MALDI interface (off-line FTIR/MALDI)	Molar masses and molar mass distribution additive identification and quantification tacticity, copolymer composition
MS with RI/UV/ELSD	Low to medium molar mass samples: absolute molar masses and molar mass distribution, end-groups

(e.g., chemical composition distribution (CCD), end-group distribution, and structural distributions).

Table 9.7 gives an overview of typical detector combinations used to investigate specific sample properties.

The following are the most commonly used SEC detectors to measure concentration profiles.

- *Refractive index (RI) detectors*: They measure the change in refractive index of the column effluent passing through the flow cell compared to that of the pure eluent. Refractive indexes are the most commonly used SEC detectors, since they allow the detection of the majority of polymeric samples with similar response (unspecific detector) strictly independent of molar mass. Only isorefractive samples cannot be detected. A well-known example is poly(dimethylsiloxane) in THF. These samples have to be measured in toluene or an alternative detector (e.g., an ELSD) has to be used.

    Refractive index detectors are used to measure concentration profiles and (if used alone) molar masses based on a calibration. In addition, they are used to measure the slice concentration when working with online light scattering detectors, viscometers, or online mass spectrometry. They are also used to measure the refractive index increment, *dn/dc*, in a single point experiment when light scattering or triple detection is used. In combination with other concentration detectors, they are used to measure the comonomer distribution and molar mass in copolymers or end-group distributions. Disadvantages are the low sensitivity, the tendency to drift (especially in solvent mixtures), and the large cell volume.

- *Variable wavelength UV/Vis detectors*: These measure the UV adsorption at a fixed wavelength for samples with chromophores. Since SEC is used to measure the molar masses and the distribution, it is sufficient to measure at one or two fixed wavelengths where the sample shows absorption. Spectra from diode array detectors (typically used for substance identification in HPLC) are only rarely needed, for example, for the analysis and identification of oligomers with special properties.

    Ultra Violet detectors are used to measure concentration profiles and (if used alone) molar masses based on a calibration curve. In addition, they are used to measure the slice concentration required when working with online light scattering detectors, viscometers, or mass spectrometers. They are the most common detectors for protein characterization. In combination with other concentration detectors, they are used to measure the comonomer distribution and molar mass in copolymers or end-group distributions.

    An advantage of UV detection is the small cell volume, its ease-of-use and sensitivity (compared to RI). A disadvantage is that it can only be applied for a limited number of polymeric samples due to missing chromophores and potentially strong changes of response factors with minor chemical changes.

- *Evaporative light scattering detectors (ELSDs)*: They measure the light scattered from the solid solute particles remaining after nebulization and evaporation of the mobile phase. This is in contrast to online light scattering detectors, which are molar mass–sensitive detectors and measure polymers in solution. Often ELSD detectors are used to measure concentration profiles and therefore molar masses based on a calibration curve, when RI detectors cannot be used (e.g., solvent mixtures, isorefractive samples, or low sample concentrations). An advantage of this detector over the RI is its higher sensitivity. Disadvantages are the low linearity, the potential danger to evaporate and therefore miss low molar mass sample parts, and the higher experimental requirements (e.g., to find volatile salts during method development).

To measure absolute molar masses or to investigate structures and branching, at least one of the detectors mentioned previously is combined with one or more of the following detectors:

- *Online light scattering detectors* (LALS, RALS, or MALS type): They measure the time-averaged intensity of light scattered by polymers in solutions at one or more fixed detection angle(s). Low-angle laser light scattering (LALS) detectors measure at 6–7° and right-angle laser light scattering (RALS) detectors at 90°. Multi-angle laser light scattering (MALS) detectors measure the scattered light intensity simultaneously at several angles. They are used to determine absolute molar masses for homopolymers and proteins and polymer structures in solution (MALS only) (Section 4.2.1). Light scattering characterization of polymers is discussed in detail in chapter 8.

- *Online viscometers* (single, dual, or four capillary type viscometers with a symmetrical or asymmetrical bridge): These measure the pressure difference between a sample path and a reference path filled with pure solvent. Viscometers are used to measure specific and intrinsic viscosity, molar masses based on Benoit's universal calibration approach [6] and Mark–Houwink coefficients.

- *Mass spectrometry (MS) detectors (see Chapter 10)*: Different MS methods have been used in macromolecular analysis. They are used to determine absolute molar masses for homo- and copolymers and to detect polymer structures. Matrix-assisted laser desorption ionization-time of flight (MALDI-ToF) and electrospray ionization (ESI) are the most common instruments used in combination with SEC (Section 9.4.2.4). A recent application summary is available [12].

A general problem in SEC is the identification of peaks. In many cases, a comprehensive analysis is possible only if the chemistry of the sample is known. Only a few chemical detectors are available in SEC, which can be used for specific applications [1].

- *Infrared (IR) detectors*: The large majority of Fourier transform infrared (FTIR) detection in polymer analysis is off-line detection. Online detection with an IR detector is mainly used for the characterization of polyolefins in high-temperature GPC (HT-GPC). FTIR signals provide information on short-chain branching, if the ratio of different wavelengths is compared. For many other SEC applications, online IR detection cannot be used due to the fact that typical solvents absorb in the same region as the investigated polymers. Far more important are off-line techniques [2]. Here, the SEC eluent is directed to a heated nozzle for evaporation of the solvent followed by deposit of the analyte on a germanium disc. The disc is then scanned in an off-line step in standard FTIR spectrometers. This elegant technique allows to separate and to detect without the influence of the solvent and is often used in additives analysis to identify unknowns, for example, in master batches.

- *Nuclear magnetic resonance (NMR) (see Chapter 10)*: Although stopped flow and online results have been published, SEC–NMR coupling is still not a standard technique [2]. The use of deuterated solvents or the need to suppress the intense solvent signals by special NMR pulse sequences has limited the applicability. An additional problem is the low sensitivity of NMR in combination with the low concentrations used in SEC.

Modern SEC instruments in R&D are often equipped with three or four detectors in series. However, this does not minimize the required user input and knowledge of the single methods applied (SEC, light scattering, and viscometry) is indispensable. Additional system parameters, such as the inter-detector delay, need to be determined carefully as they influence the results [13]. In addition, band broadening due to use of several detectors can be a problem [14–16].

9.4.1 Detection Principles (Classification of Detectors)

Size exclusion chromatography detectors are classified according to the dependence of the signal intensity on the physical properties. Two often-used terms are concentration detector and molar mass sensitive detector.

The signal intensity of concentration detectors (RI, UV/ DAD/PDA, ELSD, IR, and others) can be described as:

$$\text{Signal intensity} = K(\text{detector}) * k(\text{sample}) * m(\text{injected}) \tag{9.13}$$

where $K(\text{detector})$ is the detector constant, $k(\text{sample})$ is the sample-related constant (e.g., refractive index increment, dn/dc for RI), and $m(\text{injected})$ is the injected mass.

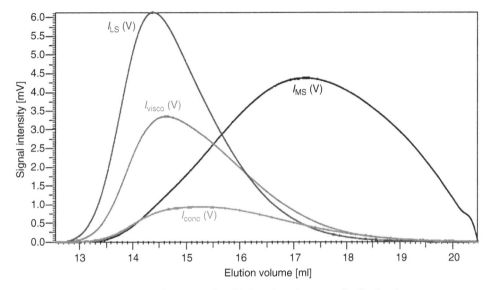

FIGURE 9.12 Difference in detector responses for a sample with broad molar mass distribution between a concentration detector and several types of molar mass sensitive detectors.

For molar mass–sensitive detectors a slightly different equation is used:

$$\text{Signal intensity} = K(\text{detector}) * k(\text{sample}) * \atop m(\text{injected}) * M^{\alpha} \qquad (9.14)$$

where M is the molar mass; $x = 1$ for light scattering detectors, while α is Mark Houwink coefficient.

The two aforementioned equations are valid for the whole peak as well as for every chromatographic slice. Many chemists replace the injected mass in these equations by the concentration, which is related to $m(\text{injected})$ and the injection volume and is more commonly used in chemistry.

Figure 9.12 shows the difference between the two detector types. While a molar mass–sensitive detector is ideal for high molar masses, a concentration detector can detect also low-molecular weights with significant concentration.

$K(\text{detector})$ and $k(\text{sample})$ can be determined using reference materials and/or dedicated instrumentation. Therefore, when working with molar mass sensitive detectors, there are two sample-related unknowns left, the mass in every slice ($m(\text{slice})$) and the molar mass ($M(\text{slice})$). To measure $m(\text{slice})$, an additional detector, the concentration detector, is used; the combination of both detectors allows to determine the unknown molar mass.

9.4.2 Application of Molar Mass Sensitive Detectors in SEC

9.4.2.1 *Light Scattering Detection* Light scattering is one of the few absolute methods for the characterization of macromolecules and biopolymers. Online light scattering

detectors are used to measure molar masses, the radius of gyration, and to identify high molar mass content (protein oligomers) at low concentrations.

Light scattering detectors measure the intensity of light scattered from the dilute sample solution. The light source is a laser or laser diode and the scattered light is measured at one or more fixed detector angles. The vast majority of light scattering detectors in SEC are static light scattering detectors; there are only a few applications (e.g., in protein analysis) where dynamic light scattering detectors are successfully applied. The term "static" here does not refer to performing an online (SEC) or batch (stand-alone) light scattering experiment, but to the fact that the time-averaged scattering intensity is measured, while "dynamic" light scattering measures the light intensity fluctuations. Light scattering theory and general practice are covered in other chapters.

There are several static light scattering detectors commercially available. The difference between these detectors is the number and the position of the angles detected. Low angle laser light scattering and RALS detectors measure intensity at only one angle, either at a low angle (e.g., 6–7°) or at right angle (90°). Multiangle laser light scattering detectors measure intensity simultaneously at several angles.

Right angle laser light scattering has the advantage of the most accurate signal with the lowest influence of stray light and dust particles that might be in the solution and disturb the signal. However, the molar mass range that can be measured accurately is limited. LS results of 90° will only be accurate for random chain linear polymers with molar masses up to 200,000 Da, globular proteins with molar masses up to 1,000,000 Da, or branched polymers over 200,000 Da (depending on the branching density) [1]. For samples with higher molar masses, LALS or MALS detectors have to be used. Low angle laser light scattering has the benefit that the observed intensities can virtually be identified with intensities at a scattering angle of $0°$. Low angles can be regarded as a direct measure of the molar mass of the species. Unfortunately, due to the complex optics, LALS presents the most experimental difficulties. An extremely clean system is required to measure the weak signals. Intense signal processing, either in the commercial detector or in the software, often make it difficult to access and visualize the true raw data. In some chromatographic systems, such as aqueous SEC systems, the requirement of a clean system can almost never be fulfilled.

Multiangle laser light scattering instruments use the extrapolation of all scattering angles to $0°$ to measure accurate molar mass and radius of gyration. The disadvantage is the high cost due to the fact that a MALS detector represents, in general, several detectors in one.

The detector angles in MALS can be, but do not have to be, the true scattering angles. This depends on the cell design and geometry:

- If the intensity is measured directly in a cell with cylindrical geometry, the scattering angle and the detector angle are always the same.
- If common glass cells are used the interface solution/ glass is responsible for the fact that all detector angles other than the $90°$ angle need a correction to obtain the true scattering angle.

Figure 9.13 shows data for a non-isotropic scatterer (polystyrene of $M_w = 254,000$ Da and PDI of 2.8), where it is obvious why a one detection angle is not sufficient and MALS is required. Depending on the detection angle, different signals are obtained.

With MALS detectors, the molar mass for every elution volume slice is obtained from a partial Zimm, Debye, or Berry plot, where the scattering intensity of all angles is plotted versus the scattering angle. Fitting the data allows to obtain the molar mass (intercept) and the radius of gyration (slope), assuming that the slice concentration is so low that extrapolation to zero concentration can be neglected.

9.4.2.2 *Viscometry Detection*

Although single and dual capillary viscometers are still in use, most commercially available detectors offered today are based on the four capillary bridge design originally developed by Haney [17].

The sample solution is split into two parts after entering the detector: a part of the sample flows through the sample path, the other part flows through the reference path. In the reference part, the sample is either delayed by hold-up columns or diluted to approximately zero concentration in a reservoir filled with pure solvent. Therefore, it is possible to measure continuously the pressure difference (delta pressure) between the sample path, filled with the viscous sample solution, and the solvent path, filled with pure solvent. In addition, most detectors measure the inlet pressure before the solution is split (Fig. 9.14).

Depending on the brand and the design of the viscometer, different calculations are required to obtain the specific viscosity from the primary information (inlet and delta pressure).

For comprehensive data evaluation, the use of a viscometer also requires the use of a concentration detector, such as RI or UV. The concentration detector measures the concentration in every elution volume slice. Division of the specific viscosity by the concentration yields the intrinsic viscosity in every elution volume slice, as well as the bulk intrinsic viscosity of the complete sample. It is assumed that the slice concentration is so low that no extrapolation to zero concentration is required.

With the capability to determine $[\eta]$ as a function of elution volume, one can compare the hydrodynamic volumes, V_h, for different polymers. V_h is related to intrinsic viscosity and molar mass through Einstein's viscosity law. Einstein's law is, strictly speaking, only valid for impenetrable spheres at highly dilute concentrations. However, it can be extended to particles of other shapes, defining the particle radius as the radius of a hydrodynamic equivalent sphere. In this case, V_h is defined as the molar volume of impenetrable spheres which would have the same frictional properties or enhanced viscosity to the same degree as the actual polymer in solution.

It has been predicted and experimentally proven by Benoit [18] and coworkers that all samples, independent of chemistry or structure, lie on one calibration curve if the logarithm of molar mass multiplied by intrinsic viscosity is plotted versus the elution volume and not the logarithm of the molar mass alone (Fig. 9.15).

If a calibration curve is established using a concentration detector, a viscometer and reference materials, it is possible to measure true molar masses for samples that are chemically and/or structurally different from the calibrants. As a consequence, using the universal calibration curve

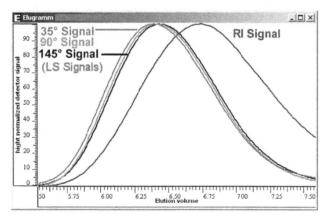

FIGURE 9.13 Raw data for a non-isotropic scatterer for three signals of a MALLS detector and an RI detector.

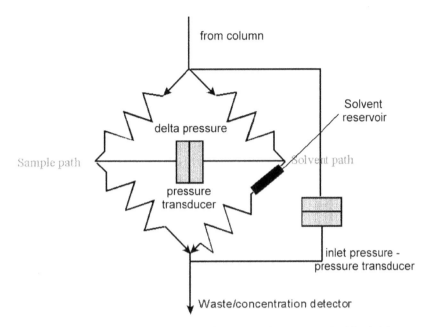

FIGURE 9.14   Scheme for the bridge design of an online viscometer cell with a holdup reservoir.

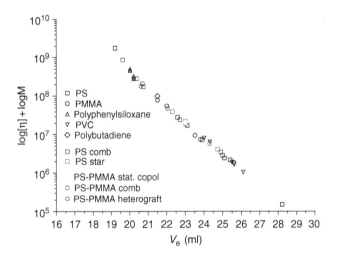

FIGURE 9.15   Universal calibration curve for different polymer types and structures established using an online viscometer and an RI detector [18].

established with known calibration standards (e.g., PS), one can obtain the SEC–molar mass calibration for an unknown polymer sample:

$$M\left(\text{sample}\right) = \frac{[\eta]\left(\text{standard}\right)M\left(\text{standard}\right)}{[\eta]\left(\text{sample}\right)} \qquad (9.15)$$

Further results can be derived from a Mark–Houwink plot, where the logarithm of the intrinsic viscosity is plotted versus the logarithm of the molar mass, for example, from the universal calibration curve. A linear fit yields

Mark–Houwink coefficient (slope) and Mark–Houwink K (intercept). Mark–Houwink coefficient, α, is a measure for the structure of the sample in solution, a change of the slope indicates a structure change. K is a measure of the segmental density and can be used to measure the effect of chemical reactions on the polymer backbone (studying polymer-analogous reactions) [19].

Mark–Houwink plots are used for the investigation of branching. Figure 9.16 shows a Mark–Houwink plot for a sample with long-chain branching. The deviation to lower intrinsic viscosities for the same molar mass indicates that branching starts.

9.4.2.3   *Triple Detection*   The term "Triple detection" has become very popular over the last years, sometimes discussed as an alternative to MALS. Unfortunately, there is not a generally accepted definition for triple detection. Sometimes, this term is used to describe the determination of intrinsic viscosity using a viscometer and molar mass using light scattering, independent of the light scattering technique applied. Sometimes it refers to the original meaning of "triple detection," used for a system composed of a viscometer, a 90° single angle light scattering and a concentration detector where the signals are processed using a special algorithm [20]. The combination of the viscometer together with the light scattering detector allows to overcome the limitations of RALS and to obtain accurate results for high molar mass samples. However, the triple algorithm includes several assumptions and therefore results in severe limitations. For each unknown sample, it needs to be known *a priori* that the sample is run under theta conditions (not SEC conditions), that the sample is linear, and that it can be

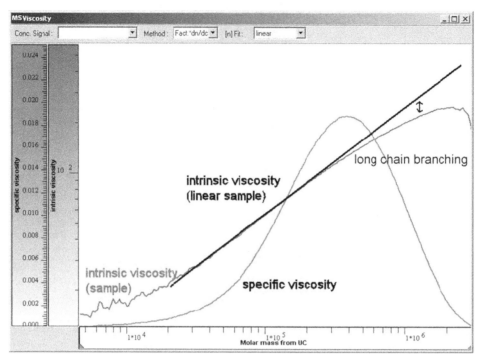

FIGURE 9.16 Mark–Houwink plot for a sample with long-chain branching. The intrinsic viscosity of the sample and the molar mass are measured online using a viscometer and an RI detector. The intrinsic viscosity of the linear counterpart is calculated using the Mark–Houwink coefficients.

represented by a simple chain model (coil, rod, and sphere). Thus, the uncritical use of triple detection might result in severe misinterpretation of experimental results [19]. If these requirements are fulfilled, triple detection allows to overcome the limitations for 90° light scattering by estimating the particle scattering function $P(q)$ from the viscometry data [20].

9.4.2.4 *MS Detection* For polymer analysis, a MS detector is a very interesting alternative to light scattering detectors, because this detector provides absolute molar masses for polymers and copolymers with high accuracy and sensitivity. Results are independent of dn/dc and since this detector is not a typical molar mass–sensitive detector, low molar mass samples can be analyzed with high precision without overloading the SEC. The advantage of the fractionation before MS analysis is that complex samples are separated into individual parts, which allows to optimize the MS device for best performance and to avoid discrimination (suppression), which can be a severe limitation with broad samples. Providing that fragmentation does not occur, the measured mass of a particular component can be correlated with chemical composition or chain length. However, the major drawback of MS is the limited mass range, preventing higher molar masses to be ionized without fragmentation.

Different gentle ionization mass spectrometric methods have been used in macromolecular analysis. MALDI-ToF

and electrospray ionization-mass spectrometry (ESI-MS) are the most common techniques in combination with SEC. For the determination of the MMD, not only the molar mass from the MS itself, but also the concentration of a species is required. Since this cannot be determined from the MS, due to the fact that MS signals cannot be quantified easily, the combination with a concentration detector is required [21].

9.4.2.4.1 *MALDI-ToF* Matrix-assisted laser desorption ionization-time of flight MS is an MS technique that allows desorption and ionization of very large molecules even in complex mixtures. In polymer analysis, it is used to perform the direct identification of mass-resolved polymer chains and for the simultaneous determination of structure and end-groups [22]. Compared to other MS techniques, the accessible mass range has been extended considerably, up to relative molar masses of about 500,000 Da. It was shown in a number of applications that functionally heterogeneous polymers can be analyzed with respect to the degree of polymerization and the type of functional groups.

Matrix-assisted laser desorption ionization-time of flight is based on the desorption of molecules from a solid surface and, therefore, the compatibility with online SEC is difficult to achieve. A number of research groups carried out off-line separations and investigated the resulting fractions with MALDI/ToF to make use of MALDI/ToF capabilities in

SEC. Although this approach is tedious, it has the advantage that many types of chromatographic separation can be combined with MS detection. The different options for using MALDI/ToF as an offline detector have been discussed by Pasch [23]. To overcome the difficulties of the off-line analysis of SEC fractions, interfaces where the SEC effluent was sprayed onto a moving matrix-coated substrate were introduced.

A new approach included an interface where the effluent from the SEC is continuously sprayed through a heated capillary nozzle on a slowly moving pre-coated MALDI target. The matrix is either deposited manually or automatically on the MALDI target. When necessary, a salt is added. Very similar equipment was also used for off-line FTIR detection.

Online coupling of SEC and MALDI-ToF was achieved in an aerosol SEC-MALDI experiment, where the effluent from the SEC column was combined with a matrix solution and sprayed directly into a ToF/MS [24]. Well-resolved MALDI-ToF spectra were obtained for PEG and poly(propylene glycol).

9.4.2.4.2   *ESI-MS*   Besides the gentle ionization, one of the major advantages of ESI is that there is a simple approach for online coupling with liquid chromatography. Shortly after the introduction of ESI-MS in the late 1990s, the first online applications for the analysis of polymers were published [25]. Since then, SEC has proven to be a highly compatible separation method for the combination with ESI-MS [12, 21]. The separation of the samples by size ensures that polymer fractions with a very narrow molecular weight distribution reach the ESI-MS. The otherwise typical mass spectrometric overlaps of different charged states due to a broad molecular weight distribution, is thus suppressed. Therefore, online coupling of SEC and ESI-MS generates easy to interpret mass spectra without overlapping masses from different charged states.

Furthermore, the sensitivity of the mass spectrometric analysis is significantly improved by SEC/ESI-MS coupling. This effect is also based on the fact that only small fractions of the narrow molecular weight distribution at a time are ionized, instead of the entire molecular weight distribution. Saturation effects have therefore less influence on the ionization efficiency. While samples in a molecular weight range of about 3 kDa can be analyzed by ESI-MS with direct infusion, the molar mass range after online connection can be extended to some 10–20 kDa (depending on the resolution of the mass spectrometer). In addition, the preliminary separation removes low–molecular weight impurities and matrix components that may otherwise interfere with the ionization process.

Modern electrospray detectors are capable of handling flows up to several 100 µl min⁻¹ and can therefore be coupled easily with existing conventional LC equipment. A parallel array of concentration detector(s) (e.g., UV, RI, and ELSD) and mass spectrometer allows operating all detectors under optimized flow rates. The bulk of the eluent is sent after the separation to the concentration detectors, and only a small amount hits the mass spectrometer. A split ratio of 9:1 has proven ideal. The ionization agent is added by a second pump, for example, syringe pump, to ensure reproducible and efficient ionization and to suppress any unwanted ionization by cationic impurities.

This type of SEC-MS hyphenation has been applied to a wide variety of homopolymers and copolymers; amongst them are polyacrylates, polyesters, polyethers, polyamides, resins, polycarbonates, proteins, and polystyrene. Synthesis optimization, investigation of the degradation behavior of polymers, polymer aging, and deformulation have been the goals of the analysis.

9.4.3   Investigation of Chemical Heterogeneity

Size exclusion chromatography reveals a multitude of molar mass information about unknown samples which are chemically simple or homogenous. However, copolymer analysis is challenging due to the effect that at least two distributions can be present simultaneously, the MMD and the CCD.

While there are molar mass sensitive detectors for the determination of the MMD of homopolymers, the applicability of these detectors for copolymers suffers from potential co-elution in SEC and (in case of light scattering) from the dependence of the *dn/dc* of composition. In addition, these detectors do not provide any insight into the CCD. There are different methods established on how to investigate the chemical heterogeneity with SEC or with related techniques.

9.4.3.1   *Determination of Chemical Composition Distribution by Dual Detection*   Based on detection with independent concentration detectors, this approach provides information on the chemical composition of copolymers or polymer blends and the variation with the molar mass [26]. An independent detector signal is required for each comonomer in the macromolecule. Many results have been published for chemically heterogeneous samples using two detectors. Obviously, there is a practical limit (four signals have been reported so far), as there are not many detectors to choose from. In addition, the approach is limited by the contribution of each detector to band broadening.

Results are obtained from the differences in detector response. To measure the chemical composition a detector response calibration is required. This is done using homopolymer samples of the comonomer with known concentrations and should not be confused with the molar mass calibration. Figure 9.17 shows two different copolymer samples with different compositions. The styrene content was calculated using the RI signal, the UV signal, and the response factors for PS (UV), PS(RI), PB(UV), and PB(RI).

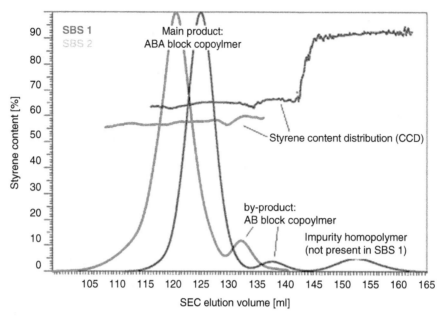

FIGURE 9.17 Analysis of two copolymers using RI/UV 254 nm dual detection, the response factors for the detectors have been determined using homopolymers of butadiene and polystyrene. (*See insert for color representation of the figure.*)

An independent task of copolymer analysis is the determination of meaningful copolymer molar masses. Obviously, results based on a single molar mass calibration generated with standards will not give accurate results, as the calibration will depend on the local composition of the species. A way to overcome this calibration dilemma is the use of online viscometers. An empirical method for samples with few hetero-contacts (e.g., block copolymers and graft copolymers) is using the multiple concentration detector approach.

From a SEC point of view, an AB block copolymer, where a sequence of comonomer A is followed by a block of B units, is a simple copolymer. The only hetero-contact in this chain is the A–B link, the A and B segments of the block copolymer will hydrodynamically behave like a pure homopolymer of the same chain length. In the case of long A and B segments, the A–B link acts as a defect position and will not change the overall hydrodynamic behavior. Consequently, the molar mass of the copolymer chain can be approximated by the molar masses of the respective segments. Similar considerations are true for ABA, ABC, and other types of block structures and for comb-shaped copolymers with low side-chain densities. In such cases, the copolymer molar mass M_c can be determined from the interpolation of the two homopolymer calibration curves $M_k(V)$ and the weight fractions w_k of the comonomers k [26]:

$$\lg M_c(V) = \sum w_k(V) \lg M_k(V) \qquad (9.16)$$

The calculation of copolymer molar mass averages and copolymer polydispersity is made as in conventional SEC

calculations using the weighted copolymer calibration curve. In cases where the number of heterocontacts can no longer be neglected, copolymer molar masses cannot be measured accurately by SEC alone. This is the case with statistical copolymers, polymers with only short comonomer sequences and high side-chain densities [26]. In such cases other methods, for example, 2D separations, have to be employed.

9.4.3.2 Determination of Chemical Heterogeneity by Interaction Chromatography

An alternative approach to measure the CCD of any kind of copolymer is the use of interaction chromatography. In such experiments, the separation is based on enthalpic interaction of the sample with the stationary phase; therefore, the name polymer HPLC (sometimes also referred to as gradient polymer elution chromatography (GPEC)) is also common.

The setup for polymer HPLC is quite similar to SEC systems with a few modifications. Liquid Adsorption Chromatography (LAC) requires the adsorption and desorption on a stationary phase. Therefore, in most cases isocratic separation is not sufficient. Gradients with respect to pH value, ionic strength, eluent composition, or temperature are applied. The most common approach is to use eluent composition gradients. In contrast to SEC where polymeric phases dominate, silica-based column packings are the most important stationary phase. Both normal phase and reversed phase separations have been described. A summary of different applications in copolymer separation is offered by Pasch [27]. Detectors used in gradient LAC are mainly UV/DAD detectors and ELSD.

When analyzing copolymers or polymer blends in LAC mode, the retention time is indicative of the average chemical composition and the peak width is a measure of the CCD. The retention axis has to be calibrated either by standards with known chemical composition or multiple detector combinations can be employed to measure the local chemical composition. The average chemical composition of a copolymer can be easily described by the moments of the CCD. The average composition, G, the width, dG, and the skew, S, of the CCD can easily be calculated. The value of the skew parameter is zero, if the composition distribution is symmetrical.

The average composition is described as:

$$\bar{G} = \mu_i(G) = \frac{\sum c_i G_i}{\sum c_i} \qquad (9.16)$$

where c_i is the local concentration in the chromatographic fraction.

The definition of dG and S can be found in the literature [4].

9.4.3.3 Comprehensive Copolymer Analysis

A fundamental problem in SEC that cannot be overcome by using any detection technique is that the method may suffer from limited chromatographic resolution.

Combination of different separation techniques into a single experiment (multi-dimensional chromatography; also called 2D chromatography, orthogonal chromatography, and cross-fractionation) has shown to overcome such types of limitations (see Section 9.5).

9.5 COMPREHENSIVE 2D CHROMATOGRAPHY

Every single separation method suffers from limited chromatographic resolution. Even powerful detection methods, such as MS techniques or light scattering, cannot resolve the fundamental separation dilemma caused by multiple overlapping property distributions [28]. These detectors rely on chromatography as a sample preparation step which reduces the complexity of the sample.

A more universal approach is the combination of different separation techniques into a single experiment (multi-dimensional chromatography; also called 2D chromatography, orthogonal chromatography, or cross-fractionation).

Two-dimensional chromatography separations can be done in planar systems or coupled-column systems. Examples of planar systems include 2D thin-layer chromatography (TLC) [29], where successive 1D TLC experiments are performed at 90° angles with different solvents, and 2D electrophoresis, where gel electrophoresis is run in the first dimension followed by isoelectric focusing in the second dimension [30].

9.5.1 Advantages of 2D Separations

Despite the fact that substantial progress has been achieved in column technology [7–9], the need and the use for multi-dimensional separation systems have increased. The main reason is due to the fact that most classes of macromolecules exhibit property distributions in more than one parameter (e.g., molar mass and chemical composition at the same time).

The separation efficiency of a single separation method is limited by the efficiency and selectivity of the separation mode, that is, the plate count of the column and the phase system selected. Adding more columns is a limited approach and does not overcome the need to identify more components in a complex sample, due to the limitation of peak capacities. The peak capacity can be described in a generalized case as:

$$n = 1 + \frac{\sqrt{N}}{4}\left(\frac{t_G}{t_0} - 1\right)\frac{1}{k_e + 1} \qquad (9.17)$$

where t_G is the gradient time, t_0 is the time for total permeation, and k_e is the retention factor at peak elution.

The corresponding peak capacity in an n-dimensional separation is higher due to the fact that each dimension contributes to the total peak capacity as a factor and not as an additive term for single dimension methods [28]:

$$n_{total} = \prod n_i \sin^{i-1} \vartheta_i \qquad (9.18)$$

where n_{total} is total peak capacity, n_i is the peak capacity in dimension i, and h_i is the separation angle between dimensions.

The "angle" between dimensions is determined by the independence of the methods. A 90° angle is obtained by two methods, which are completely independent of each other and will, for example, separate two properties solely on a single parameter without influencing each other [28].

In the case of a 2D system with isocratic methods (e.g., LCCC × SEC) the peak capacity is given by:

$$\begin{aligned} n_{2D} &= n_1 n_2 \sin \vartheta \\ &= \left(1 + \frac{\sqrt{N_1}}{4}\ln\frac{V_{p,1}}{V_{0,1}}\right)\left(1 + \frac{\sqrt{N_2}}{4}\ln\frac{V_{p,2}}{V_{0,2}}\right)\sin \vartheta \end{aligned} \qquad (9.19)$$

In 2D LC systems, an aliquot from a column is transferred into the second separation method in a sequential and repetitive manner by an inline transfer injector. Sample transfer between dimensions can be either complete (i.e., 100% of the effluent from the first dimension is transferred into the second dimension; this is also called comprehensive 2D), or partial, or simple "heart cuts" takes the eluent from a first dimension peak or a few peaks into another column during the first dimension elution process.

The use of different modes of liquid chromatography facilitates the separation of complex samples selectively with respect to different properties like molar mass, chemical composition, architecture, or functionality. A review by Kilz [28] and 2 books on 2D separation [31, 37] offer a good overview on different 2D techniques and applications.

9.5.2 Experimental Aspects of 2D Chromatography

Setting up a 2D chromatographic separation system is not difficult, if separation methods exist for the single dimensions. In most cases, the experimental implementation can be handled quite easily.

The transfer of fractions eluting from the first dimension into the second dimension is the most important aspect. Several transfer options are available:

a. *Manual transfer*: These simple off-line systems just require a fraction collection device and something or someone who reinjects the fractions into the next chromatographic dimension. However, such a setup is not comprehensive and is tedious.

b. *Comprehensive transfer*: In online 2D systems, the transfer of fractions is preferentially done by automatic transfer valves with two injection loops working in tandem mode. While one fraction is analyzed from one loop, the next fraction is collected simultaneously in the second loop (Fig. 9.18).

A strong advantage of the online 2D chromatography separations over off-line fractionation is that only online methods allow differentiating between samples that show identical chromatograms in both individual methods. This situation is illustrated in Figure 9.19a–c. Despite the fact that the sample contains two or four very different compounds, the HPLC interaction chromatogram (left-hand side) and the SEC elution profile (shown at the bottom of the 2D contour plot) are identical for all three cases. Only the online analysis allows correct identification and mapping.

Figure 9.20 shows a generalized setup for an automated 2D chromatography system.

A comprehensive 2D analysis requires total mass transfer from the first to the second dimension which can be done easily with transfer method (b) by proper selection of flow rates in both dimensions. This is a very beneficial situation as compared to heart-cut transfers, since by-products and trace-impurities can be separated even if they are not visible in the first dimensional separation. Table 9.8 presents an overview of transfer options and their benefits and limitations.

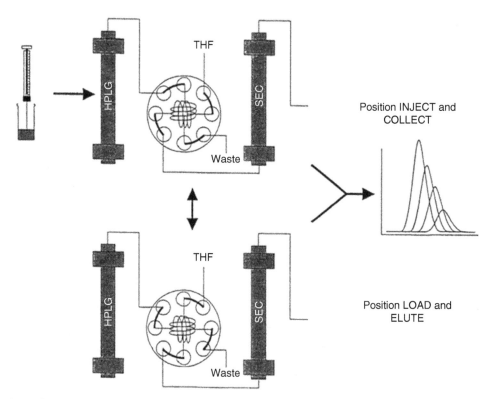

FIGURE 9.18 Automatic fraction transfer in 2D separations with an eight-port 2 position dual-loop tandem valve.

Several additional experimental aspects have to be considered for optimum 2D separations.

1. Detectability and sensitivity in the second dimension

    The consecutive dilution of the sample components requires sensitive detectors. If by-products and trace-impurities have to be determined, only the most sensitive and/or selective detection methods can be employed. ELS detectors are most commonly used in 2D experiments, due to their high sensitivity and good baseline stability even when large amounts of co-solvents from the first dimension elute. Fluorescence and diode array detectors are also sensitive but not as universal as the ELS detectors. MS detectors have a high potential in

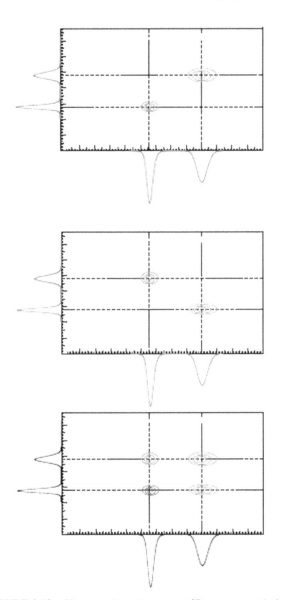

FIGURE 9.19 2D separations (shown as a 2D contour map) show clear sample differences while LC and SEC chromatograms are identical in each case. (*See insert for color representation of the figure.*)

this respect too, but have not been used in online 2D analyses yet. As a general rule, the higher the inject band dilution of a given separation method, the more sensitive the subsequent detection method must be.

2. Eluent transfer

    The miscibility of solvents which are transferred from the first to the second dimension is also important when setting up 2D experiments. Incomplete miscibility over the whole composition range will influence the separation in the second dimension and/or fractions can be trapped in the transfer valve.

    In SEC separations, the transfer of mixed mobile phases can affect molar mass calibration. In order to get proper molar mass results, the calibration curves have to be measured using the extremes of mobile phase composition and tested for changes in elution behavior and pore-size influence in the SEC column packing. The better the thermodynamic properties of the SEC eluent, the less influence is expected on the SEC calibration, when the transfer of mobile phase from the previous dimension occurs. It is advantageous to use the SEC eluent as one component of the mobile phase in the previous dimension to avoid potential interference and mobile phase incompatibility.

3. Time consumption

    Time is an important issue when designing 2D experiments. If existing techniques for each dimension are already available, then the method combination is straightforward and can be implemented within about a day. More time is required if methods are not available and have to be developed. SEC is often the simplest method to develop, despite the fact that it can be difficult to have true size-exclusion behavior for some sample/solvent systems (this applies mainly to aqueous SEC separation, e.g., starch and cellulose). Gradient LC methods are rather quick and easy to establish, especially if simple gradients raging from pure good to pure poor solvents are employed. More time could be required for optimized LC gradients to investigate subtle differences or trace impurities in otherwise very similar samples. LC at critical conditions is certainly the method that needs most time, if no other information is available. A very good review [32] on critical LC and chromatography at limiting conditions can be used as a valuable starting point for method development.

    The time needed for the 2D separations themselves can be considerable. This is especially true for 2D separations using quantitative mass transfer via dual-loop transfer valves. Heart-cut experiments require much less time and are often sufficient to check out the applicability of the approach. Cutting down on time consumption for 2D experiments is currently a heavily investigated topic. The application of HighSpeed SEC columns (Section

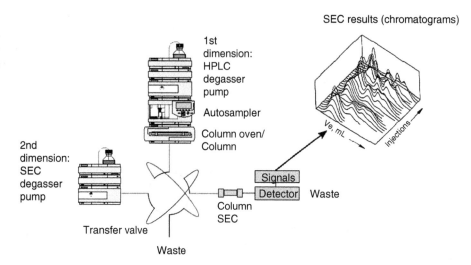

FIGURE 9.20   Generalized view of a comprehensive LC × SEC system: modules of first dimension are aligned horizontal, second dimension devices are shown in horizontal direction. (*See insert for color representation of the figure.*)

TABLE 9.8   Comparison of Transfer Options in 2D Separations and Their Requirements

Transfer	Mode	Advantages	Disadvantages	Example
Manual	Off-line	Very simple, fast setup, inexpensive	Time-consuming, not for routine work, not precise, no correlation of fraction elution to transfer time, not quantitative	Test tube
Automatic	Off-line	Simple, easy, fast setup	Not precise, no correlation of fraction elution to transfer time, not quantitative	Fraction collector, storage valve
Single loop	Online	Correct concentrations, correct transfer times, automation	Transfer not quantitative, setup time	Injection valve (with actuation)
Dual loop	Online	Correct concentrations, correct transfer times, quantitative transfer, automation	Setup time, special valve	Actuated 8-port valve, actuated 10-port valve

9.3.3.3 and Reference [9]) can reduce the analysis time in the SEC dimension by a factor of about 10. This allows 2D experiments with 60 transfer injections completed in about 1 h without loss of resolution.

Another time requirement in 2D separations is the time needed for data processing and presentation. With several dozen transfers between dimensions, data reduction, and presentation can be very time-consuming [33]. There is a clear need for specialized 2D software, which does all the data acquisition, fraction transfer, valve switching, data reduction, data consolidation, and presentation of results. Currently, only one 2D chromatography data system is commercially available [34].

9.5.3   Selection of Separation Techniques

The analytical goal governs the separation techniques to be employed in 2D experiments.

As there is no limitation on what method combinations can be employed, virtually, all samples can be analyzed by 2D separations.

Schure [35] compared different chromatographic method combinations on the basis of efficiency, sample dilution, and detectability. He investigated Capillary Electrophoresis (CE), Gas Chromatography (GC), Liquid Chromatography (LC), Supercritical Fluid Chromatography (SFC), and field flow fractionation (FFF) in detail, while

TABLE 9.9 Favorable 2D Combinations and Method Sequence

| Technique Applied in | | Typical Applications |
One Dimension	Two Dimension	
Gradient LC	SEC	Composition and molar mass of copolymers
LCCC	SEC	Functionality and molar mass of resins
Gradient LC	CE	Protein and or peptide mapping
TREF	SEC	Branching and molar mass of polyolefins
RPLC	NPLC	Deformulation of surfactants

omitting other methods, which are potential candidates for method hyphenation, for example, SFC and temperature rising elution fractionation (TREF). Obviously, destructive methods such as GC, which destroy the chromatographic phase system, play a limited role in 2D separations as they can be used only in the second dimension.

Despite early work in which the slowest technique was used as the first dimension separation, application of the method with the highest selectivity for one property should be used as the first dimension. This ensures highest purity of eluting fractions being transferred into the subsequent separation.

Best adaptability is required for the first chromatographic dimension. Interaction chromatography offers the best flexibility and is the most adjustable choice, because of the following:

- many conditions (mobile phase, mobile phase composition, mobile phase modifiers, stationary phase, temperature etc.) can be adjusted to optimize the separation based on the chemical nature of the sample;
- easier optimization in interaction chromatography allows for more homogeneous fractions from the first dimension;
- overloading of columns is less likely as, for example, for SEC columns.

Favorable 2D method combinations based on Schure's calculations and the experience of the authors are summarized in Table 9.9.

9.5.4 Current State of 2D Chromatography

There are several review publications [28, 2] and a book [31] which offer a good overview of 2D separations. A major boost for 2D separations was the first fully automated 2D system developed by Kilz et al. [33, 36]. The column-switching device is automatically driven by the software,

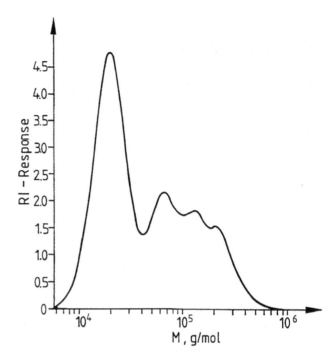

FIGURE 9.21 Apparent SEC molar mass distribution of four-arm block copolymer (see text for details).

which at the same time performs the data collection from the detectors, the data reduction, and the calculation of the results.

Only two examples of 2D separations will be presented in this chapter to illustrate the benefits and advantages of 2D separations.

As a first comprehensive 2D experiment, Kilz et al. [36] described the characterization of complex styrene–butadiene star polymers. The four-arm star polymers were prepared by anionic polymerization to give samples with perfect composition and molecular weight control. The polymerization was designed to yield a mixture of linear (of molar mass M), two-arm (2M), three-arm (3M), and four-arm (4M) species. Four samples with varying butadiene content (about 20, 40, 60, and 80%) were prepared. The 2D verification sample was a mixture of four samples with different chemical composition and molar masses. Therefore, this verification sample consisted of 16 components representing the wide range in chemical composition and molar mass.

Size exclusion chromatography separation of this 16-component star-block copolymer revealed four partially resolved peaks (Fig. 9.21). They correspond to the four molar masses of the sample consisting of species with one to four arms. The molar masses are defined by the number of arms and are M-2M-3M-4M. Despite appropriate resolution, the SEC chromatogram did not give any indication of the complex chemical structure of the sample.

The same sample mixture run in gradient HPLC mode gave poorly resolved peaks, which might suggest different composition, but gave no clear indication of different molar mass and topology as shown in Figure 9.22.

The combination of the two methods in the 2D setup dramatically increased the resolution of the separation system and gave a clear picture of the complex nature of the 16-component sample. Based on the composition of the sample, a contour map with the coordinates chemical composition and molar mass is expected to show 16 spots, equivalent to the 16 components. Each spot would represent a component which is ideally defined by a single composition and molar mass. The experimental evidence of the improved resolution in the 2D analysis is given in Figure 9.23. This contour plot was calculated from experimental data based on 28 transfer injections.

The contour plot clearly revealed the broad chemical heterogeneity (y-axis, chemical composition) and the wide MMD (x-axis) of the mixture. The relative concentrations of the components were represented by colors. Sixteen major peaks were resolved with high selectivity. These correspond directly to the components. For example, peak **1** corresponds to the component with the highest butadiene content (80%) and the lowest molar mass (molar mass 1M) whereas peak **13** relates also to a molecule with 80% butadiene content but a molar mass of 4M. Accordingly, peak **16** is due to the component with the lowest butadiene content and a molar mass of 4M, representing a four-arm star block copolymer with a styrene–butadiene content of 80:20.

A certain molar mass dependence of the HPLC separation in the first dimension is indicated by a drift of the peaks for components of similar chemical composition (e.g., with peaks **1**, **5**, **9**, and **13**). This behavior is normal for polymers, because pores in the HPLC stationary phase lead to size-exclusion effects which overlap with the enthalpic interac-

tions at the surface of the stationary phase. Consequently, 2D separations of this type will not, in general, be orthogonal but skewed, depending on the pore-size distribution of the stationary phase and the nature of the sample. The quantitative amount of butadiene in each peak could be determined via an appropriate calibration with samples of known composition. The molar masses could be calculated based on a conventional molar mass calibration of the second dimension. Kilz, Krüger, Much, and Schulz [36] analyzed aliphatic polyesters with

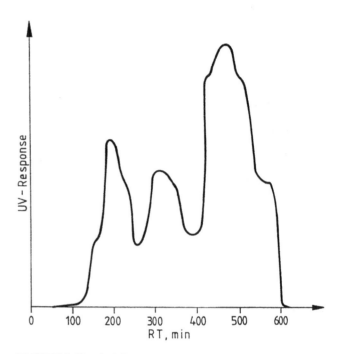

FIGURE 9.22   GPEC separation of four-arm block copolymer as observed in the UV detector.

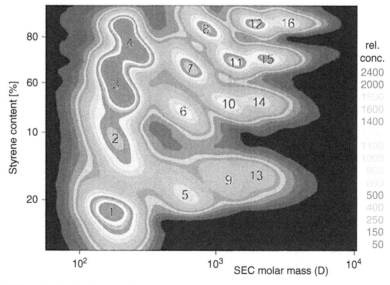

FIGURE 9.23   2D HPLC-SEC analysis of a four-arm block copolymer consisting of 16 components with four different styrene compositions and four molar masses, respectively (contour map shown; peak annotations, see text). (*See insert for color representation of the figure.*)

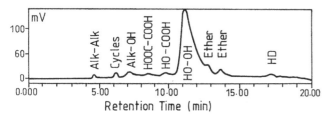

FIGURE 9.24 Separation of adipic acid–1,6-hexanediole oligo-ester end-groups by critical chromatography.

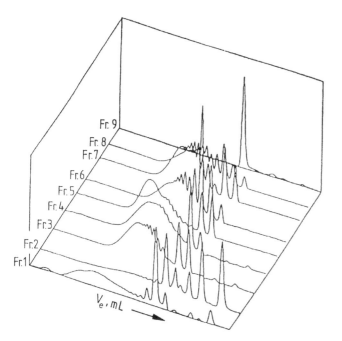

FIGURE 9.25 3D view of SEC transfer injections of oligo-ester sample build up from adipic acid and 1,6-hexanediol.

respect to end-group and MMD. Polyesters from adipic acid and 1,6-hexanediol are intermediates for the manufacture of polyurethanes and their functionality-type distribution is a major parameter in affecting the quality of the final products. In particular, nonreactive cyclic species are responsible for the "fogging effect" in polyurethane foams.

The separation of the polyesters with respect to functionality was done by LCCC. The critical point of adsorption of the polymer chain corresponded to an eluent composition of acetone–hexane 51:49 (v/v) on silica gel. The critical chromatogram of an aliphatic polyester sample together with the functionality fraction assignment is given in Figure 9.24. The "ether" peaks could be attributed to the formation of ether structures in the polyester samples by a condensation reaction.

The MMDs of the functionality fractions could be determined by preparatively separating the fractions and subjecting them to SEC. The SEC chromatograms of fractions 1–9 are summarized in Figure 9.25. For a number of fractions oligomer separations were obtained, which could be used to calibrate the SEC system.

The complex nature of the sample could be verified in a 2D experiment using LCCC as the first dimension to separate for functionality and SEC in the second dimension to determine molar masses. The contour plot in Figure 9.26 reveals separation according to end-group functionality from LC-CC as well as oligomer distribution from SEC. The sample was prepared from an adipic acid–rich reaction mixture resulting in an acid number of about five. The high content of dicarboxylic acid end-groups is clearly visible in the contour map. Quantification of the contour plot yielded quantitative information on both end-group distribution and MWD.

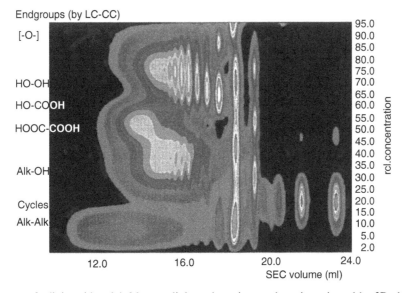

FIGURE 9.26 Contour map of adipic acid and 1,6-hexanediol condensation products investigated by 2D chromatography (see text for details). (*See insert for color representation of the figure.*)

REFERENCES

[1] Striegel AM, Yau WW, Kirkland JJ, Bly DD. *Modern Size-Exclusion Liquid Chromatography: Practice of Gel Permeation and Gel Filtration Chromatography*. 2nd ed. Hoboken: John Wiley & Sons, Inc.; 2009.

[2] Kilz P, Pasch H. Coupled LC techniques in molecular characterization. In: Myers RA, editor. *Encyclopedia of Analytical Chemistry*. Chichester: Wiley; 2000.

[3] (a) Belenkii BG, Gankina ES, Tennikov MB, Vilenchik LZ. Dokl Akad Nauk 1976;231:1147; (b) Entelis SG, Evreinov VV, Gorshkov AV. Functionality and molecular weight distribution of telechelic polymers. Adv Polym Sci 1986;76:129–175.

[4] Schröder E, Müller G, Arndt K-F. *Polymer Characterization*. Munich: Hanser; 1998.

[5] SEC Standard ISO 13885 Part 1 (THF), Part 2 (DMAc), Part 3 (aqueous). Geneva: International Organization for Standardization; 2008.

[6] Benoit H, Rempp P, Grubisic Z. A universal calibration for gel permeation chromatography. J Polym Sci B: Pol Lett 1967;5:753–759.

[7] Kilz P. Design, properties, and testing of SEC columns and optimization of SEC separations. In: Chi-san Wu, editor. *Column Handbook for Size Exclusion Chromatography*. San Diego: Academic Press; 1999. p 267.

[8] Kilz P. Optimization of GPC/SEC separations by proper selection of stationary phase and detection. In: Kromidas S, editor. *HPLC Made to Measure. A Practical Handbook for Optimization*. Weinheim: Wiley; 2006.

[9] Kilz P. Methods and columns for high-speed size exclusion chromatography separations. In: Chi-san Wu, editor. *Handbook of Size Exclusion Chromatography and Related Techniques*. 2nd ed. New York: Dekker; 2003. p 561.

[10] LaCourse WR. Column liquid chromatography: equipment and instrumentation. Anal Chem 2002;74:2813–2832.

[11] Meyer VR. *Practical High-Performance Liquid Chromatography*. Chichester: Wiley; 2010.

[12] Barner-Kowollik C, Gründling T, Falkenhagen J, Weidner S, editors. *Mass Spectrometry in Polymer Chemistry*. Weinheim: Wiley-VCH; 2011.

[13] Held D, Kilz P. Qualification of GPC/GFC/SEC data and results. In: Kromidas S, Kuss H-J, editors. *Quantification in LC and GC: A Practical Guide to Good Chromatographic Data*. Weinheim: Wiley; 2009.

[14] Mader C, Schnöll-Bitai I. Pulsed laser polymerization of styrene in microemulsion: determination of band broadening in size exclusion chromatography with multimodal distributions. Macromol Chem Phys 2005;206:649–657.

[15] Konkolewicz D, Taylor II JW, Castignolles P, Gray-Weale A, Gilbert RG. Toward a more general solution to the band-broadening problem in size separation of polymers. Macromolecules 2007;40:3477–3487.

[16] Meira G, Netopilík M, Potschka M, Schnöll-Bitai I, Vega J. Band broadening function in size exclusion chromatography of polymers: review of some recent developments. Macromol Symp 2007;258:186–197.

[17] Haney MA. The differential viscometer. II. On-line viscosity detector for size-exclusion chromatography. J Appl Polym Sci 1985;30:3037–3049.

[18] Benoit H, Grubisic Z, Rempp P, Decker D, Zilliox J-G. Gel permeation chromatography measurements on linear and branched polystyrenes of known structures. J Chim Phys 1966;63:1507–1514.

[19] Radke W. Chromatography of polymers. In: Matyjaszewski K, Gnanou Y, Leibler L, editors. *Macromolecular Engineering: Precise Synthesis, Materials Properties, Applications*. Volume 3. Weinheim: Wiley-VCH; 2007.

[20] Haney MA, Jackson C, Yau WW. *International GPC Symposium Proceedings*. 1991. p 49–63.

[21] Gruendling T, Guilhaus M, Barner-Kowollik C. Fast and accurate determination of absolute individual molecular weight distributions from mixtures of polymers via size exclusion chromatography–electrospray ionization mass spectrometry. Macromolecules 2009;42:6366–6374.

[22] Pasch H, Schrepp W. *MALDI-TOF Mass Spectrometry of Synthetic Polymers*. Berlin: Springer; 2003.

[23] Pasch H, Rode K. Use of matrix-assisted laser desorption/ionization mass spectrometry for molar mass-sensitive detection in liquid chromatography of polymers. J Chromatogr A 1995;699:21–29.

[24] Murray KK, Russell DH. Aerosol matrix-assisted laser desorption ionization mass spectrometry. J Am Soc Mass Spectrom 1994;5:1–9.

[25] Prokai L, Simonsick Jr. WJ. Electrospray ionization mass spectrometry coupled with size-exclusion chromatography. Rapid Commun Mass Spectrom 1993;7:853–856.

[26] Gores F, Kilz P. Copolymer characterization using conventional SEC and molar mass-sensitive detectors. In: Provder T, editor. *Chromatography of Polymers*. ACS Symposium Series 521. Washington, DC: American Chemical Society; 1993.

[27] Pasch H. Liquid adsorption chromatography. In: Held D, Kilz P, Pasch H, Trathnigg B, editors. *HPLC of Manuscript GPC/SEC*. Berlin: Springer; 1998.

[28] Kilz P. Two-dimensional chromatography as an essential means for understanding macromolecular structure. Chromatographia 2004;59:3–14.

[29] Grinberg N, Kalász H, Han SM, Armstrong DW. Special Techniques. In: Grinberg N, editor. *Modern Thin-Layer Chromatography*. New York: Marcel-Dekker Publishing; 1990.

[30] Celis JE, Bravo R. *Two-Dimensional Gel Electrophoresis of Proteins*, New York: Academic Press; 1984.

[31] Cohen SA, Schure MR, editors. *Multidimensional Liquid Chromatography. Theory and Applications in Industrial Chemistry and the Life Sciences*. Hoboken: John Wiley & Sons, Inc.; 2008.

[32] Macko T, Hunkeler D. Liquid chromatography under critical and limiting conditions: a survey of experimental systems for synthetic polymers. Adv Polym Sci 2003;163:61–136.

[33] Kilz P, Krüger R-P, Much H, Schulz G. Characterization of aliphatic polycondensates by online two-dimensional chromatography. Polym Mater Sci Eng 1993;69:114.

[34] PSS WinGPC 2D User Documentation, Mainz; 2011.

[35] Schure MR. Limit of detection, dilution factors, and technique compatibility in multidimensional separations utilizing chromatography, capillary electrophoresis, and field-flow fractionation. Anal Chem 1999;71:1645–1657.

[36] Kilz P, Krüger R-P, Much H, Schulz G. In: Provder T, Urban MW, Barth HG, editors. *Chromatographic Characterization of Polymers: Hyphenated and Multidimensional Techniques.* Advances in Chemistry Series 247, Washington, DC: American Chemical Society; 1995.

[37] Pasch H. Trathnigg, B. Multidimensional HPLC of Polymers, Berlin: Springer; 2014.

10

MASS SPECTROSCOPY: ESR AND NMR APPLICATIONS TO POLYMER CHARACTERIZATION AND POLYMERIZATION MONITORING

Zifu Zhu and Alina M. Alb

10.1 INTRODUCTION

The high demand for the synthesis of new materials with specific properties led to the development and diversification of the polymer characterization methods, from average molecular characterization to detailed investigation on chemical structure and composition.

Spectroscopy techniques [1] such as ultraviolet (UV), visible, infrared (IR), Raman, nuclear magnetic resonance (NMR), electron spin resonance (ESR), and mass spectroscopy (MS) are widely recognized and intensively used in structure determination and analysis.

Among them, MS has become a valuable tool in studying polymer structure, composition, molecular weight distribution (MWD), and other important physical quantities [2–5]. Different MS techniques have been employed, with different degree of applicability and success, in providing comprehensive quantitative information on the sample analyzed. Traditional approaches such as ionization techniques [2, 5–10] or modern ones such as matrix-assisted laser desorption ionization (MALDI) MS [4, 11–13] are used to obtain structural and compositional information.

A powerful and well-established analytical technique with broad applicability in all areas of synthetic chemistry is NMR [14–16]. During the last half century, the technique has become an increasingly complex and effective structural tool, being the most direct and general method for identifying the structure of pure compounds and mixtures, in solid or liquid state.

The analysis of NMR spectra of polymers in solution offers better understanding of the synthesis and structure-property relations. Thus, monomer sequence distribution in copolymers determined by NMR is used in computing reactivity ratios. The search for architectural details in the structure and end-groups has undergone major development and has reached the point where the information provided is used in establishing polymer reaction mechanisms and kinetics. Recently, significant advances have been made in the field of high resolution solid-state NMR, which not only provides high chemical resolution, but also offers key insights into polymer chain packing and morphology.

Despite limitations, NMR remains one of the most versatile techniques in studying polymers in any state [17–25].

Another important spectroscopic technique, electron spin resonance (ESR) [1, 26–28], is a powerful tool for studying chemical species with unpaired electrons. By offering information about the presence, number, and distribution of the unpaired electrons, ESR provides useful structure information. All chemical, photochemical, and electrochemical reactions which proceed through free-radical mechanism can be studied by ESR.

The advances in technology in the last years led to the development and optimization of new ESR methods such as high field/multi-frequency ESR [29–31], double resonance [32, 33], and pulse methods [34–36]. The improvements in resolution brought by these methods have made possible analysis and interpretation of more complex systems and detection and characterization of transient paramagnetic intermediates inaccessible before.

Monitoring Polymerization Reactions: From Fundamentals to Applications, First Edition. Edited by Wayne F. Reed and Alina M. Alb.
© 2014 John Wiley & Sons, Inc. Published 2014 by John Wiley & Sons, Inc.

10.2 MASS SPECTROSCOPY

Recent advances, such as increase in sensitivity, specificity, and improved data acquisition features, have made mass spectrometry a valuable tool in polymer analysis.

A main requirement for the applicability of the MS techniques to polymer analysis is that the polymer be converted to gas-phase ions. After ionizing the sample into a beam of ions in vapor phase, by separating the ions according to their mass to charge ratio (m/z) and recording the mass spectrum as a plot of ion abundance against m/z, a mass spectrometer provides the molecular mass and indicates the presence of certain structural units, thus offering information about the structure of the molecule analyzed. Oftentimes, the samples to be analyzed are fractionated by gas or liquid chromatography techniques, allowing mass spectra to be obtained for individual components and, hence, a more thorough characterization.

A brief summary of different ionization methods and their characteristics is presented next, followed by a more detailed overview of the MS methods most commonly employed in the polymer characterization and reaction monitoring.

10.2.1 Ionization Methods

10.2.1.1 *Ionization of Volatile Materials* Volatile materials are generally ionized by interaction of their vapors with electrons, ions, or strong electric fields. Relevant ionization methods included in this category are (i) electron ionization and (ii) chemical ionization.

Electron ionization. A high energy electron beam (typically 70 eV) is applied on the sample as soon as it enters the vapor phase. The most probable process to occur as a result of the collision between a molecule M with an electron, e, is that the electron ionization produces molecular cations M^+, by ejection of an electron from the highly energized sample [6, 7]:

$$M + e \rightarrow M^{+\bullet} + 2e \qquad (10.1)$$

Alternatively, but less likely, processes, are the capture of an electron by a molecule to give a radical anion, M^{-}:

$$M + e \rightarrow M^{\bullet} \qquad (10.2)$$

or to give multiple charged ions:

$$M + e \rightarrow M^{n+} + (n+1)e \qquad (10.3)$$

Chemical ionization. In this case, the sample to be analyzed is ionized by ion–molecule interactions with reagent ions (RH). Thus, when the sample is introduced in the ionization chamber with a large excess of a reagent gas, the reagent molecules, ionized by electron impact, react with other reagent molecules to form reactant ions RH^+, which protonate the sample [2, 5, 7, 8]:

$$RH^{+\bullet} + RH \rightarrow RH_2^+ + R^{\bullet} \quad \text{reagent ion formation}$$
$$RH_2^+ + M \rightarrow RH + MH^+ \quad \text{proton transfer}$$
$$RH_2^+ + M \rightarrow \left[M + RH_2\right]^+ \quad \text{electrophilic addition}$$
$$(10.4)$$

The reaction exothermicity, due to the larger proton affinity of M of RH, is controlled by the choice of the reagent. Endothermic proton transfer occurs with far less frequency and in such cases, electrophilic addition is more likely [7]. Typical reagents are methane, isobutene, or NH_3.

An advantage of this method is that it causes less fragmentation and allows a more facile interpretation of the mass spectra.

10.2.1.2 *Desorption/Ionization Methods for Nonvolatile Materials* In addition to the aforementioned ionization methods, the nonvolatile or thermally unstable compounds can be desorbed into the gas phase by fast atom bombardment (FAB), field desorption (FD), or spray ionization techniques.

Field desorption methods are based on the same mechanism as that of the field ionization ones, but in this case, the sample is deposited directly on the surface carrying the field. The method has been successfully applied to polymers with masses up to $10,000$ g mol^{-1} [37].

Secondary ion mass spectroscopy (SIMS) can be used in the case of organic materials, deposited as a film on a metal foil, oftentimes in the presence of a salt [9, 38]. A primary ion beam is applied on the sample through this high energy process, causing extensive fragmentation and release of secondary ions from the surface.

In FAB methods, extensively used for compounds with molecular weight $M_w < 5000$ g mol^{-1} (e.g., polyglycols) [39, 40], the sample is dissolved in a matrix, which is a usually a viscous liquid with low volatility (e.g., glycerol). By bombarding the liquid matrix with fast beams of atom, ions characteristic to the matrix and analyte are produced. The liquid matrix provides continuous surface renewal and thus allows primary beams to be used to produce intense and long-lasting spectra. One limitation of the method is the requirement for polymers to be miscible with the polar liquid matrix.

Matrix-Assisted Laser Desorption Ionization is among the most promising desorption methods for synthetic macromolecules and biopolymers [4, 11]. MALDI is frequently used in the analysis of the compositions, end-groups, and polymer MWDs. An advantage of the method is the capability to analyze broad molecular weight polymers (up to $\sim 10^6$ g mol^{-1}) [13]. The polymer sample, dissolved in an appropriate solvent, is added into a matrix usually at a ratio between 1:100 and 1:50,000 to which an auxiliary ionization agent was added and loaded on a target surface. After the evaporation of the solvent, the dry mixture is irradiated by laser light, which leads to different types of molecules to be desorbed upon irradiation [11, 12]. In UV-MALDI [12], the macromolecules are not energized directly upon radiation; the light is rather absorbed by the matrix which is ionized and dissociated. The process changes the matrix to a super-compressed gas in which charge transfer reactions take place. The analyte ions and molecules are transported by the gas expansion from the surface into the gas phase, where additional charge transfer reactions to neutral molecules occur.

Thermospray (TSP) ionization has been used mainly in biological and pharmaceutical areas [2, 8, 41], with very few applications in synthetic polymer analysis. A mixture of analyte and an auxiliary electrolyte enters a pumped ion chamber (1–10 Torr) after passing through a heated capillary. A supersonic beam of charged droplets emerges at the entrance of the chamber, heated to facilitate the desolvation of the droplets. Both negatively and positively charged droplets are produced, the charge being determined by the electron ion contained statistically in excess. As the droplet size decreases, analyte molecules and electrolyte ions are evaporated, yielding through their interactions analyte-indicative quasi-molar ions.

Closely related to TSP is the electrospray ionization (ESI) method [2, 42, 43]. In this case, a strong electric field is applied to the capillary containing the analyte solution and the spray is produced at atmospheric pressure. Highly charged droplets are produced, their charged depending on the polarity of the field applied to the capillary. A counter-current flow of warm nitrogen facilitates the desolvation of the droplets. Electrospray Ionization applicability to synthetic polymers is limited by the challenges in deconvoluting the mass spectra to oligomer distributions, especially for large polymers.

10.2.2 Instrumentation

A typical mass spectrometer configuration is shown in Figure 10.1 [1]. The main components are briefly discussed in the following [1]:

1. Inlet—allows introduction of the sample vapors into the ionization chamber.
2. Ionization chamber (under high vacuum, 10^{-5}–10^{-6} Torr)—the vapors of the samples are ionized by a beam of electrons (typically, 70 eV) and accelerated to their final velocities by applying a large potential.
3. Mass analyzer—separates the accelerated positively charged ions according to their mass-to-charge ratios (m/z). Certain analyzers separate the ions simultaneously while others are scanned to transmit a narrow m/z range to detector at a given time.

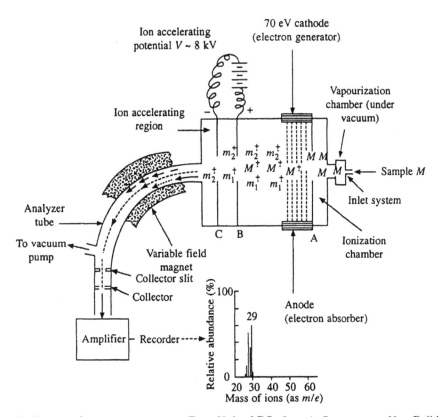

FIGURE 10.1 Schematic diagram of a mass spectrometer. From Yadav LDS. *Organic Spectroscopy*. New Delhi: Anamaya Publishers; copublished by Boston: Kluwer Academic Publishers; 2005. With kind permission from Springer Science+Business Media B.V.

4. Detector and spectrum recorder—records the relative abundances of ions versus mass of the ions (as *m/z*).

Depending on the resolution of the mass analyzer [3], a polymeric ion is either resolved into its individual isotopes or observed as an unresolved peak at the average *m/z* value of all isotopes.

Among the different types of mass analyzers, commonly found are magnetic and electric sensors, which use with fast ion beams (keV). Each ion from the ionization chamber is accelerated by a potential *V*, the energy given, *zeV*, being equal to its kinetic energy:

$$\frac{mv^2}{2} = zeV \tag{10.5}$$

where *m* and *v* are the mass and velocity of the ion, respectively, and *e* is the electron charge.

When the accelerated ion enters the mass analyzer, whose uniform magnetic field, *B*, is perpendicular to the ions velocity, it is forced into a circular trajectory of radius *r*:

$$\frac{mv^2}{r} = BzeV \tag{10.6}$$

Combining Equations 10.5 and 10.6 gives the mass analysis equation of the magnetic sector:

$$\frac{m}{z} = \frac{B^2 r^2 e}{2V} \tag{10.7}$$

The mass spectra are generally obtained by magnetic scanning, that is, increasing *B*, at constant *V*. Thus, the ions of progressively higher *m/z* reach the required value of radius to pass through the collector slit sequentially.

These types of detectors have a limited resolving power in discriminating between small mass differences. In such cases, double focusing mass spectrometers are used. Fourier-transform ion cyclotron resonance (FTICR) analyzers interfaced with ESI or MALDI are used in the high molecular weight polymer characterization [44].

10.3 APPLICATIONS OF MS METHODS TO POLYMER CHARACTERIZATION AND REACTION MONITORING

Various studies on polymer analysis and characterization, as well as mechanistic and kinetic studies of polymers using MS methods have been reported. While in the past, mass spectrometry of synthetic polymers was extremely challenging, requiring polymers to be degraded thermally or chemically prior to mass spectrometric analysis, modern soft ionization techniques, such as ESI and MALDI, have made

an enormous impact on the analysis of biopolymers and the latter has been successfully applied to synthetic polymer analysis. In this case, however, the following factors have to be taken into account: polymer MWD, different end-group chemistries, chemical composition distribution in the case of random copolymers and additional sequence, and block-length distribution in the case of block copolymers, together with architecture distribution (linear, branched, etc).

Information on the microstructure of synthetic copolymers is often obtained by NMR spectroscopy [14]. However, the sequence of individual components of the MWD cannot be obtained, NMR providing only an average of MWD. Liquid chromatographic techniques, such as gradient polymer elution chromatography (GPEC), offer information about the chemical composition of the oligomers while encountering challenges in the determination of the sequence of copolymers [45].

Information about the monomeric composition and structure can be obtained with pyrolysis MS but sequence information is lost [46]. The method was used in several applications, such as structural identification of homopolymers, differentiation of isomeric structures, copolymer composition and sequential analysis, identification of oligomers formed in the polymerization reactions, and identification of volatile additives contained in polymer samples [47]. One of the main challenges of the technique is the identification of the products in the spectrum of the multicomponent mixture produced by thermal degradation.

Fast atom bombardment has also been employed to obtain sequence information on copolymers. Montaudo et al. [48] reported the FAB-MS characterization of copolymer sequence by identifying the oligomers produced in the photolytic and hydrolytic degradation of copolyamides. A major disadvantage of FAB-MS in the characterization of the MWD is the fragmentation of parent ions during ionization [49].

10.3.1 Electrospray Ionization

10.3.1.1 *Biopolymer Characterization* Electrospray Ionization Mass Spectrometry (ESI-MS) has become a popular method in biophysical and biochemical communities as new high-resolution techniques have been developed to detect a wide range of physiologically relevant noncovalent protein complexes [50, 51]. Thus, Painter et al. [50] used Nano-ESI-MS approach for monitoring reactions of protein complexes in real time, allowing simultaneous determination of both structural and dynamical properties of the proteins studied.

Formation of protein aggregates impairs protein activity both directly and indirectly and is a particularly serious concern in the biopharmaceutical industry, where its occurrence in protein drugs not only affects the production process but also raises safety concerns. Therefore, special attention was given to biopharmaceutical applications, ESI-MS being used to monitor interactions between protein therapeutics and their physiological partners [52, 53], as well as aggregation of protein drugs under stress conditions [54, 55].

Monitoring protein aggregation with ESI-MS allowed the evolution of small soluble oligomers to be observed and characterized in real time. However, a limitation of this approach is the uncertainty regarding the relationship between the fractional concentrations of oligomeric species in solution and the relative abundance of ionic signal in ESI.

The suitability of ESI-MS for direct profiling of soluble glycoprotein aggregates was assessed by Wang et al. [55], who found a remarkable agreement between methods used, by comparing the abundance distribution of oligomeric species derived from ESI-MS to that obtained with SEC.

10.3.1.2 *Synthetic Polymer Characterization*
With major impact on the characterization of biopolymers, ESI applicability in the analysis of synthetic polymers has been limited. This is largely due to the tendency to form multiply charged ions that complicate the analysis of polymers, especially for high molecular masses. Nevertheless, the use of high resolution FTICR mass spectrometry coupled with ESI has allowed the complex charge states to be unambiguously assigned and the spectra deconvoluted.

Takáts et al. [56] showed that, under conventional conditions (low or medium cone voltage), the quantitation of distributions based on the molecular mass envelope using ESI was not useful in the case of polyethylene–polypropylene glycol copolymers, but overcame this problem by using high cone voltages and monitoring low-mass fragment ions and obtain reproducible signals, the intensity of the low-mass ions showing good correlation with the sample amount.

Koster et al. [57] used ESI as ionization technique to study the molecular structure of (co)polyesters on an FTICR mass spectrometer, which offered the advantage of a high mass resolving power and high mass accuracy. Studies presenting ESI mass spectral data along with mass spectrometry/mass spectrometry (MS/MS) have been reported for poly(propylene imine) dendrimers for selected charge states using triple quadrupole tandem mass spectrometry [58]. Particular emphasis was given to the dissociation behavior of singly protonated ions for each generation of dendrimer.

10.3.1.3 *Combined SEC/ESI-MS*
An advantage of ESI is that it can be easily interfaced with liquid separation methods. Because ESI uses a very small fraction of the eluent, conventional size exclusion chromatography (SEC) detectors can be operated in parallel with MS. With a separation of analytes according to their hydrodynamic volume, coupling of SEC with ESI-MS adds an additional dimension to the MS analysis and increases the amount of information provided. In addition, the chromatographic separation reduces the competition for the electrical charges during the

formation of gas-phase ions in the electrospray interface and improves the ionization of analytes. The separation of inorganic sample constituents from the organic compounds should also increase the ionization efficiency by decreasing the complexity of the spectra obtained.

Electrospray Ionization and MALDI mass spectrometry are especially versatile tools for the analysis of synthetic oligomers. By rendering the unfragmented macromolecule amenable to mass separation, these techniques enable accurate molecular weight determination of the individual oligomers. However, due to the fact that synthetic polymers do not exhibit uniform chain length, accurate MWD can only be obtained in limited cases. Gruendling et al. [59] combined SEC with ESI-MS (RI detection) for deriving accurate molecular weight information on synthetic polymers. The approach makes use of the complementary strengths of the two methods, in separately measuring molecular weight and abundance/concentration. The high accuracy of individual molecular weights obtained by mass spectrometry was applied to determine the concentration profiles of oligomers eluting from the chromatographic column. A computational algorithm based on the maximum entropy principle was used to process the data collected for poly(methyl methacrylate) standards (molecular weights up to 10,000 g mol^{-1}), chosen as case study. Additional information with regard to the relative ionization efficiencies and competition between the charge states of polymer molecules in ESI, as well as information on the dependence of these parameters on the polymer molecular weight was also included in the work presented in Reference [59].

Electrospray Ionization–Mass Spectrometry (in combination with online pre-separation of the polymer sample in a low molecular weight SEC) has been reported in a recent study on the reversible addition fragmentation chain transfer (RAFT) polymerization of methyl acrylate mediated by cumyl dithiobenzoate [60]. Even though MALDI-TOF-MS offers a larger accessible mass range than for ESI-MS, ESI-MS allowed for polymer sample ionization with very limited fragmentation. The authors were able to identify in Reference [60] the products corresponding to the polymeric RAFT agent, as well as disproportionation and recombination termination products, although no evidence of products associated with termination of intermediate radical species such as 3- or 4-armed polymer stars was found.

Further investigations in the mechanistic details of the RAFT process, particularly the potential side reactions of the intermediate radical species, were made using the coupled SEC/ESI-MS to map the product spectrum of a series of acrylate free-radical polymerizations mediated via the RAFT [61]. The mass spectroscopic results were compared to modeling estimations made to predict the concentrations of termination products of the intermediate species in comparison to polymeric material generated by recombination of two propagating macroradicals.

A novel methodology based on SEC-ESI-MS was developed in the study reported by Schmidt et al. [62] for the qualitative and quantitative analysis of arsenic interactions with peptides and proteins. While the suitability of the new method has to be tested for the investigation of other protein–ligand bindings, the approach was successfully used by the authors to study reactions of phenylarsine oxide with cysteine residues of biomolecules of different mass (300–14,000 g mol^{-1}) and compared equilibrium constants calculated from SEC/ESI-MS with ESI-MS-binding experiments without SEC coupling.

10.3.1.4 High-Throughput Screening

Another area of ESI applicability is in the high-throughput screening of organometallic catalysts [63, 64]. ESI mass spectrometry was employed by Hinderling and Chen [63] to compute reaction rates during homogeneous Ziegler–Natta polymerization catalyzed by a Pd(II) diimine complex. The rates were used to calculate the average molecular weight and the extent of polymerization as a function of time.

10.3.2 Matrix-Assisted Laser Desorption Ionization

One of the advantages of MALDI [4] as a desorption/ionization method is that little or no fragmentation occurs during either desorption or ionization, and at least for lower molecular weight oligomers, only singly charged ions are observed. The development of MALDI allowed the detection of intact, soluble molecules of substantially higher masses ($> 1 \times 10^6$ g mole^{-1}), while minimizing fragmentation.

Additional advantages are offered by extending the applicability of the MALDI methods to insoluble polymers [65] and saturated polyolefins [66]. One of the key factors that influence the success and the quality of MALDI mass spectrometric analysis is the sample preparation. Thus, prerequisites for traditional sample preparation method require soluble analytes and matrices and the compatibility of the solvents for both matrix and analyte systems, subsequent solvent removal, and the homogeneity of the crystallized analyte/matrix mixture on the surface of the MALDI sample holder [67–71]. These factors can introduce severe mass discrimination effects. The advantages of solvent-free preparation in comparison to conventional solvent-based MALDI-MS were recently reported [65, 72]. The MALDI mass spectra obtained by the solvent-free sample preparation method were compared by Trimpin et al. [72] with mass spectra obtained by conventional solvent-based MALDI-MS and by laser desorption (LD)-MS under identical experimental conditions, stressing the capability of solvent-free MALDI-MS to characterize insoluble samples.

10.3.2.1 End-Groups. Determination of the Rate Coefficients

Although polymer properties are determined largely by the backbone structure, the end-groups influence significantly the bulk properties of the polymers. In terms of end-groups and fragmentation, additional information was brought by the use of MALDI in combination with other spectroscopic methods [73–75]. Methods like NMR, UV-Vis, and infrared (IR) spectroscopies are commonly employed to determine the mode of chain termination in free-radical oxidative polymerization. A combination of MALDI, Fourier transform infrared (FTIR) spectrometry, and thermochemical analysis was used by Nanda et al. [73] to accurately determine the end-groups of two polyperoxide polymers. The use of MALDI-TOF to determine the propagation rate coefficients in the case of MMA pulsed laser polymerization was reported by Kowollik et al. [74].

10.3.2.2 Determination of Molecular Weight and Molecular Weight Distribution

The ability to ionize a broad range of materials, high sensitivity, large mass range, fast sample preparation, and the absence of fragmentation make MALDI-TOF mass spectrometry an effective tool in polymer characterization. Recently, a major focus was on optimizing the ability of the technique to provide accurate molar mass measurements. Determination of molar mass using MALDI-MS has been reported in numerous studies on polymer characterization, despite the fact that accurate values were only obtained for relatively narrow polydisperse polymers [56, 76, 77]. Several problems were encountered with MALDI measurements on polydisperse samples, such as: the signal being spread over many molecular weight species giving rise to a low signal to noise ratio, the different chain lengths require different intensities of laser power, and the possibility of overlap of spectra of the dimers and trimers along with multicharged species produced by the ionization process.

10.3.2.3 Combined SEC-MALDI. Mark–Houwink–Sakurada (MHS) Parameters

As mentioned earlier, combination of the quantitatively reliable LC techniques with the identification power of MS was used to overcome the limitations inherent in the use of either LC or MS alone. In addition to ESI-MS/SEC, combined applications already mentioned, several approaches to SEC/MALDI time-of-flight (TOF) MS coupling have also been reported [78–83]. In the case of polydisperse polymers, determination of molecular mass by SEC/MALDI-TOF involves the fractionation of the samples through an analytical SEC. Selected fractions are then analyzed by MALDI-TOF and the mass spectra of these nearly monodisperse samples allow the determination of the MWD moments. Montaudo et al. [81] have tested the reliability of the method for different polydisperse samples, such as poly(methylmethacrylate) (PMMA), poly(dimethylsiloxane) (PDMS), and copolyesters.

While SEC-MALDI studies have previously been conducted to determine polymer MWD and its moments, an attractive approach, utilizing the MALDI technique along with viscosity measurements, to generate Mark–Houwink–Sakurada (MHS) parameters was recently reported by Tatro et al. [84]. The authors used MALDI in generating molecular weight spectra in order to determine MHS parameters for a series of PMMA and PS samples. A method to define the width of the spectrum by the polymer spread (PSp) without bias caused by the magnitude of the molecular weight of the polymer was developed for narrow polymer standards as a basis for evaluating the broadness of MWDs in MALDI samples.

10.3.2.4 *Copolymers. Determination of Molecular Weight and Composition*

In addition to the molecular information desired for polymers (e.g., molar mass distribution, repeat unit and end-group composition, branch points), additional information is required for a complete characterization in the case of copolymers (e.g., molar fraction of each type of repeat unit, the average length of chains of each polymer type, the variation of composition as a function of molecular weight).

Matrix-assisted laser desorption ionization MS is well represented in this area [56, 85–94]. Cox et al. [88] used MALDI-MS to characterize low molecular weight polyolefin copolymers of isobutylene and paramethylstyrene and to extract composition information from spectra. Comparison of experimental oligomer distributions to a Bernoullian statistical model revealed severe overrepresentation of oligomers with higher relative amounts of paramethylstyrene. However, good agreement between the model and experimental data was obtained by the introduction of an ionization efficiency term in the model used. The differences between average composition determined by MALDI and NMR were discussed.

Matrix-assisted laser desorption ionization analysis of amphiphilic copolymers still remains a challenge since each comonomer requires its own set of experimental conditions for sample preparation [90, 91]. To circumvent this issue, Girod et al. [91] proposed a method consisting of hydrolyzing a targeted function in the junction group between the two segments of a block copolymer to produce two homopolymers which could then be independently mass-characterized using conventional MALDI protocols.

However, such a strategy can be applied only for block copolymers. Molecular weight characterization of random amphiphilic copolymers currently poses its own analytical challenges. Giordanengo et al. [92] used a methylation reaction to transform methacrylic acid (MAA)/methyl methacrylate (MMA) copolymer samples that could not be properly ionized by MALDI into PMMA homopolymers. Weight average molecular weight (M_w) parameters of the copolymers were then derived from M_w values obtained for the methylated MAA-MMA molecules by MALDI, which were also validated by NMR.

MALDI-TOF imaging mass spectrometry (IMS) has become an important tool to monitor the 2D and 3D distribution of components in biological tissues [93], although had few applications for investigating synthetic polymers. A new approach using MALDI-TOF IMS as a new detector for polymer chromatography was presented by Weidner et al. [94] in a study in which the simultaneous detection of discrete copolymer series in achromatographic run was achieved. The individual retention behavior of single structural units of PEO/PPO copolymers and changes in the copolymer composition were monitored. This approach provides new insights in the chromatographic principles of copolymer separation.

10.3.2.5 *Monitoring Polymerization Reactions*

Reproducibility and controllability of a high-throughput approach for the rapid screening and optimization of polymerization reactions was investigated by Fijten et al. [95] during automated MALDI-TOF MS measurements. RAFT polymerization of MMA utilizing an automated robotic synthesizer was used as case study. The polymerization reactions were monitored by automated MALDI-TOF MS and SEC, MALDI samples being prepared using a multiple layer spotting technique.

For illustration purposes, an example of reaction monitoring in time by MALDI-TOF MS is given in Figure 10.2 [95], which shows spectra obtained from the automated spotting of the synthesized PMMA polymers at different reaction times. In the inset to figure, peak molecular weight (M_p) computed from the obtained MALDI-TOF MS spectra are plotted versus reaction time and illustrate the growth of the polymer chains during the reaction. Even though the ionization of higher molecular weight polymers proved to be difficult and not all molecular weights and MWD were accurately calculated by MALDI, both SEC and automated MALDI-TOF MS revealed the increase in molecular weight with reaction time in a highly reproducible and comparable way. In addition, MALDI-TOF MS end-group analysis confirmed the presence of the dormant polymer chains, which, together with the increase in molecular weight and the narrow polydispersity indices, proved the controlled character of the polymerization.

In another study [96], MALDI-TOF MS end-group characterization was carried out at different time points for the cationic polymerization of oxazoline monomers containing thioethers in order to verify a proposed mechanism of the chain transfer events occurring during the polymerization. The use of time-dependent MALDI-TOF mass spectral analysis provided strong evidence to support the prevalence of chain transfer via nucleophilic addition/elimination of the thioether functional group, confirming theoretical predictions.

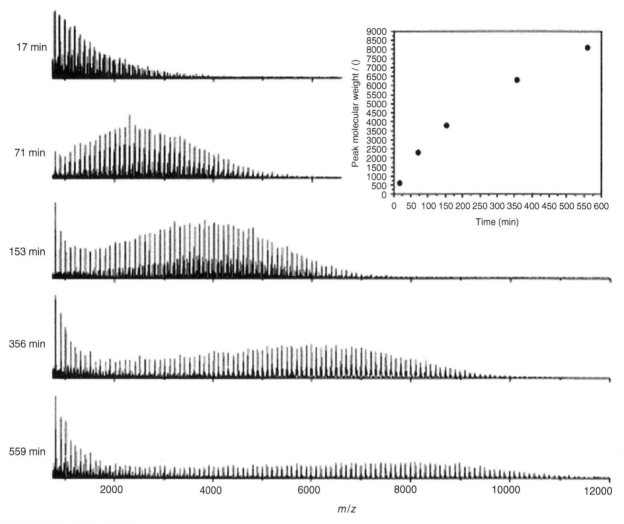

FIGURE 10.2   MALDI TOF mass spectra obtained from the automated spotting of the synthesized poly(methyl methacrylate) polymers at different reaction times. The insert (right top) shows the peak molecular weight (M_p) (obtained from MALDI TOFMS spectra) as function of the reaction time. Matrix: DCTB. This material is reproduced with permission of John Wiley & Sons, Inc. from Fijten MWM, Meier MAR, Hoogenboom R, Schubert US. Automated parallel investigations/optimizations of the reversible addition-fragmentation chain transfer polymerization of methyl methacrylate. J Polym Sci A Polym Chem 2004;42:5775–5783.

Monitoring and evaluation of the crossover reaction in ring-opening metathesis polymerization (ROMP) via MALDI methods were reported recently by Binder et al. [97]. The crossover reactions as well as the polymerization kinetics of the various monomers were studied in the presence of different catalysts by kinetic analysis and MALDI MS. Whereas the classical kinetic analysis enabled the qualitative monitoring of the chain-growth reaction, MALDI method allowed the monitoring of the reaction directly at the point of the crossover reaction, thus enabling a better evaluation of the different monomers and catalysts employed in the polymerization reactions.

An earlier work involved the use of MALDI to study the organolithium-initiated ring-opening polymerization

(ROP) of hexamethylcyclotrisiloxane in a mixed solvent system [98]. The authors' goals were to establish a quantitative way of representing the mass spectral peak populations in order to make empirical comparisons between spectra and to evaluate the changes in peak intensities as functions of reaction conditions, such as initiator concentration, polymerization time, and reaction temperature.

Another illustration of the MALDI-TOF MS applicability in monitoring polymerization reactions was presented by Xu et al. [99] in the case of the enzymatic polymerization of 4-phenylphenol. Horseradish peroxidase catalyzed polymerization reactions were carried out at room temperature in 50/50 acetone-0.01 M sodium phosphate buffer mixture. The effects of the oxidant

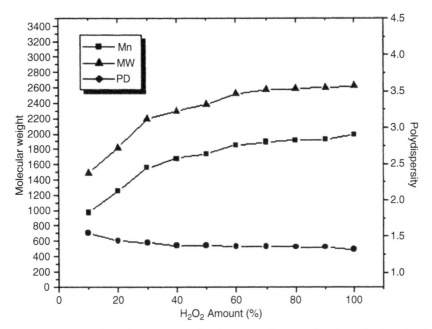

FIGURE 10.3 Molecular weight and polydispersity changes during the course of enzymatic polymerization of 4-phenylphenol as a function of the amount of H_2O_2 added. Reprinted (adapted) with permission form Xu P, Kumar J, Samuelson L, Cholli AL. Monitoring the enzymatic polymerization of 4-phenylphenol by matrix-assisted laser desorption ionization time-of-flight mass spectrometry: a novel approach. Biomacromolecules 2002;3:889–893. © 2002 American Chemical Society.

(H_2O_2) and the organic solvent on formation of dimer, trimer, and oligomers throughout the reaction were studied. A method was designed by the authors for successive sampling by MALDI-TOF MS during the reaction period due to the fact that real time *in situ* experiments were not suitable and additional sample handling was required. Variations in the molecular weight and the polydispersity during the course of the reaction are shown in Figure 10.3 [99] as functions of H_2O_2 added to the stoichiometric amount. As seen in the figure, the number average molecular weight, M_n, and weight average molecular weight, M_w increase quickly during the initial stage and continue up to 60% of the total amount of H_2O_2 when a plateau is reached. The trends in M_n and M_w of the polymer at the addition of H_2O_2 in the reaction were explained by the authors and the reaction conditions were optimized.

A final example of MALD-TOF MS applicability is in the study of alternating conjugated copolymers, whose design and synthesis allow novel means to design materials with functional, optical, and electrical properties. Alternating copolymers are generally prepared via palladium-catalyzed polycondensations by reacting either arylbis(boronic acid) derivatives (Suzuki method) or aryldistannyl derivatives (Stille method) with dihaloarenes. Mechanistic aspects of the Suzuki polycondensation of thiophenebisboronic derivatives were also studied using MALDI-TOF MS [100].

10.3.2.6 *Monitoring Postpolymerization Reactions* Oft
entimes, postpolymerization modifications are employed to optimize the performances of the polymeric materials synthesized by different polymerization methodologies. Head-to-tail coupled, regioregular poly-(alkylthiophenes) (HT-PATs) belong to an important class of conducting polymers, with well-defined primary structure and improved electronic and photonic properties. Most of the approaches made to tune their properties were focused on side-chain functionalization of the conjugated polymers. In a recent study, Liu et al. [101] used MALDI MS to monitor the Vilsmeier electrophilic reaction performed to introduce formaldehyde groups on each end of the polymer chain and compared MALDI with NMR performances in following each step of the end-group modification of HT-PATs.

10.3.3 Summary

The large number of reported studies indicates that mass spectrometry is growing in importance as a valuable technique for the analysis and characterization of synthetic polymers. While publications relating to synthetic polymer/copolymer characterization seem to favor MALDI over other mass spectrometry methods, ESI is also used consistently, in numerous studies, to follow polymer reaction kinetics and to study polymer chain growth on active catalyst substrates.

10.4 APPLICATIONS OF ESR SPECTROSCOPY TO POLYMER CHARACTERIZATION AND POLYMER REACTION MONITORING

Electron spin resonance spectroscopy or electron paramagnetic resonance (EPR) has become an important tool in understanding the kinetics and mechanism of radical polymerizations [102]. Recent advances in the ESR spectroscopic studies allowed detection of various radical species, such as primary radicals from initiators, primary propagating radicals, and propagating radicals. Propagating radicals from different types of monomers have been quantified during polymerizations in homogeneous systems. The absolute values of the rate coefficients of propagation and termination have been determined from ESR quantification of the propagating radicals. ESR techniques have been used to determine the rate coefficients from low to high conversion in bulk polymerizations of methyl methacrylate and styrene. The values of the propagation rate coefficients for copolymerization reactions have also been determined by ESR methods and the validity of the penultimate model has been supported by the dependence of the apparent values on comonomer composition. ESR spectroscopic studies of emulsion polymerization have also been employed to quantify the propagating radicals and to determine the corresponding rate coefficients.

The examples in the following highlight some of the most relevant applications of ESR spectroscopy in the study of radical polymerizations.

10.4.1 Reaction Monitoring. Determination of Rate Coefficients

Direct quantification of propagating radicals by ESR spectroscopy is known as one of the most promising techniques for the determination of the rate constants [103]. An example

is the study reported by Shen et al. [104] who have determined the rate coefficients of propagation, k_p, and termination, k_t, and their evolution with conversion for the polymerization of MMA in bulk. Changes in the monomer conversion and in the concentration of propagating radicals are shown in Figure 10.4 [104] for several MMA polymerization reactions, at different temperatures, monitored by ESR spectroscopy.

The k_p value was calculated directly from the simultaneously measured data of the concentration of radicals, [M·], and polymerization rate, R_p:

$$k_p = \frac{R_p}{[M^\cdot][M]} \tag{10.8}$$

and was found to remain constant up to ~70% conversion, followed by a fast decrease at higher conversion [104, 108]. This behavior was explained by the authors in terms of solidification of the polymerization mixture. k_t was determined under non-steady-state conditions by using the aftereffect technique in the ESR measurements.

The use of ESR for k_p determination for different monomers was reported in numerous studies [105–110]. Yamada et al. [107] have used ESR spectroscopy to study the dependence of the propagation and termination rate constants on conversion in the case of polymerization of styrene. Polymerization reactions in bulk were followed by ESR measurements over a wide range of conversion in order to determine k_p and k_t and their evolution with conversion. The concentration of the polystyrene radical, together with the polymerization rate, R_p, and the polymer molecular weight were determined throughout the polymerization. While in the case of the MMA polymerization, a change was observed in the evolution of k_p after 70% conversion, interpreted by the authors due to restricted diffusion of the polymer radical

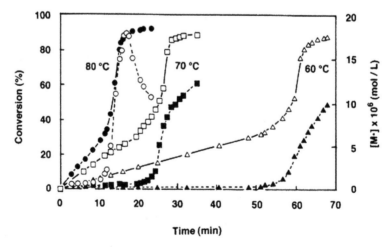

FIGURE 10.4 Changes in conversion ($\bigcirc$, $\square$, $\triangle$) and concentration of propagating radical ($\bullet$, $\blacksquare$, $\blacktriangle$) in bulk polymerization of MMA monitored by ESR spectroscopy at different temperatures: [MAIB]=0.109 mol dm^{-3}. From Shen J, Tian Y, Wang G, Yang M. Modelling and kinetic study on radical polymerization of methyl methacrylate in bulk. 1. Propagation and termination rate coefficients and initiation efficiency. Makromol Chem 1991;192:2669–2685. ©Wiley-VCH Verlag GmbH & Co. KGaA. Reproduced with permission.

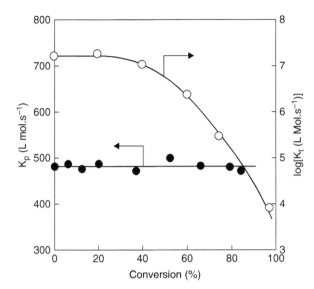

FIGURE 10.5 Decreases in k_p (●) and k_t (○) for the bulk polymerization of styrene with an increase in conversion. Reprinted (adapted) with permission from Yamada B, Kageoka M, Otsu T. Dependence of propagation and termination rate constants on conversion for the radical polymerization of styrene in bulk as studied by ESR spectroscopy. Macromolecules 1991;24:5234–5236. © 1991 American Chemical Society.

and monomer arising from lowering the T_g [111], in the case of styrene polymerization reaction studied by Yamada and coworkers, a constant k_p value was obtained at each stage during styrene polymerization. The k_t value was obtained based on the decay curve of the ESR spectrum in the presence of polystyrene.

Figure 10.5 [107] shows both k_p and k_t as functions of conversion. k_p is seen in the figure to be constant up to high conversion, signifying that styrene propagation is chemically controlled independent of conversion. A slow evolving k_t was found until ~20% conversion, followed by a steep decrease, corresponding to the increase in the viscosity of the polymerization mixture.

The properties of radicals in radical polymerizations, for example, effects of chain lengths, dynamics, and reactivity (hydrogen transfer) of propagating radicals, were investigated in a recent study by Kajiwara [112] during ESR measurements for a series of well-defined radicals generated from oligomers prepared by atom transfer radical polymerization (ATRP). Thus, radical precursors of oligo-and poly(meth)acrylates were prepared and model radicals with given chain lengths were generated by reaction with an organotin compound; the radicals were observed by ESR spectroscopy. The study allowed the estimation of the effects of chain length on propagating radicals for different monomers and, additionally, the detection of chain transfer reactions on propagating radicals in the polymerization of *tert*-butyl acrylate.

In another study, Mizuta et al. found k_p of methacrylic esters independent of chain length, in the polymerization of trans-4-*tert*-butylcyclohexyl methacrylate [113].

Electron spin resonance detection of the propagating and mid-chain radicals involved in the polymerization of phenyl acrylate (PhA) and the determination of the rate constants for this monomer using the ESR method were reported by Azukizawa et al. [114]. Absolute values of $k_p[M^\cdot]$ were determined at infinitely low conversion, to minimize the factors affecting the radical concentration.

10.4.2 Combined ESR-PLP Applications

Electron spin resonance is a mutually complementary method to the pulsed laser polymerization (PLP) in providing means to accurately obtain propagation rate coefficients. A study by Willemse et al. [115] reports the use of PLP experiments made in conjunction with ESR for detection of terminal and mid-chain radical concentrations. The fractions of the two types of free radicals occurring during PLP of *n*-butyl acrylate have been measured via ESR spectroscopy at temperatures between −50 and 70 °C. The authors found that the fraction of mid-chain radical species was negligible at the lowest experimental temperature but increased significantly with temperature. Thus, at room temperature, about 60% of the stationary radical concentration was given by mid-chain radical species. This finding is in agreement with other studies reported [116, 117] and could explain the limitations for PLP-SEC experiments in the case of *n*-butyl acrylate at temperatures of and above 30 °C [118] since the large fraction of mid-chain radicals distorts the narrow size distribution of propagating radicals and gives rise to broad structure-less SEC curves for the PLP-produced polymeric products. The fraction of mid-chain radicals is highly relevant for modeling overall propagation and also for understanding termination and transfer behavior in acrylate free-radical polymerizations.

Another example of the successful application of PLP in combination with ESR in the determination of the termination rate coefficient, k_t, can be found in the work of Barth et al. [119]. The techniques for determination of k_t, as well as the difficulties involved in deducing reliable termination rate coefficients have been recently reviewed [120, 121].

Methods based on the control of radical chain length by either using RAFT agents or by applying single pulse (SP)-PLP techniques were among the routes proposed for investigations into chain-length-dependent termination. Thus, single pulse-pulsed laser polymerization-electron spin resonance (SP-PLP-ESR) technique was used by Barth et al. [119] to study the chain-length dependence of k_t for bulk homopolymerizations of *n*-butyl methacrylate (*n*-BMA) and *tert*-butyl methacrylate (*t*-BMA) at temperatures between−30 and 60 °C. The decay of radical concentration, c_R, after laser single pulse initiation was monitored with a high time resolution of microseconds by ESR. Time-resolved measurements of c_R after laser single pulse initiation provide direct access to termination rate coefficients and should allow for model-independent determination. However, due to the fact that the signal-to-noise quality of SP-PLP-ESR was not sufficient to provide k_t from experimental $c_R(t)$ data, fitting of kinetic expressions to the measured c_R versus time data were used for adequate estimates of chain-length-dependent k_t [122]. The experimental results allowed the authors to confirm the validity of the so-called composite model for representing the decrease of k_t with radical chain length.

10.4.3 Copolymerization Kinetics

While quantification of propagation, k_p, and termination rate coefficients, k_t, in free-radical polymerization by means of ESR and PLP techniques has seen significant progress in recent years [123], there are fewer studies focusing on the middle and high conversion ranges, especially for cross-linking systems. Since the ESR method is based on direct quantification of the propagating radical concentration and may be employed over the entire conversion range, it has the additional advantage of being applicable to cross-linking systems. Thus, in the case of MMA copolymers and for various dimethacrylates it was found that higher cross-link density leads to a reduction in k_t, due to physical trapping in the polymer network [124] or "short-long" termination in an entangled system [125].

Another study on copolymerization monitored by ESR was reported by Zetterlund et al. [126], in which the bulk copolymerization of styrene (Sty) and divinylbenzene (DVB) at 70 °C has been followed by ESR and FT NIR measurements. Sty/DVB copolymers are widely employed for the preparation of cross-linked copolymer gels for applications such as SEC and ion exchange resins. The propagating radical concentrations, [P] versus conversion, obtained from ESR measurements for polymerization reactions with different DVB ratios are shown in Figure 10.6 [126]. As seen in the figure, [P] stays approximately constant up to about 20% conversion, after which it begins to increase. This trend was explained by the authors based on the dependence of k_t on conversion and the fact that a matrix with higher cross-link density (i.e., higher DVB content) would be expected to result in higher diffusion resistance for propagating radicals, therefore reducing k_t and causing the radical concentration to rise. A maximum in the radical concentration is seen in the figure at approximately 70% conversion, after which, it increased again as the reactions went to completion.

With the total concentrations of polymerizable double bonds and propagating radical species and R_p obtained as functions of conversion from ESR and FT-NIR data, k_p was estimated over the entire conversion range according to Equation 10.8 and was found to remain constant

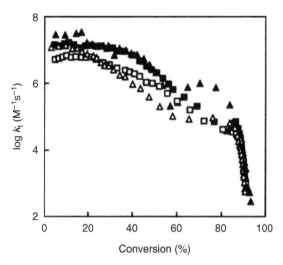

FIGURE 10.6 Free-radical concentration versus conversion for the styrene (Sty)/divinylbenzene (DVB) copolymerization initiated with 0.10 M MAIB, at 70 °C. $[DVB]_0 = 0.20$ ($\triangle$), 0.10 ($\square$), 0.05 ($\blacksquare$), and 0 ($\blacktriangle$) M. Reprinted from Zetterlund PB, Yamazoe H, Yamada B. Propagation and termination kinetics in high conversion free radical copolymerization of styrene/divinylbenzene investigated by electron spin resonance and Fourier-transform near-infrared spectroscopy. Polymer 2002;43:7027–7035. © 2002, with permission from Elsevier.

through the gel-point (which occurred at conversion lower than 50%) until ~ 80% conversion. The increase in k_p with decreasing DVB content at higher conversion was explained by the authors to be a reflection of higher rates of monomer diffusion in a more loosely cross-linked network. Under the experimental conditions of the study, k_t was computed based on:

$$k_t = \frac{2fk_d[I]_0 \exp(-k_d t)}{[M\cdot]^2} \tag{10.9}$$

where k_d is the rate constant for initiator decomposition, $[I]_0$ is the initial initiator concentration, and f is the initiator efficiency.

Computed k_t exhibited an initial plateau region, then decreased by approximately two orders of magnitude between approximately 20 and 60% conversion before reaching a plateau value in the conversion range 60–80%. The decrease was more significant and the onset of the decrease occurred at lower conversion, with increasing DVB content. A similar trend as in the case of k_p was observed at high conversion range (>80%): a decrease with increasing DVB content, consistent with reaction diffusion being the dominant termination mechanism.

Copolymerization kinetics has been investigated by numerous theoretical and experimental studies. Comonomer–copolymer composition relations and comonomer reactivity ratios have been determined, several models of copolymerization been proposed. An important role in the kinetic analysis of copolymerization was given to the quantification of propagating radicals.

The impact of the penultimate effect on the ESR spectra of the radicals and hence on the rate constants of propagation, k_p, and termination, k_t, was discussed by Sato et al. [127] in the case of the copolymerization of N-cyclohexylmaleimide and bis(2-ethylhexyl) itaconate at 50 °C, in benzene. Comparison of the ESR spectrum of the propagating polymer radicals with the corresponding homopolymers spectra in terms of any superposition of spectra revealed that penultimate monomeric unit led to a change in the ESR spectrum, hence in the structure of propagating polymer radical. The total concentration of propagating radicals was determined by a procedure similar to the one for quantification of propagating radicals of individual homopolymerizations and apparent k_p and k_t were estimated. Due to penultimate effect, higher k_p values (1.5–50 l mol^{-1} s^{-1}) than those estimated on the basis of the terminal model were obtained; k_t (1.8–5.4 × 10^3 l mol^{-1} s^{-1}) was found to reflect diffusion-control of termination reactions. Other studies on the evolution of the propagating radicals during copolymerization reactions were reported [128–132].

Gilbert et al. [133] employed EPR spectroscopy to monitor the initial stages in the free-radical copolymerization of carbohydrates and methacrylic acid, initiated by a metal–peroxide redox couple. The authors aimed to understand the behavior of mixed carbohydrate–monomer systems in the presence of radical initiators by gaining knowledge of the rate constants and their dependence on radical structure and different substitution patterns.

10.4.4 Controlled Radical Polymerization. Reaction Kinetics

Nitroxide-mediated polymerization (NMP) is one of the CRP methodologies to which ESR spectroscopy was successfully applied to study reaction mechanisms and kinetics [134–136]. Thus, MacLeod and coworkers [134] have monitored by ESR the free nitroxide level for styrene polymerization at 135 °C, initiated by TEMPO-BPO, with a large excess of initiator used relative to nitroxide (TEMPO/BPO=0.5:1).

The quantitative ESR measurements reported later by Bon et al. [135] proved to be a rather straightforward and accurate means to determine the rate coefficients for the homolytic alkoxyamine C–O bond cleavage over a range of temperatures for 2-tert-butoxy-1-phenyl-1-(1-oxy-2,2,6,6-tetramethylpiperidinyl)ethane and a polystyrene-TEMPO. The acquired kinetic data for alkoxyamines were compared by the authors with data obtained via nitroxide-exchange experiments using HPLC and with literature data [137].

Furthermore, EPR spectroscopy was used to investigate the chemistry of paramagnetic metal complexes. EPR can potentially yield information on the local structure, coordination structure, symmetry, concentration, and

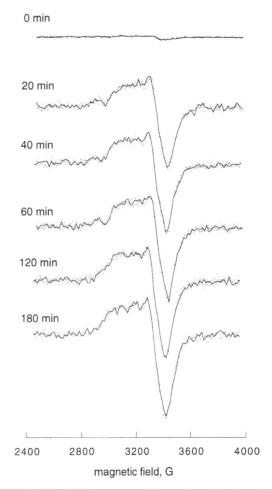

FIGURE 10.7 EPR spectra of the polymerization mixture measured at 25 °C after 0, 20, 40, 60, 120, and 180 min at 110 °C. Styrene/1-phenylethyl bromide/CuBr/dNbipy (=100/1/1/2) in toluene (50 vol%). Reprinted (adapted) with permission from Matyjaszewski K, Kajiwara A. EPR study of atom transfer radical polymerization (ATRP) of styrene. Macromolecules 1998;31: 548–550. © 1998 American Chemical Society.

aggregated structure of paramagnetic copper(II) species [138]. The fact that initiating and propagating radicals, as well as copper(II) species are paramagnetic led to the application of EPR spectroscopy to ATRP in order to investigate the mechanism of the controlled radical polymerization system.

Thus, direct determination by EPR of copper(II) species was reported by Matyjaszewski and coworkers [139] in the case of styrene ATRP. The polymerization proceeds by monomer addition to free radicals reversibly generated by an atom transfer process from dormant polymer chains with halide end-groups. In these reactions, a small amount of copper(II) species was used as a deactivator which moderates rates and keeps low polydispersity. An example of time-dependent EPR signals of copper species in the ATRP of styrene in toluene, initiated by 1-phenylethyl bromide (styrene/1-phenylethyl bromide/CuBr/dNbipy=100/1/1/2) at 110 °C is shown in Figure 10.7 [139].

Preliminary studies of the structure of copper(II) species in the presence of copper(I) indicated interactions between them, possibly via halogen bridging. The concentration of copper(II) species was estimated by double integration of these signals. Even though, in principle, all of these species could be observed by EPR spectroscopy, in the case studied, only the paramagnetic copper(II) species were observed due to the high concentration relative to the organic radicals. The computed time-dependent concentrations of copper(II) species for ATRP initiated by 1-phenylethyl bromide and benzyl bromide in toluene solution and bulk polymerization initiated by 1-phenylethyl bromide were compared, showing that benzyl bromide is a slower initiator. A system with copper(II) added before polymerization (10 mol% $CuBr_2$ relative to CuBr) was also examined.

10.4.5 Emulsion Polymerization

Recently, ESR techniques have been applied to study polymerization reactions in heterogeneous phase, for example, emulsion polymerization. The development of the direct measurements of radical concentration by ESR represents a major advance in obtaining reliable data on important parameters in emulsion copolymerization modeling, such as propagation rate coefficients or termination rate coefficients as functions of chain length and conversion [140, 141].

Ballard and coworkers [142] are among the first groups reporting direct observation of propagating free radicals in an emulsion polymerization. ESR was used in the identification and measurement of the concentration of propagating free radical in the emulsion polymerization of MMA. The ESR analysis was extended by Lau et al. [143] to semicontinuous emulsion polymerization of acrylic acid. The evolution of the propagating free-radical species as function of temperature and particle size was investigated. k_p values were calculated from the experimental feed rate, and the observed steady-state monomer and radical concentration for polymerization reactions in which 50 and 500 nm particle diameter, respectively, were produced. Linear Arrhenius plot of propagation rate constants was observed for the 50 nm particle size polymerization but not for the higher particle size polymerization, the latter case attributed by the authors to inhomogeneous polymerization.

Latex particle size is a key parameter for emulsion polymerization, with strong impact on the final properties of the polymers synthesized and, in general, on the polymerization kinetics in heterogeneous systems. An experimental methodology which combines a continuous-flow system and an ESR time sweep experiment was extended by Parker et al. [144] to study the effects of varying latex particle size on polymerization kinetics and radical distribution in the case of MMA emulsion

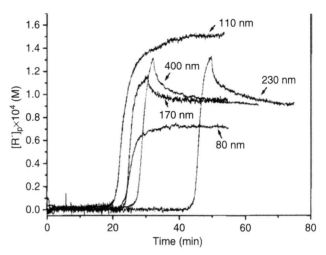

FIGURE 10.8 $[R]_p$ versus time for various particle size polymerizations. Reprinted (adapted) with permission from Parker H-Y, Westmoreland DG, Chang H-R. ESR study of MMA batch emulsion polymerization in real time: effects of particle size. Macromolecules 1996;29:5119–5127. © 1996 American Chemical Society.

polymerization initiated by a redox initiator system, at 45 °C. The system allowed continuous monitoring of the propagating free-radical concentration during the reaction. The authors aimed to investigate how the particle size influences the development of particle inhomogeneity during emulsion polymerization, which was found to have a significant impact on the evolution of several kinetic parameters, such as rate coefficients of propagation and termination.

Figure 10.8 [144] shows PMMA propagating radical concentration, $[R]_p$, as a function of the reaction time for various particle sizes.

As seen in the figure, similar $[R]_p$ trends were observed for the 80 and 110 nm particle size polymerizations: an increase after 20 min (40% conversion), with a higher rate as the reactions approach completion (80–90% conversion), followed by a plateau at maximum value within the 1 h observation time. A different $[R]_p$ behavior was observed for larger particle sizes (170, 230, and 400 nm). Although the maximum $[R]_p$ is similar as in the smaller particle size cases, a rapid decrease in the propagating radical concentration is seen shortly after the maximum $[R]_p$ value was reached, followed by a second region of more gradual decrease. The authors suggested that the decrease in the radical signal is a result of termination and explored possible factors that caused it. Thus, inhomogeneous distribution of propagating radicals in the large particles and negligible inhomogeneity in the small particles were offered as possible causes of the differences in various particle size polymerizations in radical termination. Radical termination rates would be accelerated in the larger particle size polymerizations due to higher local $[R]_p$ caused by the higher inhomogeneity of the propagating radical distribution. Inhomogeneous radical distribution has been discussed in other studies on emulsion polymerization systems [141–143].

Results of the studies reported by Ballard and Lau and their coworkers ([142, 143], discussed in the preceding text) confirmed the possibility that occluded radicals observed by ESR during the MMA emulsion polymerization could influence the termination step. At high conversion, some radical centers could become so occluded within dead polymer molecules that they would be protected from termination and lead to a high residual radical content within the latex. A "residual termination" would occur later.

These observations were extended by Cutting et al. [145], who used the cryogenic quenching procedure developed by Ballard and coworkers to record ESR spectra periodically during the course of MMA emulsion polymerization in order to monitor occluded radical concentrations, aiming to provide better insight into radical termination at very high conversions. Quantitative ESR measurements have enabled values for the polymeric radical–radical termination rate constant, k_t, to be determined. The effects of the particles size and temperature on the computed k_t values were studied. In agreement with previous studies, it was found that k_t increased as particle size decreased, possibly as a consequence of a higher rate of diffusion in the smaller particles. In terms of temperature effects, as the reaction temperature was decreased, the lower rate of initiator decomposition led to larger particle size and lower k_t values, thus leading to an increase in the occluded radical population.

Electron spin resonance spectroscopy has been employed, together with the cryogenic quenching technique, by the same group [146] to monitor occluded radical concentrations during methyl methacrylate (MMA)/butyl acrylate (BA) core-shell polymerization reactions.

10.4.6 Summary

The advances in the molecular design of new polymeric materials with targeted properties require advanced molecular characterization of the polymers. ESR techniques are among the methods under continuous development in the quest for more comprehensive physical and chemical information that could correlate microscopic properties with materials performance. ESR spectroscopy has been used in various areas of polymer science, with different goals, such as to study mechanisms of chemical reactions in polymerization and radiation effects, to identify intermediate species, to observe decay and conversion of different species, or to investigate relaxation phenomena of polymer chains by observing temperature-dependent ESR spectra of radical species trapped in solid and liquid polymers.

10.5 APPLICATIONS OF NMR SPECTROSCOPY TO POLYMER CHARACTERIZATION AND POLYMER REACTION MONITORING

With a high spectral selectivity, NMR spectroscopy offers comprehensive information about chemical composition, end-group structures, branching, comonomer composition, and sequence.

One of the main uses of NMR in polymer characterization is to correlate the structure with material properties. Even though there are various isotopes for which NMR spectroscopy can be used to yield important structural information, the most common are ^1H NMR and ^{13}C NMR. While they are both based on same principles, ^1H NMR has an enhanced spectral sensitivity due to the higher gyromagnetic ratio of hydrogen relative to ^{13}C. On the other hand, the higher spectral selectivity due to the larger dispersion of the chemical shifts confers ^{13}C NMR more detailed structural characterization.

The resolution of the available methods depends on several factors, such as the type of nuclei studied, chain dynamics, sample concentration, solvent, and temperature. Another important variable, linewidth, is related to the spin–spin relaxation time and hence, to the correlation time. Thus, large rigid molecules are expected to have broad lines. However, the flexibility of the polymer chains in solution allows high-resolution structures to be observed. Especially for the side chains, less constrained than the main ones, sharper resonances are obtained.

Online NMR spectroscopy is a powerful technique for reaction and process monitoring, being used to obtain both qualitative and quantitative information from complex reacting multicomponent mixtures [147–150]. Flow NMR spectroscopy was used to investigate reaction kinetics almost in real time under a wide range of conditions [17, 151, 152] and in high throughput screening [153]. Low field online NMR spectroscopy [154] was also applied in the reaction monitoring. Detailed overviews of NMR applications to polymer characterization are available [14–16, 155, 156].

Typical Online Setup for NMR Studies
Different setups for direct coupling of reaction and separation equipment with online NMR spectroscopy are described in a study by Maiwald et al. [17], in which the following applications of the high-resolution online NMR spectroscopy were presented:

- Formaldehyde-containing systems—due to the chemical similarity of the functional groups of individual oligomer species, formaldehyde-containing systems were chosen as a well-suited example to show the benefits of the high spectral dispersion of NMR spectroscopy.
- Rapid evaporation of formaldehyde solutions—coupling the evaporator to the NMR spectrometer to monitor the species distribution in the highly concentrated products under various process conditions.

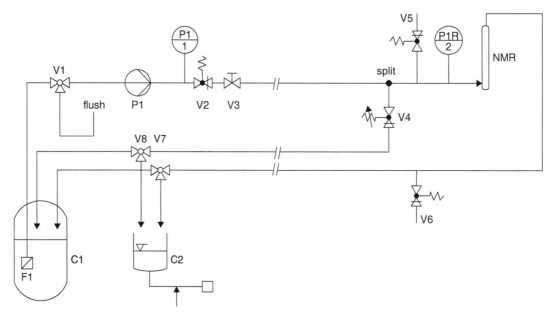

FIGURE 10.9 Typical setup for online NMR measurements. C1, laboratory reactor; F1, inlet filter; V1, (tee) purging valve; P1, thermostated dosing pump; V2, back pressure regulator; V3, shut-off valve; V4, variable back pressure regulator for split adjustment; V5, V6, pressure relief valves; PI1, PIR2, pressure transducer; NMR, thermostated flow probe of NMR spectrometer; V7, V8, tee valves, C2 container on balance for mass flow control. All tubing 1 = 1600 OD. Reprinted from Maiwald M, Fischer HH, Kim Y-K, Albert K, Hasse H. Quantitative high-resolution on-line NMR spectroscopy in reaction and process monitoring. J Magn Reson 2004;166:135–146. © 2004, with permission from Elsevier.

- Homogeneously and heterogeneously catalyzed esterification kinetics: butyl acetate esterification—online ^{1}H NMR spectroscopy monitoring of rapidly changing compositions offered a fast and reliable analysis of all reaction components.

A typical online setup for NMR studies of reacting systems is shown in Figure 10.9 [17]. A dosing pump P1 (0.1–20 ml min[−1]) was used to transport the sample from the reactor C1 to the NMR probe. Before entering the NMR spectrometer, the flow was split, allowing a quantitative flow rate in the NMR probe (0.1–1.5 ml min[−1]) while the flow rate in the transfer line was high enough to allow a rapid sample transfer.

The authors have shown, in the applications discussed in Reference [17], that online NMR spectroscopy is an important technique for physicochemical and process engineering studies, yielding reliable quantitative information on mixtures in a fast noninvasive way.

A few selected examples of the application of NMR spectroscopy to polymer analysis are presented next, to illustrate the broad applicability and diversity of approaches.

10.5.1 Living Polymerization: End-Group Structure, Reaction Kinetics, and Mechanism

The chemical nature of the polymer end-groups is determined by the initiation and termination steps and therefore offers a history of the polymerization reaction. High-field NMR spectroscopy allows the precise analysis of polymer chain ends. Better results are obtained for lower molecular weight polymers as the concentration of end-groups is higher. ^{1}H NMR measurements of the chain-end functionality of polymers prepared by CRP have already been reported [18, 157–159].

Lutz and Matyjaszewski [18] have followed the evolution of the bromine end functionality during the bulk ATRP of styrene, in the presence of the CuBr/4,4′-di-(5-nonyl)-2,2′-bipyridine catalyst. The retention of the bromide chain-end functionality was monitored through the withdrawal of aliquots at given times from the polymerization mixture and their analysis by ^{1}H NMR (600 MHz). A decrease in the functionality with conversion was observed, significant at high monomer conversion (90%). The experimental data allowed, by comparison with a kinetic model of styrene ATRP, better understanding of the side reactions that led to the loss in catalyst functionality and helped in the design of the most suitable reaction conditions in order to optimize the reaction kinetics and end-product properties.

In the context of living polymerization reactions monitored by NMR, ^{1}H NMR spectroscopy was also employed in following reaction kinetics in the case of the (ROP). Thus, a study of Li et al. [160] investigated reaction mechanism for ROP of L-lactide (LLA) in bulk and toluene, respectively, with sodium bis(2-methoxyethoxy)aluminum hydride (Red-Al) as the catalyst. The mechanism of the ROP was postulated to follow the coordination type, based on the analysis of ^{1}H NMR spectral data of the polymers formed at different reaction times. Quantitative estimations of the degree of transesterification and stereoselectivity were also made by means of ^{13}C NMR characterization.

In another recent study on the synthesis of semiconductor polymers by a combination of CRP methods, Lang et al. [161] reported the use of ^{1}H NMR and SEC to monitor the kinetics of NMP of trimethylsilyl propargyl acrylate and investigating side reactions leading to increase in polydispersity. Subsequent "click" reactions with an azide functionalized perylene bisimide derivative overcome these limitations, as proven by ^{1}H NMR and FTIR analysis.

10.5.2 Branching Detection and Characterization

A quantitative characterization of branching by the means of ^{13}C NMR was reported by Spěváček [162] in an early study in the case of high density polyethylene (HDPE) and ethylene copolymers of high density prepared by different methods. Correlations between the type of branches and the crystallinity of polyethylene were made.

Polymerizations of multifunctional monomers of suitable unequal reactivity have been proved to be feasible for preparing hyperbranched polymers. Wu et al. [163] have reported *in situ* ^{13}C NMR monitoring of the synthesis of hyperbranched poly(aminoester)s prepared via the Michael addition polymerization of a triacrylate, trimethylolpropane triacrylate (TMPTA) (A_3-type monomer) with a double molar 1-(2-aminoethyl)piperazine (AEPZ) (BB$'$B$''$-type monomer) in chloroform at room temperature.

A more recent study [164] reports the use of multidimensional solution NMR (^{19}F, ^{1}H, and ^{13}C) to determine chain-ends and backbone branching points in poly(vinylidene fluoride) (PVDF). The multidimensional NMR methods employed in the study enabled the resonance assignments of the last monomer of the chain and provided assignments for the last three monomer units of chain-end structures. 2D-NMR analysis was used to assign resonances of chain branching points along the backbone of the polymer. The results of the study confirmed previous findings reported using 1D-NMR and empirical chemical shift calculations and provided assignment of resonances from chain-ends and chain branching structures which could help in understanding the polymerization mechanism.

^{13}C-NMR is commonly employed in the detection and quantification of branching, with solid NMR offering highest resolution [165–167]. Furthermore, in a recent work, Hlalele and Klumperman [168] have used ^{1}H NMR to detect the secondary and tertiary nitroxide-capped species in the NMP of *n*-butyl acrylate.

10.5.3 Cross-Linking Polymers and Polymeric Gels

Because of their chemical and physical properties that allow their use in multiple applications, cross-linking polymers and polymeric gels prepared by different methods have been used extensively.

Nuclear magnetic resonance spectroscopy and imaging procedures were employed by Ahuja et al. [169] during polymerization of methacrylamide with N,N$'$-methylenebisacrylamide as cross-linking agent, in water at 25 °C and 35 °C, to determine the variation in the spin-lattice relaxation time and whether the polymerization is homogeneous. The authors aimed to assess the utility of NMR imaging for a nondestructive evaluation of voids and flaws in material components and for correlation of the physical properties, such as the degree of polymerization, viscosity, and strength with measurable NMR parameters, for example, the spin–lattice relaxation time and the spin–spin relaxation time.

Cryogelation is another method used to synthesize hydrophilic macroporous materials with interconnected pores, with potential bioapplicability. In this context, NMR seems to be an ideal technique to study the behavior of water and the solutes in the nonfrozen water phase. Thus, the cross-linking polymerization of dimethylacrylamide (DMAAm) with PEG diacrylate as the cross-linker in a semi-frozen aqueous solution was studied by Kirsebom et al. [170] with the use of ^{1}H NMR to monitor the amount of nonfrozen phase and the progress of the polymerization reaction *in situ* in samples frozen at −12 °C directly in the spectrometer.

10.5.4 Copolymerization Mechanism: Composition, Reactivity Ratios

Online ^{1}H NMR measurements were reported by Abdollahi et al. [171] in a recent study of the free-radical copolymerization of vinyl acetate (VA) and Me acrylate (MA) in benzene-d$_6$ at 60 °C, with benzoyl peroxide (BPO) as initiator. A significant composition drift in the comonomer mixture was observed as the reaction progressed. The comonomer reactivity ratios were calculated using the data collected only from one sample via online monitoring of the comonomer mixture and copolymer compositions at different reaction time intervals up to medium overall monomer conversions. The results were in good agreement with the literature data reported for this system, indicating the accuracy of the monomer reactivity ratios calculated by the procedure developed in the study.

A considerable interest has been shown lately to the synthesis and characterization of amphiphilic copolymers, random, star, graft, or block copolymers. Choosing the right components and conditions required to achieve self-assembled morphologies with well-defined structural order and desired dimensions allows to design materials with specific properties.

Investigations of the polymerization kinetics of 2-dimethylaminoethyl methacrylate (DEAEMA) copper mediated living radical polymerization were reported by Even et al. [172] using *in situ* ^{1}H NMR spectroscopy. Polymerizations reactions were carried out at 60° and 90 °C using a poly(ethylene butylenes)-based di-initiator as a macroinitiator in NMR

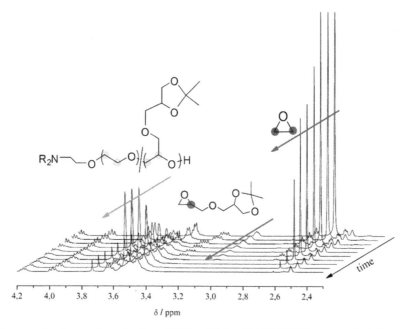

FIGURE 10.10 ^{1}H NMR spectra for the copolymerization of IGG and EO at 50 °C in DMSO-d$_6$ recorded after 0, 0.1, 2, 13, 21, 26, 87, 115, 190, and 264 min. Reprinted (adapted) with permission from Mangold C, Wurm F, Obermeier B, Frey H. "Functional poly(ethylene glycol)": PEG-based random copolymers with 1,2-diol side chains and terminal amino functionality. Macromolecules 2010;43:8511–8518. © 2010 American Chemical Society.

tubes. ^{1}H NMR spectral analysis allowed monomer conversion to be followed by integration of the vinyl peaks (6–5 ppm) relative to combined values of the CH$_2$ α to OC=O from the monomer and polymer (4.1 ppm). Data analysis led to first order kinetic plots, ln[M$_0$]/[M] versus time, allowed more data points to be obtained than from a sampled reaction, and avoided the possibility of contamination during the sampling procedure.

Mangold et al. [173] used NMR spectroscopy to monitor copolymerization kinetics during the synthesis of poly(ethylene glycol-co-isopropylidene glyceryl glycidyl ether), (PEO-co-IGG) copolymers. The random incorporation of the comonomers was verified during anionic ring-opening copolymerizations of the protected 1,2-isopropylidene glyceryl glycidyl ether (IGG) comonomer with ethylene oxide (EO) via real-time ^{1}H NMR spectroscopy and by characterization of the triad sequence distribution, relying on ^{13}C NMR analysis.

Shown in Figure 10.10 are ^{1}H NMR spectra recorded at different intervals of time during the ICG/ EO copolymerization at 50 °C in DMSO-d$_6$. Incorporation of the two comonomers and the growth of the polyether chain were studied by following the decrease of the epoxide signals located at 2.61 ppm for the four protons of the symmetric EO monomer and at 3.09 ppm for the methine proton of IGG, respectively. The spectra in the figure show the decreasing intensity of the monomer signals and the growing backbone signal.

The NMR experiments were carried out not only to verify the reaction kinetics but also to estimate the reactivity of the two comonomers, by studying the copolymer composition. The most relevant information obtained by the authors from the NMR spectra is that the comonomers were incorporated equally into the copolymer chains, a random distribution being obtained, independent of the EO/IGG ratio and for all temperatures studied. ^{13}C NMR analysis allowed to investigate the monomer triad sequence distribution and supported the authors' conclusions on the random incorporation of the comonomers.

Nuclear magnetic resonance monitoring of the synthesis of amphiphilic copolymers has also been reported by Larazz et al. [174] for the copolymerization of a methacrylic macromonomer with amphiphilic character derived from Triton X-100 (MT) with acrylic acid (AA). *In situ* ^{1}H NMR analysis was used to monitor comonomer consumption throughout the copolymerization reactions, initiated by AIBN in deuterated dioxane, at 60 °C. The results from two different approaches used by the authors to estimate the reactivity ratio of the macromonomer indicate that AA is less reactive than the macromonomer MT and a model monomer with lower molecular weight but same structure, suggesting that methacrylic double bond reactivity was not affected by poly(oxyethylene oxide) chain length.

Increased attention was shown recently to fluorinated polymers, due their unique high thermal and chemical stability, which makes them suitable in numerous applications. All these applications, however, require the availability of fluorine-based (co)polymers with well-defined structures and functionality.

^{1}H NMR spectroscopy was used by Grignard et al. [175] to prove the controlled features of the RAFT (co)polymerization of 1H,1H,2H,2H-heptadecafluorodecyl acrylate (AC8) with 2-hydroxyethyl acrylate (HEA). Monomer conversion, polymer molecular weight, and the copolymer composition (in the case of the random copolymers) versus time were determined by ^{1}H NMR analysis of the polymerization medium. The livingness of the PAC8 chains was confirmed by the synthesis of PAC8-b-PHEA diblock copolymers.

10.5.5 Coupled HPLC/GPC-NMR

Coupled HPLC-NMR measurements performed at slow flow rates in fully deuterated solvents and at room temperature have been made in several studies to determine polymer MWD, to analyze the end-groups and the copolymer chemical composition distribution, and to assess the chemical structure and the degree of polymerization of all oligomer species [176–178]. Gradient HPLC-NMR was used in the analysis of the chemical composition distribution of random poly(styrene-co-ethyl acrylate) copolymers [179]. A major drawback in most of these studies is that the measurements could only be conducted at ambient or slightly elevated temperatures, which limits the method applicability, since many polymers, such as polyethylene, polypropylene, and polyolefin copolymers are soluble at high temperatures.

The online coupling of GPC and ^{1}H NMR operating at temperatures up to 130° was recently reported by Hiller et al. [180]. A new high temperature NMR flow probe allowing flow experiments in the desired temperature range was developed by the authors. In addition, a novel interface, to be coupled to the NMR and GPC systems and enable on-flow and stop-flow experiments at high temperature conditions, was developed. The capabilities of the novel high temperature GPC-NMR system were tested in the analysis of several blends of polyethylene, poly(methylmethacrylate), and ethylene-methyl methacrylate and their separation according to the molar masses of the components. Moreover, the chemical composition of copolymers of ethylene and MMA was investigated. Differences between results of on-flow and stop-flow measurements were discussed. The setup of the high temperature GPC-NMR presented in Reference [180] opens new possibilities for the analysis of (co)polymers and polymer blends, especially polyolefins.

10.5.6 Emulsion Polymerization

Polymer synthesis in a heterogeneous environment is a complex process, depending on multiple reaction variables. The corresponding reactor models are often complicated and require numerous parameters and calibration constants. Time-resolved measurements of monomer, particle, and radical concentrations are needed to validate theoretical models and to provide experimentally determined coefficients used in the models.

The advantages of the online monitoring of chemical and physical properties during polymerization reactions have been discussed in previous chapters and several methods have been reviewed. Existing online methods used to achieve online control of conversion, molar mass, composition, or particle size rely on the accuracy of the calibration and on the many parameters involved in the models used, which are affected by the environment. Among the noninvasive techniques, calorimetry and spectroscopic techniques, such as ESR, UV/Vis, Raman, and NIR spectroscopy, are widely used in monitoring monomer/polymer properties, each with its own challenges. ACOMP for the simultaneous monitoring of the evolution of colloids polymer/monomer characteristics during emulsion polymerization was recently reported as a versatile characterization tool that offers absolute, model-independent determination of reaction features (see Chapter 12).

Due to the high selectivity to details of chemical structure and molecular dynamics, NMR spectroscopy is a powerful method to provide relevant information about morphological and dynamic properties of the polymer [172, 181–183].

The successful applicability of NMR in monitoring emulsion polymerization kinetics was recently reported by Vargas et al. [25], who followed the reaction progress by low-field ^{1}H NMR spectroscopy. The emulsion polymerization of butyl acrylate (BA) in seeded batch polymerization reactions with water and D$_2$O, respectively, as continuous phase was studied by the authors by using a low-field NMR (20 MHz) spectrometer and a flow system which pumped the reaction liquid in a closed loop bypass directly through the NMR spectrometer and back into the reactor. The polymerization rate and the rate coefficients were determined by following online the decreasing signal of the olefinic double bond in BA, as well as the growing signal of the newly formed saturated polymer main chain, in addition to the polymer chain dynamics (line width analysis). Typical emulsion polymerization three-phase conversion trends were observed for both reactions. Zero-one and pseudo-bulk kinetics were used to interpret and explain experimental data for the rate parameters, found in agreement with model predictions and existent literature data, within the conditions studied.

While ^{1}H NMR and ^{13}C NMR spectroscopy were proved to be useful tools for the online monitoring of polymerization reactions, their limitations, such as the intrinsic low sensitivity of the measurements and the time-consuming signal averaging, the use of expensive deuterated solvents (^{1}H NMR) that limits their application in industrial processes, led to the development of the so-called hyperpolarization methods which overcome the lack of sensitivity of NMR spectroscopy, allowing for a better time resolution in dynamic measurements [184–186].

Thus, Duewel et al. [186] have reported the real-time monitoring of styrene polymerization in miniemulsion using a continuous flow of hyperpolarized ¹²⁹Xe through the reaction mixture. The hyperpolarization of the ¹²⁹Xe via spin-exchange optical pumping allowed the signals to be enhanced and hence the measurements rate and resolution to be increased.

Figure 10.11 [186] shows a series of ¹²⁹Xe NMR spectra recorded during a miniemulsion polymerization of styrene reaction at 100C, initiated with 2,2′-azobis(4-methoxy-2,4-dimethylvaleronitrile (V70). The shift of the ¹²⁹Xe peaks to higher chemical shift values seen during the polymerization reaction allowed the reaction progress to be directly observed from the evolution of the ¹²⁹Xe NMR signal.

The results, in terms of conversion rates, were successfully validated by comparison with calorimetry data proving that hyperpolarized ¹²⁹Xe NMR can be a more suitable method to investigate polymerizations exhibiting complicated ¹H and ¹³C NMR spectra, due to the simplicity of the ¹²⁹Xe spectra and the simple relationship of the ¹²⁹Xe chemical shift with the composition of the reaction mixture.

10.5.7 Solid-State NMR

Solid-state NMR spectroscopy represents a powerful tool for identification and quantification of different chemical species. Techniques such as magic angle spinning (MAS), high-power proton decoupling, and cross polarization (CP) have made possible, in the last half century, to routinely measure high-resolution NMR spectra of solid samples [187–189]. Temperature-resolved experiments have been carried out previously on solid-state phase transformations [190] and solid-state reactions [191, 192], demonstrating that detailed information can be derived from these types of experiments. High-resolution solid-state NMR studies are particularly sensitive to the chemical and local structural changes associated with solid state reactions, and thus represent a vital addition to the range of *in situ* probes used to characterize the solid-state reactions in alkali halogenoacetates.

Aliev et al. reported the use of isothermal *in situ* solid-state NMR spectroscopy at 120, 130, and 140 °C to follow the synthesis of poly(hydroxyacetic acid) (polyglycolide) through the thermally induced solid-state polymerization reactions in sodium chloroacetate and sodium bromoacetate. Solid-state ²³Na-NMR spectra recorded at 130 °C (in sodium bromoacetate) and 140 °C (in sodium chloroacetate) were used to probe the sodium halogenoacetate parent phase and the sodium halide (NaCl, NaBr) formed in the reactions. The experiments tracked the sodium ions as

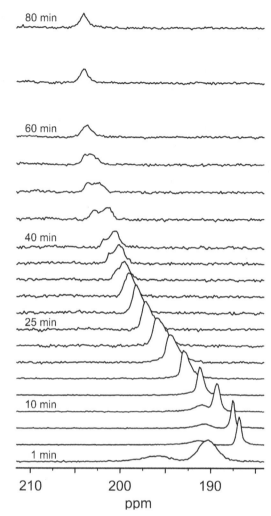

FIGURE 10.11   Time series of 129Xe NMR spectra recorded during a miniemulsion polymerization reaction at 343 K (initiated with V70). The plot shows the chemical shift range of the dissolved Xe. The insertion of the sample tube into the heated magnet was used as the starting time $t=0$ min. The spectra depict a strong dependence of the Xe chemical shift on the progress of the polymerization. Reprinted (adapted) with permission from Duewel M, Vogel N, Weiss CK, Landfester K, Spiess H-W, Münnemann K. Online monitoring of styrene polymerization in miniemulsion by hyperpolarized ¹²⁹Xenon NMR spectroscopy. Macromolecules 2012;45:1839–1846. © 2012 American Chemical Society.

they change their local environment from halogenoacetate anions to halide anions. To assess the polymerization reaction more directly, the authors have carried out *in situ* ¹³C-NMR studies of the polyglycolide reaction in sodium bromoacetate, concluding that the reaction either occurred in a single step or involved intermediates below the limits of detection. The mechanistic and kinetic information obtained from the *in situ* solid-state NMR studies was compared with data from other *in situ* probes of the polymerization reactions in these materials.

10.5.8 Summary

The improvements in technology in the last decade allowed NMR spectroscopy to emerge as one of the most important methods for polymer characterization. High resolution solution NMR, solid-state NMR, and the introduction of multidimensional NMR offer more detailed information, at higher resolution and in diverse conditions, on the polymer structure and polymerization mechanisms, chain conformation, and molecular dynamics of the polymers, blends, and multiphase polymer systems.

10.6 CONCLUSIONS

From theoretical and fundamental aspects to recent advances and novel developments in characterization and analysis of polymers, spectroscopy has, over the years, proved itself to be the most popular family of techniques in providing information at molecular levels. MS, ESR, and NMR applications highlighted in this chapter belong to the numerous approaches in the spectroscopic characterization and analysis of polymer systems.

REFERENCES

[1] Yadav LDS. *Organic Spectroscopy*. New Delhi: Anamaya Publishers; copublished by Boston: Kluwer Academic Publishers; 2005.

[2] de Hoffmann E, Charette J, Stroobant V. *Mass Spectrometry: Principles and Applications*. Chichester: Wiley; 1996.

[3] Montaudo G, Lattimer RP, editors. *Mass Spectroscopy of Polymers*. New York: CRC Press; 2001.

[4] Liang L, editor. *MALDI Mass Spectrometry for Synthetic Polymer Analysis*. Hoboken: John Wiley & Sons; 2010.

[5] Harrison AG. *Chemical Ionization Mass Spectroscopy*. 2nd ed. Boca Raton: CRC Press; 1992.

[6] McLafferty FW, Tureček F. *Interpretation of Mass Spectra*. 4th ed. Mill Valley: University Science Books; 1993.

[7] Chapman JR. *Practical Organic Mass Spectroscopy*. 2nd ed. Boca Raton: CRC Press; 1992.

[8] Johnstone RAW, Malcom ER. *Mass Spectroscopy for Chemists and Biochemists*. 2nd ed. Cambridge: Cambridge University Press; 1996.

[9] Benninghoven A, editor. *Ion Formation from Organic Solids*. New York: Springer-Verlag; 1983.

[10] Barber M, Bordoli RS, Elliot GJ, Sedgevick RD, Tyler AN. Fast atom bombardment mass spectrometry. Anal Chem 1982;54:645A–657A.

[11] Hillenkamp F, Karas M, Beavis RC, Chait BT. Matrix-assisted laser desorption/ionization mass spectrometry of biopolymers. Anal Chem 1991; 63:1193A–1203A.

[12] Cotter RJ. *Time-of-Flight Mass Spectrometry*. Washington, DC: American Chemical Society Professional Reference Books; 1997.

[13] Schriemer DC, Li L. Detection of high molecular weight narrow polydisperse polymers up to 1.5 million Daltons by MALDI mass spectrometry. Anal Chem 1996;68:2721–2725.

[14] Ibbett RN. *NMR Spectroscopy of Polymers*. Glasgow: Blackie Academic & Professional, an imprint of Chapman & Hall; 1993.

[15] Bovey FA, Mirau P. *NMR of Polymers*. San Diego: Academic Press, 1996.

[16] Cheng HN, English AD, editors. *NMR Spectroscopy of Polymers in Solution and in the Solid State*. ACS Symposium Series 834, Washington, DC: American Chemical Society; 2002.

[17] Maiwald M, Fischer HH, Kim Y-K, Albert K, Hasse H. Quantitative high-resolution on-line NMR spectroscopy in reaction and process monitoring. J Magn Reson 2004;166:135–146.

[18] Lutz J-F, Matyjaszewski K. Nuclear magnetic resonance monitoring of chain-end functionality in the atom transfer radical polymerization of styrene. J Polym Sci A Polym Chem 2005;43:897–910.

[19] Czelusniak I, Szymanska-Buzar T. Ring-opening metathesis polymerization of norbornene and norbornadiene by tungsten(II) and molybdenum(II) complexes. J Mol Catal A Chem 2002;190:131–143.

[20] Meinhold RH, Rothbaum HP, Newman RH. Polymerization of supersaturated silica solutions monitored by silicon-29 nuclear magnetic resonance. J Colloid Interface Sci 1985; 108:234–236.

[21] Larraz E, Elvira C, Gallardo A, San Roman J. Radical copolymerization studies of an amphiphilic macromonomer derived from Triton X-100. Reactivity ratios determination by in situ quantitative ^{1}H NMR monitoring. Polymer 2005;46:2040–2046.

[22] Moffat KA, Hamer GK, Georges MK. Stable free radical polymerization process: kinetic and mechanistic study of the thermal decomposition of MB-TMP monitored by NMR and ESR spectroscopy. Macromolecules 1999;32:1004–1012.

[23] Gill MT, Chapman SE, DeArmitt CL, Baines FL, Dadswell CM, Stamper JG, Lawless GA, Billingham NC, Armes SP. A study of the kinetics of polymerization of aniline using proton NMR spectroscopy. Synth Met 1998;93:227–233.

[24] Okada S, Hayamizu K, Matsuda H, Masaki A, Minami N, Nakanishi H. Solid-state polymerization of 15,17,19,21,23,25-tetracontahexayne. Macromolecules 1994;27: 6259–6266.

[25] Vargas MA, Cudaj M, Hailu K, Sachsenheimer K, Guthausen G. Online low-field ^{1}H NMR spectroscopy: monitoring of emulsion polymerization of butyl acrylate. Macromolecules 2010;43:5561–5568.

[26] Wertz JE, Bolton JR. *Electron Spin Resonance: Elementary Theory and Practical Applications*. New York: Chapman and Hall; 1986.

[27] Poole Jr. CP. *Electron Spin Resonance: A Comprehensive Treatise on Experimental Techniques*. 2nd ed. Mineola: Dover Publications; 1997.

[28] Rieger PH. *Electron Spin Resonance: Analysis and Interpretation*. Cambridge: Royal Society of Chemistry; 2007.

[29] Liang Z, Freed JH. An assessment of the applicability of multifrequency ESR to study the complex dynamics of biomolecules. J Phys Chem B 1999;103:6384–6396.

[30] Andersson KK, Schmidt PP, Katterle B, Strand KR, Palmer AE, Lee S-K, Solomon EI, Gräslund A, Barra A-L. JBIC Examples of high-frequency EPR studies in bioinorganic chemistry. J Biol Inorg Chem 2003;8:235–247.

[31] Krzystek J, Zvyagin SA, Ozarowski A, Fiedler AT, Brunold TC, Telser J. Definitive spectroscopic determination of zero-field splitting in high-spin cobalt(II). J Am Chem Soc 2004;126:2148–2155.

[32] Elsässer C, Brecht M, Bitt R. Pulsed electron-electron double resonance on multinuclear metal clusters: assignment of spin projection factors based on the dipolar interaction. J Am Chem Soc 2002;124:12606–12611.

[33] Sinnecker S, Reijerse E, Neese F, Lubitz W. Hydrogen bond geometries from electron paramagnetic resonance and electron-nuclear double resonance parameters: density functional study of quinone radical anion–solvent interactions. J Am Chem Soc 2004;126:3280–3290.

[34] Van Doorslaer S, Bachmann R, Schweiger A. A pulse EPR and ENDOR investigation of the electronic and geometric structure of cobaltous tetraphenylporphyrin (pyridine). J Phys Chem A 1999;103:5446–5455.

[35] Epel B, Goldfarb D. Two-dimensional pulsed TRIPLE at 95 GHz. J Magn Reson 2000;146:196–203.

[36] Hagen WR, van den Berg WAM, van Dongen WMAM, Reijerse EJ, van Kan PJM. EPR spectroscopy of biological iron–sulfur clusters with spin-admixed S=3/2. J Chem Soc, Faraday Trans 1998;94:2969–2973.

[37] Lattimer RP, Harmon DJ, Hansen GE. Determination of molecular weight distributions of polystyrene oligomers by field desorption mass spectroscopy. Anal Chem 1980;52;1808–1811.

[38] Day RJ, Unger SE, Cooks RG. Molecular secondary ion mass spectrometry. Anal Chem 1980;52:557–572.

[39] Watson JT. Fast atom bombardment. In: Matsuo T, Caprioli RM, Gross ML, Seyama Y, editors. *Biological Mass Spectrometry – Present and Future*. Chichester: John Wiley & Sons; 1994. p 23.

[40] Selby TL, Wesdemiotis C, Lattimer RP. Dissociation characteristics of [M+X]⁺ ions (X=H, Li, Na, K) from linear and cyclic polyglycols. J Am Soc Mass Spectrom 1994;5: 1081–1092.

[41] Fink SW, Freas RB. Enhanced analysis of poly(ethylene glycols) and peptides using thermospray mass spectrometry. Anal Chem 1989;61:2050–2054.

[42] Smith RD, Loo JA, Edmonds CG, Barinaga CJ, Udseth HR. New developments in biochemical mass spectrometry: electrospray ionization. Anal Chem 1990;62: 882–899.

[43] Gaskell SJ. Electrospray: principles and practice. J Mass Spectrom 1997;32:677–688.

[44] Buchanan MV, Hettich RL. Fourier transform mass spectroscopy of high-mass biomolecules. Anal Chem 1993;65:245A–259A.

[45] Philipsen HJA, Claessens HA, Bosman M, Klumperman B, German AL. Normal phase gradient polymer elution chromatography of polyester resins. Chromatographia 1998; 48:623–630.

[46] Garozzo D, Giuffrida M, Montaudo G. Primary thermal decomposition processes in aliphatic polyesters investigated by chemical ionization mass spectrometry. Macromolecules 1986;19:1643–1649.

[47] Sedgwick RD. In: Dawkins JD, editor. *Developments in Polymer Characterization – I*. London: Applied Science; 1978. p 41–70.

[48] Montaudo G, Scamporrino E, Vitalini D. Characterization of copolymer sequences by fast atom bombardment mass spectrometry. 2. Identification of oligomers contained in alternating and random copolyesters with photolabile units in the main chain. Macromolecules 1989;22:627–632.

[49] Montaudo MS, Puglisi C, Samperi F, Montaudo G. Structural characterization of multicomponent copolyesters by mass spectrometry. Macromolecules 1998;31:8666–8676.

[50] Painter AJ, Jaya N, Basha E, Vierling E, Robinson CV, Benesch JLP. Real-time monitoring of protein complexes reveals their quaternary organization and dynamics. Chem Biol 2008;15:246–253.

[51] Heck AJR. Native mass spectrometry: a bridge between interactomics and structural biology. Nat Methods, 2008;5: 927–933.

[52] Kaltashov IA, Bobst CE, Abzalimov RR, Berkowitz SA, Houde D. Conformation and dynamics of biopharmaceuticals: transition of mass spectrometry-based tools from academe to industry. J Am Soc Mass Spectrom 2010;21: 323–337.

[53] Kaltashov IA, Bobst CE, Abzalimov RR, Wang G, Baykal B, Wang S. Advances and challenges in analytical characterization of biotechnology products: mass spectrometry-based approaches to study properties and behavior of protein therapeutics. Biotechnol Adv 2012;30:210–222.

[54] Kukrer B, Filipe V, van Duijn E, Kasper PT, Vreeken RJ, Heck AJR, Jiskoot W. Mass spectrometric analysis of intact human monoclonal antibody aggregates fractionated by size-exclusion chromatography. Pharm Res 2010;27:2197–204.

[55] Wang G, Johnson AJ, Kaltashov IA. Evaluation of electrospray ionization mass spectrometry as a tool for characterization of small soluble protein aggregates. Anal Chem 2012;84:1718–1724.

[56] Takáts Z, Vékey K, Hegedüs L. Qualitative and quantitative determination of poloxamer surfactants by mass spectrometry. Rapid Commun Mass Spectrom 2001;15:805–810.

[57] Koster S, Duursma MC, Boon JJ, Nielen MWF, de Koster CG, Heeren RMA. Structural analysis of synthetic homo- and copolyesters by electrospray ionization on a Fourier transform ion cyclotron resonance mass spectrometer. J Mass Spectrom 2000;35:739–748.

[58] de Maaijer-Gielbert J, Gu C, Somogyi Á, Wysocki VH, Kistemaker PG, Weeding TL. Surface-induced dissociation of singly and multiply protonated polypropylenamine dendrimers. J Am Soc Mass Spectrom 1999;10:414–412.

[59] Gruendling T, Guilhaus M, Barner-Kowollik C. Quantitative LC–MS of polymers: determining accurate molecular weight distributions by combined size exclusion chromatography and electrospray mass spectrometry with maximum entropy data processing. Anal Chem 2008;80:6915–6927.

[60] Ah Toy A, Vana P, Davis TP, Barner-Kowollik C. Reversible addition fragmentation chain transfer (RAFT) polymerization of methyl acrylate: detailed structural investigation via coupled size exclusion chromatography-electrospray ionization. Mass spectrometry (SEC–ESI–MS). Macromolecules 2004;37:744–751.

[61] Feldermann, Ah Toy A, Davis TP, Stenzel MH, Barner-Kowollik C. An in-depth analytical approach to the mechanism of the RAFT process in acrylate free radical polymerizations via coupled size exclusion chromatography–electrospray ionization mass spectrometry (SEC–ESI–MS). Polymer 2005;46:8448–8457.

[62] Schmidt A-C, Fahlbusch B, Otto M. Size exclusion chromatography coupled to electrospray ionization mass spectrometry for analysis and quantitative characterization of arsenic interactions with peptides and proteins. J Mass Spectrom 2009;44:898–910.

[63] Hinderling C, Chen P. Mass spectrometric assay of polymerization catalysts for combinational screening. Int J Mass Spectrom 2000;195–196:377–383.

[64] Chena R, Li L. Lithium and transition metal ions enable low energy collision-induced dissociation of polyglycols in electrospray ionization mass spectrometry. J Am Soc Mass Spectrom 2001;12:8332–839.

[65] Trimpin S, Rouhanipour A, Az R, Rader HJ, Mullen K. New aspects in matrix-assisted laser desorption/ionization time-of-flight mass spectrometry: a universal solvent-free sample preparation. Rapid Commun Mass Spectrom 2001; 15:1364–1373.

[66] Bauer BJ, Wallace WE, Fanconi BM, Guttman CM. "Covalent cationization method" for the analysis of polyethylene by mass spectrometry. Polymer 2001;42:9949–9953.

[67] Cohen SL, Chait BT. Influence of matrix solution conditions on the MALDI-MS analysis of peptides and proteins. Anal Chem 1996;68:31–37.

[68] Chou CW, Williams P, Limbach PA. Matrix influence on the formation of positively charged oligonucleotides in matrix-assisted laser desorption/ionization mass spectrometry. Int J Mass Spectrom 1999;193:15–27.

[69] Vandell VE, Limbach PA. Polyamine co-matrices for matrix-assisted laser desorption/Ionization mass spectrometry of oligonucleotides. Rapid Commun Mass Spectrom 1999;13: 2014–2021.

[70] Jensen C, Haebel S, Andersen SO, Roepstorff P. Towards monitoring of protein purification by matrix-assisted laser desorption ionization mass spectrometry. Int J Mass Spectrom Ion Proc 1997;160:339–356.

[71] Tang XD, Dreifuss PA, Vertes A. New matrices and accelerating voltage effects in matrix-assisted laser desorption/ionization of synthetic polymers. Rapid Commun Mass Spectrom 1995;9:1141–1147.

[72] Trimpin S, Keune S, Räder HJ, Müllen K. Solvent-free MALDI-MS: developmental improvements in the reliability and the potential of MALDI in the analysis of synthetic polymers and giant organic molecules. J Am Soc Mass Spectrom 2006;17:661–671.

[73] Nanda AK, Ganesh K, Kishorea K, Surinarayanan M. End-group analysis of vinyl polyperoxides by MALDI-TOF-MS, FT-IR technique and thermochemical calculations. Polymer 2000;41:9063–9072.

[74] Barner-Kowollik C, Vana P, Davis TP. Laser-induced decomposition of 2,2-dimethoxy-2-phenylacetophenone and benzoin in methyl methacrylate homopolymerization studied via matrix-assisted laser desorption/ionization time-of-flight mass spectrometry. J Polym Sci A Polym Chem 2002;40:675–681.

[75] Jackson AT, Bunn A, Chisholm MS. Utilising matrix-assisted laser desorption/ionisation techniques for the generation of structural information from different end-group functionalised poly(methyl methacrylate)s. Polymer 2008; 49:5254–5261.

[76] Guttman CM, Wetzel SJ, Blair WR, Fanconi BM, Girard JE, Goldschmidt RJ, Wallace WE, VanderHart DL. NIST-sponsored interlaboratory comparison of polystyrene molecular mass distribution obtained by matrix-assisted laser desorption/ionization time-of-flight mass spectrometry: statistical analysis. Anal Chem 2001;73:1252–1262.

[77] Mincheva Z, Hadjieva P, Kalcheva V, Seraglia R, Traldi P, Przybylski M. Matrix-assisted laser desorption/ionization, fast atom bombardment and plasma desorption mass spectrometry of polyethylene glycol esters of (2-benzothiazolon-3-yl)acetic acid. J Mass Spectrom 2001;36: 626–632.

[78] Kassis CE, DeSimone JM, Linton RW, Remsen EE, Lange GW, Friedman RMA. Direct deposition method for coupling matrix-assisted laser desorption/ionization mass spectrometry with gel permeation chromatography for polymer characterization. Rapid Commun Mass Spectrom 1997;11: 1134–1138.

[79] Nielen MWF, Buijtenhuijs FA (Ab). Polymer analysis by liquid chromatography/ electrospray ionization time-of-flight mass spectrometry. Anal Chem 1999;71:1809–1814.

[80] Esser E, Keil C, Braun D, Montag P, Pasch H. Matrix-assisted laser desorption/ionization mass spectrometry of synthetic polymers. 4. Coupling of size exclusion chromatography and MALDI-TOF using a spray-deposition interface. Polymer 2000;41:4039–4046.

[81] Montaudo MS, Puglisi C, Samperi F, Montaudo G. Application of size exclusion chromatography matrix-assisted laser desorption/ionization time-of-flight to the determination of molecular masses in polydisperse polymers. Rapid Commun Mass Spectrom 1998;12: 519–528.

[82] Falkenhagen J, Jancke H, Krüger R-P, Rikowski E, Schulz G. Characterization of silsesquioxanes by size-exclusion chromatography and matrix-assisted laser desorption/ionization time-of-flight mass spectrometry. Rapid Commun Mass Spectrom 2003;17:285–290.

[83] Yeung B, Marecak D. Molecular weight determination of hyaluronic acid by gel filtration chromatography coupled to matrix-assisted laser desorption ionization mass spectrometry. J Chromatogr A 1999;852:573–581.

[84] Tatro SR, Baker GR, Fleming R, Harmon JP. Matrix-assisted laser desorption/ionization (MALDI) mass spectrometry: determining Mark-Houwink-Sakurada parameters and analyzing the breadth of polymer molecular weight distributions. Polymer 2002;43:2329–2335.

[85] Storey RF, Brister LB, Sherman JW. Structural characterization of poly(ε-caprolactone) and poly(ε-caprolactone-b-isobutylene-b-ε-caprolactone) block copolymers by MALDI-TOF mass spectrometry. J Macromol Sci Pure Appl Chem 2001;38:107–122.

[86] van der Hage ERE, Duursma MC, Heeren RMA, Boon JJ. Structural analysis of polyoxyalkyleneamines by matrix-assisted laser desorption/ionization on an external ion source FT-ICR-MS and NMR. Macromolecules 1997;30: 4302–4309.

[87] Maciejczek A, Mass V, Rode K, Pasch H. Analysis of poly(ethylene oxide)-b-poly(propylene oxide) block copolymers by MALDI-TOF mass spectrometry using collision induced dissociation. Polymer 2010;51:6140–6150.

[88] Cox FJ, Johnston MV, Qian K, Peiffer DG. Compositional analysis of isobutylene/p-methylstyrene copolymers by matrix-assisted laser desorption/ionization mass spectrometry. J Am Soc Mass Spectrom 2004;15:681–688.

[89] Pasch H, Schrepp W. *MALDI-TOF Mass Spectrometry of Synthetic Polymers.* Berlin, Heidelberg, New York: Springer-Verlag; 2003.

[90] Yu D, Vladimirov N, Fréchet JMJ. MALDI-TOF in the characterizations of dendritic – linear block copolymers and stars. Macromolecules 1999;32:5186–5192.

[91] Girod M, Mazarin M, Phan TNT, Gigmes D, Charles L. Determination of block size in poly(ethylene oxide)-b-polystyrene block copolymers by matrix-assisted laser desorption/ionization time-of-flight mass spectrometry. J Polym Sci A Polym Chem 2009;47:3380–3390.

[92] Giordanengo R, Vielb S, Hidalgo M, Allard-Breton B, Thévand A, Charles L. Analytical strategy for the molecular weight determination of random copolymers of poly(methyl methacrylate) and poly(methacrylic acid). J Am Soc Mass Spectrom 2010;21:1075–1085.

[93] Seeley EH, Caprioli RM. MALDI imaging mass spectrometry of human tissue: method challenges and clinical perspectives. Trends Biotechnol 2011;29:136–143.

[94] Weidner SM, Falkenhagen J. LC-MALDI-TOF imaging MS: a new approach in combining chromatography and mass spectrometry of copolymers. Anal Chem 2011;83: 9153–9158.

[95] Fijten MWM, Meier MAR, Hoogenboom R, Schubert US. Automated parallel investigations/optimizations of the reversible addition-fragmentation chain transfer polymerization of methyl methacrylate. J Polym Sci A Polym Chem 2004;42:5775–5783.

[96] Cortez MA, Grayson SM. Application of time-dependent MALDI-TOF mass spectral analysis to elucidate chain transfer mechanism during cationic polymerization of oxazoline monomers containing thioethers. Macromolecules 2010;43:10152–10156.

[97] Binder WH, Pulamagatta B, Kir O, Kurzhals S, Barqawi H, Tanner S. Monitoring block-copolymer crossover-chemistry in ROMP: catalyst evaluation via mass-spectrometry (MALDI). Macromolecules 2009;42:9457–9466.

[98] Hawkridge AM, Gardella Jr. JA. Evaluation of matrix-assisted laser desorption ionization mass spectrometry for studying the sec-butyllithium and n-butyllithium initiated ring-opening polymerization of hexamethylcyclotrisiloxane (D3). J Am Soc Mass Spectrom 2003;14: 95–101.

[99] Xu P, Kumar J, Samuelson L, Cholli AL. Monitoring the enzymatic polymerization of 4-phenylphenol by matrix-assisted laser desorption ionization time-of-flight mass spectrometry: a novel approach. Biomacromolecules 2002;3:889–893.

[100] Jayakannan M, van Dongen JLJ, Janssen RAJ. Mechanistic aspects of the suzuki polycondensation of thiophenebisboronic derivatives and diiodobenzenes analyzed by MALDI-TOF mass spectrometry. Macromolecules 2001;34:5386–5393.

[101] Liu J, McCullough RD. End group modification of regioregular polythiophene through postpolymerization functionalization. Macromolecules 2002;35:9882–9889.

[102] Kamachi M. ESR studies on radical polymerization. Adv Polym Sci 1987;82:207–275.

[103] Yamada B, Westmoreland DG, Kobatake S, Konosu O. ESR spectroscopic studies of radical polymerization. Prog Polym Sci 1999;24:565–630.

[104] Shen J, Tian Y, Wang G, Yang M. Modelling and kinetic study on radical polymerization of methyl methacrylate in bulk. 1. Propagation and termination rate coefficients and initiation efficiency. Makromol Chem 1991;192:2669–2685.

[105] Tonge MP, Kajiwara A, Kamachi M, Gilbert RG. ESR measurements of the propagation rate coefficient for styrene free radical polymerization. Polymer 1998;39:2305–2313.

[106] Zhu S, Tian Y, Hamielec AE, Eaton DR. Radical trapping and termination in free-radical polymerization of methyl methacrylate. Macromolecules 1990;23:1144–1150.

[107] Yamada B, Kageoka M, Otsu T. Dependence of propagation and termination rate constants on conversion for the radical polymerization of styrene in bulk as studied by ESR spectroscopy. Macromolecules 1991;24:5234–5236.

[108] Carswell TG, Hill DJT, Londero DL, O'Donnell JH, Pomery PJ, Winzor CL. Kinetic parameters for polymerization of methyl methacrylate at 60 °C. Polymer 1992;33:137–140.

[109] Moad G, Rizzardo E, Solomon DH, Beckwith ALJ. Absolute rate constants for radical-monomer reactions. Polym Bull 1992;29:647–652.

[110] Buback M, Kowollik C, Kamachi M, Kajiwara A. Free-radical propagation rate coefficients of dodecyl methacrylate deduced from electron spin resonance experiments. Macromolecules 1998;31:7208–7212.

[111] Sack R, Schulz GV, Meyerhoff G. Free radical polymerization of methyl methacrylate up to the glassy state. Rates of propagation and termination. Macromolecules 1988;21:3345–3352.

[112] Kajiwara A. Studying the fundamentals of radical polymerization using ESR in combination with controlled radical polymerization methods. Macromol Symp 2007;248:50–59.

[113] Matsumoto A, Mizuta K. Detailed kinetic analysis of the radical polymerization of trans-4-tert-butylcyclohexyl methacrylate in benzene based on the rate constants determined by electron spin resonance spectroscopy. Macromolecules 1994;27:5863–5870.

[114] Azukizawa M, Yamada B, Hill DJT, Pomery PJ. Radical polymerization of phenyl acrylate as studied by ESR spectroscopy: concurrence of propagating and midchain radicals. Macromol Chem Phys 2000;201:774–781.

[115] Willemse RXE, van Herk AM, Panchenko E, Junkers T, Bubackm M. PLP-ESR monitoring of midchain radicals in n-butyl acrylate polymerization. Macromolecules 2005; 38:5098–5103.

[116] Arzamendi G, Plessis C, Leiza JR, Asua JM. Effect of the intramolecular chain transfer to polymer on PLP/SEC experiments of alkyl acrylates. Macromol Theory Simul 2003;12:315–324.

[117] Sato E, Emoto T, Zetterlund PB, Yamada B. Influence of mid-chain radicals on acrylate free radical polymerization: effect of ester alkyl group. Macromol Chem Phys 2004;205:1829–1839.

[118] Asua JM, Beuermann S, Buback M, Castignolles P, Charleux B, Gilbert RG, Hutchinson RA, Leiza JR, Nikitin AN, Vairon JP, Herk AM. Critically evaluated rate coefficients for free-radical polymerization, 5. Macromol Chem Phys 2004;205: 2151–2160.

[119] Barth J, Buback M, Hesse P, Sergeeva T. Chain-length-dependent termination in n-butyl methacrylate and tert-butyl methacrylate bulk homopolymerizations studied via SP-PLP-ESR. Macromolecules 2009;42:481–488.

[120] Smith GB, Russell GT, Heuts JPA. Termination in dilute-solution free-radical polymerization: a composite model. Macromol Theory Simul 2003;12:299–314.

[121] Barner-Kowollik C, Buback M, Egorov M, Fukuda T, Goto A, Olaj OF, Russell GT, Vana P, Yamada B, Zetterlund P. Critically evaluated termination rate coefficients for free-radical polymerization: experimental methods. Prog Polym Sci 2005;30:605–643.

[122] Buback M, Müller E, Russell GT. SP–PLP–EPR study of chain-length-dependent termination in free-radical polymerization of n-dodecyl methacrylate, cyclohexyl methacrylate, and benzyl methacrylate: evidence of "composite" behavior. J Phys Chem A 2006;110:3222–3230.

[123] Kamachi M, Yamada B. Propagation and termination constants in free radical polymerization. In: Brandrup J, Immergut EH, Grulk EA, editors. *Polymer Handbook.* 4th ed. New York: Wiley; 1999. p II/77.

[124] Anseth KS, Anderson KJ, Bowman CN. Radical concentrations, environments, and reactivities during crosslinking polymerizations Macromol. Chem Phys 1996;197:833–848.

[125] O'Shaughnessy B, Yu J. Non-steady state free radical polymerization kinetics at high conversions: entangled regimes. Macromolecules 1998;31:5240–5354.

[126] Zetterlund PB, Yamazoe H, Yamada B. Propagation and termination kinetics in high conversion free radical copolymerization of styrene/divinylbenzene investigated by electron spin resonance and Fourier-transform near-infrared spectroscopy. Polymer 2002;43:7027–7035.

[127] Sato T, Kawasaki S, Seno M, Tanaka H, Kato K. Kinetic and ESR studies on radical polymerization. ESR and kinetic evidences for the penultimate effect in the radical-initiated copolymerization of N-cyclohexylmaleimide and bis(2-ethylhexyl) itaconate in benzene. Makromol Chem 1993;194: 2247–2256.

[128] Nakamura H, Seno M, Tanaka H. Sato T. Kinetic and ESR studies on the radical polymerization. Radical-initiated copolymerization of the diethyl itaconate–SnCl$_4$ complex and styrene. Makromol Chem 1993;194:1773–1783.

[129] Sato T, Shimooka S, Seno M, Tanaka H. Kinetic and electron paramagnetic resonance studies on radical polymerization. Radical copolymerization of p-tert-butoxy-styrene and dibutyl fumarate in benzene. Macromol Chem Phys 1994; 195:833–843.

[130] Cheetham PL, Tabner BJ. A spin-trap study of the aqueous heterogeneous copolymerization of acrylonitrile and vinyl acetate. Eur Polym J 1993;29:451–454.

[131] Otsu T. Reactivities of acrylic monomers and their radicals in radical polymerization. Makromol Chem Macromol Symp 1987;10–11:235–254.

[132] Carswell TG, Hill DJT, Kellman RD, Londero I, O'Donnell JH, Pomery PJ, Winzor CL. Mechanisms of polymerization of methacrylate copolymers of biological interest. Makromol Chem Macromol Symp 1991;51:183–191.

[133] Gilbert BC, Smith JRL, Ward SR, Whitwood AC, Taylor P. Kinetic EPR studies of the addition of carbohydrate-derived radicals to methacrylic acid. J Chem Soc Perkin Trans 1998;2:1565–1572.

[134] MacLeod RP, Veregin N, Odell PG, Georges MK. Electron spin resonance studies of the stable free-radical polymerization of styrene. Macromolecules 1998;31:530–531.

[135] Bon SAF, Chambard G, German AL. Nitroxide-mediated living radical polymerization: determination of the rate coefficient for alkoxyamine C–O bond homolysis by quantitative ESR. Macromolecules 1999;32:8269–8276.

[136] Zarycz N, Botek E, Champagne B, Sciannaméa V, Jérôme C, Detrembleur C. Joint theoretical experimental investigation of the electron spin resonance spectra of nitroxyl radicals: application to intermediates in in situ nitroxide mediated polymerization (in situ NMP) of vinyl monomers. J Phys Chem B 2008;112:10432–10442.

[137] Fukuda T, Terauchi T, Goto A, Ohno K, Tsujii Y, Miyamoto T, Kobatake S, Yamada B. Mechanisms and kinetics of nitroxide-controlled free radical polymerization. Macromolecules 1996;29:6393–6398.

[138] Hathaway B, Duggan M, Murphy A, Mullane JC, Power A, Walsh B. The stereochemistry and electronic properties of fluxional six-coordinate copper(II) complexes. Coord Chem Rev 1981;36:267–324.

[139] Matyjaszewski K, Kajiwara A. EPR study of atom transfer radical polymerization (ATRP) of styrene. Macromolecules 1998;31:548–550.

[140] van Herk AM III. Heterogeneous polymerization modeling of emulsion polymerization, will it ever be possible? Part-2: Determination of basic kinetic data over the last ten years. Macromol Symp 2009;275–276:120–132.

[141] Gilbert RG. *Emulsion Polymerization: A Mechanistic Approach*. London: Academic Press; 1995.

[142] Ballard MJ, Gilbert RG, Napper DH, Pomery PJ, O'Donnell JH. Application of electron spin resonance spectroscopy to emulsion polymerization. Macromolecules 1984;17:504–506.

[143] Lau W, Westmoreland DG, Novak RW. Direct measurement of propagating radical concentration in a semicontinuous emulsion polymerization. Macromolecules 1987;20:457–459.

[144] Parker H-Y, Westmoreland DG, Chang H-R. ESR study of MMA batch emulsion polymerization in real time: effects of particle size. Macromolecules 1996;29:5119–5127.

[145] Cutting GR, Tabner BJ. Radical termination and radical concentrations during the batch emulsion polymerization of methyl methacrylate studied by electron spin resonance spectroscopy. Macromolecules 1993;26:951–955.

[146] Cutting GR, Tabner BJ. Radical concentrations and reaction temperature profiles during the (batch) core–shell emulsion polymerization of methyl methacrylate and butyl acrylate, studied by electron spin resonance spectroscopy. Eur Polym J 1997;33:213–217.

[147] Keifer PA. NMR tools for biotechnology. Curr Opin Biotech 1999;10:34–41.

[148] Neudert R, Strofer E, Bremser W. On-line NMR in process engineering. Magn Res Chem 1986;24:1089–1092.

[149] Woelk K, Bargon J. High-pressure NMR probes for the in situ investigation of gas/liquid reactions. Rev Sci Instrum 1992;63:3307–3310.

[150] Keifer PA. High-resolution NMR techniques for solid-phase synthesis and combinatorial chemistry. Drug Discov Today 1997;2:468–478.

[151] Haner RL, Llanos W, Mueller L. Small volume flow probe for automated direct-injection NMR analysis: design and performance. J Magn Reson 2000;143:69–78.

[152] Albert K, editor. *On-Line LC-NMR and Related Techniques*, Chichester: Wiley; 2002.

[153] Hamper BC, Synderman DM, Owen TJ, Scates AM, Owsley DC, Kesselring AS, Chott RC. High-throughput H-1 NMR and HPLC characterization of a 96-member substituted methylene malonamic acid library. J Comb Chem 1999;1: 140–150.

[154] McDermott R, Trabesinger AH, Muck M, Hahn EL, Pines A, Clarke J. Liquid-state NMR and scalar coupling in microtesla magnetic fields. Science 2002;295:2247–2249.

[155] Koenig JL. *Spectroscopy of Polymers*. 2nd ed. New York: Elsevier Science Inc.; 1999.

[156] Hatada K, Kitayama T. *NMR Spectroscopy of Polymers*. Berlin, Heidelberg: Springer-Verlag; 2004.

[157] Matyjaszewski K, Jo SM, Paik H-J, Shipp DA. An investigation into the CuX/2,2'-bipyridine (X=Br or Cl) mediated atom transfer radical polymerization of acrylonitrile. Macromolecules 1999;32:6431–6438.

[158] Percec V, Barboiu B. "Living" radical polymerization of styrene initiated by arenesulfonyl chlorides and cui(bpy)nCl. Macromolecules 1995;28:7970–7972.

[159] Yoshida E. Polymerization of 1-chloro-2-β-naphthylacetylene by Mo catalysts and polymer properties. J Polym Sci Part A Polym Chem 1996;34:2937–2943.

[160] Li H, Wang C, Bai F, Yue J, Woo H-G. Living ring-opening polymerization of L-lactide catalyzed by red-Al. Organometallics 2004;23:1411–1415.

[161] Lang AS, Neubig A, Sommer M, Thelakkat M. NMRP versus "Click" chemistry for the synthesis of semiconductor polymers carrying pendant perylene bisimides. Macromolecules 2010;43: 7001–7010.

[162] Spěváček J. Short chain branching in high density polyethylene: 13C n.m.r. study. Polymer 1978;19:1149–1152.

[163] Wu D, Liu Y, Jiang X, Chen L, He C, Goh SH, Leong KW. Evaluation of hyperbranched poly(amino ester)s of amine constitutions similar to polyethylenimine for DNA delivery. Biomacromolecules 2005;6:3166–3173.

[164] Twum EB, Gao C, Li X, McCord EF, Fox PA, Lyons DF, Rinaldi PL. Characterization of the chain-ends and branching structures in polyvinylidene fluoride with multidimensional NMR. Macromolecules 2012;45:5501–5512.

[165] Lovell PA, Shah TH, Heatley F. Chain transfer to polymer in emulsion polymerization of n-butyl acrylate studied by carbon-13 NMR spectroscopy and gel permeation chromatography. Polym Commun 1991;32:98–103.

[166] Gaborieau M, Koo SPS, Castignolles P, Junkers T, Barner-Kowollik C. Reducing the degree of branching in polyacrylates via midchain radical patching: a quantitative melt-state NMR study. Macromolecules 2010;43:5492–5495.

[167] Pollard M, Klimke K, Graf R, Spiess HW, Wilhelm M, Sperber O, Piel C, Kaminsky W. Observation of chain branching in polyethylene in the solid state and melt via ^{13}C NMR spectroscopy and melt NMR relaxation time measurements. Macromolecules 2004;37:813–825.

[168] Hlalele L, Klumperman B. Reversible nitroxide trapping of the mid-chain radical in n-butyl acrylate polymerization. Macromolecules 2011;44:5554–5557.

[169] Ahuja S, Dieckman SL, Gopalsami N, Raptis AC. ^{1}H NMR imaging and spectroscopy studies of the polymerization of acrylamide gels. Macromolecules 1996;29:5356–5360.

[170] Kirsebom H, Rata G, Topgaard D, Mattiasson B, Galaev IY. In situ ^{1}H NMR studies of free radical cryopolymerization, Polymer 2008;49:3855–3858.

[171] Abdollahi M, Sharifpour M. A new simple procedure to calculate monomer reactivity ratios by using on-line ^{1}H NMR kinetic experiments: copolymerization system with greater difference between the monomer reactivity ratios. Polymer 2007;48:25–30.

[172] Even M, Haddleton DM, Kukulj D. Synthesis and characterization of amphiphilic triblock polymers by copper mediated living radical polymerization. Eur Polym J 2003;39: 633–639.

[173] Mangold C, Wurm F, Obermeier B, Frey H. "Functional poly(ethylene glycol)": PEG-based random copolymers with 1,2-diol side chains and terminal amino functionality. Macromolecules 2010;43:8511–8518.

[174] Larraz E, Elvira C, Gallardo A, San Román J. Radical copolymerization studies of an amphiphilic macromonomer derived from Triton X-100. Reactivity ratios determination by in situ quantitative ^{1}H NMR monitoring. Polymer 2005;46:2040–2046.

[175] Grignard B, Jérôme C, Calberg C, Detrembleur C, Jérôme R. Controlled synthesis of carboxylic acid end-capped poly(heptadecafluorodecyl acrylate) and copolymers with 2-hydroxyethyl acrylate. J Polym Sci Part A Polym Chem 2007;45:1499–1506.

[176] Ute K, Niimi R, Hongo S, Hatada K. Direct determination of molecular weight distribution by size exclusion chromatography with 750 MHz ^{1}H NMR detection (On-line SEC–NMR). Polym J 1998;30:439–443.

[177] Hatada K, Ute K, Kitayama T, Yamamoto M, Nishimura T, Kashiyama M. On-line GPC/NMR analysis of block and random copolymers of methyl and butyl methacrylates prepared with t-C_4H_9MgBr. Polym Bull 1989;21:489–495.

[178] Pasch H, Hiller W. Analysis of a technical poly(ethylene oxide) by on-line HPLC/^{1}H NMR. Macromolecules 1996;29:6556–6559.

[179] Krämer I, Hiller W, Pasch H. On-line coupling of gradient-HPLC and 1H NMR for the analysis of random poly[(styrene)-co-(ethyl acrylate)]s. Macromol Chem Phys 2000;201:1662–1666.

[180] Hiller W, Pasch H, Macko T, Hofmann M, Ganz J, Spraul M, Braumann U, Streck R, Mason J, Van Damme F. On-line coupling of high temperature GPC and ^{1}H NMR for the analysis of polymers. J Magn Res 2006;183:290–302.

[181] Reis MM, Araujo PHH, Sayer C, Giudici R. In situ near-infrared spectroscopy for simultaneous monitoring of multiple process variables in emulsion copolymerization. Ind Eng Chem Res 2004;43:7243–7250.

[182] Haddleton DM, Perrier S, Bon SAF. Copper(I)-mediated living radical polymerization in the presence of oxyethylene groups: online ^{1}H NMR spectroscopy to investigate solvent effects. Macromolecules 2000;33:8246–8251.

[183] Gramm S, Komber H, Schmaljohann D. Copolymerization kinetics of N-isopropylacrylamide and diethylene glycol monomethylether monomethacrylate determined by online NMR spectroscopy. J Polym Sci Part A Polym Chem 2005; 43:142–148.

[184] Münnemann K, Spiess HW. The art of signal enhancement. Nat Phys 2011;7:522–523.

[185] Appelt S, Baranga AB, Erickson CJ, Romalis MV, Young AR, Happer W. Theory of spin-exchange optical pumping of ^{3}He and ^{129}Xe. Phys Rev A 1998;58:1412–1439.

[186] Duewel M, Vogel N, Weiss CK, Landfester K, Spiess H-W, Münnemann K. Online monitoring of styrene polymerization in miniemulsion by hyperpolarized ^{129}Xenon NMR spectroscopy. Macromolecules 2012;45: 1839–1846.

[187] Fyfe CA. *Solid State NMR for Chemist*. Guelph: CRC Press; 1984.

[188] Engelhardt G, Michel D. *High-Resolution Solid-State NMR of Silicates and Zeolites*. Chichester: Wiley; 1987.

[189] Blümich B. Solid-state NMR of heterogeneous materials. Adv Mater 1991;3:237–245.

[190] Harris RK, Sünnetcioglu MM, Fischer RD. Variable-temperature ^{13}C solid-state NMR spectra: mobility of trimethyltin groups in coordination polymers of the type $[(Me_3Sn)_4M(CN)_6]\infty$ (M=Fe, Os, Ru). Spectrochim Acta A 1994;50:2069–2078.

[191] Brough AR, Dobson CM, Richardson IG, Groves GW. In situ solid-state NMR studies of Ca_3SiO_5: hydration at room temperature and at elevated temperatures using ^{29}Si enrichment. J Mat Sci 1994;29:3926–3940.

[192] Ferguson DB, Haw JF. Transient methods for in situ NMR of reactions on solid catalysts using temperature jumps. Anal Chem 1995;67:3342–3348.

[193] Aliev AE, Elizabé L, Kariuki BM, Kirschnick H, Thomas JM, Epple M, Harris KDM. In situ monitoring of solid-state polymerization reactions in sodium chloroacetate and sodium bromoacetate by ^{23}Na and ^{13}C solid-state NMR spectroscopy. Chem Eur J 2000;6:1120–1126.

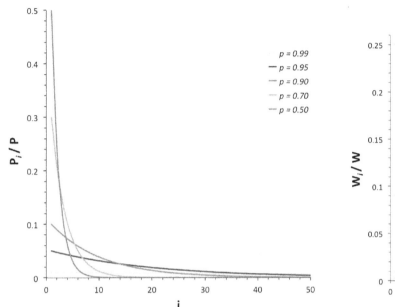

FIGURE 1.4 Geometric chain length distribution at different conversions.

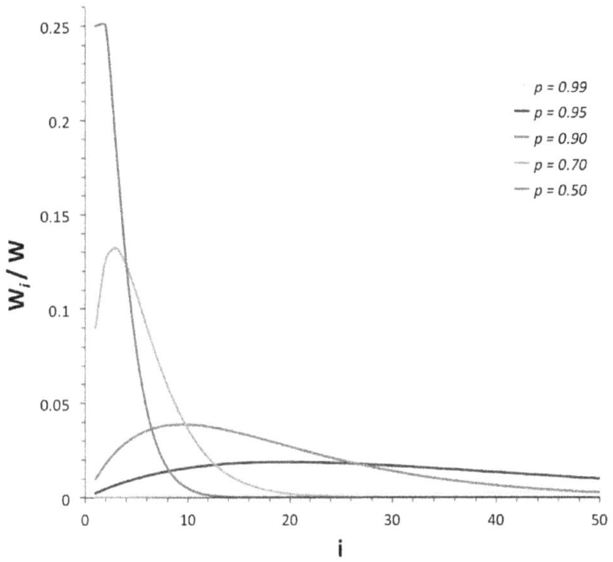

FIGURE 1.5 Geometric molecular weight distribution at different conversions.

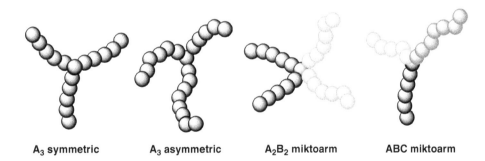

A₃ symmetric **A₃ asymmetric** **A₂B₂ miktoarm** **ABC miktoarm**

SCHEME 2.28 Examples of star polymers obtained from various arm-first and core-first methodologies

Monitoring Polymerization Reactions: From Fundamentals to Applications, First Edition. Edited by Wayne F. Reed and Alina M. Alb.
© 2014 John Wiley & Sons, Inc. Published 2014 by John Wiley & Sons, Inc.

Graft polymer   **Comb polymer**   **Bottle brush**

SCHEME 2.29   Architectural possibilities involving long-chain branching

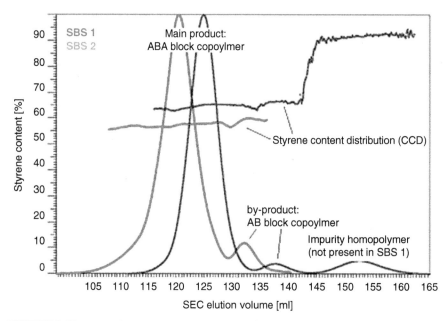

FIGURE 9.17   Analysis of two copolymers using RI/UV 254 nm dual detection, the response factors for the detectors have been determined using homopolymers of butadiene and polystyrene.

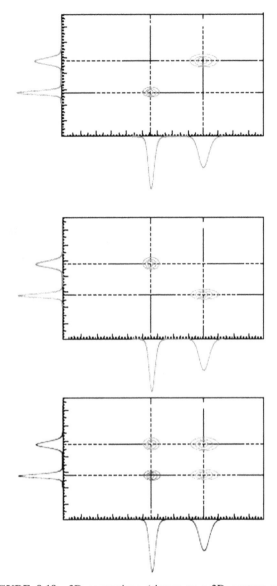

FIGURE 9.19 2D separations (shown as a 2D contour map) show clear sample differences while LC and SEC chromatograms are identical in each case.

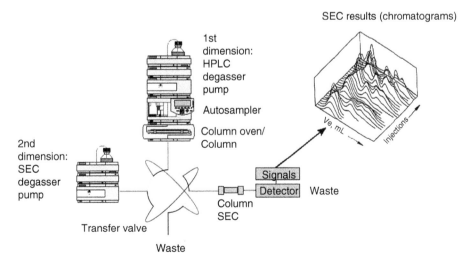

FIGURE 9.20 Generalized view of a comprehensive LC×SEC system: modules of first dimension are aligned horizontal, second dimension devices are shown in horizontal direction.

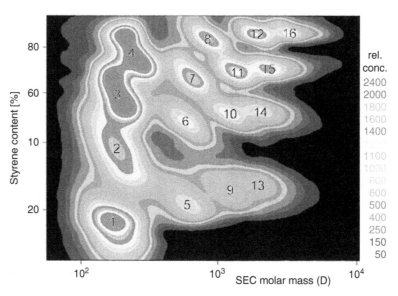

FIGURE 9.23 2D HPLC-SEC analysis of a four-arm block copolymer consisting of 16 components with four different styrene compositions and four molar masses, respectively (contour map shown; peak annotations, see text).

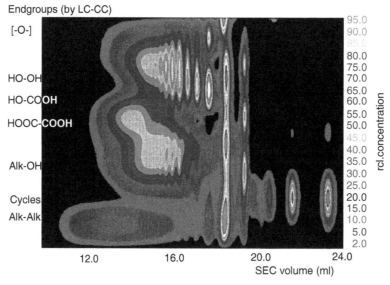

FIGURE 9.26 Contour map of adipic acid and 1,6-hexanediol condensation products investigated by 2D chromatography (see text for details).

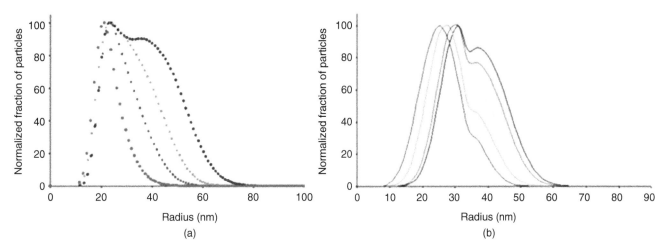

FIGURE 18.4 Evolution of the PSD at 72 °C in styrene polymerization—samples taken at half-hourly intervals (semibatch, monomer federate at 1.6×10^{-4} mol s^{-1}). (a) Experimental data; (b) model simulations. From Zeaiter J. A framework for advanced/intelligent operation of emulsion polymerization reactors [PhD Thesis]. Sydney: University of Sydney; 2002.

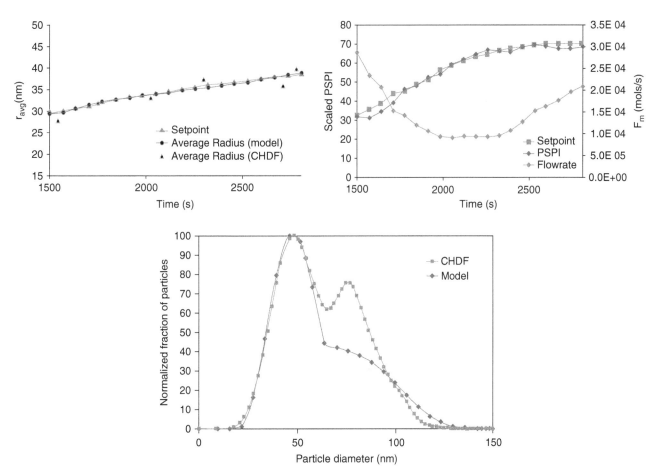

FIGURE 18.9 Experimental validation of online constrained MPC implementation. From Zeaiter J. A framework for advanced/intelligent operation of emulsion polymerization reactors [PhD Thesis]. Sydney: University of Sydney; 2002.

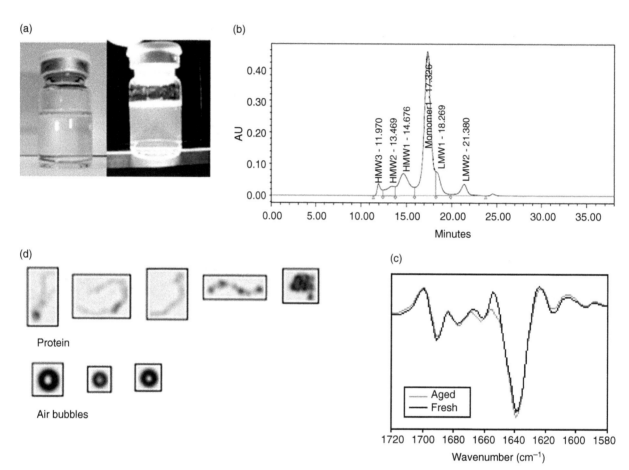

FIGURE 20.1 Analysis of a highly stressed monoclonal antibody drug product aged for ~6 years at room temperature. (a) Optically clear colorless solution under ambient room light (left); opalescent solution when viewed in a light box (right). (b) Size exclusion HPLC chromatogram showing sample degradation. (c) Second derivative FT-IR spectra of the degraded and aggregated sample (red) compared to that of an unstressed control (black). (d) Particles detected in the sample (protein aggregates and air bubbles for comparison).

SECTION 3

AUTOMATIC CONTINUOUS ONLINE MONITORING OF POLYMERIZATION REACTIONS (ACOMP)

11

BACKGROUND AND PRINCIPLES OF AUTOMATIC CONTINUOUS ONLINE MONITORING OF POLYMERIZATION REACTIONS (ACOMP)

Wayne F. Reed

11.1 OVERVIEW OF ACOMP PRINCIPLES AND APPROACH

The first work on automatic continuous online monitoring of polymerization reactions (ACOMP) was published in 1998 [1] and since then it has become an important addition to the arsenal of online monitoring methods. The principle of ACOMP is to continuously withdraw a small stream from the reactor, typically on the order of 0.050 ml min^{-1}, and dilute and condition this stream to very low polymer concentration, such that multiple fundamental properties of the reaction and polymers being produced can be characterized by a chosen combination of flow-through detectors [2]. Typical detectors include multiangle light scattering (MALS), Mie scattering, differential refractive index (RI), ultraviolet/visible (UV/VIS) absorption, viscosity, conductivity, and virtually any other desired instrument equipped to handle a flowing sample. These then yield a continuous record of kinetics, comonomer conversion, average composition drift and distribution, evolution of weight average molecular weight, M_w, and intrinsic viscosity, $[\eta]_w$, and other features, such as unexpected phenomena like microcrystallization, microgelation, premature reaction termination, and so on.

The ACOMP "front end" is the ensemble of pumps, mixing chambers, filters, and conditioning steps that prepare the continuous highly dilute and conditioned stream and deliver it to the detector train. Lag times between withdrawal and detection are typically from 10s to 100s of seconds, with similar response times, which are controlled largely by the mixing chambers' volumes and flow rates. The level of dilution is typically such that the concentration of polymer in the detector train ranges from 10^{-5} to 10^{-3} g cm^{-3}. While ACOMP is a continuous rather than discrete chromatographic method, this level of dilution is such that a periodically actuated Gel Permeation Chromatography (GPC) detection system can optionally be attached directly to the conditioned stream so that chromatographic detection can also be obtained.

Advantages of ACOMP include its versatility as a generalized approach, its ability to make fundamental measurements without recourse to empirical models and calibration, its capacity for providing a data-rich stream of complementary information from multiple independent detectors, yielding multifaceted characteristics of polymerization reactions, and its use of the "front end" to extract, dilute, and condition a sample stream that allows sensitive detectors to provide reliable data without exposing them to harsh reactor or sample conditions. Disadvantages include the mechanical complexity of the "front end," the delay time between a continuous fluid element's extraction from the reactor and downstream measurement by the detector train, and a small but continuous waste stream. ACOMP is more "invasive" than probes that can be placed at an outside reactor window, but are no more invasive than *in situ* probes, in that in either case access to the reactor contents is required.

Monitoring Polymerization Reactions: From Fundamentals to Applications, First Edition. Edited by Wayne F. Reed and Alina M. Alb.

Some guiding principles of ACOMP are listed in the following:

1. The monitoring is always adapted to the chemistry and reaction processes. The reaction processes are never interfered with or changed to suit the monitoring.

2. The quality of data obtained by each instrument is optimized through proper online sample conditioning.

3. Measurements are made at the most fundamental level possible (single scattering events, dilute regime viscosity, spectroscopy, etc.) and these are designed to obtain *model-free* primary quantities, such as conversion, composition, molar mass, intrinsic viscosity, and so on. The use of empirical and inferential models and calibration schemes is thereby avoided.

4. Obtaining high-quality data with model-free primary quantities allows the richness of the ACOMP results to be used for building chemical, physical, and mechanistic models to any degree of elaboration desired, and for potential full feedback control of reactions.

11.1.1 Comparing ACOMP with *In Situ* Methods Such As Near Infrared and Raman

While ACOMP gives the comonomer conversions, which those techniques also do, ACOMP additionally and simultaneously evolution of weight average molecular weight M_w and $[\eta]_w$, average polymer properties of critical importance in the ultimate characterization, and utilization of the polymers. Additionally, ACOMP can provide immediate detection of unforeseen or unwanted phenomena such as microgelation, runaway reaction, dead end reaction, onset of turbidity, and so on.

A seeming advantage of Raman and near infrared (NIR) compared to ACOMP is that probes for the former can often be put inside the reactor, avoiding ACOMP's complex withdrawal, dilution, and conditioning steps. The *in situ* probes also eliminate the delay times inherent to ACOMP, which are typically in the range of 50–300 s. On the other hand, whether a probe is inserted into a reactor or a tube for withdrawal is inserted for ACOMP, access into the reactor is required in either case, and hence all techniques are "invasive" to this degree. Furthermore, probes inside of reactors can easily foul and lead to erroneous data. Working at high concentrations in the reactor normally requires that empirical models and calibrations be used to interpret data [3] and other phenomena can intervene (e.g., scattering effects of emulsions) that dominate the detector's response over the desired phenomenon (e.g., monomer conversion). In fact, calibration difficulties with Raman are well known, and whole articles are devoted to them [4]. In contrast, the ACOMP "front end" (extraction, dilution, and conditioning, such as filtration, debubbling, phase inversion, etc.) is a flexible platform specifically designed to deal with the conversion of real, often impure and nonideal reactor contents into a highly conditioned, dilute, and continuous sample stream on which absolute, model-independent measurements can be made.

11.2 ACOMP INSTRUMENTATION

11.2.1 The ACOMP "Front End"

This refers to the ensemble of pumps, mixing stages, and conditioning elements that ultimately produce the diluted, conditioned stream, which continuously feeds the detection train. Extraction of liquid from the reactor typically ranges from 0.010 to 0.500 ml min^{-1}, depending on the application. Many different approaches have been taken for the front end. One system uses two high-performance liquid chromatography (HPLC) pumps, one for reactor extraction, the other for dilution, and a high pressure mixing chamber. This arrangement supplies a reliable, diluted, conditioned stream up to reactor viscosities of only about 300 centipoise (cP).

To deal with very high reactor viscosities, up to 10^6 cP, a two-stage mixing system was introduced, which consists of (i) a reactor extraction pump capable of withdrawing high viscosity fluids, such as a gear pump; (ii) a mixing stage at atmospheric pressure, which also allows any bubbles created by exothermicity or other processes in the reaction to be exhaled and excluded from the detector stream; and (iii) a high pressure mixing stage, which allows for further dilution after the low pressure stage. This normally involves five pumps [5].

Another popular front-end style uses a recirculation loop from a reactor outlet back into an inlet, powered by a gear pump or other suitable type. A mass flow controller (MFC) is introduced in a T-junction in the loop and the desired flow rate through the MFC can be set to a small fraction of the recirculation flow rate; for example, extracting 0.10 ml min^{-1} through the MFC from a 20 ml min^{-1} flow through the recirculation loop is common. The MFC itself has a feedback controlled valve for regulating the amount of reactor content that passes through it. Recirculation ensures that fresh reactor liquid is continuously sampled and reduces possible problems of spatial inhomogeneity from intermittent dip-tube sampling in a batch reactor.

11.2.2 The ACOMP Detector Train

There are no inherent limitations on which detectors can be used in ACOMP, and detector selection is made according to the needs of each monitoring situation. A standard configuration involves MALS, a differential RI detector, a UV/VIS detector, and a single capillary viscometer. Infrared, fluorescence, and conductivity detectors are other examples of instruments that can be incorporated.

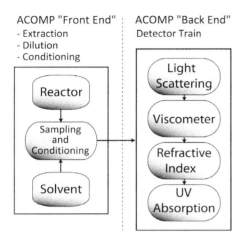

ACOMP "Front End"
- Extraction
- Dilution
- Conditioning

ACOMP "Back End"
Detector Train

Reactor

Sampling
and
Conditioning

Solvent

Light
Scattering

Viscometer

Refractive
Index

UV
Absorption

FIGURE 11.1 A simplified schematic of an ACOMP system, showing the front end used for extraction, dilution, and conditioning of reactor content, and a typical series of detectors through which the dilute, conditioned sample continuously flows.

The Tulane group developed a MALS instrument specifically to meet the demanding environment of ACOMP, and Brookhaven Instruments Corporation (BIC, Holtsville, NY) took a license from Tulane for this design and currently produces the BI-MWA (Molecular Weight Analyzer) for ACOMP, size exclusion chromatography (SEC), and batch applications (http://www.bic.com/BI-MwAmw.html).

A simplified ACOMP schematic is shown in Figure 11.1.

11.2.2.1 *ACOMP Delay Time and Response Time* Because ACOMP involves continuous withdrawal, dilution, and conditioning of reactor liquid, there is inevitably both (i) a delay time between when a fluid element is withdrawn and when a measurement of its properties is made by the detector train, and (ii) a system response time associated with the various mixing processes involved.

Delay times in ACOMP typically run from 10s to 100s of seconds, and depend on the flow rates, degree of dilution, the type of reaction, and the tubing and mixing chambers employed. In some cases, where flow rate from the reactor is variable, the detector signals can be used to continuously compute changing delay times [1].

The signal measured by a detector $D(t)$ is related to the actual physical signal in the reactor, $S(t)$, such as monomer concentration, via the ACOMP system response function $R(t)$ according to:

$$D(t) = \int_{-\infty}^{\infty} R(t-\tau) S(\tau) d\tau \qquad (11.1)$$

$S(t)$, the true concentration or other physical signal, is recovered through deconvolution. One approach to deconvolution is the convolution theorem which states that for two functions f and g under suitable conditions, a Fourier transform of the convolution of f and g, denoted as $f * g$, is a product of Fourier transforms, that is:

$$F\{f * g\} = F\{f\} \cdot F\{g\} \qquad (11.2)$$

In principle, a signal can be by taking the reverse Fourier transform of the ratio of Fourier transforms of the measured (or detected) signal $D(t)$ and the response function $R(t)$:

$$S(t) = F^{-1}\left\{ \frac{D(t)}{F\{R(t)\}} \right\} \qquad (11.3)$$

However, this approach is extremely sensitive to noise and can fail to produce good results for experimental data with noise. One successful approach was a Bayesian-based iterative method [6], which treats point spread functions as probability–frequency functions and applies Bayes's theorem [7].

The following data, illustrated in Figure 11.2, show an embodiment of this approach [8].

The inset to Figure 11.2b shows the ACOMP system response function $R(t)$, which was determined experimentally by rapidly injecting a small volume of acrylamide (Am) into the reactor to cause a step function in concentration, shown in Figure 11.2a (measured by the UV absorbance in this case):

$$S(t) = \begin{cases} 1 & t \geq t_0 \\ 0 & t < t_0 \end{cases} \qquad (11.4)$$

The corresponding $D(t)$ for this is shown in Figure 11.2b, and $R(t - t_0)$ is computed from the derivative of $D(t)$. t_0 is the lag time between withdrawal of a fluid element from the reactor and its detection and is determined directly from the time between Am injection and the leading edge of the detector signal. In the limit of instantaneous system response, $R(t - t_0)$ becomes the delta function $\delta(t - t_0)$:

$$\frac{dD(t)}{dt} = \int_{-t_0}^{\infty} \frac{d}{dt} R(t-\tau) d\tau = R(t-t_0) \qquad (11.5)$$

The deconvolved signal for Am conversion in Figure 11.2b, taken from an Am polymerization reaction, is not greatly different from the measured signal, but it eliminates the initial curvature in the measured data, providing a sharp point at the beginning of the reaction and high enough resolution to resolve fine kinetic details.

Except in cases where such kinetic resolution is required, it is not normally necessary to apply a deconvolution method, and $D(t)$ can be taken to equal $S(t)$ to a very good approximation.

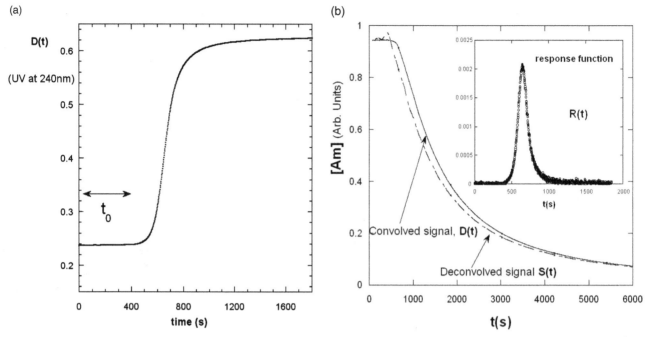

FIGURE 11.2 (a) The measured function $D(t)$ caused, via Equation 11.1, by a step function change $S(t)$, given by Equation 11.4, due to a very rapid change in the concentration of a solute in the reactor. A small amount of acrylamide (Am) was quickly injected into the reactor to cause the step change in the reactor. $D(t)$ is measured by the increase in UV absorbance at 240 nm. (b) The convolved signal $D(t)$ during an Am polymerization reaction. $D(t)$ is the concentration of Am in the reactor as the polymerization proceeds, obtained from the UV signal at 240 nm. In the inset: response function of the system, from the step function response in Figure 11.2a computed according to Equation 11.5. The deconvolved signal $S(t)$ is also shown, from Equations 11.1–11.5. Reprinted with permission, from Kreft T, Reed WF. Direct monitoring of the cross-over from diffusion controlled to decomposition controlled initiation in free radical polymerization. Macromol Chem Phys 2008;209:2463–2474.

11.3   MEASURING SPECIFIC CHARACTERISTICS

11.3.1   Computing Cumulative and Instantaneous Values of Reaction Characteristics

Whenever a reaction is monitored, the signal measured at any moment, whether viscosity, spectroscopic, SEC on a withdrawn aliquot, and so on, yields the cumulative value of the quantity being measured. For example, when a static light scattering (SLS) measurement is made at any instant, t, whether by MALS or at a single angle, the cumulative weight average molecular weight, $M_w(t)$, of the entire reactor content is determined. It is possible, nonetheless, to determine instantaneous values of a quantity via derivatives of cumulative measurements.

It is easily shown that any instantaneous quantity $Q(x)$ is related to its cumulative value $q(x)$ via:

$$Q(x) = \frac{d[xq(x)]}{dx} = q(x) + x\frac{dq}{dx} \qquad (11.6)$$

where the dependent variable x is usually either time t, or a measure of monomer or comonomer conversion, f.

The instantaneous value of weight average molecular weight, $M_{w,inst}(f)$, at any instant of conversion (or time, substituting t for f) is [9]:

$$M_{w,inst}(f) = M_w(f) + f\frac{dM_w(f)}{df} \qquad (11.7)$$

Similarly, the instantaneous average fraction of comonomer j in a copolymer chain, $F_{inst,j}(f)$, is given by:

$$F_{inst,j}(f) = y_j(f) + f\frac{dy_j(f)}{df} \qquad (11.8)$$

where y_j is the cumulative mass fraction of copolymer composed of comonomer j:

$$y_j = \frac{c_j}{c} \qquad (11.9)$$

The method can be extended to any other measured quantity, such as instantaneous reduced viscosity, hydrodynamic radius, and so on.

The main problem with the procedure is that it requires taking derivatives of data, and real data always contain some level of noise. Hence, for example, if the sampling rate of the ACOMP detectors is 1 Hz, taking differences, and hence derivatives in measured quantities at this rate $\Delta Q/\Delta t$ can lead to meaningless results. There are several strategies for circumventing this problem. One is simply to take sampling intervals over which the change in the detector values change measurably more than the inherent signal/noise level of the detector. An additional step to this procedure can be to first smooth the data, by averaging over a moving window of a fixed number of points, and then taking the derivative.

Another method that generally produces even more satisfactory results is to use a fitting procedure over segments of the data. Both analytical and numerical fits are possible. The simplicity of using analytical fits and the ability to take exact derivatives is sometimes offset by adding a systematic bias to the data by the assumed nature of the fit. Hence, unless there is a compelling reason to choose a specific functional form (e.g., a first- or second-order fit, polynomial, sigmoid, power law, etc.), it is possible to introduce artifactual structure into the data and the interpretation.

Numerical methods for interpolating between data points or over data point intervals are often better, since no *a priori* assumptions are made. There is a very wide variety of interpolation methods available, such as cubic splines, which guarantee continuity and differentiability of the interpolated data. Some programs, such as Mathematica®, have a wide variety of such menus built into their powerful data handling and representation algorithms.

Figure 11.3 shows an example of $M_{w,inst}$ found from the cumulative M_w by Equation 11.7. A smoothing type interpolation to the M_w data was used, on which the derivative was computed in Equation (11.7).

The data are from a free radical polymerization of butyl acrylate (BA) in butyl acetate. When fractional monomer conversion reached 0.4, an extra amount of azobisisobutyronitrile (AIBN) initiator was added (initiator boost). Its effect can be immediately seen in the rapid drop of $M_{w,inst}$, as the quasi-steady state approximation (QSSA) predicts for kinetic chain length (e.g., see Chapters 1 and 5) which is the molecular weight of chains being produced at any instant in a free radical reactions. It is proportional to the concentration of monomer to that of initiator. Hence, the addition of initiator causes the instantaneous chain length, and hence $M_{w,inst}$ to fall.

11.3.2 Use of Refractometer (RI) for Reactant Concentration Monitoring

In traditional SEC applications, the difference in index of refraction between a polymeric solute and the pure solvent

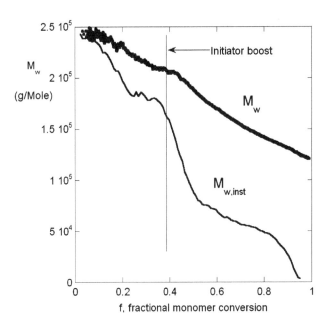

FIGURE 11.3 Instantaneous weight average molecular weight $M_{w,inst}$ derived from cumulative weight average molecular weight via Equation 11.7. The steep drop in $M_{w,inst}$ occurs rapidly after addition of an AIBN initiator "boost" to the free radical polymerization of butyl acrylate in butyl acetate. Adapted with permission from Alb AM, Drenski MF, Reed WF. Simultaneous continuous, non-chromatographic monitoring and discrete chromatographic monitoring of polymerization reactions. J Appl Polym Sci 2009; 13:190–198.

is dissolved in acts as a measure of the solute's concentration, if the differential refractive index increment, $\partial n/\partial c$, is known. If the refractometer yields a net voltage above the solvent baseline of V, and its calibration factor is CF (Δn volt^{-1}), then the solute's concentration is given by:

$$c = \frac{V * CF}{\partial n / \partial c} \qquad (11.10)$$

In the reaction monitoring context, typically, the refractive index of a polymer is greater in a given solvent than that of its corresponding monomer. This allows the use of refractometry to follow conversion, as some examples in Chapters 12 and 13 will demonstrate.

Most flow-type refractometers use a balanced, split photodiode design to measure the angular deflection of a light ray passing through a split cell, one half of which contains the pure solvent and the other half contains the solvent with polymer. Refractometers usually require special care for stabilization. Their flow cells are often too fragile to withstand much back pressure and so are often placed at the end of a detector train.

11.3.3 UV/VIS Spectrophotometer for Multiple Reactant Concentrations

This widespread method is used for concentration detection, since, for a given chemical species in a given solvent there is normally a well-defined, repeatable UV/VIS absorption spectrum, characterized by an absorption cross-section at each wavelength, usually termed an "extinction coefficient," $\varepsilon(\lambda)$, (which can be expressed in either molar or mass concentration terms), related to solute concentration, c, via

$$A(\lambda) = \varepsilon(\lambda)cl \qquad (11.11)$$

where l is the pathlength of the cell and $A(\lambda)$ is the absorption related to the intensity of light emerging from the cell $I(\lambda)$, for an incident intensity of $I_0(\lambda)$ as follows:

$$I(\lambda) = I_0(\lambda)10^{-A(\lambda)} \qquad (11.12)$$

Most common monomers have little or no VIS absorption, but frequently have measurable absorbance in the 190–300 nm UV range. It should be noted that since the transmitted intensity is measured, measured, $\varepsilon(\lambda)$ contains all terms that remove photons from the beam traversing the sample, including both absorption *per se* and scattering. This can lead to necessary corrections in UV-based concentration measurements during polymerization, since, although $\varepsilon(\lambda)$ for the converting monomer is a constant, the total $\varepsilon(\lambda)$, due to monomer and polymer contributions, can be increasing as more polymer is produced and scatters more light. This effect is generally small but can sometimes be significant. When particulate populations appear during polymerization reactions, for example, microgels, their effect on UV signals must be taken into account.

Modern UV/VIS spectrophotometers at moderate prices are available with diode array detection that offer rapid sampling and readout (>1 Hz) of the entire 190–800 nm spectrum using flow equipped cells.

11.3.4 Viscometry for Bulk, Reduced, and Intrinsic Viscosity Monitoring

The physical meaning of viscosity was discussed in Chapter 5. In the online monitoring context, a very simple and robust means of measuring total dilute solution viscosity, η, is with a single capillary viscometer. With a measurement of the pure solvent viscosity, η_s, and knowledge of the polymer concentration, c, the reduced viscosity, η_r, is given by

$$\eta_r \equiv \frac{\eta - \eta_s}{\eta_s c} = [\eta] + k_p[\eta]^2 c + O(c^2) \qquad (11.13)$$

where k_p is a constant related to the hydrodynamic interactions between polymer chains, usually around 0.4 for neutral,

coil polymers [10], and $O(c^2)$ represents terms of order c^2 and higher.

Polymer intrinsic viscosity, $[\eta]$, is the extrapolation to zero concentration and zero shear rate of the reduced viscosity, η_r. In both SEC and ACOMP, the concentration is usually low enough to a good approximation $[\eta] \cong \eta_r$, where η_r is measured by combining viscometer and concentration detector data.

Several geometries can be used to create velocity gradients in fluids for the computation of η. One of the most common arrangements is the capillary viscometer, for which the Poiseuille solution to the Navier–Stokes equation, which has a parabolic flow profile, is used:

$$\eta = \frac{\pi R^4 \Delta P}{8LQ} \qquad (11.14)$$

where Q is the flow rate of solution through the capillary (in $cm^3 \, s^{-1}$) of radius R, across whose length L there is a pressure drop ΔP.

The average shear rate in the capillary is found by integration over the capillary cross-section to be:

$$|\dot{\gamma}_{ave}| = \frac{8Q}{3\pi R^3} \qquad (11.15)$$

for which a typical value is 860 s^{-1}, given $Q = 1$ ml min^{-1}, and $R = 0.0254$ cm. Much lower average shear rates can be obtained by using capillaries of larger diameter (and correspondingly much greater length if the same pressure difference is to be kept).

Single capillary viscometers are remarkably easy and inexpensive to construct using a modern differential pressure transducer with a double T-network and capillary of given L and R.

Capillary bridge viscometers, based on the Wheatstone bridge principle of null measurement, are inherently more sensitive than single capillary viscometers, and can largely eliminate temperature and pump pressure fluctuations. The latter are most useful for SEC but are normally not suitable for long-term, continuous flow measurements, however, since a hold-up volume is used in one or more bridge arms, which leads to zeroing out of the differential signal with time. Single and bridge capillary viscometer results have been exhaustively compared [11].

11.3.5 Light Scattering Detection for Molecular Weight, Dimensions, and Other Characteristics

Light scattering was discussed in detail in Chapter 8, and some important notions were also provided in Chapter 5. Some of the key notions used in practical online light scattering are distilled here.

Most light scattering determinations of polymer mass and other characteristics assume a single scattering event of an incident photon that is subsequently detected. There is none-theless a growing literature on light scattering used directly for concentrated solutions in which multiple scattering occurs. Turbidity measurements are probably the simplest example of the latter. For both forward and backscattered light more sophisticated approaches to analyzing multiply scattered light exist [12–15]. For example, in diffusing wave spectroscopy (DWS), light scatters so many times that photons essentially undergo a random walk (diffusion) in the sample medium before reaching the detector, in which case DWS measurements can yield the mean free photon path-length as well as self-diffusion coefficients, and hence average particulate concentration and estimates of particle sizes [16, 17], in addition to micro-rheological characteristics [18–20].

For dilute polymer solutions Rayleigh–Debye theory is often used (sometimes referred to as Rayleigh–Gans–Debye theory, as in Chapter 8), with its very useful extensions due to Zimm. The underlying assumption for this theory to be applicable is that the optical pathlength differences in light scattered from different regions of a given particle (e.g., polymer chain) are due overwhelmingly to the geometrical path differences involved; that is, that the contribution to optical path difference due to the different index of refraction of the particle compared to the supporting medium is negligible. This holds particularly well for most polymers, since these can be considered as threadlike molecules in a solution, so that when scattering occurs from any element of the polymer the light then traverses just the solvent in reaching the detector (in single scattering events). In contrast, when nano- and microscopic solid or semisolid particles are involved, the scattered light can traverse a considerable distance within the scattering particle itself and hence accrue optical path difference due to the difference in particle and solvent indices of refraction.

The condition for the Rayleigh–Debye and Zimm theories to apply can be summarized as:

$$\frac{2\pi |\Delta n| a}{\lambda / n_s} \ll 1 \qquad (11.16)$$

where $\Delta n = n_p - n_s$ (n_p and n_s are the real part of the indices of refraction of the particle and supporting medium, respectively) and λ is the vacuum wavelength of the incident light.

When this condition is met, one of Zimm's most useful expressions is applicable [21]:

$$\frac{Kc}{R(c,q)} = \frac{1}{MP(q)} + 2A_2 c$$
$$+ \left[3A_3 Q(q) - 4A_2^2 MP(q)(1 - P(q)) \right] c^2 + O(c^3) \qquad (11.17)$$

where $R(c, q)$ is the excess Rayleigh scattering ratio, that is, the scattering from the polymer solution at concentration c (g cm^{-3}) measured at q minus the scattering from the pure solvent R_s.

$R(c, q)$ is obtained by dividing the scattered intensity per unit volume occupied by scatterers by the incident intensity, and multiplying by r^2 (r is distance from scatterer to detector) to eliminate the dependence on the detectors' distance from the scattering volume. $R(c, q)$, which has units of cm^{-1}, can be interpreted as the fraction of the incident intensity scattered per steradian of solid angle per centimeter of scattering media traversed. The Rayleigh scattering ratio R, for several pure liquids at a variety of wavelengths and temperature, is known to high precision; that is, R and hence calibration of light scattering detectors with such liquids is "absolute," not empirical. For example, $R = 1.069 \times 10^{-5}$ (cm^{-1}) for toluene at $T = 25°C$ when light of $\lambda = 677$ nm is incident.

This means, in practical terms, that R for any polymer solution can be determined simply by comparing the ratio of the scattering detector voltage from the polymer solution to the voltage found by scattering from pure toluene, and making any geometrical optical corrections that might be required due to cell geometry and refractive effects [22–24]. Hence, a solvent such as toluene firmly anchors light scattering measurements to absolute values of R. In turn, R is related to fundamental polymer properties.

In Equation 11.17, M is the scatterer's molar mass, $P(q)$ is the particle form factor, which sums the phase differences over all scattering centers of the particle. A_2 and A_3 are the second and third virial coefficients, respectively, and the magnitude of the scattering vector, q, is given by

$$q = \frac{4\pi n_s}{\lambda} \sin\left(\frac{\theta}{2}\right) \qquad (11.18)$$

where θ is the detection angle, K is an optical constant, given for vertically polarized incident light by

$$K = \frac{4\pi^2 n_s^2 (\partial n / \partial c)^2}{N_A \lambda^4} \qquad (11.19)$$

λ is the vacuum wavelength of the incident light, N_A is Avogadro's number, and $Q(q)$ involves a sum of Fourier transforms of the segment interactions that define A_2. $\partial n / \partial c$ is the differential refractive index for the polymer in the solvent and embodies the Clausius–Mossotti equation for a dilute solution of particle density N, which relates α, the molecular polarizability, to the index of refraction n of the polymer solution,

$$n^2 - n_s^2 = 4\pi N \alpha \qquad (11.20)$$

Most water soluble polymers have a positive value of $\partial n / \partial c$, chiefly because $n_s = 1.33$ for water is low compared to most

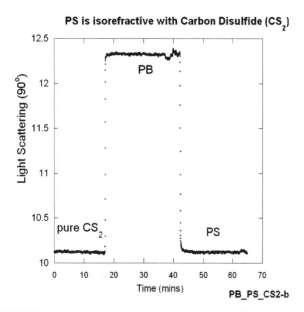

FIGURE 11.4 Isorefractivity of polystyrene (PS) in carbon disulfide (CS$_2$); that is there is no excess scattering due to PS when dissolved in CS$_2$. In contrast, polybutadiene (PB) shows strong scattering in CS$_2$ since its refractive index differs significantly from that of CS$_2$.

organic substances, whereas organo-soluble polymers in organic solvents frequently have negative values of $\partial n/\partial c$. When $\partial n/\partial c = 0$, the polymer/solvent pair is termed as "iso-refractive" and there is no excess scattering due to the polymer. $\partial n/\partial c$ also allows computation of the polymer concentration, c, in Equation 11.10 using a differential refractometer (RI).

An example of isorefractivity is given in Figure 11.4. The scattering from a carbon disulfide (CS$_2$) solution containing a mixture of 0.5 mg ml^{-1} polystyrene (PS) and 0.5 mg ml^{-1} polybutadiene (PB) reaction endproducts is shown.

As seen, the PB scatters strongly, whereas the PS does not scatter at all; that is, PS is isorefractive with CS$_2$. Hence, CS$_2$ can be used in ACOMP if one is interested in knowing the evolution of just the PB molecular weight in a PB/PS copolymerization reaction. It was also found that PB is nearly isorefractive with orthoxylene, so that PS molecular weight in the PB/PS copolymerization could be followed. Because ACOMP typically uses large dilutions, both PB and PS molecular weights could be monitored simultaneously. This would be accomplished by splitting the extracted reactor stream into two parts, one diluted by CS$_2$ and the other by orthoxylene. Since the ACOMP dilutions are normally quite large, any solvent used in the reactor would be diluted to insignificance in the process. For bulk reactions, there would be no reactor solvent. An example of this is the production of high impact polystyrene (HIPS), where bulk styrene monomer is polymerized in the presence of PB chains.

Another issue related to $\partial n/\partial c$ in the light scattering context is when polyelectrolytes are involved. $\partial n/\partial c$ should be determined at chemical equilibrium for all species, including the polyelectrolyte counterions, since this equilibrium condition will apply when making scattering measurements. Hence, in making the dilutions of the polyelectrolyte to determine the value of $\partial n/\partial c$, ionic solution in dialysis equilibrium with the polyelectrolyte should be used, in principle [25, 26]. In practice, it is often found that the errors in polymer concentration introduced by swelling in dialysis membrane tubes are larger than the effect being corrected for, so that use of the same ionic strength solution as that in which the polyelectrolyte is dissolved is expedient.

Equilibrium characterization of polymers usually takes place in dilute solution where $1 \gg 2A_2cM_w \gg 3A_3c^2M_w$ and over an angular range such that $q^2\langle S^2\rangle < 1$. In this case, the Zimm equation reduces to one of its most frequently used forms:

$$\frac{Kc}{R(c,q)} = \frac{1}{M_w}\left(1 + \frac{q^2 \langle S^2 \rangle_z}{3}\right) + 2A_2c \quad (11.21)$$

A typical batch experiment involves measuring $R(c, q)$ over a series of angles for several polymer solutions at different concentration. If for each concentration, the angular data are extrapolated to $q = 0$, and these points are then fit, the slope will yield $2A_2$, and the y-intercept will be $1/M_w$. Similarly, if the concentration points at each angle are extrapolated to $c = 0$ and these are then fit, the slope will be $\langle S^2\rangle_z/3M_w$ and the intercept is again $1/M_w$. Hence, the "Zimm technique" allows determination of M_w, A_2, and $\langle S^2\rangle_z$. The root mean square radius of gyration is often simply called the "radius of gyration" and abbreviated as $R_g \equiv \langle S^2\rangle^{1/2}$. Nothing impedes the use of the Zimm approach in time-dependent situations, as long as one is still operating within the range of validity of the approximations.

In practical terms, $R(c, q)$ can be computed from the scattering voltage $V(c, q)$ at any concentration c of solute and scattering vector amplitude q according to:

$$R(c,q) = \frac{V(c,q) - V_s(q)}{V_{ref}(q_r) - V_d(q_r)} N(q)R_{ref}F \quad (11.22)$$

where $V_{ref}(q_r)$ and $V_d(q_r)$ are the scattering voltages of the calibration reference solvent (e.g., toluene) and the dark voltage, respectively, at the reference amplitude vector q_r (corresponding to reference angle θ_r, usually chosen as 90°).

In the aforementioned equation, F is an optical constant which accounts for two effects: First of all, there is a reflection loss at each interface, and, second, because of refraction effects in two dimensions at the sample cell interface, the luminosity of the sample volume will appear smaller for solvents of higher index of refraction. Hence, if samples are measured in solvents different from that of the absolute calibration solvent, a correction for this effect must be made, which depends both on the cell geometry and the solvents used.

While $R(c, q)$ could be determined at each value of q with respect to the reference solvent scattering at that q, practical considerations concerning the higher probability of stray light at angles away from 90° make use of a preferable "normalization" procedure, where the normalization factor $N(q)$ is defined as:

$$N(q) = \frac{V_n(q_r) - V_s(q_r)}{V_n(q) - V_s(q)}$$ (11.23)

where $V_n(q_r)$ is the scattering voltage from the normalization solution at the scattering vector q_r that corresponds to the reference angle θ_r, $V_s(q_r)$ is the scattering voltage at q_r from the pure solvent the normalization solution is made in, and $V_n(q)$ and $V_s(q)$ are the normalization solution and pure solvent scattering voltages, respectively, at an angle θ. A normalization solution contains Rayleigh scatterers, small, preferably massive particles that scatter isotropically in the scattering plane.

For aqueous measurements, small latex spheres, for example, of $R_g < 10$ nm are suitable, whereas in organic solvents PS of $M < 50,000$ g mol^{-1} is suitable.

In practice, some workers use polymer molecular weight standards instead of the aforementioned procedure. Using standards has the virtue of simplicity and making it often possible to use the same solvent for the calibration standards as that in which the polymers to be measured are dissolved. This avoids the use of the optical correction factor, F. Its drawbacks are that polymer standards can sometimes degrade, leading to unreliable calibration, and that by using standards the calibration itself is no longer absolute as it is in the case of solvents with known R.

When Equation 11.17 no longer applies, as is often the case for many emulsions, latex particles, and other colloids (e.g., a typical latex particle in water 100 nm or larger), the difference in index of refraction of a particle and the surrounding medium can contribute significantly to optical path lengths of scattered light. In this case, exact solutions of the scattered waves are found using Maxwell's equations and associated boundary conditions for specified particle morphologies. Such solutions were first obtained for the dielectric spheres by Mie [27, 28], and other solutions by Lord Rayleigh [29], and this approach to scattering analysis is often termed "Mie scattering" and involves infinite series of special kinds of Bessel functions. Computation of the scattering solution is computationally intensive, but quickly performed by even modest desktop or notebook computers. Some modern particle sizers, for example, by Malvern and Shimadzu, use Mie scattering analysis for particle size analysis.

Dynamic light scattering (DLS), in the single scattering limit, is frequently used as a means of determining particle sizes [30, 31]. In its most widely used form, it is based on autocorrelating the scattered intensity of light, and relating the autocorrelation function to the mutual diffusion coefficient of the particles. DLS is typically robust over the range of 1 nm to several microns. In this sense, it is complementary to Mie scattering which usually has poor resolution below 100 nm, but is robust up to hundreds of microns. The autocorrelation function is fundamentally very sensitive to flow and attempts to adapt DLS to flow cells require either that they have a stop-flow mechanism or that the flow be low enough and the particles small enough that they not be affected by flow terms in the autocorrelation analysis. For example, Brookhaven Instruments Corporation's Nano-DLS uses an automatic stop-flow device with a bypass valve to ensure that the DLS measurements are made periodically on nonflowing solutions, thus avoiding the serious effects of flow on the results.

Whether DLS, DWS, Mie scattering, or other applications in which unfractionated samples are analyzed, the resulting "distributions" produced by modern instruments, while frequently facile to obtain and neat in appearance, must be treated with caution, as there is usually a large amount of data smoothing, fitting, and assumptions applied in using inverse Laplace transform and several other commonly employed methods. The best means of finding distributions of size and mass continue to be fractionation methods, such as SEC [32–34], field flow fractionation (FFF) [35–37], capillary electrophoresis [38], capillary hydrodynamic fractionation [39], and so on.

11.3.5.1 *Light Scattering and Determination of Copolymer Molar Mass*
It was recognized long ago that the Zimm approximation used to obtain molecular weight of homopolymers by light scattering [40] can fail dramatically for copolymers, when the value of $\partial n/\partial c$ of the different copolymer constituents is great and there is considerable compositional polydispersity [41, 42]. Light scattering determination of weight average molecular mass M_w of a two component copolymer has historically required that measurements be made in at least three separate solvents of different index of refraction. This traditional method quickly becomes intractable for terpolymers and higher order copolymers since the number of solvents required grows as $N(N + 1)/2$, where N is the number of comonomers and the number of auxiliary measurements of differential index of refraction $\partial n/\partial c$ for each copolymeric component in the solvents is $N(N + 1)^2/2$; for example, for a terpolymer, six solvents are needed and 24 differential indices of refraction. Further treatment of copolymers with different values of $\partial n/\partial c$ for its comonomeric components is given in Section 8.3 of Chapter 8.

Automatic continuous online monitoring of polymerization reactions surmounts this problem and allows M_w to be measured in a single solvent because the instantaneous composition of copolymers can be determined during synthesis, which, together with the light scattering data stream yields M_w at each instant.

The detailed derivation of the appropriate relationships between scattering, refractive index increments, and true and apparent molecular weight has been recently presented, including the generalization to an arbitrary number of comonomers N [43]. A brief outline of the major results are mentioned in the following section.

Following the homopolymer approach of extrapolating scattering data to zero concentration and zero scattering angle, one obtains an apparent mass M_{ap} that is related to the true weight average mass M_w of the copolymer via the differential refractive increments $v_j = \partial n / \partial c_j$, of each comonomeric component $j = 1, 2, \ldots, N$, where c_j is the mass concentration (g cm^{-3}) of monomer j, and the respective weight average of each comonomer in copolymeric form, $M_{w1}, M_{w2}, \ldots, M_{wN}$.

A convenient form of the Zimm equation for copolymer analysis, which is only slightly different than Equation 11.21, is:

$$\frac{K'v^2c}{R(q,c)} = \frac{1}{M_w}\left(1 + \frac{q^2 <S^2>_z}{3}\right) + 2A_2c \quad (11.24)$$

where K' is an optical constant, which for vertically polarized light is given by (this form differs from that of Equation 11.19 by the absence of the term $(\partial n/\partial c)^2$):

$$K' = \frac{4\pi^2 n^2}{N_A \lambda^4} \quad (11.25)$$

where v is the $\partial n/\partial c$ of the copolymer, which can be measured directly.

For copolymers, $\partial n/\partial c$ is simply the mass weighted sum of the refractivities of the corresponding homopolymer of each comonomer constituent:

$$v = \frac{\partial n}{\partial c} = \sum_{j=1}^{N} y_j v_j \quad (11.26)$$

The quantity $K'v^2c/R(q, c)$ extrapolated to $c = 0$ and to $q = 0$ yields only an apparent molecular weight M_{ap}:

$$\frac{1}{M_{ap}} = \begin{array}{c} \lim \\ c \to 0 \\ q \to 0 \end{array} \frac{K'v^2c}{R(q,c)} \quad (11.27)$$

M_{ap} is related to the true copolymer M_w as discussed in the following equation. Benoit's expression for the $N = 2$ case is:

$$M_{ap}v^2 = v_1 v_2 M_w + v_1(v_1 - v_2)y_1 M_{w,1} + v_2(v_2 - v_1)(1 - y_1)M_{w,2} \quad (11.28)$$

where $M_{w,j}$ is the weight average of the portion of copolymer constituted by comonomer j, defined with the ACOMP Δc_i notation as:

$$M_{w,j} = \frac{\sum F_{inst,j,i}^2 \Delta c_i M_i}{y_j c} \quad (11.29)$$

where the summation is over all measurement intervals i during the ACOMP monitored reaction. M_w has the usual definition:

$$M_w = \frac{\sum \Delta c_i M_i}{c} \quad (11.30)$$

11.3.5.2 Use of Average Comonomer Composition Parameters

Benoit provided a useful alternative representation of M_{ap} in terms of mass weighted moments of the copolymer composition distribution. Generalizing his method to N comonomers, $F_{inst,j,i}$ is expressed as:

$$F_{inst,j,i} = y_j + \delta_{j,i} \quad (11.31)$$

where $\delta_{j,i}$ is the deviation of the fractional composition of comonomer j in the element i from the average fraction y_j of comonomer j in the whole copolymer population. This yields an alternative to Equation 11.28:

$$M_{ap} = M_w + 2\left(\frac{v_1 - v_2}{v}\right)P_1 + \left(\frac{v_1 - v_2}{v}\right)^2 Q_1 \quad (11.32)$$

where P_1 is the mass-weighted first moment of the composition distribution for comonomer 1:

$$P_1 = \frac{\sum c_i M_i \delta_{1,i}}{c} \quad (11.33)$$

and Q_1 is the mass-weighted second moment of the composition distribution for comonomer 1:

$$Q_1 = \frac{\sum c_i M_i \delta_{1,i}^2}{c} \quad (11.34)$$

For the case where $N = 2$, $P_2 = -P_1$, and $Q_2 = Q_1$, Benoit and Bushuk have noted that P_1 and Q_1 vary between the limits:

$$-y_1 M_w \leq P_1 \leq (1 - y_1)M_w \quad (11.35a)$$

$$0 \leq Q_1 \leq [1 - y_1(1 - y_1)]M_w \quad (11.35b)$$

If the mass distribution is monodisperse (i.e., M_i = constant) then $P_1 = 0$, although $Q_1 \neq 0$ in this case, unless there is also no compositional heterogeneity. In the case where the copolymer sample consists of a mixture of two homopolymers, then $P_1 = y_1(1 - y_1)[M_{w,1} - M_{w,2}]$ and $Q_1 = y_1(1 - y_1)[(1 - y_1)M_{w,1} + y_1 M_{w,2}]$.

11.3.5.3 *Generalization to N Comonomers* Equation 11.28 was recently generalized to arbitrary N:

$$v^2 M_{ap} = \sum_{j=1}^{N} y_j v_j^2 M_{w,j} + 2M_w v_1 v_2$$
$$+ \frac{2}{c} \sum_i \Delta c_i M_i \left\{ \Omega_i v_1 v_2 + \sum_{k=j+2}^{N} \sum_{j=1}^{k-1} F_{\text{inst},j,i} F_{\text{inst},k,i} v_j v_k \right\}$$
(11.36)

where

$$\Omega_i = -F_{\text{inst},1,i} - F_{\text{inst},2,i} - 2\sum_{j=3}^{N} F_{\text{inst},j,i} + F_{\text{inst},2,i} \sum_{j=3}^{N} F_{\text{inst},j,i}$$
$$+ F_{1,i} \sum_{j=2}^{N} F_{\text{inst},j,i} + \sum_{j=3}^{N} F_{\text{inst},j,i}^2 + 2 \sum_{k=j+1}^{N} \sum_{j=3}^{k-1} F_{\text{inst},j,i} F_{\text{inst},k,i}$$
(11.37)

As mentioned, the minimum number N of different solvents in which scattering measurements on the copolymers must be made increases at the alarming rate of $N(N + 1)/2$ and the number of refractive increments needed grows at the even higher rate of $N(N + 1)^2/2$. For the simplest case, that of $N = 2$, nine different differential refractive index values must be known. For a terpolymer, this becomes 24 values, and for $N = 5$, 90 values! Clearly an alternative method is needed.

ACOMP provides a direct means of determination of M_w for N comonomers in a single solvent. Only $N + 1$ values of $\partial n/\partial c$ are needed; for example, only 4 are needed for a terpolymer characterization versus 24 values for the multisolvent method.

The relationship between the true M_w and M_{ap} at any instant of conversion f is given by:

$$M_W(f) = \frac{1}{f} \int \frac{d\left[f' M_{ap}(f') \left(\sum_j y_j(f') v_j \right)^2 \right]}{\left(\sum_j F_{\text{inst},j}(f') v_j \right)^2} \quad (11.38)$$

Automatic continuous online monitoring of polymerization reactions allows all the required variables to be measured at each point in time; total comonomer conversion f, apparent

molecular weight M_{ap}, total fraction y_j of comonomer j in polymeric form, and instantaneous composition of j in copolymer form $F_{\text{inst},j}$. The $N + 1$ values of v_j are constants determined in separate auxiliary measurements.

While this integral may seem formidable to integrate, especially in an experimental situation, it is actually very convenient and simple to perform numerically. The independent variable during an ACOMP experiment is time, against which *all* ACOMP values are *automatically* parametrized. Hence f, y_j, M_{ap}, and y_j, once computed from the relevant ACOMP detector signals, are automatically parametrized against t, and any combination of values can be parametrized against any other combination. Hence, the very peculiar differential $d[v(f')^2 f' M_{ap}(f')]$ in Equation 11.36 can be formed, once $F_{\text{inst},j}$ is also determined and the denominator in the integrand, $v'(f')^2$, also formed, and the numerical integration of the latter against the former can be carried out immediately.

Once $M_w(f)$ is obtained, other characteristics of the copolymer can be computed during synthesis, including instantaneous values $M_{w1}, M_{w2}, ..., M_{wN}$, as well as P_1, Q_1, and any mass averaged or other moments of the composition distribution.

Thus, ACOMP both avoids the use of multiple solvents and provides continuous values of composition and molar masses. This latter allows the possibility of controlling reactions to provide copolymers of desired composition and mass distributions, and also means the copolymer is "born characterized," which can avoid tedious and costly "postmortem" analyses of the endproduct.

11.3.5.4 *Some Important Limiting Cases for N = 2* As pointed out, $M_{ap} = M_w$ if there is no composition drift, or if $v_1 = v_2$. An important case concerns water soluble polymers. The vast majority of water soluble polymers in aqueous solutions (without extremely high ionic strength) typically range from 0.1 to 0.2 cm^3 g^{-1}; for example, most proteins have $\partial n/\partial c \sim 0.15$. M_{ap}/M_w of most aqueous copolymers will hence have values not far from unity for cases of small to moderate drift.

Another case involves acrylic monomers, where, in general $\partial n/\partial c$ are quite similar in a given solvent. Hence, except in the cases where $\partial n/\partial c$ is close to zero, or the values straddle zero, $M_{ap} \sim M_w$. Table 11.1 summarizes some limiting cases.

11.3.5.5 *Interpolymer Interactions: Angular Dependent Scattering* In the case where the polymer size is not constrained to be much smaller than λ and $\partial n/\partial c$ of one or more comonomers is close to 0, the angular dependence of the Zimm plot (Kc/R vs. q^2) becomes distorted and the $\langle S^2 \rangle_z$ obtained is now an apparent value with a complicated relationship to the copolymeric contributions to both polymer radius of gyration and refractivity [44, 45].

The relationship between scattering and intermolecular excluded volume is likewise more complex than in the

TABLE 11.1 Some Limiting Cases for $N = 2$

Type of System	Conditions on v_1, v_2	Restrictions on Composition Drift	Restrictions on M_{ap} Drift	Result
Any	$v_1 = v_2$	None	None	Homopolymer analysis is valid
Any	v_1, v_2 arbitrary	Very low drift	None	Homopolymer analysis approximately valid
Acrylics	$v_1 \cong v_2$	None	None	
Aqueous	$0.1 \leq v_1, v_2 \leq 0.2$	Low to moderate	Low to moderate	

homopolymer case, so that the A_2 measured by Equation 11.21 is also only an apparent value. The excluded volume that determines A_2 is a function of the size of the whole copolymer chain, whereas the different comonomers in the copolymer chain scatter light differently, depending on their respective values of $\partial n / \partial c$.

11.3.6 Infrared Spectroscopy and Raman Scattering

Whereas in Rayleigh scattering there is essentially no frequency shift in the scattered light, Raman scattering differs in that excitation or de-excitation of molecular vibrational energy states during the scattering event can either add to (anti-Stokes shift) or subtract from (Stokes shift) the incident photon's energy, so that the scattered light has a different frequency than the incident light. This allows different types of monochromator and cut-off optical filter schemes for measuring this relatively weak type of scattering.

Similarly, IR spectroscopy has great analytical power because the photon energies in this spectral range correspond to the quantized vibrational energy levels of covalent and ionic bonds between atoms in molecules. These levels arise from the binding energy "well" between atoms that lead to quantized (harmonic and anharmonic) oscillations. Raman scattering and IR absorption follow different quantum mechanical selection rules, making them very complementary analytical methods.

In the context of monitoring polymerization reactions, IR spectroscopy has a long history, with early reports of its use related to polymerization reaction monitoring stretching back to at least the 1950s [46, 47], with a subsequent report of continuous IR measurements at 805 cm^{-1} to monitor ethacrylate polymerization in toluene [48]. IR methods have been extended to offline copolymer sequence length analysis [49, 50], curing reactions [51], picosecond time-resolved studies of initiation events [52], and other aspects of polymer characterization.

In the past 20 years, IR monitoring has become increasingly common for polymerization reaction monitoring [53,54–57], therefore, a variety of near and mid-IR *in situ* probes are now commercially available. While IR is well suited to homopolymerization, copolymerization monitoring, while often quite feasible [58–61], can present

particular challenges due to broad overlapping of bands for comonomers and associated copolymer. A notable area for fruitful application of IR monitoring is in the field of polymerization in emulsions [62–64]. Near IR monitoring has also been coupled to the ACOMP approach [65].

Raman has been used to follow monomer concentrations [2, 66–68]. However, it has been shown that Raman spectra could be affected by particle size in emulsion polymerization [69, 70]. Many of the foregoing referenced works, and others, combine IR and Raman measurements [71, 72]. Raman has not yet been applied to ACOMP although such application is expected soon.

11.3.7 Conductivity and Other Measurements

Any detector that has a flow cell can, in principle, be used in ACOMP. Conductivity and pH probes are simple, inexpensive examples of this. In fact, these latter probes can often be inserted into the reactor, so that they need not be part of the ACOMP detector train and hence do not require flow cells. In such cases, it can be very valuable to add *in situ* data from these probes to the ACOMP data for determining other reaction characteristics.

For example, *in situ* conductivity probes were used to monitor incorporation of ionic comonomers in aqueous copolymerization reactions [73, 74], and to directly monitor counterion condensation during copolymerization [75]. Conductivity probes immersed in flow-coupled mixing chambers were used on ACOMP-extracted streams during organic phase polymerization, and for aqueous copolymerization where the reactor ionic strength was too high for direct conductivity measurements. *In situ* pH probes have been coupled with ACOMP data to monitor hydrolysis during postpolymerization modifications.

While ACOMP is inherently a nonchromatographic method, the level of dilution and conditioning of the continuous ACOMP stream is ideally suited for direct GPC (SEC) analysis. Chapter 12 reports experiments on simultaneous use of ACOMP for the usual continuous flow determination of average characteristics coupled to an automatic injector valve to a multidetector GPC system in series with the continuous flow [76].

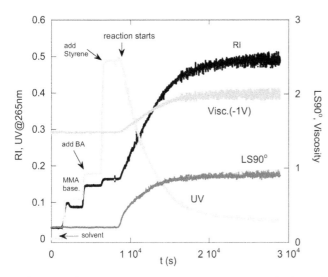

FIGURE 11.5 Automatic continuous online monitoring of polymerization reactions raw data signals from a free radical terpolymerization of methyl methacrylate (MMA), butyl acrylate (BA), and styrene in butyl acetate at 66°C. Of the many ACOMP raw signals collected a subset is shown here; one UV wavelength (from the full UV/Visible spectrum), one light scattering angle (from seven), and the single RI and viscometer signals. Reprinted with permission from Alb AM, Drenski MF, Reed WF. Automatic Continuous Online Monitoring of Polymerization reactions (ACOMP). Polym Int 2008;57:390–396 (77).

Polarimetric detectors are commonly fitted with flow cells. This can be a valuable detector when working with chiral molecules, both for detecting their concentrations, and changes in their optical activity. ACOMP has recently incorporated polarimetric detection for analysis of biopolymer modification reactions. At this time, ACOMP has not incorporated flow-cell detectors for NMR, ESR, and certain other potentially important methods.

11.3.8 Typical ACOMP Raw Data

Figure 11.5 shows raw data signals from a free radical terpolymerization of methyl methacrylate (MMA), BA, and styrene in butyl acetate, at 66°C. The total concentration of monomer plus polymer in the detector train throughout the reaction was 0.01 g cm^{-3}. The various signals can be understood as follows: During the first 2000 s pure solvent (butyl acetate) flows through the detectors. The three monomers are then added sequentially and the strong increments of RI and UV response for each are seen, whereas the LS (the 90° signal is shown of the seven angles collected) and viscometer do not respond to the dilute monomer solution. At 9000 s the initiator is added (AIBN) and the reaction begins. The decrease in the UV marks the conversion of the monomers into terpolymer, whereas the RI signal increases because *dn/dc* is higher for the polymeric form than the

dn/dc of the corresponding monomer. Complete UV spectra were acquired at the detector sampling rate of 2 s between each full detector train reading, and Figure 11.5 shows only the signal for 265 nm. The increase in LS and viscosity corresponds to the buildup of polymer in the reactor as conversion proceeds. These signals allow model-independent calculation of comonomer conversion kinetics, weight average molar mass, intrinsic viscosity, and average composition drift and distribution. Results of ACOMP analyses on this type of raw ACOMP data for different reaction scenarios are presented in Chapter 12.

11.4 SUMMARY

The background and principles of ACOMP have been discussed in this chapter, with a special focus on how important polymerization reaction characteristics are obtained from the rich data stream furnished by the ACOMP detector stream. Chapters 12 and 13 give examples of the very wide range of specific applications ACOMP has already been adapted to. Chapter 15 gives perspective on the outlook of transforming ACOMP from laboratory R&D instrumentation to a robust platform for monitoring and controlling industrial scale reactions.

REFERENCES

[1] Florenzano FH, Strelitzki R, Reed WF. Absolute online monitoring of polymerization reactions. Macromolecules 1998;31:7226–7238.

[2] Reed WF. Automatic mixing and dilution methods for online characterization of equilibrium and non-equilibrium properties of solutions containing polymers and/or colloids. US Patent 6,653,150. 2003.

[3] Elizalde O, Leal P, Leiza JR. Particle size distribution measurements of polymeric dispersions: a comparative study. Part Syst Char 2000;17:236–242.

[4] Reis MM, Araújo PH, Sayer C, Giudici R. Development of calibration models for estimation of monomer concentration by Raman spectroscopy during emulsion polymerization: facing the medium heterogeneity. J Appl Polym Sci 93, 1136–1150.

[5] Mignard E, Guerret O, Bertin D, Reed WF. Automatic Continuous Online Monitoring of Polymerization reactions (ACOMP) of high viscosity reactions. Polym Mat Sci Eng 2003;88:314–316.

[6] Richardson WH. Bayesian-based iterative method of image restoration. J Opt Soc Am 1972;62:55–59.

[7] Parzen E. *Modern Probability Theory and its Applications.* New York: John Wiley & Sons, Inc.; 1960.

[8] Kreft T, Reed WF. Direct monitoring of the cross-over from diffusion controlled to decomposition controlled initiation in free radical polymerization. Macromol Chem Phys 2008;209:2463–2474.

[9] Reed WF. A method for online determination of polydispersity during polymerization reactions. Macromolecules 2000;33:7165–7172.

[10] Huggins ML. The viscosity of dilute solutions of long-chain molecules. IV. Dependence on concentration. J Am Chem Soc 1942;64:2716–2718.

[11] Norwood DP, Reed WF. Comparison of single capillary and bridge viscometers as size exclusion chromatography detectors. Int J Polym Anal Charact 1997;4:99–132.

[12] Akkermans E, Wolf PE, Maynard R, Maret G. Optical coherent backscattering by random media: an experimental study. J Phys (Paris) 1988;49:63–98.

[13] Pine DJ, Weitz DA, Chaikin PM, Herbolzheimer E. Diffusing-wave spectroscopy. Phys Rev Lett 1988;60:1134–1137.

[14] Scheffold F, Skipetrov SE, Romer S, Schurtenberger P. Diffusing-wave spectroscopy of nonergodic media. Phys Rev E: Stat Nonlinear Soft Matter Phys 2001;63:061404/061401–061404/ 061411.

[15] Viasnoff V, Lequeux F, Pine DJ. Multispeckle diffusing-wave spectroscopy: a tool to study slow relaxation and time-dependent dynamics. Rev Sci Instrum 2002;73:2336–2344.

[16] Alexander M, Rojas-Ochoa LF, Leser M, Schurtenberger P. Structure, dynamics, and optical properties of concentrated milk suspensions: an analogy to hard-sphere liquids. J Coll Interface Sci 2002;253:35–46.

[17] Scheffold F. Particle sizing with diffusing wave, spectroscopy. J Disp Sci Technol 2002;23:591–599.

[18] Breedveld V, Pine DJ. Microrheology as a tool for high-throughput screening. J Mater Sci 2003;38:4461–4470.

[19] Narita T, Knaebel A, Munch JP, Zrinyi M, Candau S. Microrheology of chemically crosslinked polymer gels by diffusing-wave spectroscopy. Macromol Symp 2004;207:17–30.

[20] Scheffold F, Romer S, Cardinaux F, Bissig H, Stradner A, Rojas-Ochoa LF, Trappe V, Urban C, Skipetrov SE, Cipelletti L, Schurtenberger P. New trends in optical microrheology of complex fluids and gels. Prog Coll Polym Sci 2004;123:141–146.

[21] Zimm BH. Apparatus and methods for measurement and interpretation of the angular variation of light scattering; preliminary results on polystyrene solutions. J Chem Phys 1948;16:1099–1115.

[22] Doty PM, Zimm BH, Mark H. An investigation of the determination of molecular weights of high polymers by light scattering. J Chem Phys 1945;13:159.

[23] Debye PJW. Molecular-weight determination by light scattering. J Phys Coll Chem 1947;51:18–32.

[24] Hermans JJ, Levinson S. Some geometrical factors in light-scattering apparatus. J Opt Soc Am 1951;41:460–464.

[25] Casassa EF, Eisenberg H. On the definition of components in solutions containing charged macromolecular species. J Phys Chem 1960;64:753–762.

[26] Ooi T. Light scattering from multi-component systems. J Polym Sci 1958;28:459–462.

[27] Van de Hulst HC. *Light Scattering by Small Particles*. New York: John Wiley & Sons; 1957. p 470.

[28] Kerker M. *The Scattering of Light and Other Electromagnetic Radiation*. New York: Academic Press; 1969. p 666.

[29] Lord Rayleigh. On the diffraction of light by spheres of small relative index. R Soc Proc 1914;90: 219–225.

[30] Chu B. *Laser Light Scattering: Basic Principles and Practice*. Boston: Dover Publications; 2007.

[31] Berne BJ, Pecora R. *Dynamic Light Scattering*. Malabar: RE Krieger; 1990.

[32] Wu C, editor. *Handbook of Size Exclusion Chromatography*. 2nd ed. Chromatographic Science Series, Vol. 69, New York: Marcel Dekker; 2003.

[33] Potschka M, Durbin PL, editors. *Strategies in Size Exclusion Chromatography*. ACS Series 635. Washington, DC: American Chemical Society; 1996.

[34] Streigel A, editor. *Multiple Detection Size-Exclusion Chromatography*. ACS Series 893. Washington, DC: American Chemical Society; 2004.

[35] Provder T, editor. *Chromatography of Polymers: Characterization by SEC and FFF*. ACS Symposium Series 521. Washington, DC: American Chemical Society; 1993.

[36] Giddings JC. Measuring colloidal and macro-molecular properties by FFF. Anal Chem 1995;67:A592–A598.

[37] Williams SKR, Lee D. Field-flow fractionation of proteins, polysaccharides, synthetic polymers, and supramolecular assemblies. J Sep Sci 2006;29:1720–1732.

[38] Horstkotte B, Cerda V. Coupling of flow techniques with capillary electrophoresis: review of operation principles, challenges, potentials, and applications. J Chromatogr Sci 2009;47:636–647.

[39] DosRamos JG. Recent developments on resolution and applicability of capillary hydrodynamic fractionation. Part Sizing Charact 2004;881:138–150.

[40] Zimm BH. The scattering of light and the radial distribution function of high polymer solutions. J Chem Phys 1948;16:1093–1099.

[41] Stockmayer WH, Moore LD, Fixman M, Epstein BN. Copolymers in dilute solution. I. Preliminary results for styrene-methyl methacrylate. J Polym Sci Part A Polym Chem 1955;16:517–530.

[42] Bushuk W, Benoit H. Light-scattering studies of copolymers: I. Effect of heterogeneity of chain composition on the molecular weight. Can J Chem 1958;36:1616–1626.

[43] Enohnyaket P, Kreft T, Alb AM, Drenski MF, Reed WF. Determination of molecular mass during online monitoring of copolymerization reactions. Macromolecules 2007;40: 8040–8049.

[44] Prud'homme J, Bywater S. Light-scattering studies on polystyrene-polyisoprene block copolymers. Macromolecules 1971;4:543–548.

[45] Benoit H, Wippler C. Angular distribution of light scattered by a solution of copolymers. J Chim Phys Physiochem Biol 1960;57:524–527.

[46] Slowinski EJ, Claver GC. On the measurement of polymerization rates by means of infrared absorption spectra. J Polym Sci 1955;17:269–273.

[47] Luther H. Spectroscopic methods as a means of solving chemical technology problems. Zeitschrift Anal Chem 1958;164:109–120.

[48] Hirsch A, Bridgland BE. Kinetics of polymerization by infrared spectrometry. Anal Chem 1966;38:1272–1275.

[49] Winston A, Wichacheewa P. Sequence distribution in 1-chloro-1,3-butadiene: styrene copolymers. Macromolecules 1973;6:200–205.

[50] Brück D, Hummel DO. Copolymere aus Vinylhalogeniden und olefinen. III. IR-spektroskopische Sequenzanalyse an Vinylchlorid/Isobuten-Copolymeren und Bestimmung der Copolymerisationsparameter aus den gemessenen Diadenkonzentrationen. Makromol Chem 1973;163:271–279.

[51] Lee GA, Doorakian GA. Simplified analysis of UV curable compositions using infrared spectroscopy. J Radiat Curing 1977;4:2–7.

[52] Colley CS, Grills DC, Besley NA, Jockusch S, Matousek P, Parker AW, Towrie M, Turro NJ, Gill PMW, George MW. Probing the reactivity of photoinitiators for free radical polymerization: time-resolved infrared spectroscopic study of benzoyl radicals. J Am Chem Soc 2002;124:14952–14958.

[53] Darcos V, Monge S, Haddleton DM. In situ Fourier transform near infrared spectroscopy monitoring of copper mediated living radical polymerization. J Polym Sci A Polym Chem 2004;42:4933–4940.

[54] Sheibat-Othman N, Peycelon D, Fevotte G. Monitoring and control of free-radical polymerizations using near-infrared spectroscopy. Ind Eng Chem Res 2004;43:7383–7391.

[55] Zhou H, Li Q, Lee TY, Guymon CA, Jonsson ES, Hoyle CE. Photopolymerization of acid containing monomers: real-time monitoring of polymerization rates. Macromolecules 2006;39:8269–8273.

[56] Rodriguez-Guadarrama LA. Application of online near infrared spectroscopy to study the kinetics of anionic polymerization of butadiene. Eur Polym J 2007;43:928–937.

[57] de Souza FG, Anzai TK, Rodrigues MVA, Melo PA, Nele M, Pinto JC. In situ determination of aniline polymerization kinetics through near-infrared spectroscopy. J Appl Polym Sci 2009;112:157–162.

[58] Long TE, Liu HY, Schell BA, Teegarden DM, Uerz DS. Determination of solution polymerization kinetics by near-infrared spectroscopy.1. Living anionic-polymerization processes. Macromolecules 1993;26:6237–6242.

[59] Puskas JE, Lanzendorfer MG, Pattern WE. Mid-IR real-time monitoring of the carbocationic polymerization of isobutylene and styrene. Polym Bull 1998;40:55–61.

[60] Quinebeche JS, Navarro C, Gnanou Y, Fontanille M. In situ mid-IR and UV-visible spectroscopies applied to the determination of kinetic parameters in the anionic copolymerization of styrene and isoprene. Polymer 2009;50:1351–1357.

[61] Reis MM, Araujo PHH, Sayer C, Giudici R. In situ near-infrared spectroscopy for simultaneous monitoring of multiple process variables in emulsion copolymerization. Ind Eng Chem Res 2004;43:7243–7250.

[62] Wu CC, Danielsen JDS, Callis JB, Eaton M, Ricker NL. Remote in-line monitoring of emulsion polymerization of styrene by short-wavelength near-infrared spectroscopy.1. Performance during normal runs. Process Control Qual 1996;8:1–23.

[63] Puskas JE, Tzaras E, Marr G, Michel AJ. Real-time fiber optic mid-IR monitoring of solution and suspension polymerizations. Abstr Papers ACS 2001;221:U315–U315.

[64] Ouzineb K, Hua H, Jovanovic R, Dube MA, McKenna TE. Monomer compartmentalisation in miniemulsion polymerisation studied by infrared spectroscopy. C R Chim 2003;6: 1343–1349.

[65] Florenzano FH, Fleming V, Enohnyaket P, Reed WF. Coupling of near infra-red spectroscopy to automatic continuous online monitoring of polymerization reactions. Eur Polym J 2005;41:535–545.

[66] Elizalde O, Leiza JR, Asua JM. On-line monitoring of all-acrylic emulsion polymerization reactors by Raman spectroscopy. Macromol Symp 2004;206:135–148.

[67] van den Brink M, Pepers M, van Herk AM, German AL. On-line monitoring and composition control of the emulsion copolymerization of VeoVa 9 and butyl acrylate by Raman spectroscopy. Polym React Eng 2001;9:101–133.

[68] Reis MM, Araujo PHH, Sayer C, Giudici R. Spectroscopic on-line monitoring of reactions in dispersed medium: chemometric challenges. Anal Chim Acta 2007;595:257–265.

[69] Ito K, Kato T, Ona T. Non-destructive method for the quantification of the average particle diameter of latex as water-based emulsions by near-infrared Fourier transform Raman spectroscopy. J Raman Spectrosc 2002;33:466–470.

[70] Reis MM, Araujo PHH, Sayer C, Giudici R. Evidences of correlation between polymer particle size and Raman scattering. Polymer 2003;44:6123–6128.

[71] Ozpozan T, Schrader B, Keller S. Monitoring of the polymerization of vinylacetate by near IR FT Raman spectroscopy. Spectrochim Acta A Mol Biomol Spectrosc 1997;53:1–7.

[72] Adamczyk M. In situ reaction monitoring with IR and Raman spectroscopy. Labor Praxis 2002;26:50–52.

[73] Garcia GG, Kreft T, Alb AM, de la Cal JC, Asua JM, Reed WF. Monitoring the synthesis and properties of copolymeric polycations. J Phys Chem B 2008;112:14597–14608.

[74] Paril A, Alb AM, Reed WF. Online monitoring of the evolution of polyelectrolyte characteristics during postpolymerization modification processes. Macromolecules 2007;40: 4409–4413.

[75] Kreft T, Reed WF. Experimental observation of cross-over from non-condensed to counterion condensed regimes during free radical polyelectrolyte copolymerization under high composition drift conditions. J Phys Chem B 2009;113:8303–8309.

[76] Alb AM, Drenski MF, Reed WF. Simultaneous continuous, non-chromatographic monitoring and discrete chromatographic monitoring of polymerization reactions. J Appl Polym Sci 2009;13:190–198.

[77] Alb AM, Drenski MF, Reed WF. Automatic Continuous Online Monitoring of Polymerization reactions (ACOMP). Polym Int 2008;57:390–396.

12

APPLICATIONS OF ACOMP (I)

Alina M. Alb

12.1 INTRODUCTION

The ability to monitor polymerization reactions as they occur allows not only quantitative determination of reaction kinetics and the evolution of various (co)polymer characteristics, but also critical assessments of physicochemical theories that lead to advances in polymer science and engineering.

Making use of multiple detectors, automatic continuous online monitoring of polymerization (ACOMP) reactions have been applied to a large variety of reactions, including step growth reactions, free radical and controlled radical (co) polymerization, in homogeneous and heterogeneous phases, and in batch, semibatch, and continuous reactors, allowing gradient and block copolymers with different architectures to be synthesized. Grafting and cross-linking reactions, and post-polymerization modifications were also followed in real time.

Based on continuous extraction and dilution of reactor content to produce a dilute stream through the detectors, the versatile ACOMP platform combines light scattering, spectroscopy, viscometry, and other methods to monitor, during polymeric materials synthesis, various features such as reaction kinetics, monomer conversion, composition drift, linear charge density, molar mass, intrinsic viscosity, and other characteristics. The examples discussed next illustrate a number of reaction contexts where fundamental measurements are used to gain a comprehensive picture of reaction characteristics.

12.2 FREE RADICAL POLYMERIZATION

12.2.1 Homopolymerization

Free radical methods are routinely used to synthesize acrylic water-soluble polymers used in water treatment applications due to their large macromolecular size and expanded configuration in aqueous solution. Polyacrylamide, a widely used and technically important water-soluble polymer, has been intensively studied [1, 2]. Therefore, acrylamide polymerization was chosen as a first case study to demonstrate ACOMP capability to provide an accurate description of reaction kinetics [3].

12.2.1.1 *Monitoring Acrylamide Polymerization* Time-dependent signatures for conversion, reduced viscosity, and molecular weight during continuous online monitoring of acrylamide polymerization under a variety of temperature, and initiator conditions were compared in Reference [3].

Zimm approximation [4], introduced in Chapter 11, valid for $q^2 < S^2 >_z << 1$, was used to compute absolute weight-average polymer mass, M_w, and z-average mean square radius of gyration, $<S^2>_z$, from multi-angle light scattering data in conjunction with the polymer concentration c determined from collected refractive index (RI) and UV data:

$$\frac{Kc}{I(q,c)} = \frac{1}{M_w}\left(1 + \frac{q^2 < S^2 >_z}{3}\right) + 2A_2c \qquad (12.1)$$

where $I(q, c)$ is the excess Rayleigh scattering ratio, A_2 is the second virial coefficient, and K is an optical constant given for vertically polarized incident light by:

$$K = \frac{4\pi^2 n^2 (dn/dc)^2}{N_A \lambda^4} \qquad (12.2)$$

where n is the solvent index of refraction, λ is the vacuum wavelength of the incident light, and q is the scattering wave vector.

Monitoring Polymerization Reactions: From Fundamentals to Applications, First Edition. Edited by Wayne F. Reed and Alina M. Alb.
© 2014 John Wiley & Sons, Inc. Published 2014 by John Wiley & Sons, Inc.

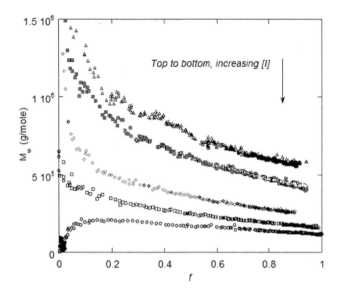

FIGURE 12.1 The cumulative weight-average polymer mass, M_w, versus conversion, f, for acrylamide polymerization reactions at 70°C in which different amounts of initiator (potassium persulfate) were used. The general trend is that at the beginning large polymer chains are produced, followed by smaller chains as conversion proceeds. Reprinted (adapted) with permission from Giz A, Çatalgil-Giz H, Alb AM, Brousseau J-L, Reed WF. Kinetics and mechanisms of acrylamide polymerization from absolute, online monitoring of polymerization reactions. Macromolecules 2001;34:1180–1191. © 2001 American Chemical Society.

Figure 12.1 [3] shows the cumulative M_w (computed based on Eq. 12.1) versus conversion, f, for polymerization reactions at 70°C in which different amounts of initiator (potassium persulfate) were used [3]. The general trend is that at the beginning large polymer chains are produced, followed by smaller chains as conversion proceeds. Two regimes are observed in the evolution of $M_w(f)$, in most cases. The first regime consists of a rapid decay at early conversion, followed by an elbow region, after which there is a linear regime, usually for the rest of the conversion. Further substantiation of the M_w trends was seen in the evolution of the reduced viscosity with conversion during the same reactions.

Despite deviations from the ideal kinetics at early and late conversion, rationalized by existing models, quasi-steady state approximation (QSSA) proved to be an appropriate approximation for analyzing the reactions and obtaining the associated rate constants. Several reaction features such as the dependence of M_w and initiator decay rates on $1/[I_2]_0^{1/2}$ and $[I_2]_0^{1/2}$, respectively, and the first-order decay of monomer during the majority of the conversion allowed verification of the ideal kinetics applicability.

Despite the fact that ACOMP is not a separation technique, in certain conditions, such as in the case of the study mentioned earlier, it is possible to quantify the broadening of the population distribution during polymerization. Two

possible routes were explored in Reference [3], each with its own advantages and drawbacks.

1. Using the slope of Kc/I versus q^2,

$$\frac{d\left(Kc_{p,d}d / I_R\right)}{dq^2} = \frac{1}{M_w}\left(\frac{<S^2>_z}{3}\right)$$
$$= \frac{A<M^\alpha>_z}{3M_w} \qquad (12.3)$$

with the use of the power law between $<S^2>$ and the polymer mass, $<S^2>=AM^\alpha$; for a random coil $\alpha=1$, $<S^2>\approx M_z/M_w$; for a coil with excluded volume $\alpha=[1.18, 1.2]$

2. Using derivatives of the online $M_w(t)$ to compute the polymer mass distributions and polydispersity indices. ACOMP measures the accumulated $M_w(t)$ at each moment; for most chain growth reactions, this involves an accumulation of dead chains, which form very quickly compared to the time scale for the total conversion of monomer.

In this case, the computation of the instantaneous weight-average molecular mass, defined as the instantaneous weight-average of chains added during an interval from f to $f+df$:

$$M_{w,inst}(f) = f\frac{dM_w}{df}\bigg|_f + M_w(f) \qquad (12.4)$$

allowed the other moments (relative) of the distribution, $M_n(f)$, $M_z(f)$, to be obtained.

12.2.1.2 *In Situ* NIR Coupling to ACOMP In some cases, it is valuable to use ACOMP in conjunction with traditional methods, such as *in situ* near infrared (NIR) spectroscopy [5] in order to either make critical comparisons of the results provided by each method or to set the stage for the coupled use of NIR and ACOMP for more complex systems, where neither technique alone would be capable of making the type of comprehensive characterization desired.

Simultaneous monitoring by *in situ* NIR spectroscopy and ACOMP was reported in the case of methyl methacrylate (MMA) polymerization reaction [6].

Keeping in mind that polymer characteristics provided by ACOMP (M_w, η_r, f) are determined from direct measurements, whereas NIR furnishes monomer conversion data via empirical calibration, an advantage of the *in situ* NIR is that it furnishes immediate information on the conversion in the reactor, whereas ACOMP relies on continuous withdrawal and dilution of a small stream of reactor fluid, so that there is a lag time of several minutes between what ACOMP reports and what is occurring in the reactor. Nonetheless, simultaneous monomer conversion data from *in situ* NIR and ACOMP, the latter derived from both RI and UV absorption, were found to be in good agreement, opening the path for the combined two methods to characterize increasingly complex systems, including co- and terpolymerization.

12.2.2 Chain Transfer

Chain transfer is a common process in polymerization reactions. The propagating radical $R_i^{\cdot}$ abstracts a weakly bonded atom from a transfer agent, T:

$$R_i^{\cdot} + T \xrightarrow{\; k_{tr} \;} T^{\cdot} + P_i \qquad (12.5)$$

where T can be any of the species present in the reaction: monomer, initiator, solvent, or other substance; k_{tr} is the chain-transfer rate constant; and P_i is the dead polymer. The new radical created, $T^{\cdot}$, can reinitiate the reaction:

$$T^{\cdot} + M \xrightarrow{\; k_{rein} \;} R^{\cdot} + P \qquad (12.6)$$

In this way, more dead chains are being formed due to the interaction of propagating chains with each other and with the chain transfer agent (CTA):

$$\frac{d[P]}{dt} = {}^{*}k_t[R^{\cdot}]^2 + k_{tr}[T][R^{\cdot}] \qquad (12.7)$$

where * is 2 in the case of termination by recombination and 1 for termination by disproportionation and takes on values between these limits when both termination mechanisms are present.

Since it is a chain breaking process, chain transfer leads to lower molecular weights. However, its effect on the polymerization rate depends on the values of the rate coefficients. In the case where the rate limiting step is a radical transfer to the CTA (T), there is a decrease in M_w but no effect on monomer conversion rate. If the radical transfer from T to monomer is much slower than the transfer to T, then inhibition can occur, which slows down the reaction and leads to formation of longer chains.

Potassium persulfate-initiated polymerization of acrylamide (AM) in water was chosen as a case study for chain transfer reactions monitored by ACOMP [7]. Chain transfer properties of ethanol (EtOH) and 2-propanol (PrOH) were investigated.

To a close approximation, the CTAs were found to obey the case expected when ideal free radical polymerization takes place and radical transfer from propagating radicals to the CTA is slower than from the CTA to monomer; that is, the polymer molar mass decreases with

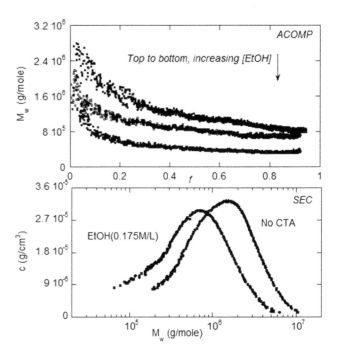

FIGURE 12.2 The effect of the chain transfer concentration, CTA, on the polyacrylamide mass: $M_w(f)$ data for reactions in which different CTA were used (top); molecular weight distribution (MWD) from SEC experiments on reaction endproducts (bottom). From Grassl B, Alb AM, Reed WF. Online monitoring of chain transfer in free radical polymerization. Macromol Chem Phys 2001;202:2518–2524. © Wiley-VCH Verlag GmbH & Co. KGaA. Reproduced with permission.

increasing CTA, but no appreciable effect occurs on the kinetics of monomer conversion.

To illustrate the effect of the CTA on the polymer mass, data from ACOMP and size exclusion chromatography (SEC) experiments are combined in Figure 12.2 [7]. Thus, $M_w(f)$ data for reactions in which different amounts of CTA were used, computed based on Equation 12.1, are shown in the top portion of the figure whereas molecular weight distribution (MWD) from SEC experiments on reaction end products are shown in the bottom part. Both methods used show similar trends in the evolution of the polymer mass, that is, the higher the concentration of CTA involved in the reaction, the lower the M_w is.

In the presence of chain transfer reactions, the number-average degree of polymerization X_n becomes:

$$\frac{1}{X_n} = \frac{1}{X_n\left([T]=0\right)} + \frac{k_{tr}}{k_p}\frac{[T]}{[M]} \qquad (12.8)$$

Hence, the instantaneous number-average mass, $M_{n,inst}$, can be expressed as:

$$\frac{1}{M_{n,inst}} = \frac{1}{X_n M_{AAm}} = \frac{1}{M_{n,inst}\left([T]=0\right)} + \frac{k_{tr}}{k_p}\frac{[T]}{[M]M_{AAm}} \qquad (12.9)$$

where $C_T = k_{tr}/k_p$ is termed as chain transfer constant, $[M] = [M]_0 e^{-kt}$ is the monomer concentration, $\kappa = k_p[R^{\cdot}] = k_p \sqrt{\dfrac{2f k_d [I_2]}{k_t}}$.

Traditionally, the cumulative M_n at the end of a reaction is determined by SEC or other molecular weight determination technique. In the QSSA regime, $M_{n,inst}(f)$ and $M_w(f)$ were computed in Reference [7] using the online technique as follows:

$$M_{n,inst} = \frac{k_p[M]M_{AAm}}{k_t[R^{\cdot}] + k_{tr}[T]} = \frac{k_p[M]_0 M_{AAm}}{k_t[R^{\cdot}] + k_{tr}[T]}(1-f) \quad (12.10a)$$

$$M_w(f) = \frac{2k_p[M]_0 M_{AAm}}{k_t[R^{\cdot}] + k_{tr}[T]}\left(1 - \frac{f}{2}\right) \quad (12.10b)$$

where $f = 1 - \dfrac{[M]}{[M]_0} = 1 - e^{-\kappa t}$ describes the monomer conversion as a first-order function.

Based on aforementioned equations, C_T was determined using different methods of computing the transfer constants, including those based on the slope and intercept behavior of the monitored cumulative weight-average molecular mass as a function of conversion, $M_w(f)$, the reduced viscosity, and corresponding SEC analysis of the reaction endproducts:

$$\frac{2}{M_w(0)} = \frac{1}{M_w(f=1)} = \frac{-1}{dM_{wt}(f)/df}$$
$$= \frac{1}{M_{AAm}}\left(\frac{k_t[R^{\cdot}]}{k_p[M]_0} + \frac{k_{tr}[T]}{k_p[M]_0}\right) \quad (12.11)$$

The agreement among data computed by various approaches proves the feasibility of ACOMP as a powerful tool to follow reaction kinetics and to offer various options in determining chain transfer parameters. It is hoped that the method will be applicable to a wider range of polymerization reactions.

12.2.3 Copolymerization

Copolymers are becoming increasingly important for high performance and new materials with specific mechanical, optical, and electrical properties. The bivariate composition and mass distribution controls many aspects of the materials behavior, such as tensile strength, processability, surface, phase stability, and so on. Nonetheless, determining the bivariate distribution can be time consuming and costly, and usually requires the use of complementary techniques, such as thermal field-flow fractionation [8], temperature rising elution fractionation (TREF) [9], Fourier transform infrared (FTIR) spectroscopy [10], and other methods [11] for determining chemical composition, SEC, and related techniques for mass distribution [12], or combinations of the preceding [13].

Light scattering is one of the standard methods for the determination of molecular weights of macromolecules. In the case of copolymers, it has historically required that measurements be made in at least three separate solvents of different index of refraction [14, 15]. Comprehensive studies are available elsewhere [16], and the ACOMP solution to the problem is given in Chapter 11.

Recently, ACOMP capabilities were expanded to follow the evolution of different copolymer features during copolymerization reactions. A first study reported the online monitoring of the copolymerization of MMA and styrene [17]. Simultaneous, model-independent evolution of the average composition and mass distributions during free radical copolymerization reactions were determined. Also, model-dependent calculations of reactivity ratios and sequence length distributions were made.

12.2.3.1 Molecular Weight
Molecular weight determination was covered in detail in Chapter 11. However, the main equations are still given here since they are part of the discussion of this chapter. As discussed in Chapter 11 and in Reference [17], copolymers are characterized by an apparent molecular weight, which is the sum of all the compositional elements constituting the copolymer population:

$$\frac{1}{M_{app}} = c_p \lim_{q \to 0} \frac{K' v^2 c_p}{I(q, c_p)}, \quad M_{app} = \frac{1}{v^2}\sum \gamma_i v_i^2 M_i \quad (12.12)$$

where γ_i is the weight fraction of the polymer population with a given pair of composition and mass values and:

$$K' = \frac{4\pi^2 n^2}{N_A \lambda^4}, \quad v = \frac{\partial n}{\partial(c_{pA} + c_{pB})} = y v_A + (1-y)v_B \quad (12.13)$$

where $c_{pA(B)}$ is the mass concentration of A (B) in polymeric form, y is the average mass fraction of A in copolymer form, v_A and v_B are the differential index of the polymer A and B, respectively, in a given solvent.

A method of finding $M_w(f)$—discussed also in Chapter 11—based on the measured values of absolute scattering $I(q, c)$ and the concentration values determined at any point using ACOMP platform was presented in Reference [17]. The evolution of the copolymer M_w was computed as follows:

$$M_w(f) = \frac{1}{f}\int \frac{v^2 m_{ap} + f M_{ap}\dfrac{dv^2}{df}}{v_A v_B + v_A(v_A - v_B)z^2 + v_B(v_B - v_A)(1-z^2)} df' \quad (12.14)$$

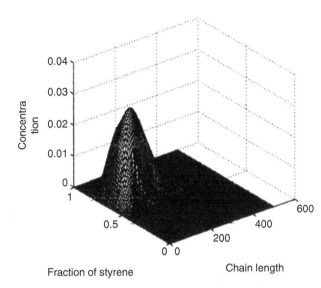

FIGURE 12.3 3D bivariate distribution for 75/25 (M/M) styrene/methyl methacrylate copolymerization reaction, using the Stockmayer bivariate distribution. Reprinted (adapted) with permission from Çatalgil-Giz H, Giz A, Alb AM, Öncül Koç A, Reed WF. Online monitoring of composition, sequence length, and molecular weight distributions during free radical copolymerization, and subsequent determination of reactivity ratios. Macromolecules 2002;35:6557–6571. © 2002 American Chemical Society.

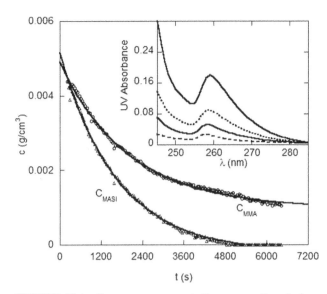

FIGURE 12.4 Comonomer concentrations versus time during a 68/32 methyl methacrylate (MMA)/N-methacryloxysuccinimide (MASI) copolymerization reaction in N,N-dimethylformamide (DMF) using the error minimization procedure developed in Reference (20). Reprinted (adapted) with permission from Alb AM, Enohnyaket P, Drenski MF, Shunmugam R, Tew GN, Reed WF. Quantitative contrasts in the copolymerization of acrylate and methacrylate based comonomers. Macromolecules 2006;39:8283–8292. © 2006 American Chemical Society.

where z is the instantaneous fraction of A incorporated into copolymer.

Once the M_w and composition were known, the aforementioned procedure for computing the Stockmayer distribution [14] was applied for one of the copolymerization reaction studied. An illustration of the method is shown in Figure 12.3 [17].

A workable and practical formalism for the determination of M_w during reactions, using a single solvent, was developed in Reference [18] for the case of N comonomers. The formalism was applied to experimental cases involving copolymerization of styrene and MMA. With the average instantaneous and cumulative compositions determined and multiangle light scattering simultaneously obtained, $M_{ap}(t)$ is computed during copolymerization reactions and thus, avoids the use of multiple solvents and provides continuous values of composition and molar masses. The integral form of $M_w(f)$ derived in Reference [18] is summarized in Equation 12.15:

$$M_w(f) = \frac{1}{f} \int \frac{d\left[f' M_{ap}(f') \left(\sum_j y_j(f') v_j \right)^2 \right]}{\left(\sum_j F_j(f') v_j \right)^2} \qquad (12.15)$$

12.2.3.2 *Comonomer Concentrations*

An important innovation made by the addition of a full spectrum UV spectrophotometer into ACOMP detector train led to the determination of the instantaneous concentration of each comonomer during the reaction, in the case of the copolymerization of comonomers with similar spectral characteristics [19, 20]. The working assumption was that a UV spectrum at any instant is a linear combination of the normalized basis spectra of the comonomers, the copolymer, and any other UV absorbing species, and that the unknown comonomer concentrations can be found by minimizing the error between measured and computed spectra over many wavelengths, even when spectral differences are small at any iven wavelength.

Figure 12.4 [20] shows comonomer concentrations versus time during a 68/32 MMA/N-methacryloxysuccinimide (MASI) copolymerization reaction in N,N-dimethylformamide (DMF) using the error minimization procedure developed in Reference [20]. In the inset to the figure, selected UV spectra from different times during the copolymerization reaction are shown.

12.2.3.3 *Reactivity Ratios*

It is important to point out that obtaining the comonomer concentrations via ACOMP is model-independent. The notion of reactivity ratios and the various formulations to express them, in contrast, are model-dependent. In some sense, ACOMP makes unnecessary their use, since it yields the average composition distribution as it evolves during the reaction, and automatically detects any unexpected or nonideal effects that may occur in a reaction, but which are not normally predicted by idealized models. Nonetheless, the reactivity ratios are still important for their predictive power, their practical use in helping to optimize reactions, their inherent theoretical interest, and for historical reasons. The values of the reactivity ratios obtained by any specific data analysis means, however, are only as meaningful as the idealized model employed in their computation.

Several ways of obtaining reactivity ratios from comonomer concentration data have been reported, including linearization, full nonlinear approaches, numerical minimization of errors, and so on. It is beyond the scope of this work to go through a detailed discussion of the existent methods and their applicability.

A straightforward approach was used in Reference [20] to compute reactivity ratios for two methacrylate-based comonomers and to compare the values obtained with the existing literature. Thus, since the comonomer conversion kinetics is known from ACOMP, the Mayo–Lewis copolymer equation:

$$\frac{d[A]}{d[B]} = \frac{[A]}{[B]}\left(\frac{r_A[A]+[B]}{[A]+r_B[B]}\right),$$
$$r_A = \frac{k_{AA}}{k_{AB}}, \quad r_B = \frac{k_{BB}}{k_{BA}} \tag{12.16}$$

offers an immediate and simple means of determining r_A and r_B from the initial conversion rates $(d[A]/dt)_{t=0}$ and $(d[B]/dt)_{t=0}$:

$$\frac{(d[A]/dt)_{t=0}}{(d[B]/dt)_{t=0}} = \left(\frac{r_A[A]_0/[B]_0+1}{[A]_0/[B]_0+r_B}\right) \tag{12.17}$$

Figure 12.5 [20] shows error contours for the values of r_A and r_B for the MASI/MMA and MASI/butyl acrylate (BA) data, using Equation 12.17, in which the ratio on the left-hand side was computed in three different ways: (i) linear fit to the initial conversion, (ii) rate constant resulting from an exponential fit to early conversion, and (iii) rate constant resulting from exponential fit to entire conversion curve, including an adjustable final baseline, to account for the experimental fact that monomer is not always fully consumed at large times. The ratios from each determination were weighted equally in the error computation. The contours were computed by forming a parameter space grid of pairs of discrete values for r_A and r_B and then computing the error in terms of the difference between experimental and theoretical points, divided by the experimental value for each point. The square root of the sum of the squares of this error for all points divided by the number of points gives

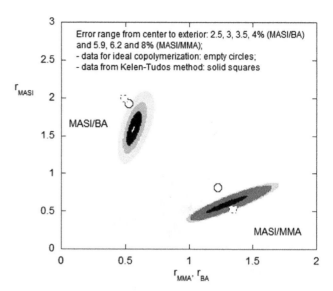

FIGURE 12.5 Error contours for the reactivity ratios for the MASI/MMA and MASI/butyl acrylate (BA) data using Equation 12.17, in which the ratio on the left-hand side was computed in different ways. Reprinted (adapted) with permission from Alb AM, Enohnyaket P, Drenski MF, Shunmugam R, Tew GN, Reed WF. Quantitative contrasts in the copolymerization of acrylate and methacrylate based comonomers. Macromolecules 2006;39: 8283–8292. © 2006 American Chemical Society.

the root mean square error, whose contours are depicted in the figure. Hollow squares shown in the figure are the results based on Kelen–Tudos method [21].

12.2.3.4 Composition Drift and Sequence Length Distributions

Automatic continuous online monitoring of polymerization continuous measurements of comonomer concentrations can be used to compute the average instantaneous molar fraction of a comonomer A in the copolymer chain formed at any given moment $F_{inst,A}$ based on:

$$F_{inst,A} = \frac{d[A]/dt}{d[A]/dt + d[B]/dt} \tag{12.18}$$

While the knowledge of the composition drift provides important information on the copolymers' compositional heterogeneity, it reveals nothing about their sequence distribution, important in determining different aspects of the copolymer behavior.

The instantaneous sequence length distribution, defined as the probability of having a sequence of k monomers in a row of type A, followed by a monomer B can be determined from ACOMP measurements at each point f [22, 23].

For illustration, shown next is the computation of the instantaneous number-average sequence lengths made by Alb et al. in the case of the synthesis of cationic copolyelectrolytes, monitored by ACOMP [24]. The instantaneous

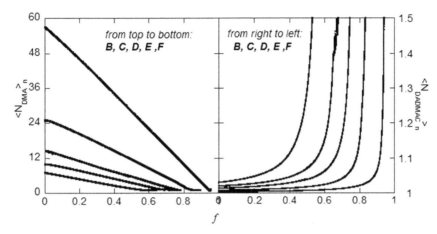

FIGURE 12.6 Computed number-average sequence length of the dimethyl acrylamide (DMA) and diallyldimethylammonium chloride (DADMAC) comonomers, $<N_{DMA}>_n$ and $<N_{DADMAC}>_n$, for several copolymerization reactions with different initial compositions. From Li Z, Alb AM. Cationic copolyelectrolytes: online monitoring of the synthesis and subsequent copolymer characterization. Macromol Chem Phys 2012;213:1397–1406. ©Wiley-VCH Verlag GmbH & Co. KGaA. Reproduced with permission.

number-average sequence length of dimethyl acrylamide (DMA) and diallyldimethylammonium chloride (DADMAC) were computed based on Equation 12.19, with $r_{DMA} = 6.3$ and $r_{DADMAC} = 0.3$ used for the comonomer reactivity ratios:

$$\langle N_A \rangle_n (f) = \frac{1}{1 - W_{AA}(f)} \qquad (12.19)$$

where W_{AA} is the probability that the propagating radical of monomer A adds to A:

$$W_{AA}(f) = \frac{r_A y_A(f)}{y_A(f)(r_A - 1) + 1} \qquad (12.20)$$

r_A is the reactivity ratio of monomer A and y_A is the mole fraction of monomer A, at any instant.

Shown in Figure 12.6 [24] are computed number-average sequence length of the comonomers, $<N_{DMA}>_n$ and $<N_{DADMAC}>_n$, for several copolymerization reactions with different initial compositions.

The addition of DADMAC monomer units into the copolymer chains is dictated by both reactivity ratios and initial comonomer composition. Thus, the evolution of $<N_{DMA}>_n$ with conversion—shown in the left side of the figure—indicates that the more DMA available, the longer sequences of DMA units are contained in the copolymer chains. The decrease with conversion corresponds to the depletion of DMA and the progressive addition of DADMAC. In the case of the reactions with less DMA in the starting composition, the addition of DADMAC is more favored at earlier conversion. The computed sequence length for DADMAC is shown in the right side of the figure. For each reaction, similar trends in the $<N_{DADMAC}>_n$ evolution are observed:

a slow increase throughout the copolymerization with DMA, followed by a significant jump, due to the DMA depletion. In these conditions, chains containing mostly or only DADMAC were obtained.

All cases presented earlier offer highlights on some of the multiple copolymer features determined during the online monitoring of (co)polymerization reactions via ACOMP for comonomers with similar or very different reactivity ratios.

12.3 HETEROGENEOUS PHASE POLYMERIZATION

A new advance with regard to the instrumentation and methods available for online monitoring of heterogeneous polymerization reactions was made by using ACOMP for monitoring the evolution of multiple characteristics during polymerization. The information-rich data collected simultaneously by multiple detectors provide absolute, model-independent determination of quantities such as conversion, composition, and molar mass distribution and avoid potentially damaging effects of the reactor environment.

Additionally, this approach offers, in the case of emulsion polymerization, a unified means of monitoring colloid/polymer characteristics giving information about both the polymer and particle evolution, and thus allowing one to make correlations between key features of the two aspects of the emulsion polymerization process. Online monitoring of both particle and polymer characteristics should allow for studies of reaction kinetics, predictive and active reaction control, and also the ability to observe deviations and unexpected phenomena.

Selected data and results are shown in the following sections in order to demonstrate the efficiency and versatility of the online method as a powerful tool for studying polymerization kinetics in heterogeneous phase polymerization reactions.

12.3.1 Inverse Emulsion Polymerization

High molecular weight acrylamide-based polymers are often synthesized through inverse emulsion polymerization processes, which allow polymerizations at high monomer concentrations while maintaining low bulk viscosities and good heat transfer throughout the reaction [25]. The locus of polymerization is in the dispersed droplets consisting of aqueous acrylamide solution, which are stabilized in the oil-continuous medium by means of emulsifying agents (surfactants) [26].

Various methods have been used to analyze aliquots from reactions [27, 28]. Online measurements by dilatometry have also been reported [29]. A significant advance in this field was the adaptation of ACOMP as a means for monitoring conversion and measures of polymer molar mass during the inverse emulsion polymerization of acrylamide, in order to understand both reaction kinetics and mechanisms and, potentially, to control them during the reaction [30].

The monomer conversion and reduced viscosity were monitored by continuously inverting and diluting the emulsion phase using a small reactor sample stream and a "breaker" surfactant solution, followed by UV absorption and viscometric detection. Sorbitan monooleate (Span 80) was used as the emulsifier, Exxsol D80 was the oil phase (aliphatic hydrocarbon), and Surfonic N-95 (alkylphenol ethoxylate), a nonionic surfactant, was chosen to make the phase inversion.

Initial efforts were focused on the optimization of the online sample conditioning and data interpretation. In order to make absolute measurements on a heterogeneous system containing oil, water, monomer, polymer, initiator, and surfactant, the monomeric and polymeric contents of the inverse emulsion or latex must be "spilled" out from the discrete phase droplet in a period of seconds and the resulting detectors signals must be interpreted to allow differentiating among the complex mixture of components.

Therefore, the first section of the study presented in Reference [30] was devoted to establishing the means of continuously inverting and conditioning the reactor liquid into a stable portion of the phase diagram, determining the type of behavior that occurs in the multicomponent system upon conditioning, and how the relevant components can be tracked with the detector train.

Two examples of preliminary experiments made by the authors in order to determine (i) the detectors response to the different components of the diluted, phase inverted emulsion, and (ii) the amount of time it takes to make the phase inversion are illustrated in Figure 12.7 [30].

As shown in the left side of the figure, the most prominent effect is the fact that, while the viscometer is insensitive to the oil droplets produced when the emulsion is broken, it responds linearly to the polymer concentration. At the same time, the light scattering (shown only at 90° in the figure, LS90°) is sensitive to the droplets and debris, in addition to the polymer, so that these data are much more complicated to analyze. It was hence decided by the authors to rely on the viscometer to provide a measure of polymer mass. The right side of Figure 12.7 shows results from polymer dissolution experiments. While the polymer-containing emulsion inverts in tens

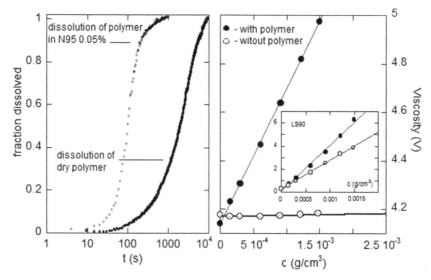

FIGURE 12.7 Automatic Continuous Online Monitoring of Polymerization Reactions detectors response (LS90° and viscosity) to the different components of the diluted, phase inverted polyacrylamide emulsion (*top*) determination of dissolution time, from RI data for polyacrylamide in emulsion and in dry form. Reprinted (adapted) with permission from Alb AM, Farinato F, Calbick J, Reed WF. Online monitoring of polymerization reactions in inverse emulsions. Langmuir 2006;22:831–840. © 2006 American Chemical Society.

of seconds when diluted with the breaker surfactant (time well within a typical ACOMP time budget), the dry polymer takes thousands of seconds to solubilize. This demonstrates another advantage of ACOMP, in that sample preparation of liquid reactor contents is orders of magnitude faster than sample preparation for postpolymerization analyses of dried polymer.

After establishing the means of making ACOMP measurements, the second part of the study was focused on resolving several reaction kinetic issues, including the evolution of molecular mass, and the simultaneous action of an "intrinsic" initiator and an added chemical initiator. The ACOMP data were used to demonstrate the determination of monomer conversion kinetics, and the detection and quantification of both surfactant-based initiators and radical scavengers. The authors found strong evidence of a radical scavenging impurity in the complex emulsion mixture. It was also shown that in one of the identical lots of stabilizer surfactant there was an "intrinsic" initiator, which influenced reaction kinetics. A model was proposed and verified with experimental data. The temperature dependence and activation energy of the intrinsic initiator were determined. The increasing trends in the computed η_r and M_w with conversion were explained as being possibly related to chain transfer to the stabilizer surfactant or other emulsion component and/or the radical scavengers.

12.3.2 Emulsion Polymerization

Emulsion polymerization was developed for producing polymers with unique properties and because of environmental considerations. Although the reaction medium in emulsion polymerization remains at low viscosity, factors such as instability of the lattices and other nonequilibrium phenomena, complexity of the materials produced, and multifaceted properties requiring multiple characterization approaches add challenges in developing accurate predictive models and/or reaction control schemes.

Therefore, it is critical to understand how these variables affect properties such as copolymer composition, MWD, branching, particle morphology, and size distribution in order to control the delicate balance among reaction components. Characterization of various properties has been carried out using different sampling procedures and instrumentation. This piecewise and tedious approach makes it difficult to assess and quantitatively evaluate the characteristics and relationships among the complex, nonequilibrium colloidal/polymer systems.

Advances in monitoring emulsion polymerization reactions can provide more accurate and integrated information, allowing thorough evaluation of different models. Details of emulsion polymerization and monitoring are given in Chapter 4. There are also several reviews about online monitoring methods for emulsion polymerization [31–33]. Existing online methods depend on the accuracy of the calibration and on the many parameters involved in the

models used, which are affected by the environment. Among the noninvasive techniques, calorimetry [34] is used for monitoring the polymerization rate by estimating the heat of polymerization and the overall heat-transfer coefficients. Spectroscopic techniques are widely used in monitoring monomer or polymer properties (see Chapter 10). Raman spectroscopy is an attractive route [35], but no direct relationship between the absolute intensity and monomer concentration can be made and calibration models are required. Multivariate calibration techniques have been developed to relate the spectra with the concentration of each monomer [36]. NIR spectroscopy has also been used to estimate monomer and polymer content and polymer particle size [37].

A large drawback of the *in situ* methods is that they work in concentrated reaction environments, and hence the signals obtained usually must be interpreted using chemometric or other empirical calibration schemes. Because emulsions also scatter electromagnetic radiation, these effects can interfere with, and often even overwhelm, the portion of the signal due to monomeric absorption. A further possible problem with the *in situ* approach is that such probes can be fouled or damaged by the reaction environment, causing large shifts of the empirical calibration schemes.

Particle size and particle size distribution (PSD) strongly affect end-product properties and applications. While there are still issues regarding accurate offline PSD determination of the latex particles, recent advances have been made [38].

In this context, the use of ACOMP for the simultaneous monitoring of the evolution of colloid phase (monomer droplets, polymer latex particles) and solution phase (polymer and monomer) characteristics during emulsion polymerization was recently reported as a versatile characterization tool that offers absolute, model-independent determination of quantities such as conversion, composition, and molecular weight [39].

By offering a unified means of monitoring colloid or polymer characteristics, the approach presented in Reference [39] allows to make correlations between key features of the two aspects of the emulsion polymerization process, leading to more accurate knowledge of reaction kinetics, predictive and active reaction control, and also improving ability to observe deviations and unexpected phenomena.

The method has been applied already to various (co)polymerizations under a variety of concentration conditions, with and without surfactant. Several features were captured by this process, such as comonomer conversion, reactivity ratios, polymer mass and composition, and correlations between monomer droplet disappearance and monomer conversion.

A few examples of data and results are presented next, together with technical details on the instrumentation.

12.3.2.1 *Instrumentation*   A novel strategy introduced in Reference [39] offered a significant advantage of ACOMP vis-à-vis the *in situ* methods by involving an ensemble of pumps, dilution, and conditioning stages to provide two streams,

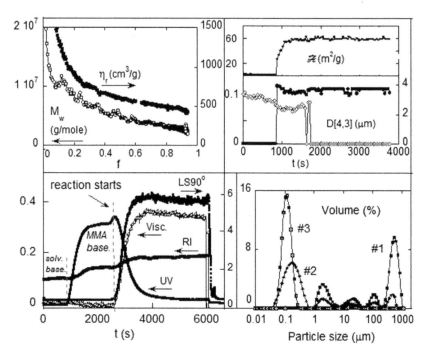

FIGURE 12.8 Simultaneous monitoring with ACOMP of particle and polymer/monomer features during the emulsion polymerization of methyl methacrylate at 70 °C. Left side: LS90°, UV, RI, and viscosity data versus T (bottom) used to compute polymer M_w and η_r as functions of conversion f (top) from the sample stream diluted by THF; right side: particle distributions from Mie scattering from the reactor content diluted with water (bottom); the increase in specific surface area, A, indicates formation of small polymer particles and furnishes a trend in particle size evolution (top). From Alb AM, Reed WF. Recent advances in Automatic Continuous Online Monitoring of Polymerization reactions (ACOMP). Macromol Symp 2008;271:15–25. ©Wiley-VCH Verlag GmbH & Co. KGaA. Reproduced with permission.

extracted continuously from the reactor in order to follow simultaneously different aspects of the polymerization reaction: (i) The evolution of the organosoluble components, including polymer and monomer characteristics, achieved by diluting the reactor content continuously and automatically throughout the reaction with an organic solvent miscible with water to create a homogeneous, dilute solution. From this, polymer weight-average molar mass, M_w, and reduced viscosity, η_r, were obtained, as well as monomer conversion. In the case of copolymers, average composition drift and distribution were measured. (ii) Monitoring the colloidal components, such as the emulsions and nucleated polymer particles, achieved by diluting the extracted emulsion stream with aqueous medium throughout the reaction and measuring with suitable particle sizing detectors. Different types of detectors were used depending on the polymer or particle features monitored.

12.3.2.2 Emulsion Polymerization in Batch Generally, it comprises of three stages: [40, 41] interval I, in which free radicals generated in the aqueous phase enter micelles and form polymer particles, interval II where particles grow as they absorb monomer from monomer droplets, until their depletion, and interval III, where polymerization continues within the particles until the monomer consumption.

An example of the comprehensive information obtainable by this method is illustrated in Figure 12.8 [42] for free

radical emulsion polymerization of MMA at 70 °C [42]. The left side of the figure shows light scattering, UV, RI, and viscosity data used to compute M_w and η_r; typical free radical homopolymerization behavior is observed in their evolution with fractional monomer conversion f (left, top) from the sample stream diluted by tetrahydrofuran (THF), containing the soluble polymer or monomer components. The right side (bottom) shows particle distributions from Mie scattering on the colloidal aspects obtained by dilution of the reactor contents with water. Distributions #1–#3 are from progressive points of monomer conversion, and show the initial large monomer droplet populations (>1 micron) are consumed and transformed into nucleated polymer particles (150 nm); the upper panel shows the evolution of large and small particle modes from Mie scattering data, and the specific particle surface area. The increase in specific surface area, A, observed at the outset of nucleation indicates formation of small polymer particles at the expense of the low surface or volume large particles and furnishes a trend in particle size evolution. Final particle sizes were cross-checked by dynamic light scattering and transmission electron microscopy and found to be in excellent agreement.

The method has been extensively used in the online monitoring of MMA and BA (co)polymerization reactions under a variety of concentration conditions, with and without surfactant [39]. A number of features were captured, such as multiphase conversion, the fast evolution of the large polymer masses produced, and correlations between monomer droplet disappearance and monomer conversion were made.

12.3.2.3 *Semibatch Emulsion Polymerization* Semibatch emulsion polymerization is a versatile process which allows polymer latexes with special particle morphology and composition to be produced [43–45].

Semibatch processes are usually carried out using two different feed types: neat monomer feed, where only monomer is flowed into reactor, or monomer emulsion feed, in which aqueous emulsifier is added with the monomer. The systems, for which a critical flow rate has been reached, such that the rate of polymerization is controlled by the rate of monomer addition, are called "starved" [46]. In this case, there are no more monomer droplets and the monomer concentration in the polymer particles is lower than the saturated concentration. If the feed rate is higher than the polymerization rate, the monomer accumulates in the reactor as monomer droplets. This is usually termed a "flooded" system. In this case, the particles are completely saturated with monomer, and reaction kinetics resembles those of a batch emulsion polymerization.

Online monitoring of different emulsion characteristics, such as conversion, comonomer composition, and particle size, have been reported in numerous studies [47–49]. In most of the cases, estimation of reaction kinetics required calibration models and algorithms to be used, which have to account for effects of competing events, such as particle nucleation, coagulation, monomer droplets, changes in concentrations, and so on.

In this context, the use of ACOMP as a versatile and absolute method to quantify in a novel manner different aspects of the reaction kinetics in semicontinuous mode offers a significant advance because it not only provides means to immediately assess the regime the reaction occurs in but also allows verification of the effects of any change in the monomer feed rates on key parameters during the synthesis.

A first study of reaction kinetics with ACOMP was made for the semibatch emulsion polymerization of MMA at 70 °C [50]. The use of continuous monitoring method described in the previous sections offered a robust means of determining the characteristic features of the starved and flooded monomer conditions and identifying them during the experiments.

Reaction Kinetics

During the reaction, the mass of monomer in the reactor, $m(t)$, is given by the mass balance equation that relates monomer and polymer mass via:

$$m(t) = m_{tot}(t) - m_p(t) \quad (12.21)$$

where $m_p(t)$ is the polymer mass in the reactor

If monomer is fed into the reactor at constant flow rate, r_{add}, from a reservoir of monomer concentration $c_{m,f}$, and reactor fluid is extracted from the reactor at rate r_{ex} for the ACOMP stream, then the total amount of monomer plus polymer in the reactor at time t, $m_{tot}(t)$, is the sum of all monomer pumped into the reactor, the amount of monomer or polymer extracted, and any initial monomer present in the reactor. Based on the aforementioned equation, this has been shown to be [51]:

$$m_{tot}(t) = \frac{m_0}{\left(1 + \frac{(r_{add} - r_{ex})}{V_0} t\right)^{r_{ex}/(r_{add} - r_{ex})}} + r_{add} c_{m,f} t \quad (12.22)$$

where m_0 and V_0 are the initial mass of monomer and initial volume of liquid, respectively, in the reactor.

Taking into account the time effects of both the polymerization kinetics and the flow of monomer into the reactor, the monomer concentration, $[M](t)$, at any time becomes:

$$[M](t) = [M]_0 \exp\left\{ -\left(\frac{r_{add} + \alpha V_0}{V_0} \right) t \right\}$$
$$+ \left(\frac{[M]_f r_{add}}{r_{add} + \alpha V_0} \right) \left(1 - \exp\left\{ -\left(\frac{r_{add} + \alpha V_0}{V_0} \right) t \right\} \right) \quad (12.23)$$

where $[M]_0$ is the initial monomer concentration in the reactor, $[M]_f$ is the concentration of bulk MMA fed in the reactor, and α is the reaction rate coefficient.

According to standard semibatch definitions, the instantaneous conversion, f_{inst}, is defined as the total mass of polymer in the reactor divided by the total mass of monomer in the reactor at the time t:

$$f_{inst}(t) = \frac{m_{tot}(t) - m(t)}{m_{tot}(t)} =$$

$$= 1 - V(t) \frac{c_{m,0} \exp\left\{ -\left(\frac{r_{add} + \alpha V_0}{V_0} \right) t \right\} + \left(\frac{[m]_f r_{add}}{r_{add} + \alpha V_0} \right) \left(1 - \exp\left\{ -\left(\frac{r_{add} + \alpha V_0}{V_0} \right) t \right\} \right)}{\dfrac{m_0}{\left(1 + \frac{(r_{add} - r_{ex})}{V_0} t\right)^{r_{ex}/(r_{add} - r_{ex})}} + r_{add} c_{m,f} t}$$

$$(12.24)$$

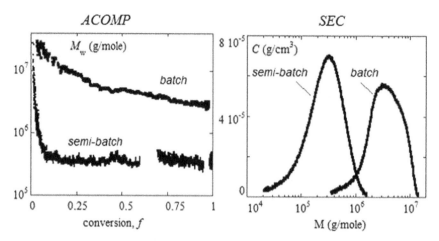

FIGURE 12.9 The effects of the monomer feed rate on M_w for reactions in semibatch mode versus batch mode (left panel); mass distributions for the same experiments, as determined from SEC (right panel). Reprinted (adapted) with permission from Alb AM, Reed WF. Online monitoring of molecular weight and other characteristics during semibatch emulsion polymerization under monomer starved and flooded conditions. Macromolecules 2009;42:8093–8101. © 2009 American Chemical Society.

Overall or total conversion, f_{tot}, is defined as the as the ratio of the polymer in the reactor to the total monomer in the recipe:

$$f_{total} = \frac{[M]_0 - [M](t)}{[M]_0} \qquad (12.25)$$

The Effect of the Feed Rate on the Polymer Molecular Weight

Particular emphasis was given in the study reported in Reference [50] to the methodology used to follow molecular weight evolution during the reaction monitoring, in an absolute fashion, based on multiangle light scattering detection.

The effects of the monomer feed rate on M_w is illustrated in Figure 12.9 [50] for MMA emulsion polymerization reactions in semibatch mode versus batch mode (left panel); mass distributions for the same experiments, as determined from SEC, are also shown (right panel). Distinct M_w kinetics was captured for a batch reaction and for a reaction in starved conditions. Thus, typical free radical behavior is observed for the M_w trend for the reaction in batch, where M_w starts with very large values and decreases with conversion nearly an order of magnitude. In contrast, for the experiment in semibatch conditions, M_w reached a plateau at around 20% conversion and has much smaller magnitude (over 20 times smaller than batch molecular mass). The shape and breadth of the MWD provided by SEC was also affected by the details of monomer addition, for example, narrower MWD for the semibatch reaction, where almost all monomer is flowed in.

The change in the monomer feed rates during reaction allowed simultaneous influence and verification of the evolution of these parameters during the synthesis, the adjustable feed rate conferring great power for controlling M_w during the reaction.

The transition from monomer starved to monomer-flooded conditions during the polymerization reaction was followed online in the work reported in Reference [50] by ACOMP

under different conditions. Figure 12.10 [50] shows an interesting M_w trend achieved during another semibatch emulsion polymerization reaction in which the flow rate at which the monomer (MMA) was added was varied during the reaction. This is an example of a "hybrid" reaction, where the polymer mass increases by an order of magnitude as the transition from monomer starved to monomer-flooded condition is made by the increase in the rate of monomer addition from 0.1 to 2 ml min^{-1}, around 50% total conversion. In the lower part of the figure, selected SEC chromatograms corroborated with ACOMP data show in their shape and breadth how the MWD is also affected by the details of monomer addition and follows the transition from low to higher molecular weight polymer. Smaller polymer mass was obtained during the slow monomer feed stage, followed by a jump in M_w as the rate of monomer addition was increased, after flooded conditions dominate the reaction kinetics.

The change in the monomer feed rates during reactions allowed simultaneous influence and verification of the evolution of these parameters during the synthesis. The adjustable feed rate confers great power for controlling M_w during the reaction. Thus, the "hybrid" reaction shown in Figure 12.10 [50] starts in semicontinuous mode, under starved conditions, achieved by slow monomer feed in the reactor, and becomes essentially a batch reaction upon the one-shot addition of MMA at 15% conversion.

The Effects of Monomer Feed Rate on the Particle Size and Number

Dynamic light scattering measurements allowed particle hydrodynamic diameter values (D_h) to be compared for reactions under different conditions.

Generally, in the case of the semibatch reactions, the particle size was proportional to the rate of monomer addition, smaller particles being produced at low flow rate. The number of particles increased at a lower rate of monomer addition.

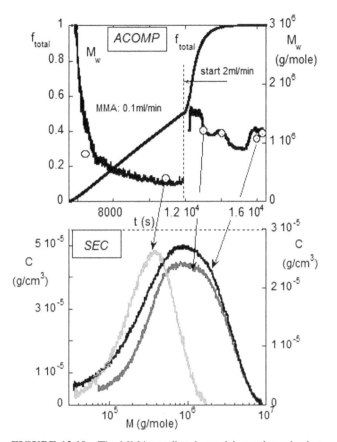

FIGURE 12.10 The MMA semibatch emulsion polymerization reaction: the polymer mass, M_w, increases by one order of magnitude as the transition from monomer starved to monomer-flooded condition is made by the increase in the rate of monomer addition from 0.1 to 2 ml min^{-1}, around 50% total conversion (top); selected SEC chromatograms corroborated with ACOMP data show in their shape and breadth how the MWD is also affected by the details of monomer addition and follows the transition from low to higher molecular weight polymer (bottom). Reprinted (adapted) with permission from Alb AM, Reed WF. Online monitoring of molecular weight and other characteristics during semibatch emulsion polymerization under monomer starved and flooded conditions. Macromolecules 2009;42:8093–8101. © 2009 American Chemical Society.

In monomer-starved conditions, particles are not saturated with monomer and grow at a rate controlled by the rate of monomer addition. In this case, the nucleation period is prolonged by a slower depletion of emulsifier micelles. Thus, lower monomer addition rates led to a decrease in the polymer latex particle size, but favored the increase in the number of the particles produced.

12.4 POSTPOLYMERIZATION MODIFICATIONS

Postpolymerization reactions, such as grafting, crosslinking, acid and base hydrolysis, quaternization, amination, and pegylation, represent attractive alternatives to using

purely synthetic routes to achieve functional polymers. Monitoring polymer characteristics during modification reactions offers means to optimize both the process and the desired properties of the end products.

An example of post polymerization modification reaction followed online by ACOMP is the hydrolysis of polyacrylamide (pAM) [52]. The evolution of polyelectrolyte characteristics during the base (NaOH) hydrolysis of polyacrylamide, which converts pAM from an electrically neutral polymer into a copolymeric polyelectrolyte, was captured in real time. Linear charge density ξ (number of elementary charges per Bjerrum length, l_B) increased during hydrolysis. The increase in ξ caused polymer coil expansion due to electrostatic repulsion, leading to an increase in the polymer intrinsic viscosity, $[\eta]$, and a suppression of the light scattering due to increased second and third virial coefficients. At the same time, the solution conductivity, σ, decreased as OH$^-$ is consumed during the reaction. ξ is proportional to the fractional degree of hydrolysis, δ (fraction of acrylamide monomers hydrolyzed) without any assumptions, as long as ξ is below the "counterion condensation threshold" [53].

Thus, the counterion condensation threshold ξ tracks $\delta(t)$ via l_B:

$$\gamma[\eta](t) = \xi(t) = \frac{l_B \delta(t)}{b} \qquad (12.26)$$

where γ is a proportionality constant and b is the monomer contour length.

For each AM monomer converted to COO$^-$ an OH$^-$ is lost:

$$\frac{d[\text{Am}]}{dt} = \frac{d[\text{OH}^-]}{dt} = -k[\text{OH}^-][\text{Am}] \qquad (12.27)$$

Solving Equation 12.27 allowed AM to be computed and with a model for the relationship between the experimentally measured decrease of $\sigma(t)$ due to loss of OH$^-$ and the buildup of charge groups on the polymer, $\delta(t)$ was computed.

While $\delta(t)$ is an important parameter to be monitored, since it allows running the reaction to any degree of conversion desired, there was no model-independent route to obtaining conversion, in contrast to typical ACOMP measurements of polymerization reactions, where model-independent conversion is directly obtained from collected data. The authors proposed the use of a chemically specific detector, such as FTIR, for the direct measurement of conversion in future studies.

Nonetheless, the ACOMP approach presented in Reference [52] provides a new means of assessing and further developing models of polyelectrolyte properties and should have numerous practical applications for optimizing product quality and process efficiency.

12.5 CONTROLLED RADICAL POLYMERIZATION

Free radical polymerization has been revolutionized in recent years by the introduction of controlled radical polymerization (CRP) methodologies [54], which confer control over the polymerization process while maintaining the versatility of conventional free radical polymerization. Synthesis of block, gradient, or other polymers of complex architecture, with predetermined molecular weight, narrow MWD, and high degrees of chain-end functionalization is no longer a difficult challenge [55–58].

Among the CRP methods, nitroxide-mediated polymerization (NMP), atom transfer radical polymerization (ATRP), and reversible addition fragmentation chain transfer (RAFT) have received special attention. CRP methods are discussed in Chapter 2.

12.5.1 Nitroxide-Mediated Polymerization (NMP)

Based on the persistent radical effect described by Fischer [59], NMP is applicable to a rather large number of monomers. The performance of a wide range of nitroxide and derived alkoxyamines has been presented in numerous experimental and theoretical studies on polymerization mechanism and kinetics [54, 60–64]. A simple selfregulation operates between transient and stable radicals; at the outset of the reaction, coupling reactions between transient radicals occur leading to an excess of stable radical. Then, bimolecular termination reactions are reduced and become negligible.

Comprehensive characterization of the controlled radical (co)polymerization of styrene and BA monomers using the ACOMP multidetector platform was recently reported [65, 66].

Monomer conversion, evolution of molecular weight, reduced viscosity, and rate constants were assessed in Reference [65] in the case of the NMP of BA using N-tertiobutyl-1-diethylphosphono-2,2-dimethylpropyl nitroxide (SG1). The reactions were carried out using both monocomponent (alkoxyamines with or without free SG1) and bicomponent (AIBN + free SG1) initiators.

One of the main findings was that conversion is roughly first order, although depends on the initial ratio of free SG1 to initiator; that is, it is zeroth order in initiator concentration. A possible explanation was given by the authors, as follows [65].

If K_{eq}, the constant for the equilibrium between active and dormant species,

$$K_{eq} = \frac{[R][SG1]}{[R-SG1]} \quad (12.28)$$

is the same no matter if SG1 is associated with a radical MA ($R = MA$) or attached to a growing PBA chain ($R = PBA$) and if the initial excess $[E]_0$ of free SG1 is

$$[E]_0 = \varepsilon[MA-SG1]_0 \quad (12.29)$$

where $[MA-SG1]_0$ is the initial, fully associated concentration of $MA-SG1$ and ε is typically a dimensionless number, less than unity.

The rate constant α of the monomer concentration decay rate:

$$\frac{d[m]}{dt} = -k_p[R][m] = -\alpha[m] \quad (12.30)$$

is controlled completely by $[R]$ since, at a fixed temperature, k_p is constant. The concentration $[SG1]$ is given by:

$$[SG1] + [R-SG1] = [MA-SG1]_0 + [E]_0$$
$$= (1+\varepsilon)[MA-SG1]_0 \quad (12.31)$$

In the presence of the free nitroxide, $[R]$ quickly reaches a quasi-steady state and with a small amount, $b[MA-SG1]_0$, during the reaction. Numerical simulations indicate $b \sim 1$, that is, only a small fraction of the initial radical concentration is terminated during the course of monomer conversion. Then,

$$[R] + [R-SG1] = b[MA-SG1]_0 \quad (12.32)$$

This leads to a quadratic form for $[R]$,

$$[R] = -\frac{(1+\varepsilon-b)[MA-SG1]_0 + K_{eq}}{2}$$
$$\pm \frac{\left\{ \begin{array}{c} <(1+\varepsilon-b)[MA-SG1]_0 + K_{eq} >^2 \\ +4bK_{eq}[MA-SG1]_0 \end{array} \right\}^{1/2}}{2}$$

$$(12.33)$$

Since the second term in the square root term is small compared to the first, this allows expansion to first order:

$$[R] \cong \frac{bK_{eq}}{(1+\varepsilon-b)} \quad (12.34)$$

and leads to an expression for the reciprocal rate constant:

$$\frac{1}{\alpha} = \frac{1}{k_p[R]} = \frac{1}{k_p}\left\{ \frac{1-b+\varepsilon}{bK_{eq}} \right\} \quad (12.35)$$

which shows that the rate constant α depends only on the excess of SG1, represented by ε, and the absolute concentration $[MA-SG1]_0$ does not appear. Equation 12.35 also allowed for determination of the equilibrium constant between active and dormant species at 120 °C ($K_{eq} = 1.53 \times 10^{-10}$ M), as well as the corresponding kinetic constant of deactivation ($k_{deact} = 2.8 \times 10^7$ l·mol^{-1} s^{-1}) and activation ($k_{act} = 4.2 \times 10^{-3}$ s^{-1}).

In terms of molecular weight behavior, typical CRP trends were obtained for weight-average and viscosity-average masses, which increased in approximately linear fashion with conversion but had finite values at zero conversion, behavior also observed in other studies [67].

Whereas the online determination of M_w, conversion, and reduced viscosity was of primary concern, the study of the time dependent light scattering signatures, $I(t)$, from different reactions was of interest since they exhibit features characteristic of the type of reaction taking place.

Both ideal CRP and free radical polymerizations are first order in $c_p(t)$:

$$c_p(t) = c_m(0)(1 - e^{-\alpha t}) \tag{12.36}$$

whereas the dependence of $M_w(t)$ is different for each type. For ideal free radical polymerization, the accumulated M_w decreases with conversion according to [68]:

$$M_w(f) = M_w(0)\left(1 - \frac{f}{2}\right) \tag{12.37}$$

while for CRP, M_w increases with conversion according to:

$$M_w(f) = a + bf \tag{12.38}$$

An example of the excellent correlation between theoretical light scattering signatures (based on Eqs. 12.1, 12.37, and 12.38) and with the experimental data is illustrated in Figure 12.11 [65], in which $I(t)$ for both free radical and NMP reactions of BA are shown, with fits using the appropriate forms of $I(t)$ for each reaction and the first-order c.

Due to the high level of end-group functionality and the lifetime of the polymer chains, NMP is an attractive methodology for synthesizing not only homopolymers but also block copolymers and well-defined gradient copolymers.

Gradient copolymers are defined as copolymers of two or more monomers, whose composition profile varies along the chain, reflecting variation in monomer concentrations as conversion proceeds. As a consequence, gradient copolymers combine the properties of the homopolymers in a way that depends on the nature of the composition profile.

In this context, the adaptation of ACOMP to online monitoring of NMP copolymerization offered the opportunity to study in more detail CRP reaction kinetics and mechanisms and to control the composition profile formation during the polymerization reaction.

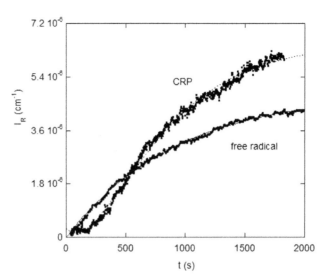

FIGURE 12.11 Excellent correlation between theoretical light scattering signatures with the experimental data for $I(t)$ for both free-radical and NMP reactions of butyl acrylate (BA), with fits using the appropriate forms of $M_w(t)$ for each reaction and the first-order c_p. From Chauvin F, Alb A, Bertin D, Tordo P, Reed WF. Kinetics and molecular weight evolution during controlled radical polymerization. Macromol Chem Phys 2002;203:2029–2041. ©Wiley-VCH Verlag GmbH & Co. KGaA. Reproduced with permission.

Styrene/BA nitroxide-mediated gradient copolymerization followed in real time by ACOMP was recently reported [66]. Using the same methodology as described in previous sections of the chapter, the simultaneous monitoring and combining signals from the continuously diluted reactor stream allowed copolymer mass, reduced viscosity, and the comonomer concentrations to be determined in a model independent fashion, based on data collected during copolymerization reactions from ACOMP light scattering, viscosity, differential RI, and UV detectors.

Both the average instantaneous composition and cumulative mole fraction of each comonomer along the polymer chains computed by the authors from continuous knowledge of the comonomer concentrations allow for a rich variety of comonomer and initiator schemes to be tried in order to control the types of profiles formed.

12.5.2 Reversible Addition Fragmentation Chain Transfer (RAFT)

Among CRP methods, much attention has been devoted to RAFT as a versatile method that allows different materials with well-defined molecular structures and architectures to be obtained [69–72].

The key feature of the reaction mechanism is the reversible transfer of the radical activity to a dormant species, making it active. Rapid equilibrium established between the active and dormant species allows all the chains to grow in a controlled fashion.

While this approach proved to be an efficient and attractive means for a wide range of monomers and reaction conditions, side reactions and deviations from ideality due to incompatibility between certain monomer types and control agents and experimental conditions are reported [73].

Despite some controversy in terminology [74], usually, the level of control is assessed based on the narrowness of the MWD and on the linear increase of the number-average molecular weight, M_n with conversion, whereas the livingness is quantified in the fraction of dormant polymer chains that can be extended by the addition of new monomer.

Automatic continuous online monitoring of polymerization joins other techniques to study and quantify RAFT polymerization kinetics under different conditions [75–77]. The utility of the technique for monitoring RAFT polymerization and to chart changes in kinetics and molar mass evolution as the concentration of RAFT agent is varied was recently demonstrated [78].

12.5.2.1 Homopolymerization

Polymerization of BA using an unsymmetrical trithiocarbonate, 2-{[(dodecylsulfanyl) carbonothioyl]sulfanyl}propanoic acid (DoPAT) was chosen as a case study [78].

Figure 12.12 [78] shows an example of ACOMP data from different detectors for one of the BA polymerization reactions performed in the work presented in Reference [78] ($T = 70\,°C$, [AIBN] = 0.0024 M, [BA] = 2.08 M, [BA]/[DoPAT] = 365.4). The increase of LS 90°, viscosity, and RI signals after the addition of AIBN follows the growth of the polymer chains, whereas the decrease in UV (260 nm) follows the monomer consumption. Additionally, UV at 305 nm was used to monitor the evolution of the trithiocarbonate (TTC) moiety during its transfer from DoPAT to a macroRAFT species and subsequent polymer growth from the latter. The inset to the figure shows UV absorption spectra at selected points during conversion. The TTC peak at 305 nm remains unchanged during the reaction, whereas the lower wavelengths dominated by BA absorbance decrease as conversion proceeds.

In terms of conversion kinetics, in agreement with previous findings [75], the rates were found to be essentially zeroth order in DoPAT, depending almost solely on the initiator (AIBN) concentration.

The evolution of M_w with conversion was monitored for a series of reactions in which the RAFT agent concentration was decreased over a wide range while all the other reaction conditions were held constant ([DOPAT]/[AIBN] ranged from 0 to 2.4). It was thus possible to follow the transition from "living"-like behavior in the CRP regime, where sufficient RAFT agent was used, to the noncontrolled radical polymerization regime as RAFT agent was progressively reduced to 0.

This is illustrated in Figure 12.13 [78]. M_w was computed on the basis of the extrapolation of Kc/I versus q^2 to $q = 0$ and

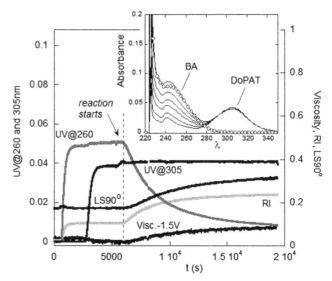

FIGURE 12.12 The ACOMP data from different detectors for a BA polymerization reaction by RAFT ($T = 70\,°C$, [AIBN] = 0.0024 M, [BA] = 2.08 M, [BA]/[DoPAT] = 365.4). The increase of LS 90°, viscosity, and RI signals after the addition of AIBN follows the growth of the polymer chains, whereas the decrease in UV (260 nm) follows the monomer consumption. UV at 305 nm monitors the evolution of the trithiocarbonate moiety during its transfer from DoPAT to a macroRAFT species and subsequent polymer growth from the latter. The inset to the figure shows UV absorption spectra at selected points during conversion. Reprinted (adapted) with permission from Alb AM, Serelis AK, Reed WF. Kinetic trends in RAFT homopolymerization from online monitoring. Macromolecules 2008;41:332–338. © 2008 American Chemical Society.

corrected for finite polymer concentration, using A_2 values determined in separate automatic continuous mixing (ACM) experiments. For high values of [DoPAT]/[AIBN] (bottom three curves in Fig. 12.13), the reactions exhibit typical CRP behavior with nearly linear increase of mass versus fractional monomer conversion, f. The downward curvature of M_w versus f in the reaction with lowest (DoPAT) (#4, second top curve in figure) indicates significant deviation from the ideal living mechanism and, as suggested by the authors, may be due to a transitional "hybrid behavior" stage where a significant fraction of polymer chains grow by a conventional uncontrolled radical polymerization mechanism [79] and/or to transfer to solvent or monomer. The top curve (#5), for a reaction with no DoPAT, shows classical uncontrolled radical polymerization behavior of M_w with conversion. All but one reaction showed a nonzero M_w at $f = 0$, in agreement with previous theoretical and experimental studies reported [80, 81].

Even though it was not possible to separate evidence of conventional chain transfer processes from the definite presence of the uncontrolled polymer radical chains with the data available at that time, deviations from living behavior allowed estimates of radical efficiency to be made.

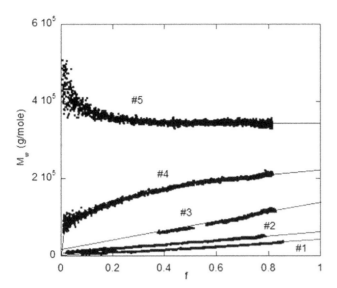

FIGURE 12.13 M_w versus conversion, f from ACOMP for several BA polymerization reactions by RAFT (#1–4) and a free radical polymerization reaction (#5). For high [DoPAT]/[AIBN] (#1–3), the reactions exhibit typical CRP behavior with nearly linear increase of mass versus f. The downward curvature in reaction #4 indicates significant deviations from the ideal living mechanism. Reaction #5 shows classical uncontrolled radical polymerization behavior of M_w with conversion. Reprinted (adapted) with permission from Alb AM, Serelis AK, Reed WF. Kinetic trends in RAFT homopolymerization from online monitoring. Macromolecules 2008;41:332–338. © 2008 American Chemical Society.

12.5.2.2 Copolymerization by RAFT Unlike in a free radical copolymerization, during the RAFT process, all the chains grow continuously throughout the reaction, allowing the composition drift to be captured within the chain structure and leading to polymer chains with similar composition.

Gradient Copolymers
Novel approaches in the online monitoring of the synthesis of the amphiphilic copolymers by Automatic continuous online monitoring of polymerization were recently reported by Alb et al. [23]. The ACOMP monitoring platform was used to follow in real time the synthesis of 2-(dimethylamino)ethyl acrylate (DMAEA)/styrene (sty) copolymers by RAFT.

The goal of the work was twofold: First, to study the feasibility of a new approach in studying the synthesis of amphiphilic copolymers by simultaneous online measurements in different solvents, and second, to investigate the degradation process of the macroRAFT agent during the copolymerization reactions, presumably caused by the tertiary-amine group of DMAEA.

Poly[2-(dimethylamino)ethyl acrylate], poly(DMAEA) is a pH-sensitive polymer in aqueous solution, as its tertiary amine can be quaternized at low pH, with wide applications in the pharmaceutical and water treatment industries. However, polymerization of DMAEA by RAFT is less studied due to the challenges imposed by the low steric hindrance of its substituents and side reactions of the polar amino group [82, 83].

The authors took advantage of the amphiphilic character of the copolymers synthesized to introduce a novel approach, consisting in the automatic, continuous withdrawal and dilution of the reactor solution with different solvents, in order to simultaneously measure different properties. This was first time when simultaneous polymer characterization in organic and aqueous environments, performed as the polymer is produced, was reported.

Automatic withdrawal of separate reactor streams and their subsequent dilution throughout the reaction with organic and aqueous solvents, respectively, allowed different features to be captured. Thus, light scattering data combined with spectroscopic and viscometric measurements in N,N-dimethylformamide (DMF), together with conductivity measurements in aqueous medium provided in real-time comonomer conversion, copolymer mass, reduced viscosity, and composition. Continuous data gathered allowed observing and quantifying the DMAEA-linked macroRAFT agent decomposition, with significant effects on the reaction kinetics. Deviations from living behavior were investigated during (co) polymerization reactions under different conditions.

Automatic continuous online monitoring of polymerization multiple detection offers flexibility in choosing the most appropriate method to compute the comonomer concentrations during a reaction. In this case, the authors used, as a novel feature, an alternate route to follow DMAEA concentration (c_{DMAEA}) by the use of the monitored conductivity of the monomer in aqueous medium. DMAEA is weakly charged in water, to a small but sufficient extent for conductivity measurements to be made.

In all experiments, excellent agreement was found between computed c_{DMAEA} from the method based on spectroscopic detection:

$$c_B = \frac{(\varepsilon_A l - \varepsilon_P l)\left[V_{RI}CF - \left(\frac{dn}{dc}\right)_D c_D - \left(\frac{dn}{dc}\right)_P c_0\right]}{(\varepsilon_P l - \varepsilon_B l)\left[\left(\frac{dn}{dc}\right)_A - \left(\frac{dn}{dc}\right)_P\right] - (\varepsilon_A l - \varepsilon_P l)\left[\left(\frac{dn}{dc}\right)_P - \left(\frac{dn}{dc}\right)_B\right]} - \frac{\left[\left(\frac{dn}{dc}\right)_A - \left(\frac{dn}{dc}\right)_P\right](V_{UV} - \varepsilon_D l c_D - \varepsilon_P l c_0)}{(\varepsilon_P l - \varepsilon_B l)\left[\left(\frac{dn}{dc}\right)_A - \left(\frac{dn}{dc}\right)_P\right] - (\varepsilon_A l - \varepsilon_P l)\left[\left(\frac{dn}{dc}\right)_P - \left(\frac{dn}{dc}\right)_B\right]}$$

(12.39)

$$c_A = \frac{V_{UV} - \varepsilon_D l c_D - \varepsilon_B l c_B - \varepsilon_P l (c_0 - c_B)}{\varepsilon_A l - \varepsilon_P l} =$$

$$= \frac{V_{RI} CF - \left(\frac{dn}{dc}\right)_D c_D - \left(\frac{dn}{dc}\right)_B c_B - \left(\frac{dn}{dc}\right)_P (c_0 - c_B)}{\frac{dn}{dc_A} - \frac{dn}{dc_P}}$$

(12.40)

and data furnished by the monitored conductivity signal:

$$c(t) = \frac{\delta(t) - \delta_f}{\delta(0)} c_0 \qquad (12.41)$$

proving the feasibility of the latter approach in following the concentration of the charged monomer.

In the aforementioned equations, $c_{A\,(B)}$ refers to the mass concentration of comonomers A (B), c_P and c_D are the mass concentrations of the polymer and of DoPAT, respectively, $\delta(t)$ is the conductivity at time t, $\delta(0)$ is the conductivity of the solution due to DMAEA before the reaction has started, and δ_f is the measured final value.

The computed concentrations from either one of the methods described earlier allowed the comonomer conversion, instantaneous composition drift, and copolymer molecular weight to be determined. It is important to point out the unique advantages ACOMP offers by providing continuously collected data that allow more detailed information about reaction kinetics to be obtained. Monitoring the evolution of molecular weight during reactions offered further means to assess deviations from ideal behavior.

Both the continuously monitored data from ACOMP and the discrete data from SEC chromatograms of samples withdrawn during the reactions studied indicated degradation of the TTC group, presumably caused by the amine group of DMAEA, a process that had a negative effect on the controlled character of the polymerization. Even though DMAEA polymerization reaction has been reported by other groups [84], as the authors were aware, this was first time when the TTC degradation was observed, within the condition studied. Therefore, the effects on the behavior of the TTC of the following factors were studied:

- *Solvent effect.* A first step in assessing the factors affecting the behavior of the TTC moiety in the polymerization of DMAEA was to eliminate the effect of other species, such as inhibitors, water on the reaction kinetics. However, the results from DMAEA polymerization reactions performed with the purified monomer in other solvents (anhydrous DMF, butyl acetate, THF, dioxane, and toluene) indicated that the degradation of the TTC still occurred.

- *Monomer effect.* A first attempt to improve the reaction livingness during DMAEA polymerization was related to its tertiary amine group. However, it was found that the use of the trimethylammonium analogue of DMAEA did not eliminate the TTC degradation.

- *RAFT agent effect.* DoPAT was replaced with other trithiocarbonates, in order to test whether the carboxylic acid in DoPAT played any role in the observed decomposition, triggered by the amine group of DMAEA. Nonetheless, TTC degradation still occurred.

- *Comonomer effect.* It was found during the copolymerization reactions studied that styrene protects the TTC moiety from decomposition and depending on the amount in the initial composition, improved the reaction controlled character.

Continuous UV data at 305 nm provided by ACOMP, shown in Figure 12.14 [23], follow the evolution of the TTC during several copolymerization reactions with different initial composition and indicate that the degradation process is slower for higher amount of styrene. Shown in the inset

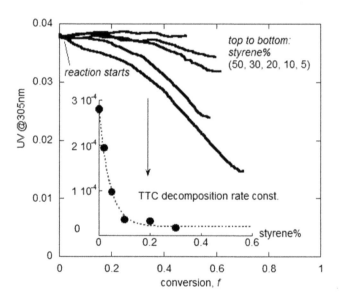

FIGURE 12.14 Continuous UV data at 305 nm from ACOMP follow the evolution of the TTC during 2-(dimethylamino)ethyl acrylate (DMAEA)/styrene (sty) copolymerization reactions by RAFT with different initial composition and indicate that the degradation process is slower for higher amount of styrene. Shown in the inset to figure are the plotted TTC decomposition rate constants versus styrene %. The rates were from exponential fits used as first-order approximations. "Reprinted from Li Z, Serelis AK, Reed WF, Alb AM. Online monitoring of the copolymerization of 2-(dimethylamino)ethyl acrylate with styrene by RAFT: deviations from reaction control. Polymer 2010;51:4726–4734. © 2010 with permission from Elsevier.

to figure are the plotted TTC decomposition rate constants versus styrene %. The decomposition rates were from exponential fits used as first-order approximations.

Size Exclusion Chromatography experiments on samples taken during reactions confirmed ACOMP findings and furnished additional evidence concerning TTC degradation and the loss of control by capturing the MWD broadening for reactions with a high amount of DMAEA.

A possible explanation offered by the authors is that increasing the amount of styrene minimizes the side effects due to the interaction of the amine group in DMAEA with the trithiocarbonate group, by removing the TTC from the vicinity of the amine group. When the last monomer unit on the chain is DMAEA, the TTC is close to the amine and the two groups can react, but when the last monomer unit is styrene, the TTC is much further away from the amine and reaction is negligible. On the other hand, the relative leaving abilities of acrylyl and styryl radicals in the fragmentation step of a RAFT-controlled polymerization have to be considered. The acrylyl radical is a much better leaving group, and so will fragment off an intermediate RAFT radical much faster than a styryl radical will, with the result that the TTC will spend most of its time on styryl-ended polymer chains and thus be protected from reaction with the amine located one or more monomer units back along the chain. This rate difference should explain why degradation of the TTC seems to be completely stopped in the case of the reaction with 50% styrene.

Despite the fact that the exact nature of the decomposition of the RAFT agent triggered by DMAEA was not elucidated, the authors found a way to slow down or, in certain conditions, to eliminate the degradation process by the addition of styrene in copolymer composition, leading to better controlled reaction kinetics. The higher the amount of styrene, the better control was achieved.

Diblock Copolymers
The fact that the thiocarbonylthio group is retained in the polymer chain allows block copolymers to be synthesized via RAFT by simply adding a second monomer.

It is known that successful preparation of block copolymers requires not only a well controlled homopolymerization of each monomer with complete conversion but also a smooth transition from one monomer type to the other at the critical covalent junctions. However, the fidelity of this crossover is often imperfect and detection of crossover inefficiency is usually not determined until SEC or other post-reaction characterization methods shows residual homopolymer from the first monomer.

The use of ACOMP in following the synthesis of diblock copolymers offers the opportunity to study and verify the efficiency of the chain extension during monitoring of the copolymerization reactions and offer means

to optimize the reaction conditions. As shown in previous sections, light scattering data collected during copolymerization allow experimental and theoretical signatures to be compared, and highlight differences between true diblock formation and formation of two homopolymers or other side reactions occurring.

12.5.3 Atom Transfer Radical Polymerization (ATRP)

Atom Transfer Radical Polymerization (ATRP), one of the most rapidly evolving CRP methodology [85, 86], has been successful in controlling polymerization of a large variety of monomers, styrenes, acrylates, methacrylates, acrylamides, and so on under various conditions [87–90].

The first use of ACOMP to follow in real-time ATRP reaction kinetics was recently reported [91]. The ACOMP multidetector platform allowed simultaneous monitoring of conversion, weight-average molar mass, M_w, and intrinsic viscosity, from which viscometric-average molar mass M_v and M_w/M_v were computed, in the case of ATRP reactions of BA, initiated by methyl 2-bromopropionate (M2BP) and catalyzed by copper bromide.

A novel feature introduced in this work was the use of a spectrophotometer at 699 nm in order to determine Cu(I) and Cu(II) concentrations, based on previous studies that showed that both $Cu^IBr/dNbpy_2$ activator and $Cu^{II}Br_2/dNbpy_2$ deactivator absorb in the visible range [92]. $Cu^{II}Br_2/dNbpy_2$ has an absorption band whose maximum is located between 650 and 750 nm; $Cu^IBr/dNbpy2$ also absorbs in the same range but with a smaller extinction coefficient. According to Reference [92], in low polar media such as BA and in the presence of an excess of Cu(I) activator, $Cu^{II}Br_2/dNbpy_2$ complexes mostly exist under a trigonal bipyramidal cationic form $[Cu(II)\text{-}(dNbpy)_2Br]^+,[Br]^-$. Hence, it was assumed that the monitored signal was only due to the active forms of each copper-based complex that is able to establish the equilibrium between activated and deactivated species, that is, $Cu^IBr/dNbpy_2$ activator and cationic $Cu^{II}Br_2/dNbpy_2$ deactivator. Preliminary ACOMP experiments were made to calibrate the visible absorption data.

The online monitored absorption at 699 nm data, allowed the concentrations of each copper complex in the reactor to be computed in real time. Generally, Cu(II) species are measured by electron spin resonance (ESR) on aliquots extracted from reactor. A good agreement with the ESR values was observed [93].

The polymerizations exhibited first-order kinetics with respect to monomer and initiator concentrations. The rate of the reaction dramatically decreased with increasing initial concentration of the deactivator $Cu^{II}Br_2$ (copper(II) bromide), as expected in ATRP kinetics since an initial excess of deactivator shifts the equilibrium toward dormant species and reduces the concentration of propagating free radicals.

The ACOMP continuously computed conversion data for experiments at 90 °C with different concentration of $Cu^{II}Br_2$ (0, 0.025, and 0.05 equiv vs. $Cu^{I}Br$) made possible to observe the persistent radical effect (PRE) kinetics at the early polymerization stages.

In the presence of initially added deactivator, the rate law for ATRP is described by

$$\log\left(\frac{c_{BA,0}}{c_{BA}}\right) = k_p K_{eq} \frac{[Cu^{I}Br]}{[Cu^{II}Br_2]}[M2BP] \times t \qquad (12.42)$$
$$= k_p[P_n^{\bullet}] \times t = k_{app} \times t$$

where $P_n^{\bullet}$ is the activated species, k_p is the absolute rate constant of polymerization, k_{app} is the apparent rate constant, and K_{eq} is the equilibrium constant between active and dormant species.

Thus, the experiments with added $Cu^{II}Br_2$ were used for calculating K_{eq} by plotting the reciprocal apparent rate constant as a function of $[Cu^{II}Br_2]0/([Cu^{I}Br]_0[M2BP]_0$ and using

$$\frac{1}{k_{app}} = \frac{1}{k_p K_{eq}} \times \frac{[Cu^{II}Br_2]_0}{[Cu^{I}Br]_0[M2BP]_0} \qquad (12.43)$$

Overall, the results presented in Reference [91] strengthen the basic understanding of ATRP kinetics and mechanisms and allow different kinetic effects and deviations from ideal behavior to be observed. Nonetheless, a more quantitative understanding of these effects and deviations has to be developed and precision in the determination of the concentrations of the copper species has to be optimized. Lately, efforts were made to adapt ACOMP to controlled copolymerization kinetics.

12.6 ACOMP WITH SIMULTANEOUS CONTINUOUS DETECTION AND DISCRETE AUTOMATIC SEC

As discussed in the previous sections, online monitoring polymerization reactions have multiple benefits, providing comprehensive, quantitative data for understanding fundamental mechanisms and kinetics, optimizing reaction processes, and, ultimately, precisely controlling reactions to yield desired products.

A novel approach consisting in coupling together non-chromatographic ACOMP and discrete chromatographic

monitoring (SEC) to follow in real-time polymerization reactions was recently introduced [94]. RAFT and free radical polymerization of BA were chosen as case study by the authors. The fact that the reactor solution is already preconditioned in the ACOMP front end to the concentration levels used in SEC made possible the addition of a multidetector SEC system to the ACOMP platform.

A main advantage of the SEC detection is that it follows the evolution of the MWD, particularly important in "living"-type reactions and in copolymerization reactions where complex mixtures of reagents make unfractionated spectroscopic resolution of comonomers difficult. On the other hand, continuous detection provides a much more detailed characterization of the reaction and, interestingly, in the case where a bimodal MWD was produced, the continuous method automatically detected the onset of the second mode in a model independent fashion, whereas SEC could only discern the bimodality by applying preconceived models.

The case of bimodal MWDs in free radical reactions is presented in the following, for illustration purposes.

Figure 12.15 [94] (left side) shows the effect of an initiator boost on both the cumulative M_w and the instantaneous weight-average molecular weight, $M_{w,inst}$, measured by ACOMP, seen in the rapid change in slope at the moment the initiator is added. $M_{w,inst}$ was computed from the cumulative M_w via Equation 12.4. The bimodal population produced was detected directly from the ACOMP data without recourse to any model and a priori assumptions about the nature of the distribution. A more clear representation of the bimodality is seen in the histogram of $M_{w,inst}$ in inset to figure. In contrast, the SEC data in the right side of Figure 12.15 show a very significant broadening of the MWD, but with no apparent bimodality. A model in which the net chromatogram is assumed to be the superposition of two single modes, each of which is best fit by a log-normal distribution was used. The result of a single log-normal fit to the early (preboost) MWD is shown in the figure, along with a double log-normal fit to an MWD toward the end of the reaction. The latter fit is good, which corroborates the assumption of the bimodal distribution, but the bimodal assumption has first to be made, and is not an immediate consequence of the direct SEC MWD analysis.

The dotted line MWD in Figure 12.15 [94] is obtained from the continuous ACOMP data, by "binning" the $M_{w,inst}$ data logarithmically. Again, the bimodality is very clear via continuous detection.

Direct comparison of kinetic and molecular weight from the continuous nonchromatographic and SEC approaches showed that addition of multidetector SEC enhances data gathering power of ACOMP by bringing valuable complementary information.

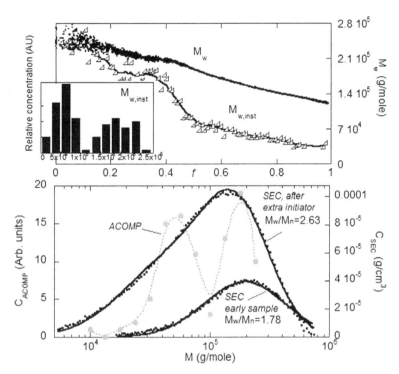

FIGURE 12.15 The effect of an initiator boost during BA free radical polymerization on both the cumulative M_w and the instantaneous weight-average molecular weight, $M_{w,inst}$, from ACOMP, illustrated in the rapid change in slope at the moment the initiator is added (top). A more clear representation of the bimodality is seen in the histogram of $M_{w,inst}$ in inset to figure. In contrast, SEC data show a very significant broadening of the MWD, but with no apparent bimodality (bottom). The dotted line MWD is obtained from the continuous ACOMP data, by "binning" the $M_{w,inst}$ data logarithmically. This material is reproduced with permission of John Wiley & Sons, Inc. from Alb AM, Drenski MF, Reed WF. Simultaneous continuous, non-chromatographic monitoring and discrete chromatographic monitoring of polymerization reactions. J Appl Polym Sci 2009;13:190–198.

12.7 SUMMARY

Recent advances in several fields of the free radical polymerization, in homogeneous and heterogeneous phase, discovery and development of controlled free radical polymerization techniques facilitate the synthesis of macromolecules with a high level of control.

At the same time, synthesis of these high performance materials requires better monitoring, control, and understanding of the reaction kinetics.

New understanding of basic aspects of the processes has become possible as a result of combining state-of-the-art polymerization chemistry with comprehensive ACOMP. The technique offers a comprehensive picture on what phenomena occurs during reactions as opposed to what is thought ideally to occur, providing monomer conversion kinetics, composition drift and distribution, evolution of molar mass, intrinsic viscosity, and other polymer and reaction characteristics.

ACKNOWLEDGMENTS

The author expresses her heartfelt gratitude to Cico for continuous support and understanding.

REFERENCES

[1] Kulicke WM, Kniewske R, Klein J. Preparation, characterization, solution properties and rheological behaviour of polyacrylamide. Prog Polym Sci 1982;8:373–468.

[2] Bamford CH. Radical polymerization. In: Mark HF, Bikales NM, Overberger CG, Menges G, editors. *Encyclopedia of Polymer Science and Engineering*, Volume 13. New York: Wiley Interscience; 1988. pp 708–867.

[3] Giz A, Çatalgil-Giz H, Alb AM, Brousseau J-L, Reed WF. Kinetics and mechanisms of acrylamide polymerization from absolute, online monitoring of polymerization reactions. Macromolecules 2001;34:1180–1191.

[4] Zimm BH. The scattering of light and the radial distribution function of high polymer solutions. J Chem Phys 1948; 16:1093–1115.

[5] Cherfi A, Fevotte G, Novat CJ. Robust on-line measurement of conversion and molecular weight using NIR spectroscopy during solution polymerization. J Appl Polym Sci 2002;85:2510–2520.

[6] Florenzano FH, Enohnyaket P, Fleming V, Reed WF. Coupling of near infrared spectroscopy to automatic continuous online monitoring of polymerization reactions. Eur Polym J 2005;41:535–545.

[7] Grassl B, Alb AM, Reed WF. Online monitoring of chain transfer in free radical polymerization. Macromol Chem Phys 2001;202:2518–2524.

[8] Dammert R, Jussila M, Vastamaki P, Riekkola M-L, Sundholm F. Determination and comparison of molar mass distributions of substituted polystyrenes and block copolymers using thermal field-flow fractionation, size exclusion chromatography and light scattering. Polymer 1997;38:6273–6280.

[9] Wignall GD, Alamo RG, Ritchson EJ, Mandelkern L, Schwahn D. SANS studies of liquid–liquid phase separation in heterogeneous and metallocene-based linear-low density polyethylene. Macromolecules 2001;34:8160–8165.

[10] Faldi A, Soares JBP. Characterization of the combined molecular weight and composition distribution of industrial ethylene/alpha-olefin copolymers. Polymer 2001;42:3057–3066.

[11] Verhelst V, Vandereecken P. Analysis of organosilicone copolymers by gradient polymer elution chromatography with evaporative light scattering detection. J Chromatogr A 2000;87:269–277.

[12] Schunk TC, Long TE. Compositional distribution characterization of poly(methyl methacrylate)-graft-polydimethylsiloxane copolymers. J Chromatogr A 1995;692:221–232.

[13] Yau WW, Gillespie D. New approaches using MW-sensitive detectors in GPC–TREF for polyolefin characterization. Polymer 2001;42:8947–8958.

[14] Stockmayer WH, Moore LD, Fixman M, Epstein BN. Copolymers in dilute solution. I. Preliminary results for styrene-methyl methacrylate. J Polym Sci 1955;16:517–530.

[15] Bushuk W, Benoit H. Light-scattering studies of copolymers: I. Effect of heterogeneity of chain composition on the molecular weight. Can J Chem 1958;36:1616–1626.

[16] Berry GC. Total intensity light scattering from solutions macromolecules. In: Pecora R, Borsali R, editors. *Soft Matter: Scattering, Manipulation and Imaging*, Volume 1. Dordrecht and London: Springer; 2007. pp 41–119.

[17] Çatalgil-Giz H, Giz A, Alb AM, Öncül Koç A, Reed WF. Online monitoring of composition, sequence length, and molecular weight distributions during free radical copolymerization, and subsequent determination of reactivity ratios. Macromolecules 2002;35:6557–6571.

[18] Enohnyaket P, Kreft T, Alb AM, Drenski MF, Reed WF. Determination of molecular mass during online monitoring of copolymerization reactions. Macromolecules 2007;40:8040–8049.

[19] Alb AM, Enohhnyaket P, Drenski MF, Head A, Reed AW, Reed WF. Online monitoring of copolymerization involving comonomers of similar spectral characteristics. Macromolecules 2006;39:5705–5713.

[20] Alb AM, Enohnyaket P, Drenski MF, Shunmugam R, Tew GN, Reed WF. Quantitative contrasts in the copolymerization of acrylate and methacrylate based comonomers. Macromolecules 2006;39:8283–8292.

[21] Kelen T, Tudos, F. Analysis of the linear methods for determining copolymerization reactivity ratios. I. A new improved linear graphic method. J Macromol Sci Part A: Chem 1975;9:1–27.

[22] Odian G. *Principles of Polymerization*. 3rd ed. New York: John Wiley & Sons; 1991.

[23] Li Z, Serelis AK, Reed WF, Alb AM. Online monitoring of the copolymerization of 2-(dimethylamino)ethyl acrylate with styrene by RAFT: deviations from reaction control. Polymer 2010;51:4726–4734.

[24] Li Z, Alb AM. Cationic copolyelectrolytes: online monitoring of the synthesis and subsequent copolymer characterization. Macromol Chem Phys 2012;213:1397–1406.

[25] Hunkeler D, Hernández-Barajas J. Inverse-Emulsion/Suspension Polymerization. In: Salamone JC, editor. *Polymeric Materials Encyclopedia*, Volume 5. Boca Raton, Florida: CRC Press; 1996. p 3330.

[26] Vanderhoff JW, Wiley RM. The Dow Chemical Co. US Patent 3,284,393. 1966 Nov 8.

[27] Hernándes-Barajas J, Hunkeler DJ. Inverse-emulsion polymerization of acrylamide using block copolymeric surfactants: mechanism, kinetics and modeling. Polymer 1997;38:437–447.

[28] Renken A, Hunkeler, D. Effect of the surfactant blend composition on the properties of polymerizing acrylamide-based inverse-emulsions: characterization by small-angle neutron scattering and quasi-elastic light scattering. Polymer 1999;40:3545–3554.

[29] Carver MT, Dreyer U, Knoesel R, Candau F. Kinetics of photopolymerization of acrylamide in AOT reverse micelles. J Polym Sci Part A: Polym Chem 1989;27:2161–2177.

[30] Alb AM, Farinato F, Calbick J, Reed WF. Online monitoring of polymerization reactions in inverse emulsions. Langmuir 2006;22:831–840.

[31] Guyot A, Guillot J, Pichot C, Rios Guerrero L. New design for producing constant-composition copolymers in emulsion polymerization. In: Basset DR, Hamielec AE, editors. *Emulsion Polymers and Emulsion Polymerization*, Volume 165. ACS Symposium Series. Washington, DC: American Chemical Society; 1981. pp 415–436.

[32] Chien DCH, Penlidis A. On-line sensors for polymerization reactors. J Macromol Sci Rev Macromol Chem Phys 1990;30:1–42.

[33] Asua JM. Emulsion polymerization: from fundamental mechanisms to process developments. J Polym Sci Part A: Polym Chem 42, 1025–1041.

[34] BenAmor S, Colombié D, McKenna T. On-line monitoring of emulsion polymerization using calorimetry. Ind Eng Chem Res 2002;41:4233–4241.

[35] van den Brink M, Hansen JF, De Peinder P, van Herk AM, German AL. Measurement of partial conversions during the solution copolymerization of styrene and butyl acrylate using on-line Raman spectroscopy. J Appl Pol Sci 2000;79:426–436.

[36] Martens H, Naes T. *Multivariate Calibration*. New York: Wiley; 1989.

[37] Vieira RAM, Sayer C, Lima EL, Pinto JC. Detection of monomer droplets in a polymer latex by near-infrared spectroscopy. Polymer 2001;42:8901–8906.

[38] Edouard D, Sheibat-Othman N, Hammouri H. Observer design for particle size distribution in emulsion polymerization. AIChE J 2005;5:3167–3185.

[39] Alb AM, Reed WF. Simultaneous monitoring of polymer and particle characteristics during emulsion polymerization. Macromolecules 2008;41:2406–2414.

[40] Harkins WD. A general theory of the reaction loci in emulsion polymerization. J Chem Phys 1945;13:381–382.

[41] Smith WV, Ewart RH. Kinetics of emulsion polymerization. J Chem Phys 1948;16:592–599.

[42] Alb AM, Reed WF. Recent advances in Automatic Continuous Online Monitoring of Polymerization reactions (ACOMP). Macromol Symp 2008;271:15–25.

[43] Elizalde O, Azpeitia M, Reis MM, Asua JM, Leiza JR. Monitoring emulsion polymerization reactors: calorimetry versus Raman spectroscopy. Ind Eng Chem Res 2005;44:7200–7207.

[44] Sajjadi S. Nanoparticle formation by monomer-starved semibatch emulsion polymerization. Langmuir 2007;23:1018–1024.

[45] Wu XQ, Schork FJ. Batch and semibatch mini/macroemulsion copolymerization of vinyl acetate and comonomers. Ind Eng Chem Res 2000;39:2855–2865.

[46] Wessling RA. Kinetics of continuous addition emulsion polymerization. J Appl Polym Sci 1968;12:309–319.

[47] Britton D, Heatley F, Lovell PA. Effect of monomer feed-rate on chain transfer to polymer in emulsion polymerization of vinyl acetate studied by NMR spectroscopy. Macromolecules 2000;33:5048–5052.

[48] Vicente M, BenAmor S, Gugliotta LM, Leiza JR, Asua JM. Control of molecular weight distribution in emulsion polymerization using on-line reaction calorimetry. Ind Eng Chem Res 2001;40:218–227.

[49] Reis MM, Araujo PHH, Sayer C, Giudici, R. In-situ NIR spectroscopy for simultaneous monitoring of multiple process variables in emulsion copolymerization. Ind Eng Chem Res 2004;43:7243–7250.

[50] Alb AM, Reed WF. Online monitoring of molecular weight and other characteristics during semibatch emulsion polymerization under monomer starved and flooded conditions. Macromolecules 2009;42:8093–8101.

[51] Kreft, T; Reed WF. Predictive control and verification of conversion kinetics and polymer molecular weight in semibatch free radical homopolymer reactions. Eur Polym J 2009;45:2288–2303.

[52] Paril A, Alb AM, Reed WF. Online monitoring of the evolution of polyelectrolyte characteristics during postpolymerization modification processes. Macromolecules 2007;40:4409–4413.

[53] Manning GS. Limiting laws and counterion condensation in polyelectrolyte solutions I. Colligative properties. J Chem Phys 1969;51:924–933.

[54] Matyjaszewski K. Comparison and classification of controlled/living radical polymerizations. In: Matyjaszewski K, editor. *Controlled/Living Radical Polymerization: Progress in ATRP, NMP, and RAFT*, Volume 768. ACS Symposium Series. Washington, DC: American Chemical Society; 2000. p 2–26.

[55] Braunecker WA, Matyjaszewski K. Progress in polymer science. Prog Polym Sci 2007;32:93–146.

[56] McCormick CL, Lowe AB. Aqueous RAFT polymerization: recent developments in synthesis of functional water-soluble (co)polymers with controlled structures. Acc Chem Res 2004;37:312–325.

[57] Moad G, Rizzardo E, Thang SH. Living radical polymerization by the RAFT process – A first update. Aust J Chem 2006;59:669–692.

[58] Patton D, Advincula, R. A versatile synthetic route to macro-monomers via RAFT polymerization. Macromolecules 2006;39:8674–8683.

[59] Fischer, H. The persistent radical effect: A principle for selective radical reactions and living radical polymerizations. Chem Rev 2001;101:3581–3610.

[60] Hawker CJ, Bosman AW, Harth E. New polymer synthesis by nitroxide mediated living radical polymerizations. Chem Rev 2001;101:3661–3688.

[61] Studer A, Schulte T. Nitroxide-mediated radical processes. Chem Rec 2005;5:27–35.

[62] Sciannamea V, Jrme R, Detrembleur C. In-situ Nitroxide-Mediated Radical Polymerization (NMP) processes: their understanding and optimization. Chem Rev 2008;108: 1104–1126.

[63] Zetterlund PB, Kagawa Y, Okubo M. Controlled/living radical polymerization in dispersed systems. Chem Rev 2008;108:3747–3794.

[64] Moad G, Rizzardo E, Thang SH. Toward living radical polymerization. Acc Chem Res 2008;41:1133–1142.

[65] Chauvin F, Alb A, Bertin D, Tordo P, Reed WF. Kinetics and molecular weight evolution during controlled radical polymerization. Macromol Chem Phys 2002;203:2029–2041.

[66] Mignard E, Leblanc T, Bertin D, Guerret O, Reed WF. Online monitoring of controlled radical polymerization: nitroxide mediated gradient copolymerization. Macromolecules 2004;37:966–975.

[67] Benoit D, Grimaldi S, Robin S, Finet JP, Tordo P, Gnanou, Y. Kinetics and mechanism of controlled free-radical polymerization of styrene and *n*-butyl acrylate in the presence of an acyclic β-phosphonylated nitroxide. J Am Chem Soc 2000;122:5929–5939.

[68] Penlidis A, MacGregor JF, Hamielec AE. Continuous emulsion polymerization reactor control. Proc Am Control Conf 1985;2:878–880.

[69] Mayadunne RTA, Rizzardo E. Mechanistic and practical aspects of RAFT polymerization. In: Jagur-Grodzinski J, editor. *Living and Controlled Polymerization: Synthesis, Characterization and Properties of the Respective Polymers and Copolymers*, Volume 65. New York: Nova Science Publishers; 2005.

[70] Moad G, Solomon DH. *The Chemistry of Radical Polymerization*. 2nd ed. Oxford: Elsevier; 2006. p 502.

[71] Lowe AB, McCormick CL. Reversible addition–fragmentation chain transfer (RAFT) radical polymerization and the synthesis of water-soluble (co)polymers under homogeneous conditions in organic and aqueous media. Prog Polym Sci 2007;32:283–351.

[72] Liu C, Hillmyer MA, Lodge TP. Multicompartment micelles from pH-responsive miktoarm star block terpolymers. Langmuir 2009;5:13718–13725.

[73] Baussard J-F, Habib-Jiwanb J-L, Laschewskya A, Mertogluc M, Storsberg, J. New chain transfer agents for Reversible Addition-Fragmentation Chain Transfer (RAFT) polymerisation in aqueous solution. Polymer 2004;45:3615–3626.

[74] Penczek, S. Terminology of kinetics, thermodynamics, and mechanisms of polymerization. J Polym Sci Part A: Polym Chem 2002;40:1665–1676.

[75] Goto A, Sato K, Tsujii Y, Fukuda T, Moad G, Rizzardo E, Thang SH. Mechanism and kinetics of RAFT-based living radical polymerizations of styrene and methyl methacrylate. Macromolecules 2001;34:402–408.

[76] Feldermann A, Ah Toy A, Davis TP, Stenzel MH, Barner-Kowollik, C. An in-depth analytical approach to the mechanism of the RAFT process in acrylate free radical polymerizations via coupled Size Exclusion Chromatography–Electrospray Ionization–Mass Spectrometry (SEC–ESI–MS). Polymer 2005;46:8448–8457.

[77] Barner-Kowollik C, Buback M, Charleux B, Coote ML, Drache M, Fukuda T, Goto A, Klumperman B, Lowe AB, McLeary J, Moad G, Monteiro MJ, Sanderson RD, Tonge MP, Vana, P. Mechanism and kinetics of dithiobenzoate-mediated RAFT polymerization. I. The current situation. J Polym Sci Part A: Polym Chem 2006;44:5809–5831.

[78] Alb AM, Serelis AK, Reed WF. Kinetic trends in RAFT homopolymerization from online monitoring. Macromolecules 2008;41:332–338.

[79] Barner-Kowollik C, Quinn FJ, Nguyen TLU, Heuts JPA, Davis TP. Kinetic investigations of reversible addition fragmentation chain transfer polymerizations: cumyl phenyldithioacetate mediated homopolymerizations of styrene and methyl methacrylate. Macromolecules 2001;34:7849–7857.

[80] Litvinenko G, Mueller AHE. General kinetic analysis and comparison of molecular weight distributions for various mechanisms of activity exchange in living polymerizations. Macromolecules 1997;30:1253–1266.

[81] Yang C, Cheng Y-L. RAFT synthesis of poly(N-isopropylacrylamide) and poly(methacrylic acid) homopolymers and block copolymers: kinetics and characterization. J Appl Polym Sci 2006;102:1191–1201.

[82] Suchao-in N, Chirachanchai S, Perrier, S. pH- and thermo-multi-responsive fluorescent micelles from block copolymers via reversible addition fragmentation chain transfer (RAFT) polymerization. Polymer 2009;50:4151–4158.

[83] Rowe, MD, Chang C-C, Thamm DH, Kraft SL, Harmon JF, Vogt AP, Sumerlin BS, Boyes SG. Tuning the magnetic resonance imaging properties of positive contrast agent nanoparticles by surface modification with RAFT polymers. Langmuir 2009;25:9487–9499.

[84] Morgan SE, Jones P, Lamont AS.; Heidenreich A, McCormick CL. Layer-by-layer assembly of pH-responsive, compositionally controlled (co)polyelectrolytes synthesized via RAFT. Langmuir 2007;23:230–240.

[85] Matyjaszewski K, Xia J. Atom transfer radical polymerization. Chem Rev 2001;101:2921–2990.

[86] Matyjaszewski K, Davis TP, editors. *Handbook of Radical Polymerization.* Hoboken: Wiley Interscience; 2002.

[87] Haddleton DM, Jasieczek CB, Hannon MJ, Shooter AJ. Atom transfer radical polymerization of methyl methacrylate initiated by alkyl bromide and 2-pyridinecarbaldehyde imine copper(I) complexes. Macromolecules 1997;30:2190–2193.

[88] Percec V, Barboiu B, Kim H-J. Arenesulfonyl halides: a universal class of functional initiators for metal-catalyzed "living" radical polymerization of styrene(s), methacrylates, and acrylates. J Am Chem Soc 1998;120:305–316.

[89] Kotani Y, Kamigaito M, Sawamoto, M. Living radical polymerization of para-substituted styrenes and synthesis of styrene-based copolymers with rhenium and iron complex catalysts. Macromolecules 2000;33:6746–6751.

[90] Teodorescu M, Gaynor SG, Matyjaszewski, K. Halide anions as ligands in iron-mediated atom transfer radical polymerization. Macromolecules 2000;33:2335–2339.

[91] Mignard E, Lutz J-F, Leblanc T, Matyjaszewski K, Guerret O, Reed WF. Kinetics and molar mass evolution during atom transfer radical polymerization of n-butyl acrylate using automatic continuous online monitoring. Macromolecules 2005;38:9556–9563.

[92] Knuehl BP T, Kajiwara A, Fischer H, Matyjaszewski, K. Characterization of Cu(II) bipyridine complexes in halogen atom transfer reactions by electron spin resonance. Macromolecules 2003;36:8291–8296.

[93] Kajiwara A, Matyjaszewski, K. EPR study of the atom transfer radical polymerization (ATRP) of (meth)acrylates. Macromol Rapid Commun 1998;19:319–321.

[94] Alb AM, Drenski MF, Reed WF. Simultaneous continuous, non-chromatographic and discrete chromatographic monitoring of polymerization reactions. J Appl Polym Sci 2009;113:190–198.

13

APPLICATIONS OF ACOMP (II)

Wayne F. Reed

13.1 KINETICS AND MECHANISTIC STUDIES

13.1.1 Initiation Kinetics

As shown in Chapter 1, the initiation step in free radical polymerization can be rate limited either by decomposition of the initiator or diffusion-controlled initiation of decomposed radicals with the first monomer. Automatic continuous online monitoring of polymerization reactions (ACOMP) was used to experimentally observe the crossover between decomposition control and diffusion control of the initiation step. The crossover was observed both among potassium persulfate initiated acrylamide (Am) free radical polymerization reactions at different concentrations of Am and in single reactions. There is no appreciable chain transfer in this particular polymerization.

The initiator I_2 thermally decomposes according to:

$$I_2 \xrightarrow{k_d} 2I^\bullet \qquad (13.1)$$

where k_d is the temperature-dependent first-order decomposition rate coefficient, and hence

$$[I_2] = [I_2]_0 \exp(-k_d t) \qquad (13.2)$$

The production of chains depends on the concentration of initiated radicals $R_1^\bullet$, which, in turn are produced from $I^\bullet$ according to the bimolecular initiation reaction:

$$I^\bullet + m \xrightarrow{k_i} R_1^\bullet \qquad (13.3)$$

where k_i is the initiation rate coefficient and m is the monomer.

Equation 13.3 is diffusion controlled because it requires $I^\bullet$ and m to encounter each other via diffusion in order to react.

Either of the reactions 1 and 3 given in Equations 13.1 and 13.3 can be rate controlling in the production of $R_1^\bullet$. If the initiator and monomer are dilute, the diffusion controlled reaction 3 in Equation 13.3 may be rate limiting, whereas if monomer concentration $[m]$ is high then the initiator decomposition reaction 1 in Equation 13.1 is rate controlling. These limits will be referred to, respectively, as the "diffusion-controlled regime" and the "decomposition-controlled regime."

In either case, once $R_1^\bullet$ forms, the propagation takes place according to:

$$R_j^\bullet + m \xrightarrow{k_p} R_{j+1}^\bullet \qquad (13.4)$$

where k_p is the propagation rate coefficient, and the propagation equation is:

$$\frac{d[m]}{dt} = -k_p[m][R^\bullet] - k_i[m][I^\bullet] \qquad (13.5a)$$

which, in the long chain limit (degree of polymerization $\gg 1$) is well approximated by:

$$\frac{d[m]}{dt} = -k_p[m][R^\bullet] \qquad (13.5b)$$

Termination proceeds according to:

$$R_j^\bullet + R_i^\bullet \xrightarrow{k_t} P_{j+i}, \text{ or}, P_j + P_i \qquad (13.6)$$

where P represents dead polymer chains formed by either recombination (the first term on the right-hand side) or disproportionation (the second term).

Monitoring Polymerization Reactions: From Fundamentals to Applications, First Edition. Edited by Wayne F. Reed and Alina M. Alb.
© 2014 John Wiley & Sons, Inc. Published 2014 by John Wiley & Sons, Inc.

The coefficients k_t and k_p are taken here as chain-length independent, although there is good evidence for chain length–dependent termination under certain reaction regimes, especially at higher monomer concentrations [1–5].

The population of radicals is governed by:

$$\frac{d[R^\bullet]}{dt} = 2Fk_d[I_2] - k_t[R^\bullet]^2, \quad \text{decomposition limited}$$

(13.7a)

$$= k_i[I^\bullet][m] - k_t[R^\bullet]^2, \text{ diffusion limited} \quad (13.7b)$$

where $[R^\bullet] = \Sigma_i[R^\bullet]_i$ is the sum of propagating radicals of all lengths, and F is the fraction of $I^\bullet$ that go on to produce $R_1{}^\bullet$, the rest being consumed by other reactions such as recombination and scavenging by impurities (e.g., oxygen).

A robust and realistic approximation is the quasi-steady-state approximation (QSSA) [6], in which it is assumed that the rate at which $R^\bullet$ population changes is very slow compared to the other processes (decomposition, initiation, propagation, and termination); that is, it is assumed that

$$\frac{d[R^\bullet]}{dt} \approx 0, \text{ QSSA} \quad (13.8)$$

With this, $[m](t)$ in Equation 13.56 can be found. For the case of decomposition-limited initiation, using Equation 13.2 for $[I_2]$, direct integration of Equation 13.3b yields an expo-exponential form:

$$[m](t) = [m]_0 \exp\left[\frac{2\alpha}{k_d}(e^{-k_d t/2} - 1)\right], \alpha = k_p\sqrt{\frac{2Fk_d[I_2]_0}{k_t}}$$

(13.9)

A further simplification results in the case where very little initiator decomposes during the entire reaction. In this case, $[I_2] \sim [I_2]_0$, where $[I_2]_0$ is the initial concentration of initiator. The aforementioned equation then reduces to:

$$[m](t) = [m]_0 \exp(-\alpha t), \text{ decomposition limited}$$

(13.10)

In this case, the polymerization rate coefficient, α (s^{-1}) is independent of $[m]_0$.

If diffusion is the rate limiting step for initiation, then:

$$[m](t) = \frac{k_d^2 e^{k_d t}[m]_0}{\left(k_d e^{\frac{1}{2}k_d t} + \sqrt{[m]_0}\,\beta - \sqrt{[m]_0}\,\beta e^{\frac{1}{2}k_d t}\right)^2},$$

$$\beta = k_p\sqrt{\frac{I^\bullet k_i}{k_t}}, \text{diffusion limited}$$

(13.11)

and if $I^\bullet$ ~ constant throughout the reaction, the aforementioned equation is reduced to:

$$[m] = \frac{[m]_0}{\left(1 + \sqrt{\frac{[m]_0}{4}}\,\beta t\right)^2}, \beta = k_p\sqrt{\frac{I^\bullet k_i}{k_t}}, \text{ diffusion limited}$$

(13.12)

In the latter case, the polymerization rate coefficient (s^{-1}) is given by $\sqrt{\frac{[m]_0}{4}}\beta$ and varies as: $\sqrt{[m]_0}$. Either mechanism can be rate controlling and the amount of radicals will depend on the monomer concentration in the following manner:

$$R^\bullet = \gamma[m]^s \qquad s = \begin{cases} 0.5 & \text{monomer diffusion} \\ 0 & \text{initiator decomposition} \end{cases}$$

(13.13)

where γ is a proportionality constant.

Deviations from first-order reactions in $[m]$ have been discussed earlier. Baer et al. [7], Ishige and Hamielec [8], and Riggs and Rodriguez [9, 10] suggested $s = 0.5$ at low monomer concentrations, a consequence of diffusion-controlled initiation. Lin [11] reported $R_{po} \propto [m]_0^{1.26}$ whereas Noyes [12, 13] explained the cage effect with $s = 0.25$.

It is also possible that at the outset of a reaction initiation is decomposition limited, and then becomes diffusion limited as monomer is depleted, and there could be a crossover regime between the two mechanisms $(0 < s < 0.5)$ during a single reaction.

In the case of general s, with $I^\bullet$ ~ constant, Equation 13.6b becomes:

$$\frac{d[m]}{dt} = -k_p\gamma[m]^{s+1} \quad (13.14)$$

$$[m] = \frac{[m]_0}{\left(1 + [m]_0^s k_p\gamma t\right)^{1/2}}, \quad \text{general case} \quad (13.15)$$

The molar mass can be addressed using the kinetic chain length concept for free radical polymerization; the instantaneous number average chain length, $N_{n,inst}$ is given in the QSSA for the aforementioned conditions by:

$$N_{n,inst} = \frac{k_p[m]}{k_t[R^\bullet]} = \frac{k_p^2}{2k_t\gamma}[m]^{1-s} \quad (13.16)$$

Instantaneous w- and z-averages depend on the type of reaction. For disproportionation, $N_z:N_w:N_n = 3:2:1$ [14]. Using this:

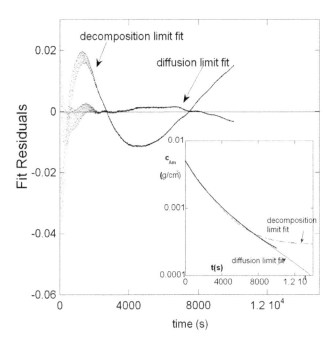

FIGURE 13.1 Residuals for fits of decomposition limited and diffusion limited initiation for Am conversion during free radical homopolymerization. The two fits from which the residuals are computed are shown in the inset. Adapted with permission from Kreft T, Reed WF. Direct monitoring of the cross-over from diffusion-controlled to decomposition-controlled initiation in free radical polymerization. Macromol Chem Phys 2008;209: 2463–2474.

$$M_w = 2M_{mon} \frac{k_p^2}{2k_t\gamma}[m]_0^{1-s}\frac{1-(1-f)^{2-s}}{f(2-s)} \qquad (13.17a)$$

$$M_{w,final} = \frac{1}{2-s}M_{w,0} \qquad (13.17b)$$

Experiments to monitor initiation in the two regimes were carried out [15] using the potassium persulfate (KPS) initiation of acrylamide (Am) in water at 60 °C, where $k_d = 1.925 \times 10^{-5}\,s^{-1}$ for KPS; that is, the approximation is fulfilled that $[I_2](t) \sim [I_2]_0$, since the reaction times were on the order of an hour and $1/k_d = 14.5\,h$.

Conversion data were deconvolved according to the method described in Section 11.2.2.1. The inset to Figure 13.1 shows the deconvolved data from Figure 11.2, which was for 0.005 g cm^{-3} Am and [KPS]=0.00078 g cm^{-3}. Fits using $c_{Am}(t)$ as predicted for the decomposition and diffusion limits are also shown, according to Equations 13.10 and 13.11. Both functions are seen to fit well to the data, with the diffusion limit data being more accurate, especially toward the end of the reaction.

The residuals to the fits are shown in the main portion of Figure 13.1. The residuals are defined as (experimental value − fit value)/(initial experimental value). There is a

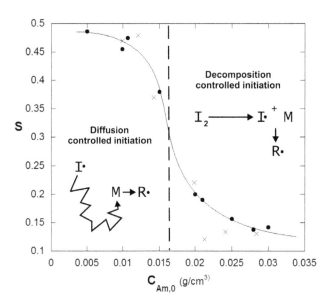

FIGURE 13.2 The scaling fit factor *s* (Equation 13.15). $s=0.5$ in the limit of diffusion controlled initiation, and $s=0$ in the limit of decomposition controlled initiation. The solid points are from fits to separate reactions under specific Am and initiator concentrations, whereas the crosses show intrareactional crossover from decomposition control to diffusion control averaged over many reactions where the crossover is prominent due to starting Am and initiator concentrations. Instantaneous molar fraction of VB, $F_{inst,VB}$ versus conversion for batch and two feed rates during VB/AM free radical copolymerization. Adapted with permission from Kreft T, Reed WF. Direct monitoring of the crossover from diffusion-controlled to decomposition-controlled initiation in free radical polymerization. Macromol Chem Phys 2008;209:2463–2474.

systematic bias in the exponential fit, seen in the quasi-oscillatory structure of the residuals, whereas the diffusion-controlled fit has much lower residuals and is nearly devoid of structure, suggesting that initiation is diffusion controlled at this low concentration of Am.

Data throughout the many experiments performed in Reference [15] consistently showed that an exponential fit to conversion kinetics was robust even when the diffusion-controlled expression gives better results and is formally a better theoretical representation of the kinetics.

Further investigation of experiments used Equation 13.15, with *s* and the cluster $k_p\gamma$ as two adjustable fitting parameters. Figure 13.2 shows this very interesting result with the solid circles. Each circle is the *s*-value obtained from individual experiments. At very low Am concentrations *s* is close to 0.5, consistent with the diffusion limited prediction. As the monomer concentration increases, *s* decreases abruptly between 0.010 and 0.020 g cm^{-3} to around 0.15 which is close to the value of $s=0$ predicted for the decomposition limited case.

Clearly, any experiment that begins at high monomer concentration will cross over from the decomposition limited regime to the diffusion limited regime as the monomer concentration decreases. This crossover from decomposition to diffusion limited regimes was found in individual experiments by fitting for s and the rate $k_p\gamma$ at just the high values of $c_{Am}(t)$ (or $[m](t)$), and then fitting again for s at just the low values of $c_{Am}(t)$. The crosses in Figure 13.2 represent the intrareactional values of s, taken as average values from multiple experiments at each concentration, as described earlier. The crosses follow very closely the solid points obtained from entire individual experiments. From the data in Figure 13.2, the crossover from diffusion limited to decomposition limited initiation occurs at $c_{Am,c} = 0.017\,g\,cm^{-3}$, or, in molar terms, $[Am]_c = 0.239\,M$.

13.1.2 Kinetics of Copolymeric Polyelectrolyte Synthesis

The field of polyelectrolyte solution properties [16] has been intensively studied theoretically and experimentally. Long-range electrostatic forces make the polymer excluded volume problem [17] very difficult, and have led to the concepts of "electrostatic persistence length and excluded volume" [18]. The electrostatic force also gives rise to such effects as angular scattering maxima of neutrons [19–21], X-rays [22], and light [23–26], including under shear [27] suggestive of liquid-like correlations, and also causes the "electroviscous effect" [28, 29], wherein reduced polyelectrolyte viscosity increases under isoionic dilution in very low ionic strength solutions. Chapter 5 provides a discussion of some of these aspects.

Automatic Continuous Online Monitoring of Polymerization Reactions can provide a unified approach to understanding copolymeric polyelectrolytes, or "copolyelectrolytes." Copolyelectrolytes combine the functional characteristics of copolymers with the unusual properties of polyelectrolytes. Copolyelectrolytes are copolymers in which at least one of the comonomers is electrically charged. The ACOMP approach allows connecting the synthesis kinetics to the resulting trivariate distribution of comonomer composition, molar mass, and linear charge density. These characteristics, in turn, control the properties of endproduct solutions, such as chain conformations, interparticle interactions, viscosity, interactions with colloids and other polymers, phase separation, etc. Unified results allow testing and improvement of existing polyelectrolyte theories, development of new quantitative physicochemical models, provide advanced characterization methods, set the stage for studying more complex copolyelectrolytes, such as hydrophobically modified ones, and provide tools for ultimately controlling and tailoring the synthesis and properties of copolyelectrolytes.

Because of counterion condensation (CC) effects, discussed in the following, the linear charge density

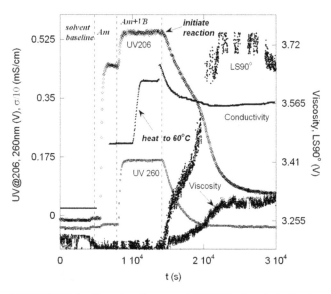

FIGURE 13.3 Raw data for several ACOMP signals during the free radical copolymerization of Am and VB. Instantaneous molar fraction of VB, $F_{inst,VB}$ versus conversion for batch and two feed rates during VB/AM free radical copolymerization reactions. Reprinted (adapted) with permission from Alb AM, Paril A, Catalgil-Giz H, Giz A, Reed WF. Evolution of composition, molar mass, and conductivity during the free radical copolymerization of polyelectrolytes. J Phys Chem B 2007;111:8560–8566. © 2007 American Chemical Society.

distribution is not simply proportional to the composition distribution of the charged comonomer. To distinguish these, a conductivity probe was inserted into the reactor and its results combined with those of the usual ACOMP detector train [30]. CC is observable, quantifiable, and does indeed cause the composition and charge density distributions to deviate from each other.

Figure 13.3 shows an example of raw ACOMP data for a copolyelectrolyte synthesis reaction involving acrylamide and vinyl benzene sulfonic acid in sodium form (VB) [31]. This case involved free radical copolymerization of a 10/90 mole percentage of [VB]/[Am]) in aqueous 0.0002 M NaCl and 0.364 M total comonomer. After solvent baseline stabilization, Am was added, and the UV206 nm signal increased. With subsequent addition of VB at 10,000 s, both UV signals and conductivity σ, increased. When the reactor temperature reached 60 °C, initiator was added (15,000 s). VB copolymerized faster than Am in first-order fashion during the first phase of the reaction, seen by the 260 nm signal. After VB was exhausted, the remaining Am homopolymerized rapidly; both phases can be seen in the 206 nm signal. Hence, inadvertently, a blend of copolyelectrolyte and pAm homopolymer was produced. σ decreased as VB was incorporated into copolyelectrolyte in the first phase, and remained constant in the second phase of Am homopolymerization. An ionic

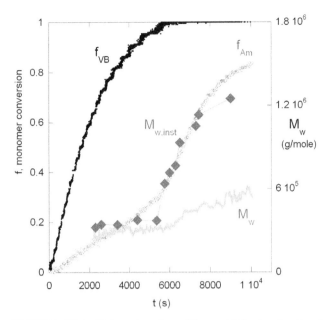

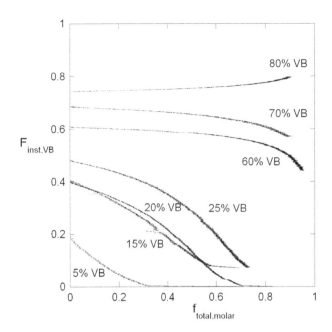

FIGURE 13.4 Using data from Figure 13.3, the fractional conversion of Am and VB are shown. A two-phase reaction occurs in which copolymer of Am and VB is produced up until 5500 s, after which homopolymers of Am form. Hence, a blend of copolyelectrolyte and Am homopolymer results. Also shown are the cumulative M_w and $M_{w,inst}$, according to Equation 11.7. The mass of the polymer chains jumps up once the final Am homopolymerization period begins. Reprinted (adapted) with permission from Alb AM, Paril A, Catalgil-Giz H, Giz A, Reed WF. Evolution of composition, molar mass, and conductivity during the free radical copolymerization of polyelectrolytes. J Phys Chem B 2007;111:8560–8566. © 2007 American Chemical Society.

FIGURE 13.5 Composition drift in Am/VB free radical copolymerization, of the type seen in Figure 13.3 and Figure 13.4. There is very high composition drift for initial VB 25% and less. Reprinted (adapted) with permission from Kreft TR, Reed WF. Experimental observation of crossover from noncondensed to counterion condensed regimes during free radical polyelectrolyte copolymerization under high-composition drift condition. J Phys Chem B 2009;113:8303–8309. © 2009 American Chemical Society.

strength of 0.1 M (from NaCl) was used in the ACOMP detector train to suppress polyelectrolyte effects.

Figure 13.4 shows the fractional conversion of each comonomer, f_{VB} and f_{Am} respectively, for the same reaction in 0.100 M NaCl, obtained from dual wavelength UV data. The two-phase behavior of f_{Am} is striking. Its polymerization rate increases when the VB is exhausted. The figure also shows the cumulative weight average mass M_w, which is seen to rise after the VB conversion phase is complete. The instantaneous M_w is also shown, and indicates that during the initial copolymerization phase a constant 3.5×10^5 g mol^{-1} copolymer is produced, whereas $M_{w,inst}$ jumps abruptly to over 10^6 g mol^{-1} during the second phase of PAM homopolymer production. $M_{w,inst}$ is obtained from the cumulative M_w by the method described in Chapter 11.

The average composition drift can be followed in terms of the instantaneous mole fraction of VB incorporated into copolyelectrolyte, $F_{inst,VB}$ determined by another method described in Chapter 11. $F_{inst,VB}$ is shown for several VB/Am reactions of different starting molar ratios in Figure 13.5 and is defined as:

$$F_{inst,VB} = \frac{d[m]_{VB}}{d([m]_{VB}+[m]_{Am})} \tag{13.18}$$

At high starting percentage of VB drift is not significant, but becomes extremely large as starting VB drops below 50%, and begins to lead to the type of blend of copolyelectrolyte and homopolymers PAM below 25% VB. Reactivity ratios were determined from ACOMP data to be $r_{VB}=2.14$ and $r_{Am}=0.18$.

13.1.3 Counterion Condensation during Synthesis of Copolymeric Polyelectrolytes

The conductivity data can be combined with the $F_{inst,VB}$ data to analyze CC effects during polymerization. CC in polyelectrolyte solutions [32, 33] has been the topic of extensive analytical and numerical investigations [34, 35], but for which far fewer detailed experimental works exist [36, 37]. CC occurs when the electrostatic potential energy between charge sites on a polymer chain and counterions at the intermonomer backbone distance exceed kT. ξ is defined in this context as the number of elementary charges per Bjerrum length l_B. In the simplest theory, CC will prevent ξ from surpassing unity.

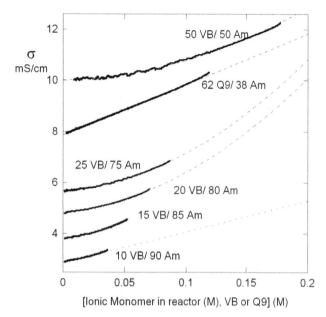

FIGURE 13.6 Increase in reactor solution conductivity σ versus ionic monomer concentration during free radical copolyelectrolyte polymerization. Where σ versus ionic monomer concentration is linear, there is no change in the degree of counterion condensation, if any, during the reaction. When it curves upward, counterion condensation occurs at changing degrees during the reaction. Reprinted (adapted) with permission from Kreft TR, Reed WF. Experimental observation of crossover from noncondensed to counterion condensed regimes during free radical polyelectrolyte copolymerization under high-composition drift condition. J Phys Chem B 2009;113:8303–8309.© 2009 American Chemical Society.

Figure 13.6 shows reactor solution conductivity versus remaining amount of ionic monomer in the reactor for a series of VB/Am copolymerization reactions at different starting compositions and also one for the copolymerization of Am with an Am derivative, cationic [2-(acryloyloxy) ethyl]-trimethylammonium chloride (Q9).

The reactivity ratios of Q9 and Am are close enough to each other ($r_{Q9} = 0.47$ and $r_{Am} = 1.10$), that there is relatively little composition drift in any synthesis reactions over a wide range of [Q9]/[Am] starting values. σ versus [Q9] is linear for all reactions, as seen for the example Q9/Am reaction in Figure 13.6; that is, the loss of σ per Q9 monomer incorporated into polyelectrolyte was independent of conversion and small amounts of composition drift. Low starting [Q9]/[Am] gives low charge density in the copolyelectrolyte chains produced, and the decrease in σ during the reaction is due solely to monomer incorporation into copolyelectrolyte chains. At higher [Q9]/[Am], the decrease in σ during the reaction was higher due to CC; that is, σ decreases due both to Q9 incorporation into the chain and a charge neutralizing condensation of a counterion onto the chain. None of the composition drifts, however, was large enough to cross through the broad noncondensed to condensed counterion transition regime, but

the slopes of the linear σ versus [Q9] data increased monotonically as [Q9]/[Am] increased in the experiments.

In the case of high drift VB/Am reactions, CC can be observed *within individual synthetic reactions*. Figure 13.6 is the primary evidence for intrareaction crossover from the noncondensed to the condensed counterion regime. σ versus [VB] for several of the high drift experiments is nonlinear for <50% starting VB. This is due to the composition drift during the reaction (see corresponding drifts in Fig. 13.5) crossing from the counterion condensed to noncondensed regimes.

In contrast, σ versus [VB] is linear for the VB/Am reactions when composition drift is low and stays within one of the regimes; for example, the 50% starting VB experiment shown in Figure 13.6 stayed within the high F_{inst} composition regime, for which all chains have condensed counterions. This linearity also demonstrated the independence of copolyelectrolyte chain electrophoretic mobility from M_w (there is large monotonic decrease in M_w vs. conversion in all the reactions of Q9/Am and for VB/Am during the VB/Am coconversion phase), and also showed that the changing viscosity of the (low polymer concentration) solution in the reactor did not measurably affect electrophoretic mobility.

The model-independent data of Figure 13.6 require an increasing amount of CC with increasing fraction of VB in the chain. If there were no CC and the specific conductivity of the copolyelectrolyte chains Σ_{cp} were constant and independent of $F_{inst,VB}$ (i.e., independent of ξ), then the slopes of σ versus [VB], $d\sigma/d$[VB], would be constant versus [VB] and the same among experiments. If there were no condensation and Σ_{cp} increased with increasing $F_{inst,VB}$, as would be expected, then $d\sigma/d$[VB] would increase versus conversion in those reactions where $F_{inst,VB}$ decreases versus conversion. $F_{inst,VB}$ decreases versus conversion for all the higher number experiments in Figure 13.3 but $d\sigma/d$[VB] decreases versus conversion for these experiments, leading to the conclusion that CC increases more rapidly than Σ_{cp} with increasing $F_{inst,VB}$. Note that the reactions occur from right (high [VB] or [Q9]) to left (low [VB] or [Q9]) in Figure 13.4.

Features concerning σ common to all reactions were (i) solution viscosity increasing during the reaction did not decrease electrophoretic mobility of polymeric and monomeric species; and (ii) molecular weight does not change mobilities.

A visualization of the crossover regime can be seen in Figure 13.7 for both the Q9/Am and VB/Am copolymerization reactions. It shows the slopes of σ with respect to change in ionic monomer; that is, the change in solution conductivity when an increment of monomer is polymerized into a copolymer chain. For the Q9 data, there is a unique slope for σ versus [Q9] since all reactions yield a linear plot, because there is no CC. The data points are averages over several experiments and the error bars reflect the spread of values obtained. For the VB data, the slope is determined over

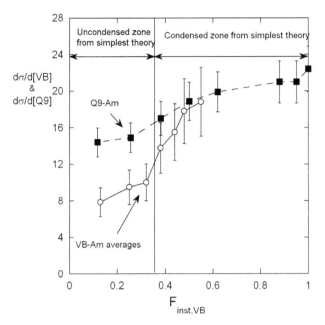

FIGURE 13.7 The slope of σ versus ionic monomer concentration, such as in Figure 13.6, provides evidence of counterion condensation for two different types of copolymerization; Am/VB (high drift) and Am/Q9 (low drift). For low drift experiments a specific, constant slope of σ versus ionic monomer concentration is obtained for each experiment at different comonomer starting ratios. For high drift experiments (Am/VB), the slope changes during the reaction, showing the crossover from counterion condensed to noncounterion condensed as VB is more rapidly consumed than Am. Reprinted (adapted) with permission from Kreft TR, Reed WF. Experimental observation of crossover from noncondensed to counterion condensed regimes during free radical polyelectrolyte copolymerization under high-composition drift condition. J Phys Chem B 2009;113:8303–8309.© 2009 American Chemical Society.

discrete conversion intervals of $F_{inst,VB}$ during individual reactions, which illustrates the effects of the nonlinear σ versus [VB].

It is seen that the slopes increase as $F_{inst,VB}$ and $F_{inst,QB}$ increase, which is due to the CC as linear charge density, ξ, increases. At low amounts of VB or Q9, the slope is low since the change in σ is due to putting ionic monomers into copolyelectrolyte chains, where they have less electrophoretic mobility than free ionic monomers, but their counterions remain free to move in an applied field. For the higher amounts of ionic monomer, counterions condense onto the polymer as the ionic monomers are incorporated into the chain, taking a larger amount of solution conductivity away. For both VB/Am and Q9/Am, the transition from noncondensed to counterion condensed is gradual, and does not involve any sharp transition as is predicted by the simplest theory. The boundary in the simplest theory is shown in the figure.

13.1.3.1 *Average Composition and Linear Charge Density Distributions* With the crossover regime mapped, it is possible to use a model to determine the average ξ distribution of the copolyelectrolyte endproduct, in addition to the average composition, molecular weight, and intrinsic viscosity distributions. If there were no CC, then ξ would simply be proportional to the composition distribution. ξ was found to be independent of the molecular weight distribution in Reference [30]. Although the deviation between the ξ and $F_{inst,VB}$ distributions can be distinguished directly by ACOMP data, a model must nonetheless be used to relate the conductivity to ξ and $F_{inst,VB}$.

There are many subtleties inherent in treating solution conductivities involving polyelectrolytes, and Kreft and Reed [30] developed a simple model for interpreting the marked nonlinearities of σ versus [VB]. Automatic Continuous Online Monitoring of Polymerization Reactions data can serve as grist for more refined and rigorous models. In Reference [30], $\phi(t)$ is the fraction of noncondensed Na^+ counterions, σ_{NaCl} is the conductivity due to added salt in the reactor at the reaction temperature, which is constant during the reaction, Σ_{VB} is the specific conductivity of free VB (it includes both the conductivities of the anionic monomer and its free Na^+ counterions), Σ_{Na} is the specific conductivity of Na^+, and Σ_{CP} is the specific conductivity of VB in the polyelectrolyte (i.e., the conductivity of a VB monomer incorporated into a copolymer chain). $\phi(t)$ was computed from $\sigma(t)$ by:

$$\phi(t) = \frac{\sigma(t) - \sigma_{NaCl} - [VB]\Sigma_{VB}}{([VB]_0 - [VB])(\Sigma_{Na} + F_{inst,VB}\Sigma_{CP})} \quad (13.19)$$

where the following parameters were determined experimentally; $\sigma_{NaCl} = 1.09$ mS cm^{-1}, $\Sigma_{Na} = 43.8$ mS^{-1}cm^{-1}mol^{-1}, $\Sigma_{CP} = 41.5$ mS^{-1}cm^{-1}mol^{-1}, $\Sigma_{VB} = 57.3$ mS^{-1}cm^{-1}mol^{-1} ±4%. $\phi(t) = 1$ when there is no condensation, and at very high VB composition there is full condensation and $\phi = 0.36$. An empirical sigmoid fit to $\phi(F_{inst,VB})$ was found:

$$\phi(F_{inst,VB}) = 0.27 + \frac{0.73}{1 + (1.969\,F_{inst,VB})^{2.843}} \quad (13.20)$$

and the charge density was computed from $\phi(t)$ by:

$$\xi(F_{inst,VB}) = \frac{l_B\,\phi(F_{inst,VB})F_{inst,VB}}{b} \quad (13.21)$$

where l_B is the Bjerrum length, which is 0.718 nm in pure water at 25 °C, and b is the contour length of both VB and Am in a polymer chain, equal to 0.26 nm.

13.2 CONTINUOUS REACTORS

Automatic Continuous Online Monitoring of Polymerization Reactions was adapted to monitoring a homogeneous continuous stirred tank reactor (HCSTR) to verify the quantitative predictions concerning f, M_w, and η_r, as a function of the flow and kinetic parameters, to determine the kinetic parameters themselves, to ascertain the ideality of mixing in the reactor, to assess the effects of feed and reactor fluctuations, and to approximate a fully continuous tube-type reactor [38].

Continuous reactors are commonly used for producing synthetic polymers. In many cases, they offer certain advantages over batch reactors in terms of product quality and ease of handling reagents and products. Because reactions can reach a steady state in continuous reactors, this approach can also be of fundamental value in studying kinetics and mechanisms of reactions.

A substantial literature exists concerning the modeling of such reactors in different contexts. A good overview is given by Dotson et al. [6]. Simulations in continuous reactors have been carried out for long-chain branching [39], copolymerization [40, 41], emulsion polymerization [4], living polymerization [42], multicomponent chain growth reactions [43], and for various other aspects of specific polymers such as nylon 6 [44], polymethyl methacrylate [45], polyvinyl acetate (modeling and experiments) [46], and polystyrene [47]. Scale-up modeling to full-scale industrial reactors has also been made [48].

In an HCSTR in which monomer and initiator are fed into the reactor at the same flow rate r, at which material is removed, a steady-state condition is reached in which the reactor contents will remain at constant values of M_w, conversion, polydispersity, etc.

13.2.1 Polymerization and Flow Considerations

It was shown that the concentration of monomer in the reactor at time t is given by:

$$[m](t) = \left\{ [m]_r - \frac{p[m]_s}{p + k_p[R]} \right\} \exp\left(-(p + k_p[R])t\right)$$
$$+ \frac{p}{p + k_p[R]}[m]_s \tag{13.22}$$

Here, $[m]_r$ is the initial concentration of monomer in the reactor, $[R]$ the concentration of radical, k_p the propagation rate constant (assumed independent of chain length), r the volume flow rate ($cm^3 s^{-1}$) from the reservoir to the reactor, and $p = r/V$ (s^{-1}) the mixing rate constant. It should be noted that $1/p$ is the average residence time of a mass element in a perfectly stirred mixing chamber. Equation 13.22 indicates that if the initial monomer concentration in the reactor is the same as in the reservoir then the monomer concentration in the reactor will exponentially decrease to a smaller value, given by the rightmost term in Equation 13.22. $[m]$ will then remain at this steady-state value for as long as the reactor is fed. At this plateau value, monomer conversion f will remain at a constant value given by:

$$f_{\text{steady state}} = \frac{k_p[R]}{p + k_p[R]} \tag{13.23}$$

Likewise, $N_{n,inst}$ will remain constant, and can be denoted by $N_{n,\text{steady state}}$:

$$N_{n,\text{steady state}} = \frac{p k_p [m]_s}{k_t[R](p + k_p[R])} \tag{13.24}$$

The polydispersity will also remain constant, with the "natural" instantaneous value characteristic of the type of termination mechanism; for example, for termination by disproportionation, the instantaneous values of the n-, w-, and z-averages of the polymer mass (M_n, M_w, and M_z) stand in the ratio 1:2:3, respectively [49].

Equation 13.22 also predicts that if the exponential prefactor on the right-hand side is set equal to 0, by properly choosing $[m]_r$ and $[m]_s$ for given p and $k_p[R]$, then there will be no exponential approach to the steady state and, in fact, the reactor will commence at and remain in the steady state.

Using $c_t(t)$ to denote the time-dependent total combined mass concentration of polymer $c(t)$, and monomer $c_m(t)$:

$$c_t(t) = c_m(t) + c(t) \tag{13.25}$$

gives the following expressions for $c_t(t)$, $c_m(t)$, and $c(t)$:

$$c_t(t) = c_{m,s} + (c_{m,r} - c_{m,s})\exp(-pt) \tag{13.26}$$

$$c_m(t) = \left(c_{m,r} - \frac{p c_{m,s}}{p + \alpha} \right) \exp\left[-(p + \alpha)t\right] + \frac{p c_{m,s}}{p + \alpha} \tag{13.27}$$

$$c(t) = \frac{\alpha c_{m,s}}{p + \alpha} + (c_{m,r} - c_{m,s})\exp(-pt) - \left(c_{m,r} - \frac{p c_{m,s}}{p + \alpha} \right)$$
$$\exp\left[-(p + \alpha)t\right] \tag{13.28}$$

where $c_{m,r}$ is the initial concentration of monomer in the reactor, $c_{m,s}$ is the concentration of monomer in the solvent reservoir, and the polymerization rate constant has been expressed as:

$$\alpha = k_p[R] \tag{13.29}$$

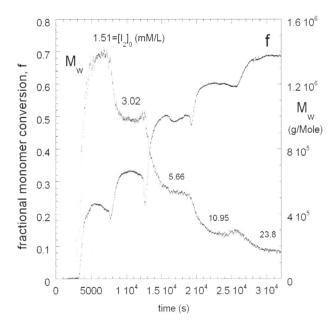

FIGURE 13.8 M_w and fractional monomer conversion f, for free radical polymerization of Am in a continuous reactor. In successive steps the initiator concentration $[I]$ is increased, while Am concentration remains constant. The kinetics of the approach to each new steady state can be seen each time $[I]$ increases. Reprinted with permission from Grassl B, Reed WF. Online polymerization monitoring in a continuous tank reactor. Macromol Chem Phys 2002;203:586–597.

The HCSTR used in Reference [38] allowed for variation of flow rates, initiator, and other reagent concentrations. The rate constant α could be determined in batch polymerization reactions monitored by ACOMP. The system was used to investigate the effects on f, M_w, and η_r of changing the different flow and concentration variables.

13.2.1.1 An Example: Varying Initiator Concentration at Constant Reactor Feed/Withdrawal Rate

A more complete study of reaction kinetics, in terms of the relation between $2Fk_d$, M_w, k_p^2/k_t, and $[I_2]_0$ (see Chapters 1 and 16 for further discussion) could be made in a single experiment by changing initiator concentration at intervals. This represents a concise, economical means of determining reaction kinetics without the need to run several independent experiments.

Figure 13.8 shows M_w and f versus t for an experiment in which p was kept constant ($7.13 \times 10^{-4}\,s^{-1}$), and the amount of initiator was increased at intervals. The steady state conversion $f_{steady\ state}$ increases with increasing amount of free radical due to increased initiator, as expected by Equation 13.23. At the same time, M_w decreases with increasing initiator as expected by Equation 13.24. This type of data allows concise determination of kinetic parameters from one single experiment.

From Equations 13.4 and 13.27, it is easily shown that the cluster of kinetic constants can be determined by plotting:

$$[I_{2,0}] = \left(\frac{k_t}{2Fk_D k_p^{\,2}} \right) \left[p \left(\frac{f}{1-f} \right) \right]^2 \qquad (13.30)$$

Figure 13.9 (top) shows this representation, from which the slope yields: $\left(\dfrac{k_t}{2Fk_d k_p^{\,2}} \right) \left(\dfrac{M \cdot s^2}{L} \right) = 9075 \qquad (13.31)$

Since the behavior of both the approach to equilibrium and the equilibrium plateau itself confirm that the reactor is perfectly mixed, for all practical purposes, and that the polymerization proceeds according to the QSSA, it is possible to exploit the M_w data in Figure 13.8 to complete the determination of the kinetic parameters. The kinetic parameter k_p^2/k_t can be computed from the aforementioned slope, together with the relation:

$$\frac{dM_w(f=0)}{d\sqrt{\dfrac{1}{[I_2]_0}}} = M_{Am} \frac{k_p[m]_0}{\sqrt{\dfrac{Fk_d k_t}{2}}} \qquad (13.32)$$

with $M_{Am} = 71\,g\,mol^{-1}$, and $[m]_0 = 0.479\,M\,l^{-1}$, via:

$$\frac{k_p^{\,2}}{k_t} = \frac{\dfrac{1}{2[m]_0 M_{Am}} \dfrac{dM_w(f=0)}{d\sqrt{\dfrac{1}{[I_2]_0}}}}{\left[\dfrac{d[I_2]_0}{d\alpha^2} \right]^{1/2}} \qquad (13.33)$$

Figure 13.9 (bottom) shows $M_w(f=0)$, from extrapolation according to $M_w(f)/(1-f)$, versus $1/\mathrm{sqrt}([I_2]_0)$. The value of k_p^2/k_t is $11.71\,M\cdot s^{-1}$. The hollow data points in Figure 13.9 (bottom) are from other experiments. All the data are in good agreement.

13.2.2 Kinetics during the Approach to Steady State

The approach to the steady state is important in industry because it controls the rate at which product grade crossovers occur in continuous reactors. For the HCSTR, Equations indicate how $c_t(t)$, $c_m(t)$, and $c(t)$ vary as functions of time for different combinations of flow, polymerization, and concentration conditions.

Figure 13.10 shows an excerpt from Figure 13.8 of one of the exponential approaches to the steady state. The exponential fit to conversion $f(t)$ is quite good, in agreement with Equation 13.22 when expressed in terms of f, which confirms two important assumptions made:

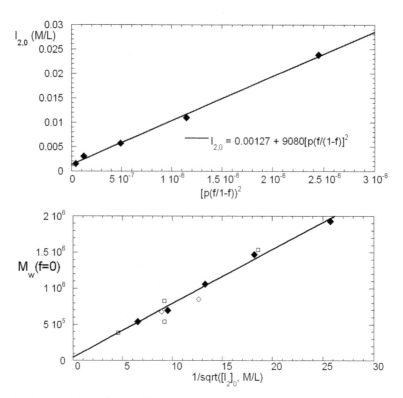

FIGURE 13.9 The conversion data (top) from Figure 13.8 are represented according to Equations 13.30 and 13.31 such that the slope yields

the cluster of constants $\left(\dfrac{k_t}{2Fk_dk_p^2}\right)\left(\dfrac{M \cdot s^2}{1}\right) = 9075$ The slope of $M_w(f=0)$ data from Figure 13.8 versus inverse square root of initiator

concentration yields, according to Equations 13.32 and 13.33, the cluster of constants k_p^2/k_t is $11.71\,m^{-1}\cdot s$ (bottom). Adapted with permission from Grassl B, Reed WF. Online polymerization monitoring in a continuous tank reactor. Macromol Chem Phys 2002;203:586–597.

First, the reaction indeed proceeds via the QSSA, and, second, the mixing of the reactor is perfect, for all practical purposes. This allows the rate constant to be determined alternatively via the rate constant of the exponential approach to equilibrium.

The approach of $M_w(t)$ to the steady state was derived in Reference [38]:

$$M_w(t) = \frac{\left\{\begin{array}{l} p\left(\dfrac{c_{m,s}}{p+\alpha}\right)^2 [\exp(pt)-1] + \dfrac{2pc_{m,s}}{\alpha(p+\alpha)}\left(c_{m,r} - \dfrac{pc_{m,s}}{p+\alpha}\right) \\[3mm] [1-\exp(-\alpha t)] + \dfrac{\left(c_{m,r} - \dfrac{pc_{m,s}}{p+\alpha}\right)^2}{p+2\alpha}[1-\exp[-(p+2\alpha)t] \end{array}\right\}}{\left(\dfrac{k_t[R]M_{Am}}{2k_p}\right)\left\{\begin{array}{l}\dfrac{c_{m,s}}{p+\alpha}[\exp(pt)-1] \\[3mm] +\dfrac{1}{\alpha}\left(c_{m,r} - \dfrac{pc_{m,s}}{p+\alpha}\right)[1-\exp(-\alpha t)]\end{array}\right\}} \tag{13.34}$$

A case often used is that $c_{m,r}=c_{m,s}$. Then $M_w(t)$ simplifies to:

$$M_w(t) = \frac{2k_p c_{m,s}}{k_t[R](p+\alpha)}$$

$$\frac{\left\{\begin{array}{l} p[\exp(pt)-1]+2p[1-\exp(-\alpha t)]+\dfrac{\alpha^2}{p+2\alpha} \\[3mm] [1-\exp[-(p+2\alpha)t] \end{array}\right\}}{\exp(pt)-\exp(-\alpha t)} \tag{13.35}$$

This latter expression implies that the approach to equilibrium of $M_w(t)$ will always be slower than for $f(t)$. Although $M_w(t)$ is not a pure exponential decay function, the time-dependent portion resembles one, in which the initial value is 1 and the final value is $p/(p+\alpha)$ and the decay rate is β. An important approximate relationship between the decay rate of the effective decay of the time-dependent part of $M_w(t)$ and the constants α and p was derived to be:

$$\beta \cong \frac{\alpha+p}{2} \tag{13.36}$$

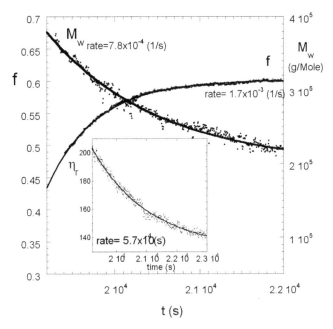

FIGURE 13.10 Approach to the steady state of M_w and f with fits according to Equations 13.22 and 13.35. The inset shows the approach to steady state of reduced viscosity η_r. Data are excerpted from one of the steady-state approaches in Figure 13.8. Adapted with permission from Grassl B, Reed WF. Online polymerization monitoring in a continuous tank reactor. Macromol Chem Phys 2002;203:586–597.

that is, the effective rate constant of the $M_w(t)$ approach to equilibrium is approximately one-half the rate constant of $f(t)$ approach to equilibrium. Figure 13.10 shows an example where this relationship is closely born out. The approach of η_r to the steady state can be found from the scaling law between viscosity and molecular weight:

$$[\eta]_w(t) = \frac{k \int_0^t M_{w,\text{inst}}(t')^\beta \, dc(t,t')}{c(t)} \qquad (13.37)$$

Since β in general can range from 0.5 to 0.8 for coil polymers, no single expression can be given for the integral in the equation. However, since it represents an average proportional to $<M\beta>$, the decay rate will be lower than for $M_w(t)$. This observation is born out in the inset of Figure 13.10, where $\eta_r(t)$ is plotted and the pseudo-exponential decay rate is found to be 73% of the $M_w(t)$ pseudo-exponential decay rate.

13.2.3 Monitoring Fluctuations in Conversion due to Drift in Reactor Conditions

Whereas the previous experiments showed how steady-state HCSTR operation is approached and can be maintained, there is considerable practical interest in being able to monitor the effects of drift in reactor conditions.

Figure 13.11 shows the results of an experiment in which a steady state was first achieved, after which the

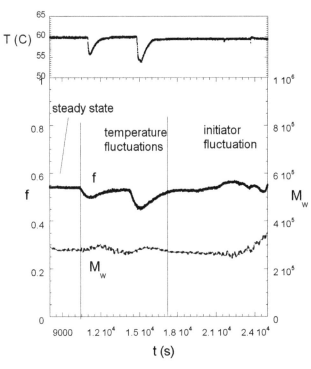

FIGURE 13.11 Effects of fluctuations of continuous reactor operating conditions on M_w and f. When temperature dips occur, f decreases and M_w increases. The effect of a small fluctuation in initiator concentration in the reservoir is seen in the later stage of the experiment. Adapted with permission from Grassl B, Reed WF. Online polymerization monitoring in a continuous tank reactor. Macromol Chem Phys 2002;203:586–597.

following manipulations were made: (i) deliberate temperature fluctuations and (ii) the effect of a change in initiator concentration. The steady state is seen up until 10,500 s at which point the temperature of the reactor was lowered from 60 to 56 °C, as seen in the temperature overset on the figure. The response of the reaction is seen in the drop in conversion f, and the increase in M_w. Conversion drops because the initiator decomposes more slowly, producing fewer free radicals. Fewer free radicals, while the monomer concentration remains steady, leads to a larger kinetic chain length, and hence the increase in M_w.

13.3 SEMIBATCH REACTORS AND PREDICTIVE CONTROL

13.3.1 Introduction

There is extensive work on measuring, controlling, and engineering more robust polymerization reactors, elaborated in Chapter 16. Richards and Congalidis have also provided a concise summary [50]. Much attention has been devoted to maintaining pressure, temperature, level, and flow (PTLF) in the reactor [51, 52]. Different online composition measurement techniques, such as IR analysis, Fourier transform infrared (FTIR), near IR (NIR), Raman spectroscopy [53], calorimetry [54], and gas chromatography (GS) are available, as discussed in Chapters 6, 9, 10, 18, and 20.

There has been a significant amount of work reported on controlling composition during copolymerization reactions. The Kalman filter method is based on a linear approximation of the nonlinear process [55] but has problems with stability and convergence [56–58]. For that reason, numerous nonlinear methods have been developed. Kravaris et al. [59] used temperature tracking as another nonlinear method to control copolymer composition. Model predictive control (MPC) [60–63], as well as nonlinear MPC (NLMPC) [64–67] algorithms have been suggested for control of nonlinear systems.

The semibatch approach, where policies are developed for selective reagent feeds to the reactor, has been extensively elaborated, especially for emulsion polymerization, and in the context of controlling composition during copolymerization reactions [68–72, 109]. Discussions are provided in Chapters 4, 7, 12, 17–19, and 21. Sun et al. developed model-based semibatch monomer feeding policies for controlled radical polymerization (CRP) [73, 74]. Vicente et al. [75, 76] controlled composition and molecular weight distribution in emulsion copolymerization in an open-loop method by maintaining the ratio of comonomers. Yanjarappa et al. [77] synthesized, via a semibatch method, copolymers with constant composition for biofunctionalization. General semibatch policies are reviewed by Asua [78].

A major issue in reactor control involves nonlinearities arising during the reaction due to gelation, cage effects, exothermicity, and large changes in kinetic coefficients; for example, propagation and termination coefficients.

The focus in this section is the control of molecular weight, conversion kinetics, and composition drift for free radical polymerization. There have been efforts to control molecular weight, such as those by Kiparissides and Morris [79]. It is frequently reported that the main difficulty in controlling the molecular weight is the lack of online sensors.

In general, the best way to control molecular weight is by manipulating the concentration of monomer, initiator, or chain transfer agent [80–85]. Due to problems with online measurements, Vicente et al. [76] controlled composition and MWD in emulsion copolymerization by an open-loop semibatch method. They computed optimal feed profiles using iterative dynamic programming. The success of this method depends on how good the mathematical models are and requires that there be no unanticipated disturbances during the process. They estimated conversion from calorimetric data. Off-line measurements after the reaction agreed with the estimations and confirmed that the mathematical model was good. The authors concluded, however, that more robust closed-loop methods involving online measurements would have to be developed for better control.

Othman et al. [86] proposed a closed-loop method to control molecular weight. They used NIR to estimate conversion. They developed a nonlinear estimator to get the reaction rate necessary for the control loop. This method relies on the quality of off-line measurements necessary for NIR calibration. They used nonlinear high-gain observers to identify model parameters and the reaction rates which were used to obtain desired monomer feeds to keep M_w constant. This feedback control produced high molecular masses which could not be achieved in open-loop cases but the approach still relies on the quality of the model.

Hur et al. [72] designed a multivariable model-on-demand predictive controller (MoD-PC) that overcomes the imperfections of other identification methods in controlling molecular weight and composition during semibatch copolymerization. In this simulation, they also used reaction temperature and the feed flow rates as variables to be manipulated and molecular weight and composition as the control outputs. The strength of the designed MoD-PC simulation lies in the fact that it identifies a model using data belonging to a small neighborhood around the current operating point rather than estimating a complex global model.

Park and Rhee [64] used a learning-based NLMPC to control semibatch copolymerization. The purpose was to linearize a nonlinear model based on previous batch data. In this way, the prediction is a function of the increment of inputs between two consecutive batches. They used an online densitometer to obtain conversion and a viscometer to calculate molecular weight. These measured properties were fed back, and, by using estimation and optimization procedures, the necessary feeds were obtained. Simulations were successful and were experimentally verified for a methacrylate/methyl acrylate copolymerization. The system did not exhibit any disturbances for any reaction. As the authors report, in general, disturbance models will have to be used for successful control.

13.3.2 Semibatch Control of M_w Using ACOMP

The precise knowledge of reaction kinetics furnished by ACOMP allows predictive control of reactions, via calculable flow rates of reagents into the reactor, to yield desired molecular weight and composition trends, with subsequent online verification of the actual reaction trajectory. In this sense, the predictive approach is a prelude to full feedback control of the reactor, where the predicted trajectory can serve as an Ansatz, and deviations from the desired trajectory can then be corrected by small changes to the feed pumps and other variables.

Figure 13.12 shows the results of predictive control of M_w for acrylamide polymerization [87]. The lower curve shows the natural tendency of M_w to monotonically decrease during

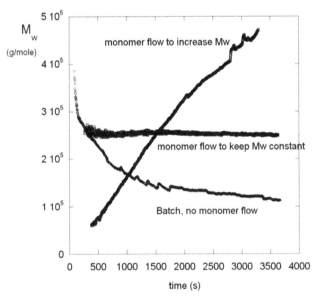

FIGURE 13.12 Results of predictive control of M_w for free radical polymerization of Am in semibatch operation. In batch mode, M_w decreases monotonically in time. By computing conditions for "isoreactivity" (Equation 13.63c) constant M_w during the reaction was achieved. By operating Am flow into the reactor in the "flooded regime" (Equation 13.63b), a predictable increase in M_w during the semibatch reaction was achieved. Adapted with permission from Kreft T, Reed WF. Predictive control and verification of conversion kinetics and polymer molecular weight in semi-batch free radical homopolymer reactions. Eur Polym J 2009;45:2288–2303.

a batch reaction. This is typical of free radical polymerization when chain transfer is negligible and the initiator decomposes slowly compared to the conversion rate. The constant M_w data was achieved by knowledge of the Am conversion rate and setting the Am reactor feed rate to maintain [Am] constant in the reactor. With a near constant concentration of initiator in the reactor, and very little change in reactor volume (a concentrated Am reservoir was used for reactor feed), the kinetic chain length remains constant. To demonstrate the robustness of the approach, the third curve shows M_w increasing versus conversion, achieved by increasing the flow rate of Am into the reactor, and hence producing the opposite trend of M_w from a batch reaction.

The way in which the different results are achieved results from a formalism to deal with the combined effects of flowing and reacting reagents in the ACOMP context, described next.

13.3.2.1 Formalism for Combined Reaction and Semibatch Flow   The following summarizes the ACOMP approach presented in [38]. Expressions were derived for the concentration of monomer and polymer in the reactor, while reactions are occurring, when N solutes in solution are allowed to flow into the reactor, each at their own rate, which need not be constant, such that solute s has caused a change in reactor volume at time t of $\Delta V_s(t)$, where $Q_s(t)$ is the instantaneous flow rate of liquid from a reservoir containing component s into the reactor. In the following, q is a constant withdrawal rate from the reactor q (cm^3 s^{-1}) that feeds the ACOMP extraction/dilution/conditioning front end.

Because volume and concentrations in the reactor are changing, mass balance of component s in the reactor, $m_s(t)$, measured in grams, is used. The rate at which $m_s(t)$ changes is:

$$\frac{dm_s}{dt} = c_s' Q_s(t) - \frac{m_s}{V(t)}q - \varepsilon_s(t) + \gamma_s(t) \qquad (13.38)$$

where c_s' is the concentration of s in the reservoir from which it is being pumped into the reactor at rate $Q_s(t)$, $\varepsilon_s(t)$ is the rate at which m_s is lost due to any reactions, and $\gamma_s(t)$ is the rate at which s is produced by any reactions. $V(t)$ is the total volume in the reactor, and is given by:

$$V(t) = V_0 - qt + \sum_{s=1}^{N} \int_0^t Q_s(t')dt' \qquad (13.39)$$

where the partial volume change due to each flowing solute s is:

$$\Delta V_s(t) = \int_0^t Q_s(t')dt' \qquad (13.40)$$

The concentration of any solute s at any time is:

$$c_s(t) = \frac{m_s(t)}{V(t)} \tag{13.41}$$

v_s is the constant relating mass concentration of solute s, c_s ($g\,cm^{-3}$), to molar concentration of s, $[s]$ ($mol\,l^{-1}$):

$$c_s = v_s[s] \tag{13.42}$$

v_s is given by $v_s = 10^{-3} \times M_s$, where M_s is the molecular weight of s. For acrylamide (Am) $v_A = 0.07108\,g^{-1}cm^{-3}mol^{-1}$, and for potassium persulfate (KPS) $v_{KPS} = 0.27032\,g^{-1}cm^{-3}mol^{-1}$. *Effect of Reactor Inflow and Outflow on Concentration of Any Solution Component* Extraction and flow of liquids into the reactor alter mass and concentrations. The molar concentration of component s, $[s]$, changes purely due to flow:

$$\frac{d[s]}{dt_{flow}} = \frac{1}{v_s V(t)} \frac{dm_s}{dt} - \frac{m_s}{v_s V(t)^2} \frac{dV(t)}{dt} \tag{13.43}$$

The rate at which m_s changes due to flow is:

$$\frac{dm_s}{dt_{flow}} = c_s' Q_s(t) - q \frac{m_s}{V(t)} \tag{13.44}$$

which leads to the expression:

$$\frac{d[s]}{dt_{flow}} = \frac{c_s' Q_s(t)}{v_s V(t)} - \frac{c_s}{v_s V(t)} \sum_j Q_j(t) \tag{13.45}$$

If q does not change the concentration of any of the components when there are no flows $Q_s(t)$ into the reactor (i.e., $Q_s(t) = 0$ for all s), pure batch formulations are used for ACOMP in which extraction rate q is the only flow in the system.

Another feature of Equation 13.45 is that the concentration of any component s will be held constant in the absence of reactions if:

$$c_s' Q_s(t) = c_s \sum_j Q_j(t) \tag{13.46}$$

If one wishes to hold the concentration of *all* components constant, then the two limiting cases for this are to use a single reservoir pumping into the reactor at rate $Q(t)$, or to use a separate reservoir for each component, each pumping into the reactor at rate $Q_s(t)$.

In the case of a single reservoir, $Q(t) = \sum_j Q_j(t)$, but since $c_s' Q_s(t) = c_s \sum_j Q_j(t)$ by Equation 13.46, $c_s' = c_s$ for all components; that is, the concentration of the reactor can only be

held constant while being fed from a single reservoir if the concentration of each component in the reservoir is the same as in the reactor itself.

For multiple reservoirs, Equation 13.46 is satisfied for each component, and each component has its own flow rate $Q_s(t)$ from its reservoir, which requires:

$$\sum_s Q_s(t) = \sum_s \frac{c_s}{c_s'} \sum_j Q_j(t) = \sum_j Q_j(t) \tag{13.47}$$

leading to the condition that:

$$\sum_s \frac{c_s}{c_s'} = 1 \tag{13.48}$$

that is, one can independently establish the flow rate Q_s from each reservoir, then use Equation 13.46 to compute c_s/c_s', which will then also satisfy Equation 13.48.

Free Radical Polymerization Kinetics and Molecular Weight Considerations

No evidence was found for changes in k_p and k_t as a function of conversion for the dilute Am concentrations in this section, but there is extensive work on how k_p depends on monomer concentration and conversion [88–91], and how k_t depends on chain length [92, 93]. There has been work devoted to acrylamide [94–96], which showed that reaction rate coefficients for Am are not constant during the polymerization reaction at high monomer concentrations (>10% monomer in solution), but are constant below 10% monomer consistent with this section.

Predictive control of molecular weight can be approached using the kinetic chain length concept for free radical polymerization; the instantaneous number average chain length $N_{n,inst}$ given by Equation 13.16.

The monomer and polymer mass balance in the reactor allows cumulative and instantaneous monomer conversion to be defined, and permits computation of M_w and $[\eta]_w$ from light scattering, viscosity, and concentration data. $m_c(t)$ is the cumulative (or total) mass of monomer plus polymer in the reactor, which is the sum of initial monomer mass in the reactor m_0 and that flowed in from the external reservoir, minus the amount flowed out for ACOMP sampling. This is found from:

$$\frac{dm_c(t)}{dt} = c_m' Q_m(t) - q \frac{m_c(t)}{V(t)} \tag{13.49}$$

For free radical reactions, where there is no elimination and condensation of products, mass balance relates monomer and polymer mass via:

$$m_c(t) = m(t) + m_p(t) \tag{13.50}$$

where $m(t)$ and $m_p(t)$ are the monomer and polymer mass in the reactor, respectively.

The ACOMP detector train provides $m(t)$, $m_c(t)$ is computed via the previous equation. $m_p(t)$ is then obtained from Equation 13.50, and $c_p(t) = m_p(t)/V(t)$. Density changes, while measurable, are usually small (negligible in this section, due to the very low Am concentration), and are not corrected for in this work. They have been addressed in the ACOMP context earlier [49, 97].

Flowing monomer into the reactor at a constant rate $Q_m(t) = \rho$ from a monomer reservoir at concentration c_m', with an initial amount of monomer m_0 in the reactor, and a constant, small ACOMP withdrawal stream at rate q, subject to $m_c(0) = m_0$ yields

$$m_c(t) = \frac{m_0}{\left(1 + \frac{(\rho - q)}{V_0} t\right)^{q/(\rho - q)}} + \rho c_m' t \qquad (13.51)$$

It is important to have a measure of fractional monomer conversion f in the reactor at any time. The simplest means, in keeping with standard batch parlance, is to define f as the total mass of polymer in the reactor divided by the total mass of combined monomer plus polymer in the reactor ($m_c(t)$). This is a "cumulative conversion," f_c, given by:

$$f_c(t) = \frac{m_c(t) - m(t)}{m_c(t)} \qquad (13.52)$$

In terms of reaction completion for a specified process, in which a certain total mass of monomer $m_{c,f}$ is to be converted completely to a corresponding amount of polymer in the reactor, the "foreseen" conversion f_v (v stands for vorausgesehen) is:

$$f_v(t) = \frac{m_{c,f}(t) - m(t)}{m_{c,f}(t)} \qquad (13.53)$$

The cumulative weight average molecular weight, $M_w(t)$, can be conveniently calculated using the mass of polymer in the reactor, $m_p(t)$:

$$M_w(t) = \frac{\int_0^t M_{w,inst}(t)\, dm_p(t)}{m_p(t)} \qquad (13.54)$$

This can be used for predicting values of $M_w(t)$, which is what the multiangle light scattering in ACOMP measures directly. On the other hand, it can be valuable to use $M_w(t)$ from ACOMP to compute $M_{w,inst}(t)$ as follows:

$$M_{w,inst}(t) = \frac{d\left(M_w(t)m_p(t)\right)}{dm_p(t)} \qquad (13.55)$$

For a constant flow into the reactor, the aforementioned formalism gives mass concentration of any component, s, present in the reactor at initial concentration $c_{s,0}$:

$$c_s(t) = \frac{c_{s,0}}{\left(1 + \frac{\rho - q}{V_0} t\right)^{\rho/(\rho - q)}} \qquad (13.56)$$

When the addition rate of component x increases linearly with time, $Q_x(t) = \sigma t$:

$$c_s(t) = \frac{c_{s,0} \exp\left\{-\frac{2q}{\sqrt{2\sigma V_0 - q^2}} \tan^{-1}\left(\frac{-q + \sigma t}{\sqrt{2\sigma V_0 - q^2}}\right)\right\}}{\left(1 - qt + \frac{\sigma}{2V_0} t^2\right)} \qquad (13.57)$$

Controlling M_w by Monomer Feed into Reactor
A powerful means of controlling M_w is to feed monomer into the reactor at a rate $Q_m(t)$ from a reservoir of monomer concentration c_m'. For this,

$$\frac{d[I_2]}{dt} = -[I_2]\left(k_d + \frac{Q_m(t)}{V(t)}\right) \qquad (13.58)$$

and the molar monomer concentration in the reactor is:

$$\frac{d[m]}{dt} = -[m]\left(k_p[R^\bullet] + \frac{Q_m(t)}{V(t)}\right) + \frac{[m]'Q_m(t)}{V(t)} \qquad (13.59)$$

where $[m]' = c_m'/v_m$ is the molar monomer concentration in the monomer reservoir. *Constant Flow Rate of Monomer into the Reactor* $Q_m = \rho$ (cm^3 s^{-1}) = *Constant* In this situation,

$$[I_2] = \frac{[I_2]_0 e^{-k_d t}}{\left(1 + \frac{(\rho - q)}{V_0} t\right)^{\rho/(\rho - q)}} \qquad (13.60)$$

Two good approximations can be made to obtain a good analytical solution. First, [Am]' in the feed reservoir can be high enough that a low flow rate Q_m and relatively little increase of V_0 occurs during the reaction. Then, $V(t) \approx V_0$. Furthermore, if the half-life of the initiator (e.g., KPS at 60 °C) is long compared to the reaction, the approximation can be used that $[I_2](t) \approx [I_2]_0$. With these approximations, the molar monomer concentration is calculated to be:

$$[m](t) = [m]_0 \exp\left\{-\left(\frac{\rho + \alpha V_0}{V_0}\right)t\right\} + \left(\frac{[m]'\rho}{\rho + \alpha V_0}\right)$$
$$\left(1 - \exp\left\{-\left(\frac{\rho + \alpha V_0}{V_0}\right)t\right\}\right) \tag{13.61}$$

This predicts that the rate coefficient controlling monomer concentration increases from α to $\alpha + \rho/V_0$, where the added term reflects the result of the added volume diluting the reactor.

Second, the concentration of monomer in the reactor, in this approximation, approaches a steady-state value, $[m]_{ss}$, given by:

$$[m]_{ss} = \left(\frac{[m]'\rho}{\rho + \alpha V_0}\right) \tag{13.62}$$

This value is approached with the same rate $\alpha + \rho/V_0$, and this steady-state concentration conveniently divides into three regimes that provide a convenient labeling for the period during which the steady state is being approached:

1. $\left(\dfrac{[m]'\rho}{\rho + \alpha V_0}\right) < [m]_0$ Starved regime   (13.63a)

2. $\left(\dfrac{[m]'\rho}{\rho + \alpha V_0}\right) > [m]_0$ Flooded regime   (13.63b)

3. $\left(\dfrac{[m]'\rho}{\rho + \alpha V_0}\right) = [m]_0$ Isoreactive regime   (13.63c)

The terms "starved regime" and "flooded regime" are commonly used for semibatch reactions, especially as regards emulsion polymerization [98, 99, 100]. In the starved regime, all monomer flowing into the reactor is being consumed in addition to the remaining $[m]_0$, whereas in the flooded condition, the inflow of monomer is fast enough that it cannot all be consumed together with the remaining $[m]_0$, so that monomer concentration builds up in the reactor.

In the isoreactive regime, the rate of monomer addition is just enough to offset the rate of monomer conversion to polymer, and to keep the concentration in the reactor constant. This condition also leads to a constant instantaneous molecular weight as long as [R·] remains approximately constant because of slow initiator decomposition.

The different approach regimes suggest that an index comparing the rates of change of $[m]$ due to the polymerization reaction and, separately, due to the flow, will be useful. With this, the "approach index" can be defined as:

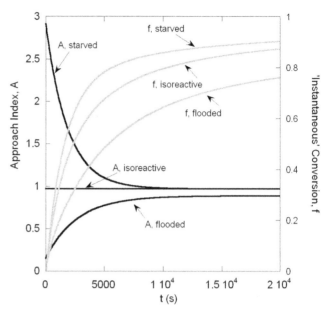

FIGURE 13.13 The computed approach index "A" (Equation 13.64) and "instantaneous conversion" versus t. Reprinted with permission from Kreft T, Reed WF. Predictive control and verification of conversion kinetics and polymer molecular weight in semibatch free radical homopolymer reactions. Eur Polym J 2009;45: 2288–2303.

$$A = \frac{(dm/dt)_{\text{reaction}}}{(dm/dt)_{\text{flow}}} \tag{13.64}$$

Some authors refer to this or a similar quantity as "instantaneous conversion." Since A can be greater than unity in this latter definition, it does not resemble a usual fractional monomer conversion, such as in a batch situation, so "approach index" term is used. For the type of reaction described earlier, and under the scenario of flowing in monomer, A becomes:

$$A = \frac{\alpha[m](t)V(t)v_{\text{m}}}{c'_{\text{m}} Q_{\text{m}}(t) - (qm(t)/V(t))} \tag{13.65}$$

Figure 13.13 shows the computed cumulative conversion f_c and the approach index A, for isoreactive and flooded cases for different scenarios related to the Am reaction. Also shown here are f_c and A for a hypothetical starved case, in which $c_{\text{Am},0} = 0.005\,\text{g cm}^{-3}$, $\rho = 0.00333\,\text{cm}^3\,\text{s}^{-1}$ $V_0 = 200\,\text{cm}^3$, $c'_{\text{Am}} = 0.054\,\text{g cm}^{-3}$, and $\alpha = 5.2 \times 10^{-4}\,\text{s}$. The steady state value of A at high conversion is $A \approx \alpha V_0/(\rho + \alpha V_0) < 1$. This is less than unity at the steady state because in order for $[m]$ to remain constant at steady state, the amount of monomer flowed into the reactor per second must compensate both the reaction that consumes monomer and the dilution of the reactor contents caused by the flow itself.

Keeping M_w Constant during the Reaction, the "Isoreactive Regime"

Keeping $[m]$ and $[R]$ constant should lead to a constant, nondrifting molecular mass which can be achieved by applying the isoreactive condition in Equation 13.63c:

$$\rho = \frac{\alpha V_0 [m]_0}{[m'] - [m]_0} \qquad (13.66)$$

Under isoreactive conditions, the cumulative weight average molecular weight equals the instantaneous weight average molecular weight, which can be predicted by:

$$M_w = M_{w,\text{inst}} = \frac{w M_{Am} k_p [m]_0}{k_t [R^\cdot]} = 1000 w c_{m,0} \frac{k_p^2}{\alpha k_t} \qquad (13.67)$$

where w is the instantaneous polydispersity M_w/M_n. For the Am reactions in Figure 13.12, the following parameters were used to compute M_w: $k_p^2/k_t = 14.7$ ($1 m^{-1} \times s$), (where $w = 2$ was used in Reference [68]), $\alpha = 0.00052 s^{-1}$ (from the batch experiment #1), $c_{m,0} = 0.005 g\ ml^{-1}$. The computation yielded $M_w = 262{,}500 g\ mol^{-1}$, which is in excellent agreement with Figure 13.12.

Increasing M_w during the Reaction

The adjustable feed rate gives substantial power for controlling M_w during the reaction. M_w can be made to increase in a predictable fashion using a higher value of the monomer flow rate ρ can be made to increa"flooded" regime. In this case:

$$c_p(t) = \frac{m_0 + \rho c_m' t}{V(t)} - [m]_0 \exp\left(-\frac{\rho + \alpha V_0}{V_0} t\right)$$
$$- \frac{[m]' \rho}{\rho + \alpha V_0} \left\{ 1 - \exp\left(-\frac{\rho + \alpha V_0}{V_0} t\right) \right\} \qquad (13.68)$$

and M_w is:

$$M_w = \frac{2000 \frac{k_p^2}{k_t}}{\alpha} \frac{1}{V_0 + \rho t} \times \left(\begin{array}{c} \dfrac{\frac{2}{\alpha}\left(\rho c_m' m_0 - \frac{\rho^2 c_m'^2}{\alpha}\right)(1 - e^{-\alpha t})}{\rho c_m' t + m_0 - \left(m_0 - \frac{\rho c_m'}{\alpha}\right)e^{-\alpha t} - \frac{\rho c_m'}{\alpha}} + \\[2em] \dfrac{\frac{1}{2\alpha}(\alpha m_0^2 - 2\rho c_m' m_0)(1 - e^{-2\alpha t})}{\rho c_m' t + m_0 - \left(m_0 - \frac{\rho c_m'}{\alpha}\right)e^{-\alpha t} - \frac{\rho c_m'}{\alpha}} + \\[2em] \dfrac{\frac{\rho^2 c_m'^2}{\alpha} t}{\rho c_m' t + m_0 - \left(m_0 - \frac{\rho c_m'}{\alpha}\right)e^{-\alpha t} - \frac{c_m'}{\alpha}} \end{array} \right)$$

$$(13.69)$$

This predicted M_w was achieved to fairly good accuracy in Figure 13.12.

Overall, fair to good predictive control of polymer molecular weight and conversion kinetics in semibatch reactions can be achieved using easily calculable reactor feed rates and reagent reservoir concentrations, and can be verified in near-real-time by online monitoring of the reactions. In the case of acrylamide, the QSSA provides a robust analytical framework for computing the reactive part of the reagent balance equations. It was thus demonstrated that M_w during a semibatch reaction can be held constant or made to increase (or decrease) in predictable fashion, once the underlying batch kinetics are quantified. Many other types of desired profiles for M_w should be attainable by predictive control.

It is recognized, however, that there is room for improvement between the predictions and the results, and that under stochastic operating scenarios, there can be potentially large deviations from the predictions, especially due to unforeseen events. The ability to monitor reactions online, however, will ultimately allow much more precise, active control of the reactions via feedback. Hence, the predictive approach may be ultimately more suited to providing an Ansatz for the reaction protocol, and incremental feedback-controlled adjustments to reactor feed and other conditions can be made automatically via the monitoring signals to keep the reaction headed toward producing the desired endproduct. Adaptation of ACOMP to feedback control will yield model independent output signals and input will rely on robust kinetic models based on basic reaction principles rather than empirical correlations. Anomalies, such as microgelation, massive cross-linking, initiator dead-end, etc., can be monitored as they occur, without reliance on inferential models [100, 101].

13.3.3 Semibatch Composition Control in Copolymerization

ACOMP has been used for approximate predictive control of average composition and molecular weight during free radical copolymerization. Using kinetics obtained from ACOMP and elementary computations as a guide, the trends in the composition and molecular weight drifts during synthesis can be approximately controlled by predetermined, constant reagent flows into the reactor. In Reference [104] work was done on (i) a comonomer pair with very different reactivity ratios, having a tendency to high composition drift (VB and Am, described earlier, $r_{VB} = 2.14$ and $r_{Am} = 0.18$), and (ii) a comonomer pair with much closer reactivity ratios (Q9 and Am, described earlier, $r_{Q9} = 0.47$ and $r_{Am} = 1.1$) resulting in relatively constant composition during batch copolymerization. Each system had high downward molecular weight drift during synthesis, typical of free radical polymerization.

Appropriate feed polices were found to control the composition and molecular weight drift tendencies at will, that is, it was possible to keep composition constant in the first case and to increase or decrease composition in the latter, while desired trends for the average molecular weight were also obtained. The approach allows practical, approximate composition and molecular weight control in free radical copolymerization with simple reactors and pumps, and without computationally intensive requirements. The results can also be useful as an Ansatz for reaction trajectory prediction, and, together with the online monitoring signals, can be used for feedback controlled corrections to the reagent flows and other reaction conditions.

The relative change in the comonomer composition can be obtained from the Mayo–Lewis equation [102]:

$$\frac{d[m_1]}{d[m_2]} = \frac{[m_1]}{[m_2]}\frac{(r_1[m_1]+[m_2])}{([m_1]+r_2[m_2])} \quad (13.70)$$

It is hence possible to integrate Equation 12.70 once the reactivity ratios are known, so that the composition drift can be predicted. The Mayo–Lewis model, or more sophisticated models (e.g., penultimate models), can be extended to the case where one or more reagents is flowed into the reactor in semibatch operation, and the resulting equations numerically integrated to predict the composition drift.

Thanks to continuous ACOMP measurements of comonomer conversion, average instantaneous molar composition of chains being produced at each point in time can be calculated from the knowledge of polymer mass created at each point. This eliminates the use of the reactivity ratios for the predictive procedure:

$$F_{inst,1}(t) = \frac{dm_{p_1}(t)/M_1}{dm_{p_1}(t)/M_1 + dm_{p_2}(t)/M_2} \quad (13.71)$$

where M_1 and M_2 are the molar masses of monomers 1 and 2, respectively, and the polymer mass increments dm_{p_1} and dm_{p_2} are computed from the derivatives of ACOMP data for each comonomer concentration according to:

$$m_{p,i}(t) = \left[c_{t,i}(t) - c_{m,i}(t)\right]V(t) \quad (13.72)$$

where $i = 1$ or 2. The total concentration (including monomer and polymer) of each component $c_{t,i}$ is approximately given by:

$$c_{t,i}(t) \approx c_{m,i}(0) + \frac{Q_{m,i}c'_{m,i}t}{V_0 + (Q_{m,i}-q)t} \quad (13.73)$$

The $c_{m,i}(t)$ are measured directly from the ACOMP UV diode array data.

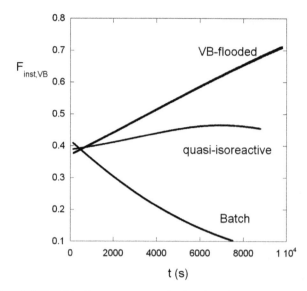

FIGURE 13.14 Instantaneous molar fraction of VB, $F_{inst,VB}$ versus conversion for batch and two feed rates during VB/AM free radical copolymerization reactions. Reprinted (adapted) with permission from Kreft T, Reed WF. Predictive control of average composition and molecular weight distributions in semibatch free radical copolymerization reactions. Macromolecules 2009;42:5558–5565. © 2009 American Chemical Society.

In order to keep the composition constant using a constant monomer inflow, the concentration of both comonomers has to stay constant during the reaction, that is:

$$\frac{d[m_1]}{dt} = \frac{d[m_2]}{dt} = 0 \quad (13.74)$$

This means that each monomer must be continuous. In Reference [103], y flowed into the reactor at its own rate Q_i from its own reservoir at concentration $[m'_i]$. Equation 13.74 was solved under certain simplifying assumptions and provided the following monomer feed rate equations:

$$Q_1 = \frac{V_0[m_1]\alpha_1}{[m'_1]-[m_1]_0} \quad (13.75a)$$

$$Q_2 = \frac{V_0[m_2]\alpha_2}{[m'_2]-[m_2]_0} \quad (13.75b)$$

Controlling High Composition Drift Comonomers VB and Am

The conditions (Eqs. 13.75 and 13.76) were implemented for the high-drift free radical copolymerization of Am and VB. With the knowledge of composition drift in batch and the conversion kinetics from separate ACOMP experiments, the authors of Reference [103] were largely able to control the composition drift. Figure 13.14 shows composition drift curves $F_{inst,VB}$ versus conversion for VB/AM copolymerization reactions made under different monomer flow schemes [103].

The quasi-isoreactive data show the results of using simultaneous VB and Am feeds. While not perfectly constant, it is much closer than an uncontrolled batch reaction, which depends on the reactivity ratios. The upper curve demonstrates how the natural batch tendency to rapidly deplete VB can be countered by flooding the reactor at a high VB addition rate from an external reservoir.

Controlling Low Composition Drift Comonomers Q9 and Am

As discussed earlier, Q9 and Am show relatively little drift in batch reactions. This is seen in Figure 13.15. The natural tendency for low drift could be countered by selective monomer feed into the reactor. $F_{inst,Q9}$ could be made to decrease when Am was continuously flowed into the reactor. By the end of the reaction, there is very little Q9 in the copolymers. In contrast, flowing Q9 into the reactor allowed $F_{inst,Q9}$ to increase. In the reaction shown, by 3000 s all the Am was depleted and continued addition of Q9 led to homopolymer production of Q9.

Hence, this type of approach can be used to tailor composition to desired distributions, including the deliberate production of blends. Virtually, any sort of taper to the composition distribution can be computed and executed in semibatch mode.

By appropriate monomer feed to the reactor, of one or both monomers, batch trends can be modulated at will and even be reversed from their natural trends. It is possible to program the feed rates so that increases, decreases, and minimal changes in composition and M_w can be achieved. This type of trend-based estimation of reaction trajectory is a simple but effective means to yield good results for synthesizing copolymers of desired average composition and molecular weight. While achieving an approximate level of control over composition and molecular weight in copolymerization reactions may often be sufficient, this approach could also be used to establish Ansatz conditions for reactions to produce on-command polymers, where automated feedback control can be implemented based on the massive ACOMP detector data stream. In this latter case, much more precise reaction control could be achieved.

13.4 SECOND GENERATION ACOMP

Whereas ACOMP discussed herein is of the "first generation" (FGA), second generation ACOMP (SGA) has the added feature of providing immediate data on how the stimuli responsiveness of polymers evolves during synthesis. There is much current work in stimuli responsive polymers, sometimes also termed "smart" materials, with potential for use in drug delivery, electronics and optics, high performance coatings, self-healing materials, etc., discussed in Chapter 3.

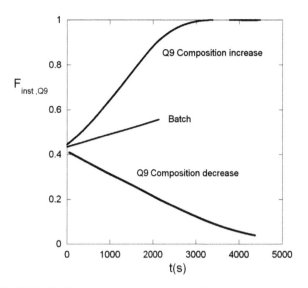

FIGURE 13.15 Low composition drift of Q9 in a batch reaction and semibatch experiments in which $F_{inst,Q9}$ was predictively arranged to either decrease or increase. In the latter case, a blend of Am/Q9 copolyelectrolyte and hompolymeric Q9 was produced. Reprinted (adapted) with permission from Kreft T, Reed WF. Predictive control of average composition and molecular weight distributions in semibatch free radical copolymerization reactions. Macromolecules 2009;42:5558–5565. © 2009 American Chemical Society.

Stimuli responsiveness includes conformational and thermodynamic phase transitions (e.g., lower critical solution temperature [LCST]), aggregation, ability to encapsulate and release other agents (e.g., drugs), ability to interact with surfactants, other polymers, etc. Stimuli producing these responses include temperature, light, pH, ionic strength, specific small molecules, surfactants, solvent type and mixtures, etc [104–109].

The goal of SGA is to monitor not only the important characteristics of the reactions, but also the onset and evolution of stimuli responsiveness of the polymers during synthesis. This will allow determination of what conditions are needed to produce specific stimuli-responsive behavior, which will help both in the fundamental understanding of these phenomena and materials, and in the practical task of optimizing their synthesis conditions.

The SGA approach involves simultaneous use of multiple sample detectors, including banks of flow cell equipped static light scattering cells with and without polarized detection, fluorescence cells, and viscometers (each being a differential transducer), either coupled to the FGA output or with separate reactor extraction and dilution stages. Depolarized scattering detection can help detect morphological changes; for example, from isotropic to anisotropic structures, whereas fluorescence detection can be used for monitoring micellization properties, encapsulation of agents, etc. Viscosity measurements have wide applicability to

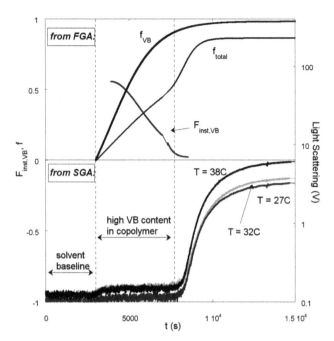

FIGURE 13.16 Second generation ACOMP used to determine how much composition of anionic SS a copolymer with NIPAM can have and still exhibit an LCST (at 10 mM ionic strength): fractional conversion of SS and total comonomers and average instantaneous SS composition $F_{inst,SS}$ (top). LS at 90° (log scale) for three detectors at different temperatures increases abruptly at the LCST when there is only about 4% SS in the copolymer (bottom). Adapted with permission from McFaul CA, Alb AM, Drenski MF, Reed WF. Simultaneous multiple sample light scattering detection of LCST during copolymer synthesis. Polymer 2011;52:4825–4833.

particle morphology, polyelectrolyte response to pH and ionic strength, etc. These detectors build on the simultaneous multiple sample light scattering (SMSLS) platform recently introduced by the authors and their colleagues [11, 110, 111], aimed at economically screening many independent samples simultaneously and continuously. Groups of cells can also be toggled between parallel and serial operation with millifluidic valves, together with multihead peristaltic pumps for simultaneous mixing and continuous detection. For example, in parallel mode, polymer behavior can be monitored under different solvent conditions by separate automatic dilution with the chosen solvents, using the multihead pump.

An example of serial operation is determination of LCST, for a polymer during synthesis by passing the diluted stream through a series of scattering detectors, each at a different temperature. *N*-isopropyl acrylamide (NIPAM) and its copolymers are a well-studied example of polymers with LCST behavior [112–116]. Below the LCST NIPAM is hydrophilic, in the form of a random coil stabilized with many hydrogen bonds. When the LCST is reached, the hydrogen bonds are sundered and the now hydrophobic NIPAM chains collapse into globular conformation, with subsequent reversible aggregation.

A recent series of experiments illustrated SGA's capability of discerning when LCST conditions are reached during the copolymerization of NIPAM and VB. The charge and hydrophilicity of VB confers resistance to the copolymer against hydrophobic collapse. As in the case of Am/VB copolymers in preceding sections of this chapter, NIPAM/VB likewise show high composition drift during copolymerization. Studied under high drift conditions in 10 mM ionic strength, it was found that VB effectively inhibited the LCST for the copolymer until the mole percentage of VB dropped to about 4%. Thus, SGA is a powerful means of automatically sweeping through many different copolymer compositions seeking the onset of LCST behavior. It is currently being used to make concise pH and ionic strength sweeps as well, for various copolymers that exhibit LCST.

Figure 13.16 shows an example of LCST results for the NIPAM/VB copolymerization. The upper panel shows the FGA results for fractional conversion of VB and total monomer (VB+NIPAM), as well as the instantaneous fraction of VB in the copolymeric chains $F_{inst,VB}$. The lower panel shows 90° light scattering intensity for three LS flow cells in series at the temperatures indicated. For the majority of VB conversion no LCST is encountered. Only when $F_{inst,VB}$ drops to 4% is the LCST finally met. This is seen in the lower panel by the enormous increase in LS, which occurred simultaneously at all temperatures monitored. The increase in LS is so large that a log scale is used.

Hence, SGA shows that the NIPAM/VB copolymer at 10 mM ionic strength can only tolerate a small fraction of VB before losing its LCST. Subsequent measurement showed that an LCST could be found for higher fractions of VB in higher ionic strength media.

13.5 SUMMARY

Monitoring polymerization reactions with detectors that make fundamental measurements of polymer properties allows a virtually model-free set of primary data concerning conversion, composition drift, and evolution of molecular weight and intrinsic viscosity. Unusual or unexpected events, such as microgelation or phase separation, can also be monitored directly.

The primary data furnished by such detection then allows interpretation of kinetics and mechanisms and can serve as the basis for building specific quantitative, mechanistic models, as well as for controlling polymerization reactions. The ACOMP platform is one means of embodying and employing any chosen set of such detectors, and these can also be complemented by widely used *in situ* detectors.

Automatic Continuous Online Monitoring of Polymerization Reactions is in a vigorous phase of widening its envelope of applications and providing many modular approaches to a very wide variety of polymerization reactions. An important goal is to combine the control potential of ACOMP with such knowledge of how macromolecular distributions correlate to endproduct properties to produce "on-command" polymers of desired properties; the producer could, in principle, select the desired end-product properties (e.g., processability, impact-resistance, rheology control, associative behavior, etc.) which are relatable to macromolecular distributions, and "dial" the distributions into a feedback control system on the reactor piloted by ACOMP to produce the desired endproduct.

REFERENCES

[1] Zhu S, Hamielec AE. Chain-length-dependent termination for free-radical polymerization. Macromolecules 1989;22, 3093–3098.

[2] Matsumoto A, Mizuta, K. Evaluation of the chain-length dependence of the termination rate-constant during radical polymerization of a methacrylic ester. Macromolecules 1994;27:1657–1659.

[3] Beuermann S. Requirements associated with studies into a chain-length dependence of propagation rate coefficients via PLP-SEC experiments. Macromolecules 2002;35:9300–9305.

[4] Moad G, Moad CL. Use of chain length distributions in determining chain transfer constants and termination mechanisms. Macromolecules 1996;29:7727–7733.

[5] Olaj OF, Vana P, Zoder, M. Chain length dependent propagation rate coefficient k_p in pulsed-laser polymerization: variation with temperature in the bulk polymerization of styrene and methyl methacrylate. Macromolecules 2002;35:1208–1214.

[6] Dotson NA, Galvan R, Laurence RL, Tirrell M. Polymerization Process Modeling. New York: VCH Publishers Inc.; 1996. p 392.

[7] Baer M, Caskey JA, Fricke AL. A kinetic study of acrylamide polymerization initiated by 4,4'-azo-bis-4-cyanopentanoic acid in aqueous solution. Makromol Chem 1972; 158:27–37.

[8] Ishige T, Hamielec AE. Solution polymerization of acrylamide to high conversion. J Appl Polym Sci 1973;17: 1479–1506.

[9] Riggs JP, Rodriguez F. Persulfate-initiated polymerization of acrylamide. J Polym Sci Pol Chem 1967;5:3151–3165.

[10] Riggs JP, Rodriguez F. Polymerization of acrylamide initiated by the persulfate–thiosulfate redox couple. J Polym Sci Pol Chem 1967;5:3167–3181.

[11] Lin HR. Solution polymerization of acrylamide using potassium persulfate as an initiator: kinetic studies, temperature and pH dependence. Eur Polym J 2001;37:1507–1510.

[12] Noyes RM. Kinetics of competitive processes when reactive fragments are produced in pairs. J Am Chem Soc 1955;77: 2042–2045.

[13] Noyes RM. Models relating molecular reactivity and diffusion in liquids. J Am Chem Soc 1956;78:5486–5490.

[14] Kim YS, Sook C, Sung P. UV and fluorescence characterization of styrene and methyl methacrylate polymerization. J Appl Polym Sci 1995;57:363–370.

[15] Kreft T, Reed WF. Direct monitoring of the cross-over from diffusion-controlled to decomposition-controlled initiation in free radical polymerization. Macromol Chem Phys 2008;209:2463–2474.

[16] Forster S, Schmidt M. Polyelectrolytes in solution. Adv Polym Sci 1995;120:51–133.

[17] Yamakawa H. Modern Theory of Polymer Solutions. New York: Harper and Row; 1972.

[18] Odijk T. Ionic-strength dependence of the intrinsic-viscosity of DNA. Biopolymers 1979;18:3111–3113.

[19] Nierlich M, Williams CE, Boue F, Cotton JP, Daoud M, Farnoux B, Janink G, Picot C, Moan M, Wolff C, Rinaudo M, de Gennes PG. Small angle neutron scattering by semi-dilute solutions of polyelectrolyte. J Phys 1979;40: 701–704.

[20] Nallet F, Jannink G, Hayter J, Oberthur R, Picot C. Observation of the dynamics a polyelectrolyte strong solutions by quasielastic neutron scattering. J Phys 1983;44: 87–99.

[21] Maier EE, Krause R, Deggelmann M, Hagenbuechl M, Weber R, Fraden S. Liquidlike order of charged rodlike particle solutions. Macromolecules 1992;25:1125–1133.

[22] Wang L, Bloomfield VA. Small-angle X-ray scattering of semidilute rodlike DNA solutions: polyelectrolyte behavior. Macromolecules 1991;24:5791–5795.

[23] Drifford M, Dalbiez JP. Light scattering by dilute solutions of salt-free polyelectrolytes. J Phys Chem 1984;88: 5368–5375.

[24] Li X, Reed WF. Polyelectrolyte properties of proteoglycan monomers. J Chem Phys 1991;94:4568–4580.

[25] Forster S, Schmidt M, Antonietti M. Static and dynamic light-scattering by aqueous polyelectrolyte solutions – effect of molecular-weight, charge-density and added salt. Polymer 1990;31:781–792.

[26] Morfin I, Reed WF, Rinaudo M, Borsali R. Further evidence for liquid-like correlations in polyelectrolyte solutions. J Phys 1994;69:1001–1019.

[27] Reed WF. Liquid-like correlations of polyelectrolytes under high shear conditions. J Phys Chem 1994;100:7825–7827.

[28] Booth F. The electroviscous effect for suspensions of solid spherical particles. Proc R Soc Lond A 1950;203:533–551.

[29] Basu S. Viscosity of sodium thymonucleate. Nature 1951; 168:341–342.

[30] Kreft TR, Reed WF. Experimental observation of crossover from noncondensed to counterion condensed regimes during free radical polyelectrolyte copolymerization under high-composition drift condition. J Phys Chem B 2009;113: 8303–8309.

[31] Alb AM, Paril A, Catalgil-Giz H, Giz A, Reed WF. Evolution of composition, molar mass, and conductivity during the free radical copolymerization of polyelectrolytes. J Phys Chem B 2007;111:8560–8566.

[32] Manning GS. Limiting laws and counterion condensation in polyelectrolyte solutions II. Self diffusion of the small ions. J Chem Phys 1969;51:934–938.

[33] Oosawa F. *Polyelectrolytes*. New York: Marcel Dekker; 1971.

[34] Deserno M, Holm C, May, S. Fraction of condensed counterions around a charged rod: comparison of Poisson-Boltzmann theory and computer simulations. Macromolecules 2000;33: 199–206.

[35] Chu JC, Mak CH. Inter and intrachain attractions in solutions of flexible polyelectrolytes at nonzero concentration. J Phys Chem 1999;110:2669–2679.

[36] Wilson RW, Bloomfield VA. Counter-ion-induced condensation of deoxyribonucleic-acid – light-scattering study. Biochemistry 1979;18:2192–2196.

[37] Hinderberger D, Spiess HW, Jeschke G. Dynamics, site binding, and distribution of counterions in polyelectrolyte solutions studied by electron paramagnetic resonance spectroscopy. J Phys Chem B 2004;108:3698–3704.

[38] Grassl B, Reed WF. Online polymerization monitoring in a continuous tank reactor. Macromol Chem Phys 2002;203: 586–597.

[39] Tobita H. A simulation-model for long-chain branching in vinyl-acetate polymerization. 2. Continuous polymerization in a stirred-tank reactor. J Polym Sci Part B: Polym Phys 1994;32:911–919.

[40] Podosenova NG, Zotikov EG. Prediction of the process kinetics and MWD of products of radical copolymerization of styrene with acrylonitrile in the presence of polybutadiene in continuous reactors. Sov Chem Ind 1991;23:15–21.

[41] Poehlein GW. Emulsion polymerization and copolymerization in continuous reactor systems. Polym Int 1993;30:243–251.

[42] Kim DM, Nauman EB. Nonterminating polymerizations in continuous flow systems. Ind Eng Chem Res 1997;36: 1088–1094.

[43] Dube MA, Soares JBP, Penlidis A, Hamielec AE. Mathematical modeling of multicomponent chain-growth polymerizations in batch, semibatch, and continuous reactors: a review. Ind Eng Chem Res 1997;36:966–1015.

[44] Plazl I. Mathematical model of industrial continuous polymerization of nylon 6. Ind Eng Chem Res 1998;37: 929–935.

[45] Ahn SM, Park MJ, Rhee HK. Extended Kalman filter-based nonlinear model predictive control for a continuous RIMA polymerization reactor. Ind Eng Chem Res 1999;38: 3942–3949.

[46] Teymour F, Ray WH. The dynamic behavior of continuous polymerization reactors. 5. Experimental investigation of limit-cycle behavior for vinyl-acetate polymerization. Chem Eng Sci 1992;47:4121–4132.

[47] Verazaluce JC, Tlacuahuac AF, Guerra ES. Steady-state nonlinear bifurcation analysis of a high-impact polystyrene continuous stirred tank reactor. Ind Eng Chem Res 2000;39: 1972–1979.

[48] Teymour F, Ray WH. The dynamic behavior of continuous polymerization reactors.6. Complex dynamics in full-scale reactors. Chem Eng Sci 1992;47:4133–4140.

[49] Flory P. *Principles of Polymer Chemistry*. Ithaca: Cornell University Press; 1953. p 688.

[50] Richards JR, Congalidis JP. Measurement and control of polymerization reactors. Comput Chem Eng 2006;30:1447–1463.

[51] Richards JR, Congalidis JP. Measurement and control of polymerization reactors. In: Meyer T, Keurentjes J, editors. *Handbook of Polymer Reaction Engineering*. Weinheim: Wiley-VCH; 2005. p 595–678.

[52] Liptak BG. *Instrument Engineers' Handbook: Process Measurement and Analysis*. New York: CRC Press; 2003. p 1920.

[53] Rodriguez F, Cohen C, Ober CK, Archer LA. *Principles of Polymer Systems*. New York: CRC Press; 2003. p 800.

[54] Kammona O, Chatzi EG, Kiparissides C. Recent developments in hardware sensors for the on-line monitoring of polymerization reactions. J Macromol Sci Polym Rev 1999;39:57–134.

[55] Hammouri H, McKenna TF, Othman S. Applications of nonlinear observers and control: improving productivity and control of free radical solution copolymerization. Ind Eng Chem Res 1999;38:4815–4824.

[56] Dochain D, Pauss, A. Online estimation of microbial specific growth-rates – an illustrative case-study. Can J Chem Eng 1988;47:327–336.

[57] Ellis MF, Taylor TW, Jensen KF. Online molecular-weight distribution estimation and control in batch polymerization. AIChE J 1994;40:445–462.

[58] Kozub DJ, Macgregor JF. State estimation for semibatch polymerization reactors. Chem Eng Sci 1992;47:1047–1062.

[59] Kravaris C, Wright RA, Carrier JF. Nonlinear controllers for trajectory tracking in batch processes. Comput Chem Eng 1989;13:73–82.

[60] Alhamad B, Romagnoli JA, Gomes VG. On-line multivariable predictive control of molar mass and particle size distributions in free-radical emulsion copolymerization. Chem Eng Sci 2005;60:6596–6606.

[61] Richalet J, Rault A, Testud JL, Papon, J. Model predictive heuristic control – applications to industrial processes. Automatica 1978;14:413–428.

[62] Garcia CE, Morari, M. Internal model control. 1. A unifying review and some new results. Ind Eng Chem Proc Des Dev 1982;21:308–323.

[63] Garcia CE, Prett DM, Morari M. Model predictive control – theory and practice – a survey. Automatica 1989;25: 335–348.

[64] Park MJ, Rhee HK. Control of copolymer properties in a semibatch methyl methacrylate/methyl acrylate copolymerization reactor by using a learning-based nonlinear model predictive controller. Ind Eng Chem Res 2004;43:2736–2746.

[65] Gattu G, Zafiriou, E. Nonlinear quadratic dynamic matrix control with state estimation. Ind Eng Chem Res 1992;31: 1096–1104.

[66] Henson MA. Nonlinear model predictive control: current status and future directions. Comput Chem Eng 1998;23: 187–202.

[67] Qin SJ, Badgwell TA. A survey of industrial model predictive control technology. Control Eng Pract 2003;11:733–764.

[68] Parouti S, Kammona O, Kiparissides C, Bousquet J. A comprehensive experimental investigation of the methyl methacrylate/butyl acrylate/acrylic acid emulsion terpolymerization. Polym React Eng 2003;11:829–853.

[69] Cao GP, Zhu ZN, Zhang MH, Le HH. Molecular weight distribution of poly(methyl methacrylate) produced in a starved feed reactor. J Polym Eng 2001;21:401–419.

[70] Aerdts AM, Theelen SJC, Smith TMC, German AL. Grafting of styrene and methyl-methacrylate concurrently onto polybutadiene in semicontinuous emulsion processes and determination of copolymer microstructure. Polymer 1994;35:1648–1653.

[71] Wu JY, Shan GR. Kinetic and molecular weight control for methyl methacrylate semi-batch polymerization. I. Modelling. J Appl Polym Sci 2006;100:2838–2846.

[72] Hur SM, Park MJ, Rhee HK. Design and application of model-on-demand predictive controller to a semibatch copolymerization reactor. Ind Eng Chem Res 2003;42:847–859.

[73] Sun X, Luo Y, Wang R, Li, B-G, Liu B, Zhu S. Programmed synthesis of copolymer with controlled chain composition distribution via semibatch RAFT copolymerization. Macromolecules 2007;40:849–859.

[74] Wang R, Luo Y, Li B, Sun X, Zhu S. Design and control of copolymer composition distribution in living radical polymerization using semi-batch feeding policies: a model simulation. Macromol Theor Simul 2006;15:356–368.

[75] Vicente M, Leiza JR, Asua JM. Simultaneous central of copolymer composition and MWD in emulsion copolymerization. AIChE J 2001;47:1594–1606.

[76] Vicente M, Sayer C, Leiza JR, Arzamendi G, Lima EL, Pinto JC, Asua JM. Dynamic optimization of non-linear emulsion copolymerization systems – open-loop control of composition and molecular weight distribution. Chem Eng J 2002;85:339–349.

[77] Yanjarappa MJ, Gujraty KV, Joshi A, Saraph A, Kane RS. Synthesis of copolymers containing an active ester of methacrylic acid by RAFT: controlled molecular weight scaffolds for biofunctionalization. Biomacromolecules 2006;7:1665–1670.

[78] Asua JM., *Polymer Reaction Engineering*. Oxford: Blackwell Publishing Ltd.; 2007. p 392.

[79] Kiparissides C, Morris J. Intelligent manufacturing of polymers. Comput Chem Eng 1996;20:S1113–S1118.

[80] Dimitratos J, Georgakis C, El-Aasser MS, Klein A. Dynamic modeling and state estimation for an emulsion copolymerization reactor. Comput Chem Eng 1989;13:21–33.

[81] Congalidis JP, Richards JR, Ray WH. Feedforward and feedback-control of a solution copolymerization reactor. AIChE J 1989;35:891–907.

[82] Clay PA, Gilbert RG. Molecular-weight distributions in free-radical polymerizations. 1. Model development and implications for data interpretation. Macromolecules 1995;28: 552–569.

[83] Ghielmi A, Storti G, Morbidelli M, Ray WH. Molecular weight distribution in emulsion polymerization: role of active chain compartmentalization. Macromolecules 1998;31:7172–7186.

[84] Othman NS, Fevotte G, Peycelon D, Egraz JB, Suau JM. Control of polymer molecular weight using near infrared spectroscopy. AIChE J 2004;50:654–664.

[85] Adebekun DK, Schork FJ. Continuous solution polymerization reactor control. 2. Estimation and nonlinear reference control during methyl-methacrylate polymerization. Ind Eng Chem Res 1989;28:1846–1861.

[86] Othman S, Barudio I, Fevotte G, McKenna TF. On-line monitoring and modeling of free radical copolymerizations: butyl acrylate/vinyl acetate. Polym React Eng 1999;7:1–12.

[87] Kreft T, Reed WF. Predictive control and verification of conversion kinetics and polymer molecular weight in semibatch free radical homopolymer reactions. Eur Polym J 2009;45:2288–2303.

[88] Kuchta FD, van Herk AM, German AL. Propagation kinetics of acrylic and methacrylic acid in water and organic solvents studied by pulsed-laser polymerization. Macromolecules 2000;33:3641–3649.

[89] Ganachaud F, Balic R, Monteiro MJ, Gilbert RG. Propagation rate coefficient of poly(N-isopropylacrylamide) in water below its lower critical solution temperature. Macromolecules 2000;33:8589–8596.

[90] Beuermann S, Buback M, Hesse P, Lacik, I. Free-radical propagation rate coefficient of nonionized methacrylic acid in aqueous solution from low monomer concentrations to bulk polymerization. Macromolecules 2006;39:184–193.

[91] Stach M, Lacik I, Chorvat D, Buback M, Hesse P, Hutchinson RA, Tang L. Propagation rate coefficient for radical polymerization of N-vinyl pyrrolidone in aqueous solution obtained by PLP-SEC. Macromolecules 2008;41: 5174–5185.

[92] Beuermann S, Buback M, Hesse P, Kuchta FD, Lacik I, Van Herk AM. Critically evaluated rate coefficients for free-radical polymerization. Part 6: Propagation rate coefficient of methacrylic acid in aqueous solution. Pure Appl Chem 2007;79:1463–1469.

[93] Beuermann S, Buback M, Hesse P, Hutchinson RA, Kukucova S, Lacik I. Termination kinetics of the free-radical polymerization of nonionized methacrylic acid in aqueous solution. Macromolecules 2008;41:3513–3520.

[94] Pascal P, Winnik MA, Napper DH, Gilbert RG. Pulsed-laser study of the propagation kinetics of acrylamide and its derivatives in water. Macromolecules 1993;26:4572–4576.

[95] Seabrook SA, Tonge MP, Gilbert RG. Pulsed laser polymerization study of the propagation kinetics of acrylamide in water. J Polym Sci A: Polym Chem 2005;43:1357–1368.

[96] Seabrook SA, Pascal P, Tonge MP, Gilbert RG. Termination rate coefficients for acrylamide in the aqueous phase at low conversion. Polymer 2005;46:9562–9573.

[97] Chauvin F, Alb AM, Bertin D, Tordo P, Reed WF. Kinetics and molecular weight evolution during controlled radical polymerization. Macromol Chem Phys 2002;203:2029–2041.

[98] Shin J, Lee Y, Park S. Optimization of the pre-polymerization step of polyethylene terephthalate (PET) production in a semi-batch reactor. Chem Eng J 1999;75:47–55.

[99] Elizalde O, Asua JM, Leiza JR. Monitoring of high solids content starved-semi-batch emulsion copolymerization reactions by Fourier transform Raman spectroscopy. Appl Spectrosc 2005;59:1270–1279.

[100] Alb AM, Mignard E, Drenski MF, Reed WF. In situ time-dependent signatures of light scattered from solutions undergoing polymerization reactions. Macromolecules 2004;37:2578–2587.

[101] Farinato RS, Calbick J, Sorci GA, Florenzano FH, Reed WF. Online monitoring of the final divergent growth phase in the stepgrowth polymerization of polyamines. Macromolecules 2005;38:1148–1158.

[102] Mayo FR, Lewis FM. Copolymerization. I. A basis for comparing the behavior of monomers in copolymerization: the copolymerization of styrene and methyl methacrylate. J Am Chem Soc 1944;66:1594–1601.

[103] Kreft T, Reed WF. Predictive control of average composition and molecular weight distributions in semibatch free radical copolymerization reactions. Macromolecules 2009;42:5558–5565.

[104] Checot F, Rodriguez-Hernandez J, Gnanou Y, Lecommandoux S. pH-responsive micelles and vesicles nanocapsules based on polypeptide diblock copolymers. Biomol Eng 2007;24:81–85.

[105] Liu SY, Billingham NC, Armes SP. A schizophrenic water-soluble diblock copolymer. Angew Chem Int Ed 2001;40:2328–2331.

[106] Skrabania K, Kristen J, Laschewsky A, Akdemir O, Hoth A, Lutz JF. Design, synthesis, and aqueous aggregation behavior of nonionic single and multiple thermoresponsive polymers. Langmuir 2007;23:84–93.

[107] Yamauchi K, Kanomata A, Inoue T, Long TE. Thermoreversible polyesters consisting of Multiple Hydrogen Bonding (MHB). Macromolecules 2004;37:3519–3522.

[108] Long KN, Scott TF, Qi HJ, Bowman CN, Dunn ML. Photomechanics of light-activated polymers. J Mech Phys Solids 2009;57:1103–1121.

[109] Lovell LG, Newman SM, Bowman CN. The effects of light intensity, temperature, and comonomer composition on the polymerization behavior of dimethacrylate dental resins. J Dent Res 1999;78:1469–1476.

[110] Drenski MF, Reed WF. Simultaneous multiple sample light scattering for characterization of polymer solutions. J Appl Polym Sci 2004;92:2724–2732.

[111] Drenski MF, Mignard E, Alb AM, Reed WF. Simultaneous in-situ monitoring of parallel polymerization reactions using light scattering: a new tool for high-throughput screening. J Comb Chem 2004;6:710–716.

[112] Liu SX, Liu X, Li F, Fang Y, Wang YJ, Yu J. Phase behavior of temperature- and pH-sensitive poly(acrylic acid-g-N-isopropylacrylamide) in dilute aqueous solution. J Appl Polym Sci 2008;109:4036–4042.

[113] Francis R, Jijil CP, Prabhu CA, Suresh CH. Synthesis of poly(N-isopropylacrylamide) copolymer containing anhydride and imide comonomers—a theoretical study on reversal of LCST. Polymer 2007;48:6707–6718.

[114] Weng Y, Ding Y, Zhang G. Microcalorimetric investigation on the lower critical solution temperature behavior of N-isopropycrylamide-co-acrylic acid copolymer in aqueous solution. J Phys Chem B 2006;110:11813–11817.

[115] Arotcarena M, Heise B, Ishaya S, Laschewsky A. Switching the inside and the outside of aggregates of water-soluble block copolymers with double thermoresponsivity. J Am Chem Soc 2002;124:3787–3793.

[116] McFaul CA, Alb AM, Drenski MF, Reed WF. Simultaneous multiple sample light scattering detection of LCST during copolymer synthesis. Polymer 2011;52:4825–4833.

14

COGNATE TECHNIQUES TO ACOMP

Michael F. Drenski, Alina M. Alb, and Wayne F. Reed

14.1 INTRODUCTION

The preceding two chapters have focused on the capabilities of automatic continuous online monitoring of polymerization reactions (ACOMP) in the area of polymerization monitoring. Related characterization challenges include time-dependent processes apart from polymerization reactions, behavior of multicomponent systems, and the issue of particulates coexisting with polymers. Recent methods for dealing with these issues are presented in this chapter.

A further characterization challenge is that of high throughput. The ACOMP is not a high-throughput method itself, since it monitors in great detail a single reaction at a time. As the name implies, high-throughput screening (HTS) is usually construed as an approach whereby many different systems, often with methodically varying parameter matrices among them, are characterized as quickly as possible, usually to determine whether or not certain criteria are met.

There are several common approaches to HTS. A central goal is to analyze as many samples per unit time as possible. There are several motivations for this. In many R&D projects, synthetic polymer chemists can produce new materials faster than they can be usefully characterized, creating an analytical bottleneck. The bottleneck can become more acute in the area of formulations where entire matrices of coexisting components are tested for products ranging from pharmaceuticals, personal care, foods and fragrances, paints, oil additives, etc.

In general, high throughput will occur either by measuring samples in parallel or rapidly in series, or in some combination of the two. Multiwell channel plates are a staple in the bioanalytical sciences. A popular configuration is the 96 well channel plate, which can require only microliters of sample per well, and on which a myriad of measurements can be made; thermographic, UV, light scattering (LS), fluorescence, etc.

An approach more suited to polymerization reactions involves multiple independent reactors run in parallel. Several manufacturers provide these, together with analytical tools for automatically sampling each reactor and making characterizing measurements.

14.2 SIMULTANEOUS MULTIPLE SAMPLE LIGHT SCATTERING

Simultaneous multiple sample light scattering (SMSLS) was recently introduced [1, 2]. It allows LS from many independent samples to be simultaneously monitored by a single instrument. Hence SMSLS is a quantitative high-throughput device. Since there are no fundamental limits to the number of independent cells in a single instrument, the number of experiments performed simultaneously can be factors of tens or even hundreds greater than individual experiments performed on single sample cell instruments. It can also be considered an HTS tool in some contexts.

Applications for SMSLS include monitoring the instability (or instability) of polymer solutions, such as therapeutic proteins, biopolymers, polymers in specific solvent, temperature, and other variable conditions, flocculants, interactions of polymers with surfactants, ions, small molecules, stimuli such as heat or light, etc. Modes of instability can include microgelation, microcrystallization, phase separation, conformational changes, specific associations, dissolution, and degradation. Examples include the

Monitoring Polymerization Reactions: From Fundamentals to Applications, First Edition. Edited by Wayne F. Reed and Alina M. Alb.
© 2014 John Wiley & Sons, Inc. Published 2014 by John Wiley & Sons, Inc.

hydrolysis of polysaccharides by specific enzymes, the kinetics of gelation of collagen, degradation of polymers by heat, LCST behavior for polyNIPAM and other semihydrophobic polymers, interaction of nanoparticles with antibodies, and much more. In flow mode, the effects of sonication, extreme thermal treatments, and other accelerated testing platforms can be used.

While most commercial LS systems are costly, SMSLS has been made possible at relatively low cost by a simplification of cell fabrication methods and materials, the ready availability of high-quality, stable diode lasers, high-sensitivity, low-noise CCD arrays, bountiful options for optical fibers, and abundant, inexpensive microcomputer power.

There are different embodiments of SMSLS. They can contain flow cells, batch cells, or both. They can use lasers with beam splitters or individual lasers per cell. Cells can be in series, for applications where there is no turbidity, in parallel when turbidity is an issue, or some combination of the two. Other features are currently being added to SMSLS. Depolarized detection, for example, can monitor processes in which there are transitions from isotropic to anisotropic morphologies or vice versa, such as nano- and micro-fibrillar modes of protein aggregation.

Depolarized operation can be achieved in different ways. An insertable half-wave plate can change incident laser polarization from vertical to horizontal to measure the depolarization. The same effect can also be achieved by placing a detection fiber perpendicular to the scattering plane, that is, aligning the fiber in the plane of the incident electric field polarization.

Other SMSLS extensions under development include turbidity measurements, by putting a detector at $q=0$ on the beam exit side of the sample cell, and adjusting sensitivity and detection electronics accordingly. Fluorescence detection over a narrow wavelength range can be had by using a laser of the desired excitation wavelength and interposing an optical notch filter between the detection fiber and the detector. The heterogeneous time-dependent static light scattering (HTDSLS) discussed in the following is also directly adaptable to the SMSLS platform, since its main design considerations are the size of the scattering volume, LS detection rate, and subsequent signal processing.

Selected examples of SMSLS capabilities are shown in the following.

14.2.1 Protein Aggregation

The Human Genome project laid the groundwork for modern proteomics in which highly sophisticated means of analyzing and modifying gene sequences have been developed. This has spawned enormous activity in the area of therapeutic proteins in the biotechnology and pharmaceutical industries.

A ubiquitous problem across this sector is the aggregation of proteins in therapeutic formulations [3–9]. This subject is treated in detail in Chapter 20. While Chaperones and the crowded physiological environment *in vivo* tend to keep proteins in their native, globular forms, there is a strong tendency for proteins *in vitro* to aggregate, often due to partial unfolding of the structure. The medicinal effects of aggregates are invariably negative; they can act as antigens, in which case the organism develops immunity to the very drug that is meant to treat it, they can reduce or eliminate the desired activity of the protein, and they prevent robust distribution of individual proteins in the organism.

A number of methods used to analyze protein aggregation have been recently compared and evaluated [10]. Currently, SEC is the most common means of detecting protein aggregation. The prevailing idea is that individual proteins, and small aggregates, for example, dimers, tetramers, will elute within the separation regime of the column(s), whereas large aggregates will elute in the exclusion volume, thus providing a means of quantifying what percentage of the protein is native and what percentage is aggregated. While widely used, SEC has several disadvantages: First, SEC is an equilibrium characterization method that assumes the protein population is unchanging. Second, it is assumed that the whole aggregate population passes through the column and arrives at the detector. Adsorption on and blockage in the column packing as well as by any prefilters or guard columns may retain some of the aggregate population leading to an underestimate of aggregate content. Third, interaction of protein aggregates under high pressure with the columns and filters can break up the aggregates, thus underestimating the aggregate content. Also, shear effects throughout the SEC components may cause aggregation of proteins; that is, the SEC method itself may create some aggregation, or accelerate aggregation that is underway. Finally, SEC does not conveniently furnish aggregation kinetics, rather "snapshots" of aggregating proteins.

Simultaneous multiple sample light scattering is emerging as a powerful high-throughput platform for monitoring protein aggregation. Figure 14.1 shows the aggregation of bovine serum albumen (BSA) at 10 mg ml^{-1} in 150 mM aqueous solution. The aggregation is expressed as M_w/M_0 versus time, where M_0 is the mass of the initial unaggregated BSA. The dependence on temperature is so strong that the timescale is logarithmic. The early phase is linear and very reproducible. Intermediate aggregation is well described by a growing exponential.

Figure 14.2 shows early-phase kinetics, represented as the first time derivative of M_w/M_0 in the linear regime. There are over three orders of magnitude increase in rate going from 53 to 75 °C. Over some of the temperature range, there is a reasonable Arrhenius fit; rate$=A\exp(-\Delta E/k_B T)$, where ΔE is the activation energy. This activation energy of 90 Kcal mol^{-1} is rather high. Pauling was one of the first to surmise that unfolding of proteins may be due to the simultaneous breaking of many weak, cooperative hydrogen bonds [11].

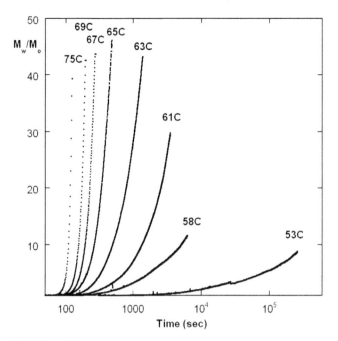

FIGURE 14.1 The SMSLS for determining kinetics of protein (BSA) aggregation at various temperatures, measured in terms of the ratio of $M_w(t)$ of the aggregates to the initial unaggregated mass M_0. The kinetics vary so markedly over the range studied that the time axis is shown on a log scale.

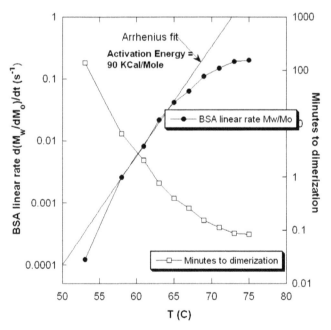

FIGURE 14.2 The initial linear rate of protein aggregation. The results of an Arrhenius fit are shown over the central temperatures. The hollow squares show the average time to an average aggregate mass equivalent to a dimer at each temperature.

Another means of conceptualizing the kinetic data in Figure 14.2 is to take the reciprocal of the linear rate constant, which is the time it takes for the protein to reach an average M_w equal to a dimer (i.e., $M_w/M_0 = 2$). This ranges from

a fraction of a minute at 75 °C to over a hundred minutes at 53 °C. The stability increases rapidly with decreasing temperature.

14.2.2 Polymer Degradation

When polymers undergo degradation, the intensity of the light they scatter diminishes. A detailed series of theoretical and applied work in this area has been published over the past 15 years [12–15]. Among many significant results is the fact that the absolute degradation rate of a polymer that resembles a random coil and undergoes random degradation can be determined by monitoring the light scattered at a single, arbitrary angle.

It was shown that for an ideal random coil undergoing random degradation, the number of bonds cleaved per second per g mol^{-1} of polymer mass is given by [16]:

$$\dot{\beta} = 2Kc\frac{d[1/I_R(t)]}{dt} \qquad (14.1)$$

The velocity normally measured in enzymology is in molar bonds cleaved per second per initial g mol^{-1} of polymer, v, which is given by:

$$v = 1000\dot{\beta}c \qquad (14.2)$$

where the factor of 1000 ensures that v is in mol $(l\text{-}s)^{-1}$.

The Michaelis–Menten–Henri model is one of the simplest for enzymatic action, and considers the velocity of enzymatic conversion at steady state, where there is a constant concentration of the enzyme–substrate complex, which itself can either dissociate or proceed to final product formation. The measured velocity is related to substrate concentration c, and the Michaelis–Menten rate constant K_M by:

$$v = v_{max}\frac{c}{c + K_M} \qquad (14.3)$$

where v_{max} is the maximum velocity obtained when the enzyme is completely saturated by substrate, and all the enzyme is in the enzyme–substrate complex.

A study on the enzymatic hydrolysis of hyaluronate by hyaluronidase by SMSLS was reported recently [2]. Several capabilities of the technique were verified: ability to make measurements of the absolute excess Rayleigh scattering ratio, allowing M_w, the monitoring of a biopolymeric degradation process (hyaluronate degradation using hyaluronidase), with subsequent determination of the Michaelis–Menten

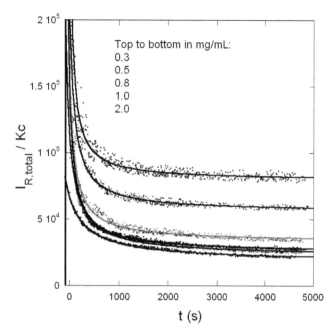

FIGURE 14.3 The decrease in light scattering intensity in time as hyaluronate is hydrolyzed by hyaluronidase. Results are shown for various concentrations of hyaluronate. Reprinted with permission from Drenski MF, Reed WF. Simultaneous multiple sample light scattering for characterization of polymer solutions. J Appl Polym Sci 2004;92:2724–2732.

constant, and an aggregation process (low-concentration gelatin solutions).

Figure 14.3 shows data for several simultaneous degradation reactions of sodium hyaluronate (HA), under the action of hyaluronidase in a buffer consisting of 0.15 M NaCl and 0.1 M sodium succinate, at pH = 5.31. There were 40 units of enzyme per milliliter in each HA solution, and the concentration of HA was different in each solution measured simultaneously by SMSLS. The fits shown yield $\dot{\beta}$ for each reaction.

v_{max} and K_M were found from a linear fit of the form:

$$\frac{1}{v} = \frac{1}{v_{max}} + \frac{K_M}{c v_{max}} \tag{14.4}$$

where the values of v were found from $\dot{\beta}$ via Equations 14.1 and 14.2. Figure 14.4 shows the data plotted in this form, for which a value of $K_M = 0.00142\,cm^3\,g^{-1}$ and $v_{max} = 1.6 \times 10^{-10}\,mol\,(l\text{-}s)^{-1}$.

14.2.3 Simultaneous Monitoring of Polymerization Reactions

Under certain conditions, SMSLS can be used to monitor parallel polymerization reactions *in situ*. Qualitatively, the SMSLS signatures immediately indicate whether reactions

occur or not, whether there is an initial lag period, and how long the reaction takes until it stops. The signatures also provide estimates of the reaction rate, weight average molecular mass M_w, and its shape can help identify mechanistic aspects; for example, controlled versus free radical polymerization, presence of impurities, etc. The method is inherently adapted to small sample volumes.

The free radical polymerization of acrylamide under varying conditions was recently reported using SMSLS [17]. Figure 14.5 shows raw scattering data, expressed as I_R (cm^{-1}) for several simultaneous polymerization reactions.

Potassium persulfate and N,N,N',N'-tetramethylethylene diamine (TMEDA) Am concentrations were from 0.00875 to 0.035 g ml^{-1} and concentrations used for KPS and TMEDA were from 0.011to of 0.1 mol l^{-1}. The ratio of the catalytic initiator system KPS/TMEDA in each reactor was 1:1 in mol. The polymerization reactions were carried out, eight at a time in the SMSLS device, at $T = 25\,°C$ under constant nitrogen flow and constant stirring. The reaction conditions are shown in Table 14.1.

In each case in Figure 14.5, the experimental TDSLS signature fit well to a class of signatures in a "library of LS signatures" computed for polymer reactions at low to medium concentration [18]. The fits themselves are shown together with the data in Figure 14.5.

All reactions except AAA in Figure 14.5, at low concentration, show a monotonic rise to a plateau value that occurs under the action of A$_2$. Reactions AAA1 and AAA2

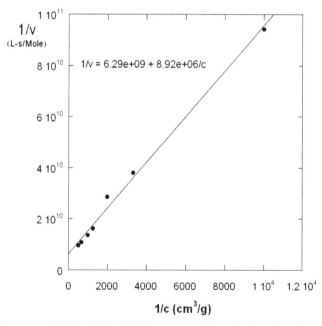

FIGURE 14.4 Michaelis–Menten analysis of the data from Figure 14.3 to determine the K_M and v_{max} of the hyaluronidase. Reprinted with permission from Drenski MF, Reed WF. Simultaneous multiple sample light scattering for characterization of polymer solutions. J Appl Polym Sci 2004;92:2724–2732.

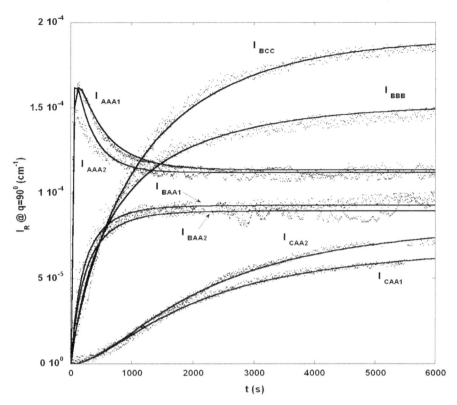

FIGURE 14.5   Absolute Rayleigh scattering ratio I_R from SMSLS during eight simultaneous Am free radical polymerization reactions. Reprinted (adapted) with permission from Drenski MF, Mignard E, Alb AM, Reed WF. Simultaneous *in situ* monitoring of parallel polymerization reactions using light scattering: a new tool for high throughput screening. J Comb Chem 2004;6:710–716. © 2004 American Chemical Society.

TABLE 14.1   Summary of Reactions in Figure 14.5

	Monomer (g ml^{-1} M^{-1})	Initiator (g ml^{-1} M^{-1})	Catalyst (g ml^{-1} M^{-1})
AAA	0.0355/0.5000	$9.03 \times 10^{-4}/0.33 \times 10^{-2}$	$3.87 \times 10^{-4}/0.33 \times 10^{-2}$
BAA	0.0088/0.5000	$9.03 \times 10^{-4}/0.\ 33 \times 10^{-2}$	$3.87 \times 10^{-4}/0.\ 33 \times 10^{-2}$
BBB	0.0088/0.1250	$2.26 \times 10^{-4}/8.33 \times 10^{-4}$	$9.68 \times 10^{-5}/8.33 \times 10^{-4}$
BCC	0.0088/0.0550	$1.00 \times 10^{-4}/3.70 \times 10^{-4}$	$4.30 \times 10^{-5}/3.70 \times 10^{-4}$
CAA	0.0039/0.5000	$9.03 \times 10^{-4}/0.33 \times 10^{-2}$	$3.87 \times 10^{-4}/0.33 \times 10^{-2}$

Some reactions were run in duplicate, in parallel.

show the characteristic maxima of I_R predicted to occur at intermediate concentration, due to the action of A_3, followed by decline to a plateau. Two fitting parameters were used in the fits shown; a M_w, assumed unchanging during conversion, and a first-order rate constant α for the conversion process. The fitted values were cross-checked with SEC and found to be in reasonable agreement. The fact that virtually all of the signatures are well fit with a single M_w suggests that M_w is largely constant throughout the reactions. The first-order rate constants α spanned the range of 6×10^{-5} to 2.8×10^{-3} s^{-1}. The higher the initiator concentration, the faster the rate, although reactions with high initiator (A) and low monomer (C) are intermediate in value, possibly because a smaller fraction of monomer is radicalized during initiator decay on

account of its lower concentration. It is noted that while SEC provides a useful cross-check on SMSLS, it is silent on kinetics, unless one were to laboriously make manual aliquot extractions during the reaction and make a SEC analysis for each.

It is notable that reactions CAA1 and CAA2 in Figure 14.5 both show the characteristic initial upturn in I_R that occurs when initial impurities compete with monomers for free radicals resulting from the initiator decomposition. This is most likely to occur when there are impurities present, such as incompletely purged O_2, and when the monomer concentration is low.

Because of second, third, and higher virial coefficient effects, SMSLS applied to polymerization reactions is restricted to rather low concentrations; the maximum in Figure 14.5 was $C_{Am} = 0.035\,\text{g cm}^{-3}$. In contrast to this type of batch mode SMSLS, the automatic dilution feature of ACOMP allows it to monitor reactions at very high concentrations and viscosities.

14.3 PARTICULATES COEXISTING WITH POLYMER POPULATIONS

A very widespread issue in polymer production is the existence of nano-, micro-, and macroscopic particulates. In some cases, these particulates are integral to the process, such as bacteria in bioreactors, cell fragments in bioextraction, or latex particles in emulsion polymerization. In most cases, however, the particulates are unwanted and can be detrimental both to the product and the reactors. Such particulates include physical aggregates, branched and crosslinked microgels, microcrystals, impurity particles, and macroscopic coagula. The issue of noncovalent protein aggregation has been discussed in the preceding. Often, the polymer producer does not know when during a process such particulates arise, what the chief factors controlling their formation are, and how the population evolves in time.

A challenge for online monitoring hence involves the early detection and characterization of particulate populations amidst the polymers being produced. An immediate solution is the use of LS and other optical techniques. Because LS is exquisitely sensitive to even tiny populations of aggregates, it should be possible to achieve very early stage particulate detection. For more advanced and dense populations of particulates, simple turbidity measurements can be useful. Another approach outlined in the following takes an inexpensive, nonoptical approach using an analytical filtration method.

14.3.1 Heterogeneous Time-Dependent Static Light Scattering [19]

Measurements of LS from heterogeneous solutions containing coexisting populations of colloids and polymers can be used to distinguish and quantify the two populations. The main requisites in this approach are a flowing solution, a small scattering volume, and a detector sampling rate sufficiently fast to detect individual large scatterers passing through the scattering volume. By appropriate recognition of scattering peaks produced by individual colloid particles and recovery of the baseline scattering due to the polymer population, the colloid particle number density can be monitored in real time while the polymer population is characterized absolutely according to weight average molecular weight M_w, and z-average mean square radius of gyration $\langle S^2 \rangle_z$. Either or both populations may also be evolving in time, which is the origin of the term "heterogeneous

time-dependent static light scattering." HTDSLS may find process-based applications, such as in biotechnology reactors, biomedical assays, and polymer reactors where gels or crystallites are produced in addition to polymer chains.

The basic HTDSLS scenario is that there will be a homogeneous population of polymer or small colloids, together with a much less numerous population of large particles (LPs). All else being equal, particle scattering at low angles increases as the sixth power of a solid particle's radius. Hence, for example, a single solid particle of 1 micron radius could scatter as much as 10^{18} enzymes of radius 1 nm. This scattering from a single cell might be much larger than the entire scattering from all the enzymes in the scattering volume. But the cell diffuses about, whereas the enzyme background remains essentially constant as long as there are a large number of enzymes in the scattering volume. Hence, when a cell diffuses into the volume there will be a large increase in scattering intensity, which will then disappear when the cell diffuses out of the volume. This increase in intensity may appear as a spike, or a more irregular shape. The profile of light scattered from the cell can be guaranteed to look like a spike if there is relative motion between the sample solution and the incident light beam used in the scattering experiment. In fact, using flowing solutions containing LPs to produce spikes has already been exploited in single particle detection.

As amply described in preceding chapters, the Rayleigh–Debye–Zimm approach is highly successful for characterizing dilute macromolecular solutions. In contrast, the characterization of the LPs in its simplest form is to merely count the particles via the scattering spikes they produce, in order to ascertain the particle number density. Beyond this, however, the actual magnitude and angular distribution of scattered light will rest upon the details of the Mie scattering involved [20]. This theory is based on the exact solution to Maxwell's equations for the particular particle geometry and composition involved. For spherical particles, the absolute, angular-dependent scattering function depends only on the ratio of λ/R, where R is the particle radius, and of n_p/n_s where n_p and n_s are the particle and solvent indices of refractions, which may be complex (i.e., involve light absorption in addition to refraction). Hence, to progress beyond particle counting for the LP requires model-dependent assumptions about the scatterers. For many scatterers, such as bacteria and yeast, the assumption of sphericity may be reasonable.

For spherical scatterers, the angular intensity in the plane perpendicular to the direction of polarization of the incident light is:

$$I_s\,(\text{horizontal plane}) = \frac{\lambda^2}{4\pi^2 r^2}\left|S_1^2\right| \qquad (14.5)$$

where

$$S_1 = \sum_{n=1}^{\infty} \frac{2n+1}{n(n+1)} \left\{ a_n \pi_n(\cos\theta) + b_n \tau_n(\cos\theta) \right\} \quad (14.6)$$

and

$$\pi_n(\cos\theta) = \frac{P_n(\cos\theta)}{\sin\theta}, \text{ and } \tau_n(\cos\theta) = \frac{dP_n(\cos\theta)}{d\theta} \quad (14.7)$$

where $P_n(\cos\theta)$ are the Legendre polynomials, and

$$a_n = \frac{\psi_n(\alpha)\psi_n'(\beta) - m\psi_n(\beta)\psi_n'(\alpha)}{\varsigma_n(\alpha)\psi_n'(\beta) - m\psi_n(\beta)\varsigma_n'(\alpha)} \quad (14.8a)$$

$$b_n = \frac{m\psi_n(\alpha)\psi_n'(\beta) - \psi_n(\beta)\psi_n'(\alpha)}{m\varsigma_n(\alpha)\psi_n'(\beta) - \psi_n(\beta)\varsigma_n'(\alpha)} \quad (14.8b)$$

where

$$\alpha = \frac{2\pi n_2 a}{\lambda}, \quad \beta = \frac{2\pi n_1 a}{\lambda}, \text{ and } m = \frac{n_1}{n_2} \quad (14.9)$$

$$\psi_n(\alpha) = \sqrt{\frac{\alpha\pi}{2}} J_{n+1/2}(\alpha) \quad (14.10)$$

where $J_{n+1/2}$ is the Ricatti–Bessel function. The $\Psi_n(a)$ obey the recursion relations [21]:

$$\psi_{n+1}(x) = \frac{2n+1}{x}\psi_n(x) - \psi_{n-1}(x) \quad (14.11a)$$

$$\psi_n'(x) = \frac{-n}{x}\psi_n(x) + \psi_{n-1}(x) \quad (14.11b)$$

Finally,

$$\varsigma_n(\alpha) = i^{(n+1)}e^{-i\alpha}, \text{ and } \varsigma_n'(\alpha) = i^n e^{-i\alpha} \quad (14.12)$$

where $i = (-1)^{1/2}$.

The assumption of single scattering—that is, that a photon only scatters once before being detected—is implicit in the aforementioned theories. Hence, there is a limit to the turbidity of a solution whose scattering can be analyzed by these theories. Being based on these theories, the HTDSLS technique presented in this work likewise requires that only single scattering occur.

14.3.1.1 Summary of Other Particle Counting and Sizing Techniques

As early as the 1930s, Moldovan [22] described a photoelectric technique to count particles flowing through a capillary. Many techniques employ light for investigation using blockage, and static and dynamic scattering [23]. These are useful over a broad range of sizes from 10 nm to ~1 mm. Other techniques include chromatography and those based on resistive or electrical measurements [24, 25]. In the latter case, particles suspended in an electrolyte pass through a sensing aperture and produce pulses, their height being proportional to the particle volume. Gear and Bednarek [26] found that cells with diameters as low as 0.45 μm can still be accurately measured. Bryant et al. ([27] found a volume of 0.62 μm^3 for *E. coli* cells, using the electrical method.

As discussed earlier, scattering from LPs requires the use of Mie theory. Therefore, particle sizing LS instruments measure scattering at angles as small as possible. For the investigation of shape and orientation, either the state of polarization of the scattered light [28] or its spatial distribution should be assessed [29]. They can be separated into different methods, which are investigations of polar scattering in a single azimuthal plane, azimuthal scattering at a single polar angle, and combinations of the two methods. The LS intensity measurements in a single plane are most commonly used. For instance, Gebhart et al. [30] developed a low angle system ($\theta < 7.5°$) with a measurement volume of 0.01 cm^3 allowing particle concentrations of 10^4 to be detected with a lower detection limit of 0.17 μm. An example for a completely automated device is the Hiac/Royco Optisizer, which measures particles with sizes from 0.5 μm to 350 μm [32]. For particles up to 1.63 μm, forward scattering is used, and above this size, extinction measurements. The system provides a constant flow of 0.01–100 ml min^{-1} and dilution from highly concentrated solutions up to 10.000:1 to avoid saturation.

Research on unicellular microorganisms by Wyatt [31] led to the development of a new device. Limited discrimination of particle shape has been shown to be possible using opposed detectors by comparing the detector output of pairs at the same angle [32]. A device allowing the measurement of LS for almost the complete 360° was described by Marshall et al. [33]. There the scattered light is intercepted by an ellipsoidal mirror which directs the light through a rotating aperture (360° in 20 ms) to a photomultiplier. Measurement errors lie within 1% for particles 11 μm > d > 0.3 μm.

Wyatt et al. [34] introduced a device that has a spherical measurement chamber with 18 small apertures for detectors on each of four azimuthal planes set equidistant apart by 45°. In addition, two large apertures at 25° and 155° are available at the same azimuthal plane. Hirst and Kaye [35] developed a system with a CCD camera, allowing 83% of the complete spatial scattering profile to be assessed.

Holve and Self [36], who developed a low angle scattering system, discuss in detail the whole range of difficulties associated with stray light, nonuniform laser beam, and scattering volume. They also analyze coincidence effects, which can also be a problem with the Coulter counter method [37].

Flow cytometry allows information to be obtained on subcellular components, and multiple methods are combined to rapidly assess cells (up to 3×10^5 min^{-1}) [38]. Analysis techniques are usually based on electrical and optical methods, such as electronic cell volume, fluorescence from stains bound to cells, LS, and extinction. In addition, cells are often separated from heterogeneous population using a cell sorting technique. Flow cytometry was used to investigate the influence of plasmids in *E. coli* cells [39]. LS measurements were used to determine the size distribution and fluorescence measurements on stained cells in order to assess their DNA content. Cross and Latimer [40] measured the angular dependence of such cells between 10 and 90° and the scattering curves were in good agreement with a modified Rayleigh–Debye approximation when the cells where modeled as coated ellipsoids. For the model, the typical dimensions for the cells ($l=2.16\,\mu$m and $d=0.74\,\mu$m) were determined by microphotography and literature values for the volume. Concentration determination of bacteria includes direct counting by light microscopy, spectrophotometry, conductivity (such as with the Coulter Counter^{tm}), and measurement of colony forming units per milliliter as determined by spreading diluted samples on agar plates and counting the colonies that form.

14.3.1.2 *Requirements and Parameters for HTDSLS*   The chief requirements for making HTDSLS feasible are the following:

1. The scattering volume V_s must be such that the average number of LP in the scattering volume per sampling period is less than one. For very sparse LP populations the scattering volume may be quite large, whereas for a high density of LP, the scattering volume must be small. The diffraction limit of focusing Gaussian laser beams will define the lowest practical scattering volume if focusing optics are used.

2. In order to form spikes from the LP, it is necessary that there be relative motion between the LP and the beam, although in some instances, pure diffusional motion of particles in and out of the scattering volume may be sufficient to form spikes. Relative motion is most easily provided by flowing the sample through a flow cell with a fixed laser beam and fixed photodetector(s).

3. Data sampling must be fast enough to resolve the LS peaks produced by LP flowing through V_s with a given velocity, v.

The requirement that there be a significant amount of time during which there are no particles in the scattering volume can be treated as follows: The time during which

there are no LP in the scattering volume is termed "clear window time" (CWT). CWT is also necessary to recover the baseline scattering from the homogeneous population of scatterers. The binomial distribution allows the fraction of time. There are k particles in the scattering detection volume, V_s, where $m=V_s/V_p$, to be computed for a given number concentration n, of randomly distributed LP, each of volume V_p, and n is the number density of LP:

$$P(k,m,n)=(nV_p)^k(1-nV_p)^{m-k}\frac{m!}{(m-k)!k!} \qquad (14.13)$$

The fraction of CWT is the probability of finding *no* randomly distributed LP in the scattering volume $P(0,n,V)$. For $m \gg 1$, which will normally hold because the LP are normally much smaller than V_s, this is approximated, based on the aforementioned equation, as:

$$P(0,m,n) \cong \exp(-nV_s), \quad m \gg 1 \qquad (14.14)$$

In other words, the fraction of CWT decreases exponentially with the size of the scattering volume for a given n, and with increasing n at fixed V_s. Detection of single LP with recovery of the background scattering level is assured only when there is an average of much less than one particle in the scattering volume, that is:

$$nV_s \ll 1 \qquad (14.15)$$

This is the criterion that dictates the range of LP concentrations measurable by the technique for an optically defined V_s for a given instrument. As long as this condition holds, each particle passing through the beam will produce an individual, resolvable spike (in fact, this condition does not have to be rigorously met, and Eq. 14.15 allows the probability to be assessed that a scattering spike is due to two or more particles simultaneously in the scattering volume). The absolute concentration of scatterers per unit volume, n, can be found from the number of LP LS spikes per time interval, with knowledge of flow velocity v and of V_s.

14.3.1.3 *Simultaneous Measurement of Bacterial and Polymer Populations*   Reference [19] used HTDSLS to simultaneously measure the increasing number density of *E. coli* bacteria in a broth at $T=38\,°$C, together with a steady population of poly(vinylpyrrolidone) polymers (PVP) of $M_w=650,000$ g mol^{-1} and concentration 0.1 mg ml^{-1} in a nutrient broth. Control tests with and without PVP showed that the *E. coli* growth rate is not measurably affected by PVP.

Figure 14.6 shows examples of 100 s swaths from the HTDSLS data continuously recorded for 26,000 s for this experiment. Data are from the 90° detector, where the

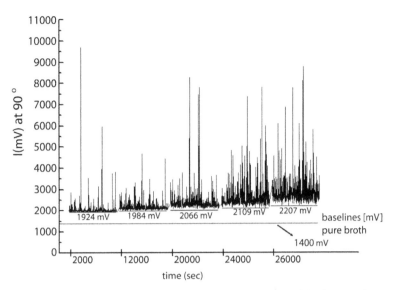

FIGURE 14.6 The HTDSLS data for the growth of *E. coli* bacteria in a nutrient broth which also contains a stable population of polymer (PVP). Examples of 100 s swaths of data are shown from the 26,000 s experiment. The scattering spikes are due to individual bacteria passing through the scattering volume. The increase in their density shows the increase in bacterial number density as the population grows. The baseline scattering remains essentially constant and corresponds to the stable PVP polymer. Reprinted (adapted) with permission from Schimanowski R, Strelitzki R, Mullin DA, Reed WF. Heterogeneous time dependent static light scattering. Macromolecules 1999;32: 7055–7063. © 1999 American Chemical Society.

sampling rate was 20 Hz and the flow rate was 0.0125 cm³ s⁻¹. The scattering level above the pure broth LS baseline was due to the population of PVP. The swaths shown are separated from each other by thousands of seconds. The spikes in the first swath are due to the original *E. coli* with which the broth was inoculated. The increasing density of spikes in the succeeding swaths shows how the bacterial population density n is increasing. The baseline due to the PVP stays constant throughout the experiment, showing that the PVP is staying intact. Were there a polymer that was either hydrolyzed or aggregated by the bacteria, then this baseline would decrease or increase, respectively, during the experiment. By the end of the experiment, the baseline has risen slightly. This is due to the fact that V_s is starting to saturate with bacteria (LP) and the criterion in Equation 14.15 begins to fail.

Figure 14.7 shows the absolute particle concentration of the *E. coli* as determined from the spike frequency count and knowledge of the flow and HTDSLS cell characteristics. There is apparently an incubation period before the cells begin to propagate. The characteristic exponential rise time, once the exponential growth region begins, was 44 min. Toward later times, the saturation of the scattering volume is apparent. At full saturation, there are no peaks to distinguish, and F plunges to 0.

The inset of Figure 14.7 shows the scattering recovered from the PVP. Using a linear fit gives $M_w = 644,000$ g mol⁻¹ and $R_g = 437$ Å, which are in good agreement with the values of PVP measured with no LP present.

A separate spectrophotometric determination of the *E. coli* growth rate was made. The doubling time was the same as determined by HTDSLS. There was a striking difference of nearly four orders of magnitude between the low particle densities that can be determined by HTDSLS and the lowest densities needed for optical density measurements. HTDSLS also gave the characterization of the coexisting polymer, to which optical density measurements are insensitive.

14.3.2 Filtrodynamics

While HTDSLS has been described in detail and other methods for particulate detection mentioned in the preceding, another practical approach which is relatively inexpensive and involves no sensitive optics is the platform recently termed "filtrodynamics." The essential notion of this is that filters act like resistors to flow and hence create pressure drops. As a filter plugs due to material being caught by it, its resistance increases which is measurable by the increased pressure. While there is nothing new in this idea, the novelty begins when one considers specific models of how a filter's resistance increases in response to particulate populations and how the mathematical signature of time-dependent pressure signals can deliver key information about the presence, characteristics, and evolution of particulate populations.

For example, as long as there are no particulates in a filtered flow, the pressure across the filter will remain constant. Only when particulates begin to appear in the flow will the

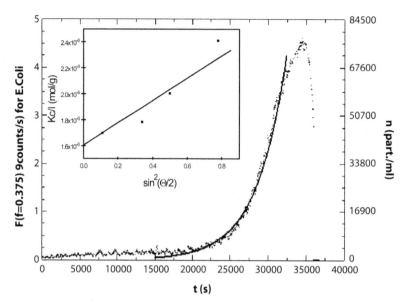

FIGURE 14.7   The SMSLS spike spectra data of Figure 14.6 analyzed to yield the bacterial population density growth. In the inset: the multiangle Zimm-type representation for the coexisting PVP polymer at fixed concentration. Reprinted (adapted) with permission from Schimanowski R, Strelitzki R, Mullin DA, Reed WF. Heterogeneous time dependent static light scattering. Macromolecules 1999;32: 7055–7063. © 2004 American Chemical Society.

pressure begin to change. Hence, monitoring transfilter pressure is the first simple step in early detection of particulates. The way the particulates interact with the specific filter will then determine the time signature of the pressure increase. Putting together multiple filters in series can lead to a histogram-type separation of the population and how it changes in time. If used with an ACOMP type front-end dilutions can be made to any level desired to balance filter sensitivity with its useful lifetime.

If filters in series are used in conjunction with a pump delivering a constant flow rate, this is equivalent to a constant current source in an electronic circuit, and the pump pressure will rise accordingly as each filter plugs at its own rate. Pressure sensors across each filter in the series will monitor the increase in pressure (resistance) for each. Similarly, if a pump at constant pressure is used in conjunction with filters in parallel, the flow rate through each filter, measured by a millifluidic flow meter, will be indicative of the increasing resistance in each.

In addition to monitoring particulate populations, such arrangements can also be used to protect downstream instrumentation from damaging particulates. Following the mathematical form of the pressure signals can allow predictions of when filter plugging will occur and changeover to another flow path will be required. In fact, using a pressure threshold, the flow can automatically be routed through successive filtration paths in parallel to ensure continuous plugging-free operation.

A wide diversity of filter types exist—membranes, sintered metallic frits, 3D filters—and many types of materials are used. Likewise, particulates can be of widely different types; soft microgels, hard microcrystals, biological cells, and aggregates

with ill-defined morphology. Hence, each filter/particulate system must be characterized as to pressure and resistance characteristics before being placed into service.

Despite the particularity of each filter/particulate system, some general characteristics can be defined and then applied to specific cases. For example, 2D (e.g., membrane filters) and 3D filters (e.g., sintered filters, cylindrical fiber filters, etc.) have distinguishable properties, and "deterministic" versus "probabilistic" mechanisms of trapping particles lead to different mathematical signatures of pressure versus time [41].

Figure 14.8 shows early data from filtrodynamic experiments. The system used $2\mu m$ latex spheres in water passing through $0.45\mu m$ PTFE membrane filters at a constant flow rate of $0.8\,ml\,min^{-1}$. The flow rate was maintained constant against the rising pressure by using a high-pressure liquid chromatography pump. The solution containing the spheres was loaded into a high-volume, 25-ml injection loop, and this plug volume passed through the filter in 1900 s.

The main part of Figure 14.8 shows the transmembrane pressure plus 1 bar versus the time that the latex sphere solution flows through the filter membrane for selected concentrations (the system outputs 1 V when there is no pressure). Since the plugging of the filters is, ideally, proportional to the number of particles stopped by the filter, the early phase of plugging should be proportional to the concentration of particles, when the pump flow rate is held constant. The early phase of the pressure increase versus time is linear and it is the slope of the early linear phase, in $bar\,s^{-1}$, that is shown in the inset to Figure 14.8.

The inset to Figure 14.8 shows that the rate of pressure change is linear over the range of 1×10^{-4} up to $2 \times 10^{-3}\,mg\,ml^{-1}$.

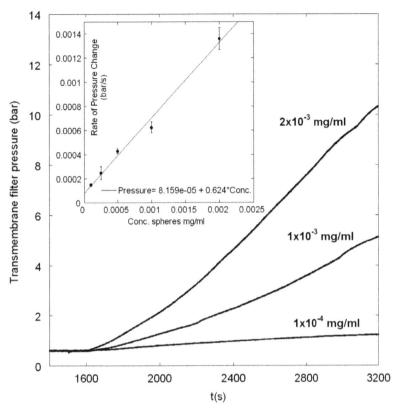

FIGURE 14.8 Pressure versus time for solutions of latex spheres in water, at three concentrations, flowing through a 0.45-µm PTFE membrane at 0.8 mg ml^{-1} (main panel). The initial linear rate of pressure increase across the membrane filters due to the spheres (inset to figure). There is good linearity between the rate of pressure increase and the concentration of spheres (data courtesy of Claiton Brusamarello).

The nonzero slope in the linear fit to the inset reflects the fact that, for this particular type of filter, pressure builds up slowly in time even when pure water flows. This is perhaps due to slow swelling of the membrane material and/or to residual "dust" and other stray particles in the nominally pure water. The experiments show good repeatability of the initial linear slope of pressure versus time, which is reflected in the error bars.

Dual function filtrodynamics, to both characterize the presence, onset, and evolution of particulates that occur in polymerization processes and protect instrumentation through which diluted polymer reactor solution flows (e.g., as in ACOMP), is currently under development. Challenges include delineating which types of filters work best with given particulate systems (e.g., microgels from natural product solutions, or from microgels occurring in polymerization reactions in emulsions and inverse emulsions, etc.)

14.4 AUTOMATIC CONTINUOUS MIXING (ACM)

Automatic continuous mixing (ACM) [42] is an important characterization technique used in studying equilibrium and quasi-equilibrium processes.

Traditional polymer characterization methods involve a labor-intensive sequence of preparing solutions of several concentrations, filtering or centrifuging them to remove dust and other scattering impurities, using several scattering cells, and gathering scattering data on each sample. The automated technique presented here has multiple advantages: reduces sample preparation to one stock polymer concentration, determines accurately the polymer concentration by using an online concentration detector, uses the same scattering chamber for all concentrations, provides many more concentration points for increased accuracy, simultaneously provides information on polymer viscosity, uses less polymer sample, and is less labor intensive.

Automatic continuous mixing includes a programmable mixing pump, operating in step gradient mode or continuous gradient mode, connected to a similar detector train as the ACOMP system, with multiangle LS, viscometry, refractometry, and UV detection. It can be used in two modes, gradient and recirculation mode, as illustrated in Figure 14.9.

14.4.1 Polymer Characterization—Automated Zimm

One of most common ACM applications is polymer characterization in terms of weight average molecular weight M_w, second- and third-order virial coefficients, A_2 and A_3, intrinsic

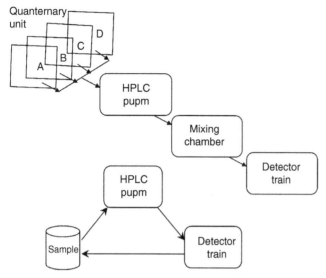

FIGURE 14.9 Automatic continuous mixing (ACM) schematic setup: gradient mode (top) and in recirculation mode (bottom).

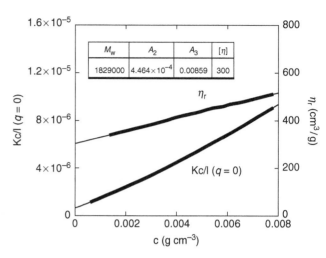

FIGURE 14.10 The change of $Kc/I(q=0)$ and reduced viscosity η_r as the polymer concentration is continuously ramped during an ACM experiment made for an aqueous polyacrylamide solution. Second-order polynomial fits to data allowed M_w, A_2, A_3, $[\eta]_w$ to be determined (unpublished data, Alina Alb).

viscosity, $[\eta]_w$, simultaneously obtained by automatic dilution of a polymer solution into the solvent [42]. Additional information on macromolecular structure, conformation, aggregation, and copolymer composition is also obtained. Compared to the traditional Zimm method, ACM offers greater accuracy (provides a continuum of points), reduces sample preparation time, and offers complete information on the polymer sample analyzed.

Equilibrium characterization of polymers usually takes place in dilute solution where $1 >> 2A_2cM_w >> 3A_3c^2M_w$, and over an angular range such that $q^2<S^2><1$. As seen in Chapters 11–13, in this case, Zimm equation reduces to one of its most frequently used forms:

$$\frac{Kc}{I(c,q)} = \frac{1}{M_w}\left(1+\frac{q^2<S^2>_z}{3}\right)+2A_2c \qquad (14.16)$$

where $I(c,q)$ is the excess Rayleigh scattering ratio, that is, the scattering from the polymer solution minus the scattering from the pure solvent, q is the scattering vector, $q=4\pi n/\lambda$ $\sin(\theta/2)$, $(<S^2>)^{1/2} \equiv R_g$ is the root mean square radius of gyration, K is an optical constant, $K=4\pi^2n^2(dn/dc)^2/N_A\lambda^4$ (for vertically polarized incident light), n is the solvent index of refraction, λ is the vacuum wavelength of the incident light, N_A is Avogadro's number, and dn/dc is the differential refractive index for the polymer.

In this regime, reduced viscosity, η_r is given by:

$$\eta_r = [\eta]+\kappa_H[\eta]^2 c \qquad (14.17)$$

where $[\eta]=\lim_{c\to0}\eta_r$ is the intrinsic viscosity, c is the polymer concentration, κ_H is Huggins' constant related to

the hydrodynamic interactions between polymer chains, usually around 0.4 for neutral, coil polymers.

An illustration of the capabilities of the method is given in Figure 14.10, which shows the evolution of $Kc/I(q=0)$ and reduced viscosity η_r as the polymer concentration is continuously ramped during an ACM experiment made for an aqueous polyacrylamide solution. Second-order polynomial fits to data allowed the determination of M_w, A_2, A_3, and $[\eta]_w$.

14.4.2 Effect of the Ionic Strength—Polyelectrolyte Effect

It is well known that dimensions of charged polymers and $[\eta]$ are affected by intramolecular electrostatic interactions [43]. In this context, the use of the ACM allowed detailed measurements of electrostatic effects in the copolyelectrolyte solutions to be made while varying ionic strength and polyelectrolyte concentration [44–47]. The evolution of second and third virial coefficients, A_2 and A_3, the angular scattering envelope, and the reduced viscosity were determined simultaneously.

In all these studies, ACM allowed, with the massive data provided, the behavior of the polyelectrolyte conformational and interaction parameters to be followed at a level of detail previously unobtainable by manual gathering of individual data points. Correlations were made between experimental data and of electrostatic persistence length and electrostatic excluded volume theories, with no adjustable parameters.

Figure 14.11 (lower part of the figure) shows viscosity and LS (90°) data versus time during a typical ACM experiment (poly(vinylbenzyl sulfonic acid), PVB) in which $NaNO_3$ was added at fixed polymer concentration (1 mg

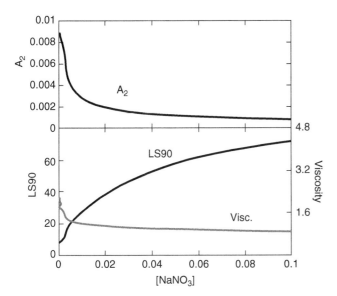

FIGURE 14.11 Effect of ionic strength for poly(vinylbenzyl sulfonic acid), PVB, in which $NaNO_3$ was added at fixed polymer concentration (1 mg ml^{-1}): LS 90° and viscosity versus [$NaNO_3$] (bottom); second-order virial coeffiecient, A_2 versus [$NaNO_3$] (top) (unpublished data, Alina Alb).

ml^{-1}). As soon as [$NaNO_3$] ramp began, expected trends were seen in the evolution of both electrostatic and hydrodynamic polymer characteristics: The LS increases since the electrostatically enhanced A_2, which suppresses scattering, decreases as ionic strength increasingly shields the charges on the polymer chains, whereas the viscosity drops as the shielding leads to a contraction in the PVB coil size. The upper part of the figure shows A_2 versus [$NaNO_3$], determined from the experiment.

14.4.2.1 *Effect of Valence and Chemical Species of Added Electrolyte*

Ion specificity is frequently encountered in charged macromolecular and colloidal structures. In particular, ion specificity is commonly found in many biological systems, where ions play a special role, apart from their mere valence [48].

An investigation on the effect of added electrolyte valence and species on polyelectrolytes studied using ACM was made in Reference [46]. First-order differences in polyelectrolyte conformations, interactions, and hydrodynamics were reported by the authors when electrolytes of different valence and symmetry were added to the polyelectrolyte solution. Each relevant characteristic, however, followed the same scaling relationship to ionic strength but the prefactors were different. Electrostatic persistence length/electrostatic excluded volume calculations without adjustable parameters suggested that the effective linear charge density is considerably lower in the presence of divalent ions than monovalent ions. Consideration of the

nonlinear Poisson–Boltzmann equation showed that charge distributions, and ultimately potential, are not merely a function of ionic strength but depended on the details of the electrolyte valence and symmetry. The asymmetric divalents gave a tighter shielding of the polyanion charge, leading to lower potentials.

Second-order effects in these properties were seen when the chemical identity of the anion was changed for a given symmetry and valence class, and yet weaker second-order effects were seen when changing the cation in a given class, effects rationalized in terms of Hoffmeister series [49].

14.4.2.2 *Copolyelectrolytes (Copolymeric Polyelectrolytes)*

Copolyelectrolytes (copolymeric polyelectrolytes) are copolymers in which at least one of the comonomers is electrically charged. They combine the functional characteristics of copolymers with the unusual properties of polyelectrolytes.

An example of the way the incorporation of a charged monomer in a copolymer chain affects the copolymer behavior, in terms of polyelectrolyte behavior and chain stiffness, is presented in the following [47].

At low ionic strength, polyelectrolytes display an extended rodlike conformation in aqueous solution due to electrostatic charge repulsion along the polymer backbone and rearrangement of counterions. A convenient means to assess the flexibility of the copolymer chains is given by the use of the apparent persistence length, L'_p [50], composed of an intrinsic persistence length and an electrostatic persistence length, which combines electrostatic stiffening and excluded volume in a single parameter, allowing chains of different mass to be compared. This topic is treated in detail in Sections 5.2.5 and 5.5.

Li and Alb studied in Reference [47] the impact that diallyldimethylammonium chloride (DADMAC) incorporation makes on the polyelectrolyte behavior and on the copolymer conformation, in the case of DADMAC/dimethyl acrylamide (DMA) copolymers. A_2 and L_p were computed from ACM experiments in H_2O and in 0.1 M NaCl made on samples withdrawn, at different stages in conversion, during copolymerization reactions with different initial comonomer composition. L'_p values were computed based on the wormlike chain expression for a polymer chain with no excluded volume (Eq. 5.26; [50]). It was found that the addition of NaCl led to the decrease of the copolymer apparent persistence length, resulting in a more compact random coil-like conformation of the polymer chains, similar to neutral polymers. This was also seen in the evolution of A_2 with increase of ionic strength, due to the shielding of electrostatic interactions as an effect of the reduction in the Debye screening length [51]. In these conditions (NaCl 0.1 M), varying the amount of DADMAC in the copolymer composition did not have significant effects on A_2 and L_p.

14.4.3   Effect of the Solvent on Polymer Conformation

Behavior of the polymer in terms of chain conformation, radius of gyration, and other properties can be influenced also by the change of solvent.

The effect of solvent on the polymer conformation is illustrated in Figure 14.12, in the case of poly(2-(dimethylamino) ethyl acrylate), poly(DMAEA) solutions in water and DMF [52]. The almost zero slope in Kc/I versus $\sin^2(\theta/2)$ for the polymer solution in DMF, expected for small, well-dissociated polymer chains, measures the small radius of gyration (R_g) of the polymer in this solvent, close to detection limits. In contrast, high R_g values were obtained for poly(DMAEA) solution in water; moreover, the upward curvature in Kc/I versus $\sin^2(\theta/2)$ indicates strong chain association. At the same time, the intercept of Kc/I data at $q=0$ gives the polymer mass, which is much larger in water than in DMF, due to the association of the amphiphilic polymer chains in aqueous media.

14.4.4   Interactions between Polymers and Surfactants

The interaction between polymers and surfactants has also been the subject of fundamental and applied research, including both neutral and charged polymers and surfactants [53–56].

There is a general consensus that in the case of the association between the two components, micelles begin to associate with the polymer around a critical aggregation concentration of surfactant, until saturation occurs and no more surfactant binds to the polymer.

In the following case study [57], interaction between neutral polymers—poly(vinylpyrrolidone (PVP) and polyethylene oxide (PEO)—and surfactants of different charge, dodecyl trimethyl ammonium chloride (DTAC) and sodium dodecyl sulfate (SDS),was investigated, with emphasis on the effects that the presence of the polymer has on the LS phenomena near the critical micelle concentration (CMC) of the surfactants.

An increase in LS as surfactant concentration approaches CMC, followed by a decrease as the concentration is raised was observed for both cationic and anionic surfactants. The increase in the LS versus surfactant concentration, sharp enough to be considered an "LS peak," occurs as the concentration of surfactant in aqueous solution approaches and passes the CMC. The angular dependence of LS showed the peaks are due to massive structures that only exist in the LS peak regime. Similar phenomena have been observed by others and different explanations were proposed, although the prevailing factor appears to be the release of hydrophobic impurities as the concentration of surfactant approaches and drops below the CMC [58, 59].

Light scattering peaks around CMC were observed for both surfactants in Reference [57]. At lower and higher concentrations than CMC, solutions were rather stable; around CMC, LS data showed a long time evolution until equilibrium was reached. The LS behavior when surfactant micelles associate with neutral polymers was investigated. While association of SDS with PVP and PEO was observed at different concentration regimes, no association of the polymers with DTAC was observed; however, "declustering" of the DTAC aggregates around CMC was suggested by LS and TEM experiments.

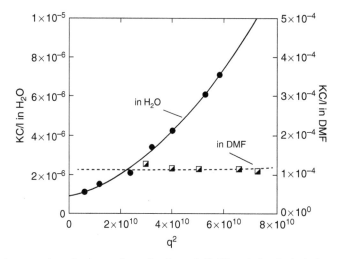

FIGURE 14.12   The effect of solvent on the polymer conformation for poly(2-(dimethylamino)ethyl acrylate), poly(DMAEA) solutions in water and in DMF: almost zero slope in Kc/I versus $\sin^2(\theta/2)$ ($\sim R_g$) for the polymer solution in DMF contrasting with high values for the solution in water; much larger mass in water than in DMF (from the intercept of Kc/I at $q=0$), due to the association of the amphiphilic polymer chains in aqueous media. From Li Z, Serelis AK, Reed WF, Alb AM. Toward amphiphilic diblock copolymers by RAFT. Kinetic study on pH responsive polymers. 239th ACS National Meeting; 2010 Mar 21–25, San Francisco; 2010.

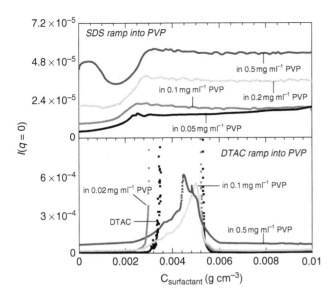

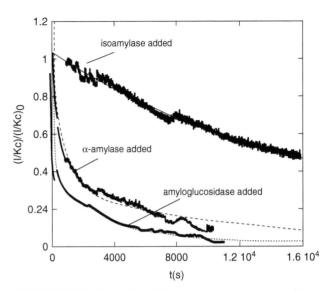

FIGURE 14.13 Contrasting features of the polymer–surfactant interaction for the PVP/DTAC and PVP/SDS systems: $I(q=0)$ versus c_{SDS} during SDS ramp (top) and during DTAC ramp (bottom), $c_{PVP} = 0.5 \, mg \, ml^{-1}$ [57].

FIGURE 14.14 Normalized I/Kc signatures as functions of time starting with the time the enzymes were introduced in the polymer solution. A two-step process is observed in the degradation process, when α-amylase and amyloglucosidase were used: a very fast rate for the first 10–15 min, followed by a slow part, which occurs during 2–3 h. In the case of isoamylase, the degradation reaction has a constant slow rate for the whole duration of the experiment (unpublished data, Alina Alb).

An example chosen to emphasize the contrasting features of the polymer–surfactant interaction for the PVP/DTAC and PVP/SDS systems is illustrated in Figure 14.13 [57], which shows $I(q=0)$ versus c_{SDS} during SDS ramp (top) and during DTAC ramp (bottom).

As seen in the top of the figure, in all cases, in the vicinity of the CMC of SDS, the addition of PVP led to the disappearance of the LS peak in the SDS solutions, fact attributable to binding of the surfactant molecules on the polymer chain. The association between SDS and PVP is seen in the evolution of the intensity of the light scattered; thus, as more SDS is introduced in the polymer solutions, an initial decrease in the LS data is observed, effect that is stronger in solutions of higher c_{PVP} ($0.5 \, mg \, ml^{-1}$). This decrease in LS, as mentioned earlier, is due to the fact that PVP becomes charged by SDS molecules which binds on the polymer chains, which increases A_2 leading to a decrease in LS as the surfactant concentration increases. Next, adding more SDS increases ionic strength of the solution and shields the charged PVP/SDS aggregates which counteracts the increase in A_2 leading to a subsequent increase in LS. On the contrary, when DTAC was added to PVP, no association effects were observed, only the LS peak around CMC was minimized due to an averaging between the large surfactant clusters and the small PVP coils or to a possible "declustering" of the aggregates.

14.4.5 Applications to Nonequilibrium Processes

The use of static LS to follow time-dependent systems includes degradation, aggregation, dissolution of dry and emulsified polymers, phase separation, etc. [60–62]. Even

though SMSLS is more suited to follow these processes, due to the high-throughput capabilities, ACM can be used, in a recirculated mode (Fig. 14.9), to monitor time-dependent processes. Two applications are presented next.

14.4.5.1 *Enzymatic Degradation* The effects of several enzymes on a modified starch (HICAP) sample were compared by monitoring the evolution of the solution recirculated through ACM detector train during the experiments. Three enzymes were added, at different concentrations, α-amylase ($50 \, U \, g^{-1}$), isoamylase ($28.5e6 \, U \, (mg \, prot)^{-1}$), and amyloglucosidase ($23,000 \, U \, g^{-1}$), into $1 \, mg \, ml^{-1}$ aqueous HICAP solutions.

Figure 14.14 shows normalized I/Kc signatures as functions of time starting with the time the enzymes were introduced in the polymer solution. A two-step process is observed in the degradation process, when α-amylase and amyloglucosidase were used: a very fast rate for the first 10–15 min, followed by a slow part, which occurs during 2–3 h. In the case of isoamylase, the degradation reaction has a slow rate for the whole duration of the experiment.

By following the polymer degradation, the experiments in the preceding allow comparison of the efficiency of the enzymes. It is also possible to obtain information about kinetics, that is, the duration of the degradation process and quantitative rate constants to be determined and structural analysis to be made.

14.4.5.2 *Polymer Dissolution*   The dissolution behavior of polymer powders is a problem of practical interest in various fields and led to extensive theoretical and experimental literature in response to the need to be able to control it [63–65].

Unusual behavior obtained when polyelectrolytes (sodium polystyrene sulfonate, NaPSS) are dissolved in pure water, in contrast with their straightforward behavior when dissolved at moderate ionic strength, was reported recently [66]. Thus, NaPSS, when dissolved in solutions of moderate ionic strength, showed dissolution behavior similar to that of neutral polymers. When dissolved in pure water, however, there was consistently a small population of aggregates that appeared at the beginning of the dissolution process, which then rapidly diminished.

Nonetheless, dissolution rates of NaPSS in both water and in the presence of ionic strength were found to be the same, and the dissolution kinetics followed the simple cubic time dependence predicted for a constant mass flux of chains from a dissolving polyelectrolyte grain, which led to the concentration of dissolved polymer $c_p(t)$ to be expressed as:

$$c_p(t) = c_0 \left[1 - \left(\frac{1 - kt}{R_0} \right)^3 \right], \quad kt \le R_0$$
$$c_p(t) = c_0, \quad kt > R_0 \tag{14.18}$$

where c_0 is the initial mass concentration of grains in solution, R_0 and ρ_0 are the grain radius and density, respectively, $k = J/\rho_0$, and J is the mass flux.

The experimentally measured $c_p(t)$ and Equation 14.18 allowed the normalized initial velocity to be determined:

$$v_{in} \equiv \frac{1}{c_0} \left(\frac{dc_p(t)}{dt} \right)_{t=0} = \frac{3k}{R_0} \tag{14.19}$$

The authors also showed evidence in the study reported in Reference [66] that aggregates seen upon dissolution in water were already present in the dry polyelectrolyte powder and represented a minor mass fraction of the entire population which, before dissolving, evolve in time from initially spherically symmetric Gaussian mass distributions into structures resembling random coils containing multiple chains.

The addition of NaCl into the solutions in water led to irreversible dissolution of the aggregates and avoids their formation when NaPSS was dissolved directly in NaCl solution.

A final comment made was that although the observations in Reference [66] may have general validity for linear polyelectrolytes and some branched ones, more complex polyelectrolyte morphologies, such as dendrimers, may have different mechanisms at work that produce slow modes.

14.5   SUMMARY

Several techniques addressing characterization challenges in studying equilibrium and quasi-equilibrium processes and in time-dependent processes apart from polymerization reactions, behavior of multicomponent systems, and the issue of particulates coexisting with polymers were introduced in this chapter, together with applications highlighting the most relevant features of the techniques.

REFERENCES

[1] Reed WF, Device and method of simultaneously measuring the light scattering from multiple liquid samples containing polymers and/or colloids. US Patent 6,618,144. 2003.

[2] Drenski MF, Reed WF. Simultaneous multiple sample light scattering for characterization of polymer solutions. J Appl Polym Sci 2004;92:2724–2732.

[3] Cromwell MEM, Hilario E, Jacobson F. Protein aggregation and bioprocessing. AAPS J 2006;8:E572–E579.

[4] Mahler HC, Friess W, Grauschopf U, Kiese S. Protein aggregation: pathways, induction factors and analysis. J Pharm Sci 2009;98:2909–2934.

[5] Chi EY, Krishnan S, Randolph TW, Carpenter JF. Physical stability of proteins in aqueous solution: mechanism and driving forces in nonnative protein aggregation. Pharm Res 2003;20:1325–1336.

[6] Roberts CJ. Kinetics of irreversible protein aggregation: analysis of extended Lumry-Eyring models and implications for predicting protein shelf life. J Phys Chem B 2003;107:1194–1207.

[7] Carpenter JF, Randolph TW, Jiskoot W, Crommelin DJA, Middaugh CR, Winter G, Fan YX, Kirshner S, Verthelyi D, Kozlowski S, Clouse KA, Swann PG, Rosenberg A, Cherney B. Overlooking subvisible particles in therapeutic protein products: gaps that may compromise product quality. J Pharm Sci 2009;98:1201–1205.

[8] Rosenberg AS. Effects of protein aggregates: an immunologic perspective. AAPS J 2006;8:E501–E507.

[9] Schellekens H. The immunogenicity of therapeutic proteins. Discov Med 2010;9:560–564.

[10] Li Y, Weiss IV WF, Roberts CJ. Characterization of high-molecular-weight nonnative aggregates and aggregation kinetics by size exclusion chromatography with inline multi-angle laser light scattering. J Pharm Sci 2009;98:3997–4016.

[11] Mirsky AE, Pauling L. On the structure of native, denatured, and coagulated proteins. Proc Natl Acad Sci USA 1936;22:439–447.

[12] Reed CE, Reed WF. Light scattering power of randomly cut random coils with application to the determination of depolymerization rates. J Chem Phys 1989;91:7193–7199.

[13] Reed CE, Reed WF. Effect of polydispersity and second virial coefficient on light scattering by randomly cut random coils. J Chem Phys 1990;93:9069–9074.

[14] Reed WF. Time dependent light scattering from single and multiply stranded linear polymers undergoing random and endwise scission. J Chem Phys 1995;103:7576–7584.

[15] Ganter JL, Sabbi J, Reed WF. Real-time monitoring of enzymatic hydrolysis of galactomannans. Biopolymers 2001; 59:226–242.

[16] Reed WF, Reed CE, Byers L. Random coil scission rates determined by time dependent total intensity light scattering: hyaluronate depolymerization by hyaluronidase. Biopolymers 1990;30:1073–1082.

[17] Drenski MF, Mignard E, Alb AM, Reed WF. Simultaneous *in situ* monitoring of parallel polymerization reactions using light scattering: a new tool for high throughput screening. J Comb Chem 2004;6:710–716.

[18] Alb AM, Mignard E, Drenski MF, Reed WF. In situ time dependant signatures of light scattered from solutions undergoing polymerization reactions. Macromolecules 2004;37: 2578–2587.

[19] Schimanowski R, Strelitzki R, Mullin DA, Reed WF. Heterogeneous time dependent static light scattering. Macromolecules 1999;32:7055–7063.

[20] Kerker M. *The Scattering of Light and Other Electromagnetic Radiation*. New York: Academic Press; 1969.

[21] Van de Hulst HC. *Light Scattering by Small Particles*. New York: John Wiley & Sons; 1957.

[22] Moldovan A. Photo-electric technique for the counting of microscopical cells. Science 1934;80:188–189.

[23] Nicoli DF, Kourti T, Gossen P, Wu J-S, Chang, Y-J, MacGregor JF. Latex particle size determination by dynamic light scattering. In: Provder T, editor. *On-Line Particle Size Distribution by Dynamic Light Scattering*. ACS Symposium Series 472. Washington, DC: American Chemical Society; 1991. p 86–97.

[24] Coulter WH. High speed automatic blood cell counter and cell size analyzer. Proc Natl Electron Conf 1956;12: 1034–1042.

[25] Kubitschek HE. Electronic counting and sizing of bacteria. Nature 1958;182:234–235.

[26] Gear ARL, Bednarek JM. Direct counting and sizing of mitochondria in solution. J Cell Biol 1972;54:325–345.

[27] Bryant FD, Seiber BA, Latimer P. Absolute optical cross sections of cells and chloroplasts. Arch Biochem Biophys 1969;135:97–108.

[28] Rochon P, Racey TJ, Zeller M. Use of light scattering to estimate the fraction of spherical particles in a mixture. Appl Opt 1988;27:3295–3298.

[29] Kaye PH. Spatial light-scattering analysis as a means of characterizing and classifying non-spherical particles. Meas Sci Technol 1998;9:141–149.

[30] Gebhart J, Bol J, Heinze W, Letschert W. A particle-size spectrometer for aerosols utilizing low-angle scattering of the particles in the laser beam. Staub-Reinhalt Luft 1970;30:5–14.

[31] Wyatt PJ, Phillips DT. New instrument for the study of individual aerosol particles. J Colloid Interface Sci 1972;39: 125–135.

[32] Diehl SR, Smith DT, Sydor M. Analysis of suspended solids by single-particle scattering. Appl Opt 1979;18:1653–1658.

[33] Marshall TR, Parmenter CS, Seaver M. Characterization of polymer latex aerosols by rapid measurement of 360° light scattering patterns from individual particles. J Colloid Interface Sci 1976;55:624–636.

[34] Wyatt PJ, Schehrer KL, Phillips SD, Jackson C, Chang YJ, Parker RG, Phillips DT, Bottiger JR. Aerosol particle analyzer. Appl Opt 1988;27:217–221.

[35] Hirst E, Kaye PHJ. Experimental and theoretical light scattering profiles from spherical and nonspherical particles. Geophys Res (Atm) 1996;101:231–235.

[36] Holve D, Self SA. Optical particle sizing for in situ measurements. Appl Opt 1979;18:1632–1652.

[37] Bader H, Gordon HR, Brown OB. Theory of coincidence counts and simple practical methods of coincidence count correction for optical and resistive pulse particle counters. Rev Sci Instrum 1972;43:1407–1412.

[38] Steinkamp JA. Flow cytometry. Rev Sci Instrum 1984;55: 1375–1400.

[39] Seo JH, Bailey JE. Cell cycle analysis of plasmid-containing *Escherichia coli* HB101 populations with flow cytometry. Biotechnol Bioeng 1987;30:297–305.

[40] Cross DA, Latimer P. Angular dependence of scattering from *Escherichia coli* cells. Appl Opt 1972;11:1225–1228.

[41] Reed WF. A device and method for monitoring the presence, onset, and evolution of particulates in chemically or physically reacting systems. US Patent Application PCT/US 0512/25041. 2003.

[42] StrelitzkI R, Reed WF. Automated batch characterization of polymer solutions by static light scattering and viscometry. J Appl Polym Sci 1999;73:2359–2368.

[43] McKee MG, Hunley MT, Layman JM, Long TE. Solution rheological behavior and electrospinning of cationic polyelectrolytes. Macromolecules 2006;39:575–583.

[44] Bayly EE, Brousseau J-L, Reed WF. Continuous monitoring of the effect of changing solvent conditions on polyelectrolyte conformations and interactions. Int J Polym Anal Char 2002;7: 1–18.

[45] Sorci GA, Reed WF. Electrostatically enhanced second and third virial coefficients, viscosity, and interparticle correlations for linear polyelectrolytes. Macromolecules 2002;35: 5218–5227.

[46] Sorci GA, Reed WF. Effect of valence and chemical species of added electrolyte on polyelectrolyte conformations and interactions. Macromolecules 2002;37:554–565.

[47] Li Z, Alb AM. Cationic copolyelectrolytes: online monitoring of the synthesis and subsequent copolymer characterization. Macromol Chem Phys 2012;213:1397–1406.

[48] Benegas JC, Cesaro A, Rizzo R, Paoletti S. Conformational stability of biological polyelectrolytes: evaluation of enthalpy and entropy changes of conformational transitions. Biopolymers 1998;45:203–216.

[49] Fixman M. Polyelectrolytes: a fuzzy sphere model. J Chem Phys 1964;41:3772–3778.

[50] Reed CE, Reed WF. Monte Carlo electrostatic persistence lengths compared with experiment and theory. J Chem Phys 1991;94:8479–8486.

[51] Landau LD, Lifschitz EM. Fluctuations. In: *Statistical Physics Part 1*. 3rd ed. Oxford: Pergamon Press; 1963.

[52] Li Z, Serelis AK, Reed WF, Alb AM. Toward amphiphilic diblock copolymers by RAFT. Kinetic study on pH responsive polymers. 239th ACS National Meeting; 2010 Mar 21–25, San Francisco; 2010.

[53] Hayakawa K, Kwak JCT. Interactions between polymers and cationic surfactants. In: Rubingh DN, Holland PM, editors. *Cationic Surfactants: Physical Chemistry*. Surfactant Science Series 37. New York: Marcel Dekker; 1991. p 189–248.

[54] Yoshida K, Dubin PL. Complex formation between polyacrylic acid and cationic/nonionic mixed micelles: effect of pH on electrostatic interaction and hydrogen bonding. Colloid Surf A 1999;147:161–167.

[55] Burke S, Palepu R, Hait SK, Moulik SP. Physicochemical investigations on the interaction of cationic cellulose ether derivatives with cationic amphiphiles in an aqueous environment. Prog Colloid Polym Sci 2003;122:47–55.

[56] Sorci GA, Reed WF. Electrostatic and association phenomena in aggregates of polymers and micelles. Langmuir 2002;18: 353–364.

[57] Zhu Z, Alb AM. Light scattering study of the contrasting interaction between ionic surfactants and neutral polymers. J Colloid Interface Sci submitted.

[58] Corti M, DiGiorgio V. Intensity-correlation study of the effect of dodecanol on sodium dodecyl sulfare aqueous solutions near the critical micelle concentration. Chem Phys Lett 1977;49:141–144.

[59] Baillet S, Grassl B, Desbrieres J. Rapid and quantitative determination of critical micelle concentration by automatic continuous mixing and static light scattering. Anal Chim Acta 2009;636:236–241.

[60] Wright LS, Chowdhury A, Russo P. Static light scattering instrument for rapid and time-resolved particle sizing in polymer and colloid solutions. Rev Sci Instrum 1996;67: 3645–3655.

[61] Norisuye T, Shibayama M, Nomura S. Time-resolved light scattering study on the gelation process of poly(N-isopropyl acrylamide). Polymer 1998;39:13, 2769–2775.

[62] Ganter JL, Reed WF. Real-time monitoring of enzymatic hydrolysis of galactomannans. Biopolymers 2001;59:226–242.

[63] Brochard F, De Gennes PG. Kinetics of polymer dissolution. Physicochem Hydrodyn 1983;4:313–322.

[64] Peppas NA, Wu JC, von Meerwall ED. Mathematical modeling and experimental characterization of polymer dissolution. Macromolecules 1994;27:5626–5638.

[65] Wang Q, Ellis PR, Ross-Murphy SB. Dissolution kinetics of guar gum powders. I. Methods for commercial polydisperse samples. Carbohydr Polym 2002;49:131–137.

[66] Michel RC, Reed WF. New evidence of the nonequilibrium nature of the "slow modes" of diffusion in polyelectrolyte solutions. Biopolymers 2000;53:19–39.

15

OUTLOOK FOR INDUSTRIAL ACOMP

MICHAEL F. DRENSKI AND ALEX W. REED

15.1 INTRODUCTION

As previously discussed in Chapters 11–13, the automatic continuous online monitoring of polymerization reactions (ACOMP) technology has been used successfully to monitor a wide range of polymerization reactions in work dating back to 1998 [1]. Numerous examples have been given in the previous three chapters (Chapters 11–13) [2–7].

ACOMP has also been used in extensive work with the polymer manufacturing industry. In the latter, ACOMP has been used to monitor bench-scale industrial reactions to assist manufacturers with quality control issues, new product development, reaction optimization, and in other advanced monitoring and characterization areas. In the laboratory, the continuous ACOMP data stream provides a clear look at the entire reaction process which can then be used to increase reaction efficiency, address quality control problems associated with the process "recipe," and to generally optimize such products for companies. Even so, the laboratory research value for ACOMP has still only been minimally exploited in terms of making an impact in the industry. Use of the technology could dramatically accelerate many areas of polymer research and product development. Looking toward using ACOMP as a polymerization reaction monitoring control tool, some work has already been done in the lab using ACOMP data to directly control reaction processes [8, 9]. The next step for ACOMP, as a mature laboratory technology, is the adaptation to the industrial setting to monitor manufacturing scale polymerization reactions with the goal of process optimization via improved process control.

This chapter discusses the outlook for the ACOMP technology's use as an advanced process control tool in polymer manufacturing. Because ACOMP has been utilized successfully in

the lab on a wide range of reaction types, the primary challenge will be proper design and engineering of the technology for continuous, reliable operation in harsh manufacturing environments. Several key areas will need to be addressed to meet this challenge including full system automation, continuous sampling, and plug-free operation, to name a few. The appeal of industrial ACOMP is that rather than seeing delayed snapshots of the reacting polymer's macromolecule by using manual sampling methods, ACOMP will provide continuous data, delayed by merely a few minutes. Once in place as a monitoring tool, ACOMP data will yield information which can then be used by the manufacturer to make process control decisions. Eventually, the vision for ACOMP includes coupling the technology to existing process control theory and developing self-correcting, predictive models for ACOMP-based feedback process control and that will allow manufacturers to maximize the efficiency of their processes with minimal human intervention.

15.1.1 Polymer Industry Overview

The global polymer manufacturing industry produces vital materials which are then turned into the products that provide the backbone of our modern way of life. The polymer industry, a subset of the over $4 trillion global chemical industry, added over $1 trillion in value in 2011 [10]. Besides direct value creation in processing raw chemical materials into polymers, these polymer materials then support dozens of other manufacturing industries through extremely broad end-use applications including commodity and engineering plastics, composite building materials, rubbers for automotive and other applications, paints, adhesives, coatings, water purification, oil recovery, agricultural, pharmaceutical, and personal

Monitoring Polymerization Reactions: From Fundamentals to Applications, First Edition. Edited by Wayne F. Reed and Alina M. Alb.
© 2014 John Wiley & Sons, Inc. Published 2014 by John Wiley & Sons, Inc.

care products, biomedical components, and "intelligent" new materials. The entire polymer industry comprises thousands of different polymer product categories and hundreds of different polymer types [11, 12]. The industry is still heavily dominated by some of the most commonly used thermoplastic resins, but there are many emerging specialty polymers used in a wide array of applications which are becoming more popular and also can be far more profitable for producers at much smaller output levels [10–13]. Other polymers are processed directly from natural sources such as guar gum from guar beans, xanthans from microbial production, alginates from seaweed, gelatin from animal sources, and many others. The processed polymers from natural sources are then used in oil recovery, agroalimentary applications, pharmaceuticals, food applications, and personal care products. Finally, new classes of "smart polymers" with stimuli responsive properties are also being developed and produced, as discussed in Chapters 2 and 3. Some of them have broad applications in many different fields and could transform the breadth of end-use applications for the entire polymer industry. As an illustrative example, a list of selected North American Industry Classification System (NAICS) codes that encompass the polymer industry can be found in the following [11]:

- **325132** (Synthetic organic dyes & pigments), for example, disperse, fluorescent dyes; color pigments; stains
- **325191** (Gums and wood chemicals), for example, hardwood distillates, oils, rosins
- **325211** (Plastics materials and resins), for example, acrylics, acrylonitrile-butadiene-styrene, alkyds; carbohydrate, casein, and cellulose acetate plastics; cellulose nitrates; elastomers; epoxy, ethylene-vinyl acetate, ion exchange, methyl cellulose, methyl methacrylate, nitrocellulose, nylon, polyacrylonitrile, polyamide, polycarbonate, polyester, polyethylene, polyethylene terephthalate, polyurethane, polyvinyl alcohol, polyvinyl chloride, silicone, urea, resins; thermoplastics, thermosets
- **325212** (Synthetic rubber), for example, acrylate, acrylate-butadiene, butyl, ethylene-propylene, chloroprene, ethylene-propylene diene, latex, neoprene, nitrile-butadiene, polyisobutylene, polysulfide, silicone, styrene-butadiene, styrene-isoprene rubber; thermoset vulcanizable elastomers; thiol rubber; urethane
- **325222** (Noncellulosic organic fibers), for example, acrylic, acrylonitrile, casein fibers and filaments; fluorocarbon, nylon fibers and filaments; polyester, polyethylene terephthalate, polyolefin, polyvinyl ester, polyvinylidene chloride fibers; protein
- **325411** (Medicinals and botanicals), for example, enzyme proteins glycosides, botanicals (i.e., for medicinal use)

- **325510** (Paints and coatings), architectural coatings; dispersions, alkyd, enamel, oil paint, lacquer, varnish, epoxy; industrial finishes, coatings; latex, marine, auto paints; plastic wood fillers
- **325520** (Adhesives), for example, adhesives, caulking, joint, pipe sealing compounds; cement, rubber cements; starch glues; tile adhesives
- **325613** (Surface active agents), textile, leather finishing, emulsifiers, oils, softeners, wetting agents
- **325910** (Printing ink), bronze, inkjet lithographic, offset
- **325998** (Miscellaneous chemical products, preparations), gasoline, concrete additives (e.g., curing, hardening); drilling muds, fire extinguisher, fire retardants; gelatin; synthetic lubricating, synthetic oils

As outlined at the beginning of this section, polymer manufacturing is a global industry, mostly concentrated in southeast Asia, United States, and Europe with increasing market share coming from the Middle East and Latin America, mainly Brazil [10, 13]. In the United States, the recent discovery and exploitation of shale gas has driven significant new investment in the chemical industry as the cost of natural gas energy has fallen and, subsequently, the polymer industry as the cost of chemical feedstocks derived from natural gas has also fallen, both driven by a glut of natural gas supply on the market [14, 15]. This phenomenon may spread to the rest of the world as shale gas is discovered and exploited in other regions around the globe [15].

Overall, global polymer demand has been increasing steadily for many years driven by consumer demand in developed economies as well as by high growth in emerging economies [10]. Increasingly, newer and smarter polymers are being synthesized with unique properties that will enable more diverse applications, further increasing demand for polymers. The advent of these new polymer applications, coupled with steadily increasing demand from both developed and developing economies, promises to keep the polymer industry an extremely relevant and vital component of the global economy. That is why it is imperative to increase the overall efficiency of the industry through technological advancements in process control, similar to what has already been achieved in many other industries. At the heart of this issue is the monitoring and control of the polymerization manufacturing process itself.

15.1.2 Common Problems during Manufacture

The polymer manufacturing industry has many documented issues with monitoring and control of reaction processes. The issue has led to government, industry groups, and collaborative panels of government, industry, and university, to comment extensively on the need for better sensor technologies and modeling capabilities for polymer reactions [16–18]. Due to regular changes to process "recipes," suppliers, raw materials, raw material quality, labor turnover, and other issues, standardizing production processes and product output is a goal of most manufacturers. One of the best ways to achieve more standardized practices is through more data and knowledge of the entire process from start to finish. Currently, a wealth of process data, ranging from supplier information for all material inputs to direct process measurements such as temperature, pressure, flow, and others, is typically compiled into historical databases for each production run or batch. These are examples of common best practices utilized by most of the industry in order to identify issues during, and more often, postproduction. Much of this activity has been enabled by the dynamic industrial automation industry and the advent of vast data storage capabilities in more modern times. However, with all of these advances, there is still one main missing piece to the puzzle of online polymer process control, and it is the capture of continuous online data of the actual reaction kinetics and intrinsic polymer properties. This has been one of the large focus points of the industry for some time [17]. The end-result of this missing piece is suboptimal process control which leads to a host of costly problems that can reduce efficiency, productivity, and profits.

Based on experience working with various different manufacturers in the industry, some examples of these problems include: (i) minutes or hours of off-spec product in a continuous process or complete batch failure or off-spec product in a batch process; (ii) inefficient polymerization reaction processes resulting in "over-cooking" or unnecessary extended cycle time in batch processes or slow grade change-over in continuous processes; (iii) suboptimal polymerization processes resulting in reduced polymer yield (i.e., finished product percent solids) per batch or run; (iv) excess emissions of controlled chemicals during manual sampling or from inefficient processes, resulting in decreased manufacturing output due to compliance with environmental regulations; (v) increased labor costs and decreased worker safety from manual sampling and quality control; (vi) excess downtime for cleanup after poor reactions and failed runs or batches; (vii) unwanted gel and particulate buildup leading to bad product or reactor shut down; and many other problems. Addressing even a few of these issues would be of tremendous benefit to manufacturers, therefore implementing an online monitoring system capable of analyzing polymerization reaction kinetics in real time which could then drive full feedback control is a primary goal of the industry.

15.1.3 Overview of Current Industrial Online Monitoring Technologies

As previously discussed, industrial polymer production presents unique problems that have so far stymied progress by manufactures to implement a continuous monitoring-based

solution which is due to several factors such as the complicated nature of polymerizations and the difficult environments of reactions themselves [19]. Nonetheless, there are numerous available technologies that currently exist to attempt to solve at least a part of this monitoring problem which are outlined in this section.

Widely used online measurements such as reactor temperature and pressure, bulk viscosity measurements of reactor contents, flow rates, and others yield valuable process information to manufacturers [19]. This data can be used for process control when it is coupled with historical knowledge of the process and, in some instances, with customized advanced process models [19]. Manufacturers also typically use online gas chromatography or other similar techniques to precisely control monomer feeds during the reaction process which can provide a measure of total polymer concentration with the proper mass-balance equations and understanding of the process flow to account for various considerations in batch, semibatch, or continuous processes [19]. In addition to these online monitoring tools, many manufacturers also perform periodic manual sampling during the reaction process to determine more information about the reaction kinetics and the polymer product by using more advanced characterization methods in the laboratory [19].

Producing an analytical result from manual sampling can take place anywhere from every 10 to 60 min or more, depending on the type of reaction, and is thus time consuming and expensive because it is repeated over and over during the process [19]. The collected samples are then typically brought to an on-site quality control laboratory for analysis. In the lab, the samples are usually prepared in some manner for analyses that oftentimes include both rheological measurements, for example, Mooney viscosity, melt flow index, tensile strength tests, and so on, and macromolecular measurements such as liquid chromatography, gas chromatography, and so on [19]. The data from these periodic lab measurements are very useful and are also commonly used for determining finished product specifications. Many of the macromolecular measurements yield data on polymer molecular weight (M_w), concentration, comonomer composition (in the case of copolymers), and other characteristics which are all of great interest to manufacturers since these measurements represent intrinsic properties of their product. Much work has also been done to correlate rheological measurements to macromolecular measurements [20].

Another online monitoring method involves the use of inferential and predictive models built on the available process data already discussed (flow, temperature, pressure, etc.), which attempt to model in real time, and/or predict into the future, the intrinsic characteristics of the polymers, as measured in the lab [19, 21]. Model-based control theory has become a very important field in polymer processing and is currently one of the best available tools. Oftentimes, however, due to the complex nature of polymerization processes, these models sometimes require input from periodic off-line manual measurements from the laboratory to correct the model throughout the polymerization process [19].

Extensive work has been done to build models around standard process data as well as other types of online data such as online melt flow index, bulk viscosity, and so on; the primary goal of these is to attempt to infer polymer properties to enable better process control [22–26]. Generally, much of the model building work is done off-line to determine optimal operating policies and optimal process variable trajectories that define end-product quality and reaction yield. Several types of advanced software have also been developed toward this goal [27, 28]. There is more discussion of process control theory and application in other chapters of this book, especially Chapters 16–18.

Additionally, online monitoring methods have been developed to adapt off-line characterization methods into *in situ* (i.e., in-reactor) probes for determination of kinetics and monomer conversion with optical methods such as mass spectroscopy (MS), ESR, FTIR, near IR, and Raman spectroscopy. However, frequently, due to high turbidity and viscosity of the polymer reaction milieu, the optical surfaces are easily fouled, leading to frequent sensor failure. Furthermore, data acquired with these probes are model dependent; the empirical and inferential calibration schemes used can be expensive and time consuming to develop and can drift and become unreliable as reactor conditions change and as sensors become fouled. Another limiting feature of these methods is that they usually measure only one characteristic of the reaction, such as monomer conversion and are not directly sensitive to polymer molecular mass and intrinsic viscosity. More detailed discussion of these techniques can be found in Chapters 6–10 of this book.

Finally, online particle sizing is another technology occasionally used in the industry that provides information on the size of the particles using dynamic light scattering or MIE scattering. This can be very useful, especially for emulsion reactions, but particle size does not give any information on comonomer composition, polymer concentration, M_w or $[\eta]$. Finally, gel permeation chromatography (GPC) is also occasionally used online with a system that prepares and injects a sample point from the reactor every 10–30 min or more, depending on the elution time required for the product to work through the gel columns. GPC, a widely accepted method for characterizing polymers, yields very useful information on comonomer composition, polymer concentration, residual monomer concentration, M_w, and more, when using the correct array of detectors and analysis methods, as discussed fully in Chapter 9. This nonexhaustive overview of current methods illustrates the complexity of the problem and highlights the fact that there is currently no

generalized approach in industrial polymerization reaction monitoring that will yield all of the real-time reaction kinetics and polymer data desired.

15.1.4 Industrial ACOMP Introduction

Using ACOMP in an industrial manufacturing process will yield a tremendous amount of valuable data on reactions which will make it ideally suited for monitoring and controlling the various process steps. This will be extremely attractive since it can complement many of the existing tools in use which were described in the previous section. The final challenge in industrial ACOMP implementation lies in the engineering of the system to address the challenges of the manufacturing environment. Ruggedness, robustness, dependability, and reliability are the focus for the industrial ACOMP technology. Addressing these will provide the polymer industry with a disruptive new technology capable of filling the gaps left by current technologies and methods.

For example, ACOMP, in contrast to *in situ* sensors, overcomes the problem of sensor fouling by interposing a high performance "front end" between the reactor and the detector train, providing a continuous, highly dilute, and conditioned sample stream on which myriad sophisticated measurements can be made without fouling detectors. Because the measurements are made on dilute solutions, model-free determinations of primary polymer reaction characteristics can be made, such as molecular weight and intrinsic viscosity, in addition to kinetics and monomer conversion. Light scattering detectors can immediately signal problems with the reaction, such as the onset of haze, gel and microgel formation, and particulate build-up, while UV or visible detectors detect side reactions that produce yellowing and other types of unwanted product degradation. The use of a modular train of multiple detectors allows cross-correlation among various characteristics, such as evolution of M_w versus conversion, changes in polydispersity as reaction proceeds, onset of branching and cross-linking reactions, conditions under which polymer phase transitions (e.g., lower critical solution temperature) occur, and so on. The signal stream from ACOMP, including monomer conversion, kinetics, M_w and $[\eta]$, as well as signals of unusual or unwanted behavior, can also be readily interfaced with existing predictive modeling and control software.

Another feature is that the ACOMP platform is fully compatible with current polymer manufacturing infrastructure. Existing manual sampling loops can be easily retrofitted with an automated ACOMP front end. In the unusual example that a sampling loop is nonexistent, the construction of such a loop could easily be carried out by in-house or contracted maintenance engineers. The ACOMP system's actual footprint at the plant is small and the system will be housed in regulation safety enclosures near reactors. Finally, the ACOMP system will be completely automated and will integrate directly with

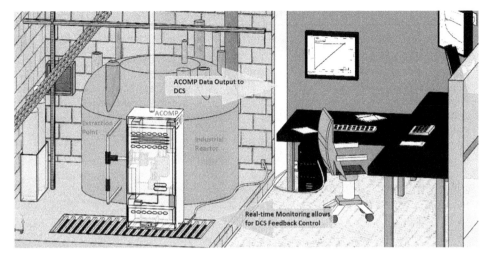

FIGURE 15.1   Artist rendering of an industrial ACOMP system attached to industrial reactor (image by Olivia Carnes).

the plant's distributed control system (DCS) for monitoring of the analyzed data output. An artist rendering of an industrial ACOMP system can be seen in Figure 15.1.

15.2   COMPARISON OF FEATURES FOR R&D ACOMP VERSUS INDUSTRIAL ACOMP

In the R&D context, one typically wants the most data possible provided by the best, most numerous, and sensitive detectors possible. The wider the range of detection methods, the more reaction characteristics can be monitored and cross-correlated. Models based on the primary monitored quantities can be elaborated to any degree of sophistication required. Highly trained personnel, typically science and engineering PhDs in industry and graduate students, postdocs, and faculty at Universities, can use all the instrumentation and software, and troubleshoot and maintain the ACOMP systems as needed. The targeted performance of the system is for periodic reaction monitoring followed by extensive analysis and modeling by the scientists for each experiment. The cycle of experiment design, execution, and analysis can take weeks or months.

In sharp contrast, in the industrial environment, ACOMP measurement repeatability, system reliability, and continuous, problem-free operation are the main requirements. The number of detectors is chosen to be the absolute minimum needed to provide the most valuable characteristics to be monitored. Oftentimes, the characteristics to be measured are a small subset of what an R&D ACOMP is aimed at; for example, an industrial system may just be needed to monitor monomer conversion and polymer reduced viscosity. In light of these goals, both the front-end and detector components are chosen with ruggedness and reliability in mind.

Another important aspect of industrial ACOMP is that its operators will not generally have the skill levels of R&D

users and, in fact, may have no understanding at all of how the ACOMP functions or of its specific components. Hence, the industrial ACOMP system will rely on an extra level of automation and "self-awareness"; that is, it requires internal flow, pressure, and temperature sensors to diagnose its own state of maintenance. Instead of relying on alert R&D operators to spot overpressure, plugging, pump failure, or detector malfunctions, all flow paths must be monitored and instruments subject to self-verification. This centralized internal monitoring of the ACOMP system will send alarms to plant operators when specific criteria are reached. The system will also include backup and redundant systems in the more problematic components to ensure that continuous operation is maintained. An illustrative table comparing the R&D ACOMP system to the industrial ACOMP system can be found in Table 15.1.

15.2.1   Filtration, Plugging, and Early Detection of Particulate Formation

Because ACOMP handles actual reactor contents, the biggest challenge to be addressed will be continuous sampling, dilution, and pumping leading into the system and within the system itself. As such, proper engineering and materials selection will be critical to achieving success in adapting ACOMP to the industrial reactor. Tying into existing reactor, sampling points will be the easiest component of the task. Many reactors have these sampling points and with the proper flow rates can interface easily with the ACOMP front end. A bigger challenge will be ensuring that the ACOMP front-end works uninterrupted in the face of gel and particulates.

The problem of gel and particulate formation is observed in many different sectors of the polymer industry and will be one of the main issues to be addressed for the ACOMP system's successful, reliable operation in the industrial environment. An important key to solving this problem will be to

TABLE 15.1   Comparison of Features between Lab ACOMP and Industrial ACOMP Systems

	ACOMP Lab R&D	Proposed Industrial ACOMP
Capability comparison		
Monomer conversion kinetics	Yes	Yes
Comonomer conversion kinetics	Yes	Yes, when needed
Evolution of M_w	Yes	Yes, when needed
Evolution of $[\eta]$	Yes	Yes, when needed
Colloid particle size distribution	Yes	Yes, when needed
SEC attachment option	Yes	Yes
Simultaneous polymer and colloid monitoring	Yes, when needed	Yes, when needed
Software/GUI	Yes	Yes
Industrial scale monitoring	No	Yes
24/7 Operation	No	Yes
Defined maintenance schedule	No	Yes
DCS integration	No	Yes
ACOMP data-driven automated feedback reaction control	No	Yes

filter the reactor contents prior to the ACOMP system. The filtration of the reactor contents will be achieved through means appropriate to the particular reaction being monitored and with minimal blockage and delay time taken into consideration. Based on pressure sensors monitoring the blockage on each filtration point, the system will automatically switch over to backup filters and send an alarm to the DCS for an operator to replace the primary filters. As an added bonus, data on filter plug rates can be recorded and will provide additional information on the process and potentially a means for early detection of particulate and gel buildup. More on this "filtrodynamics" technique is discussed in Chapter 16 of this book. The filtration systems may be placed within the sampling loop itself or immediately at the pick-off point prior to entering the ACOMP front end. Each ACOMP system installation may employ one or more types of filter configurations depending on the type of polymerization, economical filter plug rates, desired level of operator intervention, and so on. The primary goal is to transfer filtered reactor contents to the ACOMP system in the most efficient and cost-effective means possible (Fig. 15.2).

Tubing size, flow rates, valves, and tees will be important to prevent plugging at critical system points. Reactor sampling loops typically operate at much higher flow rates than currently in use in the laboratory and the tubing sizes also differ significantly. The placement of the filtration system in relation to the sampling loop will require careful consideration of flow reduction, filter pore size, and flow rate. Large volume cascading filters will be necessary if the filtration system is contained in the sampling loop where there is likely to be significantly high flow and much larger raw mass of filter plugging particulates. A special reduction tee and possibly needle valves will be required to significantly reduce the flow rate of filtered reactor contents to

integrate with the ACOMP front end. At this point, the tubing will also convert to appropriately smaller diameter to be compatible with the ACOMP front end. Much smaller volume filters will be necessary should the filtration system be placed after the reduction tee. Within the ACOMP system, similar liquid handling and filtration issues will be addressed. Once the reactor contents are roughly filtered and sent to the front end, a continuous, known quantity of reactor contents is fed into the ACOMP system's primary mixing chamber and sample conditioning stages. The sample conditioning will be customized to each particular reactor and polymerization being monitored but some commonality will be shared in each ACOMP front end as a series of pumps and mixing chambers dilute and condition the polymer stream in preparation to be sent to the detectors. As previously discussed, the polymer stream up to this point has only been filtered sufficiently to ensure no blockage occurs in the sampling and conditioning front end. This "rough" filtering may or may not be sufficient when considering the tubing size and flow rates maintained in the detector "back end" of the ACOMP system. For added security and final sample preparation, a smaller pore size duplex filtration system will be in line between ACOMP front end and the detectors. This will be the last step to ensure that the sample solution is properly conditioned prior to entering the detector train. The predetector filtration system will also have the capability of plug monitoring and automatic switchover to a clean filter with an accompanying alarm to the DCS.

15.2.2   Maintenance and Cleaning Cycles

Other safety and reliability precautions will be implemented such as the development of cleaning cycles that can be initiated automatically at specified process intervals or manually

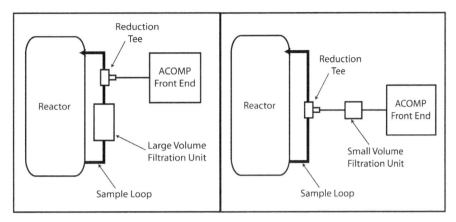

FIGURE 15.2   Schematic of possible configuration of industrial ACOMP filtration system.

by an operator. Automatic "quick-clean" cycles, executed to maintain operation during the process, can also be triggered by criteria determined by inline pressure sensors or the torque and current draw on motors associated with pumps or mixing chambers as drastic changes in pressure, torque, or current draw are often indicators of an impending blockage or a depriming event. Other automatic cleaning cycles can be simply associated with dead time during batch changeover and implemented as a precaution or in preparation for the next reaction that will take place. Cleaning cycles will likely vary depending on the implementation, but will typically consist of a rapid back-flushing the front-end dilution and conditioning system with a cleaning solution to keep gel and other particulate buildup from forming over time. The back-end detector train of the system is much less prone to blockage due to the fact that filtration and dilution of the polymer stream have removed any particulates and gel as well as solubilized the polymer in diluent. Nonetheless, at the end of each reaction, as the system is being prepared for the next batch or run, a continuous flow of pure solvent will flush through the detector train to remove all trace of polymer and other reactive materials.

15.2.3 ACOMP System Automation and Plant Integration

Automation of the entire ACOMP system will be an important focus of the industrial version. Monitoring and controlling the many different parts of the front end and detectors will be achieved through an on-board programmable logic controller (PLC) or other type of hybrid controller. This on-board PLC will monitor and control the entire system. Table 15.2 describes to some extent the PLC control of the system.

The PLC will be integrated directly with the plant DCS, into its own module, where it will be monitored by an operator along with all other plant processes (Fig. 15.3). Although

the PLC will be monitoring and recording all internal signals for the ACOMP system, only those signals which are required for reaction monitoring and system alarms will be sent to the plant DCS. Of course, any or all signals monitored by the ACOMP's internal PLC can be sent to the plant DCS if deemed necessary. In some instances, this module will also have control over various components in the system. Most of the self-sensing processes will be automated for optimum operation, but manual control of some components is envisioned, for example, pump flow rates, wavelength selection on UV detectors, and so on. Separately from the basic PLC signals' monitoring and controlling the ACOMP system's various internal components, a signal containing all of the analyzed data from the detectors will be transmitted to the DCS.

15.2.4 Simplified Data Analysis

Data analysis is a critical part of the ACOMP system. As such, the analysis procedures will be programmed into an onboard computer or microcomputer. The raw data will be collected continuously from the detectors, and will be analyzed continuously as it is collected using preprogrammed software incorporating the analysis procedure for the specific application. The results of the analysis will then be sent to the DCS module as continuous data readout from the ACOMP system. These results will usually be as simple as an analyzed number output from each detector (e.g., UV, viscosity, LS, etc.), which will be correlated to specific events in the reaction, properties of the polymer, acceptable ranges for a specified grade, monomer conversion, residual parts per million of monomer, and so on. The ACOMP system PLC data yielding system health information, filter plugging rate data, and so on, and the analyzed ACOMP data results will be added to the other reaction information from the DCS and saved in the manufacturer's historical database for each reaction.

TABLE 15.2   Description of PLC Control of the Industrial ACOMP System

Component	Monitor	Control	Alarm
Front-end filtration system	Pressure buildup across filter	Switch to backup filter(s) once elevated pressure criteria are met	Replace primary filter cartridge
Back-end filtration system	Pressure buildup at filter	Switch to backup filter(s) once elevated pressure criteria are met	Replace primary filter cartridge
Mass flow controller	Power consumption	Control flow setting	Abnormal power consumption, signaling potential problems
ACOMP system pumps	Power consumption	Flow rates	Abnormal power consumption, signaling potential pump problems
Agitator in mixing chamber(s)	Torque and power consumption	Mixing speed. Possible inferential data on polymer can be derived from torque measurements	Unusual torque and/or power consumption
Solvent reservoir(s)	Fluid levels	None	Low solvent
Clean cycle pump(s) and reservoir(s)	Fluid levels and power consumption	On/off for clean cycle	Low solvent or pump anomalies
Temperature sensors	Temperature	Temperature of various temperature controlled components, for example, tubing, mixing chambers, enclosure, etc.	Abnormal temperature increases

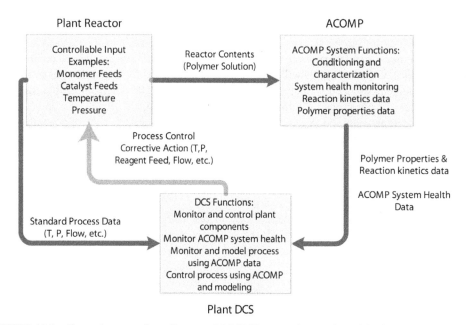

FIGURE 15.3   General process flow diagram of ACOMP system integration with plant reactor and DCS.

With this continuous real-time data stream on the polymer properties and reaction kinetics, it is anticipated that ACOMP will allow for immediate benefits with tighter operator control. Eventually, the goal will be to use the continuous stream of process data on polymer properties and reaction kinetics yielded by ACOMP to create a complete feedback control closed loop system. This will be achieved by developing low-error process control models using ACOMP and general process data. This union of ACOMP with process models will be interfaced directly into the existing plant DCS and will enable the automatic detection of faults in the polymerization reaction which will then trigger automatic corrective actions using manipulable process variables such as temperature, pressure, monomer, initiator, and other reagent feeds, to drive toward the specific reaction set point, delivering full feedback control. ACOMP-based predictive modeling and closed-loop operation will automatically monitor and control the reaction process, achieving the most efficient reaction path and yielding a superior polymer product with consistent properties. Furthermore, statistical analysis of historical process data consisted of standard process data, and the newly available ACOMP data will determine optimal operating ranges for specific process variables and will improve early fault detection and model predictive control applications.

15.3 BENEFITS OF ACOMP, OUTLOOK, AND CONCLUSION

As previously discussed, current online technologies focus on measuring basic parameters such as temperature, pressure, fluid flow rates, and concentrations of small molecules. While useful, these parameters do not directly characterize the many important features of polymerization reactions. Other approaches include off-line measurements with techniques such as liquid chromatography, filtration tests, rheological measurements, and so on. These methods are labor intensive, can be hazardous because of manual sampling during the process, and do not provide data rapidly enough for online reaction control. The other *in situ* techniques can provide some useful information but are model dependent and can foul in the harsh industrial polymer reactor environments. The results of these methods are inefficiencies resulting in millions of dollars of waste per manufacturing reactor annually. Because of these inefficiencies, the introduction of industrial ACOMP with automated feedback control will offer many benefits to manufacturers which will be primarily centered on the reduction in process time and raw material inputs and the increase in overall yield.

15.3.1 Potential Impact on Efficiency and Other Benefits

- *Monitoring grade changeover in continuous reactors* This can reduce large amounts of wasted time, energy, and feedstocks. Grade changeover is still usually monitored by manual sampling with large delay times between sample collection, obtaining results, and making reactor process decisions. ACOMP continuously monitors characteristics of polymers intermediate between grades and hence can quickly detect changeover via feedback control and determine when steady state operation for a grade is reached.

- *Monitoring conversion in batch and semibatch reactors* Multistep polymer reaction processes often move from step to step as preset monomer conversion levels are met; for example, in the production of block copolymers or for grafting reactions, such as used in high impact polystyrene. Currently, chemical conversion is

frequently gauged by empirical measures such as historical data, torque buildup on agitators, or manual sampling and subsequent analysis in an at-plant quality control laboratory. ACOMP detects monomer and comonomer conversion automatically and continuously, providing a solid basis for process step-to-step decisions.

Residual monomer chasing is of critical importance in polymer products that are used in consumer applications, such as pharmaceuticals, food and other packaging, water treatment, and personal care. Parts per million of specific monomers, for example, acrylamide and bisphenol-A, are tightly regulated by the government to very low levels, as low as 5 ppm. Hence, reactions must be run until theses trace levels are achieved. Frequently, this means adding extra initiators and "cooking" to scavenge monomers down to low levels. Many polymer manufacturers currently have no online methods for this and resort to manual sampling and plant analytical labs. ACOMP can, by a combination of detectors, directly monitor monomers down to low parts per million.

- *Copolymer composition optimization* Frequently, a manufacturer of a copolymer wants to keep a certain fraction of comonomers incorporated throughout production. Determination of copolymer composition "postmortem" (i.e., after polymerization) is a challenging and tedious task and as such there is great potential for online monitoring and control. ACOMP has used several approaches for monitoring and controlling copolymer composition to ensure desired products can be produced. Targeted copolymer products can hence be produced according to specifications, also eliminating the labor intensive postproduction analyses.

- *Optimizing trajectory of batch, semibatch, and continuous reactions* By controlling reaction processes to follow the optimum path to desired product properties, time is minimized, similar to following an optimized road route using a GPS.

- *Predictive modeling and control* With extensive reaction data and predictive models, it is envisioned that an ACOMP control software package can be created to interface the ACOMP-based models with the DCS which will allow for direct automated feedback control of reactor servomechanisms that control reagent flow rates, temperature, pressure, and so on. ACOMP control interface package will monitor the polymer reaction characteristics, plug the data in tested predictive models, and adjust process variables automatically to achieve the optimal reaction process. The design of the control interface package will be based on mechanistic and process history models. The former requires development of complex process models, whereas the latter do not assume any form of model information and rely only on process data history. Both approaches and embedding knowledge provided by either modeling method into the ACOMP platform will be explored.

- *On-command polymer properties* Product quality and product yield are much more complex issues in polymerization than in more conventional small molecule reactions. The key polymer properties are dependent on the polymer particle size distribution (PSD) and molecular weight distribution (MWD). These properties are sensitive to the process and kinetic history. The ACOMP platform, by directly monitoring these properties, will enable the production of polymers with desired properties "on-command."

- *"One shot" blending* Because ACOMP is able to both monitor and control evolution of molecular weight and copolymer composition, it can be used to produce multimodal populations of blended products by monitoring computed reaction process trajectories and flowing comonomers, initiators, and so on, into the reactor as needed (semibatch operation), rather than relying on the expensive, energy and time intensive mechanical blending of different product lots.

- *Improved product quality and consistency* It is a big focus for many manufacturers, especially in specialty polymers where there can be high variance in production lots. Even within a company, different plants can produce the same product with very different end properties which then creates negative feedback from customers. Using feedback control, predictive models, and historical databases, manufacturers will be able to share reaction data between their plants and will be able to build correlations to polymer properties based on process and ACOMP data, thereby creating the most efficient process to manufacture a uniform product.

- *Reduced need for endproduct QA/QC* will be another benefit provided by ACOMP since polymers are "born characterized" during the monitoring and control process. This will eliminate or vastly reduce the expensive, time consuming and labor intensive postproduction analytical laboratory work required for plant quality control. The technical personnel alleviated from manual sampling and laboratory analysis chores can be retrained to higher level, more productive tasks, for example, new product development, new applications.

Beyond direct near-term potential benefits of industrial ACOMP, widespread use of the ACOMP platform in the R&D laboratory and on industrial reactors will assist in the accelerated development of next generation "smart polymers," which can be stimuli-responsive (UV/Vis light, pressure, sound, stress, etc.) and can "turn-on" physical characteristics at various stimuli thresholds. Polymers with these types of properties will enable many new applications in medicine for drug delivery, construction, automotive, aircraft, machinery industries with "self-repairing" polymers and polymers that directly show stress via a characteristic change (e.g., material turns red when it is worn out), and many more exciting application areas. ACOMP at the R&D lab would enable more rapid development, design, and synthesis of these polymers and ACOMP at the industrial scale would ensure that lab-scale processes are properly scaled and implemented.

Given these clear prospective benefits of ACOMP's use in polymer manufacture, it is expected that the potential financial benefits to manufacturers and to the industry as a whole can be quite substantial. As previously mentioned, these polymer industries are responsible for many modern day products. The industry is also vital in supporting many other manufacturing sectors and ongoing polymer R&D ensures that new polymers will be developed, thereby prompting their use in more applications and further benefiting the public good.

15.3.2   Potential Impact on Environment and Worker Productivity

The U.S. chemical industry is the second largest industrial manufacturing consumer of energy, utilizing 24.4% of manufacturing energy [29]. The chemical industry (NAICS 325) has several polymer related subcategories. Energy usage in several of these target sectors, per the U.S. Energy Information Administration's (2006) Manufacturing Energy Consumption Survey, totals 2713 Trillion Btus consumed annually, equivalent to 467.7 million barrels of crude oil energy, valued at $ 50 billion ($107 per barrel). Additionally, the U.S. chemical industry's carbon emissions were 311 million metric tons in 2002, the highest of any manufacturing industry, at 22.2% of total manufacturing emissions [30]. Globally, the energy and emissions footprint for the industry is similar to what is seen in the United States. This industry's size and energy usage make it a prime target for efficiency gains to reduce overall energy consumption.

Other interesting and growing environmentally friendly industries which ACOMP may be applied to are biofuels and "green" polymers made from renewable feedstocks such as ethanol-derived ethylene and other processes. ACOMP may eventually be used to monitor fermentation processes to gauge conversion of biomass to the active biofuel or chemical feedstock. This is a completely novel and untested application for the technology, but could be of substantial commercial and environmental benefit.

In addition to the major economic impacts cited, industrial ACOMP, coupled with existing tools and automation, will yield important environmental and safety advantages. The higher efficiency will automatically ensure less emissions and pollution per kilo of polymer product. The ACOMP mode of millifluidic sampling (c. 5 ml h^{-1} of reactor content) will mean that less toxic materials will be sampled, with less releases to the atmosphere (e.g., sampling synthetic rubber in a hexane carrier releases hexane to the atmosphere). Automatic sampling will also eliminate exposure of plant personnel to toxic reactor contents from manual sampling. These new tools will allow for reductions in process steps, cycle time, batch failure, off-spec, and so on. It will enable accurate planning and execution of the most efficient processes for new and existing products. Therefore, efficiency gains from ACOMP use are expected to reduce consumption of energy and nonrenewable feedstocks, solid and liquid reactor waste, and other environmental pollutants such as volatile organic solvents and monomers, which will assist in reducing the energy consumption and environmental footprint of the polymer industry.

15.4 CONCLUSION AND OVERALL OUTLOOK FOR INDUSTRIAL ACOMP

The ACOMP technology platform is patented under a series of United States and PCT patents and applications owned by Tulane University [31–34]. The patents were licensed by a start-up company and commercialization activities are currently under way. By addressing all of the critical issues surrounding the development of industrial ACOMP, it is expected that successful implementation is achievable. Eventually, online monitoring continues to be enhanced by coupling ACOMP with existing knowledge in process control theory and with advanced process control software to create a fully compatible package for integration into each plant's DCS. Once feasibility of the ACOMP technique has been demonstrated in this industrial environment, it is expected that the clear overall efficiency benefits to the industry will drive adoption of the technology. The grand vision is for industrial ACOMP to have a transformational impact in the way that polymers are produced in a manner that will benefit manufacturers and society alike.

15.5 ACKNOWLEDGMENTS

The authors would like to thank their research team members for all of their hard work and Dr. Bill Bottoms for his mentorship and guidance over the years.

REFERENCES

[1] Florenzano FH, Strelitzki R, Reed WF. Absolute online monitoring of polymerization reactions. Macromolecules 1998;31: 7226–7238.

[2] Alb AM, Reed WF. Simultaneous monitoring of polymer and particle characteristics during emulsion polymerization. Macromolecules 2008;41:2406–2414.

[3] Alb AM, Reed WF. Online monitoring of molecular weight and other characteristics during semibatch emulsion polymerization under monomer starved and flooded conditions. Macromolecules 2009;42:8093–8101.

[4] Alb AM, Farinato R, Calbick J, Reed WF. Online monitoring of polymerization reactions in inverse emulsions. Langmuir 2006;22:831–840.

[5] Chauvin F, Alb AM, Bertin D, Tordo P, Reed WF. Kinetics and molecular weight evolution during controlled radical polymerization. Macromol Chem Phys 2000;203: 2029–2040.

[6] Alb AM, Serelis AK, Reed WF. Kinetic trends in RAFT homopolymerization from online monitoring. Macromolecules 2008;41:332–338.

[7] Mignard E, Lutz JF, LeBlanc T, Matyjaszewski K, Guerret O, Reed WF. Kinetics and molar mass evolution during atom transfer radical polymerization of butyl acrylate using automatic continuous online monitoring. Macromolecules 2005;38:9556–9563.

[8] Kreft T, Reed WF. Predictive control and verification of conversion kinetics and polymer molecular weight in semibatch free radical homopolymer reactions. Eur Polym J 2005;45:2288–2303.

[9] Kreft T, Reed WF. Predictive control of average composition and molecular weight distributions in semibatch free radical copolymerization reactions. Macromolecules 2009;42:5558–5565.

[10] American chemistry council research of global chemical industry. Available at http://www.americanchemistry.com/Jobs/EconomicStatistics/Industry-Profile/Global-Business-of-Chemistry. Accessed 2012 Sept 6.

[11] United States Bureau of the Census. 2007. *Economic Census*. Available at http://www.census.gov/econ/census07/. Accessed 2012 Sept 6.

[12] Journal staff. Facts and figures of the chemical industry. Chem Eng News 2011;90:37.

[13] Davis N. ICIS top 100 chemical companies. ICIS Chem Bus 2011;280:29–35.

[14] Chang J. US boosts cracker capacity. ICIS Chem Bus 2011;281:10.

[15] Yergin D. *The Quest: Energy, Security, and the Remaking of the Modern World*. New York: Penguin Press; 2011.

[16] Klipstein DH, Robinson S. *Vision 2020: Reaction Engineering Roadmap*. New York: Center for Waste Reduction Technologies, American Institute of Chemical Engineers & DoE of Industrial Technologies; 2001.

[17] Chemical Manufacturers Association, Council for American Research, Synthetic Organic Chemical Manufacturers Association. *Technology Vision 2020: The US Chemical Industry*. Washington, DC: American Chemical Society/American Institute of Chemical Engineers; 1996.

[18] Doyle III, FJ, Garrison A, Johnson B, Smith WD. Process measurement and control: industry needs. Workshop Report. New Orleans: National Institute of Standards and Technology; 1998.

[19] Fonseca GE, Dubé MA, Penlidis A. A critical overview of sensors for monitoring polymerizations. Macromol React Eng 2009; 3:327–373.

[20] Bremner T, Rudin A, Cook DG. Melt flow index values and molecular weight distributions of commercial thermoplastics. J Appl Polym Sci 1990;41:1617–1627.

[21] Crowley TJ, Choi KY. Experimental studies on optimal molecular weight distribution control in a batch-free radical polymerization process. Chem Eng Sci 1998;53:2769–2790.

[22] Richards JR, Congalidis JP. Measurement and control of polymerization reactors. Comput Chem Eng 2006;30: 1447–1463.

[23] Villa C. Reactor modeling for polymerization processes. Ind Eng Chem Res 2007;46:5815–5823.

[24] Sharmina R, Sundararaja U, Shah S, Griend LV, Sun YJ. Inferential sensors for estimation of polymer quality parameters: industrial application of a PLS-based soft sensor for a LDPE plant. Chem Eng Sci 2006;61:6372–6384.

[25] Alhamad B, Romagnoli JA, Gomes VG. Inferential conversion monitoring in emulsion co-polymerization through calorimetric measurements. Int J Chem React Eng 2006;4:A7, 1–10.

[26] McAuley KB, MacGregor JF. On-line inference of polymer properties in an industrial polyethylene reactor. AIChE J 1991;37:825–835.

[27] Krallis A, Pladis P, Kanellopoulos V, Kiparissides C. Development of advanced software tools for computer-aided design, simulation, and optimization of polymerization processes. Macromol React Eng 2010;4:303–318.

[28] Krallis A, Pladis P, Kanellopoulos V, Saliakas V, Touloupides V, Kiparissides C. Design, simulation and optimization of polymerization processes using advanced open architecture software tools. Comput Aided Chem Eng 2010;28:955–960.

[29] 2006 U.S. (EIA) MECS, Table 1.2: Consumption of energy for all purposes: by mfg. Industry & Region (trillion Btu). Available at http://www.eia.doe.gov/emeu/mecs/mecs2006/2006 tables.html. Accessed 2012 Aug 5.

[30] Energy-related CO_2 emissions in US manufacturing M Schipper, EIA. Table 1: CO_2 emissions from manufacturing by industry, 2002. Available at http://www.eia.doe.gov/oiaf/1605/ggrpt/pdf/industry_mecs.pdf. Accessed 2012 Aug 5.

[31] Reed WF. Miniature, submersible, versatile, light scattering probe for absolute equilibrium and nonequilibrium characterization of macromolecular and colloidal solutions. US Patent 6,052,184. 2000.

[32] Reed WF. Automatic mixing and dilution methods for online characterization of equilibrium and non-equilibrium properties of solutions containing polymers and/or colloids. US Patent 6,653,150. 2003.

[33] Reed WF, Alb AM. Methods and devices for simultaneously monitoring colloid particles and soluble components during reactions. US Patent 7,716,969. 2010.

[34] Reed WF. Automatic sampling and dilution apparatus for use in a polymer analysis system. US Patent 8,322,199 B2. 2012.

SECTION 4

REACTOR CONTROL AND DESIGN: BATCH, SEMI-BATCH, AND CONTINUOUS REACTORS

16

MATHEMATICAL MODELING OF POLYMERIZATION REACTORS

F. Joseph Schork

16.1 MATHEMATICAL MODELING TECHNIQUES

16.1.1 Introduction

If one accepts the presumption that the function of the polymerization reaction engineer is to design a *process* which will produce a given polymer structure, it is instructive to look at the magnitude of defining polymer structure [1]. To do this, consider one of the most common polymers, polyethylene. This material also has one of the simplest molecular structures (**1**, Scheme 16.1).

Polyethylene is manufactured by opening the double bond in ethylene to create a linear polymeric chain. Such a linear polyethylene is known as high density polyethylene (HDPE), and is polymerized using a coordination polymerization process. The density is high because the linear chains lead to high crystallinity, and hence high density. It is used for blow-molded products such as bottles. If the same ethylene monomer is polymerized by a free radical mechanism, a highly branched material results. The branching causes a lower crystallinity and hence a lower density. Not surprisingly, this material is known as low density polyethylene (LDPE, **2**, Scheme 16.1). If ethylene is copolymerized with butene or other short-chain alkenes using coordination polymerization, the product consists of linear polymer chains with randomly spaced short branches (of the length of two carbons in the case of butene). The short-chain branches reduce the crystallinity, and hence the density. This product is known as linear low density polyethylene (LLDPE). The primary end use for LDPE is in films. Within each density material, it is possible to produce high or low molecular weight polymers. In addition, the molecular weight distribution can be broad or narrow. Thus, a monomer

containing only two carbons and four hydrogen can be used to produce polymers of widely varying properties.

If one considers a monomer of slightly more complexity, the permutations of possible structures are even larger. Consider for example, the monomer chloroprene (2-chloro-1,3-butadiene, **3**, Scheme 16.2). Chloroprene is polymerized by free radical emulsion polymerization to form polychloroprene, or neoprene rubber. Neoprene is one of the oldest synthetic rubbers, and is used when higher performance is needed than can be provided by the lower cost styrene butadiene rubber (SBR). Being a butadiene derivative, chloroprene contains two conjugated double bonds. Polymerization takes place by the opening of one double bond; the second is less reactive. Polymerization results in one of four possible structures. The *trans*-1,4 (**4**, Scheme 16.2) structure accounts for approximately 87% of the polymer units at 40 °C [2].

The *cis*-1,4 structure (**5**, Scheme 16.2) accounts for approximately 10% of the polymer units under the same conditions. Both 1,4 structures occur when one double bond is opened by reaction with a propagating free radical, allowing the monomer unit to enter the polymer chain. The other double bond moves to the 2–3 position.

The *cis*-1,2 (**6**, Scheme 16.2) polymerization occurs when the 1–2 double bond is opened by reaction with a propagating free radical, leaving a pendant ethylene group and an allylic chlorine. This structure accounts for approximately 2% of the polymer units at 40 °C.

The *cis*-3,4 structure (**7**, Scheme 16.2) results from the opening of the 3–4 double bond, and results in a substituted ethylene pendant group. This structure makes up the last 1% of the polymer units.

Monitoring Polymerization Reactions: From Fundamentals to Applications, First Edition. Edited by Wayne F. Reed and Alina M. Alb.
© 2014 John Wiley & Sons, Inc. Published 2014 by John Wiley & Sons, Inc.

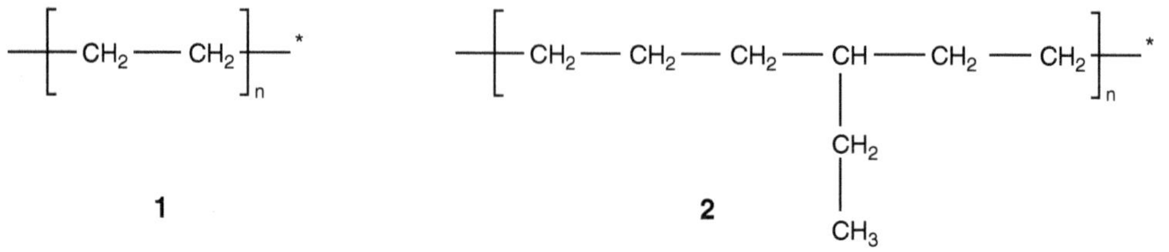

SCHEME 16.1   Polyethylene, **1**; low-density polyethylene (LDPE), **2**

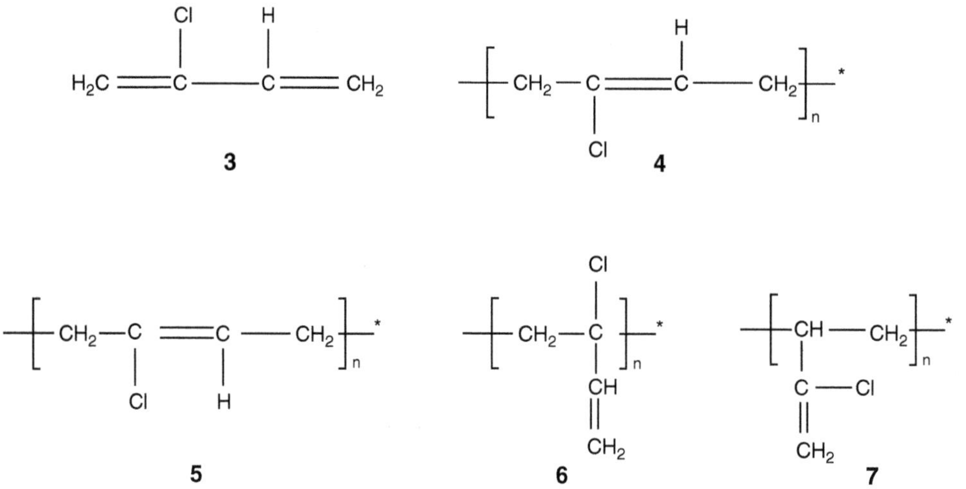

SCHEME 16.2   2-Chloro-1,3-butadiene, **3**; *trans*-1,4 structure, **4**; *cis*-1,4 structure, **5**; *cis*-1,2 structure, **6**; *cis*-3,4 structure, **7**

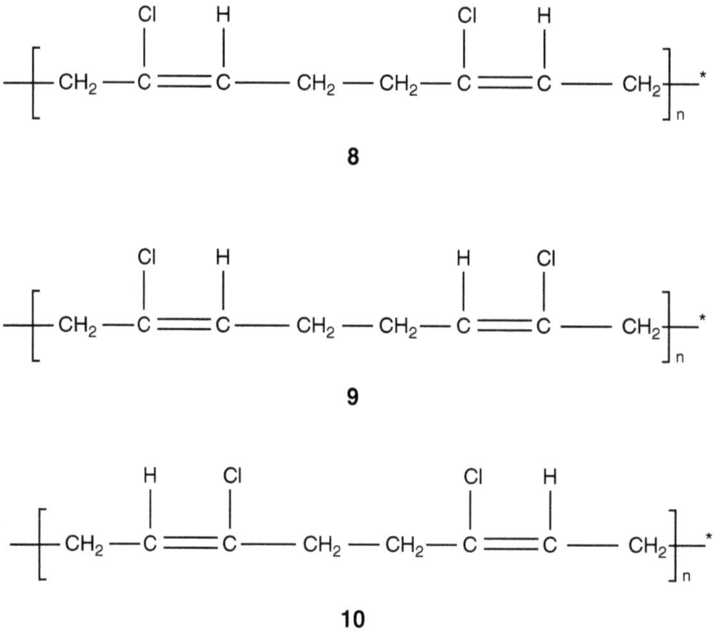

SCHEME 16.3   *Head-to-tail* configuration, **8**; *head-to-head* configuration, **9**; *tail-to-tail* configuration, **10**

Perhaps because of the fact that approximately 13% of the polymer units are structures other than *trans*-1,4, neoprene polymerized at 40 °C has approximately 12% crystallinity at 25 °C.

In addition to the four ways in which a single monomer unit can be incorporated into the polymer chain, it is possible to incorporate consecutive units *head-to-tail* or *head-to-head/tail-to-tail*. The *head-to-tail* configuration (**8**) is shown in Scheme 16.3. In the *head-to-head* configuration, sequential chlorine-containing carbons are separated by four carbons (**9**, Scheme 16.3). In the *tail-to-tail* configuration, sequential chlorine-containing carbons are separated by two carbons (**10**, Scheme 16.3). Approximately 20–30% of the units are incorporated in the *head-to-head/tail-to-tail* configuration.

Chloroprene is often copolymerized with sulfur or 2,3-dichloro-1,3-butadiene, leading to additional degrees of freedom in molecular structure in both the copolymer composition distribution (CCD—the distribution of copolymer compositions among chains), and the copolymer sequence distribution (CSD—the distribution of sequence lengths of comonomer within the chains). Copolymerization with sulfur is effected to provide sulfur–sulfur bonds in the molecular backbone which can be cleaved in postpolymerization reactions to control molecular weight. Copolymerization with 2,3-dichloro-1,3-butadiene is also carried out, as way of disrupting the regularity of the polymer chains, and thus of reducing crystallinity.

Finally, structure on a slightly larger scale is introduced by long chain branching and cross-linking. Since long chain branching is more prevalent at high monomer conversion, the polymerization is often stopped below full conversion to minimize branching. Long-chain branching can also be minimized by low polymerization temperature.

It becomes evident that with a reasonably simple monomer (chloroprene being a substituted butadiene), one must consider four structures for the incorporation of monomer units, two types of interunit connections, two potential comonomers, long-chain branching, and cross-linking, as well as molecular weight distribution. The sum of all of these structural variables determines, for the most part, the end-use properties of the polymer and all are strong functions of polymerization conditions.

The challenge in modeling polymerization reactions is the fact that the *structural properties* to be modeled (molecular weight distribution, CCD and sequence distribution, as well as the other structural properties described in the preceding) are not only functions of reaction time, but are also *distributions* of properties. To completely characterize polymer structure, evolution requires extremely complex models. The discussion in the following will serve to introduce some of the simplest models.

16.1.2 Mass and Energy Balances

As with the modeling of any chemical process system, the modeling of polymerization reactors starts with mass balances [3]. For instance, for the simplest case of living anionic polymerization in a batch reactor:

$$A^- + M \xrightarrow{k_i} P_1 \qquad (16.1)$$

$$P_n + M \xrightarrow{k_i} P_{n+1}; \quad n \geq 1 \qquad (16.2)$$

A balance on monomer, M can be written as:

$$\frac{dM}{dt} = -k_p MP \qquad (16.3)$$

For a batch reactor, the monomer is added once and can be lost only by polymerization. Similarly, a balance on the total concentration of live polymer chains, P, can be written as:

$$\frac{dP}{dt} = -k_i MA^- \qquad (16.4)$$

For a living polymerization in a batch reactor, living chains (P) are created by initiation, but are not destroyed (no termination reaction).

Likewise, an enthalpy balance for the reactor can be written as:

$$\rho C_p V \frac{dP}{dt} = V(-\Delta H_p)k_p PM - UA(T - T_j) \qquad (16.5)$$

Here, heat is added via the enthalpy of polymerization (ΔH_p) and removed through heat transfer to a jacket or cooling coil. More complex chemistries and/or reactors other than simple batch will add complexity to these equations, as well as adding an additional species balance for each additional chemical species. More complex models will be developed later in this text.

16.1.3 Population Balance Techniques

While mass balance modeling will result in the values of scalar properties (monomer concentration, etc.), properties of a polymerization system such as molecular weight, copolymer composition, and copolymer sequence length are, due to the nature of the polymerization processes, *distributions* of properties. Thus, for example, there will be a distribution of molecular weights within the product of most polymerization reactors. In these cases, *population balances*, rather than species balances must be written for these properties. Three examples are given in the following. More complete treatments of the technique can be found elsewhere [3, 4].

16.1.3.1 *Molecular Weight Distribution* The integration of the aforementioned equations allows one to calculate the conversion of monomer to polymer as a function of reaction time. However, they tell nothing about the evolution of the molecular weight. To model the development of molecular weight, a population balance approach is required. In this approach, a balance is written over every value of P_n, where P_n is the concentration of living polymer chains containing

exactly n monomer units. The total concentration of live chains, P, is the summation over n of all P_n:

$$P = \sum_{n=1}^{\infty} P_n \qquad (16.6)$$

A balance over P_n, where $n > 1$ gives:

$$\frac{dP_n}{dt} = -k_p M (P_n - P_{n-1}) \qquad (16.7)$$

Here, living chains of length n are formed when a chain of length $n-1$ adds a monomer unit and removed when a chain of length n adds a monomer unit to become a P_{n+1}.

The preceding equation is valid over all values of n except one. The balance on P_1 must contain the rate of initiation of new chains:

$$\frac{dP_1}{dt} = -k_i M A^- - k_p M P_1 \qquad (16.8)$$

Here, chains of length one are formed by initiation and removed via propagation.

Equations 16.7 and 16.8 constitute a set of *differential difference equations*. Since the value of n can range from zero to infinity, the differential difference equation can be expanded into an infinite set of ordinary differential equations.

One could conceivably write a species balance for P_n for each value of n. By solving these equations sequentially (and numerically), one could derive the chain length distribution of live chains, P_n. However, one does not know the maximum chain length (maximum value of n). For polymers with an average molecular weight of one million, n could range from one to 10,000. Simultaneous (or sequential) integration of millions of species balances is cumbersome at best, and one never knows *a priori* the value of n at which P_n becomes insignificant. Clearly, a better solution is needed in the analysis of the population balance. The most common solution is the method of moments, described later in this text.

16.1.3.2 Copolymer Composition Distribution

A simple copolymerization, with no termination, is shown in the following:

$$
\begin{aligned}
P_{n,m} + A &\xrightarrow{k_{p11}} P_{n+1,m} \\
P_{n,m} + B &\xrightarrow{k_{p12}} Q_{n,m+1} \\
Q_{n,m} + A &\xrightarrow{k_{p21}} P_{n+1,m} \\
Q_{n,m} + B &\xrightarrow{k_{p22}} Q_{n+1,m}
\end{aligned}
\qquad (16.9)
$$

Here, a $P_{n,m}$ is a living chain containing n A units and m B units and ending in an A unit. Likewise, a $Q_{n,m}$ is a living chain containing n A units and m B units and ending in a B unit. Balances on $P_{n,m}$ and $Q_{n,m}$ can be written as:

$$
\begin{aligned}
\frac{dP_{n,m}}{dt} &= -k_{p11} A P_{n-1,m} - k_{p11} A P_{n,m} + k_{p21} A Q_{n-1,m} - k_{p12} B P_{n,m} \\
\frac{dQ_{n,m}}{dt} &= -k_{p22} B Q_{n-1,m} - k_{p22} B Q_{n,m} + k_{p12} B P_{n-1,m} - k_{p21} A Q_{n,m}
\end{aligned}
$$

$$(16.10)$$

To solve these equations directly, one would need to integrate $(n_{max} \times m_{max})$ simultaneous differential equations, where n_{max} and m_{max} are the highest values of n and m expected. A better approach is to use the method of moments in two dimensions. The aforementioned equations, with some simplifications, can be used to derive the *Mayo–Lewis* or *copolymer equation*, which gives the instantaneous composition of the polymer formed from a specific comonomer composition [5].

16.1.0.1 Copolymer Sequence Distribution

Similarly, a population balance approach can be developed for the copolymer sequence distribution [6]. The chemistry can be represented as:

$$
\begin{aligned}
L_g^* + A &\xrightarrow{k_{p11}} L_{g+1}^* \\
L_g^* + B &\xrightarrow{k_{p12}} Q_1^* \\
Q_g^* + A &\xrightarrow{k_{p21}} L_1^* \\
Q_g^* + B &\xrightarrow{k_{p22}} Q_{g+1}^*
\end{aligned}
\qquad (16.11)
$$

Here, L_g^* is an active (still growing) sequence of g monomer units of type A in a row.

A balance on L_g^* can be written as:

$$\frac{dL_1^*}{dt} = -k_{p11} A L_1^* - k_{p12} B L_1^* + k_{p21} A Q^*, \quad Q^* = \sum_{g=1}^{\infty} Q_g^*$$

$$\frac{dL_g^*}{dt} = k_{p11} A L_{g-1}^* - k_{p11} A L_g^* - k_{p12} B L_g^*, \quad n \geq 2$$

$$(16.12)$$

Here, a sequence of length $g-1$ adds a unit of A to become a g sequence, a sequence of g adds an A unit to become a $g+1$ sequence, and a sequence of g units adds a B unit. The addition of the B unit terminates the sequence L_g^*. (In this notation, the asterisk denotes an active, or growing sequence. Once a B unit is added, the A sequence is terminated.) Again, the range of values of the index (g) is unknown. Similarly, a balance on O_g^* (an active sequence of *g* monomer units of type B in a row) can be written. Due to the exact symmetry, it is not given here.

16.1.4 Method of Moments

The problem inherent in population balance modeling of an infinite set of differential equations traditionally has been addressed by the method of moments, as described in the following.

16.1.4.1 *Definition of Moments*   Following the nomenclature given in the preceding and that of Ray [7], the *k*th moment of the number chain length distribution (NCLD) can be defined as:

$$\lambda_k = \sum_{n=1}^{\infty} n^k P_n, \quad k = 0, 1, 2, \ldots \tag{16.13}$$

The number average chain length (NACL) is the number average of the NCLD and can be written as the ratio of the first to zeroth moments:

$$\mu_n = \frac{\lambda_1}{\lambda_0} \tag{16.14}$$

The variance of the NCLD can be written as:

$$\sigma_n^2 = \frac{\lambda_2}{\lambda_0} - \left(\frac{\lambda_1}{\lambda_0}\right)^2 \tag{16.15}$$

The weight average chain length (WCLD) can be written as:

$$\mu_w = \frac{\lambda_2}{\lambda_1} \tag{16.16}$$

The number average molecular weight (NAMW) can be found by multiplying the NACL by the molecular weight of a monomer unit:

$$m_n = \mu_n W \tag{16.17}$$

Likewise, the weight average molecular weight (WAMW) can be found by multiplying the WACL by the molecular weight of a monomer unit:

$$m_m = \mu_m W \tag{16.18}$$

Finally, the *polydispersity* is defined as the ratio of the WACL to the NACL (or the WAMW to the NAMW):

$$D = \frac{\mu_w}{\mu_n} = \frac{m_w}{m_n} = \frac{\lambda_0 \lambda_2}{\lambda_1^2} \tag{16.19}$$

The polydispersity is the most common measure of the breadth of the molecular weight distribution because it is easily obtained from gel permeation chromatography analysis, and because, as will be shown later, it is a normalized measure of breadth of distribution (unlike variance).

16.1.4.2 *Example: Population Balance to Moment Equations*   The method of moments can be used to convert the differential difference equation (or infinite set of ordinary differential equations) represented by Equations 16.7 and 16.8 into a finite set of ordinary differential equations in the moments of the distribution. Then, using Equations , the important statistics of the NCLD can be derived.

As an example, consider the batch anionic polymerization represented by Equations 16.1 and 16.3 for very fast initiation and no terminations, the system can be described by Equations . However, for very rapid termination, Equation 16.4 may be replaced by:

$$\frac{dP}{dt} = 0 \tag{16.20}$$

The initial conditions for the system are as follows:

$$M(0) = M_0; \quad P(0) = 0; \quad P_1(0) = P_{10}; \quad P_n(0) = 0, \quad n \geq 2 \tag{16.21}$$

Equation 16.7 may be multiplied by n^k and added to Equation 16.8, and then summed over all values on *n*, resulting in:

$$\sum_{n=1}^{\infty} n^k \frac{dP_n}{dt} = -k_p M \sum_{n=1}^{\infty} n^k P_n + k_p M \sum_{n=2}^{\infty} n^k P_{n-1},$$
$$\sum_{n=1}^{\infty} P_n = P_{10} \tag{16.22}$$

Here,

$$\sum_{n=1}^{\infty} n^k P_n = \lambda_k, \quad \sum_{n=1}^{\infty} n^k \frac{dP_n}{dt} = \frac{d\lambda_k}{dt} \tag{16.23}$$

The final term in Equation 16.22 may be evaluated for $k = 1$, 2, 3 as follows:

$$k = 0, \quad \sum_{n=2}^{\infty} n^k P_{n-1} = \lambda_0 \tag{16.24}$$

$$k = 1, \quad \sum_{n=2}^{\infty} n^k P_{n-1} = \lambda_0 + \lambda_1 \tag{16.25}$$

$$k = 2, \quad \sum_{n=2}^{\infty} n^k P_{n-1} = \lambda_0 + 2\lambda_1 + \lambda_2 \tag{16.26}$$

Thus, the differential equations for the first three moments become:

$$\frac{d\lambda_0}{dt} = 0, \quad \lambda_0(0) = P_{10} \tag{16.27}$$

$$\frac{d\lambda_1}{dt} = k_p M \lambda_0, \quad \lambda_1(0) = P_{10} \tag{16.28}$$

$$\frac{d\lambda_2}{dt} = k_p M \lambda_0 + 2k_p M \lambda_1, \quad \lambda_2(0) = P_{10} \tag{16.29}$$

Integration of Equations 16.26–16.29 along with the monomer balance will describe the monomer conversion and molecular weight distribution as a function of time for the batch anionic polymerization. Addition of the enthalpy balance will allow the simulation to be done nonisothermally.

16.1.5 z-Transform Techniques

The previous example of living, anionic polymerization with very rapid initiation is, admittedly, the simplest application of the method of moments to polymerization problems. While the moment equations could be derived by clever algebra (Eqs. 16.24–16.26), it is often the case that the proper manipulation of summations is not obvious. In this case, it is possible to augment the method of moments with the method of z-transforms.

16.1.5.1 *Properties of z-Transforms* The z-transforms are the digital equivalent of Laplace transforms. They are often used in digital process control and digital signal processing. In these cases, the discreteness is in the time variable (sampled data systems). Polymerization processes are by nature discrete. However, the discreteness is in the chain length variable, rather than in time. So, while z-transforms are used in digital signal processing to convert difference equations in time into algebraic equations, they will be used here to convert differential difference equations in time (continuous) and chain length (n, discrete, since it is not possible to add a fraction of a monomer unit) into a differential equation in the continuous time domain. Additional details are available elsewhere [3]. For the present purposes, the z-transform of P_n can be defined as:

$$F(z,t) = \sum_{n=0}^{\infty} z^{-n} P_n(t), \quad P_n(t) = 0 \text{ for } n \le 0 \tag{16.30}$$

Two properties of z-transforms are important for the applications given in the following. Discussion of these properties may be found in most texts on digital control and/or digital signal processing. If $F(P_n)$ is defined as in Equation 16.30, the translation theorem is given by:

$$F(P_{n-k}) = z^{-k} F(P_n) \tag{16.31}$$

The moments of the NCLD may be determined from the moment generating equation for z-transforms [7]:

$$\lambda_k(t) = \lim_{z \to 1} \left\{ (-1)^k \frac{\partial^k F(z,t)}{\partial (\ln z)^k} \right\} \tag{16.32}$$

The advantage of using a z-transform approach lies in the fact that very large tables of z-transforms exist, allowing one to take the z-transform of rather complex summations. The simple example of living anionic polymerization with very rapid initiation in a batch reactor will be used to illustrate the technique.

16.1.5.2 *Application of z-Transform Techniques to Polymerization Modeling* Consider the polymerization system described by Equations 16.1 and 16.2 and modeled by Equations 16.3–16.8, 16.20, and 16.21. For simplicity, we will assume isothermal polymerization. In addition, one can make the following variable transformation which has the effect of removing monomer concentration from the live chain equations:

$$\tau = \int_0^t k_p M(t') dt' \tag{16.33}$$

The resulting equations are:

$$\frac{dP_1}{d\tau} = -P_1, \quad P_1(0) = P_{10} \tag{16.34}$$

$$\frac{dP_n}{d\tau} = -(P_n - P_{n-1}), \quad P_n(0) = 0, \quad n \ge 2 \tag{16.35}$$

$$\frac{dM}{d\tau} = -P = -P_{10}, \quad M(0) = M_0 \tag{16.36}$$

Taking the z-transform of Equations 16.34 and 16.35 results in:

$$\frac{dF(z,\tau)}{d\tau} = (z^{-1}-1)F(z,\tau), \quad F(z,0) = z^{-1}P_{10} \qquad (16.37)$$

The equation in the preceding is variables separable and can be easily solved to give:

$$F(z,\tau) = z^{-1}P_{10}\exp(-\tau)\exp(z^{-1}\tau) \qquad (16.38)$$

Using the moment generating equation (Eq. 16.32), the leading moments of the NCLD can be calculated:

$$\lambda_0 = P_{10} \qquad (16.39)$$

$$\lambda_2 = P_{10}(\tau^2 + 3\tau + 1) \qquad (16.40)$$

$$\lambda_1 = P_{10}(\tau + 1) \qquad (16.41)$$

The polydispersity can then be calculated from Equation 16.19:

$$D = \frac{\lambda_0\lambda_2}{\lambda_1^2} = \frac{\tau^2 + 3\tau + 1}{\tau^2 + 2\tau + 1} \qquad (16.42)$$

For large values of τ (large times or high monomer conversion), the polydispersity approaches unity. This is characteristic of living polymerizations. To see this more clearly, Equations 16.15 and 16.19 can be combined to give:

$$D = 1 + \frac{\sigma_n^2}{\mu_n^2} \qquad (16.43)$$

But from Equation 16.15,

$$\mu_n = (1+\tau) \approx \tau = \sigma_n^2 \qquad (16.44)$$

so that Equation 16.43 becomes:

$$D = 1 + \frac{\sigma_n^2}{\mu_n^2} = 1 + \frac{\tau}{\mu_n^2} = 1 + \frac{\tau}{(\tau+1)^2} \approx 1 + \frac{\tau}{\mu_n\tau} = 1 + \frac{1}{\mu_n} \qquad (16.45)$$

Thus, once again, for high NACL (large times and high monomer conversion for a living polymerization), the polydispersity approaches unity.

Alternatively, Equation 16.37 can be expanded in a power series:

$$F(z,t) = P_{10}\exp(-\tau)\sum_{n=1}^{\infty}\frac{(\tau)^{n-1}}{(n-1)!}z^{-n} \qquad (16.46)$$

Equating the coefficients gives:

$$P_n(\tau) = P_{10}\exp(-\tau)\frac{(\tau)^{n-1}}{(n-1)}, \quad n \geq 1 \qquad (16.47)$$

This is a Poisson distribution with mean $(\tau + 1)$ and variance τ (as shown earlier with the method of moments).

16.1.6 Common Distributions for Polymer Property Description

As was seen earlier, the NCLD for batch living anionic polymerization can be described by a Poisson distribution. Since most distributed properties of polymer systems can be described by standard statistical distributions, the common distributions are discussed in the following. The following distributions are written in as generic a form as possible; many molecular weight distributions are described by these distributions with additional constraints on the parameters of the distribution. In each case, the distribution is presented with the left-hand side of the equation representing the number fraction of chains of length n (P_n/P), where P_n is the concentration of chains of length n and P is the total concentration of chains of all lengths as defined by Equation 16.6. The mean and variance given are for the NCLD.

16.1.6.1 *Poisson Distribution*   The Poisson distribution is described by the following equations. For polymerization problems, Equation 16.49 is appropriate since chains of length zero are not considered; Equation 16.48 is included for completeness only. The Poisson distribution is a one-parameter distribution with the mean approximately equal to the variance. It is a symmetric distribution when plotted as NCLD. The Poisson distribution describes the MWD of simple batch anionic polymerization kinetics:

$$\frac{P_n}{P} = e^{-a}\frac{a^n}{n!}, \quad n \geq 0; \quad \mu_n = a; \quad \sigma_n^2 = a \qquad (16.48)$$

or

$$\frac{P_n}{P} = e^{-a}\frac{a^{(n-1)}}{(n-1)!}, \quad n \geq 1; \quad \mu_n = a+1; \quad \sigma_n^2 = a \qquad (16.49)$$

16.1.6.2 *Gaussian (Normal) Distribution*   The Gaussian or normal distribution is described by the following equations. It is a two-parameter distribution with the mean being a set independent of the variance b^2. The Gaussian

distribution describes the MWD of simple batch anionic polymerization kinetics when the continuous variable approximation is made. The approximation implies large values of n and in this case, the Gaussian distribution with equal mean and variance is a good approximation to the Poisson distribution:

$$\frac{P_n}{P} = \left(\frac{1}{2\pi b^2}\right)^{1/2} \exp\left[\frac{-(n-a)^2}{2b^2}\right], \quad -\infty \leq n \leq \infty; \quad (16.50)$$
$$a = \text{mean}; \quad b^2 = \text{variance}$$

16.1.6.3 *Log Normal Distribution* The log normal distribution is described by the following equations. It is a two-parameter distribution derived by using the natural logarithm of n rather than n in the normal distribution. Although the log normal distribution is not the exact solution to any of the simple polymerization kinetic schemes, it is often used to fit experimental molecular weight distribution data:

$$\frac{P_n}{P} = \left(\frac{1}{2\pi n^2 b^2}\right)^{1/2} \exp\left[\frac{-(\ln n - \ln a)^2}{2b^2}\right]; \quad 0 \leq n \leq \infty$$
$$\text{Mean} = \exp\left[\ln a + \frac{b^2}{2}\right] \quad (16.51)$$
$$\text{Variance} = (\exp[b^2] - 1)\exp[\ln a + b^2]$$

16.1.6.4 *Most Probable (Flory) Distribution* The most probable or Flory distribution is described by the following equations. For polymerization problems, Equation 16.53 is appropriate, since chains of length zero are not considered; Equation 16.52 is included for completeness only. It is a single-parameter distribution, with the shape of the distribution dependent only on the value of α. The Flory distribution describes the NCLD for a number of kinetic schemes, including batch condensation polymerization and anionic polymerization in a CSTR:

$$\frac{P_n}{P} = \alpha^n (1-\alpha); \quad n \geq 0; \quad \alpha < 1$$
$$\text{Mean} = \frac{\alpha}{1-\alpha}; \quad \text{variance} = \frac{2\alpha^2}{(1-\alpha)^2} \quad (16.52)$$

or

$$\frac{P_n}{P} = \alpha^{n-1}(1-\alpha); \quad n \geq 1; \quad \alpha < 1$$
$$\text{Mean} = \frac{\alpha}{1-\alpha}; \quad \text{variance} = \frac{\alpha}{(1-\alpha)^2} \quad (16.53)$$

16.2 KINETIC ANALYSIS OF POLYMERIZATION MECHANISMS

16.2.1 Chain-Growth Chemistries

16.2.1.1 *Anionic Polymerization* Anionic living polymerization has been considered earlier since it is perhaps the simplest kinetic scheme from which a rigorous analysis can be derived. The complete anionic polymerization kinetics can be written as follows:

$$\begin{aligned}
AC &\overset{K}{\leftrightarrow} A^- + C^+ &&\text{Initiation} \\
A^- + M &\overset{k_i}{\longrightarrow} AM^- &&\text{Initiation} \\
AM_n^- + M &\overset{k_p}{\longrightarrow} AM_{n+1}^- &&\text{Propagation} \\
AM_n^- + B &\overset{k_f}{\longrightarrow} AM_n + B^- &&\text{Chain transfer}
\end{aligned} \quad (16.54)$$

The nomenclature can be simplified to:

$$\begin{aligned}
AC &\overset{K}{\leftrightarrow} A^- + C^+ &&\text{Initiation} \\
A^- + M &\overset{k_i}{\longrightarrow} P_1 &&\text{Initiation} \\
P_n + M &\overset{k_p}{\longrightarrow} P_{n+1} &&\text{Propagation} \\
P_n + B &\overset{k_f}{\longrightarrow} M_n + B^- &&\text{Chain transfer}
\end{aligned} \quad (16.55)$$

To illustrate the essential features of an ionic polymerization (i.e., livingness), instantaneous initiation is assumed, and no termination. In this case, the population balances describing the development of molecular weight are Equations 16.6–16.8. The complete solution for this kinetics is given in the examples in previous sections. The key result is that for living polymerization, the theoretical minimum molecular weight polydispersity is unity. In addition, simple kinetic analysis yields the following for the rate of polymerization:

$$R_p = k_p PM = \frac{k_i k_p A^- M^2}{k_f B} = \frac{K k_i k_p (AC) M^2}{k_f C^+ B}; \quad P = \sum_{n=1}^{\infty} P_n \quad (16.56)$$

The instantaneous degree of polymerization, in the absence of termination, can be derived as:

$$\bar{x} = \frac{k_p M}{k_f B} \quad (16.57)$$

Note that the kinetic chain length is controlled by the extent of chain transfer. Also note that this is the *instantaneous* degree of polymerization, whereas the method of moments gives the information on the cumulative distribution.

16.2.1.2 *Cationic Polymerization*

The chemistry of cationic polymerization may be given as follows:

$$A + RH \overset{K}{\leftrightarrow} H^+AR^- \qquad \text{Initiation}$$
$$H^+AR^- + M \overset{k_i}{\longrightarrow} HM^+AR^- \qquad \text{Initiation}$$
$$HM_n^+AR^- + M \overset{k_p}{\longrightarrow} HM_{n+1}^+AR^- \qquad \text{Propagation}$$
$$HM_n^+AR^- \overset{k_t}{\longrightarrow} M_n + H^+AR^- \qquad \text{Termination}$$
$$HM_n^+AR^- + M \overset{k_f}{\longrightarrow} M_n + HM^+AR^- \qquad \text{Chain transfer}$$

$$\text{(16.58)}$$

As earlier, the nomenclature can be simplified to:

$$A + RH \overset{K}{\leftrightarrow} H^+AR^- \qquad \text{Initiation}$$
$$P_n + M \overset{k_p}{\rightarrow} P_{n+1} \qquad \text{Propagation}$$
$$P_n \overset{k_t}{\rightarrow} M_n + H^+AR^- \qquad \text{Termination}$$
$$P_n + M \overset{k_f}{\rightarrow} M_n + P_1 \qquad \text{Chain transfer}$$

$$\text{(16.59)}$$

A simple analysis will yield the rate of polymerization as:

$$R_p = k_pPM = \frac{Kk_ik_pA(AH)M^2}{k_t} \qquad P = \sum_{n=1}^{\infty} P_n \qquad \text{(16.60)}$$

The instantaneous degree of polymerization is given by:

$$\overline{x} = \frac{k_pPM}{k_tP + k_fPM} = \frac{k_pM}{k_t + k_fM} \qquad \text{(16.61)}$$

A rigorous analysis of the MWD via moment techniques is easily done, but is omitted here for brevity. In many cases, cationic polymerization can be made to be essentially living (i.e., termination is negligible). The comments on polydispersity for anionic polymerization apply here as well.

16.2.1.3 *Free Radical Polymerization*

Free radical polymerization is one of the most widely used polymerization chemistries owing to its ability to polymerize a wide variety of monomers (those with carbon–carbon double bonds) and, unlike ionic polymerization, the rather mild conditions under which it can be carried out. The complete kinetic scheme for free radical polymerization can be described as follows:

$$\text{(16.62)}$$
$$I \overset{k_d}{\longrightarrow} 2R \qquad \text{Initiation}$$
$$M + R \overset{k_i}{\longrightarrow} P_1 \qquad \text{Inititation}$$
$$P_n + M \overset{k_p}{\longrightarrow} P_{n+1} \qquad \text{Propagation}$$
$$P_n + P_m \overset{k_{t_c}}{\longrightarrow} M_{n+m} \qquad \text{Termination by combination}$$
$$P_n + P_m \overset{k_{t_d}}{\longrightarrow} M_n + M_m \qquad \text{Termination by disproportionation}$$
$$P_n + M \overset{k_{f_m}}{\longrightarrow} M_n + P_1 \qquad \text{Chain transfer to monomer}$$
$$P_n + S \overset{k_{f_s}}{\longrightarrow} M_n + S^\bullet \qquad \text{Chain transfer to solvent}$$
$$P_n + T \overset{k_{f_t}}{\longrightarrow} M_n + T^\bullet \qquad \text{Chain transfer to transfer agent}$$
$$P_n + M_m \overset{k_{f_p}}{\longrightarrow} M_n + P_m \qquad \text{Chain transfer to polymer}$$
$$R + \text{In} \overset{k_{in}}{\longrightarrow} Q \qquad \text{Inhibition}$$

A complete analysis will start with balances on monomer and initiator:

$$\frac{dM}{dt} = -k_pPM; \quad M(0) = M_0 \qquad \text{(16.63)}$$

$$\frac{dI}{dt} = -k_dI; \quad I(0) = I_0 \qquad \text{(16.64)}$$

In addition, balances on primary radicals (R) and total live chains (P) can be written as follows:

$$\frac{dR}{dt} = 2k_dfI - k_iRM \qquad \text{(16.65)}$$

$$\frac{dP}{dt} = k_iRM - k_tP^2 \qquad \text{(16.66)}$$

One can make the quasi-steady state approximation (QSSA) for radicals (R and P). This assumes that radical reactions are fast compared with other reactions and so can be considered to be always at steady state; thus the left-hand sides of Equations 16.65 and 16.66 may be set to zero. Solution of Equation 16.66 for P and substitution of Equation 16.65 into the result gives the concentration of live chains:

$$P = \left[\frac{2fk_dI}{k_{t_c} + k_{t_d}} \right]^{1/2} \qquad \text{(16.67)}$$

With this result, the rate of polymerization may be written as:

$$R_p = -\frac{dM}{dt} = k_p M P = k_p M \left[\frac{2 f k_d I}{k_t} \right]^{1/2} \qquad (16.68)$$

In the case where inhibition is significant, Equation 16.68 becomes:

$$P = \left[\left(\frac{2 f k_d I}{k_t} \right) \left(\frac{1}{1 + (k_{in} In)/(k_i M)} \right) \right]^{1/2} \qquad (16.69)$$

The instantaneous degree of polymerization gives a measure of the *instantaneous* kinetic chain length. It may be written as the ratio of the rate of propagation to the rate of production of dead chains by various mechanisms:

$$\bar{x} = \frac{k_p M P}{(1/2)k_{t_c} P^2 + k_{t_d} P^2 + k_{fm} M P + k_{fs} P S + k_{ft} P T} \qquad (16.70)$$

The method of moments may be applied to the kinetic scheme in Equation 16.62 (neglecting chain transfer to polymer) to analyze the MWD. Following studies reported by Ray [7] and Schork et al. [3], balances on living chain of length 1 and n may be written as follows:

$$\frac{dP_1}{dt} = k_i R M_1 - k_p P_1 M + (k_{fs} S + k_{fm} M)(P - P_1)$$
$$- (k_{t_c} + k_{t_d}) P P_1; \quad P_1(0) = 0 \qquad (16.71)$$

$$\frac{dP_n}{dt} = k_p M (P_{n-1} - P_n) - (k_{fs} S + k_{fm} M) P_n$$
$$- (k_{t_c} + k_{t_d}) P P_n; \quad P_n(0) = 0; \quad n > 1 \qquad (16.72)$$

If the QSSA is made for P_1, Equation 16.71 may be solved as:

$$P_1 = (1 - \alpha) P \qquad (16.73)$$

where the probability of propagation is defined as:

$$\alpha = \frac{k_p M}{k_p M + k_{fm} M + k_{fs} S + (k_{t_c} + k_{t_d}) P} \qquad (16.74)$$

Making the QSSA for P_n results in:

$$P_n = \alpha P_{n-1} \qquad (16.75)$$

Equations 16.73 and 16.75 can be combined to give:

$$P_n = (1 - \alpha) P \alpha^{n-1} \qquad (16.76)$$

Note that this is a Flory or "most probable" distribution. Equation 16.76 can also be derived from a z-transform approach [3]. Equation 16.76 and the definition of the moments of the live chain distribution can be used to derive the first three moments of the live chain distribution:

$$\lambda_0 = P \qquad (16.77)$$

$$\lambda_1 = \frac{P}{(1 - \alpha)} \qquad (16.78)$$

$$\lambda_2 = \frac{P(1 + \alpha)}{(1 - \alpha)} \qquad (16.79)$$

The NACL, WACL, and polydispersity can be expressed in terms of the moments:

$$\mu_n = \frac{\lambda_1}{\lambda_0} = \frac{1}{(1 - \alpha)} \quad \text{(live polymer)} \qquad (16.80)$$

$$\mu_w = \frac{\lambda_2}{\lambda_1} = \frac{(1 + \alpha)}{(1 - \alpha)} \quad \text{(live polymer)} \qquad (16.81)$$

$$D = \frac{m_w}{m_n} = \frac{\mu_w}{\mu_n} = (1 + \alpha) \quad \text{(live polymer)} \qquad (16.82)$$

For long chains, the probability of propagation approaches unity, so that the polydispersity becomes approximately:

$$D = (1 + \alpha) \approx 2 \quad \text{(live polymer)} \qquad (16.83)$$

In free radical polymerization, the product is made up of dead, rather than live chains, since the lifetime of a live chain in only 1–10 s. Therefore, the statistics of the MWD of the *dead* chains is critical. An approach based on population balances for M_n, the concentration of dead chains containing n monomer units can be used. This approach is given in detail in Reference [3]. It requires the use of z-transform techniques, and contains a number of assumptions. Here, we will take a simplified approach. If termination occurs solely by disproportionation, the MWD of the dead chains will be identical to that of the live chains. However, since live chains exist for only very short times, while dead chains remain forever, the live chain distribution is essentially an instantaneous distribution, while the dead chain distribution is a cumulative one. This distinction may be important in modeling a specific polymer product. If it is, then the moment equations for the dead polymer NCLD must be integrated numerically. The derivation in Reference [3] implicitly makes the assumption that the instantaneous and cumulative distributions are identical. This gives a minimum value of 2.0 for the polydispersity for the product. The approach given here suffers from the same limitations, and gives the same lower bound on polydispersity.

As stated earlier, if termination is exclusively by disproportionation, the dead chains formed over a narrow period of time will have the same MWD as the live chains from which they are derived. Thus, Equations apply as well to the dead chain distribution. Therefore, for terminations solely by disproportionation, the minimum value of polydispersity attainable is 2.0. However, very few, if any, free radical monomers terminate solely by disproportionation. While the methacrylates have significant levels of disproportionation (as well as combination), most other monomers terminate almost exclusively by combination. For termination by combination, it is easy to argue that the NACL of the dead chains should be twice that of the live chains:

$$\frac{\mu_n(\text{dead})}{\mu_n(\text{live})} = 2 \quad (\text{termination by combination}) \quad (16.84)$$

Predicting the WACL is not as easy. However, one can make use of well-known theory from mathematical statistics on combining two distributions. If elements are drawn N at a time from a distribution, the distribution of the *sums* of the N elements can be related to the original distribution by:

$$\left(\frac{\sigma^2}{\mu_n^2}\right)_{\text{sum}} = \frac{1}{N}\left(\frac{\sigma^2}{\mu_n^2}\right)_{\text{original}} \quad (16.85)$$

Then, the polydispersity of the new distribution can be written as:

$$D_{\text{sum}} = 1 + \left(\frac{\sigma^2}{\mu_n^2}\right)_{\text{sum}} = 1 + \frac{1}{N}\left(\frac{\sigma^2}{\mu_n^2}\right)_{\text{original}} = 1 + \frac{1}{N}(D_{\text{original}} - 1)$$

$$(16.86)$$

For $N=2$ (termination by combination), Equation 16.86 gives the polydispersity of the dead chains:

$$D_d = 1 + \frac{1}{2}(D_1 - 1) = 1 + \frac{1}{2}(2 - 1) = 1.5 \quad (16.87)$$

Substituting Equations 16.87 and 16.80 into Equation 16.82 and solving for the WACL gives:

$$\mu_w = D\mu_n = \frac{1.5}{(1-\alpha)} \quad (16.88)$$

(dead polymer – termination by combination)

Thus, for termination by combination, the polydispersity of the dead chains will have a minimum polydispersity of 1.5. For systems with both disproportionation and combination, the minimum polydispersity will be somewhere between 1.5 and 2.0, depending on the relative rates of the two termination mechanisms. In both cases, the minimum values of polydispersity are absolute minima, and not likely to be attained in practice. The derivations of the minima assume constant temperature, constant monomer concentration (clearly not possible in a batch reactor), constant radical flux, and perfect mixing. For these reasons, actual polydispersities will always be greater than the appropriate minimum.

16.2.1.4 *Controlled Radical Polymerization* Controlled radical polymerization (CRP) involves the use of one of a number of special chemistries to suppress termination during free radical polymerization. The most common chemistries are nitroxide-mediated polymerization (NMP), atom transfer radical polymerization (ATRP), and reversible addition-fragmentation chain transfer polymerization (RAFT). Each has advantages and disadvantages, and each is particularly successful at polymerizing certain monomers. The common theme is that bimolecular termination of living polymer chains is suppressed. In the limit, this gives a living character to polymer produced under mild, free radical polymerization conditions. Molecular weight polydispersities significantly below the theoretical limit of 1.5–2.0 for free radical polymerization, and sometimes approaching the theoretical limit of 1.0 for truly living polymerization have been reported. Mathematical modeling of these systems have been done by Zargar and Schork [6, 8], Ye and Schork [9, 10], and many others. For the purposes of this text, it is sufficient to note that the character of the polymerization and the character of the product approach are that of truly living ionic polymerizations.

The real advantage of CRP lies in the ability to synthesize block copolymers from free radical monomers without the very stringent polymerization conditions necessary for ionic polymerization. Since the chains remain living, one monomer can be completely polymerized and then another monomer added to form a second block on the same (living) chains. This process can be repeated to form multiblock polymers. This aspect of CRP will be addressed in the following as part of the modeling of copolymerization reactions.

16.2.2 Step-Growth Polymerization

Addition polymerization occurs through the sequential addition of monomer units (via ionic or free radical chemistries) to a growing polymer chain. Growth may or may not be followed by chain termination. Because of this, high polymer exists at very low conversion of monomer to polymer, and residual monomer exists at very high conversion. Step-growth polymerization proceeds by a very different set of kinetics. In step-growth, monomer units form dimers via a number of reactions, the most common being condensation. The dimers then combine with other monomers, or dimers, or longer chain oligomers. Because of this very different chemistry, high-molecular weight polymer does not exist until the reaction has come to near completion and monomer exists only at the very beginning of the reaction. There are two types of step-growth kinetics, A–B, and AA/BB. We will consider each one in the next sections.

16.2.2.1 A – B Step Growth
Step-growth polymerization takes place through the reaction of two functional groups to form a linkage. We will denote these functional groups as A and B. In general, an A functional group will not react with another A functional group, but only with a B. Likewise, a B will react only with an A. The most common step-growth reactions involve condensations; therefore, step-growth polymerization is sometime referred to as condensation polymerization. It should be stressed that condensation polymerization reactions form only a (large) subset of all step-growth polymerization. The simplest step-growth polymerization reaction takes place when both functional groups are on the same monomer unit. This is referred to as A–B polymerization. A general kinetic scheme for A–B polymerization is shown in the following:

$$(A-B)_n + (A-B)_m \underset{k_{p^-}}{\overset{k_p}{\rightleftarrows}} (A-B-A-B)_{n+m} + W \tag{16.89}$$

The rate expression can be written from mass–action kinetics. Since the stoichiometry dictates that concentration of A groups must be equal to the concentration of B groups, we can write the kinetics in terms of A only. Condensation reactions are catalyzed by acids. If the acid concentration is assumed constant and the reverse reaction can be eliminated (most often by continuously removing the condensation byproduct W, often water), Equation 16.90 gives:

$$R_p = -\frac{dA}{dt} = k_p A B H^+ = k' A^2 \tag{16.90}$$

Since the average chain length will depend on the progress of the reaction, the progress of the reaction can be quantified by introducing the *extent of reaction*, p, defined as the fraction of A or B groups which has reacted at time t. The NACL is then the total number of monomer molecules initially present divided by the total number of molecules present at time t, which can be further related to the extent of reaction as:

$$\mu_n = \frac{N_0}{N} = \frac{N_0}{N_0(1-p)} = \frac{1}{(1-p)} \tag{16.91}$$

A more in-depth analysis can be carried out based on the method of z-transforms. Equation 16.89 can be rewritten in the nomenclature previously used:

$$P_n + P_m \underset{k_{p^-}}{\overset{k_p}{\rightleftarrows}} P_{n+m} + W \tag{16.92}$$

Assuming irreversibility, the population balances can be written as follows:

$$\frac{dP_1}{dt} = -k_p P_1 P; \quad P_1(0) = P_{10} \tag{16.93}$$

$$\frac{dP_n}{dt} = \frac{1}{2} k_p \sum_{r=1}^{n-1} P_r P_{n-r} - k_p P_n P; \quad P_n(0) = P_{n0}; \quad n > 1 \tag{16.94}$$

$$P = \sum_{n=1}^{\infty} P_n \tag{16.95}$$

Multiplying Equations 16.93 and 16.94 by z^{-1} and z^{-n}, respectively, and summing over all values of n gives:

$$\frac{dF(z,t)}{dt} = \frac{1}{2} k_p F^2(z,t) - k_p P(t) F(z,t);$$
$$F(z,0) = f_0(z) = \sum_{n=1}^{\infty} z^{-n} P_{n0} \tag{16.96}$$

$P(t) = F(1, t)$ is given by

$$\frac{dP(t)}{dt} = \frac{1}{2} k_p P^2(t); \quad P(0) = f_0(1) = \sum_{n=1}^{\infty} P_{n0} \tag{16.97}$$

Transforming the time variable as earlier, using Equation 16.33 gives:

$$\frac{dF(z,\tau)}{d\tau} = F^2(z,\tau) - 2P(\tau)F(z,\tau); \quad F(z,0) = f_0(z) \tag{16.98}$$

$$\frac{dP(\tau)}{d\tau} = -P^2(\tau); \quad P(0) = f_0(1) \qquad (16.99)$$

Equation 16.99 may be solved as

$$P(\tau) = \frac{f_0(1)}{f_0(1)\tau + 1} \qquad (16.100)$$

The following substitution may be made into Equation 16.98:

$$y = \frac{F(z,\tau)}{P(\tau)} \qquad (16.101)$$

Then, Equation 16.98 may be written as:

$$\frac{dy}{dP} = \frac{1}{P}y(1-y); \quad y[f_0(1)] = \frac{f_0(z)}{f_0(1)} \qquad (16.102)$$

This may be solved (separation of variables) as:

$$F(z,\tau) = Py = \frac{f_0(z)[(P)/(f_0(1))]^2}{1 - [1 - ((P)/(f_0(1)))]((f_0(z))/(f_0(1)))} \qquad (16.103)$$

Assuming pure monomer is fed to the reactor, $f_0(z) = z^{-1}P_{10}$, Equation 16.103 can be expanded as:

$$F(z,\tau) = P_{10}\sum_{n=1}^{\infty}\left(\frac{P}{P_{10}}\right)^2\left(1 - \frac{P}{P_{10}}\right)^{n-1}z^{-n} \qquad (16.104)$$

Comparing this term by term with the definition of the z-transform yields the NCLD:

$$P_n(\tau) = P_{10}\left(\frac{P}{P_{10}}\right)^2\left(1 - \frac{P}{P_{10}}\right)^{n-1} = \frac{1}{P_{10}\tau^2}\alpha^{n+1}; \quad n \geq 1$$

$$\alpha = \frac{P_{10}\tau}{(P_{10}\tau + 1)}$$

$$(16.105)$$

This is the Flory or most probable distribution. Substituting Equation 16.100 into 16.105 gives:

$$P_n(\tau) = \frac{P_{10}(P_{10}\tau)^{n-1}}{(P_{10}\tau + 1)^{n+1}}; \quad n \geq 1 \qquad (16.106)$$

The moment generating equations (Eq. 16.32) can be applied to Equation 16.103 to give the leading moments of the NCLD, from which the statistics of the NCLD can be found:

$$\mu_n = \frac{\lambda_1}{\lambda_0} = \frac{2-\alpha}{1-\alpha} \qquad (16.107)$$

$$\mu_w = \frac{\lambda_2}{\lambda_1} = 1 + \frac{2}{(1-\alpha)(2-\alpha)} \qquad (16.108)$$

$$D = \frac{\mu_w}{\mu_n} = \frac{(1-\alpha)}{(2-\alpha)} + \frac{2}{(2-\alpha)^2} \qquad (16.109)$$

At high extent of reaction, $\alpha \approx 1$, and the polydispersity takes on a (minimum) value of 2. The mole fraction of polymer of chain length n is:

$$\frac{P_n(\tau)}{P(\tau)} = \frac{\alpha^n}{P_{10}\tau} \qquad (16.110)$$

Writing the extent of reaction as

$$p = \frac{P_{10} - P}{P_{10}} \qquad (16.111)$$

and substituting Equation 16.100 for P, assuming pure monomer initial charge, yields

$$p = \frac{P_{10}\tau}{P_{10}\tau + 1} = \alpha \qquad (16.112)$$

This confirms that a very high extent of conversion is required to form high polymer.

16.2.2.2 A–B Step-Growth Polymerization: Effect of Capping In an A–B step-growth polymerization, the addition of a small amount of monofunctional monomer can be used to control the molecular weight. This technique is known as capping. The reactions occurring (assuming due to removal of the condensation byproduct) can be written as:

$$P_n + P_m \xrightarrow{k_p} P_{n+m} \qquad (16.113)$$

$$M_n + P_m \xrightarrow{k_c} M_{n+m}$$

where P_n is the concentration of bifunctional polymeric species of length n and M_n is the concentration of monofunctional polymeric species of length n.

Balances over these two species can be written as:

$$\frac{dP_n}{dt} = \frac{1}{2} k_p \sum_{r=1}^{n-1} P_{n-r} P_r - k_p P_n P - k_c P_n M \quad (16.114)$$

$$\frac{dM_n}{dt} = k_c \sum_{r=1}^{n-1} P_{n-r} M_r - k_c M_n P \quad (16.115)$$

where, once again,

$$P = \sum_{n=1}^{\infty} P_n \quad (16.116)$$

$$M = \sum_{n=1}^{\infty} M_n \quad (16.117)$$

Initial conditions for the case of pure monomers at time zero are:

$$P_n(0) = \begin{cases} P_{10} & n=1 \\ 0 & n \neq 1 \end{cases}$$
$$\quad (16.118)$$
$$M_n(0) = \begin{cases} M_{10} & n=1 \\ 0 & n \neq 1 \end{cases}$$

This system can be solved via z-transform techniques, as was done by Kilkson [11]. Only the results are given here; however, by inspection, we can see that the concentration of P goes down, while the concentration of M remains constant. Then the initial concentration of monofunctional monomer (M_{10}) controls the ultimate NACL.

According to Kilkson,

$$r \equiv \frac{P_{10}}{P_{10} + M_{10}} \quad (16.119)$$

$$\mu_n = \frac{1}{1-rp} \quad (16.120)$$

$$\mu_w = \frac{1+rp}{1-rp} \quad (16.121)$$

$$D = 1 + rp \quad (16.122)$$

Note that the Kilkson derivation includes monomer in NACL and WACL calculations. The results in the preceding can be compared with the case with no capping agent:

$$\mu_n = \frac{(2-p)}{(1-p)} \text{(no capping)} \approx \frac{1}{1-p} \text{(with } r=1) \quad (16.123)$$

$$\mu_w = 1 + \frac{2}{(1-p)(2-p)} \text{(no capping)} \approx \frac{1+p}{1-p} \text{(with } r=1) \quad (16.124)$$

The results can be summarized as follows: The distribution with capping agent is a Flory distribution as in the case of no capping agent, but with a lower NACL and WACL due to the presence of the capping agent. The MWD is a function of temperature since k_p and k_c do not necessarily change proportionately with temperature since they might quite possibly have different activation energies. Capping, then, reduces the NACL and WACL sharply. The distribution remains a Flory distribution and, therefore, the variance of the NCLD decreases as well. The polydispersity, however, remains close to two.

16.2.2.3 A–A/B–B Step Growth If the two functional groups are on different monomer units, the kinetics of step-growth polymerization are more complex. If two monomers are used, one with two "A" functional groups and one with two "B" functional groups, the polymerization is described as A–A/B–B step-growth polymerization. An example is nylon 66. The polymerization may be represented as:

$$A-A + B-B \underset{k_{p^-}}{\overset{k_p}{\rightleftharpoons}} A-A-B-B + W \quad (16.125)$$

The kinetic derivations look very similar to the A–B case in the preceding, with one very large difference. Unless the two monomers are in exact 1:1 stoichiometric ratio, the excess monomer will limit the molecular weight development. Once polymerization has begun, there are three species present in the reaction (in addition to the by-product, W, which is removed to prevent depolymerization). The structures of these three species and the symbols to be used for them are as follows:

$$\begin{aligned} (A-A-B-B)_{n-1} \, A-A & \quad A_n \\ (A-A-B-B)_n & \quad M_n \quad (16.126) \\ (B-B-A-A)_{n-1} \, B-B & \quad B_n \end{aligned}$$

Given these conditions, three polymerization reactions may occur:

$$A_n + M_m \underset{k_{p^-}}{\overset{k_p}{\rightleftharpoons}} A_{n+m} + W \quad (16.127)$$

$$A_n + B_m \underset{k_{p^-}}{\overset{k_p}{\rightleftarrows}} M_{n+m-1} + W \qquad (16.128)$$

$$B_n + M_m \underset{k_{p^-}}{\overset{k_p}{\rightleftarrows}} B_{n+m} + W \qquad (16.129)$$

$$M_n + M_m \underset{k_{p^-}}{\overset{k_p}{\rightleftarrows}} M_{n+m} + W \qquad (16.130)$$

where W is the (condensation) by-product.
Species balances give:

$$\frac{dA_n}{dt} = 2k_p \sum_{r=1}^{n-1} A_r M_{n-r} - 4k_p A_n \sum_{r=1}^{\infty} B_r - 2k_p A_n \sum_{r=1}^{\infty} M_r r \qquad (16.131)$$

$$\frac{dB_n}{dt} = 2k_p \sum_{r=1}^{n-1} B_r M_{n-r} - 4k_p B_n \sum_{r=1}^{\infty} A_r - 2k_p B_n \sum_{r=1}^{\infty} M_r \qquad (16.132)$$

$$\frac{dM_n}{dt} = 4k_p \sum_{r=1}^{n-1} A_r B_{n-(r+1)} + 2k_p \frac{1}{2} \sum_{r=1}^{\infty} M_r M_{n-r} \\ - 2k_p M_n \sum_{r=1}^{\infty} [M_r + A_r + B_r] \qquad (16.133)$$

If one assumes the reactor is initially filled with A–A and B–B monomers only, the initial conditions can be written as:

$$A_n(0) = \begin{cases} A_{10} & n = 1 \\ 0 & n \neq 1 \end{cases} \qquad (16.134)$$

$$B_n(0) = \begin{cases} B_{10} & n = 1 \\ 0 & n \neq 1 \end{cases} \qquad (16.135)$$

$$M_n(0) = 0 \qquad (16.136)$$

This set of population balances can be solved by z-transform methods [11]. In the interest of brevity, only the solution will be given here. Defining:

$$r = \frac{B_{10}}{A_{10}} \quad \text{(A and B chosen so that } r \leq 1) \qquad (16.137)$$

$$q = \frac{r(1-\upsilon)^2}{(1-r\upsilon)^2} \qquad (16.138)$$

$$\Delta = A_{10} - B_{10} \qquad (16.139)$$

$$\upsilon = \exp[-(A_{10} - B_{10})\tau] \qquad (16.140)$$

$$\tau = \int_0^t 2k_p t' \, dt' \qquad (16.141)$$

the NCLDs for each of the polymerizing species may be derived as:

$$A_n(\tau) = \frac{\Delta}{(1-r\upsilon^2)}(1-q)q^{n-1} \qquad (16.142)$$

$$B_n(\tau) = \frac{r\Delta\upsilon^2}{(1-r\upsilon^2)}(1-q)q^{n-1} \qquad (16.143)$$

$$M_n(\tau) = 2\Delta\left[\frac{1}{(1-r\upsilon)} - \frac{1}{(1-r\upsilon^2)}\right](1-q)q^{n-1} \qquad (16.144)$$

Each of these is a Flory distribution. The NACLs can be evaluated as:

$$\mu_n^a = \mu_n^b = \mu_n^m = \frac{1}{1-q} \qquad (16.145)$$

where the superscripts represent A, B, and M species, respectively. Now,

$$\mu_n = 2\mu_n^a = 2\mu_n^b = 2\mu_n^m = \frac{2}{1-q} \qquad (16.146)$$

since Kilkson defines a monomer unit as A–A–B–B, in deriving μ_n we have defined a monomer unit as an A–A or B–B. For τ large, p goes to 1, υ goes to 0, and q goes to r (Eq. 16.138). Therefore,

$$\mu_n = \frac{2}{1-r}; \quad D = 2 \qquad (16.147)$$

In summary, the NACL is fixed by the extent of reaction and by the stoichiometric ratio of monomers. The extent of reaction must be high to give a reasonably high NACL. The NACL is independent of temperature at fixed extent of reaction. As the stoichiometric ratio becomes less than one, the NACL and WACL decrease sharply. The distribution remains a Flory distribution and, therefore, the variance of the NCLD decreases as well. The polydispersity, however, remains close to two.

16.2.3 Copolymerization

16.2.3.1 Chain-Growth Copolymerization Chain growth copolymerization can be described by Equation 16.9. Here we have ignored chain initiation and termination, since they have little effect on the CCD. Rigorous population balances can be written as in Equation 16.10. They can then be integrated numerically with the rest of the balances making up the complete model. Note that, since the species are denoted by double subscripts, two-dimensional moment equations will be required to model the system. A much simpler derivation will give the Mayo–Lewis equation, or copolymer equation. The instantaneous relative rates of consumption of the monomers A and B can be written as:

$$\frac{d\mathrm{A}}{d\mathrm{B}} = \frac{k_{\mathrm{p11}}P\mathrm{A} + k_{\mathrm{p21}}Q\mathrm{A}}{k_{\mathrm{p12}}P\mathrm{B} + k_{\mathrm{p22}}Q\mathrm{B}} \quad (16.148)$$

Here, P is the total concentration of living chains ending in an A monomer unit and Q is the total concentration of live chains ending in a B monomer unit. If the QSSA is made for P and Q, it follows that the rate of conversion of P to Q equals the rate of conversion of Q to P. Using this assumption and defining the reactivity ratios:

$$r_1 = \frac{k_{\mathrm{p11}}}{k_{\mathrm{p12}}}; \quad r_1 = \frac{k_{\mathrm{p22}}}{k_{\mathrm{p21}}} \quad (16.149)$$

the relative consumption of A and B can be written as:

$$\frac{d\mathrm{A}}{d\mathrm{B}} = \frac{\mathrm{A}(r_1\mathrm{A} + \mathrm{B})}{\mathrm{B}(\mathrm{A} + r_2\mathrm{B})} \quad (16.150)$$

This may be further simplified to:

$$F_1 = \frac{r_1 f_1^2 + f_1 f_2}{r_1 f_1^2 + 2 f_1 f_2 + r_2 f_2^2} \quad (16.151)$$

where F_1 is the mole fraction of A in the copolymer being formed at the current instant (instantaneous copolymer composition), and f_1 and f_2 are the instantaneous mole fractions of A and B, respectively, in the monomer mix.

F_1 represents the average instantaneous copolymer composition. This is adequate since the distribution of instantaneous copolymer compositions has been shown to be quite narrow.

If $r_1 = r_2$, the structure is strictly alternating since each monomer wants only to add the other. If $r_1 r_2 = 1$, the structure is random, since P and Q chains have an equal probability of adding either monomer. If r_1 and r_2 are both greater than unity, a block copolymer results, since cross polymerization is unlikely. Equation 16.151 describes the *instantaneous* copolymer composition. If the monomers are not consumed at the same rate, there will be significant compositional drift over the course of the polymerization.

This is shown in Figure 16.1. Consider, for example, an initial monomer composition of $f_1 = f_2 = 0.5$. For $r_1 = 0.1$, $F_1 \approx 0.1$. Since the polymer is very rich in monomer 2, the monomer composition will move in the direction of higher f_1. This will continue until the end of the polymerization, when the polymer will be almost completely made up of monomer 1. Thus, the first chains polymerized will have a very low composition of monomer 1 ($f_1 = 0.1$) while the last chains polymerized will be homopolymer of monomer 1 ($f_1 = 1.0$). For living polymerization, the same argument can be made for compositional drift *within a single chain*. Thus, for free radical polymerization, compositional drift will take the form of a wide distribution of copolymer composition *among the chains*. The compositional drift can be described by the *integrated* copolymer equation for batch polymerization [12].

$$1 - \frac{M}{M_0} = 1 - \left[\frac{f_1}{f_{10}}\right]^{\alpha} \left[\frac{f_2}{f_{20}}\right]^{\beta} \left[\frac{f_{10} - \delta}{f_1 - \delta}\right]^{\gamma} \quad (16.152)$$

$$\alpha = \frac{r_1}{(1 - r_2)} \qquad \beta = \frac{r_1}{(1 - r_1)}$$
$$\gamma = \frac{(1 - r_1 r_2)}{(1 - r_1)(1 - r_2)} \qquad \delta = \frac{(1 - r_2)}{(2 - r_1 - r_2)} \quad (16.153)$$

The symbols f_{10} and f_{20} are the initial values of f_1 and f_2. The right-hand side of Equation 16.151 is the overall fractional conversion of both monomers, where M is the concentration of both monomers taken together, and M_0 is the initial value of M. This equation can be solved implicitly, or values of monomer mole fractions (f_1 and f_2) can be assumed and the total monomer fractional conversion computed directly.

CCD describes the distribution of compositions of chains. It tells nothing about the distribution of monomers *within* each chain as noted. Particularly, it does not provide any information about the blockiness of the chains. For instance, is the polymer an *alternating* copolymer? A *random* copolymer? Are there long sequences of A monomer interspersed with sequences of B monomer? The CSD describes the distribution of A (or B) sequence lengths within the polymer, *independent* of the question of on which chain each sequence resides. The CSD can be developed for monomer A and monomer B. There is no reason, in general, that the two should be the same. The CCD can be derived from Equations 16.11 and 16.12 as follows. The two equations in Equation 16.12 can be multiplied by g^k and summed over all values of k to give:

$$\frac{d}{dt}\sum_{g=1}^{\infty} g^k L_g^* = k_{\mathrm{p11}}\mathrm{A}\sum_{g=1}^{\infty} g^k L_{g-1}^* - k_{\mathrm{p11}}\mathrm{A}\sum_{g=1}^{\infty} g^k L_g^*$$
$$- k_{\mathrm{p12}}\mathrm{B}\sum_{g=1}^{\infty} g^k L_g^* + k_{\mathrm{p21}}\mathrm{B}\sum_{g=1}^{\infty} g^k O_g^* \quad (16.154)$$

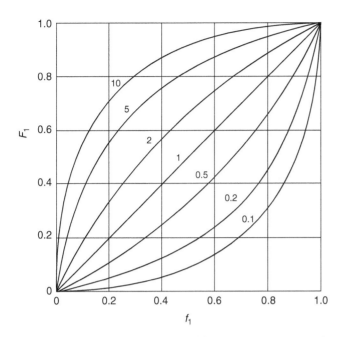

FIGURE 16.1 Copolymer composition. Instantaneous copolymer composition (F_1) as a function of monomer composition (f_1) for random copolymerization with $r_1 = 1/r_2$ as indicated. This material is reproduced with permission of John Wiley & Sons, Inc from Billmeyer FW. *Textbook of Polymer Science.* 3rd ed. New York: Wiley; 1984.

$$\sum_{g=1}^{\infty} L_g^* = L^* = \gamma_0; \qquad \sum_{g=1}^{\infty} O_g^* = O^* \qquad (16.155)$$

The moments of the CSD for monomer A may be defined as:

$$\lambda_k = \sum_{g=1}^{\infty} g^k L_g^* \qquad k = 0, 1, 2, \ldots \qquad (16.156)$$

Applying the QSSA, the left-hand side of Equation 16.154 may be set to 0. At steady state:

$$k_{p12} B L^* = k_{p21} A O^* \qquad (16.157)$$

With these simplifications and proceeding almost identically as in deriving Equations 16.27–16.29 from Equations 16.22 and 16.23, the statistics of the CSD for monomer A may be derived:

$$R = \frac{A}{B} \qquad (16.158)$$

$$\upsilon_n = 1 + r_1 R \qquad (16.159)$$

$$\upsilon_w = 1 + 2 r_1 R \qquad (16.160)$$

$$D_s = \frac{1 + 2 r_1 R}{1 + r_1 R} \qquad (16.161)$$

These results agree well with Ray's [7] derivation of sequence length based on probabilistic arguments. The advantage of the approach shown here is that the sequence population balances for a complex system can be integrated numerically with the rest of the mass and energy balances. In that case, one may also wish to consider balances on *inactive* sequences. These are sequences that are buried within a chain and can therefore no longer grow. In the simple result derived in the preceding, there is no difference between the distribution of active and inactive sequences.

16.2.3.2 Controlled Radical Polymerization For free radical polymerization, compositional drift will take the form of a wide distribution of copolymer composition *among the chains*. For controlled (living) radical polymerization (CRP), it will take the form of a wide distribution of copolymer composition *along each chain*. It should be noted that the compositional drift *within each chain* for living polymerization will not be reflected in the CCD. The CCD describes compositional drift *among* chains. For a living polymerization, the CCD might be expected to be very narrow, in spite of the fact that the compositional drift *within each chain* might significantly impact the end-use properties of the product. For instance, random or alternating copolymers can be made by CRP in much the same way as in free radical polymerization. However, since the chains remain living, one monomer can be fully polymerized and then another can be introduced into the reactor to form diblock copolymer. In a similar way, triblock copolymers (for thermoplastic elastomers) are made.

Controlled radical polymerization allows the synthesis of block copolymers from addition monomers without the kinetic constraint of the reactivity ratios. Thus, for CRP, it is perhaps the CSD rather than the CCD that is important in characterizing the polymer. Zargar and Schork [6, 8] and Ye and Schork [9, 10] have modeled the various chemistries for CRP. Their models include the CSD. Information about the CCD comes naturally from their analysis of the CSD. Mathematical modeling of CRP copolymerization would take (to a first approximation) the same form as the living (ionic) polymerization analysis given earlier.

16.2.3.3 Step-Growth Copolymerization As discussed in the preceding, step-growth polymerization must be carried out to a very high extent of reaction in order to obtain a usable molecular weight. Because of this, the average copolymer composition will approach that of the initial monomer composition. The CCD will be narrow. The CCD then is not as important for this polymerization chemistry as for others. Kilkson [11] has done rigorous analyses of various step-growth copolymerization schemes. The CSD, however, may be critically important since the properties of many of these polymers depend on controlled blockiness.

For copolymerization of monomer A–B with A′–B′, the sequence distribution may be analyzed in a manner similar to that in Section 16.2.3.1, although the analysis will be more difficult since step-growth polymerization proceeds by the combination to two existing chains.

A simpler approach is the probabilistic approach of Ray [7]. Since the reactivity ratios tend to unity for step-growth polymerization, the copolymer is likely to be random.

The copolymerization of A–A and B–B with A′–A′ and B′–B′ (or with A′–A′ only) offers additional degrees of freedom. An alternating polymer may be synthesized by prepolymerizing A–A–B–B–A–A. The low extent of reaction to produce the trimer can be effected by short reaction time and an excess of A–A. The trimer can then be reacted with added B′–B′ to produce [–A–A–B–B–A–A–B′–B′–]. If a random copolymer is desired, the four (or three) monomers can be mixed and copolymerized. The CSD then is controlled by the reactivity ratios, which, as noted, tend to unity. Since all uncapped step-growth polymers are living in the sense of being capable of additional polymerization, block copolymers can be (and are) made by sequentially polymerizing the monomers in the same way as described for living or controlled chain polymerization.

REFERENCES

[1] Billmeyer FW. *Textbook of Polymer Science*. 3rd ed. New York: Wiley; 1984.

[2] Eliseeva VI, Ivanchev SS, Kuchanov SI, Lebedev AV. *Emulsion Polymerization and its Applications in Industry*. New York: Consultants Bureau; 1981.

[3] Schork FJ, Deshpande PB, Leffew KW. *Control of Polymerization Reactors*. New York: Marcel and Dekker Inc.; 1993.

[4] Dotson NA, Galván R, Laurence RL, Tirrell M. *Polymerization Process Modeling*. New York: VCH Publishers; 1996.

[5] Mayo FR, Lewis FM. Copolymerization. I. A basis for comparing the behavior of monomers in copolymerization: the copolymerization of styrene and methyl methacrylate. J Am Chem Soc 1944;66:1594–1601.

[6] Zargar A, Schork FJ. Copolymer sequence distributions in controlled radical polymerization. Macromol React Eng 2009;3:118–130.

[7] Ray WH. On the mathematical modeling of polymerization reactors. Polym Rev 1972;8:1–56.

[8] Zargar A, Schork FJ. Design of copolymer molecular architecture via design of continuous reactor systems for controlled radical polymerization. Ind Eng Chem Res 2009;48: 4245–4253.

[9] Ye YS, Schork FJ. Modeling and control of sequence length distribution for controlled radical (RAFT) copolymerization. Ind Eng Chem Res 2009;48:10827–10839.

[10] Ye YS, Schork FJ. Modeling of sequence length and distribution for the NM-CRP of styrene and 4-methylstyrene in batch and semi-batch reactors. Macromol React Eng 2010;4: 197–209.

[11] Kilkson H. Effect of reaction path and initial distribution on molecular weight distribution of irreversible condensation polymers. Ind Eng Chem Fundam 1964;3:281–293.

[12] Odian GG. *Principles of Polymerization*. 3rd ed. New York: Wiley; 1991.

17

DESIGN AND OPERATION OF POLYMERIZATION REACTORS

F. Joseph Schork

17.1 REACTOR TYPES

The design of a polymerization reactor begins with the selection of the type of reactor (batch, semibatch or continuous) and then proceeds to the sizing and details of the reactor configuration. Only then can the details of operation and control be addressed. To this end, we will begin with a discussion of the basic types of reactors. Ultimately, it will be clear that the choice of reactor type is determined not only by practical considerations such as scale of production and propensity for fouling, but also by the specific polymerization kinetics. More complete discussions of reactor types and their residence time distributions may be found in references [1, 2].

17.1.1 Batch Reactor

A batch reactor (BR) is just that: All reactants are added at the beginning of the reaction. The reaction (polymerization) is then allowed to take place, either for a fixed time, or until some property of the product (monomer conversion, specific gravity, residual monomer, etc.) has been reached. The polymerization temperature can be held constant, the reactor can be operated in a semiadiabatic mode in which a fixed coolant load is applied and the temperature is allowed to vary (usually to rise due to the exothermic nature of polymerization), or a predetermined temperature trajectory can be followed.

The residence time distribution (RTD) for any reactor is a distribution giving the fraction of total material in the reactor that remains in the reactor for any fixed time. The RTD for a BR is not a distribution at all, but a spike. This means that all the material remains in the reactor for the same time (the reaction time or kettle time). There is no distribution of residence times.

Kinetic analyses for the various polymerization chemistries in a BR have been given in Chapter 16 and are not repeated here.

17.1.2 Semibatch/Semicontinuous Reactor

If some of the reactants are added over the course of the polymerization, while others remain in the reactor for the full kettle time, the polymerization is termed a semibatch or semicontinuous. The terms are interchangeable, and from here on, only semibatch will be used. The semibatch addition may be a small amount of material (small in volume, but not in effect) such as the addition of free radical initiator according to a predetermined trajectory (or as part of a feedback temperature control loop). It may entail the addition of various comonomers according to predetermined trajectories in order to manipulate copolymer composition distribution (CCD) or copolymer sequence distribution (CSD). In the first case, the RTD is not significantly affected by the semibatch feed (although the course of the reaction could be changed dramatically), and so from a practical point of view, this is still considered a BR. In the second case, the RTD is complex, with a fraction of the material remaining in the reactor for the total kettle time and other fractions having shorter residence times. Material added near the end of the kettle time will have a residence time approaching zero. The removal of the condensation product during condensation polymerization qualifies this type of reactor as semibatch, although, practically, such processes are considered to be batch processes.

Monitoring Polymerization Reactions: From Fundamentals to Applications, First Edition. Edited by Wayne F. Reed and Alina M. Alb.
© 2014 John Wiley & Sons, Inc. Published 2014 by John Wiley & Sons, Inc.

17.1.3 Plug Flow Reactor

The first type of continuous reactor to be considered is a tubular reactor. This is generally a tube (pipe) in which all of the reactants flow into the inlet and all of the products flow out of the outlet. It consists of a long length of straight pipe, or, more likely, is configured in a coil or "trombone" configuration. The high surface to volume ratio inherent in a tube greatly facilitates the removal of the heat of polymerization. If the contents of the reactor are in turbulent flow, then all of the fluid (from the tube wall to the centerline) has the same velocity. In this case, there is no axial mixing as an element of fluid moves down the reactor. Perfect mixing in the radial direction is assumed. Under these conditions, the tubular reactor is known as a plug flow reactor (PFR). The RTD of a PFR is the same as that of a BR: All materials remain in the reactor for exactly the same time. The RTD is a spike and the residence time is calculated as the volume of the reactor divided by the volumetric flow rate. Since the RTD is that of a batch, the kinetics of a PFR is exactly those of a BR. If desired, time in the reactor can be replaced with length from the reactor inlet divided by fluid velocity.

Tubular reactors are particularly subject to plugging. In addition, for viscous reactants (polymerization), the assumption of plug flow may not be valid, since the material near the center of the tube will move with a higher velocity than material near the walls.

The continuous analog to a semibatch reactor can be designed as a PFR with material feeds located at fixed positions along the reactor axis. Kinetic analyses for the various polymerization chemistries in a PFR are not given since they are identical to those for a BR.

17.1.4 Continuous Stirred Tank Reactor

A continuous stirred tank reactor (CSTR) is a continuous reactor as well, but whereas the PFR consists of a tube, the CSTR consists of a tank. Reactants flow into the tank continuously, and product is continuously removed. The inlet flow and outlet flow are equal so that the volume of material in the tank remains constant. One or more agitators are used to ensure that the contents of the reactor are well mixed. Thus, the concentrations of all species in the reactor are equal at all locations in the reactor and also equal to the concentrations in the outlet. Due to the well-mixed nature of the reactor, the RTD is not a spike, but a distribution. Due to the mixing, an element of fluid that has just entered the reactor has a high probability of leaving the reactor. From probability considerations, the fraction of material having a fixed residence time goes down as the value of that residence time goes up. The average residence time (generally designated as Θ) for a CSTR is calculated as the volume of the reactor divided by the volumetric flow rate. The polydispersity of the RTD (defined for the residence time distribution in the same way

as it is for the chain length distribution) for a CSTR is 2.0. A number of CSTRs in series will approach the RTD characteristics of a PFR. In fact, trains of 5–15 CSTRs in series have been used in the rubber industry to give the kinetic characteristics of a PFR without the danger of plugging due to fouling that is always a concern with a tubular reactor. The polydispersity of the RTD of a train of n equal-sized CSTRs in series is [3]:

$$D_{\text{RTD}} = \frac{1}{1+n} \qquad (17.1)$$

It can be seen that as n becomes large, D_{RTD} approaches the D_{RTD} of a PFR (unity). Figure 17.1 shows the RTD for both the PFR and the CSTR.

The segregated CSTR (SCSTR), while not discussed earlier, is included to indicate the effects of less than perfect mixing. A practical example of a SCSTR is suspension polymerization in a CSTR in which the suspension beads are well mixed within the reactor and within each bead, but in which there is no exchange of material between the various beads.

Since the RTD of the CSTR is radically different from that of the BR or the PFR, and since the RTD could have

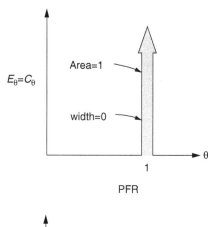

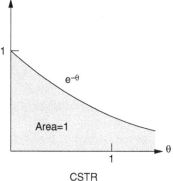

FIGURE 17.1 Residence time distributions for plug flow and continuous stirred tank reactors residence time versus fraction of the outlet stream having that residence time. This material is reproduced with permission of John Wiley & Sons, Inc. from Levenspiel O. *Chemical Reaction Engineering*. 3rd ed. New York: Wiley; 1999.

profound effects on reactor performance, analyses of the kinetics of the various polymerization chemistries in a CSTR are given in the following.

17.1.4.1 *Chain Growth Polymerization in a CSTR* *Anionic polymerization.* Recall that the idealized living anionic polymerization can be represented by the following (repeated from Chapter 16):

$$A^- + M \xrightarrow{k_i} P_1 \qquad (17.2)$$

$$P_n + M \xrightarrow{k_i} P_{n+1}; \quad n \geq 1 \qquad (17.3)$$

If one assumes rapid initiation and no termination or chain transfer (as implied in equations and following the general approach of Dotson et al. [4], the species balances and population balances are as follows:

$$V\frac{dM}{dt} = Q_f M_f - QM - Vk_p PM; \quad M(0) = M_0 \quad (17.4)$$

$$V\frac{dP}{dt} = Q_f P_f - QP; \quad P(0) = P_0; \quad P = \sum_{n=1}^{\infty} P_n \quad (17.5)$$

$$V\frac{dP_1}{dt} = Q_f P_{1f} - Vk_p MP_1; \quad P_1(0) = P_{10} \qquad (17.6)$$

$$V\frac{dP_n}{dt} = Q_f P_{nf} - QP_n - Vk_p M(P_n - P_{n-1});$$
$$P_n(0) = 0; \quad n \geq 2 \qquad (17.7)$$

Here, Q is the volumetric flow rate and the subscript f indicates feed.

If the inlet and outlet flow rates are set equal, and the feed is assumed to contain only P_1, the following steady-state balances are obtained by setting the time derivatives to *zero* in the aforementioned equations:

$$P = P_f = P_{1f} \qquad (17.8)$$

$$Q(M_f - M) = Vk_p MP \qquad (17.9)$$

$$Q(P_{1f} - P_1) = Vk_p MP_1 \qquad (17.10)$$

$$QP_n = Vk_p M(P_n - P_{n-1}) \qquad (17.11)$$

Equation 17.9 may be solved directly; Equations 17.10 and 17.11 may be solved by z-transform techniques to give:

$$M = \frac{M_f}{1 + \Theta k_p P_{1f}}; \quad \Theta = \frac{V}{Q} \qquad (17.12)$$

$$P_n = \frac{P_{1f}}{(1 + \Theta k_p M)}\left[\frac{\Theta k_p M}{1 + \Theta k_p M}\right]^{n-1} \qquad (17.13)$$
$$= P_{1f}(1 - \phi)\phi^{n-1}$$

$$\mu_n = \frac{1}{(1 - \phi)} \qquad (17.14)$$

$$D = 1 + \phi \qquad (17.15)$$

$$\phi = \left[\frac{\Theta k_p M}{1 + \Theta k_p M}\right] \qquad (17.16)$$

This is a Flory distribution or most probable distribution and much broader ($D > 1$) than the Poisson distribution ($D = 1$) resulting from batch polymerization. The increase in molecular weight polydispersity is due to the fact that the chains are living, and so are directly impacted by the RTD. Recall that the polydispersity of the RTD for a CSTR is 2. For long residence time, ϕ approaches 1 and D becomes 2. Thus, the number chain length distribution (NCLD) takes on the breadth characteristics of the RTD, due to the living nature of the polymer chains. For comparison, one should recall that the lifetime of a free radical is 1–10 s. This is insignificant in comparison with the RTD and so that RTD has little to no effect on D in free radical polymerization.

Free radical polymerization
The free radical polymerization mechanism described in Equation 16.62 in the previous chapter can be carried out in a CSTR. Inhibition is ignored here, but could easily be accounted for. Isothermal operation is assumed. Chain transfer to polymer and to chain transfer agent is ignored. Species balances on monomer and initiator can be written as:

$$V\frac{dM}{dt} = Q(M_f - M) - k_p PMV;$$
$$M(0) = M_0 \qquad (17.17)$$

$$V\frac{dI}{dt} = Q(I_f - I) - k_d I; \quad I(0) = I_0 \qquad (17.18)$$

As before, Q is the volumetric flow rate and the subscript f indicates the feed condition.

Species balances may also be written for primary radical and live chains. As before, the QSSA is made on radicals and live chains and, following the derivation from the previous chapter, the concentration of living chains can be written as (Eq. 16.67 in Chapter 16):

$$P = \left[\frac{2fk_d I}{k_{tc} + k_{td}}\right]^{1/2} \qquad (17.19)$$

Since the derivation of Equation 17.19 was from a BR, the washout of live radicals has been ignored. Ray [5] has shown this to be insignificant. As before, balances on P_1, P_n, and M_n can be written. Since flow terms were not found in Chapter 16, derivation must be added.

$$V \frac{dP_1}{dt} = k_1 R M_1 V - k_p P_1 M_1 V$$
$$+ (k_{fs} S + k_{fm} M_1)(P - P_1)V \qquad (17.20)$$
$$- (k_{tc} + k_{td})PP_1 V - QP_1$$
$$P_1(0) = 0$$

$$V \frac{dP_n}{dt} = k_p M_1 (P_{n-1} - P_n)V$$
$$- (k_{fs} S + k_{fm} M_1)P_n V \qquad (17.21)$$
$$- (k_{tc} + k_{td})PP_n V - QP_n$$
$$P_n(0) = 0; \quad n > 1$$

$$V \frac{dM_n}{dt} = (k_{fs} S + k_{fm} M_1)P_n V + k_{td} P_n P V$$
$$+ \frac{1}{2} V k_{tc} \sum_{m=1}^{n-1} P_m P_{n-m} - QM_n \qquad (17.22)$$
$$M_n(0) = 0; \quad n > 1$$

If the QSSA is made for P_1,

$$P_1 = (1 - \alpha)P \qquad (17.23)$$

where the probability of propagation is defined slightly differently (inclusion of $1/\Theta$) from the batch case:

$$\alpha = \frac{k_p M}{k_p M + k_{fm} M + k_{fs} S + (k_{tc} + k_{td})P + 1/\Theta} \qquad (17.24)$$

and the average reactor residence time is defined as:

$$\Theta = \frac{V}{Q} \qquad (17.25)$$

Now, making the QSSA for P_n gives:

$$P_n = \alpha P_{n-1} \qquad (17.26)$$

Solving Equations 17.23 and 17.26 by z-transform techniques results in:

$$P_n = (1 - \alpha)P \alpha^{n-1} \qquad (17.27)$$

Note that this is once again the Flory or most probable distribution; only the definition of α has changed.

Equations 17.20 and 17.21 can be solved by z-transform techniques to give the same results for the moment of the live chain NCLD as in the batch case, with α defined by Equation 17.24 (including the residence time):

$$\lambda_0 = P \qquad (17.28)$$

$$\lambda_1 = \frac{P}{(1 - \alpha)} \qquad (17.29)$$

$$\lambda_2 = \frac{P(1 + \alpha)}{(1 - \alpha)^2} \qquad (17.30)$$

The number-average chain length (NACL), WACL, and polydispersity for the living chains can now be calculated as follows:

$$\mu_n = \frac{\lambda_1}{\lambda_0} = \frac{1}{(1 - \alpha)} \qquad (17.31)$$

$$\mu_w = \frac{\lambda_2}{\lambda_1} = \frac{(1 + \alpha)}{(1 - \alpha)} \qquad (17.32)$$

$$D = \frac{m_w}{m_n} = \frac{\mu_w}{\mu_n} = (1 + \alpha) \approx (1 + 1) = 2 \qquad (17.33)$$

Analysis of dead polymer NCLD can be carried out from Equation 17.22 analogous to the derivation for a BR in Chapter 16. The definition of α is given by Equation 17.24, as aforementioned, and the flow terms remain:

$$\frac{d\eta_0}{dt} = (k_{fm} M + k_{td} P + k_{fs} S)PV\alpha$$
$$+ 0.5 k_{tc} P^2 - \frac{1}{\Theta}\eta_0; \quad \eta_0(0) = 0 \qquad (17.34)$$

$$\frac{d\eta_1}{dt} = \frac{[(k_{fm} M + k_{td} P + k_{fs} S)(2\alpha - \alpha^2) + k_{tc} P]P}{(1 - \alpha)}$$
$$- \frac{1}{\Theta}\eta_1; \quad \eta_1(0) = 0 \qquad (17.35)$$

$$\frac{d\eta_2}{dt} = \frac{[(k_{fm} M + k_{td} P + k_{fs} S)(\alpha^3 - 3\alpha^2 + 4\alpha) + k_{tc} P(\alpha + 2)]P}{(1 - \alpha)^2}$$
$$- \frac{1}{\Theta}\eta_1; \quad \eta_2(0) = 0 \qquad (17.36)$$

Since the polymerization is taking place in a CSTR, Equations 17.34–17.36 can be solved by setting the time derivatives to zero and solving for the moments. Then, the NACL and WACL of the dead chain distribution may be written as:

$$\mu_n = \frac{\eta_1}{\eta_0} = \frac{[(k_{fm}M_1 + k_{td}P + k_{fs}S)(2\alpha - \alpha^2) + k_{tc}P]}{(1-\alpha)[(k_{fm}M_1 + k_{td}P + k_{fs}S)\alpha + 0.5k_{tc}P]}$$

(17.37)

$$\mu_w = \frac{\eta_2}{\eta_1}$$

$$= \frac{\left[(k_{fm}M_1 + k_{td}P + k_{fs}S)(\alpha^3 - 3\alpha^2 + 4\alpha) + k_{tc}P(\alpha+2)\right]}{(1-\alpha)[(k_{fm}M_1 + k_{td}P + k_{fs}S)(2\alpha - \alpha^2) + k_{tc}P]}$$

(17.38)

To get the complete NCLD for dead chains, substituting Equations 17.23 and 17.26 into Equation 17.22 gives:

$$\frac{dM_n}{dt} = [k_{fs}S + k_{fm}M_1 + k_{td}P](1-\alpha)P\alpha^{n-1}$$

$$+ \frac{1}{2}k_{tc}P^2(1-\alpha)^2\alpha^{n-2}(n-1) - \frac{1}{\Theta}M_n;$$

$$M_n(0) = 0; \quad n > 1$$

(17.39)

Assuming steady state and solving for M_n gives:

$$M_n(t) = \Theta\Big\{[k_{fs}S + k_{fm}M_1 + k_{td}P](1-\alpha)P\alpha^{n-1}$$

$$+ \frac{1}{2}k_{tc}P^2(1-\alpha)^2\alpha^{n-2}(n-1)\Big\}$$

$$M_n(0) = 0; \quad n > 1$$

(17.40)

Recall that in the earlier batch the right-hand side of Equation 17.40 was assumed to be constant. This was strictly incorrect and resulted in the *minimum* polydispersity. For the CSTR we can assume steady state: the right-hand side of Equation 17.39 really is constant. Thus, what was a *minimum* polydispersity for the batch case is a *realistic* polydispersity for the CSTR case.

If termination by combination can be ignored:

$$M_n(t) = \eta_0(1-\alpha)\alpha^{n-2}$$

(17.41)

The power of $n - 2$ arises because this treatment excluded monomer from consideration as polymer. This is again a Flory or most probable distribution. Thus, for termination by combination, the dead polymer will have the same distribution as the living polymer. (This can be seen by inspection of the mechanism of termination by disproportionation.) The polydispersity of the dead polymer is then:

$$D = \frac{m_w}{m_n} = \frac{\mu_w}{\mu_n} = (1+\alpha) \approx (1+1) = 2$$

(17.42)

From Equations 17.27 and 17.41, it can be seen that:

$$\frac{\mu_n^{(dead)}}{\mu_n^{(live)}} = (2-\alpha) \approx 2$$

(17.43)

It should be noted that the NCLD, weight-average of the NCLD (WCLD), and polydispersity are only weak functions of residence time (through α). This can be rationalized by considering the fact that, for free radical polymerization, the lifetime of a radical is so small as to be insignificant when compared with the average residence time. Thus, individual chains are not strongly impacted by residence time.

Controlled radical polymerization.
Detailed mathematical models of the three chemistries for controlled radical polymerization (CRP) have been published by various researchers [6–10], but for the current purposes it is sufficient to approximate these systems as living polymerizations. A simple kinetic treatment would be similar to that for living ionic polymerization as in Section "Anionic Polymerization." The utility of CSTR systems for these reactions will be addressed later by considering them as living radical systems. There are significant differences between free radical polymerization and CRP. Among them is the difference in lifetime, which for a living chain can be on the order of hours rather than 1–10 s as in free radical systems. This means that the NCLD, WCLD, and polydispersity will be strongly affected by RTD. In fact, the polydispersity of the MWD will approach that of the RTD in a train of CSTRs (see Eq. 17.1). Another difference is that while in a free radical polymerization new chains are initiated, grow to length, and terminate throughout the polymerization. In a CRP system, most of the chains initiate near the beginning of the reaction and grow at approximately the same rate. This means that at some point in the reaction, there will be large numbers of living oligomeric chains. This phenomenon can lead to colloidal instability [11] in emulsion and miniemulsion systems and is described as superswelling. It can also lead to changes in the propensity for chain transfer to polymer and subsequent branching [12]. In thinking about kinetics, product properties, and reactor design for these systems, it is important to realize that they are very different from free radical systems, in spite of similar names and similar monomer types.

17.1.4.2 Step-Growth Polymerization in a CSTR As noted in Chapter 16, there are two types of step-growth kinetics: A–B and A–A/B–B. The difference is whether both functional groups are on the same monomer (A–B), or if two distinct monomers (A–A and B–B) are necessary to carry out the polymerization. The A–A/B–B kinetics are, obviously, more complex and have the additional complexity induced by the possibility of nonstoichiometric ratios of the two monomers as discussed in Chapter 16. These complexities carry over into the analysis of CSTR polymerization but will not be addressed here.

A–B step-growth polymerization. One recalls that the simplest kinetic representation for A–B step-growth polymerization (as given in Chapter 16) is:

$$(A - B)_n + (A - B)_m \underset{k_{p^-}}{\overset{k_p}{\rightleftarrows}} (A - B - A - B)_{n+m} + W \tag{17.44}$$

Again, the system is assumed to be irreversible due to the removal of the condensation product, W. Balances on living chains can be done as previously, with the flow terms added. The time derivatives have been set to zero in making the steady-state assumption:

$$P_{1f} - P_1 = k_p \Theta P_1 P \tag{17.45}$$

$$P_{nf} - P_n = \frac{1}{2} k_p \Theta \sum_{r=1}^{n-1} P_r P_{n-r} + k_p \Theta P_n P; \quad n > 1 \tag{17.46}$$

$$P = \sum_{n=2}^{\infty} P_n \tag{17.47}$$

Assuming the feed consists only of P_1, the solution to Equations can be derived by z-transform techniques to give the following:

$$\lambda_0 = \frac{-1 + [1 + 2k_p \Theta P_{1f}]^{1/2}}{k_p \Theta} \tag{17.48}$$

$$\lambda_1 = P_{1f} \tag{17.49}$$

$$\lambda_2 = P_{1f} \left(1 + k_p \Theta P_{1f}\right) \tag{17.50}$$

The analysis also shows that the NCLD is a binomial distribution. The NACL, WACL, and polydispersity can then be calculated as:

$$\mu_n = \frac{k_p \Theta P_{1f}}{\left[-1 + (1 + 2k_p \Theta P_{1f})^{1/2}\right]} \tag{17.51}$$

$$\mu_w = 1 + k_p \Theta P_{1f} \tag{17.52}$$

$$D = \frac{(1 + k_p \Theta P_{1f})\left[-1 + (1 + 2k_p \Theta P_{1f})^{1/2}\right]}{k_p \Theta P_{1f}} \tag{17.53}$$

The *extent of reaction* in a CSTR is defined as:

$$p = \frac{P_{1f} - P}{P_{1f}} = \frac{1}{2} \frac{k_p \Theta P^2}{P_{1f}} \tag{17.54}$$

Equations 17.51, 17.52, and 17.53 can be written in terms of the extent of reaction as follows:

$$\mu_n = \frac{1}{1 - p} \tag{17.55}$$

$$\mu_w = 1 + \frac{2p}{(1-p)^2} = \frac{1 + p^2}{(1-p)^2} \tag{17.56}$$

$$D = \frac{1 + p^2}{1 - p} \tag{17.57}$$

In a CSTR, as was previously shown to be true in batch, the extent of reaction must be very close to unity before significant NACL can be achieved. High conversion products are difficult to achieve in a single CSTR, because this requires a reactor of very large volume. More importantly, as the extent of reaction approaches unity, the polydispersity goes to infinity. Therefore, if CSTR polymerization were desired for this chemistry, a train of CSTRs would be used to allow the extent of reaction to approach unity with an economical reactor configuration. In practice, such chemistries (i.e., nylon) are often carried out commercially in PFRs, avoiding this problem.

Because the lifetime of a growing polymer chain is equal to the residence time, the effect of RTD is extreme broadening of the MWD. This was seen to a lesser degree in living and controlled polymerization (where the lifetime of a living chain is also equal to the residence time). The difference is that in living/controlled systems, the polydispersity mirrored the RTD, whereas in step-growth polymerization, the polydispersity goes to infinity. This is due to the geometric nature of chain growth.

17.1.4.3 Copolymerization
A CSTR has the advantage in copolymerization of having a constant composition in the reactor. This includes constant monomer concentrations, and more importantly, a constant ratio of monomer concentrations. For free radical, CRP or ionic polymerization this results in a constant CCD. The compositional drift that occurs in a batch reaction (described in Chapter 16) does not occur. If a varying copolymer composition is desired, a train of CSTRs can be used, with intermediate feeds between reactors. For nonliving chain-growth polymerizations, this will result in a broadening of the CCD as chains polymerized in one monomer ratio environment may differ from those polymerized in a downstream reactor with different monomer compositions. Since for living chain-growth polymerization, each chain continues to polymerize through the entire reactor train, intermediate monomer feed will result in changes in the copolymer composition *within* each chain. Thus, it may be possible, with specific intermediate monomer feeds, to produce blocky, gradient, or other copolymer structures. With truly living systems, it is possible to design the specific molecular structure by design of a train of CSTRs and PFRs with intermediate feeds, as demonstrated by Zargar and Schork [8].

A single CSTR should give the narrowest possible CSD, since all concentrations within the reactor remain constant. Designed variations in CSD are possible by the use of intermediate feeds in a reactor train; however, as shown by Ye and Schork [13], the CSD and CCD cannot be independently manipulated.

Because the MWD polydispersity goes to infinity in a CSTR, step-growth copolymerization is rarely done in a CSTR or CSTR train. Since the reactivity ratios for step-growth monomers vary only slightly, there is little compositional drift in batch step-growth copolymerization, so there is little advantage, in terms of CCD, to step-growth CSTR polymerization.

17.1.4.4 *Reactor Train* The possibility of using reactor trains has entered the aforementioned discussion under a number of topics. Some final remarks are made here. First, it is difficult to achieve high conversion in a single CSTR. High conversion is necessary though, since there are rigid controls on residual monomer. As noted earlier, a train of CSTRs will approximate PFR behavior. Since a PFR gives kinetics identical to those of a BR, high monomer conversion is attainable. A train of CSTRs is often desirable over a single PFR since reactor fouling is not a critical in a CSTR train. Small amount of insoluble polymer may completely plug a PFR, but may be entirely acceptable if it deposits on the wall of a CSTR. Plug flow reactors are often used as part of a reactor train but mostly as prerectors or postreactors. For instance, a tubular reactor can be used for the initial stages (low monomer conversion) of bulk free radical polymerization where the viscosity is low. The low conversion "syrup" is then used to fill a mold and the polymerization is carried out to completion. In emulsion polymerization, PFRs are often used as *seeder* reactors. In this case, the polymer particles are nucleated in a PFR, then transferred to a CSTR train where the particles are grown to full size and complete monomer conversion. The PFRs have also been used as finishing reactors in free radical chemistry. In this case, the temperature is raised and additional initiator is added to polymerize the residual monomer. Since step-growth kinetics is impractical in a single CSTR, step-growth polymers (specifically the nylons) are often polymerized in a single PFR.

One of the most widely used examples of polymerization in a CSTR train is that of synthetic rubber. Styrene–butadiene rubber (SBR) and neoprene are free radical polymerized, most often in an emulsion polymerization system. These rubbers are highly susceptible to branching. Excessive branching will cause the rubber to be unusable. Branching is accelerated by a high ratio of polymer to monomer. For this reason, the monomer is often polymerized to 50–70% conversion, then the polymerization is stopped and the monomer is steam-stripped from the product and recycled. In this case, very high conversion is not necessary and a train of 5–15 CSTRs is often used. This number of CSTRs in series will approximate PFR kinetics. Intermediate monomer addition is not used, since adding all of the monomer at the beginning of the train keeps the polymer-to-monomer ratio low and suppresses branching. As before, a single CSTR often cannot achieve sufficient monomer conversion at a reasonable reactor size.

17.2 SELECTION OF REACTOR TYPE

With the analysis of the various polymerization chemistries in the various reactor configurations completed, it is possible to draw some general conclusions about the types of reactors appropriate to each of the chemistries. Since polymers have always been characterized as products by process, it is not surprising that the quality of the product (as defined by the molecular structure of the polymer molecule) would depend on reactor design and operations. This section will discuss reactor design and the next one will describe strategies for reactor operation.

The choice of reactor type for various polymerization chemistries is summarized in Table 17.1, first developed by Gerrens [14, 15]. Following Gerrens, polymerization kinetics have been divided into three categories: monomer linkage with termination (as in free radical polymerization); monomer linkage without termination (as in ionic living polymerization); and polymer linkage (step-growth polymerization). Table 17.1 lists three categories of reactors: the BR or PFR (since the kinetics is identical for the two), the homogeneous CSTR (HCSTR), and the SCSTR. The HCSTR is defined to be the well-mixed CSTR previously discussed. The SCSTR, also discussed earlier, is included to indicate the effects of less than perfect mixing.

17.2.1 Monomer Linkage with Termination

For a polymerization featuring monomer linkage with termination, the narrowest MWD is developed in the HCSTR. In this case, the constancy of the reaction environment (at steady state) dominates over the distribution of residence times in the reactor. This is due to the fact that the lifetime of a single live polymer chain is far less than the average reactor residence time. Due to the constantly changing monomer concentration in a BR, the MWD will be wider for a BR. The distribution will be widest for the SCSTR where the effects of reaction environment and residence time variation combine to cause a broadening of the MWD. Copolymerization with the objective of producing a narrow CCD is best carried out in a HCSTR since the ratio of the comonomer concentrations stays constant at steady state.

17.2.2 Monomer Linkage without Termination

For monomer linkage without termination, the narrowest distribution occurs in the BR, since all chains are growing throughout the reaction and the effect of constant reaction

TABLE 17.1 Molecular Weight Distribution Characteristics as a Function of Chemistry and Reactor Type

| | Monomer Linkage[a] | | |
Reactor Type	With Termination	Without Termination	Polymer Linkage
Batch or PFR	Wider than Flory[b] ($D > 1.5$–2.0)	Poisson ($D = 1.0$)	Flory[c] ($D = 2.0$)
HCSTR	Flory ($D = 1.5$–2.0)	Flory ($D > 1.0$)	Wider than Flory ($D \gg 2.0$)
SCSTR	Wider than batch or PFR	Between batch and homogenous CSTR	Between batch and homogenous CSTR

[a]An example of monomer linkage with termination is free radical polymerization. An example of monomer linkage without termination is ionic living polymerization or (in the ideal case) controlled radical polymerization.
[b]Batch or PFR with termination: $D = 1.5$ for termination by combination; $D = 2.0$ for termination by disproportionation; $1.5 < D < 2.0$ for mixed termination.
[c]For polymer linkage, extent of reaction above 0.99 is assumed.

environment is moot. The HCSTR will exhibit a broader distribution due to the effect of the variations in residence times among the growing chains. The SCSTR will have an MWD whose breadth is intermediate between the BR and the HCSTR since the segregation (batch character) will cause a narrowing of the MWD while the residence time variations among the segregated beads will cause a broadening. It should be noted that, since the MWD is dependent on RTD, tubular reactors with less than perfect plug flow will result in a broadening of the MWD, due to a broadening of the RTD. This has been seen recently in CRP [16].

17.2.3 Polymer Linkage

For polymer linkage, the narrowest distribution is found in the BR for similar reasons as in monomer linkage without termination. The HCSTR causes a broadening of the MWD since the lifetime of each growing chain is equal to its residence time in the reactor and the distribution of residence times broadens the MWD. In fact, as the extent of reaction approaches unity (necessary to produce high molecular weight product), the polydispersity increases without bound due to the geometric nature of the molecular weight increase. Therefore, the HCSTR is not recommended for polymer linkage polymerizations. The SCSTR develops polymer MWD between the BR and the HCSTR since, once again, the segregation narrows the distribution while the RTD broadens it.

17.2.4 Final Comments on Reactor Selection

Of course, it is not always necessary or even desirable to produce monodisperse MWD. A certain broadening may be designed into the product by the choice of reactor type. Or, the engineer (chemist) may choose to produce a broad or bimodal MWD by blending two or more narrowly distributed products. As with most aspects of polymerization reaction engineering, the possibilities are endless. In addition, consideration of copolymer composition (uniformity, blockiness, etc.) may be more important in a given application than narrowness of MWD.

17.3 REACTOR DYNAMICS

If polymers are, in fact, "products by process," then the mode of operation, as well as the choice of reactor type will have a direct impact on polymer quality. Various types of dynamic behavior possible with polymerization reactors will be explored in this section and the impact of these types of behavior on safe and efficient operation will be discussed. By the dynamic behavior of a polymerization reactor is meant the time evolution of the states of the reactor. The states are those fundamental dependent quantities which describe the natural state of the system. A set of equations which describe how the natural state of the system varies with time is called the set of state equations. Temperature, pressure, monomer conversion, and copolymer composition could be considered states of a polymerization reactor. Independent variables such as coolant temperature in a jacketed reactor or initiator addition rate are not states but (controlled or uncontrolled) inputs. For various reactor types, different modes of dynamic behavior are observed. These can range from stable operation at a single steady state to instability, multiple steady states, or sustained oscillations.

17.3.1 CSTR Dynamics

The widest spectrum of dynamic behavior is observed in the CSTR. As we have seen, the use of a CSTR or CSTR train for polymerization reactions may be justified in some cases by kinetic considerations. However, before implementing CSTR polymerization, the engineer should be aware of the unique dynamics associated reactions in a CSTR which are exothermic and/or autocatalytic or involve nucleation phenomena.

Consider an irreversible first-order exothermic reaction in a CSTR. Figure 17.2 shows the rate of thermal energy release by reaction (curve Q_g in the figure) plotted versus temperature. At low temperature, the reaction rate is low, and the slope of Q_g is small. At high temperatures, the reactor is already operating at a high level of conversion (low reactant concentration) and additional increases in temperature result in negligible increase in reaction rate and heat evolution. If the reactor is jacketed, the rate of heat removal (for fixed jacket temperature) is linear with reaction temperature. Thus, depending on operating conditions, the rate of heat removal can be represented by the various heat removal lines marked Q_r in the figure.

Since at steady state the rate of heat generation must equal the rate of heat removal, steady-state conditions can exist only at the intersection of the Q_g and Q_r curves. Depending on operating conditions (the slope and position of the Q_r line), there may be one or three steady states. In the case of three steady states, it can be easily seen that the upper and lower steady states are stable since perturbations in temperature will result in the system returning to its original position when the perturbation is removed. The middle steady state, however, is shown to be unstable since any perturbation will drive the system away from the middle steady state and toward the upper or lower steady state (depending on the direction of the perturbation).

This type of heat balance multiplicity is common in CSTR polymerization, due to the highly exothermic nature of polymerization reactions. The presence of a gel effect will augment the potential for multiplicity. This phenomenon can be observed in free radical polymerization (nonisothermal), again, due to the exothermic nature of the polymerization reaction. However, it is also observed in some isothermal free radical polymerizations in a CSTR due to the gel effect [18].

Figure 17.3 shows the rate of polymerization plotted versus monomer conversion for the free radical solution polymerization of methyl methacrylate. Unlike a more common reaction in which the rate of reaction falls monotonically with conversion, the rate of reaction rises with conversion due to the onset of the gel effect. Thus, the system can be thought of as autocatalytic. At high conversion, the polymerization becomes monomer starved and the rate of polymerization falls to zero. At a fixed residence time, there must be a specific rate of polymerization to produce a given monomer conversion. The mass balance is represented by the dotted lines in Figure 17.3. The slope of the mass balance line will vary with operating conditions but it will always pass through the origin since at zero reaction rate the monomer conversion is zero. Inspection of Figure 17.3 reveals that for mass balances (operating lines) with slopes between the two dotted lines, three steady states exist since an intersection of the reaction rate curve and the operating line defines a steady state.

This may be seen better by referring to Figure 17.4, where monomer conversion has been plotted versus reactor residence time. (A similar plot will result from the heat balance multiplicity in a nonisothermal CSTR.) It may be seen that over a range of residence times, three values of monomer conversion are possible. As before, the upper and lower steady states are usually stable, while the middle steady state is not.

The same phenomenon is seen in emulsion polymerization. The phenomenon of multiple steady states arises in emulsion polymerization for much the same reason it appears in solution polymerization: The autocatalytic nature of the polymerization (due to the gel effect) combined with the mass balance results in the possibility of steady-state multiplicity.

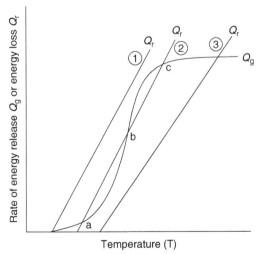

FIGURE 17.2 Heat balance multiplicity during exothermic reaction. With permission from Reference 17, Hill CG. *Chemical Engineering Kinetics and Reactor Design.* New York: Wiley; 1977.

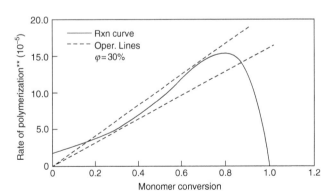

FIGURE 17.3 Rate of polymerization versus monomer conversion for free radical solution polymerization of methyl methacrylate. With permission from Reference 19, Kwalik KM. Bifurcation characteristics in closed-loop polymerization reactors [PhD thesis]. Atlanta: School of Chemical Engineering, Georgia Institute of Technology; 1988.

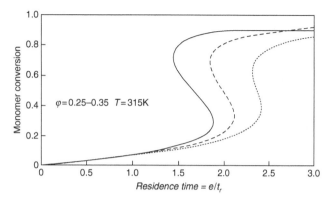

FIGURE 17.4 Multiplicity during isothermal CSTR polymerization: effect of volume fraction. With permission from Reference 19, Kwalik KM. Bifurcation characteristics in closed-loop polymerization reactors [PhD thesis]. Atlanta: School of Chemical Engineering, Georgia Institute of Technology; 1988.

Steady-state multiplicity can be an operational problem for a number of reasons. If one wishes to operate at an intermediate level of monomer conversion (perhaps to minimize viscosity or prevent excessive chain branching), one may be forced to operate in the unstable region, relying on closed-loop control to stabilize the operating point. This is tricky at best. Additionally, the steady state (upper or lower) to which the system goes on start-up will depend on how the start-up is implemented. A careful start-up policy may be needed to assure that the system arrives at the desired steady state. In general, a conservative start-up, with the temperature and initiator concentration brought to steady-state values, slowly will result in operation on the lower branch, while aggressive start-up (high temperature and/or high initiator concentration during start-up) will result in steady-state operation on the upper branch. Finally, large upsets in the process may cause ignition or extinction. This could lead to loss of temperature control in the case of ignition, or loss of reactor productivity in the case of extinction. A system designed to operate at the upper steady state will be operating well below the design product yield at the lower steady state. Additionally, the product quality (MWD, CCD, etc.) will be different for the two operating points. The polymerization reactor designer should be aware of the potential for multiplicity, and, if possible, design the system to operate outside the region of multiplicity.

Continuous stirred tank reactor polymerization reactors can also be subject to oscillatory behavior. A nonisothermal CSTR free radical solution polymerization can exhibit damped oscillatory approach to a steady state, unstable (growing) oscillations upon disturbance, and stable (limit cycle) oscillations in which the system never reaches steady state and never goes unstable, but continues to oscillate with a fixed period and amplitude. However, these phenomena are more commonly observed in emulsion polymerization. High-volume products such as styrene–butadiene rubber (SBR) often are produced by continuous emulsion polymerization. As noted earlier, this is

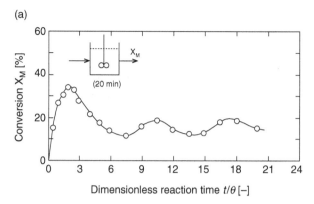

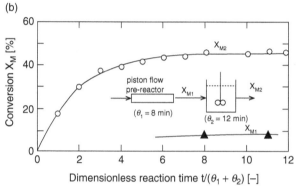

FIGURE 17.5 Steady-state oscillations during the emulsion polymerization of methyl methacrylate in a single CSTR. (a) Single CSTR; (b) tubular reactor followed by a single CSTR. Reprinted (adapted) with permission from Nomura M, Harada M. On the optimal reactor type and operations for continuous emulsion polymerization. In: Bassett DR, Hamielec AE, editors. *Emulsion Polymers and Emulsion Polymerization*. Washington, DC: p 121–144. © 1981 American Chemical Society.

most often done in a train of 5–15 CSTRs in series. Sustained oscillations (limit cycles) in conversion, particle number, and free emulsifier concentration have been reported under isothermal conditions in continuous emulsion polymerization systems. This limit cycle behavior leaves its mark on the product in the form of disturbances in the molecular weight distribution and particle size distribution, which cannot be blended away. Figure 17.5 shows evidence of a sustained oscillation (limit cycle) during emulsion polymerization of methyl methacrylate in a single CSTR [20]. Such oscillations come from the interaction of the nucleation of new particles with the growth of existing particles in competition for the available surfactant. Damped oscillations upon start-up have been noted for a large number of monomer systems.

Damped oscillations will result in lost productivity since the product during these transients may be zero quality. Unstable oscillations will, of course, preclude continued operation. Limit cycle oscillations, while not unstable, will result in a product whose quality (MWD, CCD, etc.) varies with time in a cyclic fashion. In most cases, this is undesirable. As in the case of multiplicity, the polymerization reactor designer must be aware of the potential for oscillatory phenomena and should attempt to specify operating conditions at which these phenomena do not exist. In emulsion polymerizations, oscillations (both damped and sustained) are undesirable since the product is not of a consistent quality and since oscillations in free surfactant concentration may induce coagulation and reactor fouling. Several methods of eliminating oscillations in emulsion polymerization have been suggested. Poehlein [21] has used a PFR upstream of a CSTR train. All polymer particles are nucleated in the PFR.

Since PFR kinetics are essentially those of a BR (and such oscillations do not occur in BRs), no oscillations occur. The CSTRs, then, are used to grow the existing particles. As clearly demonstrated in Figure 17.5, by segregating particle nucleation from particle growth, oscillations are eliminated. Oscillations are also observed during nonisothermal solution polymerization in a CSTR, where interactions between the heat evolution and reaction rate cause instability.

17.3.2   Other Reactor Dynamics

No other reactor configuration (batch, semibatch, and plug flow) exhibits the range of dynamic behavior of the CSTR. However, over a finite time interval, other configurations can sometimes exhibit a more narrow range of dynamic behavior. If a semibatch reactor is operated such that the rate of reaction is just balanced by the rate of dilution from the feed, a pseudo-steady state may exist. In this case, the concentration of reactants and products in the reactor will remain constant over the time interval necessary to fill the reactor. This may be exploited to provide constant polymer properties during the filling and start-up of a CSTR or CSTR train. BRs do not exhibit multiplicity or limit cycle behavior. However, thermal instability is often a major problem. Consider a batch polymerization reactor in which cooling is accomplished by a heat transfer jacket with constant coolant flow. If, for some reason, the reactor temperature increases slightly, the rate of polymerization will increase. Since polymerization is exothermic, this will cause a rise in the temperature of the reactor. If this rise is offset by the increased heat transfer to the coolant brought on the increase in temperature driving force, the system is selfregulating and will return to its original rate of polymerization. If, on the other hand, the rate of heat removal is less than the new rate of heat evolution by the polymerization, the reactor temperature will increase further. Without intervention by an automatic control system or an alert operator, the temperature will continue to increase with catastrophic results (explosion, boil-out, or, at least, unscheduled shutdown). If rapid increase in temperature causes fouling of the heat transfer surfaces or if the polymerization is autocatalytic, the situation is exacerbated. There is a possibility of thermal instability in all exothermic reactors but due to the high heat of polymerization, the autocatalytic nature of some reactions, and to the increased propensity for fouling at high temperatures, polymerization reactors are especially prone to such instability. Temperature control systems must be designed with this in mind. Plug flow reactors are subject to the same sorts of instabilities as s batch reactors (since the kinetics is essentially those of a BR). In addition, the problem of reduced heat transfer due to polymer build-up on the inside surface of the reactor is compounded by the fact that polymer near the wall may be more viscous (due to low temperature or high conversion). The resultant velocity gradient will result in continuing build-up on the walls as the viscous polymer near the walls has a longer residence time than the less viscous material near the center of the tube. This effect will lead to reduced heat transfer and an even greater tendency toward thermal instability.

17.4   REACTOR CONTROL AND OPTIMIZATION

17.4.1   Introduction

The first consideration in establishing a strategy for controlling polymerization reactors is to categorize all system inputs and outputs into those, which are to be controlled those, which may be adjusted to achieve this control, and those which are beyond the control of the designer. The cause-and-effect relationships in a polymerization reactor are depicted in Figure 17.6.

The system outputs can be divided into three categories: end-use properties, controlled variables affecting product quality, and controlled variables specifying operating conditions. In order for the final product to meet the required specifications for a given application, the polymer must have certain end-use properties. End-use properties determine the suitability of a polymer for a specific application. These properties may be well defined (e.g., tensile strength) or they may be empirical measures of suitability in a given application (e.g., scrubability for architectural paints). End-use properties such as solubility, bulk density, extrudability, and so on determine

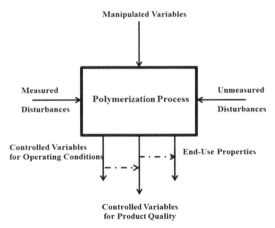

FIGURE 17.6 A view of the polymerization process from the process control perspective.

the salability of the polymer. Polymer appearance factors such as color, refractive index, particle size, and shape are also important in some cases. In most cases, the end-use properties are not measurable online. Other outputs must be controlled to produce a polymer with desired end-use properties.

From a control standpoint, the most important variables are those which ultimately affect the end-use properties of a polymer. They will be referred to as controlled variables affecting product quality. The most important are M_w, MWD, monomer conversion, CCD, CSD, and degree of branching. Most of these variables are not measurable online. The common approach is to control the variables, which are measurable, to estimate those which are estimable and control based on the estimates, and to fix those which cannot be estimated by controlling the inputs to the process. Closed-loop control involves the adjustment of some manipulated variable(s) in response to a deviation of the associated control variable from its desired value. The purpose of closed-loop control is to bring the controlled variable to its desired value and maintain it at that point. Those variables which are not controllable in a closed-loop sense are maintained at their desired values (as measured by laboratory or other off-line measurement) by controlling all identifiable inputs in order to maintain an unmeasured output at a constant value.

The next category of process outputs is called controlled variables specifying operating conditions. Some examples of these are temperatures, pressures, and flow rates associated with the process. These variables are most often measurable and are closed-loop controlled. The inputs to the polymerization system can be categorized as manipulated variables or disturbance variables. The manipulated variables are those which are adjusted, either automatically or manually, to maintain the controlled variables at their desired values. Common manipulated variables in polymerization processes include coolant or heating medium flow rates, gas or liquid flow rates for pressure control, feed rate of monomer, solvent, or initiator, and agitator speed.

The disturbance variables are those over which the control engineer has no control. Disturbances may be stochastic (random) or deterministic. Stochastic disturbances arise from the natural variability of the process. Examples are short-term variations in flow rates caused by mechanical inaccuracies. Deterministic disturbances arise from known causes, and they usually occur at longer intervals. Examples are lot-to-lot variations in feedstock quality and changes in production rates mandated by the operation of some upstream or downstream process. Although the cause of such disturbances may be known, the disturbances themselves cannot be eliminated because of constraints external to the system. Some disturbances, stochastic and deterministic, may be measurable, but by definition, they cannot be eliminated. However, the effect of such disturbances on the final product can be eliminated by compensating for them by adjusting the manipulated variables. This is the function of regulatory control.

The process variables do not always fall into such neat categories. For instance, temperature may be manipulated to adjust average molecular weight. In this case, temperature is the manipulated variable for an M_w control loop, but may at the same time be the controlled variable for a temperature control loop which uses the flow rate of a coolant as a manipulated variable. In this case, the value of the manipulated variable for the M_w control loop (temperature) is the desired value for the temperature control loop. This, of course, is the notion of cascade control. Not all process variables are measurable online. Most end-use properties are not. Some controlled variables affecting product quality are measurable or they can be estimated; many are not. Even when the technology to measure these variables online exists, the cost of such sensors may be prohibitive. Online gel permeation chromatography for the determination of MWD is an example. Almost all controlled variables specifying operating conditions are measured. Most manipulated variables are known because they are either measured or set by a control system (or both). As discussed earlier, disturbances may be measured or unmeasured.

One final point about closed-loop process control: Economic considerations dictate that to derive optimum benefits, processes must invariably be operated in the vicinity of constraints. A good control system must drive the process toward these constraints without actually violating them. In a polymerization reactor, the initiator feed rate may be manipulated to control monomer conversion or M_w; however, at times when the heat of polymerization exceeds the heat transfer capacity of the kettle, the initiator feed rate must be constrained in the interest of thermal stability. In some instances, there may be constraints on the controlled variables as well. Identification of constraints for optimized operation is an important consideration in control system design. Operation in the vicinity of constraints poses problems because the process behavior in this region becomes increasingly nonlinear. In many cases, the capability to control polymerizations is severely limited by the state of the art in measurement instrumentation. In other cases, the dynamic response of the instruments dictates the design strategy for the process.

17.4.2   State Estimation and Sensor Fusion

There are several levels to the control problem with increasing degrees of measurement difficulty. At the highest level is the overall control objective to regulate end-use properties. A unique measurement, which leads to this objective may or may not be available. This objective is often addressed by setting release specifications based on end-use properties. These are related to the controlled variables affecting product quality (e.g., M_w, monomer conversion, etc.) which again are not often measured online.

Changes in these variables are transmitted to the plant by adjusting the controlled variables specifying operating conditions (e.g., composition or temperature setpoints). And finally, the regulation of the plant requires control of temperature, pressure, and flow. The unique measurement challenges in polymerization then may be summarized as follows: vague control and measurement objectives, difficult sampling, discrete sensor output, large dead times and sampling intervals, off-line analyses, large measurement errors, and high noise levels.

The first step in reactor control (and optimization) is to accurately measure what is happening in the reactor. Other sections of this book give details of polymerization process measurements, both off-line and online. If a critical measurement is not available, it is often possible to estimate that value from other measurements plus a mathematical model. In addition, it is often possible to develop an optimal estimate of the states of the reactor from the available measurements. One is sometimes faced with very easily and frequently available measurements (such as temperatures, flow rates, etc.). However, these measurements often do not have high information content about the critical states of the process (molecular weight distribution, CCD, etc.). Other measurements, such as molecular weight via gel permeation chromatography, have high information content about the critical variables, but are only available infrequently, or even off-line. By using a mathematical model to combine the high frequency, low information content measurements with the low frequency, high information content measurements, one can develop an optimal estimate of the state of the reactor, including estimates of states for which there are no measurements. This idea of combining the results of all of the sensors into an optimal estimate of all of the states is sometimes referred to as *sensor fusion*.

There are a variety of estimation schemes ranging from the more or less ad-hoc approaches to observers and estimators which have desirable convergence properties. Perhaps the most useful of these is the Kalman Filter. The Kalman Filter is an optimal estimator for the estimation of the states of a dynamic system from a set of measurements, which are a subset of the set of states (or linear combinations of states). As such it can be used for noise filtering, estimation of unmeasured states, rectification of multiple sensors for the same property, and prediction of future values of states. Kalman filtering has been used in a number of polymerization applications [22]. A drawback is that it incorporates the process model into the filter structure. In many cases a simple observer will suffice. An observer requires model predictions of the process outputs, but the model is not incorporated into the observer structure. The process model can be updated without changing the observer. Details of the various observers and estimators can be found elsewhere [23].

17.4.3 Control Strategies

Depending upon the quantity of polymer to be produced and on the kinetics of polymerization, polymers may be produced in either a batch (or semibatch) reactor or a CSTR. Operational strategies for a BR differ from those for a CSTR even if the desired product is the same. However, the goal of the control system for either batch or continuous reactors is to maintain certain outputs within specifications by adjusting suitable inputs, subject to constraints, and in the presence of disturbances. Based on the understanding of polymerization principles and in the light of the classification of polymerization process variables, a description of current batch and continuous polymerization reactor control strategies will be presented. Following that, selected examples of polymerization control will be discussed in an attempt to show the types of strategies which may be employed. Reviews of polymerization reactor control are given by MacGregor et al. [24] and Schork et al. [22].

17.4.3.1 *Batch/Semibatch Polymerization Control Strategies* Batch or semibatch polymerization control systems commonly include preprogrammed recipe addition and start-up and shutdown procedures. Batch polymerizations are commonly carried out under temperature control either isothermally or following some predetermined temperature profile. Cooling to remove the heat of polymerization is normally provided by a coolant flowing through a jacket and possibly through external heat exchangers in a pump-around loop. In the case of condensation polymerization, cooling may be provided by vaporization of the condensation product (e.g., water). In other systems, vaporized solvent may be condensed and returned to the reactor. If a very broad MWD is desired (as in adhesives), the reactor may be operated adiabatically, although care must be taken to avoid runaway. If any of the monomers or solvents has high vapor pressures, reactor pressure may be controlled as well. This is often done with a nitrogen blanket in the head space of the reactor, removing oxygen which may retard initiation.

During batch or semibatch polymerization, temperature and pressure are usually regulated. The setpoints for these loops may be constant, or may be the result of trajectory optimization. Complications arise due to the fact that the system is never at a steady state. For this reason, a single set of tuning parameters may not be adequate for the entire batch cycle. It is often necessary to do gain scheduling (adjusting the controller tuning parameters according to a predetermined schedule through the batch cycle). This is especially important in polymerization systems with a strong gel effect, which may cause autoacceleration of the polymerization rate.

Reactor temperature is often controlled by manipulating the temperature or flowrate of coolant in the reactor cooling jacket. This scheme is hindered by the slow dynamics of heat removal and the nonlinear nature of the heat evolution process (especially in the gel effect region of free radical polymerization). If this strategy is used, it is often necessary to detune the controller in order to not become overly aggressive during times of high process gain. Some control performance can be recovered by employing a cascade strategy in which the reactor temperature (master) controller sets the setpoints for coolant temperature. The coolant temperature is then controlled at these setpoints by a slave controller, which manipulates coolant flow rate into the cooling jacket.

All of the initiator or catalyst may be added at the beginning of the reaction or some may be added over the course of the polymerization. If some of the initiator is added during polymerization, the addition may follow a predetermined trajectory or it may be adjusted to maintain a constant rate of reaction over most of the course of polymerization.

Alternatively, if an online MWD measurement is available, initiator addition may be used to regulate M_w, although this is rarely done. In any scheme in which initiator is added to achieve some closed-loop control objective, the fundamental problem of semibatch control arises: the controller can add initiator, but cannot remove it from the polymerization medium, which results in "one-sided" control action and poor control. The solution is to use an initiator with a very short half-life so that the initiator decomposes rapidly upon entering the reactor, giving "two-sided" control, since the initiator level will drop rapidly if initiator addition is stopped. This point must be emphasized, since the initiation systems for some old polymers were designed for batch systems, where all the initiator is added at the beginning of the polymerization. Initiation systems for batch polymerizations are often chosen with a long half-life in order to maintain a reasonably constant rate of polymerization over the entire kettle time. These initiation systems have often been retained when the reactor was converted to manual semibatch, and finally to closed-loop semibatch control. Therefore, if a kettle is to be converted to semibatch initiation (manual or automatic), the initiation system should be examined and possibly reformulated, according to the new mode of operation.

Often, batch polymerization control is a matter of specifying an open-loop policy for the introduction of various materials. Due to the fact that there is no steady state around which to regulate and to the fact that by the time a problem is detected, the batch may have progressed beyond the point of possible corrective action, it is often useful to track the behavior of a polymerization from batch to batch and reoptimize the current batch based on the experience gained in preceding batches. The presumption in this approach is that there is some underlying disturbance (e.g., monomer inhibition), which persists from batch to batch. Because of its simplicity and intuitive approach, this strategy has been adopted in some industrial polymerizations. However, the determination of open-loop trajectories for temperature and for the

addition of monomers and/or initiators in batch or semibatch polymerizations is mostly done off-line. These trajectories may be the result of operating experience or may be developed from mathematical models of the polymerization system.

Advanced control of a semibatch reactor involves driving the reaction from an initial state to a specified final state in some manner which is judged to be the "best" in terms of productivity or product quality. Nonlinear model predictive control may be employed to drive the reaction along a trajectory, which maximizes some predetermined objective functional [25]. Such a procedure presumes high-quality online sensors for polymer properties (monomer conversion and number-average molecular weight), an accurate mathematical model, and an optimization-based controller which allows control objectives rather than just setpoints to be included.

In copolymerization, the more reactive monomer may be added to the reactor over time to produce a more uniform CCD or to control the rate of polymerization, and hence heat evolution. One exception to this is rubber polymerization. Due to the high rate of branching in rubber which can be detrimental to the end-use properties, all of the monomer is added at the beginning of the reaction in order to suppress branching by keeping the monomer-to-polymer ratio as high as possible throughout the reaction. CCD control may be done by feeding comonomer at fixed rates by adding various comonomers at predetermined times or by following a complex monomer addition policy determinedly off-line optimization of a mathematical model of the polymerization process. If copolymer composition is measured or estimated online, the reactive monomer can be added in a closed-loop fashion. In any of these cases, the same problem of one-sided control is present. Other than by reformulating the recipe to accommodate semibatch operation or living with sluggish control brought on by the need to detune the one-sided loops, the only other solution is to adopt a "targeting" approach (as described earlier) in which a product quality variable is not regulated at each sampling period but specified at the end of the polymerization (targeted). This requires a detailed model of the polymerization process and a nonlinear predictive controller, which will adjust operating conditions at the current time in order to meet a control objective (target, in terms of product quality variables) at the end of the reaction [25].

17.4.3.2 *Statistical Process Control*   In most cases of batch and semibatch polymerizations, the only measurements of the end-use properties are at the end of the batch. As discussed previously, these are often off-line, laboratory analyses. The same is true for the product quality variables. An example from the synthetic rubber industry is the Mooney viscosity of the rubber. Neither the Mooney viscosity (end-use property) nor its underlying product quality variables (MWD and degree of branching or cross-linking) is measured online. In fact, in some cases, the polymer quality variables are not measured at

all on regular basis, since release specifications are written in terms of Mooney viscosity. In this situation, closed-loop control of Mooney viscosity is impossible. However, since the measurements of variables to be controlled (end use or polymer quality) are substantially uncorrelated in time, the use of statistical process control (SPC) is appropriate.

Likewise, in continuous polymerizations, the end use and polymer properties variables may be measured very infrequently, and most often, off-line. In addition, the laboratory analysis time adds a large delay to the measurement. Under these circumstances, the measurements may be substantially uncorrelated in time, and again, SPC may be an appropriate approach for the control of polymer quality or end-use properties.

The philosophy of SPC is to monitor the output of a process and determine when control action is necessary to correct deviations of the output from its setpoint. The most common tool for accomplishing this is the *Shewhart Chart*. In the discrete parts manufacturing industries, multiple samples are taken at fixed intervals. Quality tests are run on these samples, and the mean is plotted on a Shewhart Chart. In the absence of a disturbance, the means should be normally distributed around the setpoint. If the upper and lower control limits (UCL and LCL, respectively) are placed at three standard deviations above and below the target (hence the *six sigma* formalization of the SPC approach), a range is defined into which all of the means should fall. The likelihood of a point outside the control limits means the process is "out of control" and some adjustment should be made to the process. If the process is "in control" (points are within the control limits), no action is taken. This is to prevent manipulations of the process based on stochastic variations in the process. (Responding to stochastic variations in the process corresponds, in the field of conventional process control, to trying to compensate for noise, something no good control engineer would do.) By definition, noise is either stochastic, or deterministic but of such high frequency that it is impossible to compensate for it. An example of these is the noise on a tank level signal coming from a differential pressure transducer which senses the hydrostatic head at the bottom of the tank, and hence the depth of liquid. The level signal will contain a stochastic portion resulting from sensor imperfections and corruption of the signal during transmission. The level signal will also contain some deterministic information about the variations in level brought on by the turbulent nature of the liquid surface under agitation. This deterministic component is "real" but is of a very high frequency, and cannot be compensated. Control engineers deal with noisy signals by filtering inputs. This is effective when the signal is analog or digital with high sampling rates. However, at very low sampling rates where SPC is applicable, filtering is not appropriate (due to the assumption of lack of correlation between samples) or effective (due to the low sampling rates). This is the rational for taking action

only when the process exceeds the upper or lower control limits. Another reason for taking control action only when the process is out of control is because there is assumed to be a cost associated with control action. This is often true in the discrete manufacturing industries where it may be necessary to shut down the assembly line in order to make adjustments to equipment. However, it might not be true in the chemical process industries, where corrective action may be "free," such as in the adjustment of the temperature of a polymerization reactor.

If the hypothesis is made that the points on the Shewhart Chart are normally distributed, the probability of a point falling outside the control limits is quite small. The likelihood of various other occurrences (9 points in a row all below or above the mean, 14 points in a row alternating up and down, etc.) can be calculated as well. If the likelihood of these patterns is slight, they can be used as additional rules to determine when the process is out of control. Such pattern-based rules are easy for a chemical plant operator to implement. The process is then judged to be in control or out of control based on various patterns as described earlier. Thus, if an individual sample for a single batch of polymer is above the UCL or below the LCL, one might suspect an error in recipe makeup. If nine batches in a row are below the mean, one might suspect raw materials contamination. In continuous polymerization, if the samples are sufficiently infrequent that the process "settles" between samples and if the samples are not autocorrelated, the procedures outlined earlier can be used. This amounts to manual steady-state control with the need for control identified by the control chart.

If the process is fast compared with the sampling interval but the samples are autocorrelated (as is often the case), controller can be developed which specifies the correction to be made to the process. A process model is needed. For a process which is fast compared with the sampling interval, this can be simply a gain between the manipulated input and the process output. In addition, a disturbance model is necessary. A simple yet effective model for continuous process systems is that of an integrated white noise sequence.

With these assumptions, a minimum variance controller has been derived by MacGregor [26].

Many continuous polymerization processes fit the assumptions used to derive the controller discussed earlier (the process is fast with respect to the sampling interval, but the samples are still autocorrelated, and an integrated white noise model is appropriate). Other noise models may be used. If the process is not fast with respect to the sampling time, the process model must capture the dynamics of the process. Similar controllers can be derived for these situations. If the process is "slow" with respect to the sampling time, then conventional process control is appropriate.

17.4.3.3 *Continuous Polymerization Control Strategies* In continuous polymerizations, temperature and pressure are controlled in much the same way as in batch systems.

Obviously, temperature trajectories are not employed in continuous reactors. Instead, the various vessels in a series of polymerization reactors may operate at different temperatures. Then, the polymerizing mixture will "see" different temperatures as it passes from one vessel to the next. Likewise, monomer trajectories are replaced in continuous systems with intermediate injection of a more reactive monomer between polymerization vessels. This strategy can be exploited to adjust the CCD. Monomer conversion can be adjusted by manipulating the feed rate of initiator or catalyst. If online MWD is available, initiator flow rate or reactor temperature can be manipulated to adjust MWD. Polymer quality and end-use property control are hampered as in batch polymerization case by infrequent, off-line measurements. In addition, online measurements may be severely delayed due to the constraints of the process flow sheet. For example, even if online viscometry (via melt index) is available every 1–5 min, the viscometer may be situated at the outlet of an extruder downstream of the polymerization reactor. The transportation delay between the reactor where the M_w develops and the viscometer where the M_w is measured (or inferred) may be several hours. Thus, even with frequent sampling, the data are old.

There are two approaches possible in this case. One is to do open loop, steady-state control. In this approach, the measurement is compared to the desired output when the system is believed to be at steady state. A manual correction to the process is then made based on the error. The corrected inputs are maintained until the process reaches a new steady state, at which time the process is repeated. This approach is especially valid if the dominant dynamics of the process are substantially faster than the sampling interval. Another approach is to connect the output to the appropriate process input(s) in a closed-loop scheme. In this case, the loop must be substantially detuned to compensate for the large measurement delay. The addition of a dead-time compensator can sometimes improve the performance in this case. Optimal policies for changing product grades without shutting down the reactor train are desirable; however, what little is done in this area is based on operating experience rather than on optimization approaches.

17.4.4 Process Optimization

Optimization of (semi)batch reactor operation will include considerations of reaction time versus conversion, reaction time versus monomer recovery cost, and the potential for variations in polymerization temperature within a batch to achieve desired product quality, and hence end-use properties. Open-loop trajectories may be determined for the addition of monomers and/or initiators. Temperature programming is often done to develop polymer of unique properties. The semibatch addition of the more reactive monomer in copolymerization is often carried out to develop a product with a uniform or gradient CCD. In emulsion polymerization, programmed addition of monomer, comonomer, initiator,

surfactant, or chain transfer agent may be desirable in a semibatch reactor for the improvement of product properties. These trajectories may be the result of operating experience, or they may be developed by calculating optimal trajectories to achieve certain goals (reduced reaction time, desired MWD or CCD) subject to constraints on heat transfer capacity.

Online optimization of operating trajectories is often not done due to the lack of detailed mathematical models, and the computational time necessary to develop an optimal trajectory. (This will change as online computational capabilities increase.) The rigorous calculation of optimal trajectories requires a reasonably accurate model of the polymerization process. Trajectory optimization is most often done by defining an objective function in terms of product quality, and then using standard optimization techniques to minimize or maximize the objective by varying the process inputs with time. Solutions can be made subject to constraints on both process inputs (e.g., negative flows are not allowed and process outputs such as temperature should never exceed a safe limit). Often an optimization problem may degenerate into a regulation problem. For instance, in free radical polymerization, if the objective is to produce a product of target average M_w and minimum polydispersity, this is best done by regulating (where possible) the instantaneous average M_w at its desired value.

Optimization in continuous processes can be applied to multiple aspects of the process. The first is optimal design, including the reactor type specification as described earlier. The second is optimal steady-state operation. The latter application will require a detailed dynamic model of the process. Finally, if multiple products are to be made in a single continuous reactor system, the *product wheel* can be optimized. This involves specifying in which order the various products are to be made, based on the production volume of each product, and the propensity of that product to leave behind material which will contaminate the following product. (For instance, it is desirable to produce a light-colored product before a dark product since contamination from the light product will not be noticeable in the dark.) Finally, the transition from a product on the product wheel to the following product may be optimized. This may involve a rapid changeover or a gradual transition to the following product. The optimization objective is to maximize rector productivity while minimizing off-spec material. This also will require a detailed dynamic model.

REFERENCES

[1] Levenspiel O. *Chemical Reaction Engineering*. 3rd ed. New York: Wiley; 1999.

[2] Schork FJ, Deshpande PB, Leffew KW. *Control of Polymerization Reactors*. New York: Marcel Dekker; 1993.

[3] Schork FJ, Smulders W. On the molecular weight distribution polydispersity of continuous living-radical polymerization. J Appl Polym Sci 2004;92:539–542.

[4] Dotson NA, Laurence RL, Tirrell M. *Polymer Process Modeling.* New York: VCH Publishers; 1996.

[5] Ray WH. On the mathematical modeling of polymerization reactors. Polym Rev 1972; 8 (1):1–56.

[6] Barner-Kowollik C, Quinn JF, Nguyen TLU, Heuts JPA, Davis TP. Kinetic investigations of reversible addition fragmentation chain transfer polymerizations: cumyl phenyl-dithioacetate mediated homopolymerizations of styrene and methyl methacrylate. Macromolecules 2001;34:7849–7857.

[7] Zargar A, Schork FJ. Copolymer sequence distributions in controlled radical polymerization. Macromol React Eng 2009;3:118–130.

[8] Zargar A, Schork FJ. Design of copolymer molecular architecture via design of continuous reactor systems for controlled radical polymerization. Ind Eng Chem Res 2009;48:4245–4253.

[9] Krywko W, McAuley K, Cunningham M. Mathematical modeling of particle morphology development induced by radical concentration gradients in seeded styrene homopolymerization. Polym React Eng 2002;10:135–161.

[10] Ma JW, Cunningham MF, McAuley KB, Keoshkerian B, Georges M. Nitroxide mediated living radical polymerization of styrene in miniemulsion – modelling persulfate-initiated systems. Chem Eng Sci 2003;58:1177–1190.

[11] Luo YW, Tsavalas J, Schork FJ. Theoretical aspects of particle swelling in living free radical miniemulsion polymerization. Macromolecules 2001;34:5501–5507.

[12] Ahmad NM, Charleux B, Farcet C, Ferguson CJ, Gaynor SG, Hawkett BS, Heatley F, Klumperman B, Konkolewicz D, Lovell PA, Matyjaszewski K, Venkatesh R. Chain transfer to polymer and branching in controlled radical polymerizations of *n*-butyl acrylate. Macromol Rapid Commun 2009; 30 (23):2002–2021.

[13] Ye YS, Schork FJ. Modeling and control of sequence length distribution for controlled radical (RAFT) copolymerization. Ind Eng Chem Res 2009;48:10827–10839.

[14] Gerrens H. How to select polymerization reactors. 2. Chemtech 1982;12:434–443.

[15] Gerrens H. How to select polymerization reactors. 1. Chemtech 1982;12:380–383.

[16] Russum JP, Jones CW, Schork FJ. Impact of flow regime on polydispersity in tubular RAFT miniemulsion polymerization. AIChE J 2006;52:1566–1576.

[17] Hill CG. *Chemical Engineering Kinetics and Reactor Design.* New York: Wiley; 1977.

[18] Jaisinghani R, Ray WH. Dynamic behavior of a class of homogeneous continuous stirred tank polymerization reactors. Chem Eng Sci 1977;32:811–825.

[19] Kwalik KM. Bifurcation characteristics in closed-loop polymerization reactors [PhD thesis]. Atlanta: School of Chemical Engineering, Georgia Institute of Technology; 1988.

[20] Nomura M, Harada M. On the optimal reactor type and operations for continuous emulsion polymerization. In: Bassett DR, Hamielec AE, editors. *Emulsion Polymers and Emulsion Polymerization.* Washington, DC: American Chemical Society; 1981. p 121–144.

[21] Greene RK, Gonzalez RA, Poehlein GW. Continuous emulsion polymerization – steady state and transient experiments with vinyl acetate and methyl methacrylate. In: Piirma I, Gardon JL, editors. *Emulsion Polymerization.* Washington, DC: American Chemical Society; 1976. p 341–358.

[22] Schork FJ, Deshpande PB, Leffew KW. *Control of Polymerization Reactors.* New York: Marcel Dekker; 1993.

[23] Ray WH. *Advanced Process Control.* McGraw-Hill Chemical Engineering Series. New York: McGraw-Hill; 1981.

[24] MacGregor JF, Penlidis A, Hamielec AE. Control of polymerization reactors: a review. Polym Proc Eng 1984;2:179–206.

[25] Peterson T, Hernández E, Arkun Y, Schork FJ. A nonlinear DMC algorithm and its application to a semibatch polymerization reactor. Chem Eng Sci 1992;47:737–753.

[26] MacGregor JF. Online statistical process-control. Chem Eng Prog 1988;84:21–31.

18

OPTIMIZATION OF THE PARTICLE SIZE IN EMULSION POLYMERIZATION

Joseph Zeaiter and José A. Romagnoli

18.1 INTRODUCTION

Optimization in general represents the task of producing best results, for a given situation, with minimum costs and efforts. This is in fact a broad definition that can cover any type of decision-making problems such as in the design, operation, and analysis of manufacturing plants and industrial processes, as well as in business management and finance. The subject, seemingly simple, is indeed a complex one; it comprises two parts: (i) model identification and validation and (ii) the problem of optimal decision taking.

In fact, the success of any optimization technique critically depends on the degree to which the model represents and accurately predicts the investigated system. For this reason, the model must capture the complex dynamics in the system and predict with acceptable accuracy the proper elements of reality. Moreover, it is important to be able to recognize the characteristics of a problem and identify appropriate solution techniques; within each class of problems there are different optimization methods which vary in computational requirements and convergence properties. These problems are generally classified according to the mathematical characteristics of the objective function, the constraints, and the controllable decision variables.

In the polymer industry, the production of more and more products involves emulsion polymerization and the control of the quality of the polymer latex is becoming increasingly economical. Polymerization processes, similar to many other production facilities, continually face increasing pressures for more stringent quality and plant operation requirements. By applying advanced process optimization techniques, however, process engineers can meet these requirements and achieve significant productivity and quality improvements. There are substantial incentives expected in developing, off-line and real-time, optimal operating schemes that result in the production of polymers with desired properties. The molecular architecture of the polymer and the particle size in the emulsion polymerization processes in particular play a major role in this context.

The control of the particle size in emulsion polymerization using closed-loop strategies is a very attractive yet challenging problem [1]. Difficulties associated with online measurement of the particle size together with the complex mechanisms involved in emulsion polymerization systems limit the options and make control implementation a formidable task. In many cases, conventional optimization strategies fail to ensure a consistent product quality with the result that industries rely on traditional "recipes" and experience.

In general, the optimization of polymerization processes [2] focuses on the determination of trade-offs between polydispersity, particle size, polymer composition, number average molar mass, and reaction time with reactor temperature and reactant flow rates as manipulated variables. Certain approaches [3] apply nonlinear model predictive control and online, nonlinear, inferential feedback control [4] to both continuous and semibatch emulsion polymerization. The objectives include the control of copolymer composition,

Monitoring Polymerization Reactions: From Fundamentals to Applications, First Edition. Edited by Wayne F. Reed and Alina M. Alb.
© 2014 John Wiley & Sons, Inc. Published 2014 by John Wiley & Sons, Inc.

branching, cross-linking, monomer conversion, average molecular weight, and optimal reaction time. Optimal trajectory profiles can be calculated for the manipulated variables, such as the monomer feed rates, reactor temperature, and the flow rate of the chain transfer agent. Nonlinear model predictive control [5, 6] is also applied to control the reactor temperature and the molecular weight of the polymer; it manipulates the feed rate of the chain transfer agent to achieve the desired molecular weight distribution.

Optimal control schemes [7] can follow a hierarchical structure in which the control of the reactant composition (e.g., monomer flow rate and catalyst) is executed at the first level, and at the second level, the reactant target values are used to regulate product properties. The control and optimization [8] of the relative particle growth using state estimators along with a (shrinking horizon) model predictive controller are common approaches. State estimators such as the extended Kalman filter (EKF) are used for online state estimation during the nonlinear control of different polymer properties [9, 10] (weight-chain-length distribution, weight average molecular weight, conversion, etc.).

Attempts to optimize the particle size [11] by manipulating the surfactant feed rate and methods [12–14] focusing on the development and application of predictive algorithms for control of particle size distributions (PSDs) described by population balance models are attracting academic and industrial experts alike. The control algorithms, subject to manipulated input and product quality constraints, can be designed on the basis of finite dimensional models for particulate processes. Closed-loop simulations demonstrated the effectiveness of such proposed control algorithms. In conjunction, high-fidelity mechanistic models are used as a soft-sensor for online feedback [15] in the control of the particle size and molecular weight.

The foregoing indicates that the implementation of process optimization for the optimal control of emulsion polymerization processes is relying on technological advancement in the areas of mathematical process modeling, soft-sensing, and model-based control. This chapter illustrates some of the most successful approaches and their application to the control of the particle size in emulsion polymerization.

18.2 PROCESS OPTIMIZATION: OVERVIEW

Optimization is concerned with finding the minimum (or maximum) of a function. Usually, the problem has three basic parts:

- An objective function: any problem investigated in an optimization study will have as objective the improvement of the system. The objective is formulated in a quantitative form and subsequently subjected to a minimization (or maximization). This is represented in mathematical form as follows:

$$\underset{x(i)}{\text{Min}}\, J(x_1, x_2, \ldots, x_n) \quad i = 1, 2, \ldots, n \qquad (18.1)$$

It is appropriate to point out that the minimization problem is equivalent to the maximization problem if the sign of the objective function is reversed; that is,

$$\underset{x(i)}{\text{Max}}\, J(x_1, x_2, \ldots, x_n) = \underset{x(i)}{\text{Min}}[-J(x_1, x_2, \ldots, x_n)]$$

$$(18.2)$$

Here, $J(x_1, x_2, \ldots, x_n)$ is a short-hand notation to denote some functional relationship between the adjustable parameters x_1, $x_2, \ldots, x_n$ and the objective J. It is a measure of the difference between the required performance and the actual performance obtained.

It should be fairly obvious that, by defining an input to the system, one can find the resulting output. If this is not the case, one cannot control the system, or in other words, optimize it.

Often, the problem is formulated as a multiobjective optimization case; the decision maker has to optimize many different objectives at once. These objectives are of different scale and, in most circumstances, they conflict with each other. Moreover, the variables that optimize one objective may be far from optimal for others. In practice, such problems are dealt with by reformulating the multiobjective problem as a single-objective case by forming a weighted combination of the different objectives (i.e., weighted-sum strategy) or else by placing some objectives in the form of constraints (i.e., ε-constraint method). The optimal control problem then takes the following form for the weighted-sum strategy:

$$\underset{x(i)}{\text{Min}}\, \sum_{i=1}^{n} \lambda_i J_i \qquad (18.3)$$

where $J_1, \ldots, J_n$ are the objectives and $\lambda_1, \ldots, \lambda_n$ ($\lambda_i \in \mathscr{R}^n$) are the corresponding weights.

Note that, because J is a vector, if any of the components of J are competing, there is no unique solution to this problem. Instead, the concept of noninferiority (also called Pareto optimality) must be used to characterize the objectives. A noninferior solution is one in which an improvement in one objective requires a degradation of another. With the ε-constraint method, only one of the objectives (mainly the primary one) is expressed in the cost function while the other objectives take the form of inequality constraints:

$$\underset{x(i)}{\text{Min}}\, J(x_1, x_2, \ldots, x_n) \quad i = 1, 2, \ldots, n \qquad (18.4a)$$

subject to

$$J_i(x_1, x_2, \ldots, x_n) < \varepsilon_i \quad i = 1, 2, \ldots, m \qquad (18.4b)$$

This approach is able to identify a number of noninferior solutions on a nonconvex boundary that are not obtainable using the weighted sum technique. A problem with this method is, however, finding a suitable selection of ε to ensure a feasible solution. A further disadvantage of this approach is that the use of hard constraints is rarely adequate for expressing true design objectives. Similar methods exist, such as that of Waltz [16], which prioritizes the objectives. The optimization proceeds with reference to these priorities and allowable bounds of acceptance. The difficulty here is in expressing such information at early stages of the optimization cycle.

The steady-state version of the optimization problem is normally termed the linear programming (LP) or nonlinear programming (NLP) problem, depending on the nature of the objective function and constraints. In such problems, the objective is usually a differential function of the optimization variables. In contrast, in dynamic optimization, the objective function is usually a functional and frequently a time integral along the trajectory as illustrated in Equation 18.5. The initial conditions on the state variables have to be specified along with the final desired state:

$$\underset{x(i)}{\text{Min}} \int_{t_0}^{t_f} J(x_1, x_2, \ldots, x_n) dt \quad i = 1, 2, \ldots, n \quad (18.5a)$$

subject to

$$\frac{dx_i}{dt} = f_i(x_1, x_2, \ldots, x_n) \quad i = 1, 2, \ldots, n$$

$$x_i(0) = x_{i0} \qquad\qquad t_0 < t < t_f \quad (18.5b)$$

The solution of dynamic optimization problems is based on the classical calculus of variations, the maximum principle of Pontryagin et al. [17], and the dynamic programming of Bellman [18]. Probably, the most commonly used member of the triad is the maximum principle, as most computational techniques are concerned with satisfying the maximum principle necessary conditions for an optimum. The solution of the optimality conditions problem is generally obtained using the quasi-linearization approach [19] and through the use of multiple shooting algorithms such as those proposed by Bulirsch [20] or Dixon and Bartholomew-Biggs [21].

Other techniques employ a discretization approach whereby the optimal control problem is converted to an NLP through the discretization of all variables. This can be done using the finite difference and orthogonal collocations methods [22, 23]. The characteristic of the discretization method is that the optimization is carried out in the full space of the descretized variables and the discretized constraints are satisfied at the solution of the optimization problem only. This is therefore called the infeasible path approach. Another

technique is to discretize the control variable only and to carry out the optimization in the space of the decision variables (i.e., feasible path). For given values of the decision variables, the objective function and the constraints can be calculated for the optimal case by integrating the underlying system. The technique, known as the control vector parameterization method [24–26], has the advantage of controlling the discretization error by adjusting the magnitude and order of the integration steps.

- The decision variables: The vector of manipulated inputs $x_1, x_2, \ldots, x_n$ are chosen to have the greatest influence on the objective function. These variables must be successively adjusted during optimization to obtain the desired maximum or minimum. Each set of adjustments to these variables is termed iteration and, in general, a number of iterations are required before an optimum is obtained. In the iterative algorithm, a first estimate of the decision variables must be supplied as a starting point.

- Constraints: Although optimization studies may be required to produce the best result from a given situation, this may not always be achievable due to the imposition of a variety of restrictions known in practice as process constraints.

There are several types of constraints which can be put on a problem:

- Algebraic equality constraints written as

$$g_i(x_1, \ldots, x_n) = 0 \quad i = 1, \ldots, m \quad (18.6)$$

This type of constraints expresses the relationships connecting the decision variables $x_1, \ldots, x_n$ and, thus, reduces the degree of freedom in the system.

- Algebraic inequality constraints: These usually specify the practical operating limits of certain variables within the process. They are written in mathematical form as

$$g_i(x_1, \ldots, x_n) \leq 0 \quad i = 1, \ldots, m \quad (18.7)$$

- Differential equality constraints: These are commonly used in chemical processes, where the rate of formation or consumption of a species is a function of the state variables. This type of constraints can be represented as

$$\frac{dx_i}{dt} = f_i(x_1, \ldots, x_n) \leq 0 \quad i = 1, \ldots, m$$

$$x_i(0) = x_{i0} \qquad\qquad t_0 < t < t_f \quad (18.8)$$

Any other form of constraints is possible; however, the three aforementioned types are the most commonly encountered in process optimization. The window of operation over which the constraints are satisfied is known as the feasible region. Most solution algorithms proceed by first finding a feasible solution, then seeking to improve upon it and finally changing the decision variables to move from one feasible solution to another. In some cases, no constraints are imposed on the system. In fact, the field of unconstrained optimization is a large and important one for which a lot of algorithms and software have been developed. However, constrained optimization is much more difficult than unconstrained optimization and, therefore, a great deal of effort has been expended to reformulate constrained problems so that constraints are avoided.

The efficient and accurate solution to the optimal problem is not only dependent on the size of the problem in terms of the number of constraints and design variables but also on the characteristics of the objective function and constraints. When both the objective function and the constraints are linear functions of the design variable, the problem is known as a LP problem. Quadratic programming (QP) concerns the minimization or maximization of a quadratic objective function that is linearly constrained. For both the LP and QP problems, reliable solution procedures are readily available. More difficult to solve is the NLP problem in which the objective function and constraints may be nonlinear functions of the design variables. A solution of the NLP problem generally requires an iterative procedure to establish a direction of search at each major iteration. This is usually achieved by the solution of an LP, a QP, or an unconstrained subproblem.

Due to the transient nature of the emulsion polymerization process and the inherent nonlinearities in the system, the nonlinear constrained dynamic optimization formulation is best adopted for such process application. The optimization can be carried out to determine optimum operation trajectories to produce polymer latex with specific product characteristics.

18.3 OPTIMIZATION OF THE POLYMER PARTICLE SIZE DISTRIBUTION

The distinct feature of emulsion polymerization is the copresence of a continuous phase and a dispersed particulate phase, and the occurrence of physicochemical phenomena such as nucleation, growth, and coagulation, which are absent in other homogeneous processes. The interplay and strong coupling among these phenomena strongly affect the shape of the PSD of the polymer particles, which in turn determines the physicochemical and mechanical properties of the polymer formed. If the particle size is small, there will be far more particles (assuming no change in the total latex volume) and less chance of free movement. As a result, if everything else is kept

constant, the viscosity of an emulsion will increase as the particle size is reduced. On the other hand, mixtures of particles tend to yield lower viscosities. In relation to product quality, the PSD is particularly important. For example, a narrow (monodisperse) size distribution is required for a glossy finish to latex paints while a broad (polydisperse) PSD results in a matt finish [27]. Adhesives also require a broad distribution to maximize their strength [28]. Here, the different size particles pack closer together, minimizing the void area and thus improving adhesive performance.

Particles born at different times but all growing at the same volumetric growth rate will produce a broad PSD. Similarly, bimodal distributions can result if nucleation takes place after the end of interval I; this may happen by homogeneous nucleation at high conversion or if new micelles are introduced through surfactant addition. If during interval II all particles grow at the same rate under no secondary nucleation, the PSD narrows and the final PSD should correspond fairly closely to that at the end of interval II. The continuous monomer addition in the semibatch process will affect the reaction rate and the duration of intervals II and III. At low addition rates, a constant conversion results. If the conversion is high (i.e., 90–100%), as is usually the case, the reactor is operated in starved-feed mode. No separate monomer phase exists in the reactor and, thus, the process is operating in interval III. Contrarily, large addition rates will keep the process in interval II where high-reaction rates and particle growth are dominant. As mentioned earlier, the particle size in an emulsion is an important factor in determining the properties of the final product and the conditions under which the reaction takes place. It is generally known to be a record of the kinetic processes that took place during a reaction. In relation to product quality, the PSD is extremely important especially for latex paints and adhesives. In addition, the particle size of the latex influences the emulsion viscosity and, thus, heat and mass transfer inside the reactor.

With the plethora of competing events, process engineers often find it difficult to operate the polymerization reactor in a way that maintains the production of a polymer with specified PSD characteristics. This is due, in some cases, to the fact that the control of the PSD is practiced by controlling its characteristic variables (e.g., the mean and variance) or other easily measured process variables (e.g., temperature, conversion and concentrations). These formulations are not enough for fine PSD control and thus fail in most circumstances when applied to the real process. Moreover, the properties of the polymer formed are influenced by the process/kinetic history. Any unprecedented disturbances in the operating conditions (e.g., temperature, pressure, flow rates, etc.) may cause drastic irreversible changes in product quality. One should add to this the fact that any reactants (e.g., monomer, initiator, and surfactant) introduced to the process cannot be removed.

Achieving optimal control and operation of emulsion polymerization reactors requires the integration of mechanistic first-principle process models with optimization procedures and strategies specific to the product and process requirements. However, the modeling of polymerization reactors with a focus on PSD is a very challenging task due the distributed nature of such system. In order to produce accurate predictions, the model must account for a complexity of physical and chemical events. Skills from a variety of disciplines, such as in the areas of colloid and surface science, particle science, physical chemistry, and reaction engineering, are needed. Herein we illustrate the modeling equations needed to capture the dynamics and evolution of the PSD in a semibatch reactor employing a zero-one emulsion polymerization process [34]. Subsequently, the optimization of this process is described in detail.

18.4 ZERO–ONE SYSTEM MODELING FOR OPTIMIZATION PURPOSES

In emulsion polymerization with relatively small latex particles (<100 nm diameter), a zero–one system is considered to more accurately describe the situation. In a zero–one system, the rate of radical termination within a latex particle is fast relative to the rate of radical entry. Or, in other words, termination is not rate-determining. Consequently, the entry of a radical into a particle, which already contains a free radical, will cause immediate termination. Therefore, under the zero–one system, the latex particles can be easily split into two groups; those with zero free radicals and those containing exactly one free radical. The rate of termination, however, for the entire process is reduced due to the compartmentalization of radicals. That is, while the radicals remain in separate particles, they cannot terminate without at least one transferring phase. This naturally takes time and thus corresponds to a higher rate of propagation compared to similar bulk systems (this is one of the many attractions of emulsion polymerization). Furthermore, in a zero–one system, the following assumptions are made:

- Polymerization occurs only within latex particles and follows the radical chain addition mechanism.
- Particles are formed through both homogeneous and micellar nucleation.
- "Competition" occurs between the processes of particle growth, homogeneous nucleation, and micellar nucleation.
- Three types of particles may be identified: (i) those containing no radicals, referred to as type n_0 particles; (ii) those containing one monomeric radical, type n_1^M particles; and (iii) those containing one polymeric radical, type n_1^P particles.
- Compartmentalization plays a crucial role in the overall kinetics, as a radical in one particle will have no "access" to a radical in another particle without the intervention of a "phase transfer event."

- Two radicals in close proximity will terminate rapidly.
- The entry of a radical into a particle that already contains a free radical will instantaneously cause termination. Thus, the maximum value of the average number of radicals per particle, $\bar{n}$, is 0.5.
- Monomer diffusion inside a particle is a function of the polymer volume fraction.
- At low monomer concentration (high monomer conversion), entry, propagation, and termination become diffusion-controlled processes.

A general description of a zero–one system is illustrated in Figure 18.1 [34].

The polymerization process begins with the decomposition of the initiator in the water phase, which then reacts with the monomer to generate oligomeric radicals according to the following reactions:

$$I - I \xrightarrow{k_d} 2I^{\bullet}$$
$$I^{\bullet} + M \xrightarrow{k_{p1}} IM_1^{\bullet}$$

(18.9)

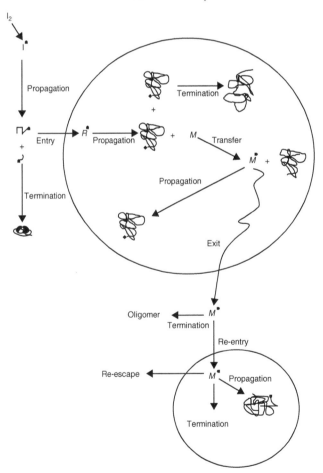

FIGURE 18.1 Mechanisms occurring in emulsion polymerization. From Zeaiter J. A framework for advanced/intelligent operation of emulsion polymerization reactors [PhD Thesis]. Sydney: University of Sydney; 2002.

These oligomeric radicals then undergo a sequence of reactions, as outlined in the following:

$$IM^{\bullet}_i + M \xrightarrow{k^i_{p,aq}} IM^{\bullet}_{i+1}$$

$$IM^{\bullet}_i + T^{\bullet} \xrightarrow{k_{t,aq}} \text{inert products}, \quad i < z$$

$$IM^{\bullet}_i + \text{latex particle} \xrightarrow{k^i_e} \text{entry}, \quad i = z, \ldots, j_{crit} - 1$$

$$IM^{\bullet}_i + \text{micelle} \xrightarrow{k^i_e} \text{new particle}, \quad i = z, \ldots, j_{crit} - 1$$

$$IM^{\bullet}_{j_{crit}} \rightarrow \text{new particle}$$

$$(18.10)$$

Mass balances on the various aqueous radical species leads to the following set of rate equations:

$$\frac{d[I^{\bullet}]}{dt} = 2k_d[I] - k_{p1}[I^{\bullet}]C_w \qquad (18.11)$$

$$\frac{d[IM^{\bullet}_1]}{dt} = k_{p1}[I^{\bullet}]C_w - k^1_{p,aq}[IM^{\bullet}_1]C_w - k_{t,aq}[IM^{\bullet}_1][T^{\bullet}] \qquad (18.12)$$

$$\frac{d[IM^{\bullet}_i]}{dt} = k^{i-1}_{p,aq}[IM^{\bullet}_{i-1}]C_w - k^i_{p,aq}[IM^{\bullet}_i]C_w \\ - k_{t,aq}[IM^{\bullet}_i][T^{\bullet}], \quad i < z \qquad (18.13)$$

$$\frac{d[IM^{\bullet}_i]}{dt} = k^{i-1}_{p,aq}[IM^{\bullet}_{i-1}]C_w - k^i_{p,aq}[IM^{\bullet}_i]C_w \\ - k_{t,aq}[IM^{\bullet}_i][T^{\bullet}] \\ + k^i_e[IM^{\bullet}_i]\frac{N_{tot}}{N_A} - k^i_{e,micelle}C_{micelle}[IM^{\bullet}_i], \\ i = z, \ldots, j_{crit} \qquad (18.14)$$

In these equations, C_w is the monomer concentration in the water phase, N_{tot} is the total number of latex particle per liter of aqueous phase, and N_A is the Avogadro's number.

Furthermore, the monomeric radicals formed by radical transfer to the monomer, $M^{\bullet}$, may desorb from the particle with such "exited" radicals denoted by $E^{\bullet}$. Once such a desorbed radical meets the surface of a particle, it will immediately penetrate the particle due to its high lipophilic nature. This adsorption–desorption process is reversible, as illustrated in the following:

$$E^{\bullet} + \text{particle} \underset{k_{dM}}{\overset{k_{eE}}{\rightleftharpoons}} \text{particle} - M^{\bullet} \qquad (18.15)$$

Hence, the total radical concentration in the aqueous phase $[T^{\bullet}]$ becomes

$$[T^{\bullet}] = \sum_{i=1}^{z-1}[IM^{\bullet}_i] + [E^{\bullet}] \qquad (18.16)$$

The rate of surfactant consumption determines the micelle concentration, $C_{micelle}$. The amount of adsorbed surfactant onto the particle surface is defined as

$$S_{ads} = \frac{4\pi r_s^2 N_{tot}}{N_A a_s} \qquad (18.17)$$

and hence,

$$C_{micelle} = \text{maximum}\left[0, \frac{[S_{added}] - S_{ads} - [cmc]}{n_{agg}}\right] \qquad (18.18)$$

where $[S_{added}]$ is the total concentration of surfactant added, a_s is the area occupied by an adsorbed surfactant molecule, n_{agg} is the average number of surfactant molecules in a micelle (known as the micellar aggregation number), and $[cmc]$ is the critical micelle concentration. From Equation 18.18, it is clear that particle formation via micelles stops when the surfactant concentration falls below the cmc. The equation also shows the critical relationship existing between the surfactant concentration and the number of particles, N_{tot}.

The population balance equations for the three particles types (n_1^P, n_0, and n_1^M) are as follows [29]:

$$\frac{\partial n_1^P(V,t)}{\partial t} = k_p^1 C_p n_1^M + \rho_{init} n_0 - \rho n_1^P \\ - k_{tr} C_p n_1^P - \frac{\partial(K n_1^P)}{\partial V} \\ + \delta(V - V_0)\left[C_{micelle}\sum_{i=z}^{j_{crit}-1} k^i_{e,micelle}[IM_i]\right] \\ + k^{j_{crit}-1}_{p,aq} C_w[IM_{j_{crit}-1}] \\ - n_1^P(V)\int_0^{\infty} B(V,V')[n_0(V') + n_1^P(V')]dV' \qquad (18.19)$$

$$\frac{\partial n_0(V,t)}{\partial t} = \rho[n_1^P + n_1^M - n_0] + k_{dM}n_1^M \\ + \int_0^{\infty} B(V, V-V')[n_0(V')n_0(V-V') \\ + n_1^P(V')n_1^P(V-V')]dV' \\ - n_0(V)\int_0^{\infty} B(V,V')[n_0(V') + n_1^P(V')]dV' \qquad (18.20)$$

$$\frac{\partial n_1^M(V,t)}{\partial t} = k_{tr}C_p n_1^P + k_{eE}[E]n_0 - (k_p^1 C_p + k_{dM} + \rho)n_1^M$$

$$(18.21)$$

$$n(V,t) = n_1^P(V,t) + n_0(V,t) + n_1^M(V,t) \qquad (18.22)$$

The coupled partial integro-differential equations in the preceding comprise the "evolution equations" for the latex particles in an emulsion polymerization process. In this mathematical description, an n_1^P type particle is formed when an oligomeric radical enters an n_0 type particle and the monomeric radical propagates in an n_1^M type particle to form a polymeric radical. Particles containing such polymeric radicals are "consumed" when an oligomeric radical enters an existing n_1^P type particle, causing instantaneous termination. The process occurs with a rate constant ρ, the pseudo first-order rate constant which describes all entry events. Alternatively, radical transfer to a monomer molecule transfers radical activity, and the resulting monomeric radical may undergo subsequent propagation or termination.

The population of n_0 type particles is increased by the entry of any radicals (oligomeric or "exited") into n_1^M and n_1^P type particles. They are also formed when monomeric radicals exit an n_1^M type particle (a process that occurs with a rate constant k_{dM}). The population of n_0 type particles is decreased when oligomeric radicals enter an existing particle (to form an n_1^P type particle), or when an exited radical enters to form an n_1^M type particle.

Finally, n_1^M type particles may be formed by chain transfer within particles containing a polymeric radical, and via the entry of exited monomeric radicals into n_0 type particles. The n_1^M particles are consumed through the propagation of radicals in existing particles by the desorption of monomeric radicals from existing n_1^M particles and by the entry of any type of radical into existing n_1^M type particles. Growth significantly affects only n_1^P type particles, as radicals propagate without changing the particle's identity.

Having established the population balances, it is important to understand how to evaluate the rate coefficients and associated parameters. First, the critical degree for micellar (z) and homogeneous (j_{crit}) nucleation can be obtained semiempirically based on free energies of hydration [30].

The entry processes and hence the rate coefficients, k_e^i and $k_{e,micelle}^i$, are assumed to be diffusion controlled (with an exponent of 1/2 describing the chain-length-dependent diffusion coefficient of small radicals). Given that oligomeric radicals of degree greater than or equal to "z" can only enter a micelle or preexisting particle, the entry coefficients can be calculated as follows:

$$k_e^i(V) = 4\pi r_s N_A e_e \frac{D_w}{i^{1/2}} \quad i \geq z; \qquad k_e^i(V) = 0 \quad i < z$$

$$(18.23)$$

$$k_{e,micelle}^i(V) = 4\pi r_{micelle} N_A e_e \frac{D_w}{i^{1/2}} \quad i \geq z;$$

$$k_{e,micelle}^i(V) = 0 \quad i < z \qquad (18.24)$$

In the aforementioned equations, D_w is the diffusion coefficient for a monomeric radical in water, e_e is the entry efficiency, $r_{micelle}$ is the radius of a micelle, and r_s is the radius of a latex particle swollen with monomer.

Note that r_s is related to the unswollen radius, r, of a particle as follows:

$$\frac{r_s}{r} = \left[\frac{d_m}{d_m - C_p M_0} \right]^{1/3} \qquad (18.25)$$

where d_m is the monomer density, M_0 is the monomer molecular weight, and C_p is the concentration of monomer in the latex particle.

The entry coefficient for exited radicals can be defined in a similar fashion, as follows:

$$k_{eE}(V) = 4\pi r_s N_A e_e D_w \qquad (18.26)$$

The rate coefficient for desorption of monomeric radicals may be written as a function of the diffusivity of monomer both in the water phase and inside the particle, the aqueous monomer concentration, the monomer concentration in the particle, and the swollen radius:

$$k_{dM}(V) = \frac{3D_w D_{mon}}{((C_p / C_w)D_{mon} + D_w)r_s^2} \qquad (18.27)$$

The overall rate for entry, ρ, is derived from the following equation:

$$\rho = k_{eE}[E^{\bullet}] + \rho_{init} \qquad (18.28)$$

where ρ_{init} is the rate of formation of "z-mers" obtained from

$$\rho_{init} = \sum_{i=z}^{j_{crit}-1} k_e^i [IM_i] \qquad (18.29)$$

The propagational growth rate, $K(V)$, is defined by

$$K(V) = \frac{k_p M_0 C_p(V)}{N_A d_p} \qquad (18.30)$$

where d_p is the density of the polymer.

The average number of radicals per particle, $\bar{n}$, is given by

$$\bar{n} = \frac{1}{\sum_{i=1}^{G} n(V_i)} \left(\sum_{i=1}^{G} n_1^P(V_i) + \sum_{i=1}^{G} n_1^M(V_i) \right) \qquad (18.31)$$

while the total number of particles per unit volume, N_{tot}, can be calculated from

$$N_{tot} = N_A \int_0^\infty n(V)dV \tag{18.32}$$

At high monomer conversion, the viscosity inside the polymer particles increases sharply and further polymerization becomes diffusion controlled. To account for these changes, the propagation rate coefficient can be expressed as follows [31]:

$$k_p = \frac{1}{1/k_{p_0} + 1/k_{diff}} \tag{18.33}$$

where k_{p_0} is the propagation rate coefficient at low conversion and k_{diff} is the diffusion controlled rate coefficient, defined as

$$k_{diff} = 4\pi\sigma N_A (D_{mon} + D_{rd}) \tag{18.34}$$

where $D_{rd} = (1/6)k_p C_p \alpha^2$.

The diffusion coefficient for the monomer, D_{mon}, is expressed as a function of the polymer volume fraction inside the particle. An additional entry phenomenon can be taken into consideration at high ϕ_p, where the entry of a z-mer into a glassy particle is impeded because of slow diffusion through the interface. The "frustrated" z-mer then desorbs back into the aqueous phase and may propagate to a j_{crit}-mer and cause secondary nucleation.

Therefore, an empirical equation for the entry efficiency, $e_e = (1 - \phi_p)^{C_{p\_sat}}$, is used to model these changes.

For a semibatch process (as being considered here), the rate of accumulation of monomer in the reaction vessel can be obtained from a molar balance equation

$$\frac{dN_m}{dt} = F_m - R_p V_r \tag{18.35}$$

where $R_p = k_p C_p (\bar{n} N_{tot}/N_A)(V_w/V_r)$ is the reaction rate and V_r is the total reaction volume, the latter being described by

$$\frac{dV_r}{dt} = M_0 R_p V_r \left(\frac{1}{d_p} - \frac{1}{d_m} \right) + \frac{F_m M_0}{d_m} \tag{18.36}$$

where F_m is the monomer feed rate.

Note that the molecular weight is still that of the monomer unit as the polymerization reaction does not completely decompose the molecule, rather it joins them together in series. Thus, for every mole of monomer consumed, the mass of polymer formed is equivalent to the mass of monomer consumed. The density of the polymer is, however, higher than that of the monomer, as the monomer units are brought closer together during polymerization. Hence, the first term in the right-hand side of Equation 18.36 represents the volume shrinkage caused by the polymerization of the monomer.

The concentration of monomer in the latex phase, C_p, which represents the amount of monomer in the latex particle, affects the rate of reaction and hence the particle growth. At high feed rates, monomer droplets accumulate inside the reactor and the monomer concentration inside the particles reaches its saturation value. Therefore, C_p may be estimated as follows:

$$V_d = \frac{N_m M_0 - C_w M_0 V_w - C_w M_0 V_p}{d_m} \tag{18.37}$$

where

$$C_p = C_{p\_sat} \quad \text{when} \quad V_d > 0 \tag{18.38}$$

and

$$C_p = \frac{N_m}{V_p} \quad \text{when} \quad V_d = 0 \tag{18.39}$$

In Equation 18.37, V_d is the droplets volume and V_p is the volume of the polymer phase, calculated by

$$\frac{dV_p}{dt} = \frac{M_0 R_p V_r}{d_p} \tag{18.40}$$

From a modeling perspective, the monomer partitions between two phases to maintain a thermodynamic equilibrium. In particular, a balance must be reached between the gain in interfacial free energy caused by an increase in surface area as a result of swelling and the loss in free energy caused by dissolving the monomer in the polymer. Based on this approach, reasonable estimates of the equilibrium concentrations can be determined. However, these estimates rely on knowing the interfacial tension and a number of Flory–Huggins interaction parameters that have not been estimated for many common double-bonded monomer groups, such as styrene [32]. Consequently, the generally accepted approach to this problem is to use empirical partition coefficients between the monomer/water phase and polymer/water phase coupled with a monomer mass balance.

The partition coefficient (K_{mwp}) for styrene between the aqueous phase and latex phase is defined as shown in Equation 18.41. Furthermore, the value of K_{mwp} can be found using the ratio of monomer concentration in the aqueous phase to that in the latex phase under saturated conditions:

$$K_{mwp} = \frac{C_{w\_sat}}{C_{p\_sat}} \tag{18.41}$$

Similarly, the partition coefficient for styrene between the monomer and polymer regions [33] is defined as

$$K_{mmp} = \frac{C_M}{C_p} = \frac{d_m K_{mwp}}{C_{w\_sat} M_0} \qquad (18.42)$$

Note that C_M is the concentration of pure monomer, which is equal to the ratio of monomer density to the monomer molecular weight. The monomer concentration in the water phase, C_w, is then given by

$$C_w = C_p K_{mwp} \qquad (18.43)$$

In the dynamic model, monomer conversion can be calculated as

$$X = 1 - \frac{N_m}{N_{m\_fed}} \qquad (18.44)$$

where N_{m_fed} is the total amount of monomer added to the reactor which can be calculated from

$$\frac{dN_{m\_fed}}{dt} = F_m \qquad (18.45)$$

The polymer volume fraction inside the particle is another indication of the particle "status" at any point in time. This can be estimated from the monomer concentration inside the particle as follows:

$$\phi_p = 1 - \frac{C_p M_0}{d_m} \qquad (18.46)$$

As discussed in depth earlier, the dynamic model completely describes the PSD for an emulsion polymerization process, given the current mechanistic knowledge. However, it is often convenient to represent the distribution by one or more measurable parameters. Such parameters can include the number average radius, defined as

$$\langle r \rangle = \frac{\sum_{i=1}^{G} n(i) r(i)}{\sum_{i=1}^{G} n(i)} \qquad (18.47)$$

This gives a good indication of the location of the distribution. The particle size polydispersity index, PSPI, on the other hand, gives an indication of the width of the distribution. The PSPI is defined as the mean squared radius divided by the mean radius squared,

$$PSPI = \frac{\langle r^2 \rangle}{\langle r \rangle^2} \qquad (18.48)$$

where the mean squared radius is defined by

$$\langle r^2 \rangle = \frac{\sum_{i=1}^{G} n(i) r(i)^2}{\sum_{i=1}^{G} n(i)} \qquad (18.49)$$

If all particles are identical, the mean squared radius and mean radius squared will be equal. This corresponds to a PSPI of 1, and the distribution is said to be monodispersed. As the distribution widens, the larger particles will more heavily influence the numerator (due to the r^2 term) and hence the index will increase.

The particle size evolution Equations 18.19–18.22 are a set of three coupled partial 2D integro-differential equations in radius and time. No analytical solution has been found for such systems and the numerical solution of such a set of equations is an extremely challenging task. One efficient method is to discretize these equations with respect to radius and convert the evolution equations into a set of coupled ordinary differential equations for each radius $r(i)$. That is, the continuous distribution can be broken down into "G" discrete groups of particles as illustrated in Figure 18.2 [34].

A useful analogy is to consider each group "i" as a box containing all latex particles with radius greater than $r(i)$ - $1/2\Delta r$ and less than $r(i) + 1/2\Delta r$, where Δr is the interval of radii covered by each box. The particle population in that box is described by its own set of three coupled ordinary, differential equations. However, these equations do, of course, depend on the particle population in other boxes where growth and coagulation are concerned.

To illustrate this fact, in Figure 18.3 [34], the total distribution has been split into the classifications of particles identity (i.e., n_1^P, n_1^M, and n_0). The diagram illustrates the transfer of particles from one box to another, although for simplicity it does not include coagulation. All movements in the vertical plane are due to the movement of radicals between the continuous phase and latex particles. The movements in the horizontal plane, on the other hand, correspond to propagation and hence growth of particles containing polymeric radicals. It is this growth which causes the distribution to continually shift to the right. Having discretized the evolution

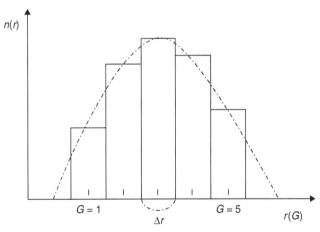

FIGURE 18.2 Discretizing the particle size distribution into $G = 5$ equally sized intervals of radius. From Zeaiter J. A framework for advanced/intelligent operation of emulsion polymerization reactors [PhD Thesis]. Sydney: University of Sydney; 2002.

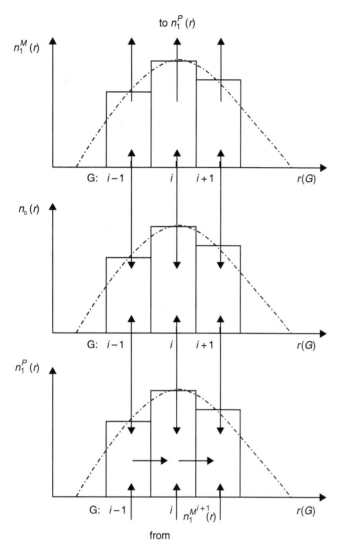

FIGURE 18.3 Transfer of radicals between boxes in the same radial interval (vertical direction) and growth across the radial distribution (horizontal arrows). From Zeaiter J. A framework for advanced/intelligent operation of emulsion polymerization reactors [PhD Thesis]. Sydney: University of Sydney; 2002.

equations, there is one ordinary differential equation (at constant radius) describing the particle population in each of the boxes illustrated in Figure 18.3. It is worth mentioning that the particle radius increases much more slowly than particle volume, hence discretizing the evolution equations by radius and not volume is computationally more efficient.

Consider a particle that doubles its radius over a period of time. The particle's volume will increase by a factor of eight, yet the radius only increases by a factor of two. Consequently, less equally sized radius boxes are required to track the particles by radius than would be required under a volume distribution. The advantage is that it dramatically reduces the number of equations to be simultaneously solved. As a result, less computational time is required.

Discretizing the partial differential evolution equations essentially makes the radius constant in each equation, by converting them to ordinary differential equations, with one equation for each radius. However, discretization also allows the integro-differential components of the equations to be expressed as finite difference approximations in equally spaced radius increments (Δr).

Integrals can be converted to the equivalent finite sum. For example, return to the contribution of entry, into preexisting particles, to the consumption of oligomeric radicals. The integral is expressed as a finite sum as shown:

$$\int_0^\infty k_e^i n(r)dr = \sum_{j=1}^\infty k_e^i(r_j)n(r_j)\Delta r \tag{18.50}$$

In terms of particle formation, the Dirac delta function $\delta(V - V_o)$ becomes $1/\Delta r$ [35]. Furthermore, the differential growth term can also be approximated by the centered finite difference method.

$$-\frac{\partial(Kn_1^P)}{\partial V} = \frac{-K}{8\pi\Delta r}\left(\frac{n_1^P(r_{i+1})}{r_{i+1}^2} - \frac{n_1^P(r_{i-1})}{r_{i-1}^2}\right) \tag{18.51}$$

For simplicity, K has been assumed to be independent of volume as the concentration of polymer in each particle (C_p) does not vary greatly with volume. The finite difference approximation is suitable for $i=2$ through to $i=G-1$; however, the followings are recommended for the endpoints [30]:

$$i = 1;\quad \frac{-K}{8\pi\Delta r}\left(\frac{n_1^P(r_2)}{r_2^2}\right) \qquad i = G;\quad \frac{-K}{8\pi\Delta r}\left(\frac{n_1^P(r_{G-1})}{r_{G-1}^2}\right)$$

$$\tag{18.52}$$

The endpoints effectively assume there is a box on either end of the discretized distribution that contains no particles. Thus, in each case, one of the terms drops out of the finite difference approximation.

In some cases, when the centered finite difference approximation (Eq. 18.51) is used, the model can predict an oscillatory response of the total particle number. Although unusual, these oscillations can be caused by successive nucleation periods of decreasing intensity [36]. In particular, it is postulated that micellar nucleation is responsible for the new particles formed at the beginning of the reaction while the oscillations in particle number were attributable to homogeneous nucleation.

In terms of reactor dynamics, micellar nucleation results in a large number of particles formed initially due to the presence of surfactant. As these particles grow, the total surface area increases substantially (as the particles are relatively small), causing a reduction in free surfactant. As a result, nucleation stops and the particle number and surface area decrease as particles exit the reactor. Under these conditions, the probability of an oligomeric radical growing to the critical length and hence of homogeneous nucleation taking place increases. Consequently, the total number of particles rises again and therefore oscillates. While this may explain the oscillations observed in continuous, it cannot be applied to a semibatch reactor model where the product is not drawn from the system. Furthermore, the associated PSD can be problematic as it is highly oscillatory and predicts particle populations less than zero which are not possible.

The oscillations do, however, point toward a problem with the growth term, as this would allow an uncertainty in one radial interval to flow quickly onto all intervals. In the discretized model the growth term is, apart from a few constants, determined by the finite difference approximation. As a result, the sensitivity of the model to both forward and backward instead of centred finite difference approximations can be tested and the most robust approach should be adopted. Notice that the forward approximation relies on populations "i" and "$i+1$" and the backward relies on populations "$i-1$" and "i." The four, in the forward and backward approximation, replaces the eight in the denominator of the centred difference approximation as the interval has been halved in size.

$$-\frac{\partial(Kn_1^P)}{\partial V} = \frac{-K}{4\pi\Delta r}\left(\frac{n_1^P(r_{i+1})}{r_{i+1}^2} - \frac{n_1^P(r_i)}{r_i^2}\right) \quad (18.53)$$

$$-\frac{\partial(Kn_1^P)}{\partial V} = \frac{-K}{4\pi\Delta r}\left(\frac{n_1^P(r_i)}{r_i^2} - \frac{n_1^P(r_{i-1})}{r_{i-1}^2}\right) \quad (18.54)$$

In general, it is found that the forward finite difference approximation produces an unstable particle number. The backward approximation, on the other hand, eliminates the oscillation and predicts a more stable PSD.

The prediction capability of the high-order mathematical model should be experimentally validated, over a wide range of operating conditions, before conducting optimization studies. Figure 18.4 [34] illustrates the evolution of the PSD with time during the emulsion polymerization of styrene [34]. Experimental measurements via capillary hydrodynamic fractionation (CHDF) are compared against the model simulations showing strong agreement.

18.5 OFF-LINE DYNAMIC OPTIMIZATION

The high fidelity model developed in the previous section can be used to conduct simulation case studies, what-if analysis, decision support, dynamic optimization, and soft-sensing with model predictive control. The off-line dynamic optimization approach is introduced here to illustrate a strategy that can be followed to develop optimum operation trajectories to achieve a specific PSD shape.

Due to the complex nature of the emulsion polymerization process and the associated set of distributed and lumped equations, the control vector parameterization approach can be adopted along with the ε-constraint optimization technique in a multiobjective form to account for the various control objectives. Several constraints should be included to define the desired final particle size and to account for different process and recipe limitations.

The systematic manipulation of the decision variables during polymerization can lead to different PSDs from the same recipe as illustrated in Figure 18.5 [34]. The question is how to achieve such customization in the product PSD and what optimum trajectory should the process follow in order to meet a specific objective?

Constraints are always present in real-life process control situations. Their importance has increased because optimal

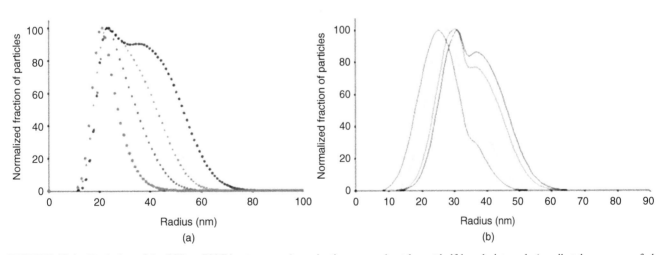

FIGURE 18.4 Evolution of the PSD at 72°C in styrene polymerization—samples taken at half-hourly intervals (semibatch, monomer federate at 1.6×10^{-4} mol s^{-1}). (a) Experimental data; (b) model simulations. From Zeaiter J. A framework for advanced/intelligent operation of emulsion polymerization reactors [PhD Thesis]. Sydney: University of Sydney; 2002. (*See insert for color representation of the figure.*)

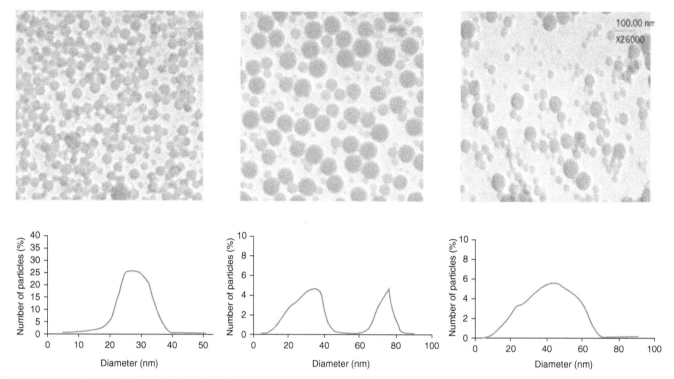

FIGURE 18.5   TEM micrographs showing different sizes of emulsion polystyrene particles. Graphs show the corresponding PSDs. From Zeaiter J. A framework for advanced/intelligent operation of emulsion polymerization reactors [PhD Thesis]. Sydney: University of Sydney; 2002.

control schemes often push the operating point toward the intersection of constraints. Generally, the objective function is set up to fulfill the decision-maker's objective, whereas the constraints which shape the feasible region usually come from the decision-maker's environment placing restrictions/conditions on achieving the desired objective. In the case of PSD control, several issues have to be considered:

- What are the objectives?
- What are the constraints?
- Should an inequality or equality type of constraint be used?
- What are the decision variables?
- What are the connections between the decision variables and the objectives?

18.5.1   Defining the Objectives

An optimization with constraints on particle concentration can be used as a prime objective but how would this result be translated and compared against measured samples (analyzed with CHDF or size exclusion chromatography, zeta-potential etc.)? Alternatively, if one uses the PSPI as the control objective then the breadth of the distribution can be defined and met. However, different shapes in the PSD can be tailored for the same PSPI. Hence, accurately defining the PSD shape would require the specification of additional polymer attributes.

The particle size (e.g., the number average radius) is an important parameter that can describe the process of particle formation and growth. During polymerization, a persistent increase in the average radius is a result of continual growth rate. On the other hand, a decrease in the average radius marks the formation of new particles via secondary nucleation mechanisms. To reach the target solution, the control problem must include the average radius along with the PSPI in the formulation. The ε-constraint method can be used to optimize the average radius profile and to force the particle size to follow the desired trajectory.

For instance, to produce a narrow PSD with specific particle size, the objectives can be placed as illustrated:

$$\text{Min}_{\substack{\text{decision} \\ \text{variables}}} [\text{PSPI}(t_{\text{final}})] \qquad (18.55)$$

subject to

$$\sum_{i=1}^{P} (r_{\text{avg}}^{\text{set point}}(k+i) - r_{\text{avg}}(k+i))^2 < \varepsilon_1 \qquad (18.56)$$

where ε_1 is the maximum value the sum of projected particle size errors can take, r_{avg} is the average radius, and $r_{\text{avg}}^{\text{set point}}$ is the set point/desired radius.

The solid content (e.g., 25%) in the reaction is another polymer attribute that has to be considered. A high solid content is obtained when the monomer is added in large amounts to the reaction, and converted consequently to polymer. Under these conditions, large particles are formed and the system resembles to bulk polymerization (caused by the dramatic increase in the viscosity of the system). The solid content also relates to the total amount of moles of added monomer in the "recipe." This specification can be included as an endpoint equality along with the monomer conversion into the set of constraints as

$$N_{m,T}(t_{final}) = \varepsilon_N$$
$$\text{Solid contents}(t_{final}) = \varepsilon_{SC} \qquad (18.57)$$
$$90\% < \text{Monomer conversion}(t_{final}) < 98\%$$

The total number of polymer particles, N_{tot}, is another criterion that needs special attention. Any variation in the operating conditions inside the reactor will affect particle growth and nucleation, hence N_{tot}. To obtain a monodisperse polymer, particle formation should only take place during interval I and the reactor should be maintained under conditions which avoid micellar formation and/or secondary nucleation. On the contrary, a polydisperse polymer can be obtained if N_{tot} is increased during reaction by micellar or secondary nucleation. Therefore, N_{tot} is considered a key performance indicator of PSD evolution. An endpoint inequality constraint should be used to constrain the particle number (and hence the particle size) within a predefined interval:

$$\varepsilon_{2low} < N_{tot} < \varepsilon_{2high} \qquad (18.58)$$

18.5.2 Selecting Appropriate Decision Variables

The effect of surfactant and initiator concentrations on the particle size and particle number is well documented in the literature. It is well known that increasing the surfactant concentration in the reactor will induce micellar formation and increase the particle number in the reactor (Fig. 18.6 [34]); smaller polymer particles are thus obtained.

To a lesser extent, the initiator concentration also affects the total particle number in the reactor. At higher concentrations, smaller polymer particles are produced. Similarly, the temperature of the reaction has a positive effect on the total particle number in the reactor as it increases. Smaller particles are produced at higher temperatures. And finally, the monomer feed rate in a semibatch reactor influences the reaction rate (under starved feed) and the particle growth. In addition, it can indirectly induce secondary nucleation at high conversions, thus reducing the particle size and increasing the total particle number. The PSPI is also affected by these four decision variables; it increases with the surfactant and initiator flow rates as well as with reactor temperature, and it decreases with the monomer feed rate.

The aforementioned decision variables (monomer, surfactant and initiator flow rates, and the reactor temperature) should be included in the optimization formulation, and subjected to

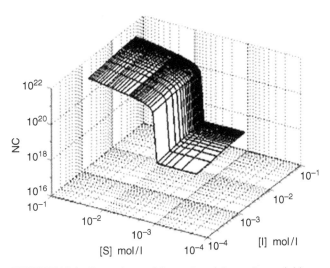

FIGURE 18.6 Dependence of the total particle number on initiator and surfactant (styrene polymerization at 50°C). From Zeaiter J. A framework for advanced/intelligent operation of emulsion polymerization reactors [PhD Thesis]. Sydney: University of Sydney; 2002.

upper and lower bounds as dictated by the process operating conditions (process limitation, equipment constraints, reaction constraints, etc.); for each variable, a path-constraint formulation should be adopted to make sure the variables are constantly maintained all the time between the high and low limits (i.e., hard constraints):

$$\varepsilon_{m,low} < F_m(t) < \varepsilon_{m,high}$$
$$\varepsilon_{s,low} < F_s(t) < \varepsilon_{s,high} \qquad (18.59)$$
$$\varepsilon_{I,low} < F_I(t) < \varepsilon_{I,high}$$
$$\varepsilon_{T,low} < T_e(t) < \varepsilon_{T,high}$$

Figure 18.7 [34] illustrates simulated and experimental results employing an optimal control strategy [34] to produce a polystyrene polymer with small particle size and a broad PSD. Monomer conversion >90% at the end of the experiment was specified. It can be clearly seen that the final shape of the experimental PSD is in good agreement with the simulation results and, also, particle nucleation (particle number) is constantly taking place due to the reduction of the monomer feed. The new particle formation mechanism causes the distribution of the PSD to broaden during the course of the reaction.

18.6 ONLINE OPTIMAL CONTROL

The off-line optimization results when applied in practice often become suboptimal due to ever-existing process disturbances and changes in process dynamics (e.g., when capacity is increased). Online optimal control can circumvent this problem and ensure optimum process operation all the time.

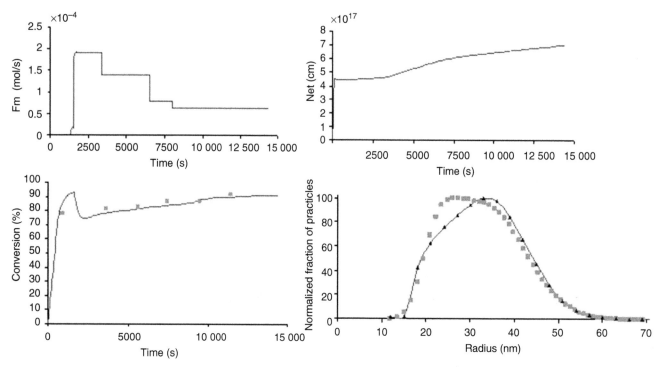

FIGURE 18.7 Schematic diagram of the optimal control strategy. From Zeaiter J. A framework for advanced/intelligent operation of emulsion polymerization reactors [PhD Thesis]. Sydney: University of Sydney; 2002.

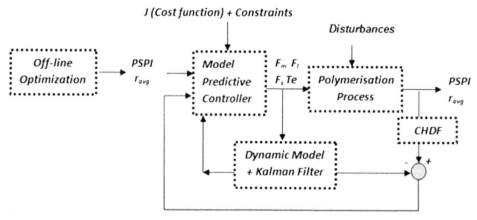

FIGURE 18.8 Optimal control results using constrained MPC with the mechanistic model acting as a soft-sensor. From Zeaiter J. A framework for advanced/intelligent operation of emulsion polymerization reactors [PhD Thesis]. Sydney: University of Sydney; 2002.

It can cope with process variability and redefine the optimal operating conditions online. Nevertheless, the control of the PSD in emulsion polymerization by means of closed-loop strategies is a challenging problem [1]. Difficulties associated with the online measurement of the PSD can limit operational options and make the control problem a formidable task. In addition, the nonlinear behaviour of the process causes conventional control strategies to fail in ensuring a consistent product quality.

Recent advances in process understanding, mathematical modeling, soft-sensing, and model-based control techniques can offer polymerization processes the chance of achieving major improvements in operation and product quality. A viable approach here would seem to involve combining these powerful tools into an effective model-based control strategy as illustrated in Figure 18.8 [34].

The closed-loop control scheme would involve a model predictive controller that receives the optimal trajectory from the off-line optimization using the dynamic model developed earlier. The MPC involves two parts: a model prediction and a control law.

The model prediction of the MPC is done using a time series model that relates the process outputs y (r_{avg}, PSPI, etc.) to the inputs u (manipulated variables: F_m, F_I, T_e, F_s),

$$y(k+1) = \sum_{i=1}^{N} a_i \Delta u(k+1-i) + y^{ss}(k+1) + d(k+1) \quad (18.60)$$

Modeling error and the impact of unmeasured disturbances are included in the last term on the right-hand side of Equation 18.60. Once the process output has been measured at a given point in time, $d(k)$ can be estimated by assuming, for the linear case, all future d values are equal, as illustrated:

$$d(k) = y^{meas}(k) - \sum_{i=1}^{N} a_i \Delta u(k-i) - y^{ss}(k) \quad (18.61)$$
$$= d(k+1) = \ldots = d(k+N)$$

In these equations, a_i are the step response coefficients obtained by model identification.

One can use the dynamic model of the emulsion polymerization to apply perturbations in the inputs u and record the model outputs y. The step response coefficients are then calculated as follows:

$$a_i = \frac{y^{step}(k+i) - y^{ss}(k+i)}{u(k+i) - u^{ss}} \quad (18.62)$$

where u_{ss} and y_{ss} are the steady-state or initial values of the inputs and outputs.

Another approach would be to obtain the step response coefficients from the optimal trajectory obtained by off-line optimization. This method helps reduce the effect of nonlinearity in the process.

$$a_i = \frac{y^{optimal}(k+i) - y^{ss}(k+i)}{u_{optimal}(k+i) - u^{ss}} \quad (18.63)$$

Furthermore, the dynamic emulsion polymerization model can be used to determine the effect of the past input moves or the future control moves on the output prediction. The step response coefficients a_i are then updated at every sequence, hence, adapting the model to the nonlinearities in the process. This is known as adaptive MPC. Another approach is to use a multimodel structure for the prediction part of the MPC (a_i model sets). As the process operating conditions change with time, the corresponding a_i model set is selected. This approach requires a bumpless mechanism to ensure a smooth transition when a new a_i model set is selected during online control execution.

To further cope with process nonlinearities, the disturbance term d can be expressed by combining two terms: d_{ext} which is the disturbance due to plant/model mismatch, and d_{nl} the disturbance due to nonlinearities. d_{nl} can be determined at every sampling time by minimizing the output prediction error between the linear and nonlinear models [5].

In the control law part, the MPC controller has to calculate the set of control moves (Δu) into the future that allows the system to follow a predefined set-point trajectory (from off-line optimization). This is done by solving for the quadratic cost function

$$J = \min_{\Delta u} \left[\sum_{i=1}^{N} \delta\left(y^{set\,point}(k+i) - y(k+i)\right)^2 + \sum_{i=1}^{N} \lambda\left(\Delta u(k+M-i)\right)^2 \right] \quad (18.64)$$

subject to constraints on the input moves: $\Delta u(k) \in [\Delta u_{min}, \Delta u_{max}]$, constraints on the input: $u(k) \in [u_{min}, u_{max}]$, and constraints on the output: $y(k) \in [y_{min}, y_{max}]$.

In Equations 18.60 and 18.61, N is the open-loop settling time of the process and in Equation 18.64, M is the control horizon (number of necessary control moves). The quantity $P = N + M$ is called the prediction horizon. In the case of a batch or semibatch process, P is a receding horizon and as such, at every control execution, the horizon is reduced by one interval.

The weight on the error, δ, and the weight on the input, λ (suppression move), are used to tune the MPC to provide tighter control for certain objectives while relaxing on others. The larger the weight, the higher is the penalty on the associated error or control move. The optimization is solved at every sampling time when a new prediction is updated by recent feedback measurements. Only the first control move is implemented to the process, the rest being discarded (although the rest could be used in case of loss of measurements).

The measurement feedback mechanism in the closed-loop strategy requires an advanced formulation similar to the MPC case. In practice, PSD measurements require sampling, dilution and off-line analysis, and data processing, typically using CHDF instruments for the PSD. The typical analysis time by CHDF is of the order 15–20 min, making this method impractical for online monitoring and control applications. To overcome this problem, a "soft-sensor" approach should be employed, with the full mechanistic model being used to provide online estimates of the polymer PSD. All relevant operating conditions (such as monomer flow rate and reactor temperature) are taken from the reactor at discrete time intervals as online measurements, and used by the dynamic model to estimate the PSD. Once off-line sample anaylsis becomes available, the measurements are used to update the model prediction of the soft-sensor. Figure 18.9 [34] shows an experimental validation of online constrained MPC implementation [34]. Here, the particle size (average radius) and the polydispersity were optimized off-line. The optimal trajectory was used as a set point trajectory for the online MPC to drive the process and to produce a polymer with a bimodal PSD. The mechanistic model was used online to provide

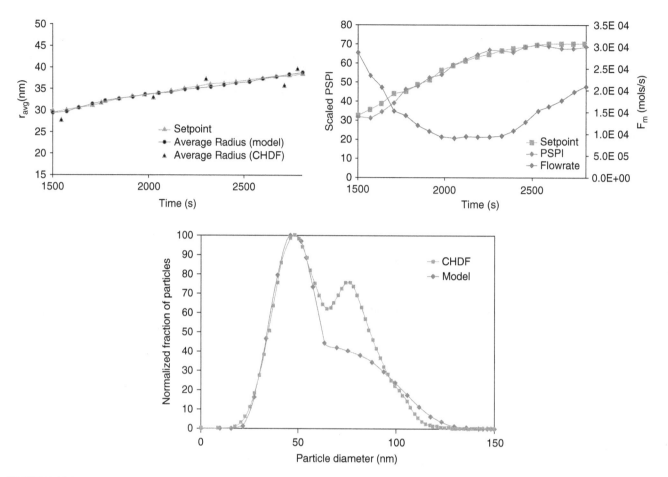

FIGURE 18.9 Experimental validation of online constrained MPC implementation. From Zeaiter J. A framework for advanced/intelligent operation of emulsion polymerization reactors [PhD Thesis]. Sydney: University of Sydney; 2002. (*See insert for color representation of the figure.*)

measurement feedback for PSPI and r_{avg} until the measurement from the CHDF becomes available.

18.6.1 State Estimator: Kalman Filter

Another successful measurement feedback/state estimation method is the EKF [37]. The Kalman filter predicts the process states and makes correction to the prediction once the measurement becomes available. The general form of the EKF equation is

$$\hat{x}_k^- = A_k(\hat{x}_{k-1}^-) + B_k u_{k-1}; \quad \text{predict the state}$$
$$P_k^- = A_k P_{k-1} A_k^T + W_k Q_{k-1} W_k^T; \quad \text{calculate covariance error}$$

$$(18.65)$$

$$K_k = P_k^- H_k^T \left(H_k P_k^- H_k^T + V_k R_k V_k^T \right)^{-1}; \text{Kalman filter gain}$$

$$\hat{x}_k = \hat{x}_k^- + K_k (z_k - H_k \hat{x}_k^-); \text{correct prediction with measurement}$$

$$P_k = \left(1 - K_k H_k\right) P_k^-; \text{update covariance error}$$

$$(18.66)$$

Equation 18.65 is to make predictions/estimations of the states (PSPI, r_{avg}) given previous filtered values and previous inputs (F_m, F_I, F_s, T_e). Once the measurement, z_k, is available, the prediction is corrected as shown in Equation 18.66. Note that matrices A and B are dependent on the sample time k, that is, the dynamic model of the emulsion process will be used to generate these at every execution interval. That way, the nonlinearity incurred in the process will be captured through the update of the process and measurement matrices. Note that H and V are the measurement Jacobians at instance k, while A, B, and W are the process Jacobians at instance k. Q and R are the process and measurement noise covariances at instant k. For the emulsion polymerization process, H is set equal to one since the states PSPI and r_{avg} are in fact measured variables (i.e., $z_k \equiv x_k$).

18.6.2 Dynamic Real-Time Optimization

In this approach, dynamic real-time optimization (DRTO) strategies directly drive the polymerization process without the intermediate MPC controller level. To do so, the embedded mechanistic dynamic model is used for optimization calculations over the whole control horizon at every control execution. The DRTO is in essence identical to an MPC with the sole difference a mechanistic first-principle model as a control engine. The advantage of such approach is the predictive power and accuracy of the DRTO associated mechanistic model, and its capability of pushing the process toward more economical operating regions usually unreachable by linear process models used in conventional MPC applications. In DRTO, the mechanistic model which is usually used for process simulation, control tuning, and design, has now become a predictor, simulator, and a controller. Given a set of desired process outputs (r_{avg}, PSPI, etc.), DRTO directly manipulates the process inputs (manipulated variables: F_m, F_I, T_c, F_s) in the most optimum manner to guarantee control, stability, and performance. Excellent references to DRTO developments are illustrated in the work of Rolandi and Romagnoli [38].

18.7 CONCLUSIONS

Nearly 30 years have passed since model predictive control (MPC) was first introduced to the chemical industry as an effective means to deal with multivariable constrained control problems. However, to this day, the online optimization, stability, and control performance are largely understood for systems described by linear dynamics but only limited progress has been made on nonlinear systems. For complex applications, especially polymerization, many questions are raised on the reliability and efficiency of these online control schemes. In particulate processes, such as emulsion polymerization, the challenges are even harder. The distributed and compartmentalized nature of the emulsion polymerization system makes conventional MPC tools a futile optimal control application. The particle size and molecular weight of the polymer represent the kinetic history of the polymerization process and are considered as the "fingerprints" of the final product. They are greatly affected by the operation of the polymerization reactor and the polymer quality/specifications can be only improved if the underlying process is well understood. Fortunately, recent technological advances in polymer colloid science, process modeling tools, and the increasing computational power are providing great opportunities to successfully tackle the optimal control problem of polymerization processes. The embedding of physico-chemical mathematical models into the model-based control architecture is rapidly becoming the favorite control approach for real-time optimization. Through these high-fidelity mathematical models, optimization strategies are continuously pushing the process to optimum operating regimes that conventional methods are incapable of achieving.

REFERENCES

[1] Dimitratos J, Elicabe G, Georgakis C. Control of emulsion polymerization reactors. AIChE J 1994;40:1993–2021.

[2] Cawthon GD, Knaebel KS. Optimization of semi-batch polymerization reactors. Comput Chem Eng 1989;13:63–72.

[3] Hidalgo PM, Brosilow CB. Non-linear model predictive control of styrene polymerization at unstable operating points. Comp Chem Eng 1990;14:481–494.

[4] Kozub DJ, MacGregor JF. Feedback control of polymer quality in semi-batch copolymerization reactors. Chem Eng Sci 1992;47:929–942.

[5] Arkun Y, Peterson T, Hernandez E, Schork FJ. A non-linear DMC algorithm and its application to a semi-batch polymerization reactor. Chem Eng Sci 1992;47:737–753.

[6] Asua JM, Echevarria A, Leiza JR, De La Cal JC. Molecular-weight distribution control in emulsion polymerization. AIChE J 1998;44:1667–1679.

[7] Ogunnaike BA. On-line modelling and predictive control of an industrial terpolymerization reactor. Int J Control 1994;59:711–729.

[8] Liotta V, Georgakis C, El-Asser MS. Real-time estimation and control of particle size in semi-batch emulsion polymerization. 15th American Control Conference Proceedings; 1997 June 4–6; New Mexico. p 1172–1176.

[9] Crowley TJ, Choi KY. Experimental studies on optimal molecular weight distribution control in a batch-free radical polymerization process. Chem Eng Sci 1998;53:2769–2790.

[10] Rhee HK, Ahn SM, Park MJ. Extended Kalman filter-based non-linear model predictive control for a continuous MMA polymerization reactor. Ind Eng Chem Res 1999;38:3942–3949.

[11] Doyle FJ, Crowley TJ, Meadows ES, Kostoulas E. Control of particle size distribution described by a population balance model of semibatch emulsion polymerization. J Proc Control 2000;10:419–432.

[12] Shi D, Mhaskar P, El-Farra NH, Christofides PD. Predictive control of crystal size distribution in protein crystallization. Nanotechnology 2005;16:S562–S574.

[13] Shi D, El-Farra NH, Li M, Mhaskar P, Christofides PD. Predictive control of particle size distribution in particulate processes. Chem Eng Sci 2006;61:268–281.

[14] Chiu T, Christofides PD. Nonlinear control of particulate processes. AIChE J 1999;45:1279–1297.

[15] Zeaiter J, Romagnoli JA, Gomes VG. Online control of molar mass and particle-size distributions in emulsion polymerization. AIChE J 2006;52:1770–1779.

[16] Waltz FM. An engineering approach: hierarchical optimization criteria. IEEE Trans Auto Control; 1967(AC-12):179–180.

[17] Pontryagin LS, Boltyanskii VG, Gamkrelidze RV, Mishchenko EF. *The Mathematical Theory of Optimal Processes*. New York: Wiley-Interscience; 1962.

[18] Bellman R. *Dynamic Programming*. Princeton: Princeton University Press; 1957.

[19] Miele A. Recent advances in gradient algorithms for optimal control problems. J Optim Theory Appl 1975;17:361–430.

[20] Bulirsch R. Die Mehrzielmethods Zur Numerischen Losung Von Nichtilnearen Randwerproblemen Und Aufgaben Der Optimalen Stewerung. Deutsche Forschungs-und Versuchsanstalt fur Luft-und Raumfahrt, *Oberphaffenhofen*. Federal Republic of Germany, Carl-Cranz Gesellschaft; 1971.

[21] Dixon LC, Bartholomew-Biggs MC. Adjoint-control transformations for solving practical optimal control problems. Optim Control Appl Methods 1981;2:365–381.

[22] Biegler LT. Solution of dynamic optimization problems by successive quadratic programming and orthogonal collocation. Comput Chem Eng 1984;8:243–248.

[23] Cuthrell JE, Biegler LT. On the optimization of differential-algebraic process systems. AIChE J 1987;33:1257–1270.

[24] Pollard GP, Sargent RWH. Off line computation of optimal controls for a plate distillation column. Automatica 1970;6:59–76.

[25] Sargent RWH, Sullivan GR. The development of an efficient optimal control package. Wurzburg: Proceedings of 8th IFIP Conference on Optimization Techniques, Part 2;1977 Sept 5–9.

[26] Sargent RWH, Sullivan GR. Development of feed changeover policies for refinery distillation units. Ind Eng Chem Process Des Dev 1979;18:113–124.

[27] Bartsch S. Kulicke WM, Fresen I, Moritz HU. Seeded emulsion polymerization of styrene: determination of the particle size by flow field-flow fractionation coupled with multi-angle laser light scattering. Acta Polym 1999;50:373–380.

[28] Collins EA. Measurement of particle size and particle size distribution. In: Lovell PA, El-Aasser MS, editors. *Emulsion Polymerization and Emulsion Polymers*. Chichester: John Wiley & Sons; 1997.

[29] Coen EM, Gilbert RG, Morrison BR, Leube H, Peach S. Modelling particle size distributions and secondary particle formation in emulsion polymerization. Polymer 1998;39:7099–7112.

[30] Gilbert RG. *Emulsion Polymerization: A Mechanistic Approach*. London: Academic Press; 1995.

[31] Russell GT, Napper DH, Gilbert RG. Initiator efficiencies in high conversion bulk polymerizations. Macromolecules 1988;21:2141–2148.

[32] Hamielec AE, MacGregor JF, Penlidis A. Multicomponent free-radical polymerization in batch, semi-batch and continuous reactors. Macromol Chem Macromol Symp 1987;10/11:521–570.

[33] Richards JR, Congalidis JP, Gilbert RG. Mathematical modeling of emulsion copolymerization reactors. J Appl Polym Sci 1989;37:2727–2756.

[34] Zeaiter J. A framework for advanced/intelligent operation of emulsion polymerization reactors [PhD Thesis]. Sydney: University of Sydney; 2002.

[35] Gilbert RG. Modelling rates, particle size distributions and molar mass distributions. In: Lovell PA, El-Aasser MS, editors. *Emulsion Polymerization and Emulsion Polymers*. Chichester: John Wiley & Sons Ltd.; 1997.

[36] Araújo PHH, de la Cal JC, Asua JM, Pinto JC. Modeling particle size distribution (PSD) in emulsion copolymerization reactions in a continuous loop reactor. In: Pierucci S, editor. *ESCAPE-10*. Amsterdam: Elsevier Science B.V.; 2000. p 565–570.

[37] Kalman RE. A new approach to linear filtering and prediction problems. Trans ASME J Basic Eng 1960;87:35–45.

[38] Rolandi PA, Romagnoli JA. A real-time optimisation engine for nonlinear model-based optimising control of large-scale systems. In: Braunschweig B, Joulia X, editors. *ESCAPE-18*, Amsterdam: Elsevier Science B.V.; 2008.

SECTION 5

INDUSTRIAL APPLICATIONS

19

WATER-SOLUBLE FREE RADICAL ADDITION POLYMERIZATIONS

WESLEY L. WHIPPLE AND HUA ZHENG

19.1 INTRODUCTION

In 2009, an estimated four billion metric tons of natural, semi-synthetic, and synthetic water-soluble polymers were consumed globally for use in the production of food, clean water, energy, for personal care, pharmaceutical, and industrial applications [1]. Synthetic water-soluble polymers synthesized by free radical addition polymerization methods account for half of this volume. They include polyacrylamides, polyacrylates, polydiallyldimethylammonium chloride (polyDADMAC), polyvinyl alcohol, and polyvinylpyrrolidones. Depending on the target applications, these polymers can have different molecular weight, charge, and architecture.

A majority of acrylamide, acrylate, and DADMAC (co) polymers produced are polyelectrolytes that carry an anionic or cationic charge in the repeating unit and function through their unique ability to adsorb to surfaces and interact with dissolved or colloidal material. Molecular weight, charge density, and structure of the polyelectrolytes are critical factors in determining the polymer applicability. For example, low molecular weight polyacrylates (<20,000 Da) with an anionic charge are used as scale inhibitors and deposit control agents in cooling tower and boiler applications, as dispersants for fillers used in the production of paper, as anti-redeposition aids in detergents, and as corrosion inhibitors [1, 2].

PolyDADMAC and polyamines are low-molecular-weight polymers (<500,000 Da) with a high cationic charge density. They are efficient at neutralizing dissolved and colloidal material and thus, are used as coagulants in municipal raw water and process water treatment. Although synthetic coagulants are more expensive than inorganic counterparts, they have greater efficiency, are effective over a broader pH range, and produce less residual sludge in dewatering applications. These polymers are also used for strength, clarification, pitch control, and in dual cationic/anionic retention programs in papermaking [3–5].

Flocculants are high-molecular-weight (>1 million Da) cationic or anionic polymers based on acrylamide or acrylic acid. They are used for water treatment in oil field, paper, mineral processing, and textile applications, for sludge conditioning and dewatering, as retention and drainage aids and strength additives in papermaking, for separations in mineral processing applications, and for enhanced oil recovery polymer augmented flood systems [6–8].

Polyvinyl alcohol is synthesized by the hydrolysis of polyvinyl acetate in a continuous process. Several grades are commercially available that differ with respect to the degree of polymerization and hydrolysis. Fully hydrolyzed samples are used primarily as warp sizes and partially hydrolyzed samples are used as adhesive components and polymerization aids [1].

During product development, measurements are established and specifications set to ensure polymer performance. At the plant scale, processes and product quality are monitored to ensure that the produced product meets specifications. Improved methods to monitor and control the reaction process variability, capture manufacturing efficiencies, and to produce a more consistent, higher quality product are continually being sought and implemented.

Monitoring Polymerization Reactions: From Fundamentals to Applications, First Edition. Edited by Wayne F. Reed and Alina M. Alb.
© 2014 John Wiley & Sons, Inc. Published 2014 by John Wiley & Sons, Inc.

19.2 POLYACRYLAMIDES

Several reviews that cover the synthesis, characterization, and application of polyacrylamides are available [6–9]. Acrylamide has the unique ability to form high-molecular-weight polymers (>1 million Da), to react with a variety of comonomers to produce cationic, anionic, nonionic, amphoteric, or zwitterionic polymers, and to produce specialized polymers available through derivatization. Monomers commonly used in the synthesis of water-soluble polymers of commercial interest are shown in Figure 19.1.

Polyacrylamides of high commercial interest are copolymers of acrylamide, **1**, with sodium or ammonium salts of acrylic acid, **4**, or 2-acrylamido tert-butylsulfonic acid (ATBS), **5**, to produce anionic polymers or with acryloyloxyethyltrimethylammonium chloride (AETAC), **6** (Fig. 19.2), to produce cationic polymers. Although not produced in large volume, copolymers of acrylamide and DADMAC, **10**, are used to produce cationic polymers for use in papermaking applications (Fig. 19.2). Polyacrylamides with a variety of charge densities

(obtained by varying monomer ratios), molecular weights, and architecture (linear, branched, and structured) are available commercially in dry and liquid form. Branched and structured polyacrylamides are obtained by including multifunctional monomers in the polymerization mix or through cross-linking following polymerization [10, 11].

Polyacrylamides are produced commercially in an aqueous environment by free radical polymerization, which follows the classical vinyl polymerization model with initiation, propagation, and termination processes. In this model, the propagation/termination ratio $(k_p/k_t)^{1/2}$ and chain transfer to monomer, polymer, initiator, and other small molecules impact the molecular weight of the formed polymer. High-molecular-weight polyacrylamides are possible because of the high $(k_p/k_t)^{1/2}$ ratio for acrylamide and the low chain-transfer activity to monomer and polymer in an aqueous environment. Polymer molecular weights can be lowered and customized through the use of chain-transfer agents. Initiation with water and oil-soluble azo compounds, redox couples, peroxides, or photochemical initiators is common [6, 12–14].

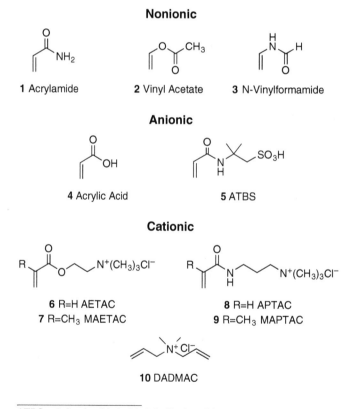

Nonionic

1 Acrylamide   **2** Vinyl Acetate   **3** N-Vinylformamide

Anionic

4 Acrylic Acid   **5** ATBS

Cationic

6 R=H AETAC
7 R=CH₃ MAETAC

8 R=H APTAC
9 R=CH₃ MAPTAC

10 DADMAC

ATBS = 2-Acrylamido tert-butylsulfonic acid
APTAC = Acrylamidopropyltrimethylammonium chloride
DADMAC = Diallyldimethylammonium chloride
AETAC = Acryloyloxyethyltrimethylammonium chloride
MAETAC = Methacryloyloxyethyltrimethylammonium chloride
MAPTAC = Methacrylamidopropyltrimethylammonium chloride

FIGURE 19.1   Common nonionic, anionic, and cationic monomers used in the synthesis of water-soluble polymers.

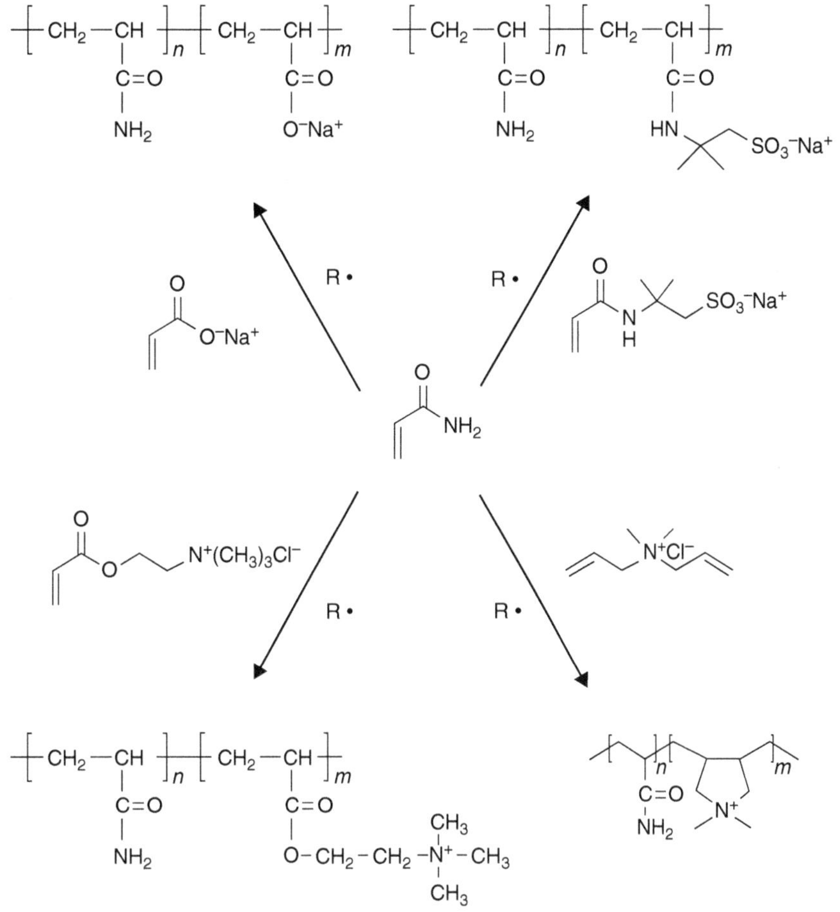

FIGURE 19.2 Synthesis scheme for the preparation of copolymers of acrylamide with sodium acrylate, sodium ATBS, AETAC, and DADMAC.

19.2.1 Cationic Polyacrylamides

Common cationic polyacrylamides produced industrially include copolymers of acrylamide, **1**, with AETAC, **6**, MAETAC, **7**, MAPTAC, **9**, or DADMAC, **10** (Fig. 19.1 and Fig. 19.2). Among them, acrylamide/AETAC copolymers are the most common due to the relatively low AETAC monomer cost and similar reactivity with acrylamide (r_1(AM)=0.61, r_2(AETAC)=0.47) producing copolymers with a roughly uniform sequence distribution of comonomers. In general, cationic methacrylates and methacrylamides are more reactive than their acrylate and acrylamide counterparts. In batch copolymerizations of acrylamide with MAETAC (r_1(AM)=0.25, r_2(MAETAC)=1.71) or MAPTAC (r_1(AM)=0.57, r_2(MAPTAC)=1.13), the more reactive MAETAC and MAPTAC monomers react faster with their own monomers than with acrylamide. This leads to compositional drift and often poor performance [6]. There are more significant differences in the reactivity ratios of AM and DADMAC ($(r_1$(AM)=6.4, r_2(DADMAC)=0.06) [14]. DADMAC, **10**, is the least expensive cationic monomer. The molecular weight of DADMAC containing polymers is

limited due to the high chain transfer activity from the DADMAC allylic moiety. The low molecular weight, compositional heterogeneity, and branching have limited the use of acrylamide/DADMAC copolymers to predominantly just a few papermaking applications (coated broke, pitch and stickies control). Semibatch strategies have been employed where the more reactive monomer is withheld from the initial reaction mixture and fed into the reactor during polymerization to obtain copolymers with similar comonomer sequence distributions [15, 16]. Extensive reactivity ratio data of acrylamide with cationic and anionic monomers under various conditions are summarized elsewhere [6, 7, 14].

Hydrolytic stability and pH have to be considered when selecting a cationic monomer. Cationic acrylate esters are susceptible to base hydrolysis above pH 6, resulting in the loss of cationic charge on the polymer. The rate of hydrolysis is concentration and temperature dependent. In contrast, cationic acrylamide copolymers containing amide monomers such as APTAC and MAPTAC are reasonably stable up to a pH of 9–10. Acrylamide/MAPTAC copolymers and acrylamide/MAPTAC/acrylate terpolymers can be found as conditioners and deposit aids in hair care formulations.

19.2.2 Anionic Polyacrylamides

Common commercial anionic polyacrylamides include copolymers of acrylamide with sodium and ammonium salts of acrylic acid, **4**, and ATBS, **5** (Fig. 19.2). Criteria for polymer selection include cost, molecular weight, polymer solubility, application pH, and salt tolerance. Acrylic acid is a rather inexpensive monomer and yields very high molecular weight acrylamide/acrylate copolymers. Sodium or ammonium acrylates are typically formed prior to polymerization by reaction of acrylic acid with sodium hydroxide or ammonia. Acrylamide/acrylate copolymers have poor solubility at low pH (acrylic acid $pKa = 4.3$) and have poor salt tolerance. The high molecular weights available for anionic flocculants render them useful in mining and papermaking applications. In contrast, acrylamide/Na-ATBS copolymers have good solubility, maintain their charge at low pH (pKa ATBS = 1.7), and have a high tolerance for salt (including many divalent cations). These features render them useful for oil field applications and as flocculants in phosphate production. Reactivity ratios of acrylamide with acrylic acid salts vary with pH. At pH 4 $r_1(AM) = 0.57$, r_2 (acrylic acid) = 0.32, and at pH 8 r_1 (AM) = 0.12, r_2 (acrylic acid) = 0.63. It was reported that a random copolymer is formed at a pH of about 5 [7].

19.2.3 Derivatized Polyacrylamides

A variety of commercially useful polyacrylamides obtained via the derivatization of polyacrylamide are reported. These include the formation of anionic polyacrylamide through hydrolysis [17], the synthesis of sulfomethylated derivatives from the reaction of polyacrylamide with formaldehyde and sodium bisulfate [18], the formation of aminomethylated polyacrylamide from the reaction of polyacrylamide with formaldehyde and dimethylamine (Mannich reaction) [19], and the generation of various derivatives by transamidation reactions including hydroxamated polyacrylamides used as a flocculant in the Bayer process [20, 21]. Fluorescent polyacrylamide or polyacrylate derivatives, useful for monitoring polymer concentration in industrial applications, are obtained by the incorporation of a small amount of fluorescent monomer into the polymer backbone [22].

19.2.4 Polyacrylamide Manufacturing Process

Liquid polyacrylamides are available as solutions, inverse (water-in-oil) emulsions, or dispersions. Dry polyacrylamides are available as powders from a dried gel or as beads from a water-in-oil suspension process. Polyacrylamides are formed from the radical chain polymerization of acrylamide with cationic or anionic monomers, which is a highly exothermic

reaction (Fig. 19.2) [7]. Reactions may be run isothermally or adiabatically. In an isothermal process, the rate of heat generation can be controlled by adjusting the initiator concentration, reaction temperature, or initiator feed rate. If the heat generation exceeds the rate at which heat can be removed from the reactor, a runaway reaction will result, which could be potentially dangerous in large-scale reactions. For an adiabatic process, temperature rise can be calculated for given monomer concentrations and the final temperature estimated based on the initial temperature. Formula adjustments can be made to avoid exceeding the pressure rating of the reaction equipment [6].

19.2.4.1 Inverse Emulsion Process
The inverse emulsion polymerization process, sometimes called an inverse-microsuspension polymerization, is a method of producing polyacrylamides and polyacrylates with extremely high molecular weight and low product viscosity since the produced surfactant stabilized polymer is suspended in an oil-continuous phase [9, 13–15]. The low viscosity (typically <2000 cP) allows for efficient heat removal during production and easy handling in the plant and at the customers site. Polyacrylamides with 25–50% polymer actives are commercially available as inverse emulsions. Removing solvent by azeotropic distillation yields concentrated products at added expense. Polymers with molecular weights that range from 1 to 50 million Da and with particle sizes from 0.2 to 2 μm are typical. The product particle size is determined by the level and type of surfactant in the formulation and the shear present in the manufacturing process, in particular during the homogenization step. The particle size and particle size distribution impact the rate of settling and oil-split in the product which impacts product shelf life.

The mechanism and kinetics of these types of heterophase acrylamide polymerization reactions have been studied [13–15]. Depending on the initiator type and oil quality the polymerization can physically and kinetically resemble either an emulsion or a suspension polymerization. With paraffinic oil phases or water-soluble initiators its mechanism resembles the one for suspension polymerization.

The commercial production of polyacrylamides by this method involves the free-radical-initiated polymerization of a water-in-oil emulsion formed from aqueous and oil phases. An aqueous phase is typically prepared by mixing together monomers and other components that could include salts, acids, bases, buffering agents, chain-transfer agents, branching agents, and chelants. An oil phase is typically prepared by mixing together oil and oil-soluble surfactants with a low hydrophilic/lipophilic balance (HLB) number in a separate vessel from the aqueous phase [23]. Common surfactants used separately or in combination include sorbitan fatty esters, sorbitan fatty ester ethoxylates, alkanolamides, linear alcohol ethoxylates, and polymeric surfactants [15, 23]. An inverse emulsion is formed by the addition of the monomer phase to the oil phase with mixing, followed by homogenization.

Free radical polymerization reactions are conducted under an inert atmosphere. Initiation with thermal or redox free-radical initiators is common [9, 14]. Industrial processes are designed to reduce residual monomer levels to below regulated limits [24, 25]. This is achieved by pushing the reaction to completion by increasing the reaction temperature or increasing the initiator levels toward the end of the reaction. Also postpolymerization strategies have been employed, such as an enzymatic treatment with amidase [26].

At the completion of the reaction, the mixture is cooled and a high HLB surfactant, sometimes referred to as an inverting surfactant, is blended into the emulsion matrix [27, 28]. Linear alcohol, tridecyl alcohol, or alkylphenol ethoxylates with an HLB number of 11–14 are examples of inverting surfactants that may be used. To prepare the polymer for use, an aqueous solution of the water-in-oil polyacrylamide is generated by vigorously mixing the desired amount of the emulsion polymer (typically about 0.25–2% polymer) with water. This process is called inversion.

Many creative methods for modifying polyacrylamides formed by inverse emulsion polymerization for enhanced activity in target applications are found in the patent literature. Flesher reported the synthesis of a cross-linked, water swellable polymer, useful for flocculation in high shear dewatering devices, by adding a small amount of a divinylic cross-linking agent [10].

Neff and Ryles describe soluble branched high-molecular-weight anionic and cation polyacrylamides formed with a combination of branching and chain-transfer agents [29, 30]. Barajas et al. [31] reported polymer synthesis with improved flocculating characteristics derived from the introduction of regularly spaced branches and/or cross-links from the continuous addition of small quantities of a chain-branching agent. To traverse the large differences in reactivity between acrylamide and DADMAC monomers, Gartner et al. [32] synthesized a soluble, linear acrylamide/DADMAC copolymer through semibatch polymerization by adding a major portion of acrylamide monomer throughout the polymerization. A chain-transfer agent was added at the end of the copolymerization to prevent branching and cross-linking reactions that would decrease polymer solubility. Cationic acrylamides with a small amount of a hydrophobic monomer, prepared by inverse emulsion polymerization aggregate or associate in solution, are useful in oily water treatment [33].

19.2.4.2 Dry Polymer Processes

Dry polymers, commercially available as powders or beads, offer better cost-efficiencies and longer shelf life than liquids, which are advantageous particularly when transporting over long distances. They have lower volatile organic compounds (VOCs) than inverse emulsions. This offers advantages in applications where VOCs may be limited due to regulation or process constraints. However, it is more difficult to handle dry polyacrylamides and it takes longer to form a fully active dilute aqueous polymer solution than inverse emulsions or dispersion polymers. If not adequately dispersed in water during dissolution, the polymer does not fully dissolve and polymer agglomerates (fisheyes) are observed, which may adversely impact the process for which they are intended.

High-molecular-weight polyacrylamide powders are manufactured using batch and continuous processes. Batch processes are run adiabatically where the adiabatic temperature rise is calculated and monomer concentrations adjusted so that the boiling point of water is not exceeded. An aqueous monomer solution is typically prepared by mixing together monomers and other components that could include salts, acids, bases, buffering agents, chain-transfer agents, branching agents, and chelants. The monomer solution is sparged with nitrogen and initiated to yield a hydrous polymeric gel that is granulated, dried, and ground to the desired particle size (typically 0–1 mm). Continuous processes for the production of dry polymers on a moving belt or an oscillating screw reactor are detailed in the patent literature [34, 35].

Dry polymer beads are manufactured using a surfactant-stabilized inverse suspension process similar to the inverse emulsion process described earlier. The final beads are obtained by removing solvent via azeotropic distillation and then centrifugation, followed by drying. The particle size of these dry beads is typically less than 0.3 mm. Polyacrylamide beads have the advantage of faster dissolution times than powders, but are more expensive to produce.

19.2.4.3 Dispersion Process

High-molecular-weight polyacrylamides dispersed in a salt matrix with 15–25% polymer are commercially available. They solubilize in water nearly instantaneously when diluted. These polymers are used in applications that prohibit the use of oil containing polymers and exclude dry polymers due to long dissolution times, or applications that have polymer feed equipment limitations. Takeda was the first to introduce a commercially viable process for the production of water-soluble cationic polyacrylamide dispersions that consisted of an acrylamide/AEDBAC (acryloyloxyethyldimethylbenzylammonium chloride) copolymer dispersed in a sodium or ammonium sulfate salt solution stabilized with AETAC or AM/AETAC copolymers [36, 37]. The polymerization reactions were carried out in custom-built reactors with ribbon type agitators designed to handle the high in-process viscosities observed (100,000–2,000,000 cP).

Subsequently, there have been many improvements to the dispersion polymerization process that have allowed for AM/AETAC and AM/DADMAC cationic, AM/sodium acrylate anionic, acrylamide nonionic, and AM/N,N-dimethyl-N-acryloyloxyethyl-N-(3-sulfopropyl) ammonium betaine polymeric dispersions [16, 38–41]. These improvements, such as the development and utilization of new dispersants, the use of polymeric seeds to facilitate precipitation, optimization of multivalent salt types and ratios, and the use of semibatch monomer addition techniques have resulted in lower in-process and final product viscosities and have allowed for production in traditional inverse emulsion type reactors and for the production of dispersions with higher polymer actives [42].

To form a dispersion polymer, a starting solution that typically consists of monomers, water, acids and/or bases for pH adjustment, buffering agents, multivalent salts (ammonium sulfate and sodium sulfate are common), chain transfer agents, the prescribed polymeric dispersant, and chelant is added to the reactor. Polymerization is accomplished with azo and/or redox initiators. As the reaction begins, the solution becomes viscous. After a short period of time, a milky dispersion is formed. Near the end of the reaction, residual monomer levels are reduced to values below the specified limits by increasing the reaction temperature and/or initiator concentration. When the polymerization reaction is complete, additional stabilizing additives may be added to the mixture.

In the polyacrylamide dispersion polymerization process, multivalent salts are used for the purpose of insolubilizing and depositing the polymer and are formulated to produce a poor solvent for the particular polyacrylamide being synthesized. At the onset of polymerization reaction, the formed polymer is soluble in the polymerization medium. As the reaction progresses, the polymer phase separates forming particles (0.1–10 μm) that are kept from agglomeration by the use of polymeric stabilizers. A mechanistic hypothesis of the formation of polyacrylamide dispersions was reported by Selvarajan [43].

19.3   POLYACRYLATES

The largest global consumption of acrylic acid is for the manufacturing of high-molecular-weight cross-linked polyacrylate super absorbents using a dry polymer process. Low-molecular-weight polyacrylates represent the second largest use of acrylic acid. They are used to control scale and deposits, as dispersants, and as corrosion inhibitors in various commercial applications. Commercial low-molecular-weight polyacrylates with an average molecular weight below 20,000 Da and 40–50% polymer are produced by solution polymerization in batch or continuous processes [44, 45].

Polymer molecular weight is controlled through the use of various chain-transfer agents such as mercaptans, alcohols, bisulfite, and phosphorous acid, through synthesis in organic solvents such as 2-propanol, and through metal-activated redox initiation [46]. Commercial polyacrylates can range from a simple homopolymer such as a sodium polyacrylate or sodium polymethacrylate to a more complex polymer derived from a combination of monomers or initiation conditions to yield a "designed" polymer with enhanced activity for a specific application or target substrate. In some

cases, polyacrylates are designed for activity in a broad spectrum of applications and conditions [2].

Common comonomers are those that place carboxylic, sulfonate, phosphite, polyethylene glycol, or hydrophobic groups along the backbone. They include maleic anhydride (subsequently hydrolyzed), sodium 2-acrylamido tert-butylsulfonic acid (Na-ATBS), sodium styrene sulfonate, 1-allyloxy-2-hydroxypropyl sulfonate, 1-allyloxy-2-hydroxypropyl phosphite, vinyl phosphonic acid, vinyl acrylate, polyethoxy acrylate, ethyl acrylate, methyl vinyl ether, and hydroxyl propyl acrylate [2, 47–50].

Strategies for adding functional groups to polyacrylates through effectively utilizing chain-transfer agents during polymerization were reported. For example, McCallum and Weinstein [51] reported a method for producing low-molecular-weight phosphonate-terminated polyacrylates by utilizing phosphorous acid as a chain-transfer agent. Polyacrylates can also effectively be modified following polymerization. Examples include the sulfomethylation and transamidation of copolymers containing carboxylic acid and (meth)acrylamide units [18].

19.4   POLYDIALLYLDIMETHYLAMMONIUM CHLORIDE

PolyDADMAC polymers are used for potable water clarification, waste water treatment, and textile processing. They are commercially available as solutions with solids as high as 70% and molecular weights ranging from 10,000 to 500,000 Da. These polymers are synthesized by the free radical solution polymerization of DADMAC monomer, obtained from the reaction of allyl chloride, dimethylamine, and sodium hydroxide in an aqueous solution. The structure of polyDADMAC consists of about 98% five-member pyrrodinium rings from an internal radical cycloaddition reaction and 2% pendant double bonds from the radical addition across one of the allylic double bonds (Fig. 19.3) [6].

19.5   POLYVINYL ALCOHOL

Polyvinyl alcohol is typically produced by the hydrolysis of polyvinyl acetate in a continuous process [52, 53]. Although specialty grades of polyvinyl acetate may be formed using a batch process, the majority of commercial polyvinyl acetate is formed by the free radical polymerization of a vinyl acetate solution in methanol using a continuous process. Polymerization reactions are typically conducted at 55–85 °C, where heat is removed from the reaction mixture by condensing the monomer. At the end of the process, residual vinyl acetate and methanol are stripped from the reactor and recycled. Polyvinyl acetate formed by this process has long and short chain

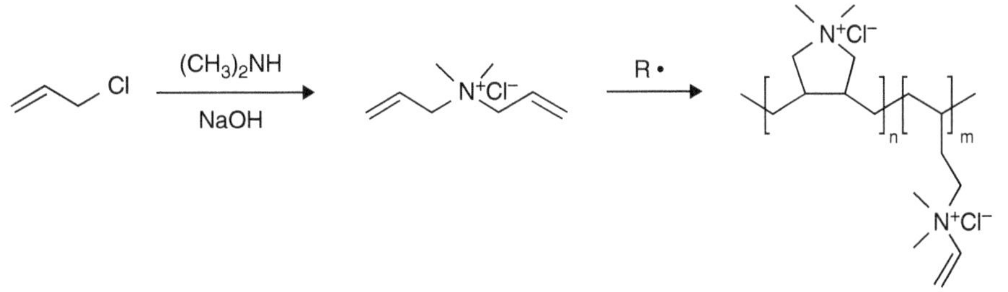

FIGURE 19.3 Synthesis scheme for the preparation of polyDADMAC. The polymer contains about 98% five-membered pyrrolidinium rings (n) from internal cycloaddition and 2% pendant double bonds (m). From Heitner HI. Flocculating agents. In: *Kirk-Othmer Encyclopedia of Chemical Technology*. New York: John Wiley and Sons; 2004, with permission.

branches. Long chain branches result from chain transfer to the acetyl group of the monomer, or polymer. The branch points, through the acetoxy group, are hydrolyzed during conversion to polyvinyl alcohol. Short four carbon–side chains are observed when the growing radical "backbites" through a six-membered ring intermediate.

Polyvinyl acetate is hydrolyzed to polyvinyl alcohol in a continuous belt or extrusion process using sodium or potassium hydroxide, methoxide, or ethoxide catalysts. The hydrolysis process is complicated on a commercial scale due to a significant increase in viscosity that accompanies conversion. Following hydrolysis, the polyvinyl alcohol gel is ground, dried, ground, sized, and packaged [53].

19.6 POLYMER CHARACTERIZATION

Although polymer molecular weight, composition, and charge density are some of the most important parameters that define water-soluble polymer activity, other parameters related to product handling, stability, and regulatory compliance are monitored during product development and production. These include product viscosity, particle size, pH, product homogeneity, and residual monomer levels. Different analytical methods are used to assess these parameters during product development in the laboratory and production, based on the instrument availability and the complexity of the analysis.

The molecular weight determination of dilute water-soluble polymer solutions by light scattering, sedimentation, and matrix-assisted laser desorption/ionization (MALDI) mass spectrometry techniques is utilized almost exclusively for product development and troubleshooting due to the specialized analytical equipment and sample preparation [54–56]. In a plant setting, viscosity and size exclusion chromatography methods are commonly used to determine the molecular weight or the average degree of polymerization. Dilute solution measurements involving polyelectrolytes are normally carried out in salt solutions to minimize the charge repulsion between ionic groups on the polymer

backbone allowing the polymer chain to adopt a random coil configuration.

Absolute polymer molecular weight and size information are available from light scattering–based measurements. Commercially available gel permeation/size exclusion chromatography (GPC/SEC) systems with refractive index and multiangle laser light scattering (MALLS) detectors are used to determine polymer concentration and to calculate molecular weight for each polymer fraction providing insight into polymer linearity and branching through the use of different scaling laws and algorithms for molecular weight and mean square radii data [54].

MALDI mass spectrometry is an established technique in the biopolymer area, particularly useful for low-molecular-weight polymers. Not only does MALDI-MS provide a polymer molecular weight distribution, but for low-molecular-weight polymers the method can provide useful information regarding end-groups incorporated into the polymer from initiators or chain-transfer agents [56]. In many instances this functionality has a substantial impact on polymer performance.

Intrinsic viscosities (IV) are correlated to a polymers weight-averaged molecular weight (M_w) through the Mark–Houwink–Sakurada relationship: $[\eta] = KM^{\alpha}$. The K and α values are derived from polymeric standards with defined molecular weights (and limited molecular weight range) under specific solvent and temperature conditions. Values obtained under various conditions are available in the literature [6, 7, 57]. In industrial quality control laboratories, reduced specific viscosities (RSV), measured with capillary viscometers, and standard viscosities (SV), measured with rotational viscometers, are used as an indicator of molecular weight for high-molecular-weight flocculants [17, 40]. Santini and Yankie [17] reported an approximate correlation between SV, IV and M_w.

Several techniques are used in determining copolymer composition. These include nuclear magnetic resonance (NMR) and infrared (IR) spectroscopy, MALDI mass spectrometry for low-molecular-weight polymers, titration for determining the degree of hydrolysis (e.g., polyvinyl alcohol) [58] and polymer composition inferred from monomer levels measured by high performance liquid chromatography (HPLC) from reaction samples obtained during polymerization (see Chapter 10). Nuclear magnetic resonance (NMR) spectroscopic techniques have proven useful not only for the determination of comonomer composition, but also for providing insight into the sequence distribution of monomers in the polymer backbone [51, 59]. Fourier transform infrared (FTIR) spectroscopic techniques are used in following conversion and monomer concentrations with or without multivariate algorithms.

Polymer charge density is usually expressed as mole percent of charged groups or in terms of milliequivalents per gram (meq g^{-1}). For copolymers with a formal charge independent of pH, such as for AM/APTAC copolymers, the charge density may be calculated from the formula weights of the comonomers. Colloid titration is a method frequently used to measure polymer charge density [60]. Using this technique, the direct determination of the cationic charge is made by titrating with an anionic colloid such as potassium polyvinyl sulfate (PPVS) to an o-toluidine blue endpoint. Anionic charges are typically measured based on the reaction of the anionic polymer with a known excess of a cationic colloid such as polyDADMAC and back-titration with PPVS. Colloid titration can also be used to provide structural insight into cross-linked and branched polyacrylamide flocculants that are useful in high-shear dewatering applications. For these types of polymers, the full polymer charge is not initially assessable by colloid titration; it increases with mixing in a high-shear environment. The degree of "ionic regain" determined by colloid titration is a parameter used to define polymer solubility and predict performance in high-shear dewatering applications [10].

The rate of polymer dissolution is an important property that impacts high-molecular-weight polyacrylamide availability and performance commercially. Polymer particle size distribution in dry powder polymers and the polymer solids content and level of activating surfactant in inverse emulsion polymers are factors that impact polymer dissolution. The rate of polymer dissolution is typically determined by following the rate at which the viscosity in a polymer solution builds during dissolution.

Particle-size distribution for inverse emulsion polymers and powders can be determined by light scattering–based methods using any of the various commercially available analyzers. The particle size distribution of powders can also be determined using sieving techniques where the dry powder is passed through a stack of standard mesh sieves of decreasing size while shaking [58].

Product stability is an important parameter. Polymer molecular weight stability is verified over time at elevated temperature. Rheological measurements have been used to predict the settling-stability of polyacrylamide inverse emulsions [61]. For inverse emulsions and dispersions, product stability may be determined by observing product separation and sediment formation over time. Ultrasonic and light scattering/back scattering devices can be used to automate and assist with these measurements. Some quality control specifications for inverse emulsion and dispersion polyacrylamides call for accelerated centrifuge sedimentation tests.

19.7 MONITORING PRODUCTION

Manufacturers of water-soluble polymers are continually striving for optimized efficiency, regulatory control, and consistent product quality. To meet these goals, improved cost-effective methods for monitoring and controlling production are continually sought. This is a challenging task because often a range of products with different processes that vary in complexity are manufactured in the same reactor. Although process monitoring techniques are available and are utilized more frequently to monitor production and to certify product quality, the reality is that a significant portion of monitoring during production deals with parameters that define the "state of the reactor" such as temperature, pressure, fluid level, flow rate, pH, density, and amperage drawn on the motor. These are captured by a reactor control system and monitored by the operator. Calorimetric techniques are frequently used to monitor reaction rates and conversion, and rates are often adjusted so the heat evolved during polymerization is balanced with the cooling capacity of the reactor to maintain reactor efficiency [62].

Product quality is usually determined and certified off-line in a quality control laboratory to ensure that specified properties of a production batch fall within the established range. Specifications related to molecular weight, charge density, composition, particle size, viscosity, rate of dissolution, solution viscosity, and residual monomer are common. Production batches that do not meet the specifications either require further processing, rework, or disposal and reduce production efficiency. As online analytical methods are employed, first pass yields (the proportion of production that meets specifications the first time) usually increase.

19.8 ONLINE PROCESS MONITORING

Several online monitoring methods for polymerization have been reviewed [63, 64]. While some of these methods are still in development or suited for laboratory use, many monitoring techniques, in particular spectroscopic techniques, have found their way into industrial processes. There are several reaction features that could be addressed through improved online monitoring techniques in the production of water-soluble polymers. These techniques are used to monitor different features, such as monomer concentration, copolymer composition, polymer molecular weight, and particle size in hetero-phase processes.

The ability to follow comonomer concentration and copolymer composition in real-time allows operators to efficiently determine the end of the reaction. Residual monomer levels are regulated, and in a majority of cases are determined through the HPLC analysis of reactor samples. Reactor sampling, sample preparation, and analysis take time and reduce productivity, particularly if the batch is held in a reactor pending analytical clearance. Additional work that may be required would further reduce productivity.

In many applications, improved performance is observed in the case of the polyelectrolytes with uniform charge distribution. If the comonomers have different reactivity ratios, these copolymers are synthesized using semibatch techniques. In this case, monitoring the monomer concentrations and composition would allow for monomer feed control. This would be particularly useful for systems where monomer reactivity changes as a function of conversion [65]. Also, this capability would help automate and standardize processes with critical in-process additions. One such process is the manufacture of dispersion polymers where in-process viscosities are sensitive to the concentrations of monomer, polymer, and various salts. Through strategic additions of various components, in-process viscosities can be significantly reduced.

The formation of gel particles and coagulum during the production of inverse emulsion and dispersion polymerization require filtration and diminish productivity. Continuous monitoring of the particle size and of the change in particle properties during production might provide insight into the onset of gel formation and could allow for specific action.

There are some unique challenges that impact online process monitoring in an industrial environment. These include keeping sensors clean and functioning, maintaining a side-stream flow for sampling or analysis, and modifying analytical equipment for use in a hazardous area or alternatively locating it outside of the hazardous zone. Water-soluble polymers have a tendency to coagulate or precipitate on reaction surfaces during production, including sensors. If fouling occurs in reactors at the laboratory scale, the reaction equipment is disassembled and cleaned. At a production scale, cleaning sensors is more involved. Sensors integrated into reaction equipment must be robust and withstand cleaning at elevated temperatures in caustic or oxidizing environments and often withstand a high pressure cleaning spray. Also, sampling may be an issue. Often it is difficult to sample reactors utilizing side stream loops due to high in-process viscosities or gels and coagulum that may form and impede flow and impact analysis.

A few approaches for the online monitoring of water-soluble polymerization reactions are highlighted in the following.

19.8.1 Automatic Continuous Online Monitoring of Polymerization Reactions

Automatic continuous online monitoring of polymerizations (ACOMP) is a method developed for monitoring polymerization reaction kinetics in real time and thus offering a complete characterization of the polymer while it is being produced. Principles and applications of the ACOMP technique are discussed in Chapters 11–13. The technique, with a complex multidetection platform, has been used for different polymerization systems, from homogeneous to heterogeneous phase polymerization reactions. ACOMP involves continuous sampling of the reaction mixture, diluting and conditioning the sample and passing through a detector train typically comprising viscometer, ultraviolet, refractive index, and multiangle light scattering detectors. The technique provides information in real time (2–15 min delay) on the evolution of the monomer and polymer concentrations, composition drift, weight average molecular weight, reduced viscosity, and other polymer characteristics [66].

19.8.2 Spectroscopic Techniques

Spectroscopic techniques are powerful process analytical tools. They are unparalleled in terms of molecular specificity, minimal sample preparation that translates to speed, and are capable of inferring physical properties [63, 67, 68]. Mid-infrared Fourier transform spectroscopy (FTIR), near-IR (NIR), and Raman spectroscopic techniques are the workhorse tools commonly used to monitor online polymerization processes.

19.8.2.1 *Fourier Transform Infrared Spectroscopy* Fourier transform infrared spectroscopy (FTIR) coupled with an *in situ* probe is a powerful technique for instantaneously monitoring chemical species including monomers in a water-soluble polymerization matrix. Monomer concentrations can be determined from a partial least square (PLS) calibration curve established from a series of reference samples with known monomer concentration generated by HPLC analysis. The determination of residual monomer is limited only by the sensitivity of the detector. Calibration curves are product and formula specific.

Several commercial vendors including Mettler Toledo offer small footprint *in situ* FTIR systems that are amenable for lab-scale monitoring as well as larger instruments for process monitoring. PLS calibrations established at the laboratory-scale can be transferred to process instrumentation for use in monitoring production. Recent advancements in immersion optics such as attenuated total reflectance (ATR) allow immersion optic probes to be installed in the reactor wall, in a side loop, or as an *in situ* probe joined by an optic fiber. Because of its short path-length (~ 100 μm), the length of the optic fiber is limited to about 6 ft which requires that a dedicated FTIR spectrometer designed for use in a hazardous environment be located next to the reactor or equipment being monitored.

19.8.2.2 *Near-Infrared Spectroscopy*   There are several commonalities between NIR and Raman spectroscopic techniques. In contrast to FTIR spectroscopic techniques where instrument location near a dedicated reactor is required, NIR and Raman spectroscopic techniques utilize long optic fibers and have a multiplexing feature that allow for multiple sensors to be monitored by a single instrument located in a remote location. Both NIR and Raman techniques are sensitive to the scattering effect of particles allowing for particle size characterization [63, 67, 68] and are capable of monitoring multiple parameters of a reaction by extracting information using multivariate techniques.

NIR spectroscopy weaker signals due to overtone and combination bands and broad spectral bands that are hard to interpret makes the method limited to molecules with OH, CH, NH, and SH functionalities. Weaker signals are compensated by the use of long optic fibers and a sensitive InGaAs (indium gallium arsenide) detector. Acrylamide monomers and polymers do not have strong absorption in the region other than weak spectral features caused by hydrogen bonding.

19.8.2.3 *Raman Spectroscopy*   Historically, Raman spectroscopy was never considered a sensitive technique because only 1 in 10^6 photons emitted from a molecule is collected. However, Raman systems have improved tremendously in the last several years. It is no longer deemed an insensitive, irreproducible, fluorescence-dominated technique. Raman is a versatile technique capable of providing information on several parameters simultaneously, such as monomer concentration and particle size. Raman is especially amenable for monomer detection in water-soluble polymers because symmetric vinylic monomer structures are good Raman scatterers and water has a weak signal. To that end, Raman is a complementary technique to FTIR and can be used to monitor monomer concentration and conversion. By employing a near-IR laser (785 nm) which removes most of the fluorescence, along with sharp monomer and polymer peaks that are often separated, monomer concentrations may be determined with univariate calibration. Additionally, since Raman is sensitive to the local molecular environment, it may be used to provide particle size information.

The installation of Raman for online process monitoring can either be by immersion optics or in a noncontact fashion through a sapphire or diamond window. The laser excitation source can be hazardous. Low voltage lasers are required to avoid sparks or ignition when hitting black particles and must be located outside of a production area unless well shielded and conforms to safety requirements. Additionally, laser power will deteriorate with time requiring periodic replacement.

19.9 SUMMARY

A spectrum of synthetic water-soluble polymers is available and utilized daily in areas such as the production of energy, clean water, and industrial and personal care products that enhance our quality of life. It is hard to imagine what our world would be like without them. Advances continue to be made in the areas of new polymer design, production, and in the instrumentation and methods utilized for monitoring polymerizations. There is unlimited opportunity for researchers in academia and industry alike to develop and utilize cost-effective monitoring tools in water-soluble polymer production. Increased use of online monitoring tools at the commercial-scale will allow for improved production efficiency, reduced product variability, and the accommodation and tighter control of more complex processes.

ACKNOWLEDGMENTS

The authors would like to thank Dr. Cathy Doucette and Dr. Shamel Shawki of Nalco for the support of this work.

REFERENCES

[1] Will RK. Water-soluble polymers. In: *Specialty Chemicals Strategies for Success*. Volume 16. Menlo Park: SRI International; 2010.

[2] Frayne C. Chemical treatments and programs for cooling water. In: *Cooling Water Treatment – Principles and Practice*. New York: Chemical Publishing Company Inc.; 1999. p 137–176. Available at www.knovel.com/web/portal/browse/display?_EXT_KNOVEL_DISPLAY_bookid = 3074&VerticalID=0. Accessed 2012 Feb 25.

[3] Bolto B, Gregory J. Organic polyelectrolytes in water treatment. Water Res 2007;41:2301–2324.

[4] Jackson LA. Applications of cationic polymers in water treatment. In: Amjad Z, editor. *Science and Technology of Industrial Water Treatment*. Boca Raton: CRC Press; 2010. p 465–479.

[5] Heitner HI. Flocculating agents. In: Kroschwitz, JI, Howe-Grant, M, editors. *Kirk-Othmer Encyclopedia of Chemical Technology*. New York: John Wiley and Sons; 2004. Available at http://onlinelibrary.wiley.com/doi/10.1002/0471238961.061 2150308050920.a01.pub2/full. Accessed 2012 Feb 25.

[6] Huang SY, Lipp DW, Farinato RS. Acrylamide polymers. In: Mark HF, editor. *Encyclopedia of Polymer Science and Technology*; 2001. New York: John Wiley & Sons Inc. Available at http://dx.doi.org/10.1002/0471440264.pst004. Accessed 2012 Feb 25.

[7] Buchholz FL. Polyacrylamides and poly(acrylic acids). In: Elvers B, Hawkins S, Schulz G, editors. *Ullmann's Encyclopedia*

of Industrial Chemistry. 5th ed. Volume A21. Weinheim: VCH Publishers, Inc.; 1992. p 143–156.

[8] Caulfield MJ, Qiao GG, Solomon DH. Some aspects of the properties and degradation of polyacrylamides. Chem Rev 2002;102:3067–3083.

[9] Hunkeler DJ, Hernandez-Barajas J. Inverse-emulsion/ suspension polymerization. In: Salamone JC, editor. *Polymeric Materials Encyclopedia.* Volume 5. Boca Raton: CRC Press; 1996. p 3322–3334.

[10] Flesher P, Farrar D, Field JR. Flocculation processes. European patent 202780. 1986.

[11] Neff RE, Pellon JJ, Ryles RG. High performance polymer flocculating agents. European patent 374458. 1988.

[12] Sarac AS. Redox polymerization. Prog Polym Sci 1999;24:1149–1204.

[13] Hunkeler D, Hamielec AE, Baade W. Mechanism, kinetics, and modeling of the inverse-microsuspension homopolymerization of acrylamide. Polymer 1989;30:127–142.

[14] Baade W, Hunkeler D, Hamielec AE. Copolymerization of acrylamide with cationic monomers in solution and inverse-microsuspension. J Appl Poly Sci 1989;38:185–201.

[15] Hernandez-Barajas J, Hunkeler DJ. Inverse-emulsion copolymerization of acrylamide and quaternary ammonium cationic monomers with block copolymeric surfactants: copolymer composition control using batch and semi-batch techniques. Polymer 1997;38:449–458.

[16] Wong Shing JB, Hurlock JR, Maltesh C, Nagarajan R. Papermaking process utilizing hydrophilic dispersion polymers of diallyldimethyl ammonium chloride and acrylamide as retention and drainage aids. US patent 6071379. 2000.

[17] Santini JJ, Yankie NA. Stabilized polyacrylamide emulsions and methods of making same. US patent 5548020. 1996.

[18] Fong DW, Kowalski DJ. Sulfomethylamide-containing polymers. US patent 5120797. 1992.

[19] Ryan M, Pawlowska L. Quaternary mannich polymer microemulsion (QMM) with rapid standard viscosity (SV) development. US patent 5789472. 1998.

[20] Domb AJ, Langer RS, Cravalho EG, Golomb G, Mathiowitz E, Laurencin CT. Method of making hydroxamic acid polymers from primary amide polymers. US patent 5128420. 1992.

[21] Rothenberg A, Flieg G, Cole R. Reduction of impurities in Bayer process alumina trihydrate. US patent 5665244. 1997.

[22] Murray PG, Whipple WL. Fluorescent water-soluble polymers. US patent 6344531. 2002.

[23] Greenshields JN. Surfactants in inverse (water-in-oil) emulsion polymers of acrylamide. In: Karsa DR, editor. *Surface Active Behavior of Performance Surfactants.* Boca Raton: CRC Press; 2000. p 66–96.

[24] Kunststoffempfehlungen des BfR. Database BfR recommendations on food contact materials. Available at http://bfr.zadi. de/kse/faces/DBEmpfehlung_en.jsp. Accessed 2012 Feb 25.

[25] 40 CFR 141.111 – Treatment techniques for acrylamide and epichlorohydrin. Available at http://cfr.vlex.com/vid/141–111-acrylamide-epichlorohydrin-19813677. Accessed 2012 Feb 25.

[26] Wetegrove RL, Kaesler RW, Bhattacharyya B. Water-in-oil emulsions of amidase. US patent 4786679. 1988.

[27] Anderson DR, Frisque AJ. Process for rapidly dissolving water-soluble polymers. US patent 3624019. 1971.

[28] Anderson DR, Frisque AJ. Rapid dissolving water-soluble polymers. US patent 3734873. 1973.

[29] Neff RE, Pellon JJ, Ryles RG. High performance cationic polymer flocculating agents. US patent 5945494. 1999.

[30] Neff RE, Pellon JJ, Ryles RG. Method of dewatering suspensions with unsheared anionic flocculants. US patent 5961840. 1999.

[31] Barajas JH, Wandrey C, Hunkeler D. Polymer flocculants with improved dewatering characteristics. US patent 6294622. 2001.

[32] Gartner HA. Process for the production of high molecular weight copolymers of diallyl dimethyl ammonium chloride and acrylamide in an aqueous dispersed phase. European patent 363024. 1988.

[33] Yang HW, Pacansky TJ. Inverse emulsion process for preparing hydrophobe-containing polymers. US patent 4918123. 1990.

[34] Chmelir M, Pauen J. Process for the continuous production of polymers and copolymers of water-soluble monomers. US patent 4857610. 1989.

[35] Patel M, Huang C-S, Reese RW, Cramm JR, Harris PJ. Horizontally flowing continuous free radical polymerization process for manufacturing water-soluble polymers from monomers in aqueous solution. US patent 6103839. 2000.

[36] Takeda H, Kawano M. Process for production of water-soluble polymer dispersion. US patent 4929655. 1990.

[37] Takeda H, Kawano M. Process for the preparation of dispersion of water-soluble cationic polymer. US patent 5006590. 1991.

[38] Ramesh M, Howland CP, Cramm JR. Dispersion polymerization process. European patent 630909. 1993.

[39] Sparapany JW, Hurlock JR. Hydrophilic dispersion polymers for treating wastewater. US patent 5938937. 1999.

[40] Hurlock JR. Aqueous dispersion of a particulate high molecular weight anionic or nonionic polymer. US patent 6265477. 2001.

[41] Carter PW, Murray PG, Brammer LE, Dunham AJ. High molecular weight zwitterionic polymers. US patent 6590051. 2003.

[42] Werges DL, Ramesh M. An improved process for the preparation of water soluble polymer dispersion. European patent 657478. 1995.

[43] Selvarajan R. Synthesis of high molecular weight anionic dispersion polymers. US patent 5605970. 1997.

[44] Goretta LA, Otremba RR. Method of controlling the molecular weight of vinyl carboxylic acid-acrylamide copolymers. US patent 4143222. 1979.

[45] Holy NL, Bortnick NM, Hughes KA. High temperature aqueous polymerization process. US patent 5263437. 1993.

[46] Hughes K, Swift G. Metal-activated redox initiation for the synthesis of low molecular weight water-soluble polymers.

In: Glass JE, editor. *Water-Soluble Polymers*. Volume 213. Washington, DC: American Chemical Society; 1986. p 145–151.

[47] Chen F, Bair KA. Water treatment polymers and methods of use thereof. US patent 4659480. 1987.

[48] Chen F. Water treatment polymers and methods of use thereof. US patent 4659481. 1987.

[49] Chen F, Bair KA, Boyette SM. Polymers for the treatment of boiler water. US patent 5242599. 1993.

[50] Alfano JC, Godfrey MR, Selvarajan R, Uhing MC. Monitoring boiler internal treatment with fluorescent-tagged polymers. US patent 5736405. 1998.

[51] McCallum TF, Weinstein B. *Process for preparing phosphonate-terminated polymers*. US patent 5866664. 1999.

[52] Marten FL, Zvanut CW. Manufacture of polyvinyl acetate for polyvinyl alcohol. In: Finch CA, editor. *Polyvinyl Alcohol-Developments*. New York: John Wiley & Sons, Inc.; 1992. p 31–56.

[53] Marten FL, Zvanut CW. Hydrolysis of polyvinyl acetate to polyvinyl alcohol. In: Finch CA, editor. *Polyvinyl Alcohol-Developments*. New York: John Wiley & Sons, Inc.; 1992. p 57–76.

[54] Wyatt PJ. Light scattering and the absolute characterization of macromolecules. Anal Chim Acta 1992;272:1–40.

[55] Munk P, Aminabhavi TM, Williams P, Hoffman DE, Chmelir M. Some solution properties of polyacrylamide. Macromolecules 1980;13:871–876.

[56] Hiemenz PC, Lodge TP. *Polymer Chemistry*. 2nd ed. Boca Raton: CRC Press; 2007. p 24–37.

[57] McCarthy KJ, Burkhardt CW, Parazak DP. Mark-Houwink-Sakurada constants and dilute solution behavior of heterodisperse poly(acrylamide-co-sodium acrylate) in 0.5 M and 1 M NaCl. J Appl Polym Sci 1987;33:1699–1714.

[58] Finch CA. Analytical methods for polyvinyl alcohol. In: Finch CA, editor. *Polyvinyl Alcohol-Developments*. New York: John Wiley & Sons, Inc.; 1992. p 751–761.

[59] Vu C, Cabestany J. Characterization of cationic water-soluble polyacrylamides. J Appl Polym Sci 1991;42:2857–2869.

[60] Kam S, Gregory J. Charge Determination of synthetic cationic polyelectrolytes by colloid titration. Colloid Surf 1999;159:165–179.

[61] Armanet L, Hunkeler D. Manifestation of polyacrylamide inverse-emulsion instabilities through oscillatory shear. Langmuir 2003;19:7164–7172.

[62] Saenz de Buruaga I, Echevarria A, Armitage PD, de la Cal JC, Leiza JR, Asua JM. On-line control of a semibatch emulsion polymerization reactor based on calorimetry. AIChE J 1997;43:1069–1081.

[63] Frauendorfer E, Wolf A, Hergeth W. Polymerization online monitoring. Chem Eng Technol 2010;33:1767–1778.

[64] Bakeev KA, editor. *Process Analytical Technology*. 2nd ed. West Sussex: John Wiley & Sons, Ltd.; 2010.

[65] Ni H, Hunkeler D. Prediction of copolymer composition drift using artificial neural networks: copolymerization of acrylamide with quaternary ammonium cationic monomers. Polymer 1997;38:667–675.

[66] Alb A, Farinato R, Calbick J, Reed WF. Online monitoring of polymerization reactions in inverse emulsions. Langmuir 2006;22;831–840.

[67] Maher S, Elizalde O, Leiza JR. Raman application in emulsion polymerization systems. In: Amer MS, editor. *Raman Spectroscopy for Soft Matter Applications*. Hoboken: John Wiley & Sons, Inc.; 2008. p 95–144.

[68] Francisco TW, Yembrick S, Leffew KW. Semi-batch polymerization analysis by multi-point Raman spectroscopy. Proc Anal Technol 2006;5:1–6.

20

PROTEIN AGGREGATION IN PHARMACEUTICAL BIOTECHNOLOGY

Mark L. Brader

20.1 INTRODUCTION

The propensity for proteins to aggregate is one of the most ubiquitous stability problems encountered in the development of protein pharmaceuticals. Aggregation can impact the safety and efficacy of protein pharmaceutical products and is a major issue within the biotechnology industry because virtually every element of the manufacture, fill-finish, storage, transport, distribution, and patient administration is fraught with potential sources of aggregation. Aggregation often originates from a clearly identifiable stress such as a freeze–thaw step or the agitation of a mixing operation. However, the cause of aggregation may be exceedingly cryptic, the kinetics may be nonlinear, and accelerated stability experiments misleading. This chapter is intended to provide context to the significance of protein aggregation for the development and commercialization of protein drug products and to place some perspective on the methodologies currently in use for its characterization and control.

The concept of pharmaceutically undesirable protein aggregation encompasses a broad range of molecular level phenomena. Viewed purely from a molecular size perspective, the troublesome species may be as small as a dimer possessing an effective hydrodynamic diameter only incrementally greater than that of the constituent monomer. At the other extreme aggregates may attain macroscopic sizes. As aggregates become prolific, visually observable physical effects may occur such as colloidal phenomena, turbidity, gelation, phase separation, and precipitation. The deleterious effects of protein aggregation on product quality often lead to an obvious and easily measurable consequence such as filter occlusion or loss of activity. But the effects can also be insidious and less predictable. The industry is acutely aware of the potential for even trace levels of protein aggregates to cause adverse immunological effects and to compromise clinical safety and efficacy [1, 2].

There are established principles of immunology that make protein aggregates a risk factor. Notably, the conceptual analogy between protein aggregates in pharmaceutical products and the large protein assemblies displaying repetitive arrays of antigens that form the basis of vaccines – provoking an immune response, is good for vaccines but bad for protein drugs! Although the risk itself is clear, establishing a rational and systematic basis for understanding and controlling the risk posed by aggregates remains extremely challenging. This challenge arises because there is no established basis to predict the extent of an immune response expected from administration of a specific quantity of protein aggregate or from consideration of its specific structural characteristics. In addition to unknown factors, the primary structure of the constituent protein is relevant, the size and shape of the aggregate is important, as is the topological molecular surface exposed on the aggregate, and the quantity of aggregate. Because it is known that the immune response to an antigen depends on molecular-level details, it is logically inferred that different protein aggregate structures will present different immunogenicity risks. Currently, there are renewed efforts to more thoroughly characterize aggregates as a function of specific stresses [3] and to make rational connections to immunogenicity [4].

While it is convenient to use the term "aggregates" as a catch-all for the undesirable large-species-content of a protein pharmaceutical product, it is important to recognize that the aggregating protein solution comprises a kinetically

Monitoring Polymerization Reactions: From Fundamentals to Applications, First Edition. Edited by Wayne F. Reed and Alina M. Alb.
© 2014 John Wiley & Sons, Inc. Published 2014 by John Wiley & Sons, Inc.

changing heterogeneous ensemble of variously sized conformers. The challenge to minimize, characterize, and control this unstable cocktail of components spanning a size range of ~10 nm to ~100 μm represents the essence of the pharmaceutical protein aggregation problem.

20.2 PROTEIN MOLECULAR ASSEMBLY AND AGGREGATE STRUCTURE

In the world of pharmaceutical protein product development, the term "aggregation" possesses overwhelmingly negative connotations and is considered synonymous with compromised product quality. The classification of aggregate types is not definitive and popular terminology relates to various aspects of protein structure and size including distinctions between so-called native and nonnative protein conformation, small oligomers versus high molecular weight polymers, covalent versus noncovalent aggregates, "soluble" versus "insoluble" aggregates, reversible and irreversible aggregates, visible and subvisible particles. Protein aggregation is also connected to a variety of macroscopic effects deleterious to product quality. Physical instability can manifest as the formation of visible particles (snow-globe-like), amorphous precipitates, clear gels, turbid gels, fibrils, and microcrystals. The kinetics of aggregation can be slow or fast on timescales relevant to production, analysis, storage, and administration of protein drugs. It should also be apparent that some of the concepts used in discussing the significance of pharmaceutical aggregates (e.g., "irreversible aggregate" or "nonnative conformation") are not necessarily fundamentally rigorous but are used in a context relating to aqueous formulation compositions, protein concentration ranges, and analyses typical to pharmaceutical biotechnology. It is common to restrict the topic of pharmaceutical protein aggregation to the boundaries of nonnative, physically and conformationally altered species possessing clearly undesirable properties. However, it is important to recognize that the full scope of protein molecular assembly phenomena can impact pharmaceutical properties and can, sometimes, be exploited in advantageous ways. The major types of species relevant to the topic of protein aggregation are considered briefly in the following sections.

20.2.1 Covalent Oligomers

Covalent aggregates form as a result of chemical degradation reactions which crosslink protein molecules, for example, disulfide-mediated reactions. Some proteins such as human serum albumin possess free thiol groups which are prone to intermolecular disulfide formation. More commonly, therapeutic protein candidates may contain one or more intramolecular disulfide group which can be prone to thiol-disulfide interchange. Although covalent oligomers,

such as dimers, may be relatively small and may remain fully soluble, they can possess enhanced toxicity and immunogenicity relative to the monomer. Because the origin of covalent bond formation is chemical reactivity, it is often possible to inhibit covalent aggregation through astute formulation design, for example, through the inclusion of antioxidants or scavenger molecules to inhibit thiol/disulfide-mediated reactions.

Many therapeutic proteins are too unstable to be commercialized as liquid drug products. In these cases, formulating the protein as a lyophilized powder becomes the configuration of choice because degradation reactions are inherently slower in the solid state. However, covalent aggregation can become more pronounced relative to other degradation mechanisms in the solid state [5]. Stresses specific to the lyophilization process become relevant and it is necessary to understand the impact of each stage of the lyophilization cycle and its interplay with formulation composition. Protein lyophilization is a complex balancing act between a large number of competing variables associated with product stability, clinical compatibility, manufacturing process robustness, and economics. For this reason, it may not be practicable to simply select a composition and lyophilization cycle based exclusively on the criterion of minimization of aggregation. Stresses that need to be understood as part of the lyophilization development process are described in Table 20.2.

From an analytical perspective, covalent aggregates are usually more straightforward to quantitate and characterize because they are less prone to dissociate under chromatography column conditions and are more amenable to mass spectrometry analysis. Because of the inherent stability of covalent species, it can be possible to prepare enriched samples for structure–activity relationship (SAR) studies in preclinical models. An example of well-characterized covalent aggregates formed by an IgG1 antibody in response to pH excursions is described in Kukrer et al. [6]. Using SE-HPLC together with native electrospray ionization time-of-flight mass spectrometry (ESI-TOF MS), the authors fractionated and characterized dimers, trimers, and tetramers as examples of mAb covalent aggregates.

20.2.2 Reversible Noncovalent Oligomers

The propensity of many proteins to form noncovalent assemblies is not necessarily a problem in itself for the development of a pharmaceutically viable product. However, such behavior can impact pharmaceutical properties and stability by bringing reactive groups into close proximity leading to the formation of covalent oligomers. The increased size of the noncovalent assembly may impact absorption kinetics and bioavailability, the extent of which is influenced by the dissociation constant. Noncovalent association can confer beneficial effects. The conformational stability of the protein

TABLE 20.1   Some Common Sources of Aggregation in Biopharmaceutical Proteins

Source	Stress	Note
Cell culture	Unfavorable intracellular environment	Intracellular accumulation during expression
Cell culture	Cell culture medium and high temperature destabilizes properly folded conformation	Cell culture medium is optimized for growth and not the conformational stability of the protein
Viral inactivation	Exposure to very low pH ~3.5–4	A viral inactivation step is an essential part of purifying therapeutic proteins harvested from cells
Ultrafiltration	Shear stresses and surface-induced unfolding	Solvent exchange of purified protein into final bulk drug substance buffer
Bulk drug substance storage	Freeze–thawing	Commercial bulk protein drug substances are often stored as frozen solutions at −70 °C
Compounding	Local pH hotspots and high excipient concentrations	pH adjustments and addition of excipients create transient regions of destabilizing conditions
Drug product filling	Pumping	Final filling into vials or prefilled syringes involve peristaltic pumps that agitate and flow solution through filling lines
Bulk drug substance container closure	Material-specific surface effects	Bulk drug substance may typically be stored in teflon, polycarbonate, polypropylene bottles, or flexible bags constructed of various types of multilaminar films
Prefilled syringes	Leachates, metals, silicone oil	Prefilled syringes may expose the protein to silicone oil, stainless steel needle, reactive tungsten species from syringe-forming process, reactive species from glue holding staked needle in place
Headspace	Air–liquid interface	Many liquid drug products possess an air or N_2 headspace which allows "sloshing" during transport or agitation of the drug product
Storage and in-use conditions	Thermally induced chemical degradation and unfolding	Protein solution drug products are typically stored for 24–36 months at 2–8 °C and from a few hours to 1 month at room temperature
UV light	UV light can promote thiol-disulfide reactivity and other degradation pathways	Manufacturing and in-use
Leachables/metals	Trace metals can leach from contact surfaces or are introduced from raw materials	Trace levels of metal ions can promote reactivity (potentially catalytic) or create particles via precipitation

may be greater in an oligomeric assembly than in the monomer, and the self-association may bury reactive groups thus providing a protective effect. These phenomena sometimes lead to the seemingly paradoxical observation that proteins can be more prone to aggregation at dilute concentrations than at high concentrations. This is relevant to protein formulations supplied for intravenous delivery, which are typically diluted 10- to 100-fold into an intravenous bag using saline or other standard infusion diluent, an effect which can be further exacerbated by the concomitant dilution of stabilizing excipients.

The ability of some proteins to order themselves into noncovalent assemblies is an essential part of their biological structure and function. Therapeutic proteins may or may not possess a self-association behavior innate to their therapeutic activity; however, their tendency to self-associate and the potential impact of this on stability and analysis needs to be understood as part of the early drug product development process. Many proteins remain happily stable and soluble as monomers even at high concentrations (~100–200 mg ml⁻¹)

and under varying aqueous solution conditions. Formulated therapeutic protein solutions may, therefore, comprise a 100% population of monomer, a 100% population of a specific oligomer (e.g., dimer), or a distribution of several oligomeric species. Many proteins can be formulated quite successfully under conditions that stabilize the protein in a form that does not correspond to the biologically active conformer. This is not an issue as long as dilution and conformational re-equilibration upon administration results in the biologically active species in the bloodstream. Reversible self-association is not necessarily a problem and can even be exploited to modulate subcutaneous absorption kinetics as exemplified by insulin and its rapid-acting analogs *vide infra*.

20.2.3   Irreversible Noncovalent Aggregates

The impact of noncovalent aggregation depends greatly on its degree of reversibility under conditions of analysis and administration *in vivo*. Noncovalent aggregation can result in soluble aggregates that are so tightly associated they do not dissociate upon injection with the result that bioavailability is diminished. Furthermore, conformational changes upon self-association can result in significantly altered physical properties such as decreased solubility causing insoluble aggregates. An example reported by Clodfelter et al. [7] characterizes the noncovalent self-association of an octanoylated glucagon-like peptide-1 analog and reports three distinctly different sets of subcutaneous absorption pharmacokinetics corresponding to different degrees of self-association and reversibility. In the most extreme of the three conditions, noncovalent aggregation was effectively irreversible and resulted in negligible bioavailability upon subcutaneous administration. Under different formulation conditions, the same molecule adopted less tightly associated states and bioavailability was restored.

A major category of irreversible noncovalent aggregation is *in vivo* deposits of misfolded proteins that are associated with neurodegenerative diseases such as Alzheimers and prion diseases. Amyloid structures have been extensively characterized as long fibrils comprising densely packed β-pleated sheets. These structures form highly insoluble plaque deposits that cannot be broken down by enzymes and appear to be responsible for the harmful manifestations of these diseases. A major realization of the research on protein deposition diseases has been that the specific types of protein aggregates associated with these diseases can also be formed by proteins that are not associated with these diseases. From a pharmaceutical biotechnology perspective, two key points from the field of protein misfolding and disease are (i) the propensity for proteins to adopt highly altered conformations leading to irreversible events is a general property and (ii) the misfolded protein can exhibit remarkably different physical properties and toxicity relative to its native conformation. The potential for these types of phenomena to occur with a therapeutic protein candidate must be kept in mind throughout its development, characterization, and long-term stability monitoring.

20.2.4 Particles

An important quality attribute and degradation mechanism for therapeutic protein solutions is the formation of particles. Particles that occur in protein pharmaceuticals can be proteinaceous (homogeneous) or nonproteinaceous (heterogeneous). Adventitious particles can sometimes find their way into products due to manufacturing and/or container closure deficiencies. For example, particles shed from rubber, glass, and stainless steel surfaces are not uncommon as are airborne particulates that inevitably find their way into products when air quality systems underperform. A major concern with heterogeneous particles is that the protein will adsorb to the surface creating an adjuvant-like presentation with increased immune response risks.

Particles have been categorized as "visible" and "subvisible." Visible particle content is generally controlled by visual inspection in accordance with a product-specific description such as "a clear colorless solution free of visible particles." Visible particles are loosely considered to be particles bigger than 125 μm. The United States Pharmacopeia (USP) compendial method for quantifying subvisible particles is USP <788> *Particulate Matter in Injections*. This defines allowable levels of particulates per drug product container. For small volume parenteral products (<100 ml), acceptable limits are defined as not more than 6000 particles per container of size >10 μm, and not more than 600 particles per container >25 μm. For biologics products, light obscuration instrumentation is used for these particle counting measurements. The acceptance limits originated from small molecule parenteral drugs and considerations around solid particles posing risks for blood vessel occlusion. However, the rationale behind this standard does not translate particularly well to biotechnology products and is a discrepancy that is currently the source of much attention. Major points at issue are (i) the numerical limits themselves as a meaningful indicator of risk and product quality, (ii) the suitability of light obscuration for measuring protein particles, and (iii) the fact that particles within the 0.1–10 μm range are not monitored, yet particles of this size are known to pose risks of immunogenicity. Adding to the difficulty of achieving a more rigorous approach to this aspect of product quality control is the fact that the immune response itself is poorly understood with respect to prediction of risk from protein particles and aggregates. A commentary article published in 2009 by representatives of the FDA and several distinguished academic groups describes these deficiencies in subvisible particle analysis and control [2]. A central thesis of this article is that while analytical technologies and quality standards have made continuous improvements since the introduction of recombinant protein drugs, the standards for particle testing have not advanced much. An industry perspective highlighting the technical limitations and uncertainties in the connection between these particles and clinical immunogenicity has subsequently been published [8].

20.3 MECHANISTIC ASPECTS

Understanding the mechanism of protein aggregation can be instrumental in its control or avoidance during biologics development. In a recent comprehensive review of the protein aggregation literature, Morris et al. [9] described five classes of postulated protein aggregation mechanisms. These are: (i) the subsequent monomer addition mechanism, (ii) the reversible association mechanism, (iii) prion aggregation mechanisms, (iv) the Finke–Watsky two-step model, and (v) quantitative SAR models. It is clear that pharmaceutically undesirable aggregation can originate from various sources and several underlying mechanisms have been encountered as recurring adversaries to the commercialization of therapeutic proteins [10]. Certain assembly mechanisms are relatively easily identifiable such as the formation of covalent aggregates attributed to an underlying chemical reaction; or the formation of reversible noncovalent oligomers from native monomeric conformers. In contrast, aggregation processes that occur slowly or result in conformationally altered aggregates can be difficult to elucidate and the intermediate states are often mysterious. For example, the formation of particles upon long-term storage of highly purified, optically clear, sterile filtered protein solutions is perhaps one of the more distressing and difficult to study protein aggregation phenomena. This type of aggregation may originate from the presence of a trace impurity (e.g., a leached metal ion) which causes the slow growth of seed nuclei which promote the subsequent rapid formation of very large aggregates in a catalytic-like fashion. Glucagon is one of the most thoroughly studied examples of this type of aggregation [11]. Because this nucleation-controlled aggregation mechanism has a significantly stochastic basis, it can be notoriously variable and difficult to predict or study on a convenient time scale using accelerated methods.

20.3.1 Conformational Stability

A central theme at the origin of many noncovalent aggregation problems is conformational instability of the folded protein. Proteins in aqueous solution tend to adopt tertiary structures that remove hydrophobic side chains from solvent exposure such that the adopted structure represents a global free-energy minimum. In addition to the covalent bonds of the primary structure, the tertiary structure is maintained by a particular network of noncovalent bonding which includes hydrogen bonding, salt bridges, and hydrophobic interactions. Anything that perturbs these bonding interactions may destabilize the overall secondary and tertiary structure potentially resulting in the exposure of hydrophobic surfaces that would otherwise be buried. These exposed hydrophobic patches provide a pathway to form tightly associated noncovalent aggregates. This mechanism is important because the process of isolating, purifying, and manufacturing a formulated drug product for long-term storage exposes the protein to many factors capable of disrupting its noncovalent bonding and thereby generating transient aggregation-prone species. This situation is described by the classic Lumry–Eyring [12] equation illustrated in Equation 20.1.

$$N \leftarrow \rightarrow U \rightarrow F \qquad (20.1)$$

It is apparent that the relationship between N (native) and U (unfolded) relates to thermodynamic stability of the native

protein conformation. U represents an ensemble of unfolded and partially unfolded states, that is, the aggregation-prone conformational variant. Consequently, identification of additives and solution conditions that thermodynamically favor the native conformation is a key formulation paradigm to minimize aggregation. Protein preformulation strategies often rely heavily on screening combinations of stabilizers, pH values, and ionic strength using techniques such as differential scanning calorimetry (DSC), circular dichroism (CD), and fluorescence spectroscopy that can monitor a protein unfolding signal in response to a denaturing stress. Conversely, formulation strategies may be based on destabilizing the unfolded form via preferential exclusion of an excipient. This exploits the principle that water–protein interactions are stronger than excipient–protein interactions. Therefore, in the presence of a preferentially excluded solute, the unfolded protein is destabilized relative to the folded state.

Identifying solution conditions that enhance stability of the native conformation represents an excellent starting point in protein formulation design. However, practical experience has shown that formulation conditions of maximal conformational stability do not always translate into formulations that are optimal for shelf life or for minimization of aggregates. An example is reported for human immunoglobulin G (IgG) [13]. This serves as a reminder that other factors including kinetics and colloidal stability are also relevant to the overall aggregation propensity of the formulated protein.

20.3.2 Kinetics

The right-hand side of the Lumry–Eyring equation indicates that the *kinetics* of formation of the irreversible aggregate will be of prime significance to the fate of the partially unfolded species (U). When the irreversible step is fast, U does not accumulate and the mechanism appears as a two-state kinetically controlled conversion from native state to aggregate. In contrast, when the right-hand side of Equation 20.1 is comparatively slow, the rate limiting step is bimolecular, giving rise to second-order kinetics.

The pharmaceutical shelf-life of a liquid biologics product is typically ~24 months at 2–8 °C, thus it is the kinetics relative to this timescale and the magnitude of aggregate accumulation relative to formal acceptance criteria that are of utmost importance. A worthy goal in the characterization of protein aggregation is to fit the aggregate formation data with a mathematical model corresponding to the proposed aggregation mechanism thereby enabling the calculation of rate constants.

20.3.3 Colloidal Stability

Protein assembly is driven by short range and long range attractive intermolecular forces arising from hydrogen bonding, hydrophobic, ionic, and van der Waals interactions.

These interactions will be influenced by the sequence and folding of the protein as well as extrinsic properties notably ionic strength, pH, and solute content of aqueous formulations. The shape of the protein, its surface charge distribution, and its unique arrangement of hydrophobic and hydrophilic surfaces are major factors affecting the degree of attraction between protein molecules. Local surface geometrical complementarity is another significant factor. Because a large degree of variation in these properties is to be expected for therapeutic protein candidates, it is not surprising that aggregation propensity and optimal formulation conditions can vary highly from one product to the next. Conformational dynamics also play a significant role because a protein structure is not static, it is dynamically sampling conformational space causing fluctuations in the attractive and repulsive interactions between molecules and transiently exposing aggregation-prone regions.

The osmotic second virial coefficient (B_{22}) represents an experimentally accessible thermodynamic parameter to evaluate colloidal stability. This parameter reflects the overall attraction or repulsion between molecules such that a positive B_{22} value represents net repulsion and vice versa. B_{22} measurements have been used extensively in the field of protein crystallization to screen for solution conditions favorable to crystallization. In contrast, good pharmaceutical stability seeks strong protein–solvent interactions (positive B_{22} values) representing conditions under which protein molecules should be less likely to associate. Measurements of B_{22} have proven helpful in developing formulation conditions to minimize aggregation. However, in cases where aggregation proceeds via the formation of a small fraction of a partially unfolded species, the B_{22} measurement is not likely to adequately reflect the aggregation propensity of the system [14].

20.4 EXAMPLES OF AGGREGATION AND ITS IMPACT

20.4.1 Rapid-Acting Insulin Analogs: Reversible Self-Association

Insulin provides a beautifully instructive case study that exemplifies several principles pertinent to the pharmaceutical significance of molecular assembly and reversible self-association of protein oligomers. Insulin is secreted by the β-cells of the pancreas where it is stored as zinc-containing granules comprising crystalline arrays of hexameric zinc insulin [15]. Upon secretion into the bloodstream, insulin dissociates to the monomer which asserts its biological activity via binding to insulin receptors. There is an interesting parallel to the formulation science here because insulin solutions *in vitro* are found to be much more stable as

zinc-insulin hexamers than as monomers. The hexameric structure serves to sequester labile groups within the core of the hexamer and inhibit the effects of interfacial stress. Insulin is thus formulated as a zinc-stabilized hexamer to attain optimal storage and in-use stability [16].

Another general implication of molecular aggregation is the principle that larger species will absorb more slowly into the bloodstream from the subcutaneous tissue. Although the zinc insulin hexamer dissociates in the subcutaneous space soon after injection, the kinetics of this dissociation significantly affect the onset of action which is a parameter of central importance to insulin therapy. This process was recognized as the basis for engineering rapid-acting insulin molecules with amino acid substitutions selected to destabilize inter-subunit contacts critical to hexameric structure [17, 18]. Although these insulin analogs generally do not exist as homogeneous populations of monomers in solution, the basic principle of rapid subcutaneous absorption from less self-associated insulin has been clearly established. The downside of forcing insulin out of its hexameric state, however, was significantly poorer chemical and physical stability. Pharmaceutical stability was restored by invoking formulation conditions that drive self-association and stabilize the rapid-acting insulin as a zinc hexamer [19]. It would seem contradictory to engineer the molecule specifically to dissociate the hexamer, then to choose formulation conditions intended to force it back into a hexameric state. However, the interesting additional principle exemplified here is that *dissociation kinetics* are important too. Although formulated as a zinc hexamer, the rapid-acting insulin, Humalog®, has been shown to dissociate much more readily than the native human insulin zinc hexamer [19]. This results in a "best of both worlds" situation where pharmaceutical stability is attained through formulation as a zinc hexamer, yet this hexamer is able to dissociate rapidly in the subcutaneous tissue resulting in a significantly faster absorption and onset of action. The example of insulin and its rapid-acting analogs emphasize that the propensity for self-association is not necessarily bad in a therapeutic protein candidate, and can in fact be used in clever ways to enhance stability and modulate pharmacokinetics.

20.4.2 Covalent Insulin Dimers

Although covalent oligomers such as dimers may be relatively small and may remain fully soluble, they can possess enhanced toxicity and immunogenicity relative to the monomer. It is not trivial to isolate and characterize individual protein oligomers to the extent of resolving toxicology in preclinical models, so published examples are not plentiful. However, the rich history of insulin as a therapeutic protein provides a well-characterized example. Covalent insulin dimers have been detected in the blood of insulin patients [20] and shown to be responsible for approximately 28% of total circulating insulin immunoreactivity in type I diabetic patients [21]. In contrast, as described previously the noncovalent dimerization of insulin represents an innate property of the molecule essential to its ability to self-associate into the physiological storage form of the hormone. This emphasizes that it is the irreversibility of the covalent linkage that is the significant detail affecting the impact on product safety.

20.4.3 Monoclonal Antibodies

Monoclonal antibodies (mAbs) represent a significant proportion of all new therapeutic proteins currently in development. Antibodies are large glycoproteins that comprise two identical heavy chains and two identical light chains connected by inter- and intra-chain disulfide linkages. There are several classes of antibodies. Characteristic properties of these molecules are their large size (~150 kDa) and "Y-"shaped structures. The IgG class is the most common for therapeutic applications. IgG molecules are further divided into subclasses IgG1, IgG2, IgG3, and IgG4 which differ in the disulfide linkages and hinge regions. Because of the structural commonality between mAbs and the large number of these molecules in development, they are considered prime candidates for "platform" production processes and analytical characterization methods.

The mAbs have been successfully commercialized as lyophilized powders and stable liquids, the latter sometimes exhibiting spectacular stabilities enabling shelf-lives longer than 2 years at 2–8 °C. Excellent stability can frequently be attainable with simple platform formulations and the general similarities between mAb molecules of a given subclass can translate into extensive physicochemical and analytical similarities from one molecule to the next. Despite these commonalities, mAbs can exhibit a wide variation in pharmaceutical stabilities and can display unpredictable aggregation and physical behaviors. The fact that disulfide bonding is different for different subclasses portends that some variation in propensity for covalent oligomer formation is to be expected. The large size of mAbs means that a significant degree of Rayleigh scatter is expected, even from stable soluble monomeric solutions. The mAbs often require relatively large doses, therefore, high concentration formulations (>~100 mg ml^{-1}) become relevant. Physical properties such as colloidal stability, opalescence, phase separation, and viscosity become significant considerations for this class of therapeutic protein. These phenomena represent technical challenges that can often be mitigated via astute formulation design. Kanai et al. [22] reported that the reversible self-association of a concentrated mAb solution resulted in high viscosity. (In contrast to the useful role in insulin pharmaceutics, this exemplifies how reversible self-association may cause undesirable product

effects and annoying analytical complications.) One unusual example of noncovalent dimerization of an mAb clinical candidate was reported by Moore et al. [23] This mAb exists predominantly as a monomer as determined by SE-HPLC, but forms noncovalent aggregates in response to changes in pH, temperature, and ionic strength. The aggregates were found to be dimers that dissociate to monomer upon dilution. Noteworthy was the finding that the time-scale of the self-association of this molecule is particularly slow compared to other characterized examples of self-associating proteins.

An anecdotal case study of aggregation by an mAb is presented in Figure 20.1. An mAb drug product sample comprising a sterile-filled 10 ml glass vial containing 5 ml of a 10 mg ml^{-1} mAb formulation was exposed to uncontrolled room temperature conditions and light for ~6 years.

It is remarkable that even after this degree of stress and aging that the drug product still appears as a clear colorless solution essentially free of visible particles. Analysis of the highly stressed sample by high resolution microcapillary electrophoresis determined its purity to be 78% intact mAb. When the solution, shown in the left side of Figure 20.1a, is observed under intense light, a degree of opalescence is apparent (shown in the right side of Fig. 20.1a). This appearance is typical of many mAb solutions, even those where aggregate levels are very low, as Rayleigh scatter is expected to be significant from the large monomer molecules. The size exclusion HPLC (SE-HPLC) chromatogram shown in Figure 20.1b indicates that the sample has degraded to a mixture of high molecular weight species and low molecular weight fragments together with the remaining monomer main peak. The total aggregate content determined from the analysis of this sample is significant, at 23%. However, it is noteworthy that the visual appearance and particle count (light obscuration) are perfectly acceptable with respect to typical acceptance criteria for parenteral protein drug products.

Aggregation is not always accompanied by distinctive conformational changes. A second derivative Fourier transform infrared (FT-IR) spectrum of this highly degraded and aggregated sample is shown in Figure 20.1c (in red) and compared with that of an unstressed control (in black). The FT-IR analysis shows that the overall secondary structure of this sample appears to be essentially unchanged relative to that of the control, despite the significant stress the sample has been exposed to and the appreciable degree of protein aggregation. This illustrates that a significant perturbation of secondary structure does not always accompany protein aggregation. Although spectroscopic techniques such as FT-IR and CD can provide useful signatures of protein aggregation, this example and others in the literature show that this is not always the case. Several images of particles in the 10–25 µm size range, presumed to be protein aggregates,

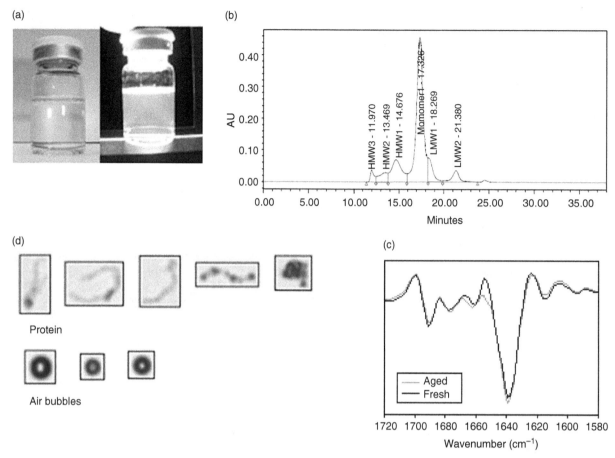

FIGURE 20.1 Analysis of a highly stressed monoclonal antibody drug product aged for ~6 years at room temperature. (a) Optically clear colorless solution under ambient room light (left); opalescent solution when viewed in a light box (right). (b) Size exclusion HPLC chromatogram showing sample degradation. (c) Second derivative FT-IR spectra of the degraded and aggregated sample (red) compared to that of an unstressed control (black). (d) Particles detected in the sample (protein aggregates and air bubbles for comparison). (*See insert for color representation of the figure.*)

are shown in Figure 20.1d for this stressed sample. The Microflow digital imaging (Brightwell DPA 4200) used shows protein aggregates and air bubbles for comparison. A particle count analysis of the sample by light obscuration yielded a result of 1200 particles/container ≥ 10 μm and 175 particles/container ≥ 25 μm. These particle counts are within USP acceptance criteria for particles in small volume parenterals (6000 and 600 per container, respectively).

20.4.4 Glucagon: β-Sheet Fibrils and Gel Formation

Glucagon is a therapeutic peptide that has been thoroughly characterized and provides an excellent example of problematic aggregation. This molecule plays an important role in glycemic metabolism and is used therapeutically to treat severe hypoglycemia. Glucagon comprises a single chain of 29 amino acids and adopts an α-helical secondary structure [24]. The conformational behavior and stability of glucagon has been characterized extensively over the years.

As a pharmaceutical product it is supplied as a lyophilized powder to be reconstituted with diluent for subcutaneous injection. The glucagon molecule possesses poor conformational stability and readily adopts conformers that result in extremely undesirable physical properties. It has been shown to form β-sheet aggregates and gels, including cytotoxic amyloid-type fibrils at high protein concentration under solution conditions similar to formulated conditions.

20.4.5 Glucagon-Like Peptide-1: Solid Precipitates and Microcrystals

Glucagon-like peptide-1 (GLP-1) is an incretin gut peptide that plays a physiological role in postprandial glucose control exerting its biological activity via several mechanisms. The biologically active forms of the peptide are GLP-1(7–37) and GLP-1(7–36)NH$_2$. This peptide is relatively stable quiescent in solution, however, it is very susceptible to the effects of agitation, forming highly insoluble β-sheet precipitates in stirred solutions [25]. The solubility and conformational behavior of this peptide are also highly sensitive to formulation composition and microcrystals form readily from saline solutions.

Aggregation can be accompanied by large changes in conformation. Figure 20.2 shows second derivative FT-IR spectra of a GLP-1 analog described in Doyle et al. [26].

Spectrum (a) was recorded on an aggregate-free solution. The prominent band 1657 cm^{-1} is indicative of a highly α-helical conformation. Spectrum (b) corresponds to a gel centrifuged from solution containing a high level of aggregate. The major bands at 1615 cm^{-1} and 1698 cm^{-1} are signatures of intermolecular β-sheet indicating that this aggregated material is highly β-sheet in nature. Spectrum (c) was recorded on a white precipitate formed by agitation of an initially aggregate-free solution. This spectrum is unlike (a) and (b) and represents a distinctly different aggregated conformational state of this molecule. These spectra illustrate the significant alteration of secondary structure that accompanies aggregation of this molecule. Such conformational instability and susceptibility to agitation present significant challenges for the manufacture and storage/in-use stability of an injectable solution product.

20.4.6 Glucagon-Like Peptide-1 Analog: Large Soluble Noncovalent Aggregates

An octanoylated analog of GLP-1 illustrates how dramatically formulation- and temperature-dependent noncovalent aggregation can impact subcutaneous bioavailability and pharmacokinetics [7]. In this study, Clodfelter et al. showed the octanoylated GLP-1 analog exhibited noncovalent aggregation highly dependent upon ionic strength and temperature. Even though the solutions remained clear with no visible evidence of precipitation, dynamic light scattering and spectroscopic measurements showed that the conformational and association states were significantly impacted by ionic strength and temperature. The measurements performed in parallel with *in vivo* subcutaneous absorption experiments showed that very low bioavailability of a 0.15 M NaCl formulation originated from noncovalent aggregation in which the peptide was so tightly associated that it was effectively irreversible. In contrast, when formulated at low ionic strength, biophysical measurements indicated a significantly lesser degree of self-association and excellent bioavailability was attained *in vivo*. The significantly different subcutaneous absorption profiles of the low and higher ionic strength formulations are shown in reference 7. These data show significantly different subcutaneous absorption kinetics attributed to a low ionic strength (5 mM phosphate buffer) formulation versus a phosphate buffered saline (PBS) formulation. The difference in pharmacokinetics is attributed

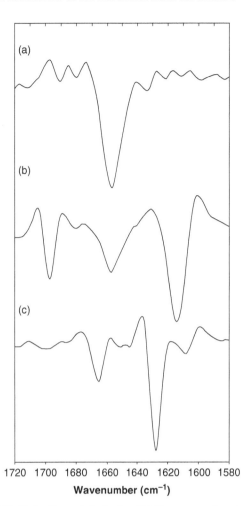

FIGURE 20.2 Second derivative FT-IR spectra of a glucagon-like peptide-1 analog. (a) Spectrum of an aggregate-free solution. (b) Spectrum of a gel centrifuged from solution with a high level of aggregate. (c) Spectrum of a white precipitate formed by agitation of an initially aggregate-free solution. This material is reproduced with permission of John Wiley & Sons, Inc. from Doyle BL, Pollo MJ, Pekar AH, Roy ML, Thomas BA, Brader ML. Biophysical signatures of noncovalent aggregates formed by a glucagonlike peptide-1 analog: A prototypical example of biopharmaceutical aggregation. J Pharm Sci 2005;94:2749–2763.

to the significantly greater noncovalent self-association characterized for the PBS formulation.

Despite the formidable behavior reported for GLP-1 and certain analogs, it is noteworthy that a C16 fatty acid (palmitic acid) acylated derivative of a GLP-1 analog has been successfully commercialized as an injectable treatment for type II diabetes (Victoza®/Liraglutide, Novo Nordisk). It has been suggested that the C16 fatty acid group facilitates reversible self-association to heptamers which helps to favorably protract time-action of this molecule and protect it from *in vivo* enzymatic degradation [27] (another example of noncovalent aggregation being exploited to enhance therapeutic effectiveness).

TABLE 20.2 Sources of Aggregation Specific to Protein Lyophilization

Lyophilization Stress	Impact
Freezing—interfacial effects	The first stage of lyophilization is controlled rate freezing. This process has a major impact on the outcome of the subsequent drying steps. The freezing process affects crystal size of ice and crystallizing excipients. Interfacial stresses associated with crystalline surface area can cause protein unfolding creating aggregation-prone species
Protein freeze concentration	Freezing can cause a significant degree of partitioning and freeze concentration of the protein itself enhancing concentration-dependent aggregation processes
Excipient freeze concentration	Freezing-induced concentration of excipients, such as salt, can expose the protein to local concentrations that are highly suboptimal for conformational stability
Primary drying (water removal by sublimation) causes protein dehydration	Protein unfolding creates aggregation-prone species
Secondary drying (water removal by desorption) causes protein dehydration	Protein unfolding creates aggregation-prone species
Moisture	Aggregation rates of lyophilized powders usually correlate directly with moisture content. High moisture levels upon storage can occur due to ingress from the stopper or simply from variability in the drying efficiency of the scaled-up process

20.5 SOURCES OF AGGREGATION

Protein structure is delicate and biopharmaceutical production, storage, and delivery inherently abusive toward the structural integrity of the molecule. When therapeutic proteins are manufactured and packaged for large scale commercial supply, the protein is exposed to conditions vastly different to its natural *in vivo* environment. These can include pH extremes, intermediate hold steps, agitation and shear stresses, freeze–thaw steps, pumping through filling lines and nozzles, agitation during shipping and use, exposure to various surfaces during clinical administration (e.g., intravenous bags, syringes, tubing, filters, and needles), and finally patient noncompliance with recommended storage and use conditions. During the manufacture of biologics, proteins are exposed to a large number of materials representing a physically diverse set of surfaces each with its own set of potential leachables. These materials include metals, glass, plastics, filters, chromatography resins, elastomers, diafiltration membranes, and silicone oil used to lubricate stoppers and prefilled syringes. A specific interfacial stress is frequently found to be the culprit in generating the partially unfolded aggregation-prone species (*U*) represented in Equation 20.1. A listing of common sources of aggregation in biologics manufacturing is presented in Table 20.1 and a list specific to lyophilization in Table 20.2.

20.6 AGGREGATE ANALYSIS AND CONTROL

20.6.1 Quality Control

Biotechnology products span a diverse set of molecular structures ranging from peptides possessing limited secondary structure through to very large protein assemblies possessing high degrees of secondary, tertiary, and quaternary structures. The fundamental purpose of quality control is to assure safety, efficacy, and consistency of the drug product to the patient. Analytically this is further differentiated as safety, identity, strength, purity, and quality. Because of the inherent inhomogeneity and complexity of biotechnology molecules, quality control comprises more than a series of tests on the final product. The complete process—from cell culture to stoppering of the filled drug product container—becomes a series of highly characterized unit operations incorporating multitudes of target parameters, operating ranges, and in-process controls. These manufacturing controls in conjunction with final testing on drug substance and drug product represent the overall control strategy for the protein pharmaceutical product.

Because protein aggregation can be caused by any one of a number of manufacturing unit operations, changes or "improvements" to that process must be closely scrutinized for potential deleterious impact to protein aggregation levels. Characterization of protein aggregation will thus be a key element of nonclinical comparability studies performed to support manufacturing or formulation changes.

The mainstay technique for the measurement and control of soluble aggregates in therapeutic protein products is size exclusion chromatography (SE-HPLC). Visible particles have been controlled by visual inspection of the final product, and subvisible particles measured using light obscuration instrumentation in accordance with USP <788> (counting particles >10 μm and >25 μm). This leaves the size range ~100 nm–10 μm unmonitored although this deficiency can be addressed in part by extending light obscuration methodology down to 2 μm. In commercial protein drug products, allowable aggregate content measured by SE-HPLC is product-specific. Acceptance limits for new molecules will be established based on no adverse effect (NAE) levels demonstrated via animal toxicology studies, human clinical trials, and batch history data showing manufacturing capability. As a general example, aggregate levels of around 1% are commonly considered acceptable in mAb products and levels may be as high as several percent upon product expiry. A therapeutic protein solution can thus be highly degraded relative to allowable aggregate specifications yet still present as an optically clear, colorless solution with no visible turbidity or solid material (see the case study presented in Fig. 20.1). Some specific examples of allowable aggregate content from monographs in the European Pharmacopoeia are given as follows: Human immunoglobulin for intravenous delivery specifies no more than 3% polymers and aggregates by chromatography. The monograph for Interferon gamma-1b concentrated solution explains that this molecule exists as a noncovalent homodimer and specifies that the content of monomer and aggregates is not more than 2%.

20.6.2 Measurement Methods

Protein aggregates cover a wide range of sizes, from dimers to visible solids. Consequently, a comprehensive approach to monitoring aggregates across the entire size range will require several techniques. For quality control applications, several major challenges exist for the implementation of more rigorous particle characterization and data interpretation approaches [8]. The particles themselves are likely to be highly fragile and susceptible to damage during the process of sampling and introduction into the instrument. Even the innocuous step of pipetting an aliquot of sample has been identified as capable of significantly altering protein particle populations. Particle measurement perhaps represents a classic example of perturbing a system by measuring it. It is also apparent that protein particles may exhibit very diverse morphologies such as thin transparent plates, long fibers, and solid chunks. Defining particle size in terms of an equivalent hard sphere or effective hydrodynamic radius thus becomes questionable. Many pharmaceutical proteins are now sold in prefilled syringes which utilize silicone oil applied to the barrels as a lubricating agent for the plunger. Application levels in the range 0.3–0.8 mg per syringe (1 ml) are typical, representing a significant quantity of silicone oil that may become dispersed as droplets within the subvisible size range. The microflow digital imaging technique provides some capability to distinguish silicone oil particles from protein particles on the basis of particle circularity. This technique offers promise as a valuable supplement to light obscuration measurements [28].

There are numerous instrumental methods capable of interrogating a protein solution for the presence of soluble aggregates; however, there are relatively few methods currently suited to implementation in a quality control laboratory for routine monitoring and lot release. For example, analytical ultracentrifugation is perhaps one of the most fundamentally rigorous and powerful methods for detecting and characterizing protein assemblies. However, it has very low throughput, a highly specialized user requirement, and instrumentation is expensive, collectively making it unsuitable for routine quality control. The most common methods for analysis of aggregates in therapeutic proteins are described briefly in Table 20.3. Methods including dynamic light scattering, static light scattering, field-flow fractionation (FFF), and nanoparticle tracking analysis (Brownian motion) are also examples of techniques that are very effective for specific applications but suffer limitations of accuracy, precision, resolution, or throughput when considered as routine analytical methods for quality control. In contrast, despite some well-known limitations [29], SE-HPLC allows high throughput, is easy to validate, and provides a simple, cost-effective, and robust method that can be run routinely for lot release in a quality control laboratory. Sedimentation velocity and FFF are often used as orthogonal methods to support the development of an SE-HPLC method. These are considered to be more "gentle" methods of analysis because there is less sample dilution and the absence of the column matrix enables a more faithful indication of aggregates in the original sample. Results from a study by Gabrielson et al. [30] comparing SE-HPLC, AUC, and FFF for the analysis of stressed and unstressed mAb samples are shown in Table 20.4

20.7 FINAL COMMENTS

It is fitting to conclude by stating that challenging aggregation problems can certainly be overcome through good product development science and the application of suitable methods for protein aggregation analysis. Glucagon and GLP-1 analogs are exemplified herein as poster children for the worst of biopharmaceutical aggregation. Nevertheless, an acylated GLP-1 analog has been successfully commercialized as a stable liquid injectable for the treatment of type II diabetes (Victoza®, Novo-Nordisk) and glucagon analogs have been designed that possess significantly improved biophysical and clinical properties [31]. The exendin-4 peptide has been commercialized successfully as a stable multi-dose injectable solution for the treatment of type II diabetes (Byetta®, Amylin/Lilly). This molecule is noteworthy because it has a substantially similar amino acid sequence as glucagon-like peptide-1 (7–36) amide (GLP-1). The major difference is the presence of an additional nine amino acids at its C-terminus which form a tryptophan cage thereby dramatically improving its conformational stability and physical properties. The mAb, antiVEGF, was shown by Moore et al. [23] to possess an unusual and problematic noncovalent dimerization behavior, yet was developed into a stable liquid drug product (Avastin®, Genentech) representing a highly successful therapy for certain cancer indications.

TABLE 20.3 Prominent Methods for Protein Aggregation Analysis

Method	Typically Applied Size Range	Separation Medium	Strength	Weakness
Size exclusion chromatography	~1–100 nm	Chromatog. column	High throughput and robust for quantitative analysis, easily validated	Column-binding effects and dilution may alter aggregate profile
AUC-sedimentation velocity	~1–100 nm	Centrifugal field	Least perturbing method. Good solvent versatility. High information content. Can quantitate low level aggregates	Low throughput, complex technique and data analysis
AUC-sedimentation equilibrium	~1–100 nm	Centrifugal field	Rigorous characterization of reversible self-association	Low throughput, complex technique and data analysis
Asymmetric flow field-flow fractionation	~1 nm–1 μm	Flow	Relatively gentle separation mechanism. Good solvent versatility	Interactions with cross-flow membrane may alter aggregate profile
Native gel electrophoresis	~1–100 nm	Electric charge	Good for screening	Semi quantitative
Dynamic light scattering	~0.1–1 μm	None	Extremely high sensitivity to trace levels of large aggregates (~0.01%)	Poor resolution of polydisperse samples. Qualitative
Static light scattering	~0.1–1 μm	None	Small sample volumes amenable to high throughput. Combines well with fractionation techniques	Qualitative
Light obscuration	~2–150 μm	None	Simple, quantitative, easily validated	Poor accuracy
Microflow digital imaging	~2–150 μm	None	Wide dynamic range, image analysis enables particle morphology categorization	Image resolution declines at low end of size range
Nanoparticle tracking analysis	0.1–1 μm	None	One of the few techniques optimal for 0.1–1 μm size range. Excellent resolution of polydisperse samples	Sensitivity depends on refractive index difference between solvent and particle, therefore, relatively poor sensitivity for protein particles in aqueous formulation

TABLE 20.4 Comparison of Results Using SE-HPLC, AF4, and AUC-SV to Analyze Aggregation in a Monoclonal Antibody Sample as Reported by Gabrielson et al. [30]

mAb Sample	Species	% by Mass of Total Protein		
		SV-AUC	SEC	AF4
Unstressed	Monomer	95.8	99.6	97.8
	Total soluble aggregate	4.2	0.4	2.4
Acidified	Monomer	84.6	86.4	80.3
	Total soluble aggregate	15.4	13.6	19.7

In conclusion, even daunting aggregation problems can be overcome by astute formulation and processing strategies combined with sound biophysical and analytical methodologies. In instances where formulation-based approaches are not sufficient, undesirable physicochemical behavior and aggregation propensity can be pinpointed to specific regions and obviated via molecular engineering approaches. Further advances are imminent in the development of quality control strategies for aggregates and particles in protein drug products as well as better insights into risk factors connecting aggregate measurements and immunogenicity.

ACKNOWLEDGMENTS

The author thanks Hao Li for her excellent technical assistance in generating the data presented in Figure 20.1 and is grateful to Olivia Henderson for providing the sample used for this case study. The author thanks Kevin Maloney for a critical review of the manuscript and helpful comments.

REFERENCES

[1] Rosenberg AS. Effects of protein aggregates: an immunologic perspective. AAPS J 2006;8(3), Article 59:E501–E507. Available at http://www.aapsj.org. Accessed 2013 Jun 11.

[2] Carpenter JF, Randolph TW, Jiskoot W, Crommelin DJA, Middaugh CR, Winter G, Fan YX, Kirshner S, Verthelyi D, Kozlowski S, Clouse KA, Swann PG, Rosenberg A, Cherney B. Overlooking subvisible particles in therapeutic protein products: gaps that may compromise product quality. J Pharm Sci 2009;98:1201–1205.

[3] Joubert MK, Luo QZ, Nashed-Samuel Y, Wypych J, Narhi LO. Classification and characterization of therapeutic antibody aggregates. J Biol Chem 2011;286:25118–25133.

[4] Fradkin AH, Boand CS, Eisenberg SP, Rosendahl MS, Randolph TW. Recombinant murine growth hormone from *E. coli* inclusion bodies: expression, high-pressure solubilization and refolding, and characterization of activity and structure. Biotechnol Prog 2010;26:743–749.

[5] Chang LQ, Pikal MJ. Mechanisms of protein stabilization in the solid state. J Pharm Sci 2009;98:2886–2908.

[6] Kukrer B, Filipe V, van Duijn E, Kasper PT, Vreeken RJ, Heck AJR, Jiskoot W. Mass spectrometric analysis of intact human monoclonal antibody aggregates fractionated by size-exclusion chromatography. Pharm Res 2010;27, 2197–2204.

[7] Clodfelter DK, Pekar AH, Rebhun DM, Destrampe KA, Havel HA, Myers SR, Brader ML. Effects of non-covalent self-association on the subcutaneous absorption of a therapeutic peptide. Pharm Res 1998;15:254–262.

[8] Singh SK, Afonina N, Awwad M, Bechtold-Peters K, Blue JT, Chou D, Cromwell M, Krause HJ, Mahler HC, Meyer BK, Narhi L, Nesta DP, Spitznagel T. An industry perspective on the monitoring of subvisible particles as a quality attribute for protein therapeutics. J Pharm Sci 2010;99:3302–3321.

[9] Morris AM, Watzky MA, Finke RG. Protein aggregation kinetics, mechanism, and curve-fitting: a review of the literature. Biochim Biophys Acta, Proteins Proteomics 2009;1794:375–397.

[10] Philo JS, Arakawa T. Mechanisms of protein aggregation. Curr Pharm Biotechnol 2009;10:348–351.

[11] Hoppe CC, Nguyen LT, Kirsch LE, Wiencek JM. Characterization of seed nuclei in glucagon aggregation using light scattering methods and field-flow fractionation. J Biol Eng 2008;2:1–11.

[12] Lumry R, Eyring H. Conformation changes of proteins. J Phys Chem 1954;58:110–120.

[13] Ahrer K, Buchacher A, Iberer G, Jungbauer A. Thermodynamic stability and formation of aggregates of human immunoglobulin

G characterised by differential scanning calorimetry and dynamic light scattering. J Biochem Biophys Methods 2006;66:73–86.

[14] Bajaj H, Sharma VK, Badkar A, Zeng D, Nema S, Kalonia DS. Protein structural conformation and not second virial coefficient relates to long-term irreversible aggregation of a monoclonal antibody and ovalbumin in solution. Pharm Res 2006;23:1382–1394.

[15] Emdin SO, Dodson GG, Cutfield JM, Cutfield SM. Role of zinc in insulin biosynthesis some possible zinc insulin interactions in the pancreatic beta cell. Diabetologia 1980;19: 174–182.

[16] Brange J, Langkjaer L. Insulin formulation and delivery. Pharm Biotechnol 1997;10:343–409.

[17] Brange J, Ribel U, Hansen JF, Dodson G, Hansen MT, Havelund S, Melberg SG, Norris F, Norris K, Snel A, Sorensen AR, Vogt HO. Monomeric insulins obtained by protein engineering and their medical implications. Nature 1988;333: 679–682.

[18] Brange J, Volund A. Insulin analogs with improved pharmacokinetic profiles. Adv Drug Deliv Rev 1999;35:307–335.

[19] Bakaysa DL, Radziuk J, Havel HA, Brader ML, Li S, Dodd SW, Beals JM, Pekar AH, Brems DN. Physicochemical basis for the rapid time-action of Lys(B28)Pro(B29)-insulin: dissociation of a protein–ligand complex. Protein Sci 1996;5:2521–2531.

[20] Robbins DC, Mead PM. Free covalent aggregates of therapeutic insulin in blood of insulin-dependent diabetics. Diabetes 1987;36:147–151.

[21] Robbins DC, Cooper SM, Fineberg SE, Mead PM. Antibodies to covalent aggregates of insulin in blood of insulin-using diabetic patients. Diabetes 1987;36:838–841.

[22] Kanai S, Liu J, Patapoff TW, Shire SJ. Reversible self-association of a concentrated monoclonal antibody solution mediated by fab–fab interaction that impacts solution viscosity. J Pharm Sci 2008;97:4219–4227.

[23] Moore JMR, Patapoff TW, Cromwell MEM. Kinetics and thermodynamics of dimer formation and dissociation for a recombinant humanized monoclonal antibody to vascular endothelial growth factor. Biochemistry 1999;38:13960–13967.

[24] Sasaki K, Dockerill S, Adamiak DA, Tickle IJ, Blundell T. X-ray analysis of glucagon and its relationship to receptor binding. Nature 1975;257:751–757.

[25] Kim Y, Rose CA, Liu Y, Ozaki Y, Datta G, Tu AT. FT-IR and near-infrared FT-Raman studies of the secondary structure of insulinotropin in the solid state: alpha-helix to beta-sheet conversion induced by phenol and/or by high shear force. J Pharm Sci 1994;83:1175–1180.

[26] Doyle BL, Pollo MJ, Pekar AH, Roy ML, Thomas BA, Brader ML. Biophysical signatures of noncovalent aggregates formed by a glucagonlike peptide-1 analog: A prototypical example of biopharmaceutical aggregation. J Pharm Sci 2005;94: 2749–2763.

[27] Russell-Jones D. Molecular, pharmacological and clinical aspects of liraglutide, a once-daily human GLP-1 analogue. Mol Cell Endocrinol 2009;297:137–140.

[28] Wuchner K, Buchler J, Spycher R, Dalmonte P, Volkin DB. Development of a microflow digital imaging assay to

characterize protein particulates during storage of a high concentration IgG1 monoclonal antibody formulation. J Pharm Sci 2010;99: 3343–3361.

[29] Carpenter JF, Randolph TW, Jiskoot W, Crommelin DJA, Middaugh CR, Winter G. Potential inaccurate quantitation and sizing of protein aggregates by size exclusion chromatography: essential need to use orthogonal methods to assure the quality of therapeutic protein products. J Pharm Sci 2010;99: 2200–2208.

[30] Gabrielson JP, Brader ML, Pekar AH, Mathis KB, Winter G, Carpenter JF, Randolph TW. Quantitation of aggregate levels in a recombinant humanized monoclonal antibody formulation by size-exclusion chromatography, asymmetrical flow field flow fractionation, and sedimentation velocity. J Pharm Sci 2007;96:268–279.

[31] Chabenne JR, DiMarchi MA, Gelfanov VM, DiMarchi RD. Optimization of the native glucagon sequence for medicinal purposes. J Diabetes Sci Technol 4:1322–1331.

21

RUBBERS AND ELASTOMERS

Jorge Soto

21.1 INTRODUCTION

A rubber or elastomer is a polymer with viscoelasticity [1–3]. Since rubbers are amorphous polymers above the glass transition temperature, considerable segmental motion is possible, making them relatively soft and deformable, with better elongation, compression, and torsional properties than other polymeric materials. Elastomers and rubbers thus come back fairly close to their original size and shape after the stress is released. The primary uses for these materials include tires, hoses, seals, adhesives, and molded flexible parts, among many others.

An important aspect of the development of rubber and elastomeric technology is the understanding of the relationship between polymerization reaction parameters and product properties [4]. Understanding the effects and interactions of the input reaction parameters in the polymerization reaction helps in the design of an elastomer with specific final properties. There are multitudes of approaches to gather such knowledge. Among them, chemical, spectrophotometric, and thermomechanical analysis are commonly used in the research and development facilities. It is in the process of scaling up that the parameters will need to be adapted to get to the desired product, as a result of different levels of impurities in the raw material feeds, the change in mass and energy transportation, etc. Given the high throughput in plant production compared to smaller scales, it is important to assess the product composition during synthesis and to implement any correction so that the produced material has the expected chemical, physical, thermal, mechanical, and rheological properties.

The global business of the rubber industry is becoming increasingly competitive decade after decade. This is related to the continuously increasing price of oil. This in itself is an incentive to make the elastomer-producing processes more efficient. Additionally, the competition from a challenging global economy and energy cost requires the polymerization processes to be optimized, debottlenecked, and as free of scrap as possible. It is thus imperative to have optimum control of the polymerization process in order to achieve maximum efficiency. The process control will provide more consistent product quality and reduced waste, both of raw materials (resulting in a reduction of scrap) and energy (reduced recycling and/or waste of monomers, for example). One way to accomplish process control is by online monitoring of the polymerization reactions.

The traditional techniques for monitoring the progress and status of a polymerization reaction are based on sample withdrawal for subsequent off-line analysis. Depending on the objective, these methods include methods based on percent solids and viscosity to analytical techniques such as infrared (IR), gel permeation chromatography (GPC), and nuclear magnetic resonance (NMR). The key for successful monitoring is to be able to get a representative sample of the process reaction mixture at the moment of the sampling and to correlate the data from analysis with the process parameters. Ideally, the online monitoring methods chosen have to allow for fast feedback to provide quick and meaningful control of the material made in the polymerization process, especially if any changes are required. Depending on the type of the analysis required, in most of the cases, by the time analytical feedback is obtained, the reaction mixture might have evolved and, if the material is off specifications, it leads to waste of raw materials. Therefore, it will be extremely beneficial if a monitoring system can be implemented and provide feedback in a matter of just a few minutes.

Monitoring Polymerization Reactions: From Fundamentals to Applications, First Edition. Edited by Wayne F. Reed and Alina M. Alb.
© 2014 John Wiley & Sons, Inc. Published 2014 by John Wiley & Sons, Inc.

Many advances have been achieved in the field of online polymerization reaction monitoring. There have been a large number of advances in the field of on-line polymerization reaction monitoring. These have been due to improved detection (sensitivity) of the instruments, improved modeling capabilities, and even the use of neural networks [5] so that the system can learn from previous mistakes, thus creating a "memory" of previous good and bad decisions by itself, and directing and maintaining the process under control from those "learnings."

Given the large breadth of the field of rubbers and elastomers, only a few types of relevant commercial materials will be treated in this chapter. Monomer concentration, polymer molecular weight, percent solids in the system, and in the case of homogeneous systems, the viscosity of the reaction mixture are among the variables that change with time during the polymerization reactions. The parameters monitored depend on the specific type of polymerization, so the online monitoring devices to be used in a given process should be chosen to provide the most valuable feedback to maintain the reaction control.

In this chapter, in-line (real-time) monitoring refers to techniques that provide analysis feedback in about 1 s–1 min, online monitoring means that feedback is available in about 1 min–1 h, and off-line analysis means the classical sampling that takes hours to a day to be completed. The terms in-line and online monitoring are used here interchangeably.

21.2   INDUSTRIALLY PRODUCED RUBBERS AND ELASTOMERS

The term elastomer, derived from elastic polymer, is often used interchangeably with the term rubber, although the latter is preferred when referring to vulcanizates. Listed in Table 21.1 are some of the commercially available rubbers and elastomers.

TABLE 21.1   Common Rubbers and Elastomers—Partial List

Rubbers	Elastomers
Acrylate rubbers	Ethylene-propylene copolymers
Acrylonitrile-butadiene	Fluorinated monomers (vinylidene fluoride)
Ethylene-propylene-diene materials	Polysiloxanes
Hydrogenated nitrile-butadiene	Styrene-butadiene triblocks
Neoprene	Thermoplastic copolyesters
Polybutadiene	Thermoplastic polyurethanes
Polyisobutylene	Polysulfide rubber
Polyisoprene	Ethyl vinyl acetate
Styrene-butadiene copolymers	Chlorosulfonated polyethylene

21.3   POLYMERIZATION REACTIONS

The general reaction chemistry used in the synthesis of common rubbers and elastomers mentioned in Table 21.1 is described in the following. The discussion covers four types of rubbers: styrene-butadiene rubbers (SBRs), polybutadiene, ethylene-propylene-diene rubbers, and thermoplastic polyurethanes.

The SBRs can be synthesized using different processes, each of them producing materials with different properties. Details about the differences among the SBRs, including the microstructural ones, are given in the next section.

21.3.1   Styrene-Butadiene Copolymers

SBR is the synthetic rubber produced in largest volume. Compared to natural rubber, SBR has better processability, heat aging, and abrasion resistance, but is inferior in terms of elongation, hot tear strength, hysteresis, resilience, and tensile strength. With over 70% being consumed in the manufacture of tires and tire products, the demand is very much dependent on the automotive sector.

SBR's share of total synthetic rubber consumption declined in the 1980s and 1990s as radial tires, which use less SBR, replaced belt tires. Currently, there is an increased interest in developing low rolling resistance tires for improved gas mileage and reduced emissions and this should keep a good balance for increased SBR demand [5].

In nontire applications, there has been a faster growth in the use of synthetic rubbers such as ethylene-propylene-diene monomer (EPDM) and nitrile rubbers, as a substitute for SBR. However, nontire uses for SBR are growing with applications including conveyor belts, gaskets, floor tiles, footwear, and adhesives.

The microstructure is the most important parameter which dictates polymer performance, so it is important to monitor the specific parameter that allows correlations with the key performance parameters. The means to modify the polymer architecture are discussed in the following.

The aforementioned styrene-butadiene materials can be produced through various processes:

1. Emulsion polymerization (ESBR)—via free radical polymerization. Makes a random copolymer. The material has some long chain branching. This is an economical way of making this type of rubber. Viscosity is not a problem, since the emulsion does not change viscosity.

2. Solution polymerization (SSBR)—via anionic polymerization. Produces a random copolymer. Its composition and monomer reactivity ratios (monomer sequence distribution along macromolecule chain) are different than in the case of the ESBR (Table 21.2).

3. Solution polymerization (SBS, SEBS)—via dianionic polymerization. Produces blocky terpolymers. These materials are very different than the ones described in the preceding, as will be explained in the following. The blocky terpolymers have styrene blocks which cocrystallize leading to materials with good mechanical properties (high tensile) and good elastomeric properties (elongation).

SBR is produced by two basically different processes: emulsion (ESBR) and solution (SSBR). The emulsion process can be carried out at two temperatures: 5 °C (cold-emulsion) and 50–60 °C (hot-emulsion). The resulting materials have significantly different properties. Thus, cold SBR process offers better abrasion resistance and provides better tread wear, while the hot emulsion SBR process gives a more

TABLE 21.2 General Styrene and Butadiene Isomer Composition for ESBR and SSBR

	Stereoisomer	ESBR	SSBR
Butadiene	cis	9–12%	23–27%
	trans	54–70%	42–44%
	Vinyl	13–18%	31–33%+
Styrene		23%	25%
Number average molecular weight, M_n		100,000	150,000
Weight average molecular weight, M_w		500,000	310,000
Polydispersity, M_w/M_n		5	2.1

branched polymer. Both emulsion processes are free radical polymerizations. The solution SBR is made via anionic polymerization. There is an increasing trend toward the use of solution SBR, due to the ability of the process to meet the increasingly stringent specifications in the manufacture of high-performance tires.

SBR is produced by copolymerizing butadiene and styrene in a 3:1 weight ratio. In the past, emulsion SBR has been favored due to better processing properties which make consumers able to switch easily between suppliers without any need to reconfigure their processing machines. Emulsion SBR is typically produced in a continuous process making it cost effective, while SSBR can be produced in both continuous and batch processes.

Given the nature of its reactivity, butadiene can be incorporated into the growing chain during polymerization in two ways, leading to three types of isomers. The 1,4-addition of butadiene leads to *trans-* and *cis-*isomers, and the 1,2-addition leads to a pendant vinyl groups from the main chain. These isomers define the performance of the final product. Scheme 21.1 shows the idealized structures of these stereoisomers.

The SBR material includes random segments of these isomers plus a random distribution of the styrene (phenyl) moiety in the chain (Scheme 21.2). Since the *cis-* and *trans-*geometric isomers occur randomly throughout the main polymer chain, they are shown in the following as a "crossed" or "twisted" double bond.

Scheme 21.2 offers a snapshot of what would be a small portion of the ever-random insertion of monomers through the length of the polymerization.

A major difference between the emulsion and solution SBR materials is the relative ratio of the insertion modes and geometric isomers derived from butadiene [6], which makes

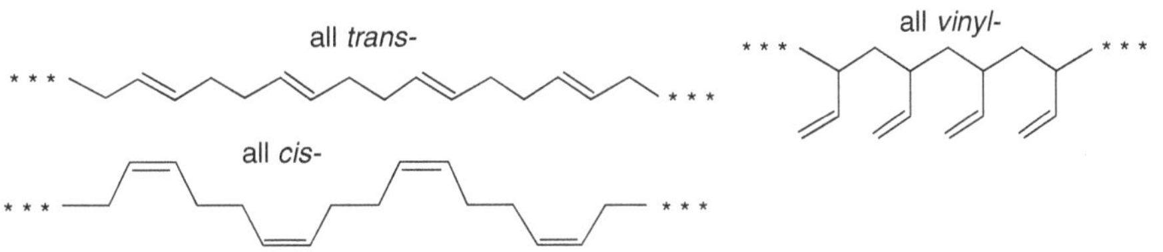

SCHEME 21.1 Idealized 100% all *trans-*, all *cis-*, and all 1,2-*vinyl* polybutadiene chains

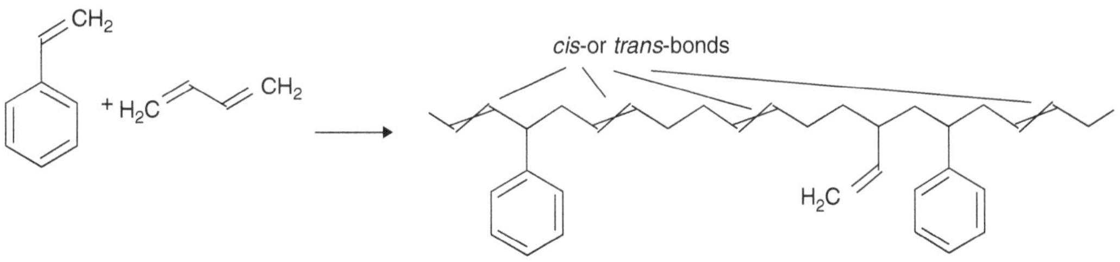

SCHEME 21.2 A schematic molecule of SBR, which includes an appreciation for the randomness of *cis-*, *trans-*, and *vinyl* insertions of butadiene and styrene

TABLE 21.3   Styrene-Butadiene Rubber Commercial Series Descriptors

Series	Description
1000	Hot polymerized polymers
1500	Nonextended cold polymerized polymers
1600	Non-oil-extended cold carbon black master batches
1700	Cold oil-extended polymers
1800	Cold oil-extended carbon black master batches
1900	Miscellaneous high styrene resin master batches

them different in their performance capabilities. This is exemplified in Table 21.2.

SBR is similar to natural rubber in its resistance to mild solvents and chemicals and, like natural rubber, can be successfully bonded to many materials. Natural rubber is polyisoprene of high 1,4-*cis* content. It has a molecular weight of 100,000–1,000,000 g mol^{-1} and has up to 5% of other materials, such as proteins (which may cause allergies), fatty acids, resins, and salts.

21.3.1.1 Styrene-Butadiene Copolymers: Emulsion Process via Free Radical Polymerization

SBR is a synthetic rubber consisting of styrene and butadiene. It has good abrasion resistance and is widely used in car tires. When the proper additive package is used, SBR has also good aging stability. This type of material is a random and, thus, amorphous copolymer. The elastomer is used widely in pneumatic tires, shoe heels and soles, gaskets, and even chewing gum. It is a commodity material which competes with natural rubber. Latex (emulsion) SBR is extensively used in coated papers, being one of the most cost-effective resins to bind pigmented coatings. It is also used in building applications as a sealing and binding agent and serves as an alternative to PVA, but is more expensive. In the latter application, it offers better durability, reduced shrinkage, and increased flexibility, as well as being resistant to emulsification in damp conditions. SBR can be used to "tank" damp rooms or surfaces, a process in which the rubber is painted onto the entire surface (walls, floor, and ceiling) forming a continuous, seamless damp-proof liner; a typical example would be a building's basement.

ESBR is predominantly used for the production of car and light truck tires, and truck tire retread compounds. A list of the uses of SBR includes houseware mats, drain board trays, shoes, chewing gum, food container sealants, tires, conveyor belts, sponge articles, adhesives and caulks, automobile mats, brake and clutch pads, hose, V-belts, flooring, military tank pads, hard rubber battery box cases, extruded gaskets, rubber toys, molded rubber goods, shoe soling, cable insulation and jacketing, pharmaceutical, surgical, and sanitary products, food packaging, etc.

The SBRs have been grouped into different categories, depending on the molecular weight of the polymer and on

the way the material is finished after being processed, for example, being extended with or without oil (Table 21.3).

It is important to understand the process that generates the free radical initiator, which is essentially the heart of the polymerization reaction: catalytic decomposition of an organic peroxide with a ferrous complex, with consequent formation of peroxide for polymerization reaction, plus oxidized ferric ion as shown in the following equation:

$$Fe(II)EDTA + ROOH \rightarrow Fe(III)EDTA + RO^{\bullet} + OH^{\bullet}$$

$$(21.1)$$

where EDTA is ethylene diamine tetra acetic acid.

In this case, it is not practical to have in-line monitoring. It is more convenient to qualify the reagents to be "within specification" limits prior to adding them to the polymerization reaction. Ergo, not everything needs to be monitored in-line, in spite of the nature of the chemical reaction happening (redox type in this case; Eq. 21.2): redox couple of ferric complex and SFS (sodium formaldehyde sulfoxylate, NaO_2SCH_2OH) to convert iron to its reduced catalytic active state.

$$Fe(III)EDTA + SFS \rightarrow Fe(II)EDTA$$
$$- \text{Reduced ferric back to ferrous}$$
$$- \text{Keeps the peroxide cycle alive} \qquad (21.2)$$

Mercaptan is added to provide free radicals and to control the molecular weight distribution by terminating existing growing chains while initiating a new chain. The thiol group acts as a chain transfer agent to prevent high molecular weight values, possible in emulsion systems. The sulfur–hydrogen bond in the thiol group is extremely susceptible to attack by the growing polymer radical and thus loses a hydrogen atom by reacting with polymer radicals (Eq. 21.3)—interaction of growing polymer free radical group with mercaptan leading to hydrogen transfer to form inactive polymer P–H, plus mercaptan radical RS$^{\bullet}$:

$$P^{\bullet} + RSH \rightarrow P-H + RS^{\bullet} \qquad (21.3)$$

The RS$^{\bullet}$ (radical) formed will continue to initiate the growth of a new chain as shown in Equation 21.4—reaction between radical RS$^{\bullet}$ and another monomer (or double bond moiety) to start a new polymer chain. The thiol prevents gel formation and improves the processability of the rubber:

$$RS^{\bullet} + M \rightarrow RS-M^{\bullet} \qquad (21.4)$$

The single polymer parameter with the greatest effect on cure rate is the polymer microstructure, since this defines the disposition of reactive double bonds for the cure. Curing (vulcanization) with sulfur occurs faster at higher *cis*- and *trans*-content. However, the residual soap from the emulsion polymerization reactions slows down the cure of such materials. The SBR made via solution polymerization does require the use of surfactant during the polymerization; not having surfactant in the final product makes it cure relatively faster than emulsion SBR. One way to partially overcome the surfactant issues is to use sulfonate emulsifiers instead of carboxylates (fatty or rosin acids); this topic is beyond the scope of the chapter.

In the ESBRs made by the cold polymerization method, the butadiene component has, on average, about 9% *cis*, 54.5% *trans*, and 13% of *vinyl* structure [6]. At a 23.5% bound styrene level, the glass transition temperature, T_g, of SBR is about −50 °C. As the styrene content in the SBR increases, the glass transition temperature also increases. This process is also referred to as ESBR cold polymerization and operates via a free radical mechanism. The reaction is typically made at 5 °C. Compared to other processes, the emulsion process involves mild reaction conditions. Moreover, the reaction medium is water, which is not tolerated in many other types of polymerizations reactions. Even though large amounts of oxygen are undesirable in the process, emulsion polymerization can still work fairly well with levels of oxygen (air) which would be unacceptable to other systems, even as high as 1% oxygen. The reaction is fairly robust to impurities and can handle a range of functionalized and nonfunctionalized monomers. An additional advantage is the fact that emulsion polymerization gives high solid contents with low reaction viscosity and is a cost-effective process. The physical state of the emulsion system makes it easy to control the process. Thermal and viscosity problems are much less significant than those encountered in the solution polymerization process.

Low-pressure reactors are required. The styrene and butadiene monomers are dispersed in a turbulent mixture of water and an emulsifier (surfactant, about 5% by weight of the total mixture), and charged into the reactor. A package containing a free radical generator is added; due to the low temperature, various activators can be added as catalysts for the radical initiator to work efficiently. This is immediately followed by the addition of a chain transfer agent such as an alkyl mercaptan, which controls the molecular weight of the growing polymer. At the end, the reaction is terminated with a short stop such as sodium dimethyldithiocarbamate, a polysulfide, or isopropylhydroxylamine.

One could follow the polymerization process by in-line monitoring of the formation of *cis*-, *trans*-, or vinyl groups, molecular weight of the polymer, the increasing percent solids, and/or the viscosity of the system. Since this is a well-known process, the microstructure is very well defined and will not change. However, the reaction can be affected by (i) impurities in the recycled monomers, (ii) activator

package components out of specifications (more of this in the next paragraph), and (iii) impurities in other components of the mixture, for example, high levels of oxygen (air) in the system.

It is important to have a quality system to ensure that the components going into the polymerization reaction meet the raw materials specification requirements.

21.3.1.2 *Styrene-Butadiene Copolymers: Solution Process via Anionic Polymerization*
The process starts with the monomers being dissolved in a hydrocarbon solvent. The reaction is initiated anionically (alkyl lithiums). The SSBR material has a narrower molecular weight distribution than ESBR. No soap is used in the reaction, therefore the level of nonrubber materials is less than 2%. The lack of soap improves the curing properties of the SSBR compared to the ESBR, which contains the soap used in the emulsion after coagulation of the material with acid and/or salts (Ca^{2+}, Mg^{2+}). One of the drawbacks of the solution process is that the viscosity of the mixture increases significantly with monomer conversion and the concurrent increasing percent of solids. The challenges to monitor the emulsion and solution processes are thus different.

SSBR is used in essentially the same applications as ESBR: shoe soles, tires, machine parts, energy-saving tires, and high-efficiency tires. The properties of these two polymers are different, so careful evaluation has to be done to choose the best material for specific applications.

The SSBR material should not be confused with the styrene-butadiene block copolymer, a thermoplastic elastomer made from the same monomers also by an anionic polymerization mechanism (see the following). This material has very different mechanical properties and applications.

21.3.1.3 *Styrene-Butadiene Triblock Copolymers: Solution Process via Anionic Polymerization (SBS, SEBS)*
Like SSBR, the block styrene-butadiene copolymers are made in solution. However, more effort is required to make the polymers have the specific sequence and microstructure of monomer segments in order to achieve the targeted polymer properties. The general approach is to start with a compound that will give a dianionic species, as shown in the following diagram, which can initiate anionic polymerization simultaneously from both ends. When the first monomer used is depleted (butadiene in the example shown), the dianionic polybutadiene is a living polymer which can further initiate polymerization with other olefins. After the addition of the desired amount of styrene, the polymer growth continues until all the styrene is consumed, leading to a polystyrene-polybutadiene-polystyrene (SBS) triblock copolymer material to be produced. Due to the ability of polystyrene to cocrystallize with itself, the crystalline PS ends give the material the ability to maintain its integrity without the need for vulcanization, while the PB amorphous center provides the elastomeric properties to the SBS. Inherently, the SBS

materials are prone to oxidation so they are hydrogenated to saturate the olefinic bonds, giving SEBS as the final product. Isoprene is also used as the elastomeric portion of the material. The SEBS block line of polymers is sold under the name Kraton®. Kraton® D (SBS and SIS) and their selectively hydrogenated versions Kraton® G (SEBS and SEPS) are the major Kraton® polymer structures. When blended with thermoplastic polypropylene, SEBS polymers improve the material ductility and the toughening mechanism.

21.3.2 Polybutadiene

Polybutadiene (BR) is the second largest volume synthetic rubber produced, next to SBR (Scheme 21.3). The major use of polybutadiene is in tires, with over 70% of the polymer produced going into treads and sidewalls. Cured BR imparts excellent abrasion resistance (good tread wear), and low rolling resistance (good fuel economy) due to its low glass transition temperature (T_g). The low T_g (lower than $-90\,°C$) is a result of the low "vinyl" content of polybutadiene. Due to the fact that the low T_g leads to poor wet traction properties, polybutadiene is usually blended with other elastomers, such as natural rubber or SBR for tread compounds.

Polybutadiene has a major application as an impact modifier for polystyrene and acrylonitrile-butadiene-styrene (ABS) resin, with about 25% of the total volume of BR going into these applications. Typically, about 7% polybutadiene

is added to the polymerization process to make these rubber-toughened resins. Most polybutadienes are made by solution polymerization, using as catalyst either a transition metal (Nd, Ni, or Co) complex or an alkyl metal, like butyllithium.

The use of the alkyl lithium and transition metal catalysts leads to very different products. The Ziegler catalysts produce stereoregular polybutadiene with high levels of *cis*-configuration product. *cis*-Content >95% gives rise to better green strength and increased cut growth resistance in the cured product. Green strength, which is the strength of the uncured rubber compound, is important for the tire-building process and cut growth resistance is necessary for tire performance. Cut growth resistance is the resistance to the propagation of a tear or crack during a dynamic operation like the flexing of a tire in use.

High *cis*-polybutadiene shows lower T_g compared to alkyllithium-based BR because it has almost no vinyl structure. As mentioned earlier, vinyl tends to increase the T_g of the polymer. The low vinyl content and low T_g makes high *cis*-polybutadiene ideal for golf ball cores. Golf ball cores are cured with peroxides, which tend to "overcure" the vinyl units making a very hard and slow golf ball. The neodymium catalyst system produces the highest *cis*-content of about 99% and also makes the most linear chain structure (no branching) producing a polymer with the best tensile and hysteresis (low heat build-up) properties of all the high

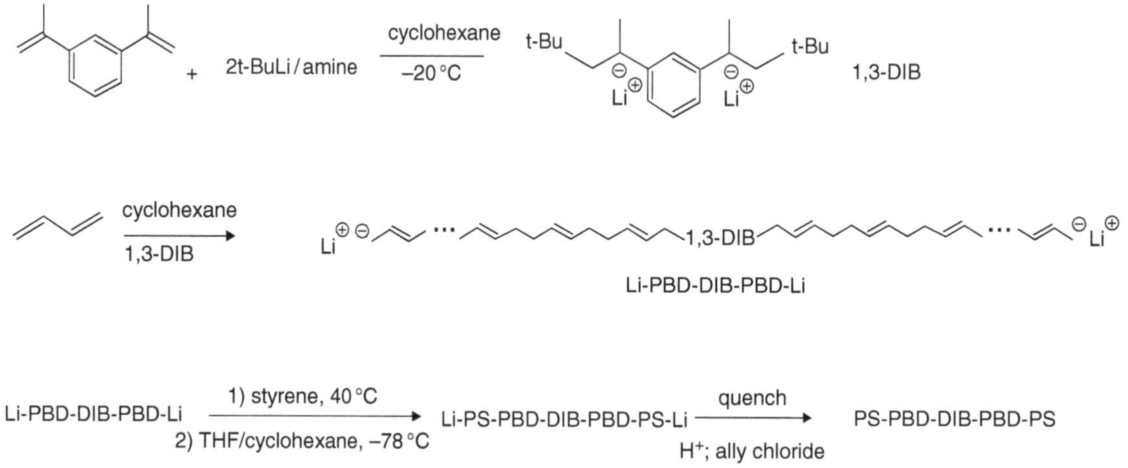

SCHEME 21.3 Block styrene-butadiene copolymers

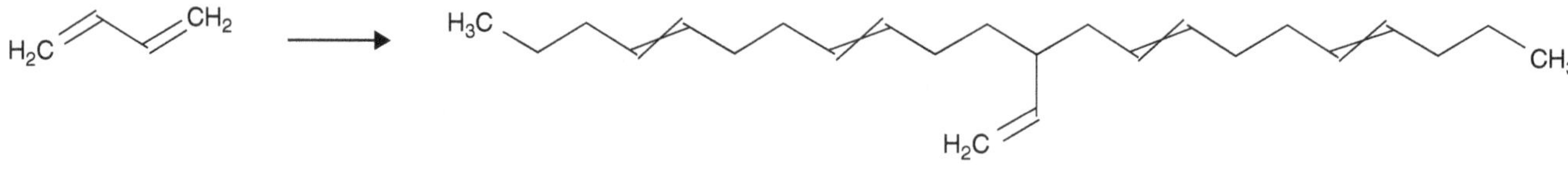

SCHEME 21.4 PBD synthesis

cis types. The cobalt system produces a highly branched BR with a low solution viscosity that makes a good polystyrene and ABS modifier. The nickel catalyst makes polybutadiene with an intermediate level of branching. The alkyl lithium (anionic) catalyst system produces a polymer with about 40% *cis*, 50% *trans*, and 10% *vinyl* when no special polar modifiers are used in the process. The alkyl lithium process is probably the most versatile since the growing chain end contains a "living" anion which can be further reacted with coupling agents or functional groups to make a variety of modified polybutadienes. Vinyl increases the T_g of the polybutadiene by creating a stiffer chain structure. It also tends to cross-link or cure under high heat conditions so the high vinyl polymers are less thermally stable than low vinyl. Polar modifiers (N and O containing compounds) direct the addition of the propagating anion on the living chain end to give a 1,2-addition of the butadiene monomer.

21.3.3 Ethylene-Propylene and Ethylene-Propylene-Diene Materials

Ethylene-propylene (EPM) and ethylene-propylene-diene (EPDM) materials are synthesized via coordination (Ziegler–Natta) polymerization typically using catalytic systems with $TiCl_4$, $ZrCl_4$, VCl_4, or $VOCl_3$ and alkyl aluminums as cocatalysts (Scheme 21.5). These have been used for over half a century and deliver materials with good mechanical properties. A more recently developed technology of metallocene chemistry has revolutionized the field of polyolefins, EPM and EPDM polymers. These metallocenes are used at levels of about 1000th or less of the Ziegler–Natta counterparts. Depending on the type of metallocene used (L_1L_2: ligand, M: Ti, Zr, Hf; thus $L_1L_2MCl_2$, *dichloride*, or $L_1L_2M(CH_3)_2$ *dimethyl*), they are differently activated. The dichlorides have to be activated with alumoxanes, which are obtained by controlled partial hydrolysis of alkyl aluminums. The dimethyl complexes are typically activated with (tris-alkyl) ammonium-(perfluoro)borate salts, which are very soluble in the hydrocarbon polymerization medium. About half of the current manufacturers of EPDM used Ziegler catalysts, the other half use metallocenes. Metallocenes are more reactive than the Ziegler catalysts, so they are used in lower amounts. They are also more expensive, which fairly balances the difference in amount. Due to the low level of

metallocene and corresponding activator used in the reaction, the residuals do not need to be washed off from the polymer. Most of the residual catalyst remnants from the Ziegler process are removed with water.

Due to the fact that the polymer composition is sensitive to changes in temperature, it is important to maintain a good temperature control, especially when the reaction is fairly exothermic. Most of the heat of reaction is used to keep the reaction mixture hot. Molecular weight is controlled with hydrogen, added in small amounts at the beginning of the reaction. There are two major parameters to be taken into account in specifications for the final product, composition and molecular weight. Therefore, online monitoring methods have to address these parameters in order to maintain the targeted polymerization features.

In the aforementioned system, metallocenes cannot be used together with the Ziegler catalyst. Either catalytic system reduces polymerization efficiency with incremental addition of diene. This reduction of activity can be as much as 80% in some cases. The molecular weight distribution of the polymers obtained with Ziegler catalysts is broad, with a M_w/M_n ~3–6; in the case of the polymers obtained with metallocenes, $M_w/M_n = 2$.

21.3.4 Thermoelastic Polyurethanes via Growth Polymerization

Polyurethanes are made by mixing a diisocyanate and a diol or mixture of diol/triol. The aromatic diisocyanates can be aromatic, alicyclic, or aliphatic (Scheme 21.6). Proper personal protective equipment is required when handling diisocyanates, especially the more volatile ones. Direct exposure (inhalation) and contact with diisocyanates can lead to sensitization effects. The diol/triol combination can be a polyether, polyester, or specialty polyol for specific applications.

The reaction proceeds via step growth polymerization (condensation). The degree of polymerization depends on conversion and is sensitive to impurities and molar (stoichiometric) imbalance. Water should be excluded from the reaction medium since it reacts with the diisocyanate and converts it to CO_2 plus an amine, which reacts with unreacted diisocyanate to give polyurea. This prevents the formation of high molecular weight and therefore leads to

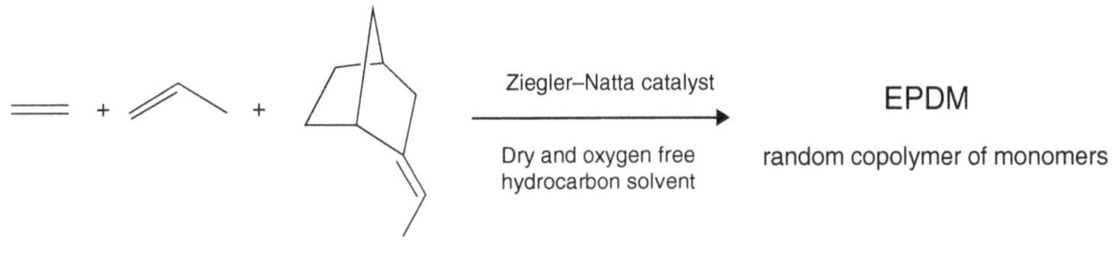

SCHEME 21.5 EPDM synthesis

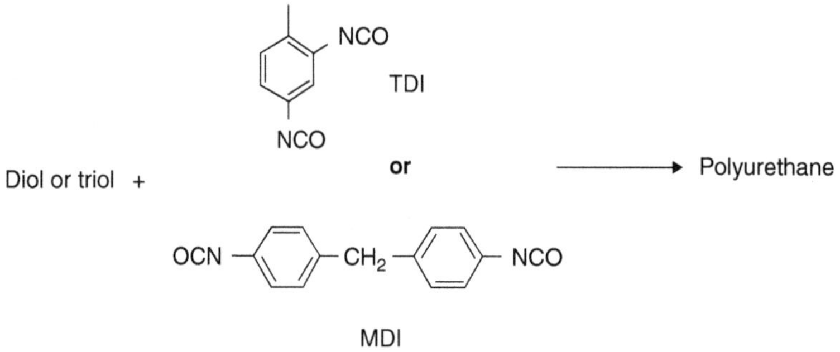

NCO

TDI

NCO

Diol or triol   +   **or**   ⟶   Polyurethane

OCN—⟨ ⟩—CH₂—⟨ ⟩—NCO

MDI

SCHEME 21.6   Polyurethane synthesis

polymers with far from targeted properties. Other consequences could be secondary reactions such as gelation, chain branching, and/or chain degradation. It is thus highly desirable to implement online monitoring systems in order to achieve the targeted polymer properties. More details will be discussed in the next section.

21.4   PROCESS MONITORING

21.4.1   Monitoring Techniques: Overview

In-process measurements represent key tools for achieving high-quality, well-controlled repeatable production:

- Process monitoring has the potential to achieve or move toward full closed-loop process control.
- Specific issues to be addressed by process measurements, for example, lowering energy consumption, reducing scrap, and increasing utilization of recycled materials.
- Wireless sensing is an emerging technology; a research project to develop applications for polymer process monitoring was suggested.
- Scale-up is seen as a particular problem, where process measurements combined with modeling could be beneficial.

The high and competitive pace to produce more cost- and energy-efficient polymers requires the control of chemical composition at molecular level. The basic foundation for the engineering performance of polymers comes from an intertwined design and interplay of composition, polymer architecture, and processing history of the product, represented by the materials science tetrahedron in Scheme 21.7. It will be beneficial if these factors can be controlled and performance evaluations could be done while the material is being made, so any changes could be made in a timely fashion to eliminate scrap and off-spec product.

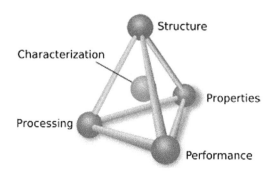

SCHEME 21.7   Materials science tetrahedron

In-line monitoring can provide key information to frame the polymer microstructural, mechanical, thermal, and rheological properties. This occurs as the polymer is being built and, thus, allows for a quick interaction with the process to adjust for disturbances of the system. A few examples are given next.

In-line monitoring devices have been used for many years to test various physicochemical properties of chemical processes. They provide fast and vital information such as temperature, pressure, viscosity [7] (percent solids), refractive index, and pH. However, this type of data provides indirect and/or limited information about the polymer or has similar disadvantages as the off-line methods, such as time-consuming sample preparation, analysis, and maintenance.

Depending on the type of polymerization performed and the number of target criteria for the control of the process, one or multiple in-line systems may be installed to achieve the expected qualified product. An excellent example of monitoring online polymerization reactions by multiple detection was reported [8]. Mostly individual methodologies are described as follows; sometimes, the reader will notice a combination of techniques used to obtain the data sought.

21.4.2   Ultrasound Doppler Velocimetry

Online monitoring of viscosity is currently used in industry. Despite the fact that useful rheological characterization of materials requires a wide range of shear rate measurements, the viscosity instruments used for measurements on processes in various fields can only operate at a single or in a limited range of shear rate.

The method of ultrasound Doppler velocimetry (UDV) [9] was proposed to measure the viscosity of non-Newtonian fluids over a wide range of shear rates and in a short period of time. This is a noninvasive, nondisturbing, quick, and accurate procedure. The distribution of shear-stress can be found by pressure drop. At a radial position, the ratio of shear-stress to shear rate, by definition, yields the viscosity at that point. Thus, for the shear rate range in the flow, viscosity values can be obtained by means of only one online experiment. This is a method known in the literature as pointwise rheological measurement [10, 11].

An interesting study [12] reported the use of a miniaturized Helmholtz RF coil combined with a capillary tube for the design of a portable online/in-line magnetic resonance imaging (MRI) based viscometer. Viscosity was successfully measured for carboxymethyl cellulose (CMC), sodium salt solutions of different polymer concentrations and molecular weights. The MRI measurements were in good agreement with measurements of the same samples performed on a conventional rotational rheometer. The range of shear rates covered by the obtained MRI viscosity data was $3–20\,s^{-1}$. The authors mention that a miniaturized magnetic resonance instrument, based on a small, low-cost, low-field, electro- or permanent magnet, would minimize the cost and safety problems. Such instrument could be readily utilized in a typical industrial environment. It was suggested by the authors that a miniaturized magnetic resonance probe for online/in-line flow studies could be microfabricated by combining an RF coil [13, 14] with microfabricated gradients [15] and electronics [16] around a small tube into one assembly. Microfabrication techniques with the features of integration, reproduction, and precision are particularly suitable for the implementation of miniaturized designs.

An interdisciplinary group at the Pacific Northwest National Laboratory in the United States recently accomplished NMR spectroscopy measurements without magnet suggesting that the concept of an NMR in the hood capability may not be that far away [17–21]. The work could lead to portable NMR spectrometers and possibly even small personalized spectrometers for medical diagnosis. The team demonstrated the performances of the technique in distinguishing between several similar hydrocarbon molecules— the first time that zero-field NMR has been successfully used for such complex chemical analysis.

21.4.3 Heat Flow and Heat Balance Calorimetry

Several reports have been published on the in-line monitoring of vinyl acetate emulsion polymerization reactions in semi-batch mode [22]. With appropriate models, this approach can provide good feedback about the polymerization reaction kinetics. Heat flow calorimetry (Hfc) is frequently used to

monitor and control batch and semibatch polymerization reactions. A major advantage of the calorimetric technique is the high sensitivity, due to the considerable difference between jacket and reactor temperatures, regardless of the jacket fluid and its properties. Nonetheless, this technique requires the determination of the overall heat transfer coefficient between the reaction medium and the jacket, which is a function of the properties of the reaction medium that changes throughout the reaction since the viscosity of the reaction medium can increase drastically and fouling at the reactor wall can occur. In semibatch polymerization processes, the strong variation in the overall heat transfer coefficient between reactor and jacket, as well as the increase in heat transfer area, have to be considered. The calorimetric method is subdivided into the two main concepts of Hfc and heat balance calorimetry (Hbc). In the case of Hfc, the heat generated is established as a difference between the reactor and the cooling jacket, and the jacket temperature is assumed to be dependent on reactor temperature and considered to have no dynamics of its own. In the case of Hbc, the energy balance is performed on the cooling fluid taking into account the dynamics of the jacket. The overall heat transfer coefficient between the reaction medium and the jacket, referred as UARJ (Hfbc), was proposed.

Esposito and his coworkers [22] showed the potential of this method and its sensitivity by introducing disturbances in the cooling fluid and thus, into the system. The approach proposed in this work shows very promising results for the in-line estimation of conversion during batch and semibatch vinyl acetate (VAc) emulsion polymerizations, even in the presence of disturbances, due to its predictive characteristic since it does not require sampling.

21.4.4 Attenuated Total Reflection UV Spectroscopy

As discussed, emulsion polymerization is an important industrial process. Miniemulsion polymerization is a rapidly emerging technology. Conversion of monomer to polymer is one of the most important process variables in batch or semibatch polymerization. This is affected by a series of variables among which impurities, oxygen in the system and poor initiator (free radical) are typical ones. Attenuated total reflection (ATR)-UV spectroscopy was evaluated as a method for monitoring emulsion and miniemulsion polymerization of acrylates with excellent results [23].

There are several methods available to monitor emulsion polymerization reactions, such as gravimetric and GC analysis, nonetheless, they are time consuming. Others such as densitometry, ultrasound velocity, and calorimetry can be applied for online analysis, but they are recipe specific and are unable to discriminate between monomers in a copolymerization. More recently, advanced analytical techniques such as Fourier transform infrared spectroscopy (FTIR) and Raman spectroscopy have been developed for online and in-line monitoring of emulsion polymerization processes. The major drawback of the near-infrared (NIR) spectroscopic applications is that the absorption peak in this region is non-specific and, additionally, most NIR absorption are overtones and combinations of the fundamental mid-infrared (MIR) molecular vibration bands and are typically weaker than their corresponding fundamental transition. In the emulsion case, the problem is compounded by the nonlinearities associated with light scattering in concentrated suspensions.

MIR spectroscopy via ATR or internal reflection is the most widely used of the spectroscopic sensors for these types of polymerizations. In this approach, the so-called evanescent wave, probing the medium outside the probe, is utilized. Thus, the intense absorption of water that covers the major part of the normal analytical region in the conventional MIR spectroscopy can be minimized, which makes possible to record the MIR spectra of monomer/polymer emulsions. However, the signal-to-noise ratios are relatively low, which leads to poor precision in the spectroscopic measurements. Although Raman spectroscopy was regarded as a very promising technique in emulsion applications because of the noninvasive, scattering measurement features as well as due to small water interference, the signal-to-noise ratios are rather low.

The UV-visible (Vis) range has found less use for a direct process analysis. Spectra in this range are broader and less informative, but the main obstacle is probably highly absorptive over short path lengths. At high concentrations of the measured species often found in liquid process streams, this leads to extremely high absorbances when using normal optical path lengths. With the introduction of an ATR probe for UV-Vis range, new possibilities for measurements on optically thick solutions have arisen.

In the UV-Vis range, the effective optical path length for ATR can be reduced to 1–2 μm per reflection. By varying the length of the probe, one can change the number of reflections and thus the sensitivity. On-line and in-line ATR probes for UV-Vis measurements are available commercially. Because of the low cost and ease of maintenance, the ATR-UV technique may provide a better alternative to ATR-NIR or ATR-FTIR methods.

It was found that changes in particle size affected the UV absorption. The spectral absorption at wavelengths above 300 nm is in general independent on the monomer and polymer absorption. The particle size effect was estimated using the spectral information from the wavelength range, and the conversion was obtained with good accuracy. ATR-UV spectroscopy was thus shown to be effective in monitoring methyl methacrylate (MMA) conversion. Good agreement between ATR-UV spectroscopy and gravimetry was demonstrated.

21.4.5   Vibrational Spectroscopy: Near-Infrared and Raman

Both of these techniques have the same physical origin: molecular vibrations. The main difference is that NIR is an absorption technique while Raman is a scattering technique.

These techniques complement each other and could become a power team if both are used in a particular reaction mixture.

21.4.5.1   *Near-Infrared Spectroscopy*   NIR spectroscopy uses the NIR region of the electromagnetic spectrum (from about 800 to 2500 nm). The molar absorptivity in the NIR region is typically quite small; however, NIR can typically penetrate much further into a sample than MIR radiation. So, in spite of the lack of sensitivity, NIR spectroscopy can be very useful in probing bulk material with little or no sample preparation.

The spectra obtain by NIR are typically broad and difficult to assign to specific chemical functional groups. It is thus important to have multivariate calibration techniques to extract the desired chemical information. Beyers [24] describes the in-line monitoring of controlled radical copolymerization reactions using NIR. This is a good case study that describes how a calibration model was built to resolve the issue of similar spectral characteristics for methacrylate and acrylamide monomers.

The control of the molecular weight of acrylic acid was reported by Othman et al. [25]. The authors describe drawbacks of other techniques, such as size exclusion chromatography (SEC), and how the delay in the analysis and the maintenance of the columns lead to significant difficulties in real-time control. Very importantly, they claim that the molecular weight values obtained give a delayed mean value and an efficient control loop should be based on a process model and an estimator of the instantaneous molecular weight. They described the model used and the importance of using the rate of transfer to solvent in the calculation of molecular weight.

Typically, the control of the molecular weight is achieved by changing the concentration of monomer, initiator, solvent, or the reaction temperature. However, some of the process conditions cannot be manipulated easily in the process under investigation. Due to poor efficiency of the initiator at low temperatures and possibility of solvent evaporation at high temperatures, temperature was not used as a variable. In order to maintain a good amount of free radicals for an efficient reaction, the concentration of the initiator was not used as a variable to control M_w either. Also, to avoid effects on mass and heat transfer phenomena in the process, the amount of solvent in the reactor should not be used to control the molecular weight.

An efficient strategy aiming at controlling molecular weight consists in the direct manipulation of the concentration of monomer through changes the inlet flow rate of monomer. Controlling the concentration of reactive monomers such as acrylic acid in the reactor was also essential to maximize the process productivity by decreasing the reaction time. Because the polymerization of acrylic acid is rapid and highly exothermic, the amount of monomer in the reactor must be calculated in such a way that the cooling system is able to remove the heat of polymerization efficiently. In order to make a good comparison with the model, the reaction temperature ought to be stable within a degree centigrade or so.

The concentration of acrylic acid was thus measured online via NIR spectroscopy. Othman et al. [25] describe the control scheme used to feedback the needed rate of polymerization and calculation of the sought molecular weight. The limitations of their model to small amounts of monomer in the system (<1700 ppm) are mentioned. A careful validation of the first model estimations for molecular weight is needed since molecular weight is inaccurate if there is a small amount of polymer in the reactor. This is expected to be addressed in new developing technologies.

Styrene-butadiene by NIR. Polybutadiene and styrene-butadiene copolymer are used extensively in the tire and rubber industries. As mentioned earlier in this chapter, there are various stereoisomers associated with the polymerization of butadiene: *cis-*, *trans-*, and *vinyl*, and their relative amounts appreciably affect the polymer properties. NMR and infrared spectroscopy can accurately determine the microstructure and composition of these materials. These methods usually require extensive sample preparation and usually, dissolving the polymer in a solvent or pressing the polymer into a thin film.

Typically, NIR spectra contain bands that arise from OH, CH, and NH groups in a sample. NIR bands of individual analytes are generally broader and heavily overlapped than bands are in MIR spectra. As a result, it is difficult to determine an analyte concentration from only one or a few NIR absorbances. Nonetheless, full-spectrum multivariate calibration methods, such as partial least squares (PLSs), classical least squares (CLS), and principal components regression (PCR) have been used to accurately determine analyte concentrations from NIR spectra.

Polyurethanes by NIR. It was discussed that the formation of polyurethane materials with good mechanical properties, that is, with high-molecular weight and no undesirable structural formations (gels, branches, etc.), is essential; yet the system is sensitive to the presence of impurities, such as water. These can affect the ultimate microstructure of the polyurethane and thus its performance. Various techniques have been used in the past to measure polymer properties, including densitometry and refractometry [26]. In spite of all the reports on NIR applications for quality control in the polymer field, it is surprising to find that only a few publications have used this technique for in-line monitoring and control of molecular weight in solution step growth polymerization. NIR has been described for addition (free radical) polymerization, where the conversion and the molecular weight are essentially decoupled, since high molecular weights are obtained even during the first seconds of the reaction. In step growth polymerization, they are intimately

intertwined and the only way to achieve high molecular weight polymers is to achieve more than 97% conversion. The feedback in-line monitoring loop to maintain good control of the polyurethane molecular weight (conversion) is shown in Figure 21.1 (from 26).

A comprehensive discussion about in-line monitoring using viscometry and SEC is made by Pinto et al. [26], pointing out that in the latter case industrial applications are missing. One of the reasons is that SEC requires sampling followed by time-consuming analysis, and consequent temperature control. Viscometry, on the other hand, has been used more successfully to monitor polyurethane and polystyrene reactions. The monomers used in the study were methylene diphenyl diisocyanate (MDI) and poly(ethylene glycol) (PEG 1500). The reaction was made in *N,N*-dimethylformamide (DMF), with 1,4-butanediol as the chain extender. The polyurethanes were made in the classical two-step solution polycondensation reaction. Controls and calibrations were accurately made. Monomer consumption was monitored in the 1800–2100 nm window during the reaction, following the level of MDI consumed. In order to develop a calibration model for the MDI concentration, several samples were collected at different times from different runs. The actual level of MDI was obtained by standard titrimetric technique with dibutylamine. The standard error of the calibration was the same as the one for the reference technique (5%). The final calibration model, presenting five factors, had a correlation coefficient of 0.97 and a standard error of calibration (SEC) of 6%. Control of the M_w is based on (i) the minimization of feed molar imbalances and feed impurities which affect M_w and on (ii) in-line monitoring and control of M_w during the chain extension step. A model was developed by the authors, and combined with the NIR, it was tested. The final polymer was essentially linear despite the formation of a few cross-links. The reaction was quenched by adding an inhibitor once the desired M_w was obtained. This work showed the potential for accurate NIR-based PLS calibration model for MDI concentrations. These models can also be used to monitor conversion and M_w during the prepolymerization step of polycondensation reactions.

Key points validated in this study were the proposed closed-loop control strategy for polyurethane reactions, as well as the fact that the relative torque signal provided by the agitator can be used to monitor the M_w of the polymer resin during the polymerization reaction. Importantly, in-line monitoring of M_w may allow the prevention of gelation by controlled addition of inhibitor feed rates.

21.4.5.2 *Raman Spectroscopy* Unlike conventional process monitoring techniques such as UV-Vis, chromatography, or NIR, Raman spectroscopy approach provides high chemical selectivity, requires no sample preparation, is unaffected by water, and is nondestructive. In the case of some of the materials of interest here, such as SBR,

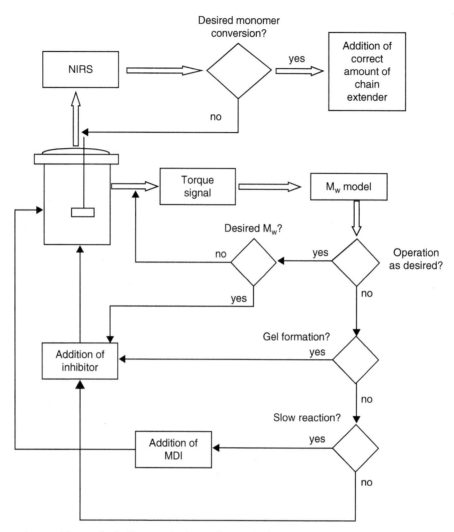

FIGURE 21.1 Proposed control layout for in-line monitoring and control of conversion of polyurethanes. From Nogueira ES, Borges C, Pinto JC. In-line monitoring and control of conversion and weight-average molecular weight of polyurethanes in solution step-growth polymerization based on near infrared spectroscopy and torquemetry. Macromol Mater Eng 2005;290:272–282. © Wiley-VCH Verlag GmbH & Co. KGaA. Reproduced with permission.

spectra show good selectivity for specific symmetrical functional groups, such as $C=C$. There is a fair amount of information on the technique being used for the monitoring of polymers containing unsaturations. Some applications are studying diene and vinyl polymers, such as polystyrene [27] and polybutadiene [28]. Nonetheless, considering the fact that spectroscopy was around for so long, there are no major uses in actual processes. The field is under development, Horiba Jobin Yvon is promoting the technology, which seems particularly promising in the areas of polymer manufacturing [29, 30] and pharmaceutical industry.

Due to the fact that water and CO_2 do not affect the spectra, Raman spectra is particularly effective for monitoring emulsion polymerizations, being used to follow the monomer consumption and the polymer formation.

On top of that, the technique offers probes which are suitable for harsh environments, such as high pressure and temperature. The convenience of little or no sample preparation or sample extraction is a plus, and very importantly, chemical and molecular information is rich with high specificity compared to NIR. What makes this even more attractive is that the data analysis is easier since Raman bands are usually sharper and interferences weaker than in the MIR and NIR. The technique can also be used remotely, with optical fibers up to 1000 ft in length.

21.4.6 Nuclear Magnetic Resonance

During the early 1990s, Borealis implemented the widespread use of the NMR technology for online analysis of polypropylene resin. The technology has been included into their Magneflow® online NMR technology [31–33]. This technology is advantageously used to monitor tacticity parameters, including xylene soluble material and ethylene content, and offers quick feedback to avoid plant shutdowns, while providing continuous monitoring for a more consistent product. Online NMR comes close to the ideal of being a noninvasive, nondisturbing analytical technique to the process. Thermostated tubing and pieces are used so samples can rapidly be transferred to the place of detection. The typical delay time of online monitoring with ^{1}H NMR spectroscopy is below 1 min compared to 5–10 min using conventional 5 mm tubes (sample preparation and shimming).

In polypropylene applications, a small sample (less than 50 ml) is collected from a transfer line, purge tank, or extruder. The sample is sent to the analyzer, and after the sample is done, the sample is returned to the process. The measurements are based on the shape of the NMR signal obtained (free induction decay (FID)). The concept was proven with the capability to be extended to the prediction of mechanical properties such as flexural modulus and Charpy.

NMR spectroscopic techniques are invaluable to obtain quantitative information on reaction processes *in situ*. A few references are provided here for online NMR analysis of small molecule synthesis for the interested reader [34–36]. The information can be used in modeling studies to enable the refinement and development of theories with predictive power which describe the relevant processes.

21.4.6.1 *Acrylate Polymerization* The ability to observe the microstructure of the reaction product makes NMR a

powerful technique to closely monitor and control a polymerization reaction. The technique has also been used to study the polymerization of MMA. The reaction progress was followed using the spatial distribution of the MMA imaged molecules.

Significant progress in the field was reported by Maiwald et al. using flow NMR spectroscopy (Fig. 21.2) [35]. Typical delay with ^{1}H-NMR monitoring is around 5–10 min. Their study was made on fast reactions with less than 1 min turnaround time. It is conceivable that this could be used for deeper NMR studies than what is currently done. As the authors mention, sensitivity of the current technology is one of the major issues. Nonetheless, they were able to collect data at a fast rate. In the case of the polymers, viscosity is the main issue in handling the samples. This needs to be resolved with the proper design of tubing, pumps, and NMR probes which can handle the polymerization mixture in such a way that the analysis is as representative as possible to the reactor instantaneous situation at the time the sample was taken. This is a challenge, especially for solution polymerization, where viscosity increases quickly with the percent solids in the sample. The temperature has to be held steady to prevent precipitation of polymer; additionally, one has to take into account the fact that the kinetics of the heated reacting mixture continue evolving and thus the composition of the sample could quickly change.

The key to the success of this monitoring process is the accurate modeling of polymer properties based on two fundamental characteristics of NMR signals: the amplitude of the signal itself, but more importantly the shape of the

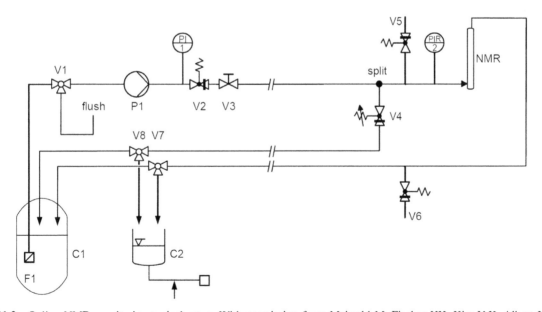

FIGURE 21.2 Online NMR monitoring typical setup. With permission from Maiwald M, Fischer HH, Kim Y-K, Albert K, Hasse H. Quantitative high-resolution on-line NMR spectroscopy in reaction and process monitoring. J Magn Reson 2004;166:135–146. © 2004 with permission from Elsevier.

NMR signal. The raw data NMR signal contains the information that provides the information of the material microstructure. The signal is called the free induction decay (FID), and it is typically Fourier transformed (mathematical operation that decomposes a signal into its constituent frequencies) to give the spectrum of the material. In the case of this technology, this is not done, but the FID is compared to modeled outcome. An example of this is shown in Figure 21.3, where the difference in the FIDs is visually obvious as the peaks, and thus the microstructure of the polymer, change [37].

Another relevant study which shows the potential of bringing the NMR monitoring technology not only closer to production, but to provide kinetic data of reactions, was published by Bart et al. [34]. His group put together a microfluidic NMR chip with high spectral resolution for the case of small molecules reaction, the acetylation of benzyl alcohol.

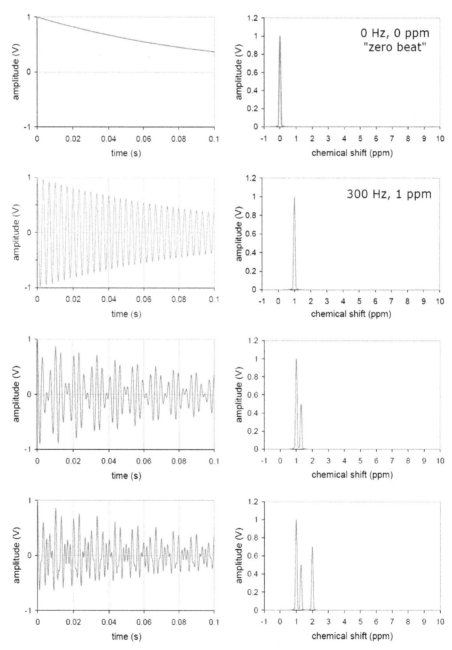

FIGURE 21.3 Examples of FID (free induction decay) curves, prior to performing FT (Fourier transform) to get the corresponding spectra shown on the right of each set [37]. Fast acquisition of FID's is important, the shape of which carries the encoded frequencies and amplitudes of the components of the reaction mixture.

NMR analytical techniques continue to develop new approaches and pulse sequences for more detailed analysis into the samples of interest. These could be used to probe deeper into the polymer molecules. In conjunction with the process, fine tuning or controlling of the polymer molecular could essentially be done as it is being built. Currently, the NMR measurements take more than 5 min, so cannot be used as in-line process monitoring techniques. Improvements in NMR sensitivity together with increased resolution and faster computing processing capabilities will make these seemingly inaccessible techniques available within the next decade. Conceivably, they can include in-line monitoring with pulse sequences such as DEPT and COSY, among others.

COSY is proton–proton correlation spectroscopy used as a quick method of establishing connectivity within a molecule. This is important in defining thermal and mechanical properties of polymers, as such stereoisomers in the polymer provide the "fine tuning" of polymer properties. As such, consistency in the amount of such isomers will provide a consistent polymer from the process.

A more sophisticated methodology that might eventually be used is heteroatomic hydrogen–carbon correlation (HETCOR), which can define deeper correlations in the polymer molecule. Conceivably, this can be used for copolymerizations of complex mixtures with monomers with very different reactivities and functional groups.

To conclude this subsection, the discussed development of an "NMR in the hood" capability will open many opportunities to apply the method for monitoring of many types of chemical reactions, including polymerization reactions, either by itself or in combination with other techniques, for the improvement of polymerization reaction in-line monitoring [20, 21].

21.4.7 Electron Spin Resonance

Electron spin resonance (ESR) spectroscopy is currently used in measuring concentration and composition of chemical species with unpaired electrons. Also called electron paramagnetic resonance (EPR) spectroscopy, this is a technique for studying chemical species that have one or more unpaired electrons, such as organic and inorganic free radicals or inorganic complexes possessing a transition metal ion. Since the basic physical concepts of EPR are analogous to those of NMR, it is conceivable that the same principle of zero-field splitting could be used. The problem of spin coupling without an applied magnetic field can be overcome by employing a technique known as "parahydrogen-induced polarization." Parahydrogen is a spin isomer form of hydrogen with the antiparallel spin alignment, forming a "singlet state." The technique transfers the polarization from the parahydrogen to the sample molecule resulting in the signal enhancement. While the phenomenon of parahydrogen-induced polarization has been known for some time, the current work is the first to successfully use it in zero-field, so far only in NMR; nonetheless, given the similarities, one can expect the ESR methodology to benefit from these findings.

In ESR spectroscopy, the electron spins are excited instead of spins of atomic nuclei. Most stable molecules have all their electrons paired, so the EPR technique is less widely used than NMR, but one can see that this could be advantageously used to determine the concentration of active free radical species or inactivated/dormant metal species in coordination chemistry (e.g., V^{2+} vs. V^{3+} in Ziegler–Natta polymerizations) in the corresponding polymerization reactions. In the case of free radical polymerizations, lack of the free radical will immediately point to such as the main cause of poor polymerization rates. In the case of Ziegler polymerization, the active catalyst is typically deactivated by reduction with cocatalyst or by other side reactions during the polymerization. ESR monitoring could detect the presence of paramagnetic species and thus, the activity of the process could be correlated accordingly.

Such off-line monitoring is common and has been reported in laboratory settings. Gueta-Neyroud et al. [38] confirmed in their study that the Al-OEt species characterized by the reaction of ethanol and methyl aluminoxane (MAO) were not the active species in the polymerization process, as a blank polymerization of propylene was performed with a 1:200 ethanol:MAO solution in toluene. The reaction yielded no polymer, thus indicating that the polymerization proceeded via titanium species. ESR technique can be implemented at pilot plant level if the parahydrogen-induced polarization becomes a reality.

21.5 COMPUTER SIMULATION AND MODELING

It is worth mentioning that there is a fair amount of information on computer simulation to achieve steady-state reaction conditions [39]. These simulations use artificial neural networks, [40] advances in computational fluid dynamics (CFD) [41], as well as combination of new experimental and modeling techniques, whence the application of these techniques can lead to improved models of polymerization systems as well as the discovery of "new" kinetic mechanisms that control polymerization rate and properties.

An example of this is discussed by Campbell et al. [42]. The frequency factors for backbiting, chain transfer to polymer, and addition fragmentation reactions were fitted using a combination of molecular weight, oligomer concentration, and terminal double bond (TDB) distribution data for three different temperature values. Using these parameter values, the model was found to adequately predict the oligomer concentrations, molecular weights, and TDB distributions over the entire range of experimental conditions examined in the first part of this work. The kinetic model indicates that the polymerization is characterized and controlled by aggressive backbiting, followed by beta-scission degradation.

REFERENCES

[1] Morton M. *Rubber Technology*. 3rd ed. Boston: Rubber Division American Chemical Society, Kluwer Academic Publishers; 1999.

[2] Mark JE, Erman B, Eirich FR. *The Science and Technology of Rubber*. 3rd ed. Elsevier Academic Press; 2005.

[3] Available at http://en.wikipedia.org/wiki/Synthetic_rubber. Accessed 2013 Jun 22.

[4] Litvinov VM, De PP. *Spectroscopy of Rubbers and Rubbery Materials*. Shawbury: RAPRA Technology Limited; 2002.

[5] Leon F, Piuleac CG, Curteanu S. Stacked neural network modeling applied to the synthesis of polyacrylamide-based multicomponent hydrogels. Macromol React Eng 2010;4 :591–598.

[6] Available at http://www.afdc.energy.gov/conserve/fuel_economy_tires_light.html. Accessed 2013 Jun 22.

[7] Available at http://www.iisrp.com/WebPolymers/09E-SBR PolymerSummaryJuly16.pdf. Accessed 2013 Jun 22.

[8] (a) Galvanic Applied Solutions. Available at www.galvanic.com. Accessed 2013 Jun 22; (b) Available at http://www.interline.nl/media/1000052/nametreviscoliner.pdf. Accessed 2013 Jun 22.

[9] Alb AM, Reed WF. Fundamental measurements in online polymerization reaction monitoring and control with a focus on ACOMP. Macromol React Eng 2010; 4:470–485.

[10] Koseli V, Zeybek S, Uludag Y. Online viscosity measurement of complex solutions using ultrasound doppler velocimetry. Turk J Chem 2006; 30:297–305.

[11] Arola DF, Barrall GA, Powell RL, McCarthy KL, McCarthy MJ. Use of nuclear magnetic resonance imaging as a viscometer for process monitoring. Chem Eng Sci 1997; 52: 2049–2057.

[12] Uludag Y, McCarthy MJ, Barrall GA, Powell RL. Polymer melt rheology by magnetic resonance imaging. Macromolecules 2001;34:5520–5524.

[13] Goloshevsky AG, Walton JH, Shutov MV, de Ropp JS, Collins SD, McCarthy MJ. Nuclear magnetic resonance imaging for viscosity measurements of non-Newtonian fluids using a miniaturized RF coil. Meas Sci Technol 2005;16:513–518. DOI:10.1088/0957-0233/16/2/025

[14] Webb AG. Radiofrequency microcoils in magnetic resonance. Prog Nucl Magn Reson Spectrosc 1997;31:1–42.

[15] Walton JH, de Ropp JS, Shutov MV, Goloshevsky AG, McCarthy MJ, Smith RL, Collins SD. A micromachined double-tuned NMR microprobe. Anal Chem 2003;75:5030–5036.

[16] Goloshevsky AG, Walton JH, Shutov MV, de Ropp JS, Collins SD, McCarthy MJ. Integration of biaxial planar gradient coils and an RF microcoil for NMR flow imaging. Meas Sci Technol 2005;16:505–512.

[17] Dechow J, Lanz T, Stumber M, Forchel A, Haase A. Preamplified planar microcoil on GaAs substrates for microspectroscopy. Rev Sci Instrum 2003;74:4855–4857.

[18] Available at http://physicsworld.com/cws/article/news/2011/may/29/nmr-spectroscopy-without-the-m. Accessed 2013 Jun 22.

[19] Theis T, Ganssle P, Kervern G, Knappe S, Kitching J, Ledbetter MP, Budker D, Pines A. Parahydrogen-enhanced zero-field nuclear magnetic resonance. Nat Phys 2011;7:571–575. DOI:10.1038/nphys1986. Available at http://www.nature.com/nphys/journal/vaop/ncurrent/pdf/nphys1986.pdf. Accessed 2013 Jun 22.

[20] Available at http://www.rsc.org/chemistryworld/News/2011/June/24061101.asp. Accessed 2013 Jun 22.

[21] Küster SK, Danieli E, Blümich B, Casanova F. High-resolution NMR spectroscopy under the fume hood. Phys Chem Chem Phys Aug 7, 2011;13(29):13172–13176. Epub Jun 23, 2011.

[22] Esposito M, Sayer C, Hermes de Araujo PH. In-line monitoring of emulsion polymerization reactions combining heat flow and heat balance calorimetry. Macromol React Eng 2010; 4:682–690.

[23] Chai X-S, Schork FJ, Oliver EM. ATR-UV monitoring of methyl methacrylate miniemulsion polymerization for determination of monomer conversion. J Appl Polym Sci 2006;99:1471–1475.

[24] Beyers CP. Technical University of Eindhoven; 2003. Available at http://alexandria.tue.nl/extra2/200312295.pdf. Accessed 2013 Jun 11.

[25] Othman NS, Févotte G, Peycelon D, Egraz JB, Suau JM. Control of polymer molecular weight using near infrared spectroscopy. AIChE J 2004;50:654–664.

[26] Nogueira ES, Borges C, Pinto JC. In-Lline monitoring and control of conversion and weight-average molecular weight of polyurethanes in solution step-growth polymerization based on near infrared spectroscopy and torquemetry. Macromol Mater Eng 2005;290:272–282.

[27] Cornell SW, Koenig JL. Laser-excited Raman scattering in polystyrene. J Appl Phys 1968;39:4883–4890.

[28] Cornell SW, Koenig JL. The Raman spectra of polybutadiene rubbers. Macromolecules 1969;2(5):540–545.

[29] Available at http://www.horiba.com/fileadmin/uploads/Scientific/Documents/HJY/hjybro.pdf. Accessed 2013 Jun 22.

[30] Available at http://www.upc.edu/pct/documents_equipament/d_210_id-499.pdf. Accessed 2013 Jun 22.

[31] Available at http://www.progression-systems.com/PDF/technical-papers/TP05_PP.pdf. Accessed 2013 Jun 22.

[32] MagModule II™ NMR analyzer: can be used in the polymer industry to measure morphological properties of polymers. Available at http://www.progression-systems.com/products/nmr-spectroscopy/magmoduleii.php. Accessed 2013 Jun 22.

[33] Product consistency and manufacturing flexibility using process NMR. Available at http://www.progression-systems.com/PDF/technical-papers/TP03_PP.pdf. Accessed 2013 Jun 22.

[34] Bart J, Oosthoek de Vries AJ, Tijssen K, Janssen JWG, van Bentum PJM, Gardeniers JGE, Kentgens APM. In-line NMR analysis using stripline based detectors. 14th International Conference on Miniaturized Systems for Chemistry and Life Sciences; 2010 Oct 3–7; Groningen.

[35] Maiwald M, Fischer HH, Kim YK, Albert K, Hasse H. Quantitative high-resolution on-line NMR spectroscopy in

reaction and process monitoring. J Magn Reson 2004;166: 135–146.

[36] Available at http://www.process-nmr.com/index.html. Accessed 2013 Jun 22.

[37] Available at http://www.chem.purdue.edu/courses/chm621/ lecture/multiplex/10.5%20FT%20NMR%20Instrumentation. pdf. Accessed 2013 Jun 22.

[38] Gueta-Neyroud T, Tumanskii B, Kapon M, Eisen MS. Synthesis and characterization of dichlorotitanium alkoxide complex and its activity in the polymerization of α-olefins. Macromolecules 2007;40:5261–5270.

[39] Soares JBP. editor. *Polymer Reaction Engineering. V. Macromolecular Symposia.* Weinheim: Wiley-VCH; 2004.

[40] Minari RJ, Stegmayer GS, Gugliotta LM, Chiotti OA, Vega JR. Industrial SBR process: computer simulation study for online estimation of steady-state variables using neural networks. Macromol React Eng 2007;1:405–412.

[41] Kolhapure NH, Fox RO. *DECHEMA*, Monographs. Volume 74; 2001, p 1089.

[42] Campbell JD, Teymour F, Morbidelli M. High temperature free radical polymerization. 2. Modeling continuous styrene polymerization. Macromolecules 2003;36:5502–5515.

22

POLYMERS FROM NATURAL PRODUCTS

Dimitrios Samios, Aline Nicolau, and Miriam B. Roza

22.1 OILS AND FATS

22.1.1 Chemical Structure of Triglycerides and Fatty Acids

Triglycerides, also called triacylglycerols, are plants, fish, and algae oils, as well as animal fats, in which three fatty acids are joined to a glycerol molecule (Fig. 22.1). These fatty acids can be equal or different. Oils are triglycerides that are liquids at room temperature, while solid triglycerides are called fats [1, 2].

Fatty acids are carboxylic acids with long hydrocarbon chains, which can be saturated or unsaturated. They differ from one another in the length of their carbon chain and in the number and position of their double bonds [3]. Fatty acids with more than one double bond are called polyunsaturated fatty acids. Vegetable oils have more than 80% of unsaturated fatty acids, while fats contain mainly saturated fatty acids. Table 22.1 presents the chemical structures of the most common fatty acids in natural oils and fats. Naturally occurring unsaturated fatty acids have all double bonds with *cis* configuration. The melting point of saturated fatty acids is higher than the unsaturated fatty acids with comparable molecular weight. This is a consequence of the *cis* configuration, which causes a bend in the molecule impeding the efficient packing of the chains.

Table 22.2 presents the fatty acid composition of the different vegetable oils commonly used in polymerization reactions [4–6]. Additionally, the fatty acid distribution of fish and algae oils are shown. As shown in the table, each oil has a specific fatty acid distribution. Oils from algae, such as microalgae, are raw materials with good perspectives for use in polymerization reactions due to the high content of arachidonic acid, with four double bonds in its carbon chain. Currently, there is abundant research related to the use of algae oil for biofuel production [7–9]; nonetheless, the use of these materials in the polymer area is incipient.

In general, unsaturated fatty acids with 18 carbons (oleic, linoleic, and linolenic) are the dominant constituents of vegetable oils. Oleic acid (C18:1) presents only one unsaturation at C9–C10 position. It is the main component of olive oil, canola oil, and rapeseed oil. Linoleic acid (C18:2) includes two unsaturated bonds at C9–C10 and C12–C13 positions, while linolenic acid (C18:3) has three unsaturations at C9–C10, C12–C13, and C15–C16 positions [4–6]. Soybean oil and linseed oil are examples of the natural oils rich, respectively, in linoleic acid and linolenic acid. Exceptions are castor oil, tung oil, and oil from *Vernonia galemensis*. Castor oil is a hydroxyl-containing oil with ~90% ricinoleic acid [5]. Tung oil has ~80% α-eleostearic acid [4], which is a fatty acid with three conjugated double bonds. *V. galemensis* is a natural epoxidized plant, which contains 60–65% vernolic acid [5].

The physical and chemical properties of the polymers produced from biological oils and animal fats depend strongly on the fatty acid distribution in the triglyceride molecules. Consequently, each oil or fat has a characteristic degree of unsaturation, which is determined by iodine value [4]. This parameter is defined as the number of grams of iodine that reacts with the unsaturated bonds contained in 100 g of oil or fat. According to their iodine value, triglycerides are classified as nondrying,

Monitoring Polymerization Reactions: From Fundamentals to Applications, First Edition. Edited by Wayne F. Reed and Alina M. Alb.
© 2014 John Wiley & Sons, Inc. Published 2014 by John Wiley & Sons, Inc.

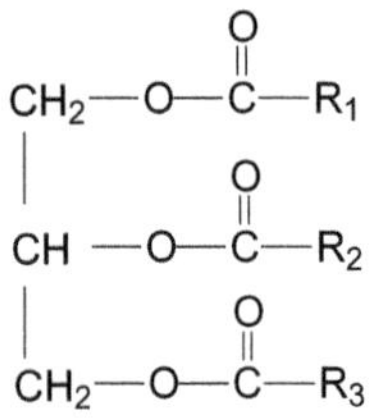

FIGURE 22.1   Structure of a triglyceride. R_1, R_2, and R_3 are fatty acid molecules.

semidrying, and drying. Nondrying oils have iodine value (IV) smaller than 90. Semidrying oils are characterized by IV between 90 and 130, while drying oils have IV higher than 130. The presence of other functionalities (such as epoxies and hydroxyls) also affects the properties of the polymers produced. In general, the molecular weight or the unsaturation degree of the polymers from triglycerides varies with the unsaturation degree of the starting oil [4–6].

TABLE 22.1   Chemical Structure of the Most Common Fatty Acids in Natural Oils and Fats

Fatty Acid	CN:DN	Structure
Caprilyc	8:0	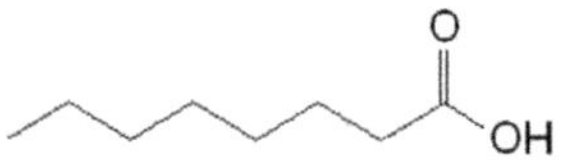
Capric	10:0	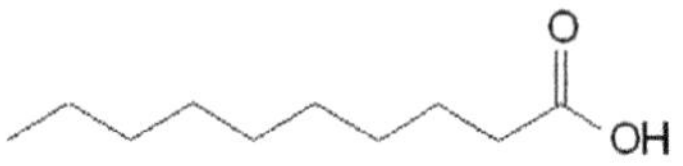
Lauric	12:0	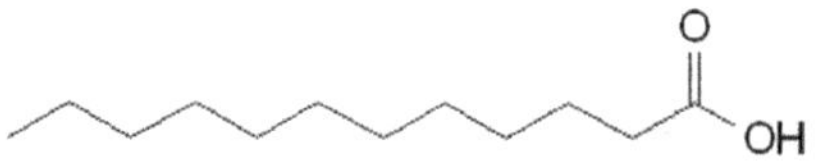
Myristic	14:0	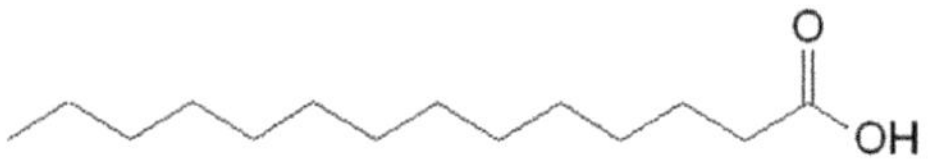
Palmitic	16:0	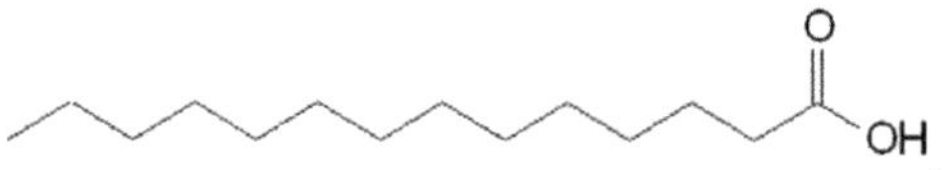
Palmitoleic	16:1	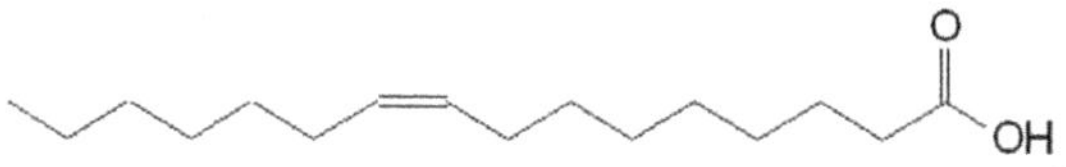
Stearic	18:0	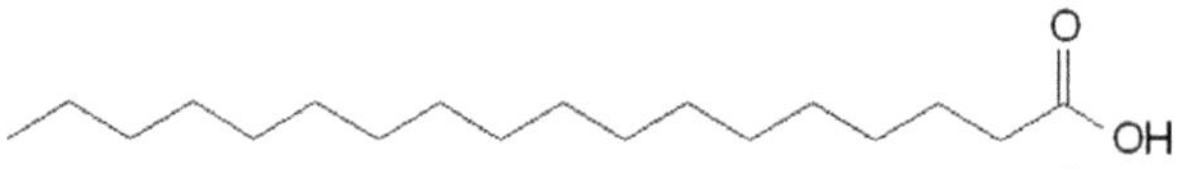
Oleic	18:1	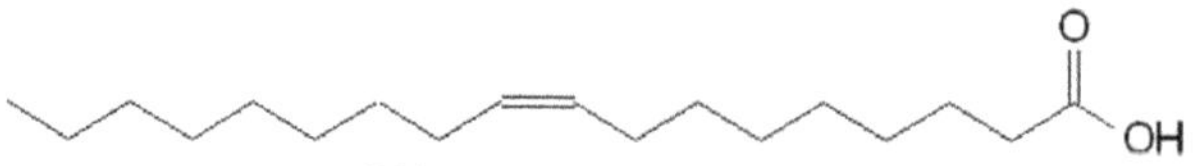
Ricinoleic	18:1	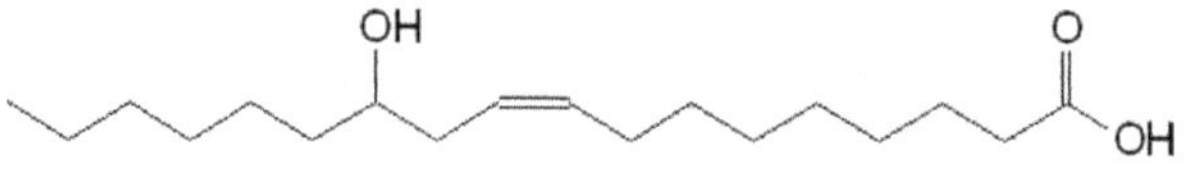
Vernolic	18:1	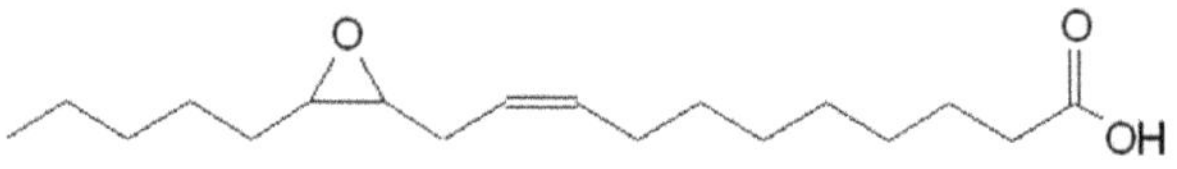
Linoleic	18:2	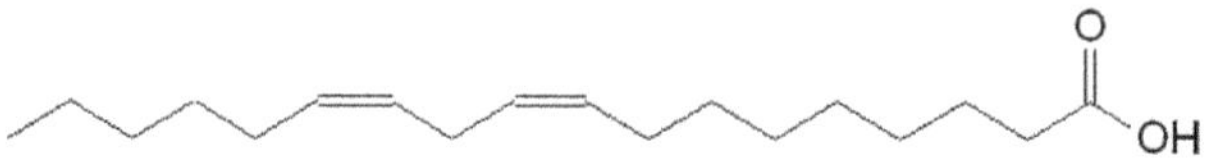
Linolenic	18:3	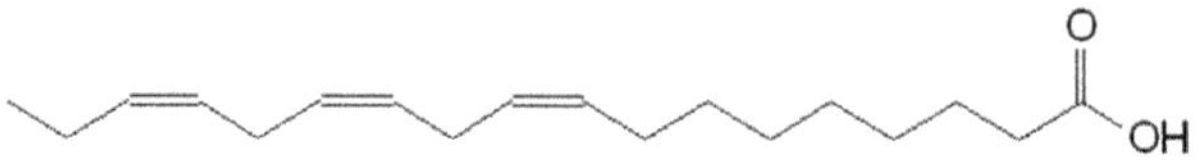
α-Eleostearic	18:3	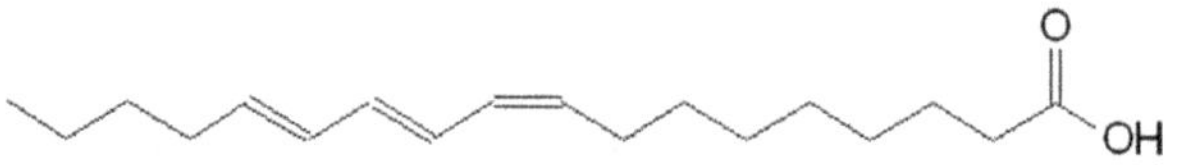
Araquidonic	20:4	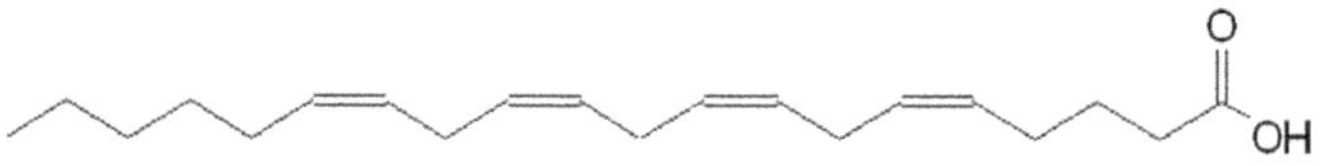

CN is the carbon number and DB is the double bonds number.

TABLE 22.2 Composition of Different Vegetable Oils

Fatty Acid	Rapeseed Oil (%)	Olive Oil (%)	Soybean Oil (%)	Sunflower Oil (%)	Linseed Oil (%)	Castor Oil (%)	Canola Oil (%)	Fish Oil (%)	Microalgae Oil (%)
Palmitic	4	13.7	12	6	5	1.5	4.1	—	13.9
Palmitoleic	—	—	—	—	—	—	—	—	4.7
Stearic	2	2.5	4	4	4	0.5	1.8	—	1.7
Oleic	56	71.1	24	42	22	5	60.9	18.20	11.8
Linoleic	26	10.0	53	47	17	4	21.0	1.10	13.2
Linolenic	10	0.6	7	1	52	0.5	8.8	0.99	11.8
Ricinoleic	—	—	—	—	—	87.5	—	—	—
Arachidonic	—	—	—	—	—	—	—	—	34
References	—	[4]	[5]	[5]	[5]	[5]	[4]	[4]	[6]

22.2 THE PRODUCTION OF BIODIESEL

22.2.1 Biodiesel as Fuel and Raw Chemical Material

Hundred years after the discovery of Rudolph Diesel's motor, scientists evaluate the possibility to obtain alternative fuels, similar to Diesel fuel. It is well known that several fuels are chemical materials normally employed in simple and complex chemical processes. Biodiesel is comprised of esters that are derived from the transesterification of oils and fats with alcohols and has emerged in recent years as a great potential complementary resource to petroleum-based fuels and derivatives. Rudolf Diesel predicted that the use of vegetable oils for engine fuels may seem insignificant during his time but could become as significant as petroleum and coal-tar products in the future [10].

Currently, vegetable oil-derived esters do not have the same importance as fossil derivatives in fuel usage; however, the growing significance of these products is undeniable, agreeing with Diesel's prediction. In this sense, biofuels, especially Biodiesel, can be used in different types of chemical processes and constitute raw material for second generation products including polymers, surfactants, and others. The growth in biodiesel production worldwide led to the development of novel processes that are more rapid, cost-effective, and efficient [11–16]. These new procedures should consider the environmental significance of biodiesel, which is synthesized from a renewable and biodegradable source, to promote green technology for all biodiesel production and distribution, supporting the sustainable development of an ever-growing product [17–21].

One of the chemical treatments of triglycerides of different vegetable and algae oils and animal fats produces the biofuel known as biodiesel. This development was based on the knowledge of the catalytic processes as well as on the evolution of instrumental techniques, which permitted the elucidation of the esterification mechanisms and consequently, the study of transesterification. Homogeneous and heterogeneous processes were studied focusing on basic, acid, and enzymatic catalysis [13, 15, 16, 22–36]. Usually,

two-step procedures were used to prepare biodiesel from different oils with high content of free fatty acids (FFAs) [12, 37–43]. In these studies, the first step is the acid catalyzed esterification, which reduces the FFA content of the oil and minimizes the soap formation in the second step (basic catalyzed transesterification). Other articles reported the biodiesel production by two-step process using supercritical conditions [44–47] or enzymes [48–50]. The alcohol mainly used in the transesterification procedure is methanol but recently, the use of ethanol in the production of ethyl-biodiesel was also reported.

The use of methanol offers the best results in the transesterification of oils and fats. Compared with other alcohols, methanol requires shorter reaction times and smaller catalyst amounts and alcohol/oil molar ratios [10, 12, 15, 16, 51, 52]. These advantages lead to reduced consumption of steam, heat, water, and electricity, and use of smaller processing equipment to produce the same amount of biodiesel. Biodiesel applications continue to expand. Thus, in addition to its use as fuel, biodiesel has been employed in the synthesis of resins, polymers, emulsifiers, and lubricants [53–64]. Concerning the range of applications, new biodiesel production processes should be considered as alternatives to the production based on methanol. Currently, methanol is primarily produced from fossil matter. Due to its high toxicity, methanol may cause cancer and blindness in humans, if they are overexposed to it. Methanol traces are not desired in food and other products for human consumption [15]. In contrast, ethanol emerges as an excellent alternative to methanol as it is mainly produced from biomass, is easily metabolized by humans, and generates stable fatty acid esters. Additionally, fatty acid ester production with ethanol requires shorter reaction times and smaller amounts of alcohol and catalyst compared to the other alcohols, except methanol, used in transesterification processes [11, 15, 16].

The first biodiesel produced was derived from palm oil and ethanol, as described in a patent of Chavanne in 1938 [65]. Since then, different procedures have been developed for the biodiesel production of fatty acid ethyl esters (FAEEs). Given the novel applications for biodiesel, fatty acid ethyl esters may be more suitable than fatty acid methyl esters.

22.2.2 Transesterification

The ideal transesterification with a primary alcohol is given by the overall reaction shown in Figure 22.2. The main feature of the reaction is the transformation of the triglyceride to glycerol and three ester molecules. These esters constitute the components of biodiesel. The reaction includes at least a catalytic component. In reality, the transesterification process never occurs in the ideal form.

A better approximation is given in Figure 22.3, where the chemical agents and products are present without consideration of stoichiometric parameters. So, at the end of the reaction, one must quantify the presence of triglycerides, diglycerides,

monoglycerides, biodiesel esters, glycerol, alcohol, and catalytic components and in the case of basic catalysis, soap moieties. There is no evidence in the literature of 100% conversion efficiency during the transesterification process.

Generally, the basic catalysis presents conversion of ~82%, the acid catalysis provides higher conversion, approximately of 98% but the process is extremely slow [13, 22–24, 27, 28].

22.2.2.1 Acid Catalysis

Acid catalysis transesterification includes the combination of three reversible reactions (Fig. 22.4). The high conversion of the acid catalyzed transesterification procedure is due to the capacity to transesterify fatty acids and fatty acid salts present in the system. The acids employed are HCl, H_2SO_4, BF_3, and sulfonic acids [15, 22–24]. Generally, acid catalysis is many times slower than basic catalysis. The rate of the biodiesel production reaction

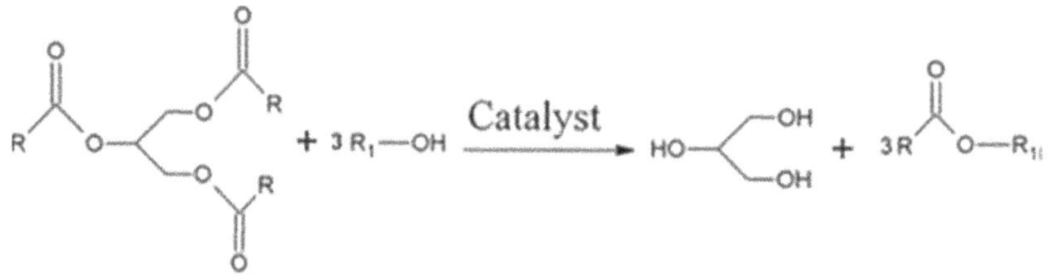

FIGURE 22.2   Ideal process of transesterification.

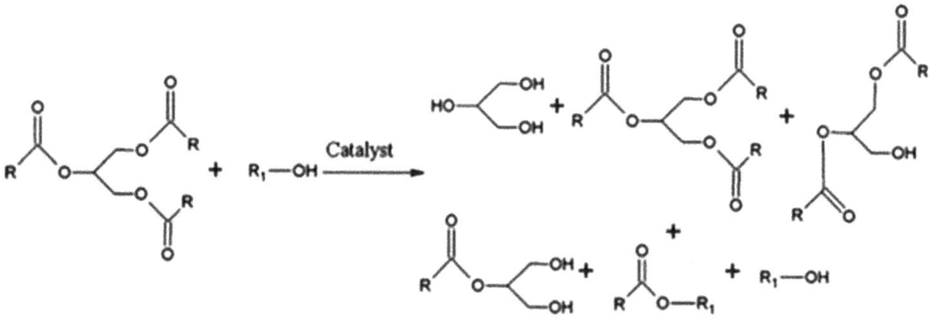

FIGURE 22.3   Real transesterification process.

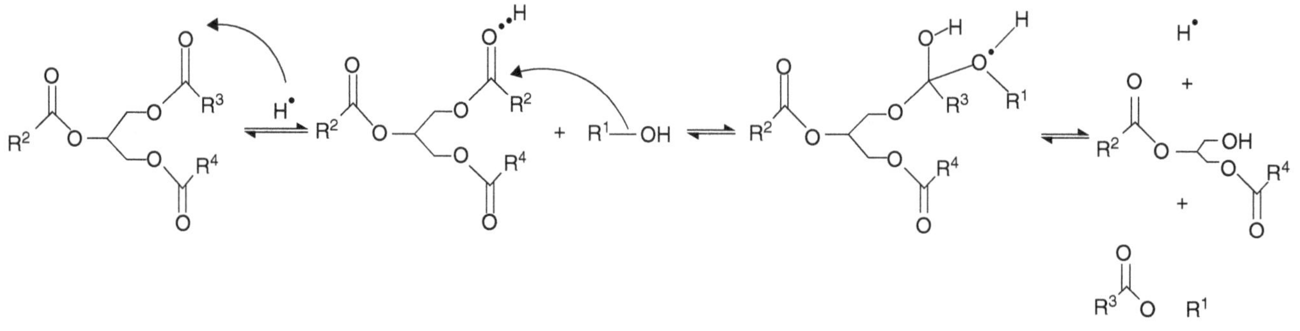

FIGURE 22.4   Mechanism of transesterification according to acid catalysis.

with acid catalyst is related to the conditions of the transesterification reaction. Alcohol/triglyceride molar ratio, temperature, concentration of catalyst, purity of reactants affect the rate and chemical equilibrium [22, 25–27]. The alcohols used in this process are methanol, ethanol, propanol, butanol (*n*, *s*, and *t*), and amyl alcohol [22]. The alcohol/triglyceride molar ratio is essentially large; generally, the optimum is ~12 alcohol molecules for 1 triglyceride molecule. The excess is used in order to shift the reaction equilibrium, to avoid the reverse reactions and to accelerate the process [22, 26, 27]. The temperature influences the velocity of the electrophilic attack; this affects the yields and consequently decreases production cost [13, 22, 26]. The concentration of catalyst is directly related to the yields; in excess, the one promotes reverse reactions, while a low catalyst concentration results in a limited process evolution [13, 25, 28, 33]. The impurities promote parallel reactions; higher temperature and pressure can eliminate this influence, but yields cannot be optimized [25].

22.2.2.2 Basic Catalysis

Basic catalysis is the industrially more used transesterification process. Despite the many advantages as a fast reaction with low alcohol/triglyceride molar ratio and good yields [15, 26], basic catalysis requires more rigid anhydrous conditions than the acid one since the presence of water leads to irreversible hydrolysis of lipids [24–27, 29]. It can form emulsion or soap if the catalyst concentration is higher than needed. Essentially, the species used in this process are potassium hydroxide, sodium hydroxide, and generally Lewis bases, with the exception of the oxides that form heterogeneous systems. As shown in Figure 22.5, the formation of the active species is observed, in a similar manner as in the case of alkoxides. The active species attack the carbonyl moiety in triglyceride, originating a tetrahedral intermediate, from which the alkyl ester and the corresponding anion of the diglyceride are formed [32, 34, 35]. Regeneration of the active species occurs after the liberation of the mono alkyl ester. The formation of the glycerol occurs toward the end of the process. Normally, a small excess of the alcohol is necessary in order to shift the equilibrium to the products and to avoid parallel reactions [26]. The reaction time decreases significantly by using temperatures near the alcohol boiling point [25, 26].

In the following section, a two-step procedure is presented, based on the combination of consecutive basic and acid catalyses [66]. In this procedure, there is no initial acid attack and acid esterification stage for elimination of FFAs as proposed by other studies. The procedure includes consecutive basic and acidic catalytic stages.

22.2.2.3 The Double Step Transesterification Process

A two consecutive steps basic–acid transesterification process (transesterification double-step process—TDSP) was proposed [66, 67] for the biodiesel production from vegetable oils. The process involves homogeneous consecutive basic–acid catalysis steps and is characterized by the formation of well-defined phases, easy separation procedures, high reaction rate, and high conversion efficiency. The proposed TDSP is different from other traditional two-step procedures which normally include acid esterification followed by basic transesterification, or enzymatic, or even supercritical transesterification conditions. The results of the analysis of the biodiesel (fatty acid methyl esters) by standard biodiesel techniques in addition to ^{1}H-NMR indicated high-quality and pure biodiesel products. The transesterification of sunflower and linseed oils resulted in oil conversion higher than 97% corresponding to yields of 85%.

- *TDSP Procedure for Production of Methyl Ester Biodiesel.*

  The procedure is a sequence of operations, performed approximately in 4 h. Initially, the alkali agent is dissolved in methanol (25 g KOH in 1 l methanol) at a temperature near 45 °C allowing the formation of the basic catalysis active species. Then, 40 ml of this solution and 100 ml of vegetable oil, under vigorous and constant agitation, are introduced in a simple reactor equipped with reflux device. The molar ratio of alcohol/oil is 10 and catalyst/alcohol is 1.78×10^{-2}. The temperature is increased until 60 °C, close to the alcohol boiling point and maintained for ~1 h, after which the system is cooled down to 25 °C. The second step of the procedure consists in the addition into the reaction mixture of 60 ml methanol and 1.5 ml sulfuric acid, 18 mol l^{-1}, followed by slow heating. The temperature is increased to 60 °C using the reflux equipment while the mixture is under constant agitation. After 1 h, the system is cooled slowly to 25 °C. At the end of this step, the formation of

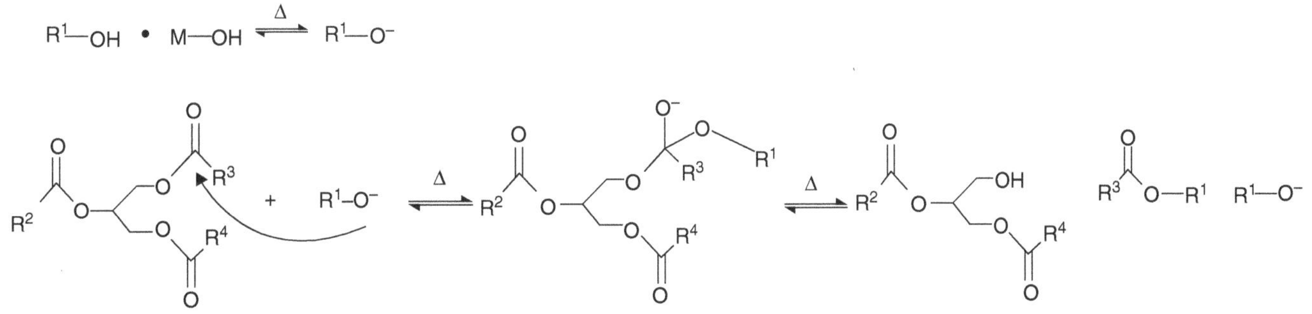

FIGURE 22.5 Mechanism of transesterification according to basic catalysis.

two phases occurs. The phases are separated and processed further. The biodiesel phase is washed with cold water and the residual alcohol is removed by evaporation under vacuum. The lower phase, with a pH ~6, is used for recuperation of the methanol excess, the glycerol, as well as other secondary products.

- *TDSP Procedure for Production of Ethyl Ester Biodiesel.* The procedure is a sequence of operations performed in ~4h [68]. Initially, 120 ml of ethanol is introduced in a simple reactor equipped with a reflux device and stabilized temperature (65 °C). Potassium hydroxide (2.0 g) is added to the ethanol and the mixture is vigorously and constantly agitated until the potassium is completely dissolved, allowing the formation of the basic catalyst active species. The agitation is maintained while 100 ml of vegetable oil is added to the reaction vessel. The alcohol/oil/catalyst molar ratio is 20:1:0.35. After 30 min, during the second main step, 4 ml of sulfuric acid (P.A., 18.77 mol l⁻¹) is added dropwise to the reaction mixture, followed by a slow heating

to 80 °C of the mixture and the addition of 60 ml of ethanol. After 2h 30 min, the reaction mixture was removed from the reflux system and was immediately filtered to remove the solid residue (K_2SO_4). The reaction mixture was then concentrated using a rotary evaporator to remove the alcohol excess. This specific procedure promoted a rapid and clear separation in two phases. The ethyl esters (biodiesel) were located in the upper phase, with some traces of nonreacted oil, monoglycerides, diglycerides and a small amount of ethanol. This phase was washed with water and dried under anhydrous sodium sulfate or through heating in a temperature range between 95 and 100 °C. The lower phase was composed of glycerol and a little amount of ethanol. The excess ethanol was recovered from vacuum evaporation and stored for further use.

- *The Mechanism for the Transesterification Double-Step Process.*

The mechanism related to this process (Fig. 22.6) can be described as a basic–acid consecutive transesterification

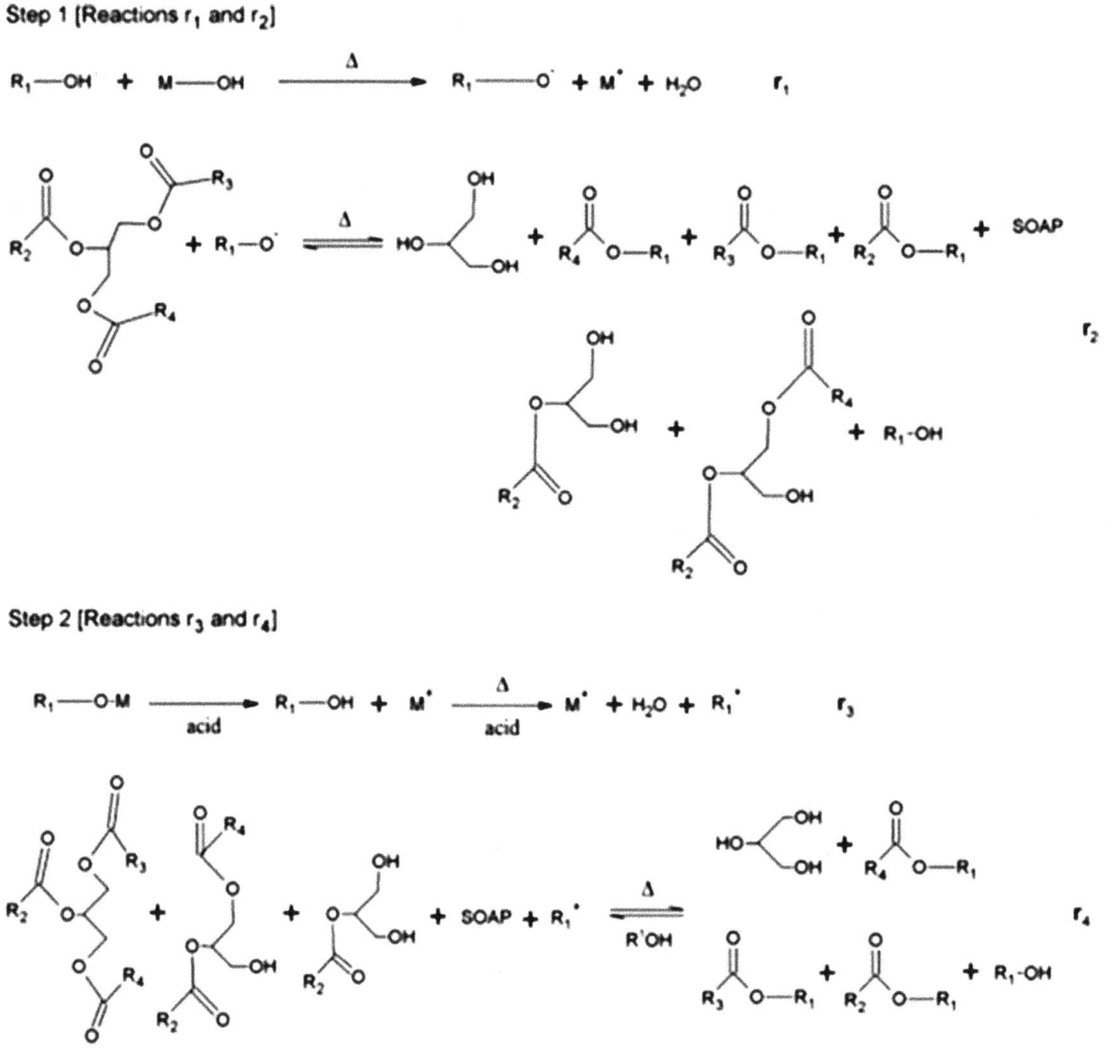

FIGURE 22.6   A suggested mechanism for the transesterification double-step process (TDSP).

mechanism [66] incorporating the main characteristics and principles of the basic and acid catalyses. The first two reactions (r_1 and r_2) represent the first step and describe the typical basic catalysis process. The reaction r_1 is related to the alkoxide preparation while the second one corresponds to the basic transesterification. The mechanism of these reactions does not require additional explanation, the basic catalysis principles being described in the preceding. The second operational step occurs with the addition of sulfuric acid and alcohol in the reaction mixture with pH between 12 and 14; in other words, in the absence of H⁺. The reactions r_3 and r_4 occur after the addition of the acid catalyst and alcohol. As described, in the reaction r_3, the H⁺ active species produced by the addition of the acid is responsible for the neutralization and simultaneous production of the R_1^+ species, which attack directly the carbonylic groups. The difference between this acid catalysis step and the classical acid one is that the acid is added in to the reaction mixture, which includes alkoxide and thus has a very high pH value. In the classical acid catalysis, described previously, the acid/alcohol solution is added into the neutral triglyceride. The very active species R_1^+ reacts immediately, according to reaction r_4, with mono-, di-, and residual triglycerides and with soap already present in the reaction mixture, leading to a fast process and very high conversion degree, proved by the purity of the final product. It is important to mention that, according to the suggested mechanism, the fatty acid methyl esters produced by the basic catalysis process do not suffer any degeneration by the acid catalysis.

22.2.3 Techniques for Characterization of Triglycerides and Biodiesel

The oils and the biodiesel products of the transesterification procedures are mainly characterized by nuclear magnetic resonance (¹H-NMR) and gas chromatography (GC) techniques. The ¹H-NMR technique provides chemical characteristics of the oils, fats, and products and the conversion degrees of the transesterification procedures. GC allows a more accurate characterization of the molecular species involved in the transesterification procedure. Additionally, the "Analysis Biodiesel Protocol" for the characterization of the methyl or ethyl biodiesel must include information of the following physicochemical techniques: kinematic viscosity, density, flash point, cloud point, pour point, cold filter plugging point, free and total glycerol, ethanol residue, sulfur content, acid number, oxidative stability, and refractive index.

22.2.3.1 The ¹H-NMR Technique ¹H-NMR spectra of linseed oil and biodiesel product are shown in Figure 22.7 and Figure 22.8, respectively. In both figures are shown general schemes, which include the characteristic NMR shift values of protons for vegetable oils and methyl esters. As shown in Figure 22.7, the methylene and the methine protons of the glyceride sequence –CH₂–CH–CH₂– can be observed in the range between 4.22–4.42 and 5.37 ppm, respectively [69]. Qualitative analysis of Figure 22.8 indicates the conversion of the triglycerides to methyl esters due to the disappearance of the resonance signal between 4.22 and 4.42 ppm and the emergence of the new signal at 3.67 ppm [21]. The same ¹H-NMR characteristics are

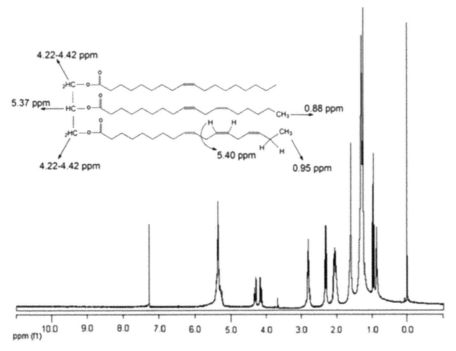

FIGURE 22.7 ¹H NMR spectrum of linseed oil and a general scheme, which includes the NMR shift values of protons for a vegetable oil.

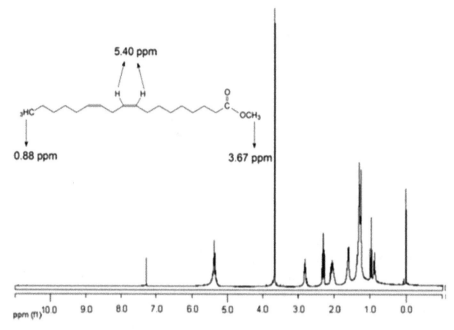

FIGURE 22.8 ¹H NMR spectrum of biodiesel from linseed oil and a general scheme, which includes the NMR shift values of protons for a methyl ester.

observed in the transesterification of different fats and oils. A reliable methodology for ¹H-NMR characterization of mono, di, and tri acylglycerides, as well as fatty acid methyl esters was established during the last decade [16, 30–32, 36].

The aforementioned literature presents fundamentals for the calculation of the conversion, C_{ME} (in %), where ME responds to the methyl ester. According to the ¹H-NMR technique, the normalized integrated intensity of the glyceridic (4.22 and 4.42 ppm) or the methyl ester hydrogens (3.67 ppm) has to be considered. The conversion C (in %) of vegetable oil to biodiesel can be calculated using the following equation:

$$C = 100 \times \left[\frac{I_{\text{i–TAG}} - I_{\text{f–TAG}}}{I_{\text{i–TAG}}} \right] \qquad (22.1)$$

where $I_{\text{i–TAG}}$ and $I_{\text{f–TAG}}$ are the initial and final normalized integrated intensities of the signal in the range between 4.22 and 4.42 ppm.

Considering the integration values of the glyceridic (TAG), the methyl ester (ME) hydrogens, and the methylene groups adjacent (α-CH$_2$) to the ester group NMR signals, Equations 22.2 and 22.3 allow the conversion calculation:

$$C_{ME} = 100 \times \left[\frac{4 \times I_{ME}}{4 \times I_{ME} + 9 \times I_{ATG}} \right] \qquad (22.2)$$

$$C_{ME} = 100 \times \left(\frac{6 \times I_{ME}}{9 \times I_{\alpha\text{-CH}_2}} \right) \qquad (22.3)$$

where I_{ME} is the integration value of the methyl ester peak and I_{TAG} is the integration value of the glyceridic peaks in the TAGs of the vegetable oil.

The factors 9 and 6 come from the fact that the three methyl ester moieties resulting from one triglyceride molecule have nine hydrogens (3.67 ppm) and the three methylene groups adjacent to the ester group in both vegetable oil and biodiesel have six hydrogens (2.31 ppm) [30]. With respect to Equation 22.2, some authors have erroneously used the factor 5, considering that the glycerol moiety of a triglyceride has five protons contributing at 4.22–4.42 ppm; however, according to literature [69], the methane hydrogens present signal at 5.37 ppm and the correct factor must be 4. Note that mono- and diacylglyccrides, which are formed as intermediates in the transesterification reaction, exhibit signals of their glyceridic hydrogens in the same region as glyceridic hydrogens of the triacylglycerides in the feedstock.

The normalization factor, NF, can be obtained by using one or the sum of some peaks that do not change after reaction. For this reason, the signal at 2.31 ppm is chosen, which corresponds to the methylene hydrogens adjacent to the ester group in both vegetable oil and biodiesel. The terminal methyl group might also be used. According to Vigli et al. [69] and Gelbard et al. [32], it appears as a superposition of triplets in the range between 1.1 and 0.88 ppm, as indicated in Figure 22.8. More specifically, the end methyl triplet of tri-unsaturated fatty acids and esters is centered at 0.95 ppm while all the other appears at 0.88 ppm.

The evaluation of the ¹H-NMR data of linseed oils and the corresponding biodiesel products indicates conversion degrees of vegetable oil to fatty acid methyl ester higher than

97%. As seen in Figure 22.8, the ¹H-NMR spectrum indicates only the presence of pure fatty acid methyl ester. There is no presence of triglyceride (4.22–4.42 ppm) and di- or mono-glycerol (small peaks at 3.4–4.1 ppm) evidenced. In other words, the proposed procedure leads to high purity fatty acid methyl ester and this is coherent with the final elevated value of methanol/triglyceride ratio used for this purpose.

The yield, α (w/w %), by using the TDSP procedure, can be calculated by Equation 22.4 taking into account the initial volume of the used oil (V_{oil}), the volume of the biodiesel (VBD), the corresponding densities (ρ_{oil} and ρ_{BD}), and the conversion degrees (X_{ME}) obtained by the ¹H-NMR technique.

$$\alpha = 100 \times \left(\frac{V_{BD} \times \rho_{BD} \times X_{ME}}{V_{oil} \times \rho_{oil}} \right) \qquad (22.4)$$

Equation 22.4 offers a good approximation of the yield with a small error, lower than 0.4%. In spite of very high conversions values, the yields for TDSP procedures were lower than 90% (the average α values for six experiments are: $\alpha_{sunflower}$ 86 ± 5% and α_l in seed 85 ± 4%). The results are related to the purity of the starting materials, the nonanhydrous conditions and the losses of material inherent to procedure handling. Nonetheless, these values are significantly higher than the basic catalysis industrial yields and lower than the acid catalysis, which makes it difficult to apply for industrial purposes.

Figure 22.9 shows the ¹H-NMR spectrum of the generated product after complete sunflower oil conversion to ethyl esters. The completely converted products appear at chemical shift (δ=4.15 ppm). The disappearance of signals at ~4.3 ppm confirms that there is no residual mono-, di-, and triacylglycerides. The integration data of some of the presented peaks were used to evaluate the oil conversion to the ethyl esters (CEE). The peak that appears at 4.15 ppm, regarding the ethyl ester CH_2 group corresponds to the biodiesel content. The peak at 2.30 ppm was chosen to represent all possible oil derivatives because it remains unchanged for the biodiesels and mono-, di-, and triacylglycerides, soaps, and all other species formed. This signal was easily identifiable, well defined, and was not superimposed with the other hydrogen signals. This region corresponds to the methylene group hydrogen atoms that are adjacent to the carbonyl (α-CH_2), identified as E in Figure 22.9. The proposed ethyl ester conversion calculation methodology is based on a reliable ¹H-NMR characterization strategy of mono-, di-, and triacylglycerides, as well as fatty acids methyl esters; this methodology was established during the last decade by numerous contributions [30–32, 36, 69–74].

Equation 22.5 describes the oil reaction conversion to biodiesel, relating the NMR integration values of the ethyl ester CH_2 hydrogen atoms (EE–CH_2, δ=4.15 ppm) to the methylene group hydrogen atoms that are adjacent to the carbonyl group (α-CH_2, δ=2.30 ppm) [68]:

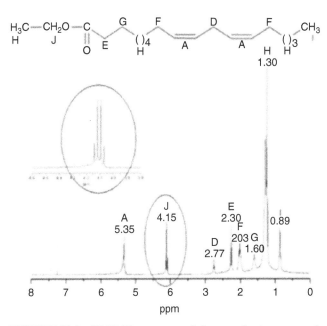

FIGURE 22.9 ¹H NMR spectrum of the completely converted sunflower oil to biodiesel.

$$C_{EE} = \left(\frac{I_{EE-CH_2}/2}{I_{\alpha-CH_2}/2} \right) \qquad (22.5)$$

where I_{EE-CH_2} is the integration value of the ethyl ester peak and $I_{\alpha-CH_2}$ is the integration value of the methylene group that is adjacent to the carbonyl. The factor of 2, which divides both integration values, is the normalization factor.

22.2.3.2 *The GC Technique* Figure 22.10 includes three chromatograms which allow the quantitative analysis of materials and products related to the transesterification, as well as to glycerolysis procedures [75].

The chromatogram A is composed by the external standards linolenic acid methyl ester, monoolein, diolein, and triolein including the internal standard tricaprin. The chromatogram B includes biodiesel and internal standard tricaprin. The chromatogram C shows a sample which includes linolenic acid methyl ester (a) appears in 10.4 min retention time. The biodiesel (in chromatogram B) includes a mixture of C16 and different C18 methyl esters (a_1, a_2, a_3) with retention times between 9 and 13 min. These peaks (a_1, a_2, a_3) appear in the chromatogram C as residual biodiesel. Monoacylglycerols (b_1, b_2) appear in chromatogram A in the range between 14 and 17 min. Small amounts of monoacylglycerols (b_1, b_2) are present in chromatogram B and a significant increase is observed in chromatogram C. The presence of tricaprin (c_1, c_2, c_3) in all three chromatograms ensures the accuracy of the retention times. In chromatogram A, diolein appears (d) after 25 min and, as it can be seen, there are no traces of diacylglycerols in bio-

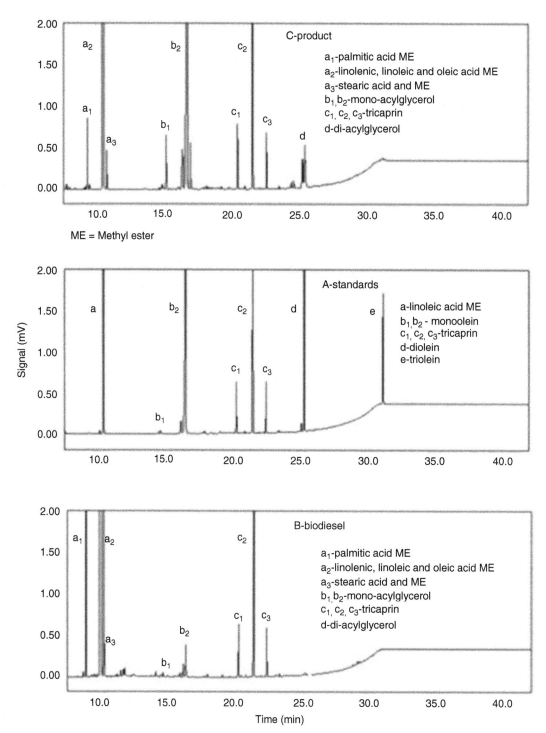

FIGURE 22.10 Three chromatograms including different tri, di, and mono acylglycerides; different fatty acid methyl esters; and tricaprin as internal standard.

diesel and a small quantity is detected in chromatogram C. The triolein (e) is detected in chromatogram A after 31 min but triacylglycerols are not present in the used biodiesel or the analyzed sample. In order to evaluate quantitatively the obtained products, the chromatography response factors (r_F) of the involved species have to be considered.

22.3 POLYMERIZATION

The use of renewable resources, such as triglyceride oils, pure fatty acids, and fatty acid esters (biodiesel), is attractive due to their great potential for replacing petrochemical derivatives. In addition to the gradual increase of

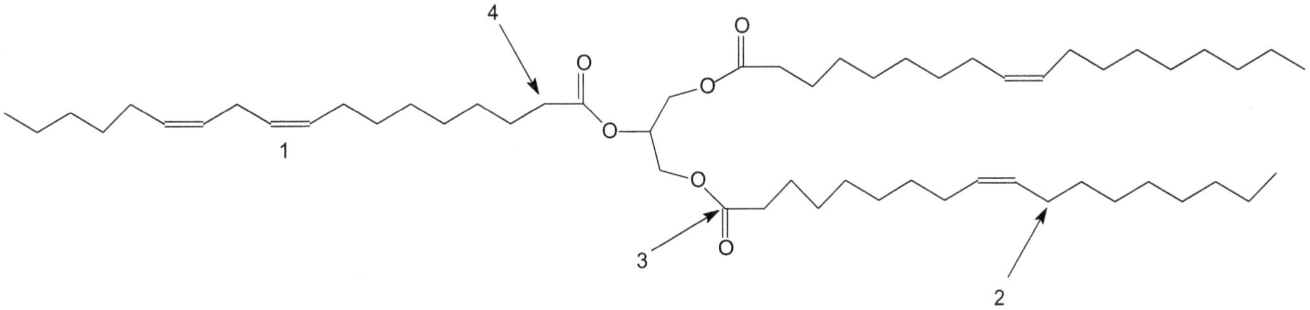

FIGURE 22.11 Different functionalities of triglyceride molecule.

petroleum prices and the decline of its reserves, the motivation for new materials from renewable resources includes lower costs, biodegradability, ready availability, and production periodicity. The diversity of vegetable oil applications highlights their importance in polymer synthesis and justifies the development of new technologies in order to promote changes in its structure. Depending on the fatty acid distribution, triglycerides and the materials produced from them have specific physical and chemical properties.

The fatty acid molecules include a variety of reactive sites for functionalization: double bonds, allylic carbons, ester groups, and α-carbon to ester group [6]. These positions are illustrated in Figure 22.11. The C=C double bonds of the biodiesel and triglycerides can be polymerized directly or after functionalization.

22.3.1 Direct Polymerization of Double Bonds

22.3.1.1 Oxypolymerized Oils Direct polymerization of double bonds can be achieved by oxidative polymerization via the free radical mechanism. Oils containing high content of unsaturation (drying and semidrying) can be oxidized with the atmospheric oxygen to form highly branched materials or cross-linked polymers [76].

The oxidative polymerization mechanism of drying oil is illustrated in Figure 22.12 [77]. In the initiation step, naturally present hydroperoxides decompose to form free radicals. In the propagation step, the free radicals can react with the fatty acid chain of the oil. The hydrogen abstraction on a methylene group located between two double bonds in polyunsatured fatty acid chain, leading to free radical as **1**. The free radical **1** can react with oxygen to form the peroxy free radical **2**, which can be conjugated or nonconjugated. Cross-linking proceeds by radical–radical termination reaction. The initiation step can be catalyzed by driers or heat. Driers are metallic salts that shorten the induction time of the drying process and accelerate the rate of oxygen absorption [77].

22.3.1.2 Polymerization of Triglycerides by Cationic Polymerization Triglycerides can be made by cationic polymerization. Boron trifluoride diethyl etherate (BF$_3$-OEt$_2$), a Lewis acid, is commonly used initiator for the cationic polymerization of alkenes.

Figure 22.13 presents the mechanism for the cationic polymerization of simple alkenes [78, 79]. According to this mechanism, the catalyst reacts with a small amount of water, producing a boron trifluoride hydrate. Next, the hydrate reacts with the alkene to form the carbocation–counterion pair complex. The olefin cationic polymerization proceeds by the attack of the electrophilic carbocation on the π system of the monomer. The triglyceride double bonds represent sites for electrophilic attack of the reactive species generated by the catalyst. In general, the homopolymerization of vegetable oils leads to low molecular weight viscous oils or weak rubbery materials [78].

The cationic copolymerization of vegetable oils with vinylic monomers such as divinylbenzene or a combination of styrene/divinylbenzene yield polymeric materials ranging from elastomers to rigid and tough plastics [80].

22.3.1.3 Copolymerization with Vinyl Polymers via Free Radical Polymerization Another option for direct modification of triglycerides, pure fatty acids, and biodiesel is the reaction of drying or semidrying oils with vinyl monomers. Styrene is an important vinyl monomer for the copolymerization of triglycerides. Styrenation process of oils is used in the production of oil-based binders. These materials have important characteristics, such as good stability toward pigments, excellent suspension medium for pigments, good water resistance, and good electrical properties [81]. The reaction between styrene and triglycerides involves free radical initiated polymerization. Peroxide catalysts are effectives for this purpose. Conjugated oils are more reactive than the nonconjugated ones.

Figure 22.14 presents the reaction mechanism [5] for the copolymerization reaction between styrene and conjugated and nonconjugated vegetable oils. In order to minimize the

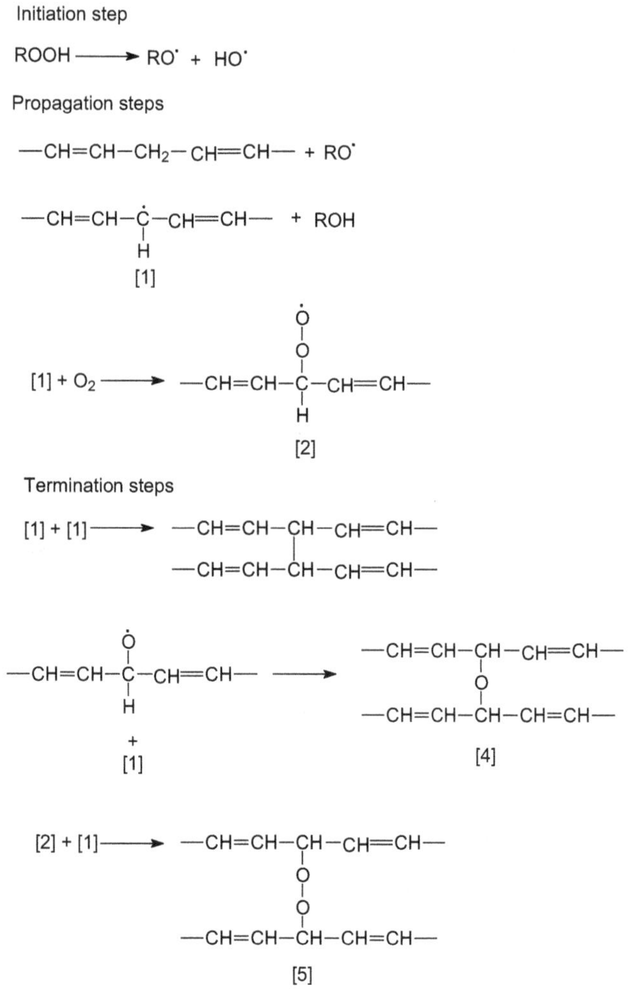

Initiation step

ROOH ⟶ RO· + HO·

Propagation steps

—CH=CH—CH₂—CH=CH— + RO·

—CH=CH—Ċ—CH=CH— + ROH
 　　　　　|
 　　　　　H
 　　　　[1]

[1] + O₂ ⟶ —CH=CH—C—CH=CH—
 　　　　　　　　　　　|
 　　　　　　　　　　　H
 　　　　　　　　　　[2]

Termination steps

[1] + [1] ⟶ —CH=CH—CH—CH=CH—
 　　　　　　　　　　　|
 　　　　　　—CH=CH—CH—CH=CH—

—CH=CH—C—CH=CH— ⟶ 　　—CH=CH—CH—CH=CH—
 　　　　　|　　　　　　　　　　　　　　|
 　　　　　H　　　　　　　　　　　　　O
 　　　　　+　　　　　　　　　—CH=CH—CH—CH=CH—
 　　　　[1]　　　　　　　　　　　　[4]

[2] + [1] ⟶ —CH=CH—CH—CH=CH—
 　　　　　　　　　　　|
 　　　　　　　　　　　O
 　　　　　　　　　　　|
 　　　　　　　　　　　O
 　　　　　—CH=CH—CH—CH=CH—
 　　　　　　　　　[5]

FIGURE 22.12 Oxidative polymerization mechanism of drying oil. With permission of Springer from Li F, Larock RC. Novel polymeric materials from biological oils. J Polym Environ 2002;10:59–67.

styrene homopolymerization, number and position of the double bonds are important aspects. The reaction can be performed with or without solvent [78].

22.3.2 Functionalization of Fatty Acid Derivatives

22.3.2.1 *Double Bonds Functionalization: Epoxidation and Hydroxylation* Fatty acid derivatives such as biological oils (plant, animal, and algae), as well as fatty acids and fatty acid esters (biodiesel) are excellent renewable sources for the preparation of different polymeric materials. However, high-strength materials can be obtained by improving the reactivity of the double bonds, introducing other functionalities to the molecules. Epoxides and hydroxyls have an important role in the functionalization of fatty acid derivatives [82–85]. The polymerization degree depends on the unsaturation degree of the oil as well as the number of the functional groups introduced in the fatty acid molecules.

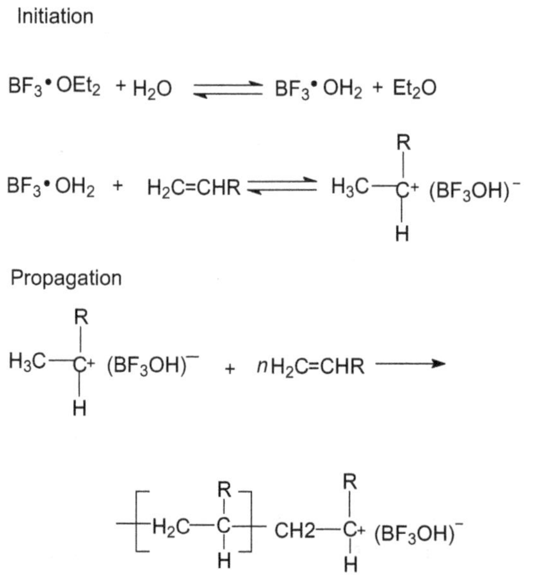

Initiation

BF₃·OEt₂ + H₂O ⇌ BF₃·OH₂ + Et₂O

 　　　　　　　　　　　　　　　　　R
 　　　　　　　　　　　　　　　　　|
BF₃·OH₂ + H₂C=CHR ⇌ H₃C—C⁺ (BF₃OH)⁻
 　　　　　　　　　　　　　　　　　|
 　　　　　　　　　　　　　　　　　H

Propagation

 　R
 　|
H₃C—C⁺ (BF₃OH)⁻ + nH₂C=CHR ⟶
 　|
 　H

 　　　　　R　　　　　　　R
 　　　　　|　　　　　　　|
—[H₂C—C]— CH2—C⁺ (BF₃OH)⁻
 　　　　　|　　　　　　　|
 　　　　　H　　　　　　　H

FIGURE 22.13 Mechanism for the cationic polymerization of simple alkenes. With permission of Elsevier from Andjelkovic DD, Valverde M, Henna P, Li F, Larock RC. Novel thermosets prepared by cationic copolymerization of various vegetable oils – synthesis and their structure–property relationships. Polymer 2005;46:9674–9685 and Kabasaka OS. Styrenation of mahaleb and anchovy oils. Prog Org Coat 2005;53:235–238.

Some plant oils have, naturally, high functionality and can be used without modifications to produce different polymeric materials. Such is the case for *V. galemensis*, a natural epoxidized plant, which contains 60–65% vernolic acid. Castor oil is a hydroxyl-containing oil with ~90% of ricinoleic cid.

Epoxides are very versatile structures in organic synthesis and are susceptible to a large number of nucleophilic, eletrophilic, acidic, alkaline, and reducing and oxidizing agents. Because of their versatility, the epoxides are susceptible to polymerize when submitted to heat in the presence of reticulation agents, which are, generally, amines and anhydrides [86].The epoxidation of olefinic compounds can be performed by organic peracids [87–91], transition metals [92–96], enzymes [97], and molecular oxygen [98].

The epoxidation of fatty acids and their derivatives using organic acids is the most used process in industry. According to this methodology, the oxidant agent is a peracid formed *in situ* from hydrogen peroxide and a carboxylic acid. The peracid reacts with the unsaturated bonds, forming the epoxy ring. Carboxylic acids, such as acetic acid and formic acid, are mostly used for this purpose. The reaction using acetic acid requires the presence of an inorganic acid, as sulphuric acid, to catalyze the peracid formation (Fig. 22.15). However, the use of strong mineral acid leads to undesirable by-products

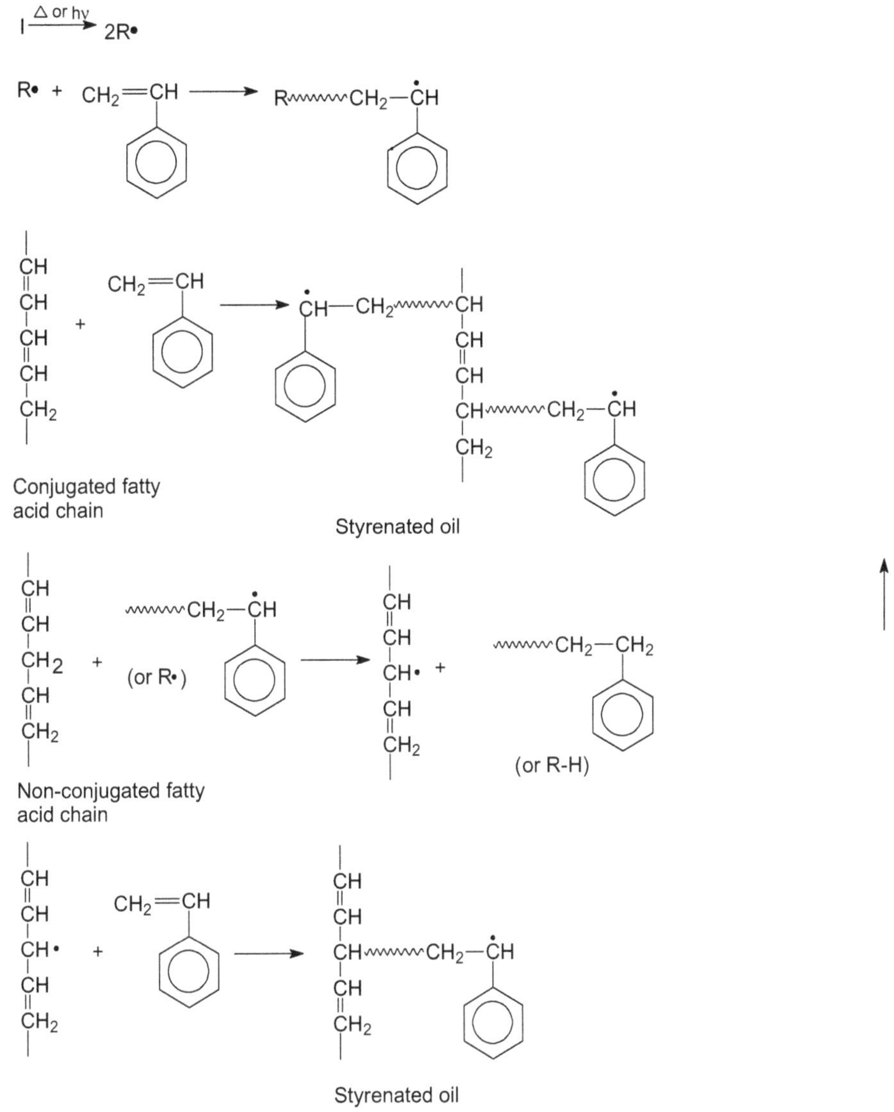

FIGURE 22.14 Copolymerization reaction between styrene and conjugated and nonconjugated vegetable oils [5].

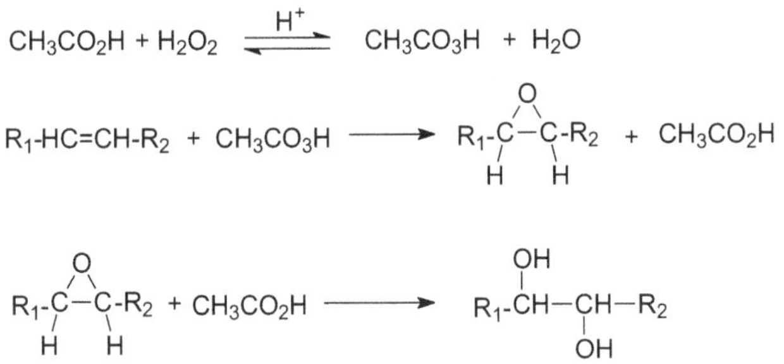

FIGURE 22.15 Epoxidation with peracetic acid.

such as diols. The formic acid minimizes the ring-opening reactions [99].

NMR spectroscopy is an important technique used to verify the epoxidation reactions. Figure 22.16a and b show, respectively, the ^{1}H-NMR spectra of the oleic acid and the epoxidized oleic acid. The complete disappearance of vinylic hydrogen at $\delta=5.3$ ppm and the appearance of epoxide hydrogen at $\delta=3.0$ ppm indicate the conversion of the double bonds to epoxides. Additionally, the disappearance of the peak at $\delta=2.02$ ppm and the appearance of the peak at $\delta=1.5$ ppm also confirm the oleic acid epoxidation.

The epoxidation reactions can also be confirmed by ^{13}C-NMR. Figure 22.17a and b show, respectively, the ^{13}C-NMR spectra of the oleic acid and the epoxidized oleic acid. The disappearance of vinylic carbons at $\delta=130$ ppm of oleic acid and the appearance of epoxidilic carbons at $\delta=54.3$–57.2 ppm in the ^{13}C-NMR spectra of epoxidized oleic acid are observed.

Similar peaks are observed in the ^{1}H-NMR and ^{13}C-NMR for the epoxidation of vegetable oils, pure fatty acids, or biodiesel.

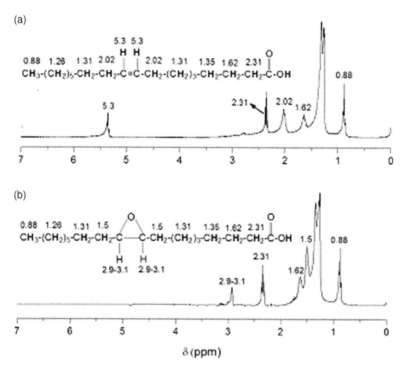

FIGURE 22.16   ¹³C NMR spectra of the (a) oleic acid and (b) epoxidized oleic acid.

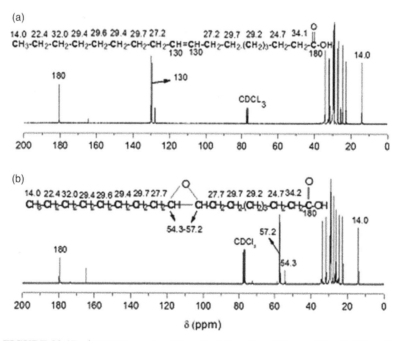

FIGURE 22.17   ¹H NMR spectra of the (a) oleic acid and (b) epoxidized oleic acid.

The most common way to introduce hydroxyl groups in unsaturated fatty acids is the epoxide ring opening with a proton donors such as organic and inorganic acids, alcohols, water, or hydrogenation (Fig. 22.18) [6, 100].

Another alternative method to hydroxylation of fatty acids is the hydroformylation, which allows the introduction

of aldehyde groups (Fig. 22.19) [6, 101]. The aldehyde functionality can be reduced to produce OH groups.

Polyols of a given triglyceride prepared via epoxidation and hydroformylation reactions are illustrated in Figure 22.20. The hydroformylation allows the introduction of an extra carbon per double bond. Additionally, the primary OH groups

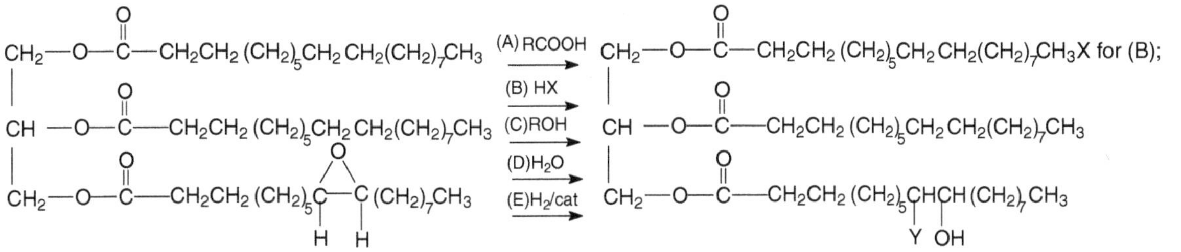

FIGURE 22.18 Hydroxylation of an epoxidized vegetable oil, where Y = –OC(O)R for (A), Cl or Br for (B), –OR for (C), –OH for (D), and –H for (E) [6].

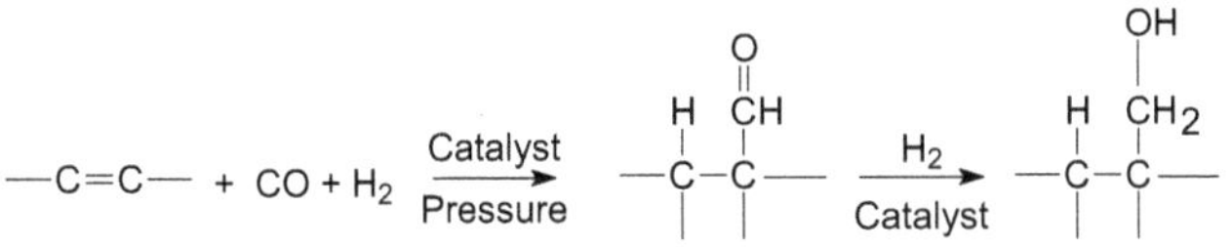

FIGURE 22.19 Hydroxylation reaction. With permission from Petrović ZS. Polymers from biological oils. Contemp Mater 2010;I–1:39–50 and Kandanarachchi P, Guo A, Petrovic Z. The hydroformylation of vegetable oils and model compounds by ligand modified rhodium catalysis. J Mol Catal A: Chem 2002;184:65–71.

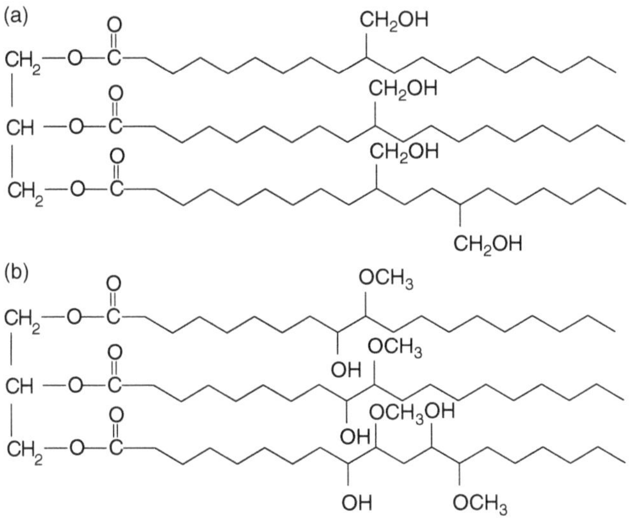

FIGURE 22.20 Polyols of a given triglyceride prepared via (a) epoxidation reaction and (b) hydroformylation reaction. With permission from Petrović ZS. Polymers from biological oils. Contemp Mater 2010;I–1:39–50.

of these polyols are more reactive than the ones obtained via epoxidation.

Other means to introduce polymerizable groups on the fatty acids is the acrylation of the epoxide rings (Fig. 22.21) [4, 103]. In this sense, the reaction of epoxidized fatty acids with acrylic acid incorporates acrylate groups on the fatty acid structure. Next, the monomers produced can be copolymerized with styrene using a free radical initiator to produce cross-linked materials.

Similarly, fatty acids can be modified using maleic anhydride [102, 103]. An example is shown in Figure 22.22 [102].

22.3.2.2 Functionalization via Glycerolysis: Monoglyceride Production
Monoglycerides or monoacylglycerols can be synthesized from glycerol using triglycerides [104], FFAs [105], pure fatty acid methyl esters [106, 107], and biodiesel [75]. The reactions can be performed by basic or acid homogeneous catalysis, using catalysts as KOH, Ca(OH)$_2$, and H$_2$SO$_4$.

Monoglycerides are widely manufactured by the glycerolysis of vegetable oils or animal fats at high temperatures (approximately ~210–250 °C) using an inorganic alkaline catalysts under nitrogen gas atmosphere [108].

A mixture of monoglycerides, diglycerides, and triglycerides is obtained in all the homogeneously catalyzed processes. Figure 22.23 shows the glycerolysis reaction for a fatty acid methyl ester. In this case, methanol is also formed in the reaction. The reaction counts for ~40–60% monoglyceride, the rest being diglyceride and triglyceride [109]. Therefore, several purification steps are required to obtain monoglyceride with high grade purity, such as neutralization of the reaction media followed by molecular distillation [110].

Monoglycerides have a hydrophilic head and a hydrophobic tail having surfactant and emulsifying properties that find application in food, detergent, plasticizer, cosmetic, and pharmaceutical formulations [111]. They are very important monomers for polymers production, such as alkyd resins and polyurethanes, being used without [112] or after modification [103]. An example is the maleinated monoglyceride (Fig. 22.24). The monomer is synthesized from the triglyceride in two steps. The first one is the glycerolysis to convert triglycerides into monoglycerides, which can be reacted with maleic anhydride [103].

22.3.3 Polymers from Functionalized Triglycerides and Functionalized Biodiesel

Different polymeric materials such as alkyd resins, polyesters, and polyurethanes can be prepared from triglycerides and biodiesel after functionalization. As the synthesis of monomers and polymers from vegetable oils has several industrial applications, research in this area is widely

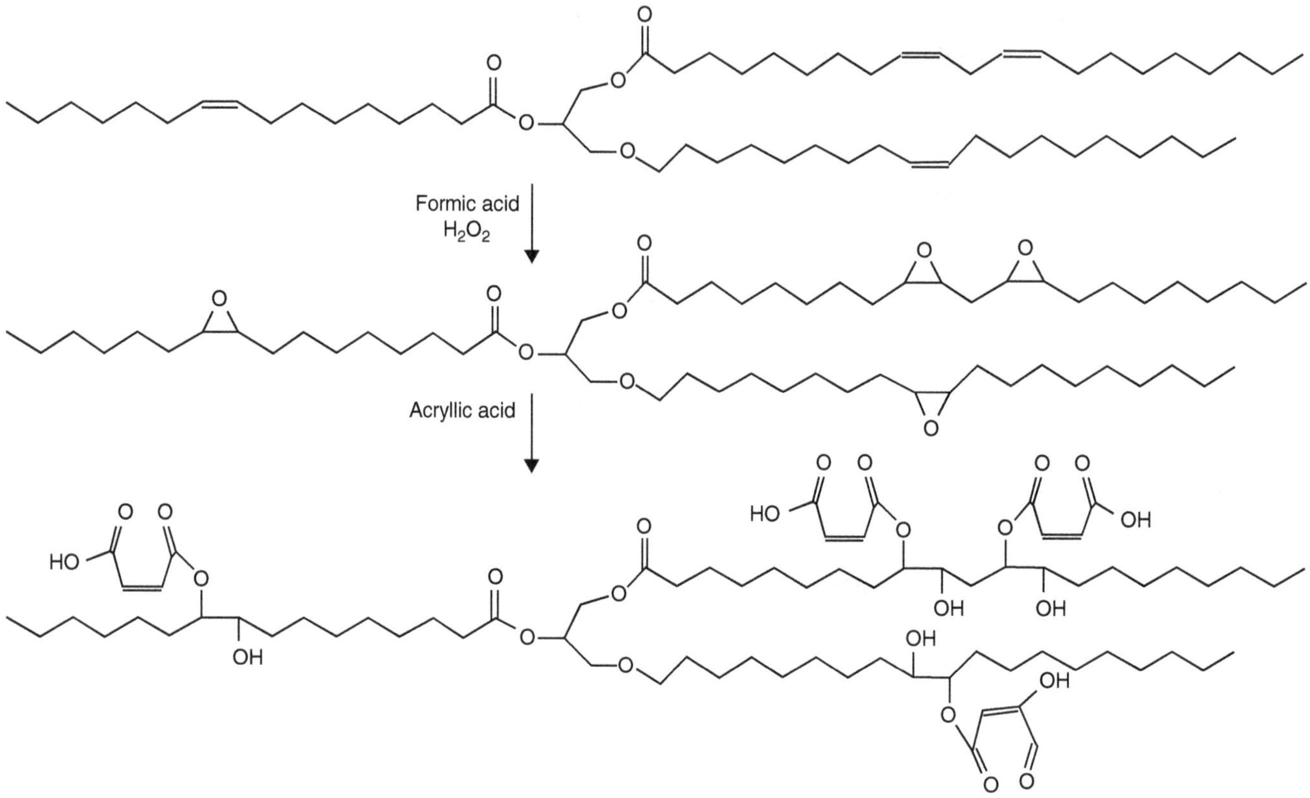

FIGURE 22.21 Acrylation of the epoxidized vegetable oil. With permission of Elsevier from Lu J, Khot S, Wool RP. New sheet molding compound resins from soybean oil. I. Synthesis and characterization. Polymer 2005;46:71–80.

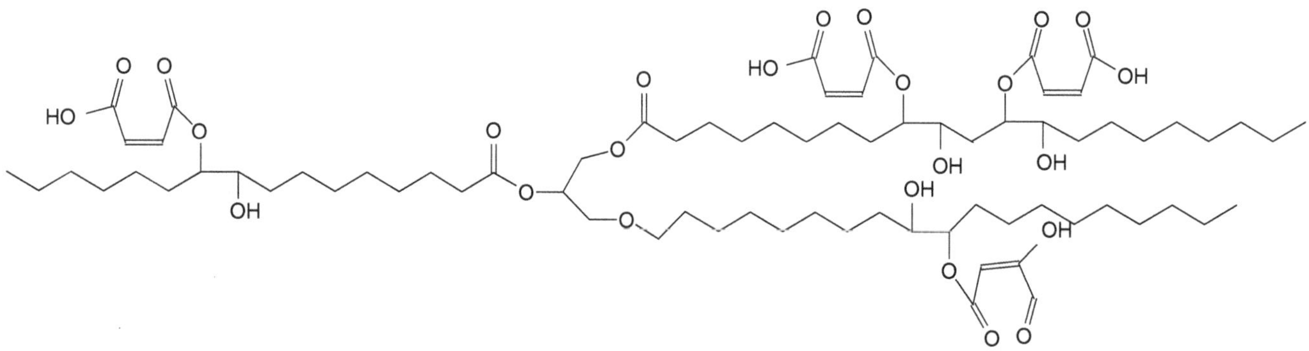

FIGURE 22.22 Maleinated vegetable oil [4].

explored. The uses of biodiesel in the fuel area as well as the research for new synthetic routes are intensively studied in recent years. Monitoring the physicochemical properties during these processes will constitute a necessary step for their optimization. In the last years, important progress has been reported [113–115].

22.3.3.1 Polymers from Functionalized Triglycerides Polyesters. One of the oldest polyesters prepared from triglyceride oils are the alkyd resins. They are widely used in

the coating and paint industries. The two main methods used in the preparation of alkyd resins are the one-stage process and the two-stage process [116]. In the one-stage process, these resins are prepared by polycondensation reaction between fatty acids, polyols (usually glycols) or hydroxyl acid, and dibasic acid or anhydride. In the two-stage process (Fig. 22.25), oil is initially converted into monoglycerides by glycerolysis and then the resins were prepared by reacting monoglycerides with an anhydride (usually phthalic or maleic anhydrides).

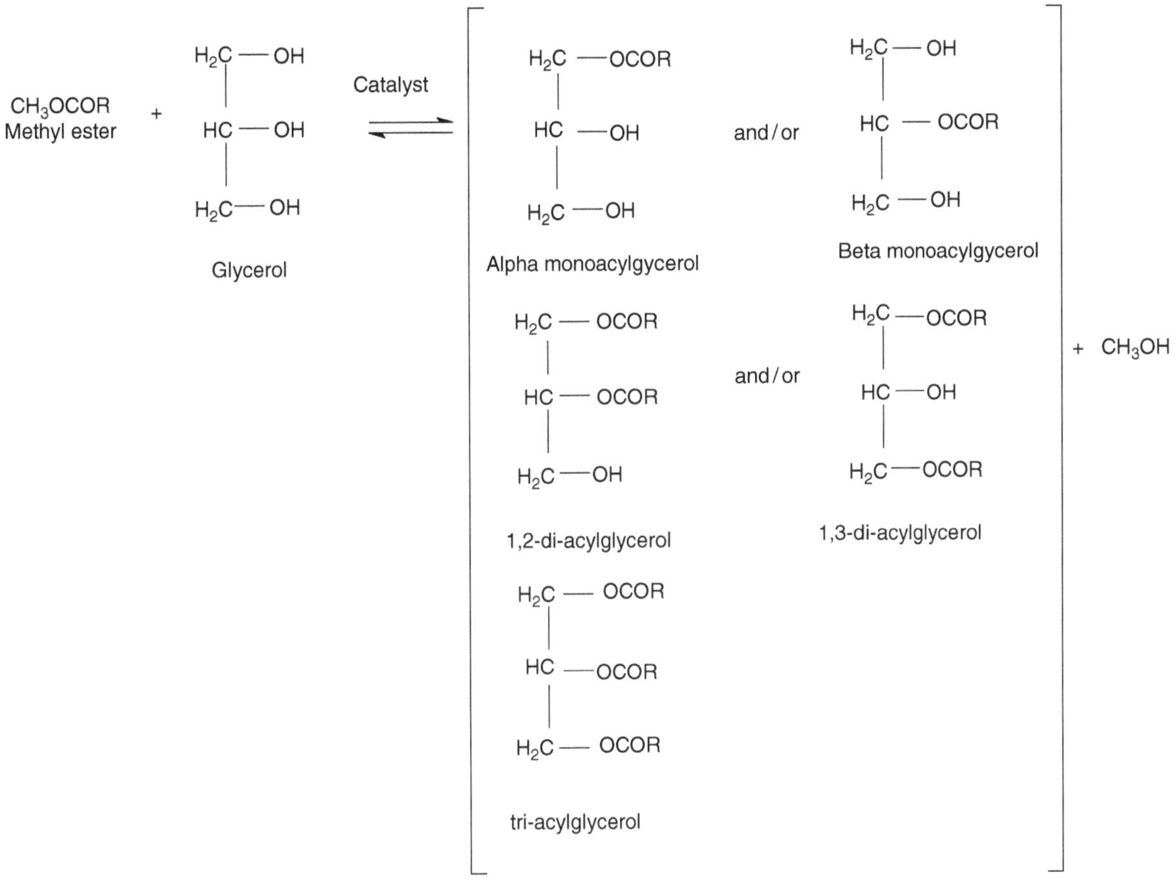

FIGURE 22.23 Glycerolysis from a fatty acid methyl ester. From Schulz GAS, da Silveira KC, Libardi DB M. do Carmo RP, Samios D. Synthesis and characterization of mono-acylglycerols through the glycerolysis of methyl esters obtained from linseed oil. Eur J Lipid Sci Technol 2011;113:1533–1540. ©Wiley-VCH Verlag GmbH & Co. KGaA. Reproduced with permission.

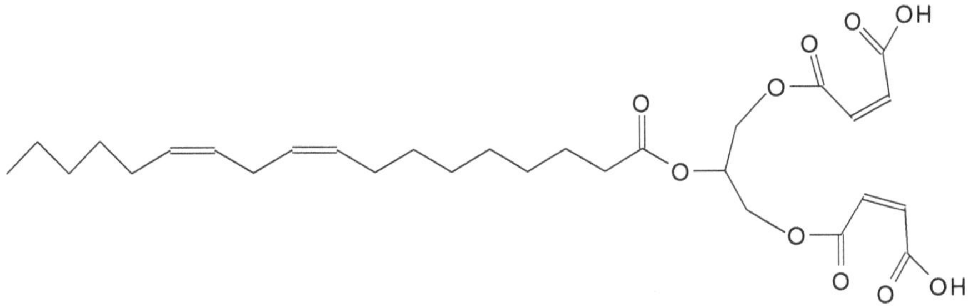

FIGURE 22.24 Maleinated monoglyceride.

Alkyd resins are classified according to their oil length, that is, the oil percent of the final resin [116]. This factor has a great impact on the properties and the applications of these resins. Alkyd resins are classified as "long oil" alkyds, because they contain above 60% oil or fatty acids. Oil amounts between 40% and 60% are characteristics of medium alkyds, while the short ones contain below 40% oil. Drying and semidrying oils, such as linseed, soya, sunflower, and castor oil, are commonly used in "long oil" applications because of their fatty acid composition and lower costs

[117]. The fatty acid portion gives many important properties to the resin, such as cross-linking potential, flexibility, compatibility with solvents, and control of solubility [117].

Polyurethanes. Polyurethanes are polymers with urethane linkage (–NHCOO–), which is formed by reacting isocyanate and hydroxyl-containing material (polyol) [6]. Polyurethanes based on triglycerides are obtained from isocyanates and oils with natural hydroxyl groups (e.g. castor oil) or oils functionalized with hydroxyl groups. Castor oil is a natural polyol with 85–95% ricinoleic acid and can be used directly

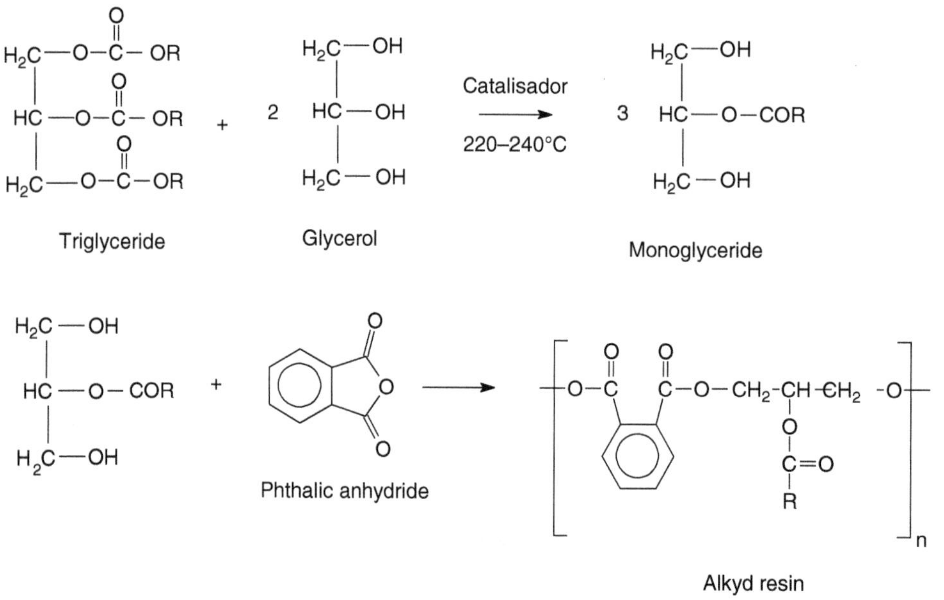

FIGURE 22.25 Alkyd resin preparation from a two-stage process.

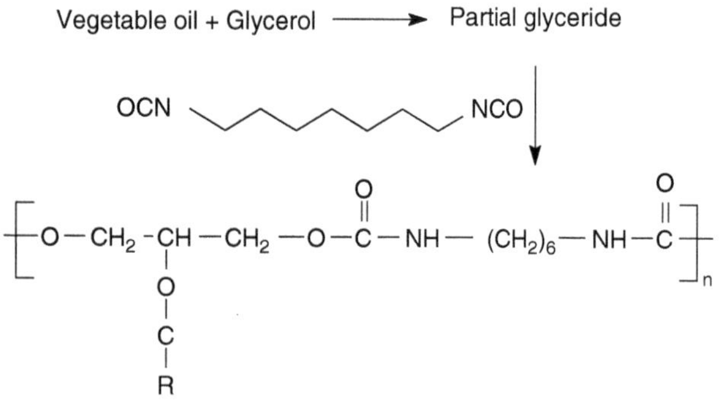

FIGURE 22.26 Preparation of polyurethanes from partial glycerides and hexamethylenediisocyanate (HMDI). With permission from Seniha Güner F, Yağci Y, Erciyes Tuncer A. Polymers from triglyceride oils. Prog Polym Sci 2006;31:633–670.

as renewable material for the preparation of polyurethanes, without any modification. In general, polyols from triglycerides are based on partial glycerides prepared from oil and glycerol or prepared from oxirane ring opening. In the case of double bonds hydroxylation from oxirane ring opening, it is expected that oils with higher unsaturation should give polyols with higher functionality. Vegetable polyol functionality results in different cross-linking density of the polyurethane networks [118]. Figure 22.26 shows the preparation of oil-modified polyurethanes from partial glycerides and hexamethylenediisocyanate (HMDI) [5].

22.3.3.2 *Polymers from Functionalized Biodiesel* Thorough knowledge about the polymerization of triglycerides was achieved in the last decades. On the other hand, there is still a

limited number of studies on biodiesel polymerization [58, 59, 119, 120].

Samios and coauthors have prepared oligoesters and polyesters from epoxidized biodiesel [58, 59, 119, 120]. To obtain these polymeric materials, the respective vegetable oils were transesterified and epoxidized with performic acid generated *in situ* (Fig. 22.27). Epoxidized biodiesel from different vegetable oils were polymerized with cyclic anhydrides using tertiary amines as initiator [58, 59, 119, 120]. Depending on the fatty acid distribution of the starting oil, as well as the anhydride used as comonomer, oligoesters, or cross-linked polyesters can be obtained.

Figure 22.28 and Figure 22.29 show, respectively, the ^{1}H and ^{13}C-NMR spectra of the oligoesters prepared from epoxidized sunflower oil methyl esters (methyl biodiesel from sunflower oil) and *cis*-1,2-cyclohexanedicarboxylic anhydride using triethylamine as initiator. These materials are soluble in common organic solvents such as acetone, ethanol, tetrahydrofurane, and chloroform, but insoluble in water. Oligoesters from epoxidized biodiesel can be used as intermediate materials for the synthesis of polyelectrolytes by saponification reactions with aqueous solution of sodium or potassium hydroxide at room temperature (Fig. 22.27). The products obtained after saponification present solubility in water. Amphiphilic materials, such as the polyelectrolytes prepared from epoxidized biodiesel, have hydrophobic and hydrophilic segments. They can spontaneously self-organize in a wide variety of structures in aqueous solution. Understanding the dynamics of the formation and transition between the various self-organized structures is important for technological applications.

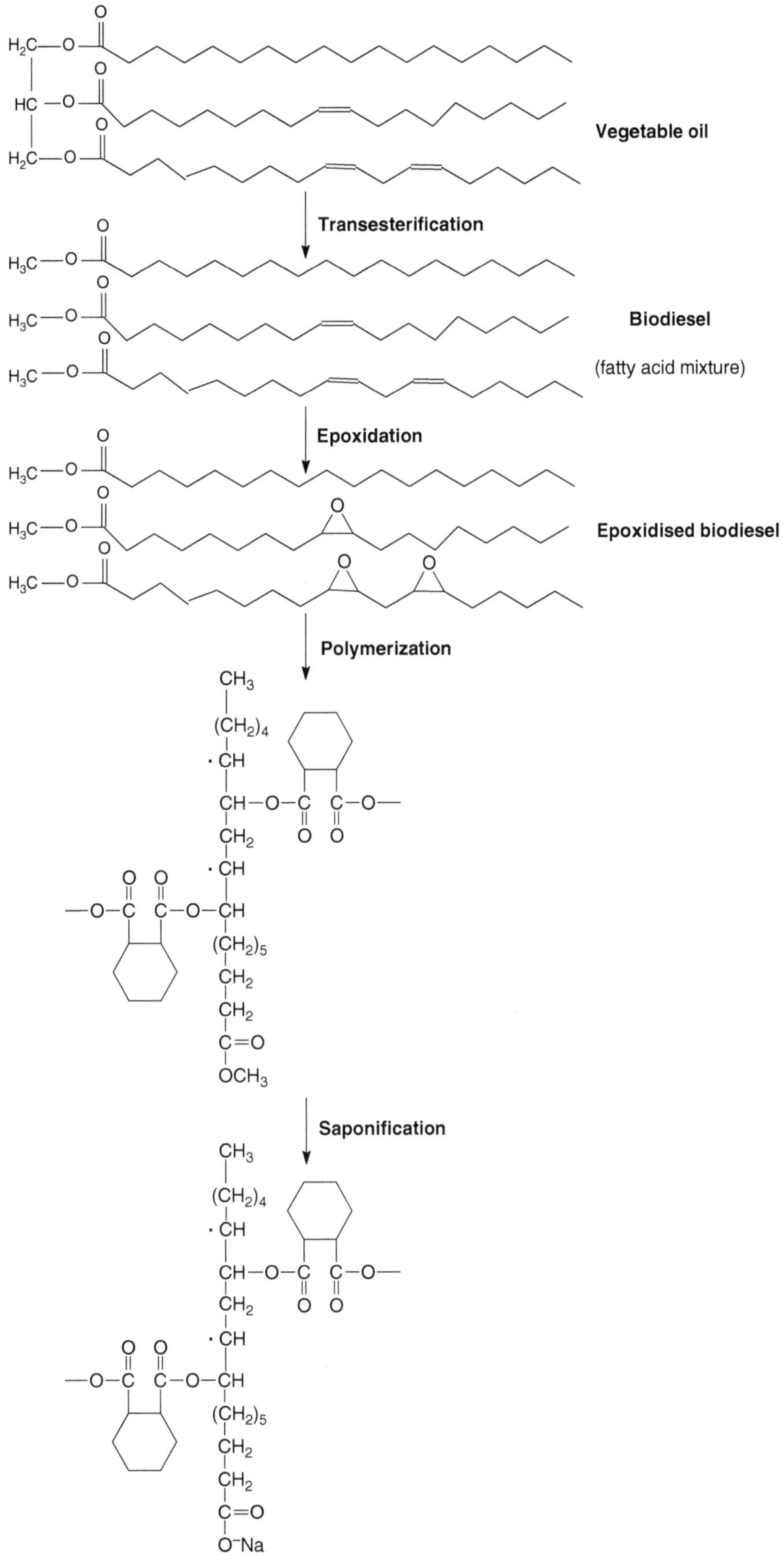

FIGURE 22.27 Polymerization of epoxidized biodiesel with subsequent saponification.

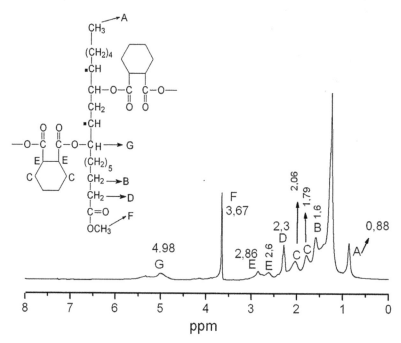

FIGURE 22.28 ¹H NMR spectrum of the oligoesters prepared from epoxidized sunflower oil methyl esters.

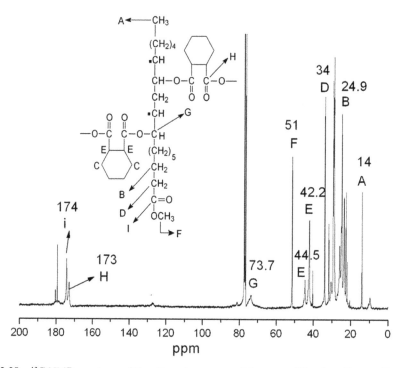

FIGURE 22.29 ¹³C NMR spectrum of the oligoesters prepared from epoxidized sunflower oil methyl esters.

REFERENCES

[1] Solomons G, Fryhle C. *Organic Chemistry*. 7th ed. New York: John Wiley & Sons Inc.; 1998.

[2] Kaynond KW. *General, Organic and Biological Chemistry: An Integrated Approach*. 2nd ed. New York: John Wiley & Sons Inc.; 2008.

[3] Yurkanis Bruice P. *Organic Chemistry*. 5th ed. New York: John Wiley & Sons Inc.; 2006.

[4] Sharma, V, Kundu PP. Addition polymers from natural oils – a review. Prog Polym Sci 2006;31:983–1008.

[5] Seniha Güner F, Yağci Y, Erciyes Tuncer A. Polymers from triglyceride oils. Prog Polym Sci 2006;31: 633–670.

[6] Petrović ZS. Polymers from biological oils. Contemp Mater 2010;I–1:39–50.

[7] Fatih Demirbas M. Biofuels from algae for sustainable development. Appl Energy 2011;88:3473–3480.

[8] Scott SA, Davey MP, Dennis JS, Horst I, Howe CJ, Lea-Smith DJ, Smith AG. Biodiesel from algae: challenges and prospects. Curr Opin Biotechnol 2010;21:277–286.

[9] Demirbas A. Use of algae as biofuel sources. Energy Convers Manage 2010;51:2738–2749.

[10] Knothe G. The history of vegetable oil based diesel fuels. In: Knothe G, Van Gerpen J, Krahl J, editors. *The Biodiesel Handbook*. Champaign: American Oil Chemists' Society Press; 2005. p 4–16.

[11] Demirbas A. Progress and recent trends in biodiesel fuels. Energy Convers Manage 2009;50:14–34.

[12] Wang Y, Ou S, Liu P, Zhang Z. Preparation of biodiesel from waste cooking oil via two-step catalyzed process. Energy Convers Manage 2007;48:184–188.

[13] Ma F, Hanna MA. Biodiesel production: a review. Bioresour Technol 1999;70:1–15.

[14] Zheng S, Kates M, Dubé MA, McLean DD. Acid-catalyzed production of biodiesel from waste frying oil. Biomass Bioenergy 2006;30:267–272.

[15] Marchetti JM, Miguel VU, Errazu AF. Possible methods for biodiesel production. Renew Sust Energy Rev 2007;11: 1300–1311.

[16] Meher LC, Sagar DV, Naik SN. Technical aspects of biodiesel production by transesterification – a review. Renew Sust Energy Rev 2006;10:248–268.

[17] Demirbas A. Political, economic and environmental impacts of biofuels: a review. Appl Energy 2009;86:108–117.

[18] Hill J, Nelson E, Tilman D, Polasky S, Tiffany D. Environmental, economic, and energetic costs and benefits of biodiesel and ethanol biofuels. Proc Natl Acad Sci USA 2006;103:11206–11210.

[19] Pinzi S, Garcia IL, Lopez-Gimenez FJ, Luque de Castro MD, Dorado G, Dorado MP. The ideal vegetable oil-based biodiesel composition: a review of social, economical and technical implications. Energy Fuels 2009;23:2325–2341.

[20] Pinto AC, Guarieiro LLN, Rezende MJC, Ribeiro NM, Torres EA, Lopes WA. Biodiesel: an overview. J Brazil Chem Soc 2005;16:1313–1330.

[21] Ahmad M, Ullah K, Khan MA, Zafar M, Tariq M, Ali S. Physicochemical analysis of hemp oil biodiesel: a promising non edible new source for bioenergy. Energy Source Part A 2011;33:1365–1374.

[22] Goodwin JG, Lotero E, Liu Y, Lopez DE, Suwannakarn K, Bruce DA. Synthesis of biodiesel via acid catalysis. Ind Eng Chem Res 2005;44:5353–5363.

[23] Otera J. Transesterification. Chem Rev 1993;95: 1449–1470.

[24] Liu Y, Lotero E, Goodwin JG. Effect of water on sulfuric acid catalyzed esterification. J Mol Catal A: Chem 2006;245: 132–140.

[25] Srivastava A, Prasad R. Triglycerides-based diesel fuels. Renew Sustain Energy Rev 2000;4:111–133.

[26] Freedman B, Pryde EH, Mounts TL. Variables affecting the yields of fatty esters from transesterified vegetable oils. J Am Oil Chem Soc 1984;61:1638–1643.

[27] Crabbe E, Nolasco-Hipolito C, Kobayashi G, Sonomoto K, Ishizaki A. Biodiesel production from crude palm oil and evaluation of butanol extraction and fuel properties. Process Biochem 2001;37:65–71.

[28] Fukuda H, Kondo A, Noda H. Biodiesel fuel production by transesterification of oils. J Biosci Bioeng 2001;92: 405–416.

[29] Canakci M, Van Gerpen J. Biodiesel production via acid catalysis. Trans ASAE 1999;42:1203–1210.

[30] Morgenstern M, Cline J, Meyer S, Cataldo S. Determination of the kinetics of biodiesel production using proton nuclear magnetic resonance spectroscopy (^{1}H-NMR). Energy Fuels 2006;20:1350–1353.

[31] Knothe G, Kenar JA. Determination of the fatty acid profile by ^{1}H-NMR spectroscopy. Eur J Lipid Sci Technol 2004;106: 88–96.

[32] Gelbard G, Bres O, Vargas RM, Vielfaure F, Schuchardt UF. ^{1}H nuclear magnetic resonance determination of the yield of the transesterification of grape seed oil with methanol. J Am Oil Chem Soc 1995;72:1239–1241.

[33] Santacesaria E, Di Serio M, Ledda M, Cozzolino M, Tesser R. Transesterification of soybean oil to biodiesel by using heterogeneous basic catalysts. Ind Eng Chem Res 2006;45: 3009–3014.

[34] Singh A, He B, Thompson J, Van Gerpen J. Process optimization of biodiesel production using alkaline catalysts. Appl Eng Agric 2006;22:597–600.

[35] Van Gerpen J. Biodiesel processing and production. Fuel Process Technol 2008;86:1097–1107.

[36] Knothe G. Monitoring a progressing transesterification reaction by fiber-optic near infrared spectroscopy with correlation to H^1-nuclear magnetic resonance spectroscopy. J Am Oil Chem Soc 2000;77:489–493.

[37] Yang F, Su Y, Li X, Zhang Q, Sun R. Studies on the preparation of biodiesel from *Zanthoxylum bungeanum* maxim seed oil. J Agric Food Chem 2008;56:7891–7896.

[38] Berchmans HJ, Hirata S. Biodiesel production from crude *Jatropha curcas* L. seed oil with a high content of free fatty acids. Bioresour Technol 2008;99:1716–1721.

[39] El-Mashad HM, Zhang R, Avena-Bustillos RJ. A two-step process for biodiesel production from salmon oil. Biosyst Eng 2008;99:220–227.

[40] Çayl G, Küsefoğlu S. Increased yields in biodiesel production from used cooking oils by a two-step process: comparison with one step process by using TGA. Fuel Process Technol 2008;89:118–122.

[41] Ramadhas AS, Jayaraj S, Muraleedharan C. Biodiesel production from high FFA rubber seed oil. Fuel 2005;84:335–340.

[42] Veljkovic VB, Lakicevic SH, Stamenkovic OS, Todorovic ZB, Lazic ML. Biodiesel production from tobacco (*Nicotiana tabacum* L.) seed oil with a high content of free fatty acids. Fuel 2006;85:2671–2675.

[43] Naik M, Meher LC, Naik SN, Das LM. Production of biodiesel from high free fatty acid Karanja (*Pongamia pinnata*) oil. Biomass Bioenergy 2008;32:354–357.

[44] Kusdiana D, Saka S. Two-step preparation for catalyst-free supercritical biodiesel fuel production. Appl Biochem Biotechnol 2004;113:781–791.

[45] D'Ippolito SA, Yori JC, Iturria ME, Pieck CL, Vera CR. Analysis of a two-step, noncatalytic, supercritical biodiesel production process with heat recovery. Energy Fuels 2007;21: 339–346.

[46] Minami E, Saka S. Kinetics of hydrolysis and methyl esterification for biodiesel production in two-step supercritical methanol process. Fuel 2006;85:2479–2483.

[47] Isayama Y, Saka S. Biodiesel production by supercritical process with crude biomethanol prepared by wood gasification. Bioresour Technol 2008;99:4775–4779.

[48] Watanabe Y, Pinsirodom P, Nagao T, Kobayashi T, Nishida Y, Takagi Y, Shimada Y. Production of FAME from acid oil model using immobilized *Candida antarctica* lipase. J Am Oil Chem Soc 2005;82:825–831.

[49] Lai CC, Zullaikah S, Vali SR, Ju YH. Lipase-catalyzed production of biodiesel from rice bran oil. J Chem Technol Biotechnol 2005;80:331–337.

[50] Shimada Y, Watanabe Y, Sugihara A, Tominaga Y. Enzymatic alcoholysis for biodiesel fuel production and application of the reaction to oil processing. J Mol Catal B: Enzym 2002;17:133–142.

[51] Meneghetti SMP, Meneghetti MR, Wolf CR, Silva EC, Lima GES, Silva LL. Biodiesel from castor oil: a comparison of ethanolysis versus methanolysis. Energy Fuels 2006;20: 2262–2265.

[52] Sanli H, Canakci M. Effects of different alcohol and catalyst usage on biodiesel production from different vegetable oils. Energy Fuels 2008;22:2713–2719.

[53] Willing A. Oleochemical esters – environmentally compatible raw materials for oils and lubricants from renewable resources. Fett/Lipid 1999;101:192–198.

[54] Salehpour S, Dubé MA. Biodiesel: a green polymerization solvent. Green Chem 2008;10:321–326.

[55] Salehpour S, Dubé MA. The use of biodiesel as a green polymerization solvent at elevated temperatures. Polym Int 2002;57:854–862.

[56] Salehpour S, Dubé MA, Murphy M. Solution polymerization of styrene using biodiesel as a solvent: effect of biodiesel feedstock. Can J Chem Eng 2009;87:129–135.

[57] Nicolau A, Mariath RM, Samios D. Study of the properties of polymers obtained from vegetable oil derivatives by light scattering techniques. Mater Sci Eng C 2009;29:452–457.

[58] Reiznautt QB, Garcia ITS, Samios D. Oligoesters and polyesters produced by the curing of sunflower oil epoxidized biodiesel with cis-cyclohexane dicarboxylic anhydride: synthesis and characterization. Mater Sci Eng C 2009;29: 2302–2311.

[59] Martini DS, Braga BA, Samios D. On the curing of linseed oil epoxidized methyl esters with different cyclic dicarboxylic anhydrides. Polymer 2009;50:2919–2925.

[60] Wehlmann J. Use of esterified rapeseed oil as plasticizer in plastics processing. Fett/Lipid 1999;101:249–256.

[61] Hu J, Du Z, Tang Z, Min E. Study on the solvent power of a new green solvent: biodiesel. Ind Eng Chem Res 2004;43: 7928–7931.

[62] Poirier MA, Steere DE, Krogh JA, inventors; Exxon Research and Engineering Co., assignee. Cetane improver compositions comprising nitrated fatty acid derivatives. US Patent 5454842. 1995 Oct 3.

[63] Suppes GJ, Goff M, Burkhart ML, Bockwinkel K, Mason MH, Botts JB. Multi-functional diesel fuel additives from triglycerides. Energy Fuels 2001;15:151–157.

[64] Suppes GJ, Chen Z, Rui Y, Mason M, Heppert JA. Synthesis and cetane improver performance of fatty acid glycol nitrates. Fuel 1999;78:73–81.

[65] Chavanne G. Procédé de transformation d'huiles végétales en vue de leur utilisation comme carburants. Belgian Patent 422877. 1937 Aug 31. Chem Abstr 1938;32:4313.

[66] Samios D, Pedrotti F, Nicolau A, Reiznautt QB, Martini DD, Dalcin FM. A Transesterification Double Step Process – TDSP for biodiesel preparation from fatty acids triglycerides. Fuel Process Technol 2009;90:599–605.

[67] Guzatto R, de Martini TL, Samios D. The use of a modified TDSP for biodiesel production from soybean, linseed and waste cooking oil. Fuel Process Technol 2012;92:197–203.

[68] Guzatto R, Defferrari D, Bülow Q, Reiznautt Í, Cadore R, Samios D. Transesterification double step process modification for ethyl ester biodiesel production from vegetable and waste oils. Fuel 2012;92:197–203.

[69] Vigli G, Philippidis A, Spyros A, Dais P. Classification of edible oils by employing ^{31}P and ^{1}H-NMR spectroscopy in combination with multivariate statistical analysis. A proposal for the detection of seed oil adulteration in virgin olive oils. J Agric Food Chem 2003;51:5715–5722.

[70] Rashid U, Anwara F, Knothe G. Evaluation of biodiesel obtained from cottonseed oil. Fuel Process Technol 2009;90:1157–1163.

[71] De Boni LAB, Lima Da Silva IN. Monitoring the transesterification reaction with laser spectroscopy. Fuel Process Technol 2011;92:1001–1006.

[72] Sharma YC, Singh B, Korstad J. High yield and conversion of biodiesel from a nonedible feedstock (*Pongamia pinnata*). J Agric Food Chem 2010;58:242–247.

[73] Sharma YC, Singh B, Korstad J. Application of an efficient nonconventional heterogeneous catalyst for biodiesel synthesis from *Pongamia pinnata* oil. Energy Fuels 2010;24: 3223–3231.

[74] Tariq M, Ali S, Ahmad F, Ahmad M, Zafar M, Khalid N. Identification, FT-IR, NMR (^{1}H and^{13}C) and GC/MS studies of fatty acid methyl esters in biodiesel from rocket seed oil. Fuel Process Technol 2011;92:336–341.

[75] Schulz GAS, da Silveira KC, Libardi DB M. do Carmo RP, Samios D. Synthesis and characterization of monoacylglycerols through the glycerolysis of methyl esters obtained from linseed oil. Eur J Lipid Sci Technol 2011;113:1533–1540.

[76] Mallégol J, Lemaire J, Gardette J-L. Drier influence on the curing of linseed oil. Prog Org Coat 2000;39:107–113.

[77] Tuman SJ, Chamberlain D, Scholsky KM, Soucek MD. Differential scanning calorimetry study of linseed oil cured with metal catalysts. Prog Org Coat 1996;28: 251–258.

[78] Li F, Larock RC. Novel polymeric materials from biological oils. J Polym Environ 2002;10:59–67.

[79] Xia Y. Larock RC. Vegetable oil-based polymeric materials: synthesis, properties, and applications. Green Chem 2010;12: 1893–1909.

[80] Andjelkovic DD, Valverde M, Henna P, Li F, Larock RC. Novel thermosets prepared by cationic copolymerization of various vegetable oils – synthesis and their structure–property relationships. Polymer 2005;46:9674–9685.

[81] Kabasaka OS. Styrenation of mahaleb and anchovy oils. Prog Org Coat 2005;53:235–238.

[82] Milchert E, Smagowicz A, Lewandowski G. Optimization of the reaction parameters of epoxidation of rapeseed oil with peracetic acid. J Chem Technol Biotechnol 2010;85: 1099–1107.

[83] Campanella A, Baltanás MA. Degradation of the oxirane ring of epoxidized vegetable oils in liquid–liquid heterogeneous reaction systems. Chem Eng J 2006;118:141–152.

[84] Gan LH, Goh SH, Ooi KS. Kinetic studies of epoxidation and oxirane cleavage of palm olein methyl esters. J Am Oil Chem Soc 1992;69:347–351.

[85] Sonnet PE, Foglia TA. Epoxidation of natural triglycerides with ethylmethyldioxirane. J Am Oil Chem Soc 1996;73: 461–464.

[86] Kangas SL, Jones FN. Binders for higher-solids coating. J Coat Technol 1987;59:99–103.

[87] Rios LA, Weckes P, Schuster H, Hoelderich WF. Mesoporous and amorphous Ti-silicas on the epoxidation of vegetable oils. J Catal 2005;232:19–26.

[88] Piazza JG, Nuñes A, Foglia TA. Epoxidation of fatty acids, fatty methyl esters, and alkenes by immobilized oat seed peroxygenase. J Mol Catal B: Enzym 2003; 21:143–151.

[89] Rangarajan B, Havey A, Grulke EA, Culnan PD. Kinetic parameters of a two-phase model for in situ epoxidation of soybean oil. J Am Oil Chem Soc 1995;72:1161–1169.

[90] Du G, Tekin A, Hammond EG, Woo LK. Catalytic epoxidation of methyl linoleate. J Am Oil Chem Soc 2004;81: 477–480.

[91] Schmitz WR, Wallace JG. Epoxidation of methyl oleate with hydrogen peroxide. J Am Oil Chem Soc 1954;31: 363–365.

[92] Bouh AO, Espenson JH. Epoxidation reactions with urea–hydrogen peroxide catalyzed by methyltrioxorhenium(VII) on niobia. J Mol Catal A: Chem 2003;200:43–47.

[93] Hermann WA, Fischer RW, Marz DW. Methyltrioxorhenium as catalyst for olefin oxidation. Angew Chem Int Ed 1991;30: 1638–1641.

[94] Adam W, Mitchell CM. Methyltrioxorhenium(VII)-catalyzed epoxidation of alkenes with the urea/hydrogen peroxide adduct. Angew Chem Int Ed 1996;35:533–535.

[95] Marks DW, Larock RC. The conjugation and epoxidation of fish oil. J Am Oil Chem Soc 2002;79:65–68.

[96] Gerbase AE, Gregório JR, Martinelli M, Brazil MC, Mendes ANF. Epoxidation of soybean oil by the methyltrioxorhenium-CH$_2$Cl$_2$/H$_2$O$_2$ catalytic biphasic system. J Am Oil Chem Soc 2002;79:179–181.

[97] Warwel S, Klaas MR. Chemo-enzymatic epoxidation of unsaturated carboxylic acids. J Mol Catal B: Enzym 1995;1: 29–35.

[98] Fischer RF. Polyesters from expoxides and anhydrides. J Polym Sci 1960;44:155–172.

[99] Raquez J-M, Deléglise M, Lacrampe M-F, Krawczak P. Thermosetting (bio)materials derived from renewable resources: a critical review. Prog Polym Sci 2010;35: 487–509.

[100] Guo A, Cho Y-J, Petrovic ZS. Structure and properties of halogenated and nonhalogenated soy-based polyols. J Polym Sci Part A: Polym Chem 2000;38:3900–3910.

[101] Kandanarachchi P, Guo A, Petrovic Z. The hydroformylation of vegetable oils and model compounds by ligand modified rhodium catalysis. J Mol Catal A: Chem 2002;184:65–71.

[102] Lu J, Khot S, Wool RP. New sheet molding compound resins from soybean oil. I. Synthesis and characterization. Polymer 2005;46:71–80.

[103] Khot SN, Lascala JJ, Can E, Morye SS, Williams GI, Palmese GR, Kusefoglu SH, Wool RP. Development and application of triglyceride-based polymers and composites. J Appl Polym Sci 2001;82:703–723.

[104] Noureddini H, Medikonduru V. Glycerolysis of fats and methyl esters. J Am Oil Chem Soc 1997;74:419–425.

[105] Felizardo P, Machado J, Vergueiro D, Correia MJN, Gomes JP, Bordado JM. Study on the glycerolysis reaction of high free fatty acid oils for use as biodiesel feedstock Fuel Process Technol 2011;92:1225–1229.

[106] Ferretti CA, Fuente S, Ferullo R, Castellani N, Apesteguía CR, Di Cosimo JI. Monoglyceride synthesis by glycerolysis of methyl oleate on MgO: catalytic and DFT study of the active site. Appl Catal A: Gen 2012;413–414:322–331.

[107] Negi DS, Sobotka F, Kimmel T, Wozny G, Schomäcker R. Glycerolysis of fatty acid methyl esters: 1. Investigations in a batch reactor. J Am Oil Chem Soc 2007;84:83–90.

[108] Damstrup ML, Abildskov J, Kiil S, Jensen AD, Sparsø FV, Xu X. Evaluation of binary solvent mixtures for efficient monoacylglycerol production by continuous enzymatic glycerolysis. J Agric Food Chem 2006;54:7113–7119.

[109] Ferretti CA, Olcese RN, Apesteguía C. R, Di Cosimo JI. Heterogeneously-catalyzed glycerolysis of fatty acid methyl esters: reaction parameter optimization. Ind Eng Chem Res 2009;48:10387–10394.

[110] Ferretti CA, Apesteguía CR, Di Cosimo JI. MgO-based catalysts for monoglyceride synthesis from methyl oleate and glycerol: effect of Li promotion. Appl Catal A: Gen 2011;399: 146–153.

[111] Stevenson DE, Stanley RA, Fumeaux RH. Glycerolysis of tallow with immobilised lipase. Biotechnol Lett 1993;15: 1043–1048.

[112] Dutta N, Karak N, Dolui SK. Synthesis and characterization of polyester resins based on Nahar seed oil. Prog Org Coat 2004;49:146–152.

[113] McFaul CA, Alb AM, Drenski MF, Reed WF. Simultaneous multiple sample light scattering detection of LCST during copolymer synthesis. Polymer 2011;52:4825–4833.

[114] Li Z, Serelis AK, Reed WF, Alb AM. Online monitoring of the copolymerization of 2-(dimethylamino)ethyl acrylate with styrene by RAFT. Deviations from reaction control. Polymer 2010;51:4726–4734.

[115] Alb AM, Reed WF. Fundamental measurements in online polymerization reaction monitoring and control with a focus on ACOMP. Macromol React Eng 2010;4:470–485.

[116] Tiwari S, Kumari B, Singh SN. Evaluation of metal mobility/ immobility in fly ash induced by bacterial strains isolated from the rhizospheric zone of *Typha latifolia* growing on fly ash dumps. Bioresour Technol 2008;99:1300–1304.

[117] Ploeger R, Scalarone D, Chiantore O. The characterization of commercial artists' alkyd paints. J Cult Heritage 2008;9: 412–419.

[118] Javni I, Petrović ZS, Guo A, Fuller R. Thermal stability of polyurethanes based on vegetable oils. J Appl Polym Sci 2000;77:1723–1734.

[119] Nicolau A, Martignago Mariath R, Agostini Martini E, dos Santos Martini D, Samios D.The polymerization products of epoxidized oleic acid and epoxidized methyl oleate with cis-1,2-cyclohexanedicarboxylic anhydride and triethylamine as the initiator: chemical structures, thermal and electrical properties. Mater Sci Eng C 2010;30:951–962.

[120] da Roza MB, Nicolau A, Angeloni LM, Sidou PN, Samios D. Thermodynamic and kinetic evaluation of the polymerization process of epoxidized biodiesel with dicarboxylic anhydride. Mol Phys 2012;110:1171–1178.

INDEX

Monitoring Polymerization Reactions: From Fundamentals to Applications, First Edition. Edited by Wayne F. Reed and Alina M. Alb.
© 2014 John Wiley & Sons, Inc. Published 2014 by John Wiley & Sons, Inc.

CPSIA information can be obtained
at www.ICGtesting.com
Printed in the USA
JSHW040824290522
26351JS00004B/14

9 780470 917381